# Magnetic Coordination Compounds and More. . . a Long and Successful Story: A Tribute to M. Julve and F. Lloret

# Magnetic Coordination Compounds and More... a Long and Successful Story: A Tribute to M. Julve and F. Lloret

Guest Editors

**Carlos J. Gómez García**
**Salah-Eddine Stiriba**

MDPI

Basel • Beijing • Wuhan • Barcelona • Belgrade • Novi Sad • Cluj • Manchester

*Guest Editors*
Carlos J. Gómez García
Department of Inorganic Chemistry
University of Valencia
Valencia
Spain

Salah-Eddine Stiriba
The Institute of Molecular Science (ICMol)
University of Valencia
Valencia
Spain

*Editorial Office*
MDPI AG
Grosspeteranlage 5
4052 Basel, Switzerland

This is a reprint of the Special Issue, published open access by the journal *Magnetochemistry* (ISSN 2312-7481), freely accessible at: https://www.mdpi.com/journal/magnetochemistry/special_issues/Q19T1CC7AR.

For citation purposes, cite each article independently as indicated on the article page online and as indicated below:

| Lastname, A.A.; Lastname, B.B. Article Title. *Journal Name* **Year**, *Volume Number*, Page Range. |
|---|

**ISBN 978-3-7258-4641-2 (Hbk)**
**ISBN 978-3-7258-4642-9 (PDF)**
**https://doi.org/10.3390/books978-3-7258-4642-9**

Cover image courtesy of Santiago Álvarez

# Contents

# About the Editors

**Carlos J. Gómez García**

Prof. Dr. Carlos J. Gómez García (Valladolid, Spain, 1964) received his PhD from the University of Valencia with the Extraordinary Doctoral Award in 1991. After taking two postdoctoral positions in Rennes and Bordeaux (France), he joined the University of Valencia as an Associate Professor in 1993. In 1998 he was promoted to Assistant Professor, and became a full Professor in 2007 in the Inorganic Chemistry Department of the University of Valencia (after a national qualification process). Since 2014 he has led the Multifunctional Modulable Molecular Materials (M4) research group, and since its foundation in 2015 he has been the Editor-in-Chief of the journal *Magnetochemistry*, where he has co-edited four other Special Issues. His main interests are magnetic and conducting MOFs and coordination compounds, including polyoxometalate-based materials for energy applications and catalysis. Prof. Gómez has co-authored almost five hundred articles in international journals, more than twenty book chapters and has presented around eighty invited and oral conferences in national and international congresses.

**Salah-Eddine Stiriba**

Prof. Dr. Salah-Eddine Stiriba (Marrakech, Morocco, 1971) received his PhD from the University of Valencia in 1998. He took three postdoctoral positions, first at Texas A&M University (CS, USA), then at the University Louis Pasteur in Strasbourg (France), and finally at the Institut für Makromolekulare Chemie at Freiburg University (Germany) as an Alexander von Humboldt fellow, where he started his independent career. He joined the University of Valencia as Senior Research Scientist in 2002, and then became Professor of Organic Chemistry in 2007. He is the Editor-in-Chief of the Macromolecules section of *IJMS* and a member of the editorial board of *Catalysts*, and has edited several Special Issues in both journals. His main research interests are in the areas of homogenous and heterogeneous catalysis, sustainable catalysis, organic polymer chemistry, and multifunctional coordination compounds with magnetic properties. He has published more than one hundred and forty articles and four book chapters.

# Preface

This Special Issue was initially intended to pay a more than deserved tribute to the long scientific careers of professors Miguel Julve and Francisco Lloret. Sadly, Professor Miguel Julve, a beloved professor, outstanding scholar and colleague, passed away in July 2024 and so transforming this reprint, unfortunately, into a posthumous homage to him. Readers will find in this reprint contributions from a long list of world-leading researchers, most of whom are former PhD students, postdoctoral fellows, collaborators, and friends of Miguel and Paco. The contributions cover a wide range of topics related to molecular magnetism of discrete and polymeric materials, mainly coordination compounds and metal–organic frameworks, discussing properties ranging from spin crossover to single-ion and single-molecule magnets.

**Carlos J. Gómez García and Salah-Eddine Stiriba**
*Guest Editors*

*magnetochemistry* MDPI

*Editorial*

# Magnetic Coordination Compounds and More... a Long and Successful Story: A Tribute to M. Julve and F. Lloret

**Carlos J. Gómez-García [1,*] and Salah-Eddine Stiriba [2,*]**

1 Departamento de Química Inorgánica, Universidad de Valencia, C/Dr. Moliner 50, 46100 Burjasot, Spain
2 Instituto de Ciencia Molecular/ICMol, Universidad de Valencia, C/Catedrático José Beltrán 2, 46980 Valencia, Spain
* Correspondence: carlos.gomez@uv.es (C.J.G.-G.); stiriba@uv.es (S.-E.S.)

This Special Issue of *Magnetochemistry*, entitled "*Magnetic Coordination Compounds and More... a Long and Successful Story: A Tribute to M. Julve and F. Lloret*", was organized to commemorate the partial retirement in 2024, institutionally speaking, rather than scientifically, of two extraordinary chemists and friends, Francisco Lloret Pastor and Miguel Julve Olcina, both faculty members of the Department of Inorganic Chemistry and the Institute of Molecular Science at the University of Valencia, after an active and successful research and teaching career.

Sadly, on 9 July 2024, while preparing this Special Issue, the chemical community lost one of its distinguished scholars, Prof. Miguel Julve. Miguel was a great friend, collaborator, and mentor for his graduates and postdoctoral associates. He always had his desk door open to anyone who wanted to ask his advice, eat some fruit, or simply listen to his last jokes. He was proud of his work and enjoyed talking about it almost as much as about his grandson Sebastian. He will be sorely missed by colleagues, friends, his wife María José, and daughters Andrea, Ingrid, and Michaela. Fortunately, we have Paco to help us fill the huge gap left by Miguel.

The two protagonists of this Special Issue had established an extraordinary scientific and personal relationship between each other for almost 48 years, since their undergraduate studies, then doctoral theses, later postdoctoral stays, and finally when they joined the Department of Inorganic Chemistry at the University of Valencia as faculty members. Francisco Lloret Pastor and the late Miguel Julve Olcina were two good friends who shared, in a *hybrid* mode, the passion for making new complex inorganic magnetic systems and creating an experimental and theoretical understanding of them.

We believe that the personal *Laudatio* entitled "*Miguel Julve and Francisco Lloret, a Friendly Pair of Two Exceptional Coordination Chemists in Molecular Magnetism*", written by their good friend and close colleague *Michel Verdaguer*, from the Sorbonne University in Paris (France), says everything about their scientific and academic careers [1]. This laudatio summarizes, in three parts, the scientific trajectory of the Coordination Chemistry team lead by Miguel and Paco at the University of Valencia, their most important results, and, finally, the strong friendly link between them (Figure 1).

Following a chronological order, the first contribution in this Special Issue by *B. Casanovas, R. Vicente, M. Font-Bardía, and M. Salah El Fallah*, from the University of Barcelona (Spain), ref. [2], nicely describes six new polynuclear compounds prepared with $Mn^{II}$, R-salicylaldehyde oximes, and 9-anthracenecarboxylato. Their crystal structures contain trinuclear $Mn^{III}_3$ units forming chains, dimers, or monomers, and their magnetic properties show antiferromagnetic interactions in three cases and a combination of antiferromagnetic and ferromagnetic interactions in the other three. The contribution includes a magneto-structural correlation between the magnetic exchange and the intramolecular $Mn^{III}$–N–O–$Mn^{III}$ torsion angle. Furthermore, two of these compounds showed slow relaxation of magnetization.

**Figure 1.** Paco (**left**) and Miguel (**right**). Two exceptional chemists, teachers and friends. Reprinted from Ref. [1].

The contribution by *A. Palii, S. Zilberg, and B. Tsukerblat*, from the Federal Research Center of Problems of Chemical Physics and Medicinal Chemistry RAS (Russia), the Ariel University (Israel), and the Ben-Gurion University of the Negev (Israel), ref. [3], reports on the theoretical foundations of the rational design of molecular cells for quantum cellular automata (QCA) devices with optimized properties. The vibronic approach was applied to address the analysis of the cell–cell response and energy dissipation, which are the two key properties of such molecular cells. This study shows that in order to reach a good compromise between low energy dissipation and a strong cell–cell response, the use of weakly interacting mixed valence molecules with weak electron delocalization as cells is recommended. Some relevant results from a recent ab initio study on electron transfer and vibronic coupling are briefly discussed with the idea of controlling the key parameters of molecular QCA cells.

*J. Löhr, M. Font-Bardia, J. Mayans, and A. Escuer*, from the University of Barcelona (Spain), ref. [4], studied the magnetic properties of a series of manganese clusters with $\{Mn^{III}{}_6Mn^{II}Na_2\}$, $\{Mn^{III}{}_3Mn^{II}Na\}$, and $\{Mn^{III}{}_3\}$ metallic cores, prepared by the reaction of manganese halides with polydentate $NO_3$ Schiff bases. Their magnetic properties show spin ground states of 19/2, 7/2, and 12/2 for the $\{Mn^{III}{}_6Mn^{II}Na_2\}$, $\{Mn^{III}{}_3Mn^{II}Na\}$, and $\{Mn^{III}{}_3\}$ clusters, respectively, including the presence of field-induced slow relaxation of the magnetization for the trinuclear complexes.

The contribution by *C. Andrei Spinu, D. O. T. A. Martins, T. Mocanu, M. Hillebrand, J.-P. Sutter, F. Tuna, and M. Andruh*, from the University of Bucharest (Romania), the Romanian Academy, the University of Manchester (UK), and the University of Toulouse (France), ref. [5], presents two new 2p–3d complexes of the type $(Et_3NH)[ML(hfac)_2]$ [M = Mn and Co], obtained with a nitronylnitroxide radical. The magnetic measurements show strong antiferromagnetic M(II)-Rad interactions, confirmed by EPR spectra.

*A. L Świtlicka*, from the University of Silesia (Poland), ref. [6], provides an updated overview of the different topologies and magneto-structural correlations in $Co^{II}$ compounds with dicyanamide as a ligand.

The contribution by *B. Gil-Hernández, S. Millan, I. Gruber, C. Janiak, C. J. Gómez-García, and J. Sanchiz*, from the University of La Laguna (Spain), Heinrich Heine University (Germany), and the University of Valencia (Spain), ref. [7], reports the X-ray single crystal structure and magnetic properties of two new mesoxalate-bridged heptanuclear copper(II) compounds, formulated as $(NH_4)_4[Cu_7(Hmesox)_6(H_2O)_8]\cdot10H_2O$ and $[Ru(bpy)_3]_4[Cu_7(Hmesox)_6Cl_2]Cl_2\cdot2CH_3CN\cdot12H_2O$. Magnetic studies show the presence of antiferromagnetic interactions in the former compound and the coexistence of ferro- and antiferromagnetic interactions in the latter that, additionally, shows luminescent properties arising from the $[Ru(bpy)_3]^{2+}$ cation, supporting its multifunctionality.

*W. X. C. Oliveira, V. G. Araújo, C. B. Pinheiro, M. Julve, and C. L. M. Pereira*, from the Federal University of Minas Gerais (Brazil) and the University of Valencia (Spain), ref. [8], investigated the synthesis, structure, and magnetic characterization of the hexanuclear copper(II) complex $[Cu_6(en)_4(eg)_2(pyox)_4]\cdot3eg\cdot en\cdot12H_2O$, with en = ethylenediamine, eg = ethylene glycol, and $H_2$pyox = 4-(1H-pyrazole-4-yl)phenylene-N-oxamic acid. This hexacopper(II) complex is built from two linear tricopper(II) complexes containing $[Cu(en)]^{2+}$ moieties connected to a [Cu(eg)] unit by two $pyox^{2-}$ ligands and a $[eg]^{2-}$ ligand. DC magnetic studies reveal a strong antiferromagnetic Cu···Cu interaction within the trinuclear subunits, mediated by the alkoxide and pyrazolate bridges.

*J. D. Tempesta, F. Faria Paiva, L. A. Ferreira, R. M. R. da Silva, L. D. G. Botelho, I. M. L. Rosa, C. Cesar Candido, A. Marcio Gomes, W. C. Nunes, G. P. Guedes, and M. Vanda Marinho*, from the Universities of Alfenas, Goiás, Rio de Janeiro, and Fluminense (Brazil), ref. [9], report the synthesis, structure, and magnetic properties of the heterobimetallic 15-MC-5 metallacrown-based Cu–Ce complex $[CeCu_5(5mpzHA)_5(NO_3)(H_2O)_7]\cdot2NO_3\cdot7H_2O$. Magnetic measurements show the presence of antiferromagnetic interactions inside the heterometallic complex.

The article by *P. Escamilla, N. Moliner, D. Armentano, E. Pardo, J. Ferrando-Soria, and T. Grancha*, from the University of Valencia (Spain) and the University of Calabria (Italy), ref. [10], reports an interesting example of the partial post-synthetic metal exchange of $Zn^{II}$ metal ions by $Co^{II}$ ones in the water-stable three-dimensional $CaZn_6$-MOF 1, containing the amino acid S-methyl-L-cysteine. This ion exchange yields two novel MOFs with increasing contents of $Co^{II}$ ions of 4 % and 8%. This post-synthetic metal exchange methodology led to MOFs that cannot be obtained by direct synthesis, featuring a modulation of the magnetic properties and field-induced slow relaxation of the magnetization.

*A. Colin, Y. Wang, F. Lambert, N. Bridonneau, N. Suaud, R. Guillot, E. Rivière, Z. Halime, N. Guihéry, S.-I. Ohkoshi, and T. Mallah*, from the University of Paris-Saclay, the University of Toulouse (France), and The University of Tokyo (Japan), ref. [11], undertook a systematic study on a trinuclear $Co^{II}$-containing complex assembled using the non-innocent hexahydroxytriphenylene bridging ligand. The characterization of this complex shows that the central ligand undergoes up to four reversible redox processes, leading to species with different optical properties. The exchange coupling among the electrons of the bridge resulted in a spin doublet ($S = 1/2$) localized close to one of the three $Co^{2+}$ ions, as demonstrated by the experimental magnetic data.

The article by *T. T. da Cunha, J. H. de Araujo-Neto, M. E. Alvarenga, F. Terra Martins, E. F. Pedroso, D. L. Mariano, W. C. Nunes, N. Moliner, F. Lloret, M. Julve, and C. L. M. Pereira*, form the Universities of Minas Gerais, São Paulo, Goiás, Rio de Janeiro, and Fluminense (Brazil) and the University of Valencia (Spain), ref. [12], reports on the synthesis, crystal structures, and magnetic properties of four air-stable mononuclear lanthanide(III) complexes with the ligand N-(2,4,6-trimethylphenyl)oxamate. The magnetic properties show the typical behavior for the ground state terms of the $Ln^{III}$ ions. AC magnetic measurements reveal the presence of slow magnetic relaxation without the presence of a DC field only for the last compound, whereas a DC field is needed to observe slow magnetic relaxation behavior in the other three compounds.

*I. F. Diaz-Ortega, Y. Ye, J. Jover, E. Ruiz, E. Colacio, and J. M. Herrera*, from the Universities of Granada, Almería, and Barcelona (Spain), ref. [13] report the synthesis of a series of nine mononuclear $Ln^{III}$ complexes with $LnN_4O_2Cl_2$ and $LnN_4O_4$ coordination spheres and rare hexagonal bipyramidal geometries exhibiting field-induced slow magnetization relaxation (SMR).

The review by *R. Rabelo, L. M. Toma, A. Bentama, S.-E. Stiriba, R. Ruiz-García, and J. Cano*, from the University of Valencia (Spain), the Federal University of Goiás (Brazil), and the University of Fès (Morocco), ref. [14], presents the current trends and future directions of the use of mononuclear six-coordinate $Co^{II}$ spin crossover and single-ion magnet complexes with opto-, electro-, or chemo-active ligands for the preparation of multifunctional and multiresponsive magnetic devices for applications in molecular spintronics and quantum computing technologies. The review gives an illuminating insight in how these spin-crossover cobalt(II) molecular nanomagnets are suited for the preparation of devices on different supports, like metal molecular junctions, carbon nanomaterials, metal–organic frameworks, and metal–covalent organic frameworks in order to measure the single-molecule electron transport and quantum coherence properties, which are two major challenges in single-molecule spintronics (SMSs) and quantum information processing.

*F. S. Delgado, L. Cañadillas-Delgado, J. Rodríguez-Carvajal, Ó. Fabelo, and J. Pasán*, from the University of La Laguna (Spain) and Institut Laue-Langevin (France), ref. [15], investigate the magnetic structure of the molecular compound $[Mn(mal)(H_2O)]_n$ (mal = dianion of malonic acid) with unpolarized neutron diffraction. The combined neutron diffraction and magnetometry results leads to a comprehensive understanding of how structural and symmetry factors influence the magnetic properties of malonate-based manganese compounds.

The contribution by *S. Benmansour, C. Cerezo-Navarrete, and C. J. Gómez-García*, from the University of Valencia (Spain), ref. [16], reports on the synthesis and characterization of two isostructural layered MOFs with the asymmetric ligand chlorocyananilato and $Eu^{III}$ or $Dy^{III}$ as metal ions. Both compounds show a (4,4)-layered square structure with square cavities. The magnetic properties show the presence of a field-induced slow relaxation of the magnetization in the $Dy^{III}$ derivative at low temperatures that follows direct and Orbach relaxation mechanisms.

*N. Marino, F. Lloret, M. Julve, and G. De Munno*, from the University of Calabria (Italy) and the University of Valencia (Spain), ref. [17], show the simultaneous synthesis of up to three copper compounds: the $Cu^{II}$-containing compound $Cu^{II}(bpm)_3](I_3)(I)$ (**1**), the mixed valence $Cu^{II}/Cu^{I}$ compound $\{[Cu^{I}(I_3)Cu^{II}(I)(bpm)_2](I_3)\}_n$ (**2**), and a fully reduced $Cu^{I}$ compound: $\{[Cu^{I}_2(\mu\text{-}I)_2(bpm)]\}_n$ (**3**). Compound **1** consists of a rare tris(2,2′-bipyrimidine)copper(II) monomeric dication, charge balanced by both iodide and triiodide anions. Compound **2** consists of a regular alternating μ-bpm/di-μ-iodide copper(I) chain. Interestingly, the mixed valence $Cu^{II}/Cu^{I}$ compound **3** consists of a rare, regular alternating mixed-valent $Cu^{II}$-$Cu^{I}$ μ bpm copper chain, with a very weak antiferromagnetic coupling between well-separated paramagnetic $Cu^{II}$ ions.

The contribution by *J. W. Maciel, L. H. G. Kalinke, R. Rabelo, M. E. Alvarenga, F. Terra Martins, N. Moliner, and D. Cangussu*, from the Federal University of Goiás (Brazil) and University of Valencia (Spain), ref. [18], presents the synthesis and characterization of three isostructural lanthanide(III) compounds with the ligand *N*-(4-carboxyphenyl)oxamic acid ($H_3pcpa$) formulated as $\{[Ln_2(Hpcpa)_3(H_2O)_5]\}_n$, with Ln = $Dy^{III}$, $Ho^{III}$, and $Er^{III}$. Their structures consist of neutral zig–zag chains of $Ln^{III}$ ions, with $Hpcpa^{2-}$ ligands acting as bridges. The $Dy^{III}$ derivative shows a field-induced slow relaxation of the magnetization with a reciprocating thermal behavior below 5 K for H = 0.25 T, whereas the $Ho^{III}$ compound shows maxima of the magnetic entropy from 3 to 6 K for ΔH > 2 T.

*V. Jornet-Mollá, C. J. Gómez-García, M. J. Dolz-Lozano, and F. M. Romero*, from the University of Valencia (Spain), ref. [19], present the synthesis and characterization of a series of isostructural coordination polymers with a homoditopic picolinato ligand ($L_1$), formulated as $[(CH_3)_2NH_2][Ln(L_1)_2]\cdot H_2O\cdot CH_3COOH$ with Ln = Eu, Gd, Tb, Dy, and Ho. Their crystal structures show 3D-anionic lattices with triangular cavities. AC magnetic susceptibility measurements indicate that the Gd, Tb, and Dy derivatives show field-induced slow relaxation of the magnetization with the Gd and Dy derivatives following direct and Orbach relaxation mechanisms. The Dy derivative also shows slow relaxation of the magnetization in the absence of an external field. The Gd derivative is one of the very few reported $Gd^{III}$ compounds showing slow relaxation of its magnetization.

The contribution by *C. Pejo, S. Valiero, C. Rojas-Dotti, G. P. Guedes, J. Cano, M. A. Novak, R. Chiozzone, M. G. F. Vaz, and R. González*, from the University of the Republic (Uruguay), of the Fluminense Federal University (Brazil), and University of Valencia (Spain), ref. [20], reports on two tetranuclear complexes formulated as $[M_4\{(py)_2C(OH)O\}_4(O_2CPh)_4]$ with M = Co and Ni, whose structures show $[M_4O_4]$ cubane-like cores. The magnetic properties, analyzed with a two-J model including magnetic anisotropy, indicate the presence of dominant ferromagnetic interactions within the tetranuclear cores. DFT calculations support the experimental magnetic properties.

The list of the contributions of this Special Issue is completed by *G. Abellán-Dumont, J. M. Clemente-Juan, and C. Giménez-Saiz*, from the University of Valencia (Spain), ref. [21], who report the new polyoxometalate $[Co_7(OH)_6(H_2O)_2(CH_3COO)_4(PW_9O_{34})_2]^{13-}$, whose crystal structure consists of two trilacunary heptadentate B-α-$[PW_9O_{34}]^{9-}$ fragments encapsulating a hepta-cobalt di-cubane-like $\{Co^{II}{}_6Co^{III}O_8\}$ core. The magnetic properties of this compound have been fitted with an anisotropic exchange model in the low-temperature regime and show the presence of ferromagnetic interactions between $Co^{II}$ ions with Co–O–Co angles close to orthogonality, whereas the $Co^{II}$ ions connected through the central $Co^{III}$ ion present a weak antiferromagnetic coupling.

The list of contributions, a total of twenty-one, in this commemorative Special Issue that pays tribute to the key contributions of Paco and Miguel in the field of coordination chemistry and magnetochemistry clearly shows that Miguel and Paco have not only left a long list of key contributions in coordination chemistry and magnetic materials but also an endless list of collaborators and friends in many different countries. The positive answers to participating in this Special Issue were immediate and enthusiastic by all contributors, as can be read in their commemorative words for each contribution. They were all impressed by the scientific exchange and friendship of Paco and Miguel. We would like to thank all of them for their high-level contributions in this Special Issue devoted to these two distinguished scholars. We are sure that Paco will enjoy and Miguel would also have enjoyed reading all of them as much as we will miss Miguel.

**Author Contributions:** C.J.G.-G. and S.-E.S. contributed to the writing of this Editorial. All authors have read and agreed to the published version of the manuscript.

**Conflicts of Interest:** The authors declare no conflicts of interest.

## References

1. Verdaguer, M. Laudatio: Miguel Julve and Francisco Lloret, a Friendly Pair of Two Exceptional Coordination Chemists in Molecular Magnetism. *Magnetochemistry* **2025**, *11*, 6. [CrossRef]
2. Casanovas, B.; Vicente, R.; Font-Bardía, M.; El Fallah, M.S. Manganese(III) Compounds Derived from R-Salicylaldoxime and 9-Anthracenecarboxylate Ligands: A Study of Their Synthesis and Structural, Magnetic, and Luminescent Properties. *Magnetochemistry* **2024**, *10*, 55. [CrossRef]

3. Palii, A.; Zilberg, S.; Tsukerblat, B. Theoretical Hints to Optimize Energy Dissipation and Cell–Cell Response in Quantum Cellular Automata Based on Tetrameric and Bidimeric Cells. *Magnetochemistry* **2024**, *10*, 73. [CrossRef]
4. Löhr, J.; Font-Bardia, M.; Mayans, J.; Escuer, A. $[Mn^{III}_6Mn^{II}Na^{I}_2]$, $[Mn^{III}_3Mn^{II}Na^{I}]$, and $[Mn^{III}_3]$ Clusters Derived from Schiff Bases: Syntheses, Structures, and Magnetic Properties. *Magnetochemistry* **2024**, *10*, 76. [CrossRef]
5. Spinu, C.A.; Martins, D.O.; Mocanu, T.; Hillebrand, M.; Sutter, J.; Tuna, F.; Andruh, M. Two New 2p–3d Metal Complexes with a Nitronyl-Nitroxide Ligand Derived from o-Vanillin: Synthesis, Crystals Structures and Magnetic Properties. *Magnetochemistry* **2024**, *10*, 86. [CrossRef]
6. Świtlicka, A. Recent Insights into Magneto-Structural Properties of Co(II) Dicyanamide Coordination Compounds. *Magnetochemistry* **2024**, *10*, 90. [CrossRef]
7. Gil-Hernández, B.; Millan, S.; Gruber, I.; Janiak, C.; Gómez-García, C.J.; Sanchiz, J. Mesoxalate-Bridged Heptanuclear Copper(II) Complexes: Structure and Magnetic Properties. *Magnetochemistry* **2024**, *10*, 93. [CrossRef]
8. Oliveira, W.X.; Araújo, V.G.; Pinheiro, C.B.; Julve, M.; Pereira, C.L. Building Up a Hexacopper(II)-Pyrazolate/Oxamate Magnetic Complex with Rare Ethane-1,2-Dioxide ($–OCH_2CH_2O–$) as a Bridge Between Copper(II) Units. *Magnetochemistry* **2024**, *10*, 94. [CrossRef]
9. Tempesta, J.D.; Paiva, F.F.; Ferreira, L.A.; da Silva, R.M.; Botelho, L.D.; Rosa, I.M.; Candido, C.C.; Gomes, A.M.; Nunes, W.C.; Guedes, G.P. Metallacrown of $Ce^{III}Cu^{II}_5$: Synthesis, Structural Characterization and Insights for Nanoparticles. *Magnetochemistry* **2024**, *10*, 96. [CrossRef]
10. Escamilla, P.; Moliner, N.; Armentano, D.; Pardo, E.; Ferrando-Soria, J.; Grancha, T. Controlled Zn(II) to Co(II) Transmetalation in a Metal–Organic Framework Inducing Single-Ion Magnet Behavior. *Magnetochemistry* **2024**, *10*, 99. [CrossRef]
11. Colin, A.; Wang, Y.; Lambert, F.; Bridonneau, N.; Suaud, N.; Guillot, R.; Rivière, E.; Halime, Z.; Guihéry, N.; Ohkoshi, S. A Trinuclear Co(II) Complex Based on the Tris-Dioxolene Triphenylene Non-Innocent Bridge: Complementary Redox, Magnetic Behavior and Theoretical Calculations. *Magnetochemistry* **2024**, *10*, 102. [CrossRef]
12. da Cunha, T.T.; Honorato de Araujo-Neto, J.; Alvarenga, M.E.; Martins, F.T.; Pedroso, E.F.; Mariano, D.L.; Nunes, W.C.; Moliner, N.; Lloret, F.; Julve, M. Calixarene-like Lanthanide Single-Ion Magnets Based on $Nd^{III}$, $Gd^{III}$, $Tb^{III}$ and $Dy^{III}$ Oxamato Complexes. *Magnetochemistry* **2024**, *10*, 103. [CrossRef]
13. Diaz-Ortega, I.F.; Ye, Y.; Jover, J.; Ruiz, E.; Colacio, E.; Herrera, J.M. Engineering Mononuclear Ln(III) Complexes with a Pseudo-Macrocyclic Hexadentate $N_4O_2$ Schiff Base Ligand Exhibiting Slow Magnetic Relaxation. *Magnetochemistry* **2024**, *10*, 104. [CrossRef]
14. Rabelo, R.; Toma, L.M.; Bentama, A.; Stiriba, S.; Ruiz-García, R.; Cano, J. Exploring Spin-Crossover Cobalt(II) Single-Ion Magnets as Multifunctional and Multiresponsive Magnetic Devices: Advancements and Prospects in Molecular Spintronics and Quantum Computing Technologies. *Magnetochemistry* **2024**, *10*, 107. [CrossRef]
15. Delgado, F.S.; Cañadillas-Delgado, L.; Rodríguez-Carvajal, J.; Fabelo, Ó; Pasán, J. Uncovering the Mechanisms of Long-Range Magnetic Order in $[Mn(mal)(H_2O)]_n$: Insights from Microscopic and Macroscopic Magnetic Analysis. *Magnetochemistry* **2024**, *10*, 109. [CrossRef]
16. Benmansour, S.; Cerezo-Navarrete, C.; Gómez-García, C.J. Slow Relaxation of the Magnetisation in a Two-Dimensional Metal–Organic Framework with a Layered Square Lattice. *Magnetochemistry* **2024**, *11*, 1. [CrossRef]
17. Marino, N.; Lloret, F.; Julve, M.; De Munno, G. Chemical and Structural Versatility in the Copper/2, 2′-Bipyrimidine/Iodide System: A Regular Alternating Mixed-Valent Cu(II)-Cu(I) Chain Showing Unusually Similar Metal Coordination Environments. *Magnetochemistry* **2025**, *11*, 20. [CrossRef]
18. Maciel, J.W.; Kalinke, L.H.; Rabelo, R.; Alvarenga, M.E.; Martins, F.T.; Moliner, N.; Cangussu, D. Slow Relaxation of Magnetization and Magnetocaloric Effects in One-Dimensional Oxamato-Based Lanthanide(III) Coordination Polymers. *Magnetochemistry* **2025**, *11*, 23. [CrossRef]
19. Jornet-Mollá, V.; Gómez-García, C.J.; Dolz-Lozano, M.J.; Romero, F.M. Lanthanoid Coordination Polymers Based on Homoditopic Picolinate Ligands: Synthesis, Structure and Magnetic Properties. *Magnetochemistry* **2025**, *11*, 31. [CrossRef]
20. Pejo, C.; Valiero, S.; Rojas-Dotti, C.; Guedes, G.P.; Cano, J.; Novak, M.A.; Chiozzone, R.; Vaz, M.G.; González, R. Cobalt(II) and Nickel(II) Cubane $\{M_4O_4\}$ Complexes Derived from Di-2-pyridyl Ketone and Benzoate: Syntheses, Structure and Magnetic Properties. *Magnetochemistry* **2025**, *11*, 34. [CrossRef]
21. Abellán-Dumont, G.; Clemente-Juan, J.M.; Giménez-Saiz, C. A Heptacobalt(II/III) Dicubane Cluster with Polyoxometalate and Acetato Ligands: Synthesis, Crystal Structure, and Magnetic Properties. *Magnetochemistry* **2025**, *11*, 48. [CrossRef]

magnetochemistry

MDPI

Article

# Laudatio: Miguel Julve and Francisco Lloret, a Friendly Pair of Two Exceptional Coordination Chemists in Molecular Magnetism

Michel Verdaguer

Institut Parisien de Chimie Moléculaire (IPCM), CNRS Unit 8232, Sorbonne Université, Campus Pierre et Marie Curie, Box 229, 4 Place Jussieu, 75252 Paris Cedex 05, France; michel.verdaguer@upmc.fr; Tel.: +33-6-87-46-13-32

**Abstract:** This *laudatio* is dedicated to Professors Miguel Julve Olcina and Francisco Lloret Pastor on the occasion of their retirement in 2024. The first part deals with the scientific trajectory of the Coordination Chemistry team at the University of Valencia, within the Department of Inorganic Chemistry on the Burjassot campus and then in the Paterna Institute of Molecular Science. The second part relates some of the more salient results of the heritage left by our two colleagues in molecular magnetism, where they developed, in their own way, a rational approach to designing, creating and understanding a wealth of brand new systems from the simplest to Multifunctional Molecule-based Magnetic Materials. The robust and friendly links between our two colleagues are emphasized in the third part.

**Keywords:** coordination chemistry; molecular magnetism; exchange interaction; anisotropy; copper(II); Co(II); Re(IV); bimetallic ferrimagnetic chains; oxalate; oxamidate; oxamate; cyanide; complex-as-ligand; porous magnets; multifunctional materials; rational approach; friendship

## 1. Introduction

When our Editor-in-Chief, Carlos Gómez, proposed that I write an article for this Special Issue of *Magnetochemistry*, dedicated to the extraordinary scientific and human story of my two good friends, Francisco Lloret Pastor and Miguel Julve Olcina, I accepted with enthusiasm.

Now, a few months later, the terrible news of Miguel's recent loss has knocked down my enthusiasm. I am sorry to tell the reader that the lines below were the most difficult to conceive and to write in my life, fed at the same time by my admiration for the accomplished work, my affection for my two friends and an immense sadness.

The story is so endearing that I nevertheless decided to try to tell it. I was happy to publish two tribute papers in the past, one celebrating Professor Juan Faus, the initiator of the Coordination Chemistry lab in Valencia [1], and the other for the 65th anniversary of Professor Miguel Julve [2]. I dare to use quotations from these papers below.

I had the privilege of meeting Francisco Lloret Pastor and Miguel Julve Olcina to work with them and become their friend. I am happy to have been, during more than 40 years of exchange, the witness and, among others, an actor of this remarkable scientific and friendly saga. The present contribution is an attempt to present the scientific and human story of my two friends, often affectionately called Miguel and Paco in the text.

## 2. The Coordination Chemistry Team at University of Valencia

We undertake here a journey of almost 50 years of scientific activity, located half and half in the XX and XXI centuries, following the track of two friends, from modest Valen-

 https://doi.org/10.3390/magnetochemistry11020006

cian students at the beginning of the Spanish democratic transition to their international scientific recognition as European Academicians.

### 2.1. At the Beginning (1976)

To begin, we go back to before the beginning of the eighties. Let us imagine Professor Juan Faus at the University of Valencia, creating a group for coordination chemistry in 1976 to explore fundamental problems in solution by potentiometric and spectroscopic studies; see, e.g., the 1977 paper on violurate complexes of transition metals [3].

To locate the date, in 1976, Spain was one year after the death of Franco, at the beginning of the reign of Juan Carlos and the so-called Spanish democratic transition. "Spain knew a sudden burst of changes and hopes in many aspects of the social life. It is significant that the new group was born at this time" [1].

Among the first students of J. Faus were J.M. Moratal and our two young promising friends, Miguel and Paco. Miguel Julve was pursuing the thermodynamical study of violurato complexes of transition metals [4–6]. A little later, Francisco Lloret was undertaking analogous studies with other ligands, including salen [7–9]. Once, attending a conference of a famous French Professor of Solid-State Chemistry, Paul Hagenmüller, Miguel Julve was looking for a post-doctoral stay in a good coordination chemistry laboratory. He was advised to contact Professor Olivier Kahn at Orsay. Olivier was heading a small enthusiastic team, recently created (also in 1976), dedicated to molecular magnetism, a scientific discipline in which Olivier Kahn was one of the pioneers [10].

### 2.2. Orsay's Interlude

When Miguel arrived at Orsay, "Olivier Kahn decided that Miguel and me (member of the "long chain" group in the lab with J.J. Girerd, and speaking some broken Spanish) should work together to check experimentally the Heitler-London model of exchange recently coined by him and Briat, thanks to specially designed compounds. (Figure 1). We started with a very simple system: an oxalate bis-chelating ligand bridging two Cu(II), spin = 1/2 and diamine or triamine chelating terminal ligands" [2]. The reader can find in this article how the conditions of hard work, exchanges and discussions with Olivier and with the senior scientists of the lab, Irène Morgenstern-Badarau, Marie-France Charlot, Jean-Jacques Girerd, Yves Mathey, allowed the achievement of rapid progress. This initial common experience of research, combining experiments and theory, enriched by confident discussions with colleagues in a friendly atmosphere, has coined for Miguel and me our life's behaviour as researchers.

**Figure 1.** The Orsay team at the arrival of M. Julve in 1981. One can recognize some of the protagonists of our narrative, O. Kahn (centre), at his left I. Morgensten-Badarau, J. Zarembowitch, Dr. Journaux (looking back), the author. © A. Kahn-Harari.

After a few months, Miguel produced various systems, which revealed "textbook examples" to illustrate the relevance of Olivier's model and its efficiency in understanding and foreseeing the magnetic properties of simple dinuclear transition metal complexes and more sophisticated bimetallic ferrimagnetic chains, another hot topic at this time. As the oxalato chains were reluctant to form single crystals, we began EXAFS at LURE, the French synchrotron source at Orsay, which was continued later by I. Castro in Valencia (at ALBA in Bellaterra).

Several papers followed [11–16]. They appear important to me since they can be considered as the first seeds of the work on molecular magnetism in Valencia: the circular close link between experiment and theory, syntheses and crystal growth thanks to Miguel, who gained here his status of "green fingers" chemist, who was open to new techniques (magnetic measurements, EPR, EXAFS), close relations and discussions with crystallographers. One of the articles, a textbook example [15], is still among Miguel's most quoted works. The last reference introduces an important new window. After the appearance of μ-oxalato binuclear copper (II) complexes, μ-oxamato and μ-oxamido ones followed [16]. A tiny change, but big developments were to come in Burjassot.

At the end of Miguel's stay in Orsay, where he did not accept a permanent position as a researcher at the CNRS, we convinced Olivier that it could be good to pursue collaboration with the Valencia team. Indeed, a few years later, in October 1986, Francisco Lloret arrived, in collaboration with a young colleague, Dr. Y. Journaux and José Antonio Real in collaboration with Jacqueline Zarembowitch, under the leadership of Olivier. José Antonio was coming from the Inorganic Chemistry Department of the Faculty of Pharmacy in Valencia.

The stay and work of Paco, involved in the problem of magnetic exchange in the Mn-Cu chains, were very fruitful as demonstrated by [17–21]. New windows appeared: oxamidate derivatives, ordered bimetallic chains, molecular magnets, deeper investigation of the physics and magnetism of materials, where Paco excelled, and closer collaboration with outstanding physicists such as Jean-Pierre Renard.

José Antonio was involved in the Spin Crossover (SCO) problem, in particular combining SCO and magnetic exchange in Fe(II)-bypyrimidine dinuclear complexes, a topic proposed by Olivier. José Antonio's stay was the source of a new burst of results. It could be, and should be, told as another amazing story.

The three post-doctoral stays, as a whole, were at the origin of several molecular magnetic seeds carefully grown by the future creative work in Valencia.

### *2.3. The Coordination Chemistry Group in the Department of Inorganic Chemistry (DIC), University of Valencia (Up to 2000)*

Thanks to a new Spanish governmental program, talented young Spanish scientists benefitted from post-doctoral fellowships abroad. Hence, J.M. Moratal and E. García-España visited the groups of Professors Bertini and Paoletti in Florence (in bioinorganic and supramolecular chemistry), along their own scientific ways.

"Their reincorporation in the mother group in Valencia allowed the achievement of a solid background in solid-state coordination chemistry, in structural studies and in magnetism" [1]. "This national and European opening-up transformed deeply Faus's group and beyond, the Inorganic Chemistry Department. It brought new blood, modernized and diversified the themes, and opened the way to publications in European and American journals" [1].

"Furthermore, some of the systems investigated by Professor Faus underwent spin changes [complexes of Co(II) and Fe(II) with violurate and its alkyl derivatives] and led him to be interested in Molecular Magnetism. New equipment, a variable-temperature Faraday balance permitted to carry out this new research avenue at home, always keeping and reinforcing the collaboration with foreign teams. Along the years, the scientific partnership transformed into reciprocal esteem and friendship" [1].

The lab of our two friends was initially located on the campus of Burjassot, a pleasant green place close to the teaching theatres and the students. During this time, several of them, attracted by the teaching talents of J. Faus and his coworkers, Miguel, Paco and others, as well as the dynamics of the team, joined the team and contributed to its expansion.

It is not easy to give a brief overview of the scientific production of our two friends in the period 1990–2000. As for the quantitative scientific production in this period, Miguel published around 200 papers and Paco published more than 120.

It is more enlightening to look at the scientific orientation of the group, not forgetting former solution studies but clearly turning to molecular magnetism. There was, in my opinion, a triple movement: (i) some continuity with the synthesis of appropriate chemical systems to study exchange interaction between two metallic ions using new bridging and peripheral ligands; (ii) the study of more and more sophisticated systems, including polynuclear, 1D chains, 2D sheets, 3D materials; new metallic ions beyond $Cu^{II}$ and $Ni^{II}$ with a particular interest in $Co^{II}$; and a look at lanthanides, coupled at the same time with a thorough and deeper interpretation of the physical properties; (iii) the opening to national and international collaborations with several laboratories, in Spain and abroad, often through bilateral actions (Picasso with France, INTAS with Russia, and European TMR Networks).

In the following sections, I select a few examples of ligands, complexes used as ligands, combined experimental and theoretical studies and new transition metallic ions, enlarging the research scope and the width of collaborations, hence preparing projects to develop at ICMol.

#### 2.3.1. New Ligands, New Systems

As an extension of the post-doctoral stays, at a time of learning independence, the research continued on the exchange interaction through different bridges, including oxalate, oxamates, oxamidates, bipyrimidine, terephtalate, dithiosquarate, azide, hydroxo, cyanato, thio-cyanato, cyanido, carbonate.

I like the way Miguel answered the question "Why other bridges?" on the example of the simplest bridging ligands, two carbon atoms surrounded simply by oxygen, oxygen and nitrogen, sulfur, with their atomic orbitals ready for tuning exchange interaction. (Figure 2).

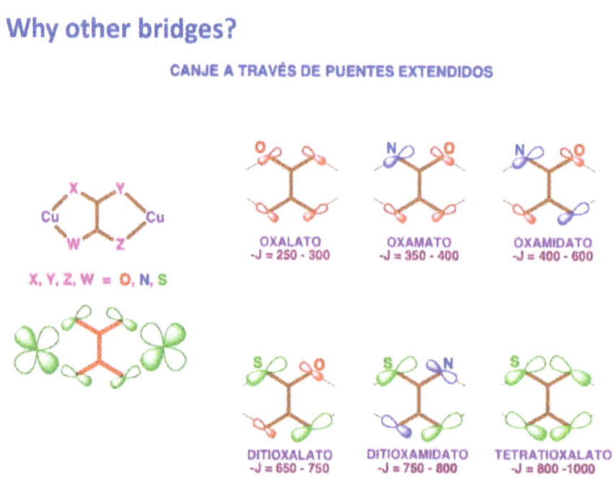


**Figure 2.** A few molecular bridges used in Valencia to study exchange interaction between two metallic ions. From [22], courtesy of M. Julve, adapted with permission from *Nouv. J. Chim.* **1985**, *9*, 325–334 [16]; *Inorg. Chem.* **1990**, *29*, 4500–4507 [23]; *J. Chem. Soc. Dalton Trans.* **1999**, 1669–1676 [24].

Paco and Miguel actively continued their exploration of new bridges and launched a series of PhD theses (I. Castro, Cu(II) complexes with Oxalate, Squarate and Croconate, 1991; R. Ruiz, Oxamidate, 1995; R. Lescouëzec, Cyanido, 2002) along with the theoretical interpretation of their magnetic properties (J. Cano, 1998).

*Oxalate*

As for oxalate $(C_2O_4)^{2-}$ derivatives, the syntheses were most often accompanied by crystallographic structures and theoretical interpretations of magnetism, up to a three-dimensional chiral cobalt network and magnet, introducing new perspectives in the group, with chirality and spin canting [23,25,26]. Oxalate was also elegantly used in networks with bipyrimidine (see below).

*Oxamate and Oxamidate*

When looking at the structure of oxamate and oxamidate in Figure 2, one discovers that nitrogen can welcome an organic tail R (N–R), offering many remarkable possibilities, structures, properties and ... publications. "Oxamidato complexes", for example, were treated in six papers and provided original structures of different dimensionalities, a combination of transition ions and lanthanides. We quote only a few [16,20,27–30].

The one-dimensional bimetallic ferrimagnetic system CuMn(obbz) [17] deserves a special comment [obbz = oxamido bis(benzoate)]. It is a nice example of a complex used as a ligand, first published at Orsay by Francisco and Olivier as a magnet at $T_c$ = 14 K, "exhibiting the highest ordering temperature reported to date" [20]. Thanks to endeavours in Valencia, the structure of its DMF solvate was solved after three years, giving the possibility of understanding the 3D magnetic ordering [29] (Figure 3).

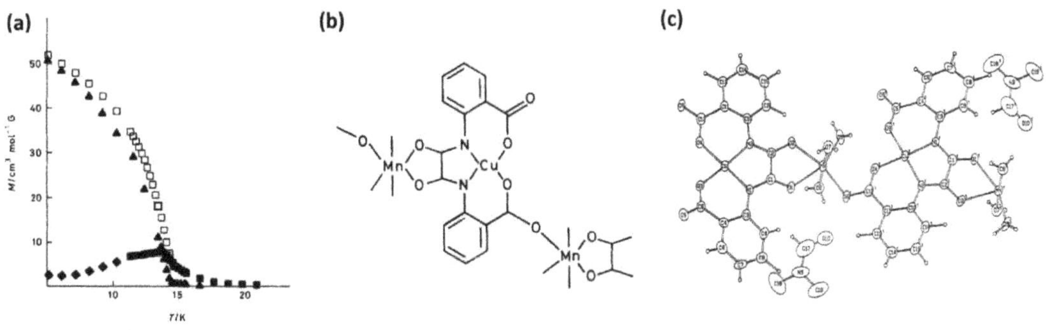


**Figure 3.** MnCu(obbz) Z (**a**) magnetization of MnCu(obbz)·$H_2O$ [17]; (**b**) postulated molecular structure [17]; (**c**) crystal structure of MnCu(obbz)$(H_2O)_3$·DMF [29]. Courtesy of F. Lloret, adapted from *J. Chem. Soc. Chem. Comm.* **1988**, *10*, 642–643 [17] and *Inorg. Chem.* **1993**, *32*, 27–31 [29].

The possibilities offered by oxamidates also appear on the original pentanuclear compound obtained with the so-called ligand apox ($H_2$apox = N,N′-bis(3-aminopropyl)oxamide) with Cu(II) and lanthanide appearing here for the first time with a ferromagnetic Gd(III)-Cu(II) interaction [30].

The first of the substantial reviews published by the group on the accomplished work was precisely on oxamidates, written by Rafael Ruiz (Rafa) after his PhD thesis [31], and was a splendid summary, gathering equilibria in solution—a pedestal of the team and a strength for synthesis—organic ligand tailored syntheses, rational synthetic strategies and coordination chemistry of oxamidates, from mononuclear complexes to bimetallic chains and 2D sheets.

*Bipyrimidine*

Among the bischelating ligands, bipyrimidine (bipym) became a new favourite, initiated with a collaboration with Giovanni de Munno at Arcavata di Rende University (Italy), his colleague Giuseppe Bruno (Univ. Messina) and then Donatella Armentano. Several papers were dedicated to 2,2′-bipyrimidine-bridged dinuclear complexes with different transition metallic ions, e.g., Cu(II) [32], Ni(II), Fe(II), Co(II) [33]. The chelating ability of bipyrimidine and its flexibility were combined with many other ligands, such as azido, cyanato, hydroxo and oxalate (Figure 4). The azido one appeared as a copper(II) 1D ferromagnetic chain [34] and a Mn(II) plane [35]. With cyanato, 1D, 2D and 3D systems were reached [36]. The oxalate ones provided systems with dimensionalities from 0 to 3 [37–39].

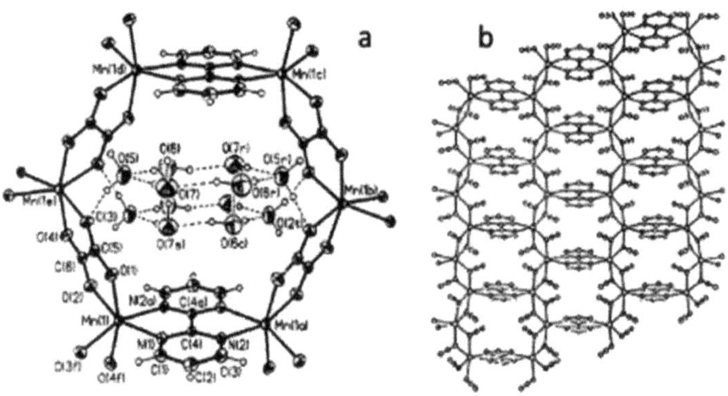


**Figure 4.** (**a**) Crystallographic structure of the Cu(II)(bipym)Ox·n$H_2O$ unit and the corresponding 2D network. Courtesy of F. Lloret, adapted with permission from *Angew. Chem. Int. Ed.* **1993**, *32*, 613–615. [37]. (**b**) 2D structure of the Cu(II)(bipym)$(N_3)_2$. Courtesy of F. Lloret.

Continuing their review of ligands, our friends explored squarate [40], croconate [41], carbonate [42,43] and malonate.

*Malonate*

Malonic acid is another flexible molecule that coordinates well with copper(II) and many other metallic ions. The malonate field was developed with the team of Professor Catalina Ruiz-Pérez, Univ. la Laguna, Canarias [44].

The study of metal–metal distances on the importance of the exchange was extended from diamines [45] to carboxylic acids [46].

To conclude on Burjassot's times, a few important domains still need to be presented.

The pioneering work on spin cross-over by J.A. Real, with M.C. Muñoz, Miguel and French groups, allowed them to find and report the beautiful Fe(II) interlocked system allowed by the catenane, 1,2-di-(4-pyridyl)-ethylene, (tvp, Scheme 1), which turned out to be the first porous metal–organic framework with spin cross-over behaviour (Figure 5) [47].

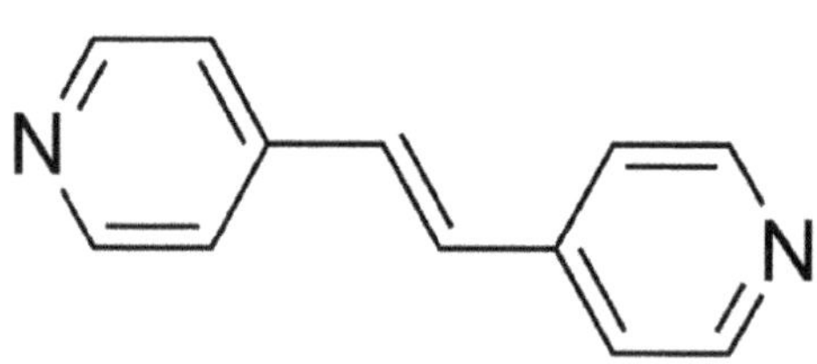


**Scheme 1.** Catenane 1,2-di-(4-pyridyl)-ethylene.

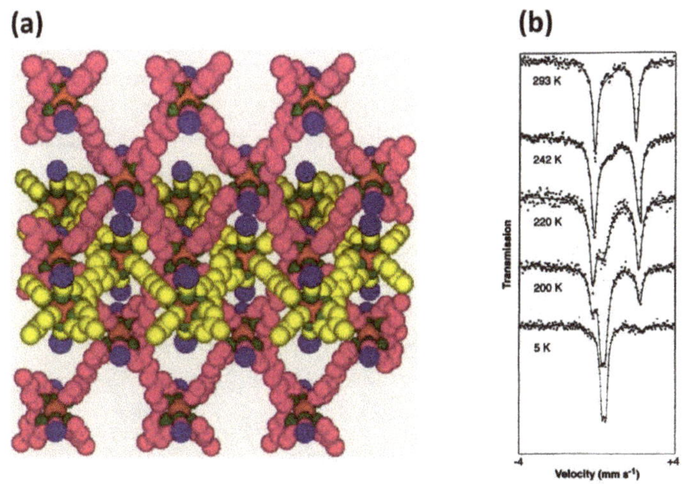


**Figure 5.** Spin cross-over. (**a**) Structure of $[Fe(tvp)_2(NCS)_2]{\cdot}CH_3OH$ (tvp = 1,2di-(4-pyridyl)-ethylene), in the [001] direction; (**b**) Mössbauer spectra from high spin to low spin. Courtesy of J.A. Real, adapted with permission from *Science* **1995**, *268*, 265–267 [47].

The 2D honeycomb-like cyanide bridged complex $[Ni(cyclam)]_3[Fe(CN)_6]_2{\cdot}nH_2O$ metamagnet opened a long-term collaboration between F. Lloret and Professor Colacio (Univ. Granada) (Figure 6) [48].

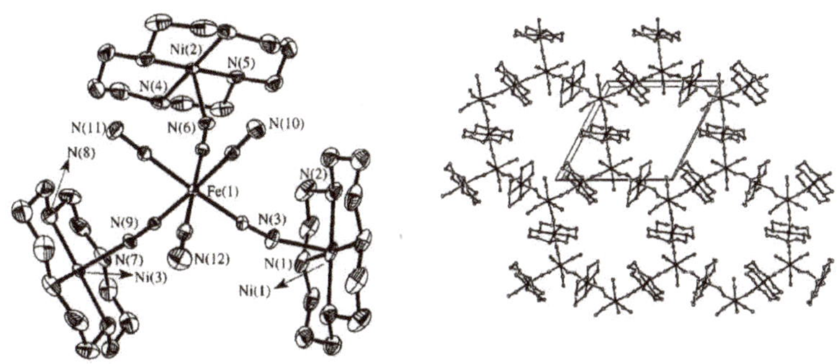


**Figure 6.** Crystallographic structure of the 2D honeycomb-like $[Ni(cyclam)_3[Fe(CN)_6]_2{\cdot}nH_2O$ (cyclam = 1,4,8,11-tetraazacyclodecane)], combination of $[Fe^{III}(CN)_6]$ units and Nicyclam. Courtesy of F. Lloret, adapted with permission from *Chem. Comm.* **1999**, *11*, 987–988 [48].

Juan Faus introduced the study of third-row transition metal ions, developing the chemistry of Re(IV), $5d^3$ ion, rare at this time, and its exchange with Cu(II), in collaboration with the group of Professor Kremer (Univ. Montevideo, Uruguay) [49].

2.3.2. New Collaborations

The preceding section demonstrates that in this phase of development, the group expanded a lot of its collaborations from Spain and France to Europe and beyond. (Figures 7 and 8) Several of these scientific links continued for many years with outstanding impact. They contributed to the scientific production in the ICMol years. Let us give a few prominent examples of close and sustainable collaborations—figures should be taken as rough approximations: Professor C. Ruiz-Pérez (145 articles from 1997 to 2020) among them malonates; Professor de Munno (118 items, from 1988 to 2022) among them bipyrimidine complexes. It even gave rise to a tribute by Miguel Julve. [50], Professor D. Armentano (71 items from 1999 to 2019); Dr. Journaux, who nourished in part the oxamidate work (66 items from 1989 to 2021); Professor J. Sletten (46 items from 1988 to 2011); Professor M. Andruh (36 items from 1998 to 2021); Professor A. Gleizes (13 from 1983 to 2018), including a 1996 INTAS project with Russia (Prof. N. Kuzmina). As for the author, 38 common articles either with Miguel, Paco, or both appeared from 1983 to 2018.

**Figure 7.** Working visit at the Department of Inorganic Chemistry, Burjassot 1995. From left to right: Pedro Gómez-Romero, Santiago Alvarez [25], Francisco Lloret, Ramdane Ouahes [51], Pascual Román [25]. © MV.

**Figure 8.** Spanish-French "Picasso's actions" times. After a working session at the department, on the way to a paella at the Albufera. From left to right: Juan Faus, Miguel Julve, Francisco Lloret, Michel Verdaguer (visitor) © MV.

*Crystallographers and Theoreticians*

Our two friends carefully cultivated their collaborations with crystallographers along the years. Their names appear in the references, but it is a pleasure, and a privilege, to quote some names, with some who are from my own laboratory: Alain Gleizes (ENSIACET Toulouse), Jorunn Sletten (Univ. Bergen), Catalina Ruiz-Pérez, Jorge Pasán (Univ. La Laguna), Yves Jeannin, Michèle Philoche-Levisalles, Claudette Bois, Jacqueline Vaissermann, Patrick Herson (CIM2 Paris), Lise Marie-Chamoreau, Kamal Boukekeur (IPCM Paris), Stoekli-Evans (Univ. Neuchatel), Xavier Solans, M. Font-Bardia (Univ. Barcelona) and M. Carmen Muñoz (Univ. Valencia), without forgetting neutron scientists Oscar Fabelo and Laura Cañadillas-Delgado (I.L.L.). The crystallographic structure, molecular and supramolecular, represents an essential piece in the work and steps of Paco and Miguel for at least two reasons: (i) to check how the changes in the synthetic process modify the structure and to orient further syntheses planned by Miguel and (ii) to obtain the most precise information on the structures in view of an exact interpretation of the physical properties, the subtle magnetic ones in particular, as requested by the sophisticated theoretical interpretations by Paco.

The second kind of privileged collaboration is with theoreticians, either at home (Joan Cano, precious in molecular orbitals and spin calculations and interpretations) or elsewhere, in particular with the Department of Inorganic Chemistry of Barcelona University, Professors Santiago Alvarez and Pere Alemany, who welcomed Joan as an ICREA researcher (Institució Catalana de Recerca I Estudis Avancats) during several years.

A final word on Professor Juan Faus, the founding father.

"Professor Faus was appointed as an Assistant Professor in 1968 at the Department of Inorganic Chemistry of the University of Valencia. He became Full Professor at the same Institution in 1980. He held this position until his retirement in 2017, after almost 50 years of continuous dedication to the University. Amazingly, he decided to continue to work in the old Inorganic Chemistry building of the university, even when the new and more functional buildings of the Institute of Molecular Chemistry (ICMol) became available. His passion for teaching induced him to stay closer to the students by remaining in the old building" [1] (Figure 9).

(a) (b) (c) (d)

**Figure 9.** From Burjassot (DQI) to Paterna (ICMol), continuity and changes. From left to right: (**a**) Professors J. Faus, (**b**) M. Julve, (**c**) F. Lloret (Coordination Chemistry Group), (**d**) Professor E. Coronado (creator and director of ICMol). © MV and © Alvarez.

## *2.4. Paterna, Molecular Science Institute (ICMol) from 2000*

"The ICMol was founded in 2000 to develop a competitive and high-quality research in materials science using a molecular approach. In fact, ICMol is the sole research centre in Spain exclusively focused on the molecular aspects of Nanoscience, with a special emphasis on the study of functional molecules and materials exhibiting useful magnetic, electrical or optical properties" [52]. It was created by a young, charismatic and ambitious Professor, Eugenio Coronado (Figure 9). A coordination chemist, he was studying EDTA bimetallic chains with M. Drillon in Strasbourg in the 1980s.

The group moved in 2005 to the premises of the Molecular Science Institute of Valencia, when the new building became available.

This was a new important impulsion and turn in the activity of the team. The coordination chemistry team at ICMol gathered the older members with our two full Professors, Miguel Julve and Francisco Lloret, who was just promoted (Figure 9), whereas Professors J.A. Real and M.C. Muñoz created their own group dedicated to spin cross-over. Along the years, Miguel and Paco became heads of the team, gathering colleagues such as Isabel Castro (Professor), Salah Eddine-Stiriba (Lecturer), Joan Cano and Rafael Ruiz (permanent researchers), hired researchers (Emilio Pardo, José Martínez-Lillo, Marta Viciano, Luminita Toma, Jesús Ferrando-Soria), one technician (Nicolás Moliner) and many doctoral students and Post Docs (Figure 10). It is frustrating to be unable to quote their names. They appear in the references and the Supplementary Materials.

**Figure 10.** 1st row, Consuelo Yuste, Julia Vallejo; 2nd row, from left to right, Isabel Castro, Paco Camallonga, Joan Cano, Jesús Ferrando-Soria, María Castellano, Luminita Toma, Thais Grancha, María Grancha, Rafael Ruiz, Emilio Pardo; 3rd row, Salah Stiriba, Miguel Julve, Marius Andruh (visitor), Francisco Lloret, Francisco Fortea-Pérez, Jeremy Chaignon © I. Castro.

To understand the evolution of the theme of the coordination chemistry team, it is perhaps appropriate to remind the reader of two papers published in *Nature* in the 1990s.

The first one, in 1993, by Sessoli and coworkers, opened the possibility to achieve molecular magnetic bistability without long range order, some kind of a revolution in molecular magnetism. [53] It provoked the exponential apparition of single-molecule magnets (SMMs), single-chain magnets (SCMs) and even single-ion magnets (SIMs). Ref. [53] highlights species very present in the production of Paco and Miguel. The second one in

1995, by Ferlay and coworkers, realized, through a rational approach, the initial dream of the protagonists, as shown in Figure 1, at Orsay (succeeding in obtaining a room temperature molecular magnet). At the same time, it made new endeavours and competition in this field less exciting [51]. One day, Paco even told me, "Michel, you killed the game". Of course, the game continued, along many other different lines, illustrated by the opus of our two friends. But, from then, isotropic exchange coexisted with anisotropy. And static magnetic measurements left a useful place for dynamic ones.

We can now travel through the tremendous work performed in ICMol thanks to the beautiful reviews published by our two colleagues and by their coworkers (Figure 11) authors); they often present an exhaustive review of the field and a perspective. Most often, the first author is the researcher, PhD student or Post Doc fellow who made the more substantial contribution. This is not insignificant since it shows the care of Miguel and Paco not only for training young scientists but also to lighten their work and to support their careers. Many of these young scientists are now full-time researchers, in Spain or abroad. The corresponding authors are frequently one of the two friends (starred * in the references) who accompanied the young researchers, giving them independence. Following a chronological order helps to demonstrate how the different concepts and topics developed over time.

**Figure 11.** A bird's-eye view of the reviews produced by the group from 1999, with first and corresponding authors. A readable version is available in the references. © MV.

The second review, after the one by Ruiz [31], was produced by R. Lescouëzec, a French student, joint PhD Paris-Valencia, dealing with cyanide derivatives. Single-chain magnets (SCMs) entered in the range of the team's realizations (Figure 12) [54].

One year later, L. Toma, Rumanian PhD student in Valencia, introduced in the game Ru(III), a 4d element, with a novel cyanido precursor, *trans*-$[Ru^{III}(acac)_2(CN)_2]^-$ [55].

Emilio Pardo, a Valencian PhD student, intended to demonstrate the general validity of the "complex-as-ligand" approach for the rational design of metallosupramolecular assemblies with increasing structural and magnetic complexity, using a novel family of aromatic polyoxamato (or apoxa) ligands (Figure 13) [56].

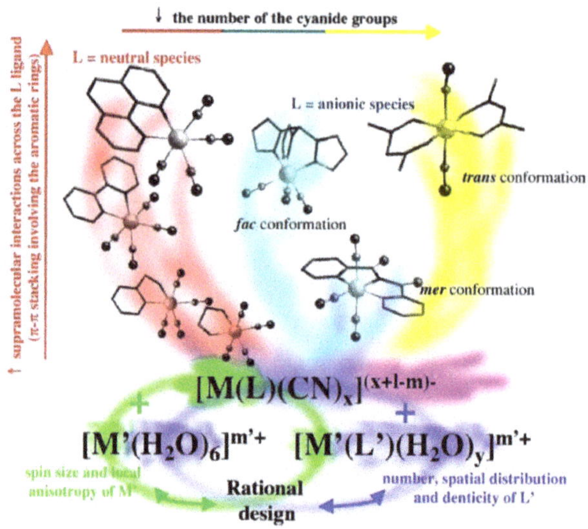


**Figure 12.** Miguel's dynamic colorful presentation of the complex-as-ligand cyanido precursors. From [22], courtesy of M. Julve. Adapted with permission from *Coord. Chem. Rev.* **2005**, *249*, 2691–2729 [54].

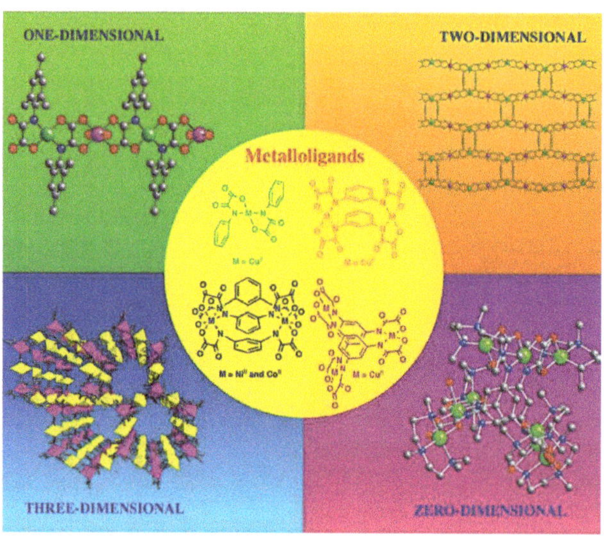


**Figure 13.** A graphical summary of Metal Organic Ligands (MOLs) from oxamate able to build materials of increasing dimensionality (D), from 0D (Metal Organic Clusters; MOCs) to 1D, 2D, 3D-Metal–Organic Polymers (MOPs). Courtesy of E. Pardo, adapted with permission from *Dalton Trans.* **2008**, 2780–2805 [56].

M.C. Dul, a French student, joint PhD Valencia-Paris, in a review dedicated to the late Professor Olivier Kahn, introduced a novel family of aromatic polyoxalamide using a supramolecular approach to multifunctional magnetic materials, with a number of international coworkers, including Brazilians colleagues. Metallacyclic complexes gave rise to metallacyclophanes or metallacryptands. (Figure 14) Devices began to be envisioned [57].

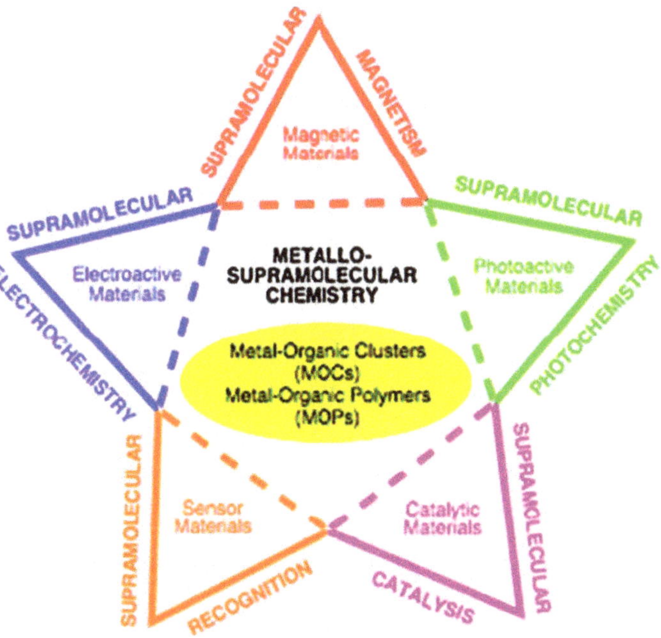


**Figure 14.** Metallosupramolecular chemistry and related molecular sciences. Courtesy of F. Lloret, adapted with permission from *Coord. Chem. Rev.* **2010**, *254*, 2281–2296 [57].

G. Marinescu and M. Andruh, a long-time coworker, proposed a new tecton, bis(oxalato)chromium (III), based on an inert $d^3$ Cr(III) ion, which allows many versatile syntheses [58].

Thais Grancha, a Spanish PhD student, enlarged the view in a feature article, introducing advances in multifunctional magnetic devices from oxamato coordination polymers (CPs). Exploratory studies reveal a wide variety of multifunctional magnetic materials (optically active chiral magnets, luminescent ones and dynamic porous magnets [59].

María Castellano, a Spanish PhD student, shows, in *Accounts of Chemical Research*, how to obtain magnetic devices, combining computation and experiments, from oxamato polymers and original copper(II) metallocyclophanes [60].

María continues the prospects offered in [60] by considering magnetic devices for molecular spintronics, an emerging discipline focusing on spin that is in constant evolution [61].

At the same time, J. Martínez-Lillo, a Spanish PhD student, focused on the use of a mononuclear complex of the anisotropic six-coordinate rhenium (IV), $5d^3$, as a metalloligand for the rational design of heterometallic magnetic compounds, a long-term project of J. Faus (Figure 15) [62].

Isabel Castro Bleda, one of the first members of the team, and Nadia Marino (Univ. Calabria) developed the chemistry of copper(II) pyrazolenophanes with peculiar attention to the ligand effects on the structures and the magnetism [63].

Jesús Ferrando-Soria, a Spanish PhD student, presented a general reflection of the team on the future of molecular magnetism. It is a wide and well-documented fresco on the history of molecular magnetism and some of its perspectives for coordination chemistry (no less than 87 pages, 115 references) (Figure 16) [64].

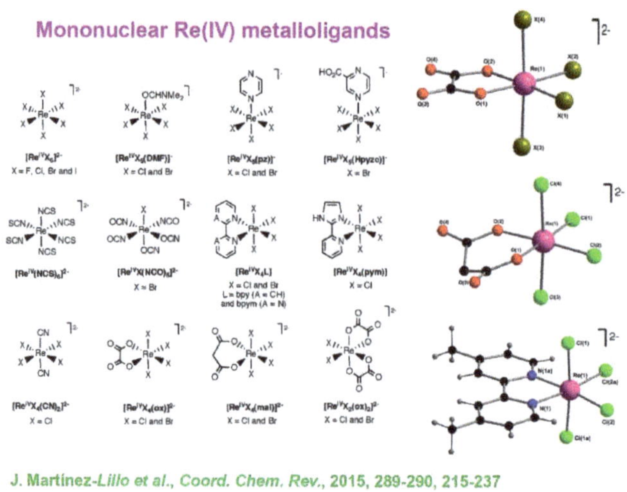


**Figure 15.** Metalloligands built from Rhenium (IV). From [22], courtesy of M. Julve. Adapted with permission from *Coord. Chem. Rev.* **2015**, *289–290*, 215–237 [62].

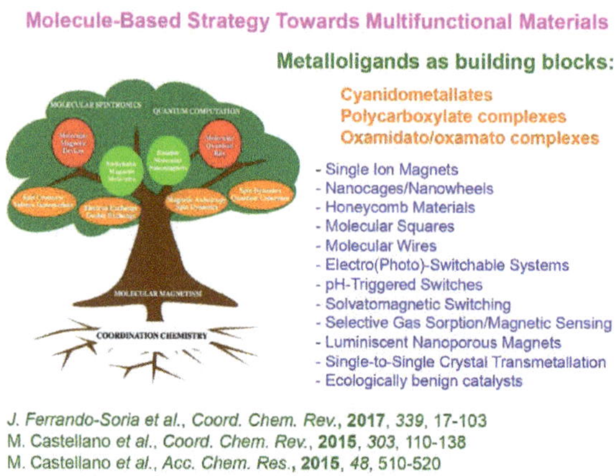


**Figure 16.** A general prospect on the future of molecular magnetism, expected fruits and suggested saps. At the roots, coordination chemistry. From [22], courtesy of M. Julve. Adapted with permission from *Coord. Chem. Rev.* **2017**, *339*, 17–103 [64]; *Coord. Chem. Rev.* **2015**, *303*, 110–138 [61]; *Acc. Chem. Res.* **2015**, *48*, 510–520 [60].

Miguel Julve collected in reference [65] the common results with Professors Greenfield and Doyle (Univ. Syracuse) and seized the opportunity to exhaustively describe the coordination chemistry of pyrophosphate, an anion with a strong biological impact.

Isabel Castro comes back in reference [66] on the many oxalate results with an original approach, considering oxalate not only as a magnetic coupler but as a polyatomic coordination centre, introducing the concepts of inverse polynuclear complexes (IPCs) and coordination polymers (ICPs).

Consuelo Yuste, a Spanish PhD student, proposes an ambitious perspective review. This perspective review, dedicated to G. de Munno, shows how the coordination chemistry approach to magnetic wires based on dinuclear copper(II) complexes with rodlike aromatic oligomers allows reaching a novel class of metallo-carbon nanostructures, possible models for single-molecule spintronics and quantum computing nanotechnologies. The quotation of famous authors and poets, road companions of Miguel and Joan, already present in preceding reviews (Molière, Pessoa, Reis-Prosas ...), is even crowned by the Raphael's Athens School famous fresco as a link between theory and experiment (Figure 17) [67].

**Figure 17.** Art and Science, Earth and Sky. A simple dinuclear copper(II) complex at the centre of experiments and theories in the molecular design of metallo-carbon nanostructures with polycyclic aromatic hydrocarbon spacers and prototypes of advanced magnetic nanodevices for single-molecule spintronics and quantum computing. Background: Part of Raphael's "The School of Athens". Courtesy J. Cano. Adapted with permission from *J. Coord. Chem.* **2022**, *75*, 2359–2383 [67].

To conclude this section, I hope that the reader is convinced that the work and leadership realized by Miguel and Francisco created among their coworkers a common culture for a rational approach to new molecular systems, combining (i) the fundamental aspects of molecular magnetism (exchange interaction, anisotropy, orbitals and spin densities' calculations and analyses); (ii) programmed synthesis; (iii) crystallography; (iv) physical measurements and their analysis; (v) and conclusions, applying them to new systems.

It is remarkable that many of the coworkers cited have now found their own way in science, securing stable positions within the team, at ICMol or in Spain, Europe and South America, with some becoming professors leading their own groups. This was a real pleasure and a source of deep satisfaction for Paco and Miguel.

## 3. The Scientific Work, an Extraordinary Harvest

### 3.1. Career

This first subsection presents the career of Miguel Julve and Francisco Lloret, their positions, publication impact and the following deserved awards. Figure 18 gathers this information, making apparent the extraordinary parallel between the two scientific lives. Beyond the individual competencies of the two friends, this tight connection is one of the sources of their common success.

**Miguel Julve Olcina**
**Born 1953**
ORCID 0000-0001-9006-8268

**Francesc Lloret Pastor**
**Born 1954**
ORCID 0000-0003-2959-0879

**CAREER**

1977 Bachelor University of Valencia
1978 Chemistry Degree University of Valencia
1981 PhD (Chemistry) University of Valencia
1981–1983 Post Doctoral Position
Paris-Sud University (Prof. Olivier Kahn)

**Positions**

1977–1982 Teaching assistant (ayudante)
1982–1983 CNRS Fellow Paris-Sud University
Department of Inorganic Chemistry, University of Valencia
1982–1985 Assistant Professor assistant (colaborador)
1985–1992 Associate Professor
1992– Full Professor University of Valencia
Department of Inorganic Chemistry & ICMol (2000–)

**Publications**

644 articles, 28269 citations
(average per item 44), h-index 85

**Awards**

1982 Extraordinary doctorate award , University of Valencia
2005 Best Senior Spanish Inorganic Chemist Prize
Spanish Royal Society of Chemistry (RSEQ)
2008 Catalan-Sabatier Prize, French Chemical Society (SCF)
2011 Election to the Academia Europaea (Chemical Science)

**Honorary Doctorate**

2014 Doctor Honoris Causa, Bucharest University

**CAREER**

1977 Bachelor University of Valencia
1979 Chemistry Degree University of Valencia
1982 PhD (Chemistry) University of Valencia
1985–1987 Post Doctoral Position
Paris-Sud University (Prof. Olivier Kahn)

**Positions**

1979–1984 Lecturer, Col. Univ. Castellón, Jaume I Univ.
1980–1984 Lecturer, CEU San Pablo, Card. Herera Univ.
1984–1986 Lecturer, Chemistry Fac., University of Valencia
1987 Associate Professor, University of Valencia
2000– Full Professor, University of Valencia
Department of Inorganic Chemistry & ICMol

**Publications**

644 articles, 29000 citations
average per item 44), h-index 85

**Awards**

2005 Best Senior Spanish Inorganic Chemist
Real Sociedad Española de Química (RSEQ)
2014 Election to Academia Europaea (Chemical Science)
2018 Catalan-Sabatier Prize, French Chemical Society
2022 Distinguished Career Award,
Real Sociedad Española de Química

**Honorary Doctorate**

2014 Doctor Honoris Causa, Bucharest University

**Figure 18.** Career, positions, publication impact and awards of Miguel Julve and Francisco Lloret. (from Supplementary Materials (Document S3)) © MV.

### 3.2. *Scientific Production*

In this second part, we try to overview and summarize the salient discoveries in the scientific production of Professors Miguel Julve Olcina and Francisco Lloret Pastor, who have an abundant crop of results (Figure 18). Together, they found their own original way by developing a rational synthetic approach to design, create, study and understand Molecule-based Magnetic Materials from the simplest systems to more complex, multifunctional ones towards molecular electronics and spintronics. Their complex-as-ligand or metalloligand approach was efficient and produced new tailor-made homo- and heterobimetallic assemblies with tunable magnetic and associated properties. One knows that the rational syntheses, fed by experience and theory, are worth the candle and are also full of unexpected surprises, often sources of new inspiration. Miguel and Paco revealed excellent in both exercises.

Unfortunately, the limited space obliges me to focus on a few particularly remarkable features, related to methodology, new systems or new phenomena.

#### 3.2.1. Research Methodology

Francisco Lloret and Miguel Julve worked following a constant rational approach: after having defined a goal to reach (compounds with requested properties or dimensionalities), they gathered the necessary ingredients for synthesis [not the simplest step, since they need to find the good metallic ion, oxidation state, spin, anisotropy, reactivity and the appropriate ligand to permit or to avoid coordination], the reaction's conditions (solvent, temperature, ...) and the full characterization of the products (analytical, structural, physical, magnetic, functional...) before performing careful analysis, interpretation and modelization of the results. The conclusions were made towards the end (and publication) or, to go further, were applied back to other syntheses. This was an apparently simple way to scientifically follow Descartes's Method Discourse, but was rigorously applied to thousands of experiments and theoretical models, intimately mixed, and systems of growing complexity, with organic and inorganic chemistries being intertwined. They always had an eye for defining future directions, from coordination chemistry to molecular electronics and spintronics. The reviews quoted in Section 2 [40,54,57,58,61–67] are nice witnesses of such steps.

Of course, as evocated before, such a rigorous step always includes an exploratory and playful side, enjoyed by Miguel and by the students, and also bad or beautiful surprises, exciting the creativity of the molecular chemists. Enthalpy H and order are faithful guides, but a touch of entropy S is always there and a welcome constitutive part of free enthalpy G.

The chemical strategy "complex-as-ligands" or "metalloligands" to design molecular, supramolecular and multifunctional materials is one of the efficient tools of this rational approach. More trendy expressions such as "molecular-programmed self-assembly methods that exploit the coordination preferences of metal ions and specifically tailored ligands" [51–60] are also used and translate the same rational and practical approach to the creation of materials.

#### 3.2.2. Continuing Previous Research

In an intermediate period, works on ligands such as malonate [44,68,69], dicyanamide [70], oxalate [71], oxamate [72] and cyanide were continued.

Special attention was paid by the team to ferromagnetism, either in an iron tetranuclear complex, the first example of a molecular species containing three peripheral low-spin iron(III) ions linked to a central high-spin iron(III) cation by single cyanide bridges [73], in a metallacyclophane by spin polarization [74]; a 2D sheet cobalt(II), in collaboration with A. Caneschi (Univ. Firenze) [75] or the ferromagnetically ordered 3D framework built from Co(II) and octacyanidotungstate(V) brick [W(V), $5d^1$, $S = 1/2$], a newcomer [76].

We do not forget the first and rare homometallic magnet ($T_c$ = 6 K), composed of Ni(II) chains with alternating $Ni_2$ units linked by hydroxo bridges, ferromagnetically coupled (S = 2), and Ni centres (S = 1) linked by fumarate bridges, antiferromagnetically coupled, resulting in the observed ferrimagnetism; a study with N.R. Chaudhuri (Univ. Kolkata) [77] on new cases of spin-frustrated systems was conducted in collaboration with J. Haasnoot (Univ. Leiden) [78].

The studies on the lanthanides, which began with gadolinium in [30], are pursued in [79].

#### 3.2.3. The Special Attention Paid to Cobalt(II) and to Anisotropy

The work on cobalt(II) began in the 1980s [3,6] and in the 1990s [75,80,81]. Cobalt was one of the favourite transition metals of the team [26,46,82], up until the PhD thesis of Renato Rabelo (2022). The total represents more than 85 articles, indeed less than the 189 ones with the friendly copper(II) with its simple spin $\frac{1}{2}$, but much more complex and exciting to treat and understand.

This interest was shared by the collaborating teams abroad, as shown by recent publications, where cobalt participates in multifunctional systems [83,84].

Reference [83] presents a new class of triply thermal-, magneto-, and electro-switchable spin cross-over cobalt(II)-PDI (PDI = pyridine-2,6-diimine) compounds with field-induced SMM behaviour, which possesses multiple metal- or ligand-based redox and spin states capable of reversible interconversion, whereas [84] presents a Co(II) triple helicate supramolecular nanomagnet, a common work with D. Cangussu (Univ. Goias).

Francisco Lloret applied himself strongly in establishing the theoretical background of Co(II) complexes. The difficulty in dealing with the magnetism and fitting the magnetic susceptibility data of this $d^7$ ion, presenting a remarkable spin–orbit coupling, led F. Lloret, J. Cano and the team of coworkers to analyze the influence of the different parameters involved in the magnetic susceptibility of six coordinated high-spin Co(II) complexes and "to propose an empirical expression to fit the magnetic susceptibility of polycrystalline samples of mononuclear Co(II) complexes with an axial distortion". The variable parameters are $\Delta$ (axial distortion), $\alpha$ (orbital reduction factor) and $\lambda$ (spin–orbit coupling). For polynuclear Co(II) complexes, with an exchange coupling J between effective spin S eff = 1/2, a perturbational approach allows to describe the magnetic susceptibility in the temperature range (2–300 K) as a function of J, $\Delta$, $\alpha$ and k. This approach is valid in the limit of weak magnetic coupling compared to spin–orbit coupling.

As in the Lines model, this expression is given under the form of a fictitious temperature-dependent Landé factor G(T). Figure 19 gives the energy diagrams of the splitting of the Co(II) $^4F$ and $^4T_1$ terms as a function of octahedral ligand field ($O_h$), spin–orbit and magnetic field. The $^4T_1$ term is further split in $^4A_{2g}$ and $^4E_1$ terms by an energy $\Delta$ when going from an $O_h$ symmetry to an elongated $D_{4h}$ one.

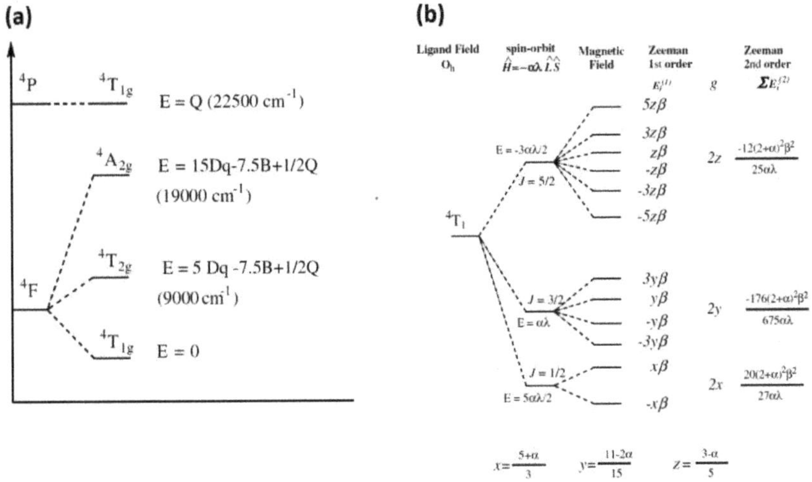


**Figure 19.** (**a**) Octahedral ligand field splitting of the $^4F$ and $^4P$ terms of the single-ion Co(II) with a $3d^7$ electronic configuration (Dq = B = 1000 cm$^{-1}$ and Q = $[225B^2 + 100Dq^2 + 180DqB]^{1/2}$). (**b**) Energy diagram showing the splitting of the ground $^4T_1$ term of a high-spin Co(II) ion in an octahedral field by spin–orbit coupling and Zeeman effects. Courtesy of F. Lloret. Adapted with permission from *Inorg. Chim. Acta* **2008**, *361*, 3432–3445 [82].

Reference [82] details the elaboration of the model and proposes equations to fit the susceptibility, such as

$$\chi_M T = \frac{N\beta^2}{4k}[G(T)]^2 \tag{1}$$

$$G(T) = \frac{\sum_{k=0}^{4}\left[\prod_{j=1}^{3}\left(\sum_{i=0}^{2} A_{i,j,k}x_j^i\right)T^k\right]}{\sum_{k=0}^{4}\left[\prod_{j=1}^{3}\left(\sum_{i=0}^{2} B_{i,j,k}x_j^i\right)T^k\right]} \tag{2}$$

and simple schemes for linear polynuclear Co(II) compounds (Scheme 2).

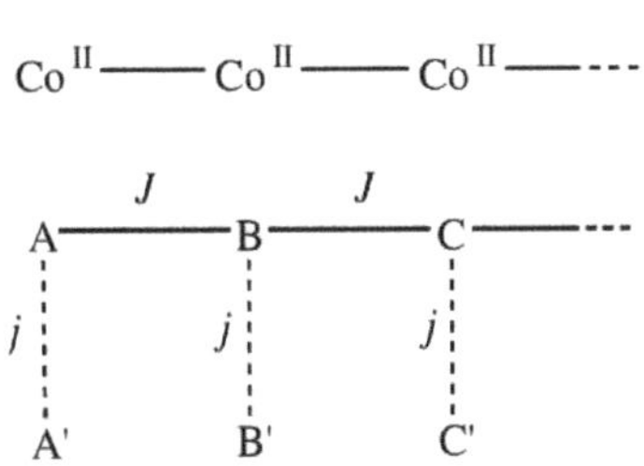


**Scheme 2.** A fictitious A-A′ dimer is used to treat orbital moment and spin-orbit coupling of Co(II) in a Spin Hamiltonian frame (example of a chain). Reproduced with permission from *Inorg. Chim. Acta* **2008**, *361*, 3432–3445 [82].

Reference [82] further discusses the model's limits and how the empirical expression accurately describes the powder magnetic susceptibility data for octahedral distorted high-spin Co(II) ion in many mono- or bimetallic systems. The model, going beyond the Lines model, has been well received. This article, dedicated to Dante Gatteschi, an international expert in anisotropy, was published in a Special Issue tribute to Dante. It is indeed the most quoted of Paco and a featured paper in the journal.

3.2.4. New Systems and New Phenomena

a. *The Oxalate Saga*

The work on oxalate really deserves the "saga" name [1–6,12]. Miguel constantly joked about his exhaustive work on the oxalate ligand, which "is a ligand that will always be on trend (ever green). After so many years of hard work on it, even my kidneys have specialized in its synthesis and the crystallization of calcium oxalate". In this way, he makes fun of his kidney stones, even if he has been following an anti-oxalate diet since his painful stay at Orsay's hospital in 1982 [2].

Oxalate (Ox) gave rise to around 90 publications, from [11,12] in 1983 to [66] in 2022, and 10 international conferences. It is present in many complex-as-ligand precursors (Figure 20) and provides systems of the three dimensionalities (0–3D).

The *fac*-$[Fe^{III}\{HB(pz)_3(CN)_3]^-$ anionic complex ($[HB(pz)_3]^-$ = hydrotris(1-pyrazolyl)borate) (Figure 20f) was first synthesized in Valencia. It is known as a member of the $HB(pz)_3$ or Tp class of scorpionates discovered by Trofimenko. Its steric and electronic properties and the ones of its substituted R-Tp make it suitable for building low-dimensional cyanido derivatives and commuting ones, and it was used often.

Miguel and Isabel Castro (a faithful and long-standing researcher on oxalates) recall some striking oxalate results of the team in [66], a review with more than 150 references. They furthermore describe them using the concepts of Inverse Coordination Chemistry (ICC), introduced by Haiduc [85]. Our usual oxalato polynuclear copper(II) systems then present oxalate as a polyatomic coordination centre for amine terminal ligands, forming inverse polynuclear complexes (IPCs) or multidimensional inverse coordination polymers (ICPs). The review then focuses on polypyrazole as a terminal ligand.

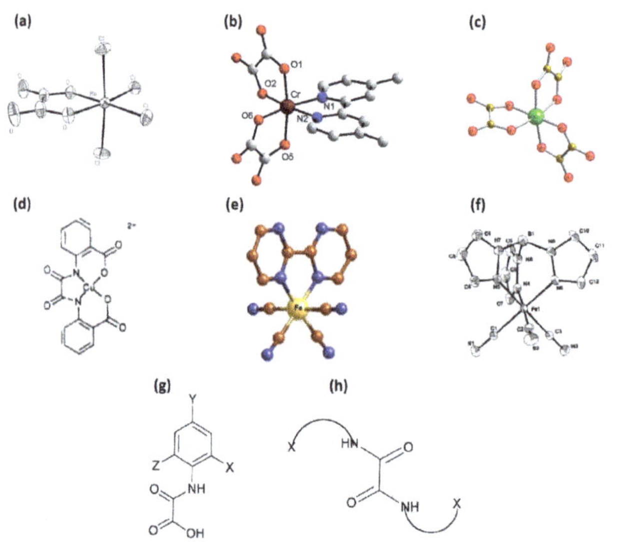


**Figure 20.** Complex used as ligands: (**a**–**c**) with 1, 2 or 3 oxalato ligands; (**d**) $Cu^{II}$(obbz); (**e**,**f**) with cyanido ligands, (**e**) $[Fe^{III}bipym(CN)_4{}^-]$; (**f**) *fac*-$[Fe^{III}\{HB(pz)_3\}(CN)_3]^-$; familiar ligands of the team: (**g**) oxamate; (**h**) oxamidate (X, substituent).

I give only one example of such an "ICP", the initial simple system [Cu–Oxalate–Cu], quoted in [66], in a completely different context. To explore all its facets, we decided 30 years ago to look at the role of the counterion around the dication $[(bipy)Cu(Ox)Cu(bipy)]^{2+}$.

We found that different anions ensured neutrality and deeply modified the structures and the magnetism, with the antiferromagnetic J coupling constant moving from $-10$ to $-380 cm^{-1}$ [2].

The first paper was submitted, rejected and forgotten for 25 years. Thanks to Miguel's determination, it was completely adapted to the new publication constraints on crystallographic data: new single crystals were grown by Miguel's green fingers, new crystallographic structures were determined by L M. Chamoreau and DFT calculations were performed by E. Ruiz. Paco welcomed me in his Valencia office to better fit the susceptibility data with his homemade programs! The paper was published roughly 30 years after the beginning of the work, realizing one of Miguel's oxalate dreams [86] (Figure 21). It reminds us of an important message: in synthesis, there is no "innocent" component. "Programming" all the details, even apparently the most trivial, is an essential step for success.

b. *Oxamate and Oxamidate, Another Fruitful Adventure*

As for the oxalate, the work on these ligands is enormous, original and fruitful, including no less than seven PhD theses (R. Ruiz 1998), E. Pardo (2006), M.C. Dul (2009), J. Ferrando-Soria (2012), M. Castellano (2013), F. Fortea-Pérez (2015), T. Grancha (2015), more than 70 articles, and 10 conferences in congresses. It started in 1985 [16] and is still going on today [84,87,88]. The first goal was to achieve molecular-based magnets mainly with the precious bridging qualities of oxamidates [16,21,27–29]. Then, it moved to more sophisticated materials of all dimensionalities, thanks to the versatile chemistry of flexible oxamates and the derived cyclophanes.

The giant synthetic work, which can be perceived in Figures 13 and 16, was accompanied by calculations and theoretical orbital and spin densities interpretations, in the intricated theoretical/experimental manner of our two friends. Figure 22 shows that previous work on spin polarization [74] allows us to complete and interpret how a simple change in the meta- or para-position on the aromatic spacer between two copper(II) oxamate moieties leads either to a singlet or to a triplet ground state.

More generally, the programmed change in the spacer between oxamate moieties in cyclophanes opened the way to porous materials. Many results on porous magnets take their origin from them (see list-item f. below) and multifunctional systems (see list-item g. below).

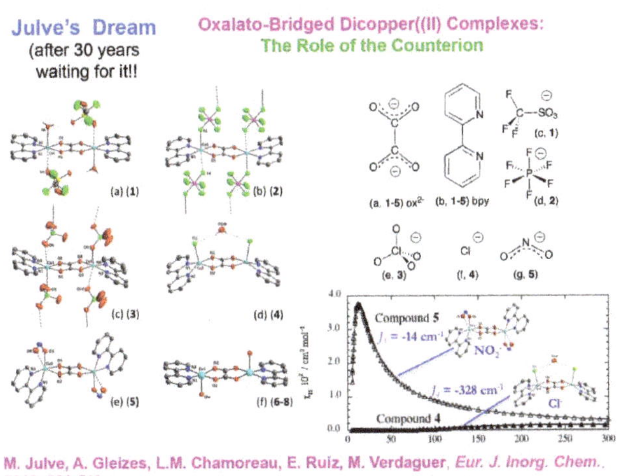


**Figure 21.** A long-waited rational resurrection miracle. Realization of one of Julve's oxalate dreams. From [22], adapted with permission from *Eur. J. Inorg. Chem.* **2018**, 509–516 [86].

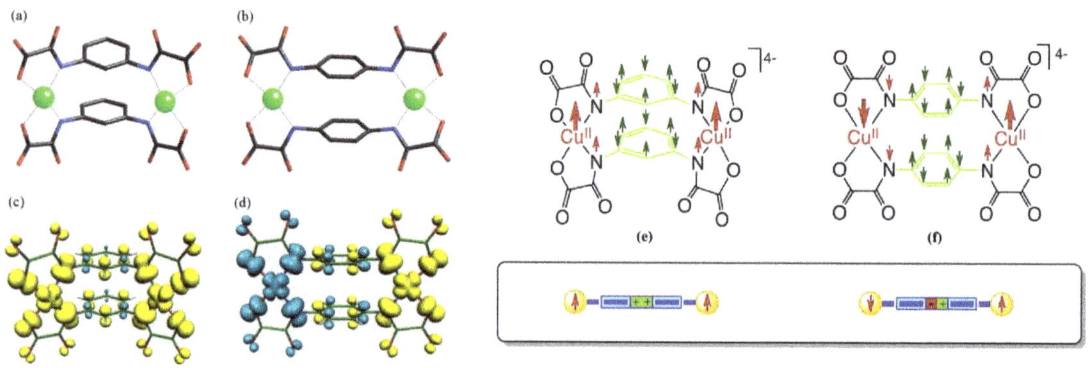

**Figure 22.** Dinuclear copper(II) cyclophane, meta- (**a**) and para (**b**). Ligand and metal are shown by sticks and balls. Spin density distributions in the ground state triplet (**c**) and singlet (**d**). Yellow and blue contours are the positive and negative spin densities. Spin polarization interpretation, triplet (**e**), singlet (**f**). Courtesy of J. Cano. Adapted with permission from *Coord. Chem. Rev.* **2010**, *254*, 2281–2296 [57]; *Coord. Chem. Rev.* **2015**, *303*, 110–138 [61].

c. *Single-Chain, Single-Molecule and Single-Ion Magnets (SCMs, SMMs, SIMs)*

Anisotropy of the metallic ion is essential in this field as predicted by Glauber. Cobalt(II) is often present. The energy barrier between the up and down magnetization ground states governs the long relaxation time for the magnetization reversal, tuned by quantum tunnelling.

Among the results obtained with cyanides, we highlight the seminal paper published on the two nanowires [{$Fe^{III}$(L)$(CN)_4$}$_2Co^{II}(H_2O)_2$]·$4H_2O$] (L = 2,2′-bipyridine) or 1,10-phenanthroline), with D. Gatteschi and W. Wernsdorfer, leaders in the field, in *Angewandte Chemie*, 2003 [89]. Figure 23 displays the structure and the magnetism.

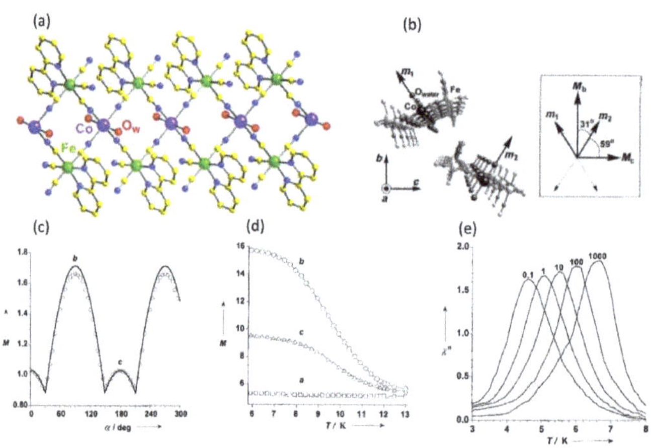

**Figure 23.** One of the first Single-Chain Magnet. (**a**) Structure of the $Fe_2Co$ chain; (**b**) double-chain magnetization vectors; (**c**) magnetization when rotating the field; (**d**) magnet behaviour at low T; (**e**) *ac* susceptibility. Adapted with permission from *Angew. Chem. Int. Ed.* **2003**, *42*, 13, 1483–1486 [89].

Similar chains were soon identified [90]. The beautiful review [54] replaced the name "nanowires" with the more trendy "single-chain magnets" and reviewed studies on stable cyanide-bearing six-coordinate complexes of formula $[M^{III}(L)(CN)_x]^{(x+l-m)-}$

(M = trivalent transition metal ion and L = polydentate blocking ligand) to build materials, including single-chain magnet (SCMs). Then, other building blocks were proposed, such as $[Fe^{III}(bipyrimidine)(CN)_4]^-$, to obtain new SCMs [91]. It is difficult for me to resist the pleasure of showing how revealing the magnetic data are when measured on a micro-SQUID (W. Wernsdorfer) and how spectacular their representation becomes (Figure 24).

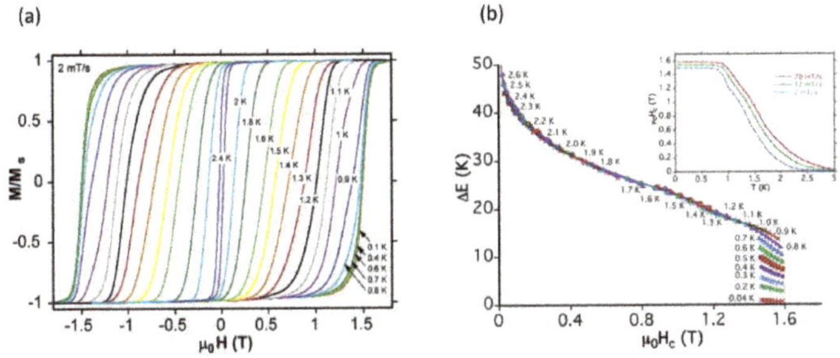

**Figure 24.** Micro-SQUID measurements on the$\{[Fe^{III}(bpym)(CN)_4]_2Co^{II}(H_2O)_2{\cdot}6H_2O]\}$ Single-Chain Magnet. (**a**) single crystal hysteresis loops at several temperatures and 0.002 T/s. (**b**) field dependence of the energy barrier and coercive field $H_c$ for the chain as a function of temperature. Adapted with permission from *J. Am. Chem. Soc.* **2006**, *12*, 4842–4853 [91].

The first oxamate Single-Chain Magnet, a bimetallic Co(II)-Cu(II) chain, with a large Ising anisotropy, was identified, and a new class of SCMs was discovered [92]. A review was written with other experts 8 years after the discovery of SCMs in Florence in a Special Issue in *Inorganic Chemistry* dedicated to molecular magnetism [93].

Single-molecule magnets were also tackled with ions such as Mn(III) or Co(II).

Mn(III) was studied with polynuclear complexes Mn6 and the famous Mn12 [53] in a Greek-Catalan collaboration with Paco [94]; also, a mononuclear complex was studied with J. Krzystek and W. Wernsdorfer. It can be named a single-ion magnet and is the first Mn(III) SIM [95].

The few other SIMs existing at this date were iron or cobalt complexes.

With Co(II), positive anisotropy was demonstrated in the first example of field-induced single-molecule magnet behaviour in a mononuclear six-coordinate Co(II) complex with a transverse anisotropy energy barrier [96].

d. *Coordination Chemistry of Heavier 4d and 5d Elements, Re(IV), Lanthanides*

The remarkable creative contribution of the group to coordination chemistry is also illustrated by the development of the chemistry of heavy transition elements of the second (4d) and third period (5d) and lanthanides (4f). Ruthenium is presented in [55].

The chemistry of rhenium ion Re(IV), which exploits the large magnetic anisotropy of this $5d^3$ ion, was initiated by J. Faus. [49], followed by further work from the group and José Martínez-Lillo. This difficult enterprise was reviewed by José and Joan [62]. Structures and magnetism are synthetically illustrated in Figures 15 and 25.

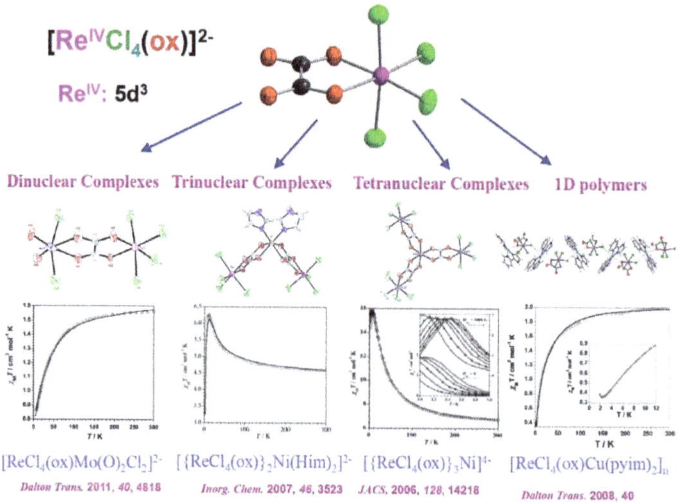


**Figure 25.** Above: mononuclear Re(IV) metalloligand. Middle: nuclearity and structure of the synthesized complexes; Below: magnetism. From [22], courtesy of M. Julve, J. Martínez-Lillo. Adapted with permission from *Dalton Trans.* **2011**, *40*, 4818–4820 [97]; *Inorg. Chem.* **2007**, *46*, 3523–3530 [98]; *J. Am. Chem. Soc.* **2006**, *128*, 14218–14219 [99]; *Dalton Trans.* **2008**, 40–43 [100].

Lanthanides were addressed on many occasions, including the thesis by Alejandro Alvarez (2017). It began with the common work with A. Gleizes and N. Kuzmina on d–f heterometallic complexes [101]. Several works dealt with the d–f exchange interaction [30,79]; familiar ligands, malonate and dicarboxylate [102,103], cyanide [79,104] or others [105,106]; the anisotropy in single-ion magnets (SIMs) [107]; or the luminescent properties (see list-item g).

Lanthanides were also sources of rarely observed trinuclear structures [108] and phenomena [109] carefully explored.

e. *Spin Cross-Over (SCO)*

The change in spin of a transition metal ion under different constraints (T, P, hν; ...) was and remains a hot topic in molecular magnetism. The domain was beautifully developed by José Antonio Real and M. Carmen Muñoz in another ICMol team and it was not specifically looked at closely by Paco and Miguel. The 1995 *Science* paper was already quoted [47]. The SCO phenomenon emerged when appropriate systems, mainly with cobalt(II), were developed, starting a long time ago with the violurate [81]. This work was conducted in collaboration with B. Machura (Univ. Silesia, Katowice] [110,111] and was even applied in a multifunctional context (electroswitching in a Co(II) SMM) [112].

f. *Porous Magnets*

This is another important issue in the scientific trajectory of Paco and Miguel. Their strategy to obtain porous magnets with predictable structures and tunable magnetic properties was similar to the strategy of Robson, Yaghi and Kahn but followed their own original approach, using, for example, ferromagnetically coupled, oxamato-based dinuclear metallacyclic complexes as ligands towards other ions to elaborate mixed metal–organic frameworks. The following figures (Figures 26–29), extracted from a presentation by Paco at a Spanish National School (in Spanish and English), illustrate the rational strategy described above, which was put into action in a very clear pedagogical manner together with the results [113]. When useful, titles have been translated to English in the captions.

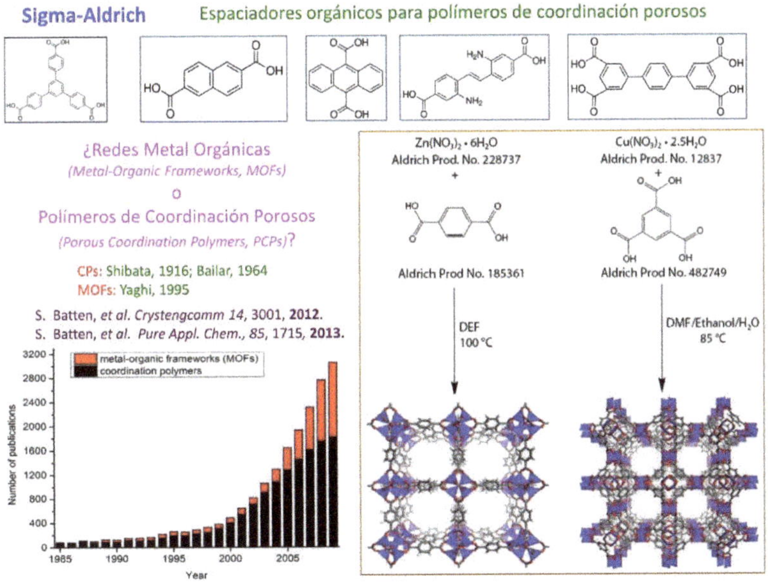


**Figure 26.** Spacers envisioned for synthesizing porous polymers; growing interest in porous systems; how to obtain various porous solids. From [113], courtesy of F. Lloret. Acknowledgement to *CrystEngComm* **2012**, *14*, 3001–3004 [114]; *Pure Appl. Chem.* **2013**, *85*, 8, 1715–1724 [115]. "Espaciadores orgánicos para polímeros de coordinación porosos" (English: "Organic spacers for porous coordination polymers").

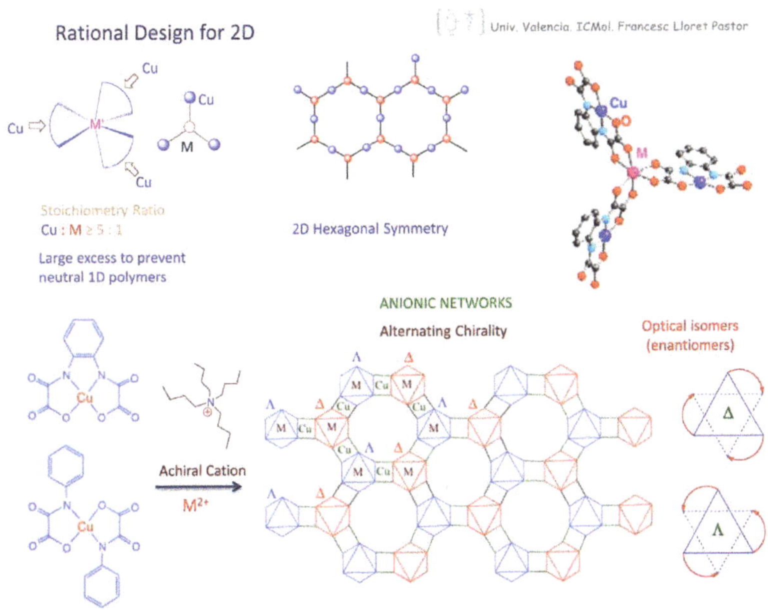


**Figure 27.** Rational design of achiral 2D networks from complexes used as ligands. From [113], courtesy of F. Lloret.

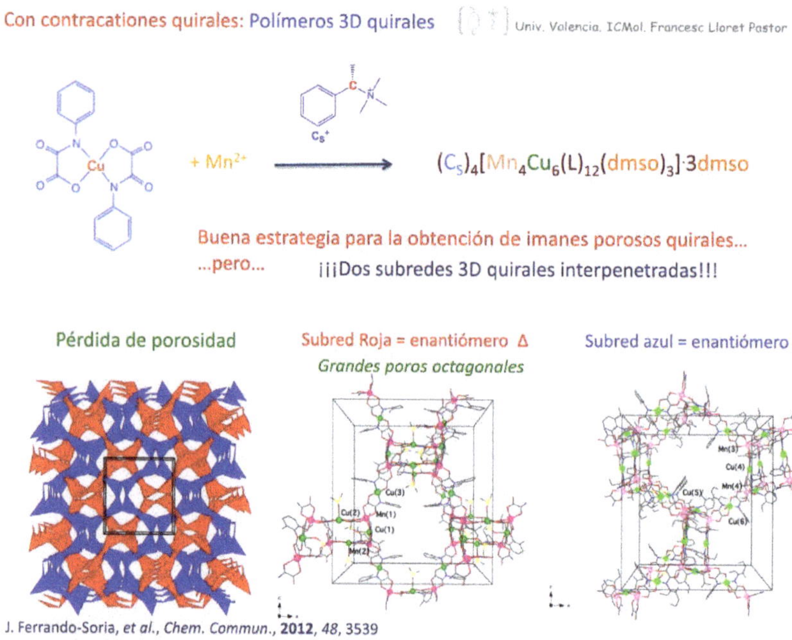


**Figure 28.** Rational design of chiral 3D networks and the obstacle of networks' interpenetration. From [113], courtesy of F. Lloret. Adapted with permission from *Chem. Commun.* **2012**, *48*, 3539–3541 [116]. "Con contracationes quirales: Polímeros 3D quirales" (English: "With chiral counterions, 3D chiral polymers").

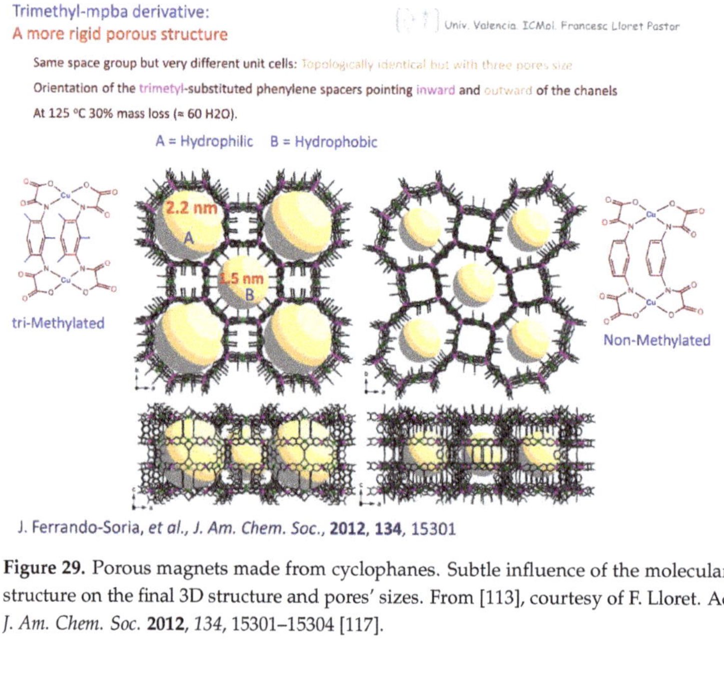


**Figure 29.** Porous magnets made from cyclophanes. Subtle influence of the molecular precursors' structure on the final 3D structure and pores' sizes. From [113], courtesy of F. Lloret. Adapted from *J. Am. Chem. Soc.* **2012**, *134*, 15301–15304 [117].

The porous magnets' realizations relied on previous works performed on metallocyclophanes, coordination polymers, MOFs and open frameworks [118,119].

The presence of pores permits the selective sorption or insertion of gases and molecules. It opens the way to other selected functions accompanying magnetism, solvatomagnetic switching [120], spin control [121] and selective gas and vapour sorption, including $CO_2$, which tunes the $T_c$ temperature of the 3D solid [117]. They can be used for applications in multifunctional materials, such as catalysts, or luminescent selective chemical sensing [122].

g. *Multifunctional Materials*

When the essentials of molecular magnetism (exchange, anisotropy, ...) are mastered, it is natural, with beautiful flexible building blocks in hand and extensive experience in building magnetic molecular and supramolecular solids, to try associating magnetic properties with one or more additional characteristics.

This is the way followed by Paco and Miguel and their team. Many of their funded projects (Valencian, Spanish, European) are devoted to this research, and examples of realized multifunctionality appear in the reviews reported above [56,59–62]. We need therefore to add only a few other references.

It is important to first highlights results with practical environmental applications, a sign that our two friends are conscious that science can help solve social challenges [123–125]. MOFs are able to welcome many guests: alkaline cations, spin cross-over complexes, big molecules [126]. Slow relaxation of the magnetization and luminescence can coexist in lanthanide complexes [107,127].

A more elaborated goal is to go beyond the coexistence of two functions by conceiving them so that they interact and one is able to control the other. Ferroicity is one of the targets, with the goal to one day be able to govern magnetism by electric means or conduction using a magnetic field. This implies the need for intricate experiments such as the study conducted on oxalate-based helix made of $(NH_4)_4[MnCr_2(ox)_6]_3{\cdot}4H_2O$ by Lloret, Pardo, and Train in a Spanish-French-Japanese collaboration (S.I. Ohkoshi, H. Tokoro, Univ. Tokyo) [128] (Figure 30) with neutron scientists [129].

**Figure 30.** Artist view of the Chiral Ferromagnetic Metal-Organic Quartz-like Framework, low-T magnet and High Proton Conductor [128]. Courtesy C. Train. Adapted from *J. Am. Chem. Soc.* **2011**, *133*, 15328–15331.

Multifunctionality also leads to further exploration in spintronics and quantum computing. This is performed in review [130], where R. Rabelo proposes proofs of concepts, relying on the hypotheses of Castellano, Lloret, and Julve [61] for the design of molecular magnetic switches for single-molecule spintronics and quantum computing. Metallacyclophanes presenting different possible chemo-, electro-, photo-, electro- and/or photoactive excitations are the privileged models. One example is given in Figure 31, with a clear and elegant design.

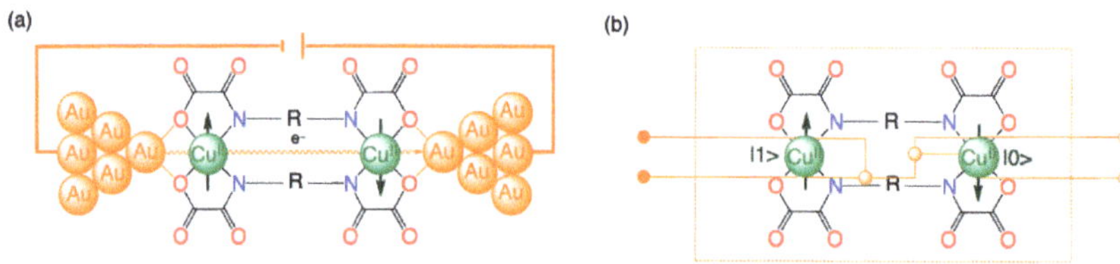


**Figure 31.** Copper(II) Metallacyclophanes as models for electrical (**a**) and logic circuits (**b**) of single-molecule spintronics and quantum computing [130]. From authors' private communication.

Students can find some other sources of inspiration in the two last references, where the work of Paco and Miguel is present. The first is the Dover edition of Olivier's book [10]. The second glorifies the electron in the molecule, electrons and molecules that Paco and Miguel elegantly showed how to master [131].

## 4. A Long Friendly Story

It is now a wonderful time to show that the harvest of the results evoked above is largely due to the marvellous friendship unifying Miguel Julve and Francisco Lloret since their students' years in Burjassot, their families and their talent for establishing a warm friendly atmosphere in the team, which was extended to external collaborations. I am lucky to have benefitted from this lesson of life. I simply add a few pictures as witnesses of those times (Figures 32–35).

**Figure 32.** The two friends and their gifts during the ICMol science day, 15 December 2023, on the occasion of their retirement. © Isabel Lloret, María José Sanz, ICMol.

**Figure 33.** ICMol tribute to Miguel and Paco on the occasion of their retirement. Left: Francisco Romero, head of the Department of Inorganic Chemistry (Univ. Valencia), Eugenio Coronado, head of ICMol. The retirees with their wives, daughters and friends. © Alvarez.

**Figure 34.** Relaxed atmosphere in the coordination chemistry group. From left to right, Giovanni de Munno (visitor), Joan Cano, Isabel Castro, Juan Faus, Francisco Lloret, Miguel Julve. © I. Castro.

**Figure 35.** After a lecture in the Department of Inorganic Chemistry, Burjassot 2009. From left to right, first row: Miguel Julve, José Antonio Real; second row: Francisco Lloret, Michel Verdaguer (visitor), Juan Faus, Santiago Alvarez (visitor), M. Carmen Muñoz. © MV.

Miguel and Paco were also united in receiving prestigious professional awards. Both members of Academia Europaea, they received Spanish and Spanish-French Awards and an Honoris Doctorate from Bucharest University (Figure 36).

(**a**) (**b**) (**c**)

**Figure 36.** (**a**,**b**) Presentation of the Spanish-French Catalan Sabatier Prize to Miguel and Paco in Paris; (**c**) Honoris Causa Doctorate for three Valencian chemists—150th anniversary of Bucharest University. © MV.

Along these hard-working years, our two friends enjoyed a nice and fulfilling family life. (Figure 37).

**Figure 37.** Family times for Paco and Miguel. "Happiness is not an option, it is a way of life". Left © Isabel Lloret; Right © MV.

The teaching qualities of our two friends have been told previously. It is time to also celebrate Miguel's activities in citizen awareness among young pupils or science festivals. Figure 38 shows Miguel in the lab or in action, happy to be working, alongside the cover of a book he just published (in Spanish) with two colleagues of the Department of Inorganic Chemistry in Valencia, a model and a very useful volume for children and teachers.

**Figure 38. Left,** Ana Isabel Terraes and Miguel Julve in the lab (© Anabel Terraes). **Centre,** Miguel performing experiments in a Valencia's Science Festivai (© Carlos Gómez). **Right,** Book just published (2024) for teachers and pupils: *Learning Chemistry: Experimenting*.

## 5. Conclusions

It is difficult to conclude. We still have so many things to tell, as we planned with Miguel. We are condemned now to continue without him.

"The consternation of all the members of the Faculty shows the personal affection of which he was worthy. Miguel was a good person. Around him, this university world was kinder, less complex, less tense" [132].

We knew of course that his health was fragile but he was so alive, so active that he appeared indestructible. We are not.

Nevertheless, it is important nevertheless to tell today that the work achieved by Francisco Lloret, Miguel Julve, Juan Faus and their coworkers of the coordination chemistry group at the University of Valencia has played a key role in the foundation of what is known nowadays as *Molecular Magnetism*. The two friends have contributed to effectively demonstrate, through their rigorous steps and their many "Premières", that coordination chemistry is an efficient tool in the design of advanced materials. They have reported exciting examples of Multifunctional Molecule-based Magnetic Materials (chiral magnets, porous magnets, photomagnetic systems, redox switches, among others). They have defined possible illuminating routes to go beyond. They have trained a countless number of young scientists, coming from different continents and now pursuing their careers.

I tried to emphasize that this was made possible in happy and peaceful times. We are facing today a world full of violence and wars. A few weeks ago, dramatic climate events in the Valencia region reminded us that nature and people need to be understood and respected.

The life-long endeavours of our two friends to develop science, better understand matter and world, spread knowledge and the scientific method and respect each other are a timely message that strengthens their outstanding scientific legacy. I wish that the present Special Issue contributes to relaying their example.

**Supplementary Materials:** The following supporting information can be downloaded at https://www.mdpi.com/article/10.3390/magnetochemistry11020006/s1, Document S1: Publications List of Professor M. Julve, private communication. Document S2: Publications List of Professor F. Lloret, private communication. Document S3: M. Verdaguer, Conference "Homenaje a los Profesores Miguel Julve Olcina y Francisco Lloret Pastor, a personal and scientific retrospective", ICMOl Scientific Day, 15 December 2023.

**Funding:** Written at retirement times, this work received no external funding.

**Data Availability Statement:** This is a Laudatio for Prof. Miguel Julve and Prof. Francisco Lloret, based solely on published data and sources related to the two honorees and their families. No new data were created in this paper, and data sharing is not applicable.

**Acknowledgments:** As a retiree and volunteer, I informed the head of my research team and of my institute who reviewed the work. This article covers 40 years of memories about the scientific trajectory of two close friends. I quote many of their coworkers and friends. Many provided information and pictures and are acknowledged. I would like to acknowledge particularly all the people and friends who made possible this article, including Francisco and our late friend Miguel; their wives Isabel and María José for documents, pictures and advice; their coworkers and friends, in particular Juan Faus, Joan Cano, Emilio Pardo, José Antonio Real and M. Carmen Muñoz; Rodrigue Lescouëzec, Santiago Alvarez and Carlos Gómez for valuable comments and careful reading of the initial manuscript; Isabel Castro, Anabel Terraes, Santiago Alvarez and the ICMol staffs for pictures; Francisco Romero and Eugenio Coronado for their welcome in Valencia; my own coworkers Sylvie

Ferlay, Talal Mallah, Cyrille Train, Valérie Marvaud and Anne Bleuzen. Their names and others appear in the references.

**Conflicts of Interest:** The author declares no conflicts of interest.

## References


1. Julve, M.; Lloret, F.; Verdaguer, M. A tribute to Professor Juan Faus Paya. *J. Coord. Chem.* **2018**, *71*, 585–589. [CrossRef]
2. Verdaguer, M. Miguel Julve, a creative chemist and scholar, a personal account. *Polyhedron* **2019**, *170*, 109–114. [CrossRef]
3. Moratal, J.; Faus, J. Violurato complexes of transition-metals. 3. Synthesis and characterization of tris(dihydrogenviolurato)cobalt(III)—Deprotonation equilibria in aqueous-solution. *Inorg. Chim. Acta* **1977**, *25*, L1–L3. [CrossRef]
4. Julve, M.; Moratal, J.; Faus, J. Spectrophotometric and Potentiometic Study of Fe(III) Complexes in DMSO. *Rev. Chim. Min.* **1980**, *17*, 138–145.
5. Faus, J.; Moratal, J.; Julve, M. Potentiometric Study of the Ni(II)-dihydrogenviolurate System in Dimethylsulphoxide (DMSO). *J. Inorg. Nucl. Chem.* **1980**, *42*, 1786–1788. [CrossRef]
6. Moratal, J.M.; Julve, M.; Faus, J. Violurato Complexes of Fe(II), Co(II) and Cu(II) in DMSO Solution. A Potentiometric Study. *Rev. Chim. Min.* **1982**, *19*, 72–79.
7. Lloret, F.; Moratal, J.; Faus, J. The formation of the 1-1 complex Fe(III)-$H_2$PI in DMSO solution. In *An. Quim. Serie B*; Real Sociedad Española de Física y Química: Madrid, Spain, 1981; pp. 202–205.
8. Lloret, F.; Moratal, J.; Faus, J. Solution chemistry of NN-ethylenebis (salicylideneiminato)iron(III). 1. Deprotonation equilibria and reversible decomposition in acid-medium of NN′-ethylenebis (salicylideneimine)—Stability constant of NN′-ethylenebis(salicylideneiminato) iron(III). *J. Chem. Soc. Dalton Trans.* **1983**, 1743–1748. [CrossRef]
9. Lloret, F.; Moratal, J.; Faus, J. Solution chemistry of NN-ethylenebis (salicylideneiminato)iron(III). 2. Formation equilibria of µ-oxo-bis (NN-ethylenebis (salicylideneiminato)iron(III)) and catecholato (NN′-ethylenebis (salicylideneiminato)) ferrate(III). *J. Chem. Soc. Dalton Trans.* **1983**, 1749–1753. [CrossRef]
10. Kahn, O. *Molecular Magnetism*; Dover: Mineola, NY, USA, 2021.
11. Julve, M.; Kahn, O. Synthesis, Magnetic Properties and EPR of $\mu$-Oxalato-tetrakis(acetylacetonato)diiron(III). *Inorg. Chim. Acta* **1983**, *76*, L39–L41. [CrossRef]
12. Julve, M.; Verdaguer, M.; Kahn, O.; Gleizes, A.; Philoche-Levisalles, M. Tunable exchange interaction in $\mu$-oxalato copper(II) binuclear complexes. *Inorg. Chem.* **1983**, *22*, 368–370. [CrossRef]
13. Verdaguer, M.; Julve, M.; Michalowicz, A.; Kahn, O. EXAFS structure and magnetic properties of a $Cu^{II}Ni^{II}$ $\mu$-oxalato mixed linear chain. *Inorg. Chem.* **1983**, *19*, 2624–2629. [CrossRef]
14. Julve, M.; Verdaguer, M.; Charlot, M.F.; Kahn, O. Interactions in Cu(II)-Cu(II), VO(IV)VO(IV), and Cu(II)-VO(IV) pairs through oxalato ligands. *Inorg. Chim. Acta* **1984**, *82*, 5–12. [CrossRef]
15. Julve, M.; Verdaguer, M.; Gleizes, A.; Philoche-Levisalles, M.; Kahn, O. Design of $\mu$-oxalato copper(II) binuclear complexes exhibiting expected magnetic properties. *Inorg. Chem.* **1984**, *23*, 3808–3818. [CrossRef]
16. Verdaguer, M.; Kahn, O.; Julve, M.; Gleizes, A. A way to increase the exchange interaction through multiatomic bridges: Magnetic properties of $\mu$-oxalato, $\mu$-oxamato, $\mu$-oxamido binuclear copper (II) complexes; crystal structure of $\mu$-oxamido (2-) (O,N:O′,N′) bis-[(N,N,N′,N′ tetramethylethanediamine N,N′) copper (II)] hexafluorophosphate. *Nouv. J. Chim.* **1985**, *9*, 325–334.
17. Lloret, F.; Nakatani, K.; Journaux, Y.; Kahn, O.; Pei, Y.; Renard, J.P. A novel molecular based ferromagnet ordering at 14K. *J. Chem. Soc. Chem. Commun.* **1988**, *10*, 642–643. [CrossRef]
18. Journaux, Y.; Vankoningsbruggen, P.; Lloret, F.; Nakatani, K.; Pei, Y.; Kahn, O.; Renard, J.P. Chemistry and Physics of molecular based ferromagnets. *J. Physique* **1988**, *49*, 851–852. [CrossRef]
19. Lloret, F.; Julve, M.; Mollar, M.; Castro, I.; Latorre, J.; Faus, J.; Solans, X.; Morgenstern-Badarau, I. Formation in solution, synthesis, and electrochemical study of oxalato complexes of *N,N′*ethylenebis(salicylideneiminato)-Chromium(III) and *N,N′*ethylenebis(salicylidene iminato)-Iron(III)—Crystal-structures of piperidinium [N,N′-ethylenebis(salicylidene iminato).(oxalato-$O^1O^2$)Chromate(III) and Ferrate(III). *J. Chem. Soc. Dalton Trans.* **1989**, 729–738.
20. Nakatani, K.; Carriat, J.Y.; Journaux, Y.; Kahn, O.; Lloret, F.; Renard, J.P.; Pei, Y.; Sletten, J.; Verdaguer, M. Chemistry and physics of the novel molecular-based compound exhibiting a spontaneous magnetization below $T_c$ = 14 K, MnCu(obbz)·1$H_2O$ (obbz = oxamidobis(benzoato)). Comparison with the antiferromagnet MnCu(obbz)·5$H_2O$. Crystal structure and magnetic properties of NiCu(obbz)·6$H_2O$. *J. Am. Chem. Soc.* **1989**, *111*, 5739–5748.
21. Lloret, F.; Julve, M.; Faus, J.; Journaux, Y.; Philoche-Levisalles, M.; Jeannin, Y. Oxamidato complexes. 1. A study of the formation of complexes between copper(II) and $H_2$Pmox ($H_2$Pmox = *N,N′*-bis(2-pyridylmethyl)oxamide)—Synthesis, crystal-structure, and magnetic-properties of the alternating-chain compound $[Cu_2(Pmox)(H_2Pmox)(H_2O)_2](NO_3)_2\cdot 8H_2O$. *Inorg. Chem.* **1989**, *28*, 3702–3706.

22. Julve, M. Tailor-made molecules as metalloligands in designing metal assemblies with interesting spin topologies. In Proceedings of the Second World Chemistry Congress and Exhibition, Valencia, Spain, 9–11 July 2018.
23. Alvarez, S.; Julve, M.; Verdaguer, M. Oxalato-bridged and related dinuclear copper(II) complexes: Theoretical analysis of their structures and magnetic coupling. *Inorg. Chem.* **1990**, *29*, 4500–4507. [CrossRef]
24. Cano, J.; Ruiz, E.; Alemany, P.; Lloret, F.; Alvarez, S. Theoretical study of the exchange coupling in copper(II) binuclear compounds with oxamidate and related polyatomic bridging ligands. *J. Chem. Soc. Dalton Trans.* **1999**, 1669–1676. [CrossRef]
25. Román, P.; Guzmán-Miralles, C.; Luque, A.; Beitia, J.I.; Cano, J.; Lloret, F.; Julve, M.; Alvarez, S. Influence of the Peripheral Ligand Atoms on the Exchange Interaction in Oxalato-Bridged Nickel(II) Complexes: An Orbital Model. Crystal Structures and Magnetic Properties of $(H_3dien)_2[Ni_2(ox)_5]\cdot 12H_2O$ and $[Ni_2(dien)_2(H_2O)_2(ox)]Cl_2$. *Inorg. Chem.* **1996**, *35*, 3741–3751. [CrossRef]
26. Hernández-Molina, M.; Lloret, F.; Ruiz-Pérez, C.; Julve, M. Weak Ferromagnetism in Chiral 3-Dimensional Oxalato-Bridged Cobalt(II) Compounds. Crystal Structure of $[Co(bpy)_3][Co_2(ox)_3]ClO_4$. *Inorg. Chem.* **1998**, *37*, 4131–4135. [CrossRef] [PubMed]
27. Lloret, F.; Sletten, J.; Ruiz, R.; Julve, M.; Faus, J.; Verdaguer, M. Oxamidato complexes. 2. Copper(II) and Nickel(II) complexes with oxamide-*N,N′*-diacetic acid: Solution study, synthesis, crystal structures and magnetic properties. *Inorg. Chem.* **1992**, *31*, 778–784. [CrossRef]
28. Real, J.A.; Mollar, M.; Ruiz, R.; Faus, J.; Lloret, F.; Julve, M.; Philoche-Levisalles, M. Oxamidato Complexes. Part 6. Complex Formation between Copper(II) and *N,N′*-Bis2-(dimethylamino)ethyl]oxamide ($H_2$dmaeoxd). Preparation, Crystal Structure, and Magnetic Properties of $[Cu_2\ (dmaeoxd)(NCO)_2(H_2O)_2]$. *J. Chem. Soc. Dalton Trans.* **1993**, 1483–1488. [CrossRef]
29. Lloret, F.; Julve, M.; Ruiz, R.; Journaux, Y.; Nakatani, K.; Kahn, O.; Sletten, J. Crystal Structure and Magnetic Properties of $CuMn(obbz)(H_2O)_3\cdot DMF$ (obbz = *N,N′*-oxamidobis(benzoato) and dmf = dimethylformamide), a Precursor of the Molecular-Based Magnet $MnCu(obbz)\cdot H_2O$ with $T_c$ = 14 K. *Inorg. Chem.* **1993**, *32*, 27–31. [CrossRef]
30. Sanz, J.L.; Ruiz, R.; Gleizes, A.; Lloret, F.; Faus, J.; Julve, M.; Borrás-Almenar, J.J.; Journaux, Y. Crystal Structures and Magnetic Properties of Novel $[Ln^{III}Cu_4^{II}]$ (Ln = Gd, Dy, Ho) Pentanuclear Complexes: Topology and Ferromagnetic Interaction in the $Ln^{III}$–$Cu^{II}$ Pair. *Inorg. Chem.* **1996**, *35*, 7384–7393. [CrossRef]
31. Ruiz, R.; Faus, J.; Lloret, F.; Julve, M.; Journaux, Y. Coordination Chemistry of *N,N′*-Bis(coordinating group substituted)oxamides: A Rational Design of Nuclearity Tailored Polynuclear Complexes. *Coord. Chem. Rev.* **1999**, *195*, 1069–1117. [CrossRef]
32. Julve, M.; De Munno, G.; Bruno, G.; Verdaguer, M. Synthesis, structure and magnetic properties of $\mu$-2,2′-bipyrimidine-*N, N′*, *N″,N‴* copper(II) complexes. *Inorg. Chem.* **1988**, *27*, 3160–3165. [CrossRef]
33. De Munno, G.; Julve, M.; Lloret, F.; Faus, J.; Caneschi, A. 2,2′-Bipyrimidine (Bipym)-Bridged Dinuclear Complexes. Part 4. Synthesis, Crystal Structure and Magnetic Properties of $[Co_2(H_2O)_8(bipym)[NO_3]_4$, $[Co_2(H_2O)_8\ (bipym)][SO_4]_2\cdot 2H_2O$ and $[Co_2(bipym)_3(NCS)_4]$. *J. Chem. Soc. Dalton Trans.* **1994**, 1175–1183. [CrossRef]
34. Thompson, L.K.; Tandon, S.S.; Lloret, F.; Cano, J.; Julve, M. An Azide-Bridged Copper(II) Ferromagnetic Chain Compound Exhibiting Metamagnetic Behavior. *Inorg. Chem.* **1997**, *36*, 3301–3306. [CrossRef] [PubMed]
35. De Munno, G.; Julve, M.; Viau, G.; Lloret, F.; Faus, J.; Viterbo, D. Azido and 2,2′-Bipyrimidine Ligands as Useful Tools in Designing Two- and Three-Dimensional Manganese(II) Networks. *Angew. Chem. Int. Ed. Engl.* **1996**, *35*, 1807–1810. [CrossRef]
36. Julve, M.; Verdaguer, M.; De Munno, G.; Real, J.A.; Bruno, G. Synthesis, crystal structure and magnetic properties of μ-bipyrimidine(cyanato) copper(II) and (thiocyanato) copper(II) complexes. *Inorg. Chem.* **1993**, *32*, 795–802. [CrossRef]
37. De Munno, G.; Julve, M.; Nicolò, F.; Lloret, F.; Faus, J.; Ruiz, R.; Sinn, E. 2,2′-Bipyrimidine(oxalato)copper(II) Complexes: From the Mononuclear Complex to the 2D Sheetlike Polymer. *Angew. Chem. Int. Ed.* **1993**, *32*, 613–615. [CrossRef]
38. De Munno, G.; Poerio, T.; Julve, M.; Lloret, F.; Faus, J.; Caneschi, A. Syntheses, Crystal Structures and Magnetic Properties of One-, Two- and Three-Dimensional 2,2′-Bipyrimidine-Containing Copper(II) Complexes. *J. Chem. Soc., Dalton Trans.* **1998**, 1679–1685. [CrossRef]
39. De Munno, G.; Armentano, D.; Julve, M.; Lloret, F.; Lescouëzec, R.; Faus, J. Título: Two-Dimensional Assembling of (2,2′-Bipyrimidine)bis(oxalato)chromate(III) Units Through Alkaline Cations. *Inorg. Chem.* **1999**, *38*, 2234–2237. [CrossRef] [PubMed]
40. Solans, X.; Aguiló, M.; Gleizes, A.; Faus, J.; Julve, M.; Verdaguer, M. Coordination modes of the squarate ligand: Synthesis and crystal structures of six copper(II) squarate complexes. *Inorg. Chem.* **1990**, *29*, 775–784. [CrossRef]
41. Castro, I.; Sletten, J.; Faus, J.; Julve, M.; Journaux, Y.; Lloret, F.; Alvarez, S. Exchange Interaction through Croconato Bridge: Synthesis, Crystal Structure and Magnetic Properties of ($\mu$-Croconato)bis{bis(2pyridylcarbonyl)amido}copper(II)]Trihydrate. *Inorg. Chem.* **1992**, *3*, 1889–1894. [CrossRef]
42. Sletten, J.; Hope, H.; Julve, M.; Kahn, O.; Verdaguer, M.; Dworkin, A. Phase transition and exchange interaction in ($\mu$-carbonato) bis-[[*N,N,N′,N″,N″*-pentaethyldiethylenetriamine] copper(II)] diperchlorate. *Inorg. Chem.* **1988**, *27*, 542–549. [CrossRef]
43. Sertucha, J.; Luque, A.; Castillo, O.; Román, P.; Lloret, F.; Julve, M. Ferromagnetic Coupling through a Carbonate Bridge in the Copper(II) Chain $[Cu(CO_3)(4\text{-}apy)_2\cdot H_2O$ (4-apy = 4-aminopyridine). *Inorg. Chem. Commun.* **1999**, *2*, 214–216. [CrossRef]

44. Ruiz-Pérez, C.; Sanchiz, J.; Hernández-Molina, M.; Lloret, F.; Julve, M. Ferromagnetism in Malonato-Bridged Copper(II) Complexes. Synthesis, Crystal Structures, and Magnetic Properties of $\{[Cu(H_2O)_3][Cu(mal)_2(H_2O)]\}_n$ and $\{[Cu(H_2O)_4]_2[Cu(mal)_2(H_2O)]\}[Cu(mal)_2(H_2O)_2]\{[Cu(H_2O)_4][Cu(mal)_2(H_2O)_2]\}$ ($H_2$mal = Malonic Acid). *Inorg. Chem.* **2000**, *39*, 1363–1370. [CrossRef] [PubMed]
45. Julve, M.; Verdaguer, M.; Faus, J.; Tinti, F.; Moratal, J.M.; Monge, A.; Gutiérrez-Puebla, E. Exchange Interaction through Extended Molecular Bridges: Magnetic Properties of *μ*-4,4′-Bipyridine and μ-Pyrazine Copper(II) Binuclear Complexes and Crystal Structures of (*μ*-4,4′-Bipyridine)bis[(diethylenetriamine)(perchlorato)copper(II). Perchlorate and Aquo(4,4′-Bipyridine)(diethylenetriamine)copper(II) Perchlorate. *Inorg. Chem.* **1987**, *2*, 3520–3527.
46. Cano, J.; De Munno, G.; Sanz, J.L.; Ruiz, R.; Faus, J.; Lloret, F.; Julve, M.; Caneschi, A. Ability of Terephthalate (ta) to Mediate Exchange Coupling in ta-Bridged Copper(II), Nickel(II), Cobalt(II) And Manganese(II) Dinuclear Complexes. *J. Chem. Soc. Dalton Trans.* **1997**, 1915–1923. [CrossRef]
47. Real, J.A.; Andrés, E.; Muñoz, M.C.; Julve, M.; Granier, T.; Bousseksou, A.; Varret, F. Spin Crossover in a Catenane Supramolecular System. *Science* **1995**, *268*, 265–267. [CrossRef]
48. Colacio, E.; Dominguez-Vera, J.M.; Ghazi, M.; Kivekas, R.; Lloret, F.; Moreno, J.M.; Stoeckli-Evans, H. A novel two-dimensional honeycomb-like bimetallic iron(III)-nickel(II) cyanide-bridged magnetic material $[Ni(cyclam)]_3[Fe(CN)_6]_2 \cdot nH_2O$ (cyclam=1,4,8,11-tetraazacyclodecane). *Chem. Commun.* **1999**, *11*, 987–988. [CrossRef]
49. Chiozzone, R.; González, R.; Kremer, C.; De Munno, G.; Cano, J.; Lloret, F.; Julve, M.; Faus, J. Synthesis, Crystal Structure, and Magnetic Properties of Tetraphenylarsonium Tetrachloro(oxalato)rhenate(IV) and Bis(2,2′bipyridine)tetrachloro(μ-oxalato)copper(II)rhenium(IV). *Inorg. Chem.* **1999**, *38*, 4745–4752. [CrossRef] [PubMed]
50. Julve, M. A tribute to Professor Giovanni De Munno. *J. Coord. Chem.* **2022**, *75*, 2278–2286. [CrossRef]
51. Ferlay, S.; Mallah, T.; Ouahès, R.; Veillet, P.; Verdaguer, M. A Room-Temperature organometallic magnet based on Prussian Blue. *Nature* **1995**, *378*, 701–703. [CrossRef]
52. ICMol Website. Available online: https://www.icmol.es/ (accessed on 10 January 2025).
53. Sessoli, R.; Gatteschi, D.; Caneschi Novak, M.A. Magnetic bistability in a metal-ion cluster. *Nature* **1993**, *365*, 141–143. [CrossRef]
54. Lescouëzec, R.; Toma, L.M.; Vaissermann, J.; Verdaguer, M.; Delgado, F.S.; Ruiz-Pérez, C.; Lloret, F.; Julve, M. Design of single chain magnets through cyanide-bearing six-coordinate complexes. *Coord. Chem. Rev.* **2005**, *249*, 2691–2729. [CrossRef]
55. Toma, L.M.; Toma, L.D.; Delgado, F.S.; Ruiz-Pérez, C.; Sletten, J.; Cano, J.; Clemente-Juan, J.M.; Lloret, F.; Julve, M. Trans-dicyanobis(acetylacetonato)ruthenate(III) as a precursor to build novel cyanide-bridged $Ru^{III}$–$M^{II}$ bimetallic compounds [M = Co and Ni]. *Coord. Chem. Rev.* **2006**, *250*, 2176–2193. [CrossRef]
56. Pardo, E.; Ruiz-Garcıa, R.; Cano, J.; Ottenwaelder, X.; Lescouëzec, R.; Journaux, Y.; Lloret, F.; Julve, M. Ligand design for multidimensional magnetic materials: A metallosupramolecular perspective. *Dalton Trans.* **2008**, 2780–2805. [CrossRef]
57. Dul, M.C.; Pardo, E.; Lescouëzec, R.; Journaux, Y.; Ferrando-Soria, J.; Ruiz-García, R.; Cano, J.; Julve, M.; Lloret, F.; Cangussu, D.; et al. Supramolecular Coordination Chemistry of Aromatic, Polyoxalamide Ligands: A Metallosupramolecular Approach Towards Functional Magnetic Materials. *Coord. Chem. Rev.* **2010**, *254*, 2281–2296. [CrossRef]
58. Marinescu, G.; Andruh, M.; Lloret, F.; Julve, M. Bis(oxalato)chromium(III) Complexes. Versatile Tectons in Designing Heterometallic Coordination Compounds. *Coord. Chem. Rev.* **2011**, *255*, 161–185. [CrossRef]
59. Grancha, T.; Ferrando-Soria, J.; Castellano, M.; Julve, M.; Pasán, J.; Armentano, D.; Pardo, E. Oxamato-based coordination polymers: Recent advances in multifunctional magnetic materials. *Chem. Commun.* **2014**, *50*, 7569–7585. [CrossRef] [PubMed]
60. Castellano, M.; Ruiz-García, R.; Cano, J.; Ferrando-Soria, J.; Pardo, E.; Fortea-Pérez, F.R.; Stiriba, S.E.; Julve, M.; Lloret, F. Dicopper(II) Metallacyclophanes as Multifunctional Magnetic Devices: A Joint Experimental and Computational Study. *Acc. Chem. Res.* **2015**, *48*, 510–520. [CrossRef]
61. Castellano, M.; Ruiz-García, R.; Cano, J.; Ferrando-Soria, J.; Pardo, E.; Fortea-Pérez, F.R.; Stiriba, S.E.; Barros, W.P.; Stumpf, H.O.; Cañadillas-Delgado, L.; et al. Metallosupramolecular approach toward multifunctional magnetic devices for molecular spintronics. *Coord. Chem. Rev.* **2015**, *303*, 110–138. [CrossRef]
62. Faus, J.; Lloret, F.; Julve, M. Towards multifunctional magnetic systems through molecular-programmed self-assembly of Re(IV) metalloligands. *Coord. Chem. Rev.* **2015**, *289–290*, 215–237.
63. Castro, I.; Barros, W.P.; Calatayud, M.L.; Lloret, F.; Marino, N.; De Munno, G.; Stumpf, H.O.; Ruiz-García, R.; Julve, M. Dicopper(II) pyrazolenophanes: Ligand effects on their structures and magnetic properties. *Coord. Chem. Rev.* **2016**, *315*, 135–152. [CrossRef]
64. Ferrando-Soria, J.; Vallejo, J.; Castellano, M.; Martínez-Lillo, J.; Pardo, E.; Cano, J.; Castro, I.; Lloret, F.; Ruiz-García, R.; Julve, M. Molecular magnetism, *quo vadis*? A historical perspective from a coordination chemistry viewpoint. *Coord. Chem. Rev.* **2017**, *339*, 17–103. [CrossRef]
65. Greenfield, T.J.; Julve, M.; Doyle, R.P. Exploring the biological, catalytic, and magnetic properties of transition metal coordination complexes incorporating pyrophosphate. *Coord. Chem. Rev.* **2019**, *384*, 37–64. [CrossRef]

66. Castro, I.; Calatayud, M.L.; Orts-Arroyo, M.; Marino, N.; De Munno, G.; Lloret, F.; Ruiz-García, R.; Julve, M. Oxalato as polyatomic coordination center and magnetic coupler in copper(II)-polypyrazole inverse polynuclear complexes and coordination polymers. *Coord. Chem. Rev.* **2022**, *471*, 214730. [CrossRef]
67. Yuste, C.; Castellano, M.; Ferrando-Soria, J.; Stiriba, S.-E.; Marino, N.; Julve, M.; Lloret, F.; Ruiz-García, R.; Cano, J. From computational design to the synthesis of molecular magnetic wires for single-molecule spintronics and quantum computing nanotechnologies. *J. Coord. Chem.* **2022**, *75*, 2359–2383. [CrossRef]
68. Ruiz-Pérez, C.; Hernández-Molina, M.; Lorenzo-Luis, P.A.; Lloret, F.; Cano, J.; Julve, M. Magnetic Coupling Through the Carbon Skeleton of Malonate in Two Polymorphs of {[Cu(bpy)($H_2O$)][Cu(bpy)(mal)($H_2O$)]}($ClO_4$)$_2$ ($H_2$mal = Malonic Acid; bpy = 2,2′-Bipyridine). *Inorg. Chem.* **2000**, *39*, 3845–3852. [CrossRef]
69. Rodríguez-Martín, Y.; Hernández-Molina, M.; Delgado, F.S.; Pasán, J.; Ruiz-Pérez, C.; Sanchiz, J.; Lloret, F.; Julve, M. Structural Versatility of the Malonate Ligand as a Tool for Crystal Engineering in the Design of Molecular Magnets. *CrystEngComm* **2002**, *4*, 522–535. [CrossRef]
70. Vangdal, B.; Carranza, J.; Lloret, F.; Julve, M.; Sletten, J. Syntheses, Crystal Structures and Magnetic Properties of Copper(II) Dicyanamide Complexes; Dinuclear, Chain and Ladder Compounds. *J. Chem. Soc. Dalton Trans.* **2002**, *4*, 566–574. [CrossRef]
71. Castillo, O.; Luque, A.; Román, P.; Lloret, F.; Julve, M. Syntheses, Crystal Structures, and Magnetic Properties of One-Dimensional Oxalated-Bridged Co(II), Ni(II), and Cu(II) Complexes with n-Aminopyridine (n = 2–4) as Terminal Ligand. *Inorg. Chem.* **2001**, *40*, 5526–5535. [CrossRef] [PubMed]
72. Cervera, B.; Sanz, J.L.; Ibanez, M.J.; Vila, G.; Lloret, F.; Julve, M.; Ruiz, R.; Ottenwaelder, X.; Aukauloo, A.; Poussereau, S.; et al. Stabilization of copper(III) complexes by substituted oxamate ligands. *J. Chem. Soc. Dalton Trans.* **1998**, 781–790. [CrossRef]
73. Lescouëzec, R.; Vaissermann, J.; Lloret, F.; Julve, M.; Verdaguer, M. Ferromagnetic Coupling between Low- and High-Spin Iron(III) Ions in the Tetranuclear Complex *fac*-{[$Fe^{III}${HB(pz)$_3$}(CN)$_2$(μ-CN)]$_3$ $Fe^{III}$($H_2O$)$_3$}.6$H_2O$ ([HB(pz)$_3$]$^-$ = Hydrotris(1-pyrazolyl)borate). *Inorg. Chem.* **2002**, *41*, 5943–5945. [CrossRef] [PubMed]
74. Fernandez, I.; Ruiz, R.; Faus, J.; Julve, M.; Lloret, F.; Cano, J.; Ottenwaelder, X.; Journaux, Y.; Munoz, M.C. Ferromagnetic coupling through spin polarization in a dinuclear copper(II) metallacyclophane. *Angew. Chem. Int. Ed.* **2001**, *40*, 3039–3042. [CrossRef]
75. Lloret, F.; De Munno, G.; Julve, M.; Cano, J.; Ruiz, R.; Caneschi, A. Spin polarization and ferromagnetism in two-dimensional Sheetlike cobalt(II) polymers: [Co($L_2$)($NCS)_2$] (L = pyrimidine or pyrazine). *Angew. Chem. Int. Ed.* **1998**, *37*, 135–138. [CrossRef]
76. Herrera, J.M.; Bleuzen, A.; Dromzée, Y.; Julve, M.; Lloret, F.; Verdaguer, M. Crystal Structures and Magnetic Properties of Two Octacyanotungstate(IV) and (V)-Cobalt(II) Three-Dimensional Bimetallic Frameworks. *Inorg. Chem.* **2003**, *42*, 7052–7059. [CrossRef] [PubMed]
77. Konar, S.; Mukherjee, P.S.; Zangrando, E.; Lloret, F.; Chaudhuri, N.R. A three-dimensional homometallic molecular ferrimagnet. *Angew. Chem. Int. Ed.* **2002**, *41*, 1561–1563. [CrossRef]
78. Ferrer, S.; Lloret, F.; Bertomeu, I.; Alzuet, G.; Borras, J.; García-Granda, S.; Liu-Gonzalez, M.; Haasnoot, J.G. Cyclic trinuclear and chain of cyclic trinuclear copper(II) complexes containing a pyramidal $Cu_3O$(H) core. Crystal structures and magnetic properties of [$Cu_3$($\mu_3$-OH)(aaat)$_3$($H_2O$)$_3$)]($NO_3$)$_2$)·$H_2O$] (aaat = 3-acetylamino-5-amino-1,2,4-triazolate) and {[Cu-$_3$($\mu_3$-OH)(aat)$_3$($\mu_3$-$SO_4$)]·6$H_2O$}$_n$ (aat = 3-acetylamino-1,2,4-triazolate): New cases of spin-frustrated systems. *Inorg. Chem.* **2002**, *41*, 5821–5830.
79. Figuerola, A.; Diaz, C.; Ribas, J.; Tangoulis, V.; Granell, J.; Lloret, F.; Mahia, J.; Maestro, M. Synthesis and characterization of heterodinuclear $Ln^{3+}$–$Fe^{3+}$ and $Ln^{3+}$–$Co^{3+}$ complexes, bridged by cyanide ligand ($Ln^{3+}$ = lanthanide ions). Nature of the magnetic interaction in the $Ln^{3+}$–$Fe^{3+}$ complexes. *Inorg. Chem.* **2003**, *42*, 641–649. [CrossRef] [PubMed]
80. Faus, J.; Julve, M.; Lloret, F.; Muñoz, M.C. Bis(dimethylviolurato)(phenanthroline)cobalt(II), a Low-Spin Octahedral Co(II) Complex. Crystal Structure of [Co(dmvi)$_2$phen]·2$CHCl_3$. *Inorg. Chem* **1993**, *32*, 2013–2017. [CrossRef]
81. Faus, J.; Julve, M.; Lloret, F.; Real, J.A.; Sletten, J. (Dimethylviolurato)bis(phenanthroline)cobalt(II), a Novel Spin-Crossover Octahedral Co(II) Complex. Synthesis, Crystal Structure, and Magnetic Properties of [Co(dmvi)(phen)$_2$]$ClO_4$·3$H_2O$. *Inorg. Chem.* **1994**, *33*, 5535–5540. [CrossRef]
82. Lloret, F.; Julve, M.; Cano, J.; Ruiz-García, R.; Pardo, E. Magnetic properties of six-coordinated high-spin cobalt(II) complexes: Theoretical background and its application. *Inorg. Chim. Acta* **2008**, *361*, 3432–3445. [CrossRef]
83. Rabelo, R.; Toma, L.; Julve, M.; Lloret, F.; Pasán, J.; Cangussu, D.; Ruiz-García, R.; Cano, J. Multielectron transfer and field-induced slow magnetic relaxation in opto-electroactive spin crossover cobalt(II) complexes: Structure–function correlations. *Inorg. Chem. Front.* **2024**, *11*, 6028–6043. [CrossRef]
84. Kalinke, L.H.G.; Rabelo, R.; Valdo, A.K.; Martins, F.T.; Moliner, N.; Ferrando-Soria, J.; Julve, M.; Lloret, F.; Cano, J.; Cangussu, D. Trinuclear Cobalt(II) Triple Helicate with a Multidentate Bithiazole bis(oxamate) Ligand as a Supramolecular Nanomagnet. *Inorg. Chem.* **2022**, *61*, 5696–5700. [CrossRef] [PubMed]
85. Haiduc, I.; Tiekink, E.R.T. *Inverse Coordination, A Novel Chemical Concept*; Sunway University Press: Selangor, Malaysia, 2020.

86. Julve, M.; Gleizes, A.; Chamoreau, L.M.; Ruiz, E.; Verdaguer, M. Antiferromagnetic Interactions in copper(II) μ-oxalato dinuclear complexes: The role of the counterion. *Eur. J. Inorg. Chem.* **2018**, *2018*, 509–516. [CrossRef]
87. Costa, A.I.; da Silva, R.M.R.; Botelho, L.D.G.; Coelho, S.F.N.; Sigoli, F.A.; Honorato, J.; Ellena, J.; de Felipe, T.; Martins, F.T.; Angelo, M.; et al. Intensity and lifetime ratiometric luminescent thermometer based on a Tb(III) coordination polymer. *Dalton Trans.* **2024**, *53*, 3994–4004. [CrossRef]
88. Oliveira, W.X.C.; Pinheiro, C.B.; Journaux, Y.; Julve, M.; Pereira, C.L.M. Effects on the magnetic interaction caused by molecular recognition in complexes of 1,2-azolebased oxamate and $[Cu(bpca)]^+$ units. *CrystEngComm* **2024**, *26*, 647–655. [CrossRef]
89. Lescouëzec, R.; Vaissermann, J.; Ruiz-Pérez, C.; Lloret, F.; Carrasco, R.; Julve, M.; Verdaguer, M.; Dromzée, Y.; Gatteschi, D.; Wernsdorfer, W. Cyanide-Bridged Iron(III)–Cobalt(II) Double Zigzag Ferromagnetic Chains: Two New Molecular Magnetic Nanowires. *Angew. Chem. Int. Ed.* **2003**, *42*, 1483–1486. [CrossRef] [PubMed]
90. Toma, L.M.; Lescouëzec, R.; Lloret, F.; Julve, M.; Vaissermann, J.; Verdaguer, M. Título: Cyanide-Bridged Fe(III)–Co(II) Bis Double Zigzag Chains with a Slow Relaxation of The Magnetisation. *Chem. Commun.* **2003**, 1850–1851. [CrossRef] [PubMed]
91. Toma, L.M.; Lescouëzec, R.; Pasán, J.; Ruiz-Pérez, C.; Vaissermann, J.; Cano, J.; Carrasco, R.; Wernsdorfer, W.; Lloret, F.; Julve, M. $Fe(bpym)(CN)_4]^-$: A new building block for designing single-chain magnets. *J. Am. Chem. Soc.* **2006**, *12*, 4842–4853. [CrossRef]
92. Pardo, E.; Ruiz-García, R.; Lloret, F.; Faus, J.; Julve, M.; Journaux, Y.; Delgado, F.; Ruiz-Pérez, C. Cobalt(II)–Copper(II) Bimetallic Chains as a New Class of Single-Chain Magnets. *Adv. Mater.* **2004**, *16*, 1597–1600. [CrossRef]
93. Miyasaka, H.; Julve, M.; Yamashita, M.; Clérac, R. Slow Dynamics of the Magnetization in One-Dimensional Coordination Polymers: Single-Chain Magnets. *Inorg. Chem.* **2009**, *48*, 3420–3437. [CrossRef] [PubMed]
94. Milios, C.; Raptopoulou, C.; Terzis, A.; Lloret, F.; Vicente, R.; Perlepes, S.; Escuer, A. Hexanuclear Manganese(III) Single-Molecule Magnets. *Angew. Chem. Int. Ed.* **2004**, *43*, 210–212. [CrossRef]
95. Vallejo, J.; Pascual-Alvarez, A.; Cano, J.; Castro, I.; Julve, M.; Lloret, F.; Krzystek, J.; De Munno, G.; Armentano, D.; Wernsdorfer, W.; et al. Field-Induced Hysteresis and Quantum Tunneling of the Magnetization in a Mononuclear Manganese(III) Complex. *Angew. Chem. Int. Ed.* **2013**, *52*, 14075–14079. [CrossRef] [PubMed]
96. Vallejo, J.; Castro, I.; Ruiz-Pérez, R.; Cano, J.; Julve, M.; Lloret, F.; De Munno, G.; Wernsdorfer, W.; Pardo, E. Field Induced Slow Magnetic Relaxation in a Six-Coordinate Mononuclear Cobalt(II) Complex with a Positive Anisotropy. *J. Am. Chem. Soc.* **2012**, *134*, 15704–15707. [CrossRef]
97. Martínez- Lillo, J.; Armentano, D.; De Munno, G.; Lloret, F.; Julve, M.; Faus, J. Synthesis, crystal structure and magnetic properties of an oxalato-bridged $Re^{IV}Mo^{VI}$ heterobimetallic complex. *Dalton Trans.* **2011**, *40*, 4818–4820. [CrossRef] [PubMed]
98. Martínez- Lillo, J.; Delgado, F.S.; Ruiz-Pérez, C.; Lloret, F.; Julve, M.; Faus, J. Heterotrimetallic Oxalato-Bridged $Re^{IV}{}_2M^{II}$ Complexes (M = Mn, Co, Ni, Cu): Synthesis, Crystal Structure, and Magnetic Properties. *Inorg. Chem.* **2007**, *46*, 3523–3530. [CrossRef] [PubMed]
99. Martínez- Lillo, J.; Armentano, D.; De Munno, G.; Wernsdorfer, W.; Julve, M.; Lloret, F.; Faus, J. A Heterotetranuclear $[Ni^{II}Re^{IV}{}_3]$ Single-Molecule Magnet. *J. Am. Chem. Soc.* **2006**, *128*, 14218–14219. [CrossRef] [PubMed]
100. Martínez- Lillo, J.; Armentano, D.; De Munno, G.; Lloret, F.; Julve, M.; Faus, J. X-Ray structure of $[ReCl_4(\mu\text{-ox})Cu(pyim)_2]$: A new heterobimetallic $Re^{IV}Cu^{II}$ ferrimagnetic chain. *Dalton Trans.* **2008**, 40–43. [CrossRef]
101. Gleizes, A.; Julve, M.; Kuzmina, N.; Alikhanyan, A.; Lloret, F.; Malkerova, I.; Sanz, J.L.; Senocq, F. Heterobimetallic d–f Metal Complexes as Potential Single-Source Precursors For MOCVD: Structure and Thermodynamic Study of the Sublimation of $[Ni(salen)Ln(hfa)_3]$, Ln = Y, Gd. *Eur. J. Inorg. Chem.* **1998**, *1998*, 1169–1174. [CrossRef]
102. Hernández-Molina, M.; Lorenzo-Luis, P.A.; López, T.; Ruiz-Pérez, C.; Lloret, F.; Julve, M. Generation of lanthanide coordination polymers with dicarboxylate ligands: Synthesis, structure, thermal decomposition and magnetic properties of the two-dimensional triaquabis(malonato)dipraseodymium(III) dihydrate $\{[Pr_2(C_3H_2O_4)_3(H_2O)_3]\cdot 2H_2O\}$. *CrystEngComm* **2000**, *31*, 1–5. [CrossRef]
103. Vallejo, J.; Cano, J.; Castro, I.; Julve, M.; Lloret, F.; Fabelo, O.; Cañadillas-Delgado, L.; Pardo, E. Slow magnetic relaxation in carbonato-bridged dinuclear lanthanide(III) complexes with 2,3-quinoxalinediolate ligands. *Chem. Commun.* **2012**, *48*, 7726–7728. [CrossRef]
104. Alexandru, M.G.; Visinescu, D.; Cula, B.; Shova, S.; Rabelo, N.; Moliner, F.; Lloret, J.; Cano, J.; Julve, M. A rare isostructural series of 3d–4f cyanido-bridged heterometallic squares obtained by assembling $[Fe^{III}\{HB(pz)_3\}(CN)_3]^-$ and $Ln^{III}$ ions: Synthesis, X-ray structure and cryomagnetic study. *Dalton Trans.* **2021**, *50*, 14640–14652. [CrossRef] [PubMed]
105. Sourav, D.; Sakiat, H.; Atanu, D.; Sourav, B.; Emilio, P.; Francesc, L.; Vadapalli, C. Heterometallic Pentanuclear $[Ni_4Ln]$ ($Ln^{III}$ = Gd, Tb, Dy, Ho) Complexes: Accidental Orthogonality Leading to Ferromagnetic Interactions. *Eur. J. Inorg. Chem.* **2014**, *21*, 3393–3400.
106. Díaz-Galliga, P.; Fabelo, O.; Pasán, J.; Cañadillas-Delgado, L.; Lloret, F.; Julve, M.; Ruiz-Pérez, C. Two-Dimensional 3d–4f Heterometallic Coordination Polymers: Synthesis, Crystal Structures, and Magnetic Properties of Six New Co(II)–Ln(III) Compounds. *Inorg. Chem.* **2014**, *53*, 6299–6308. [CrossRef] [PubMed]

107. Vaz, R.C.A.; Esteves, I.O.; Oliveira, W.X.C.; Honorato, J.; Martins, F.T.; Marques, L.F.; Dos Santos, G.L.; Freire, R.O.; Jesús, L.T.; Pedroso, E.F.; et al. Mononuclear Lanthanide(III)-oxamate complexes as new photoluminescent field-induced single-molecule magnets: Solid-state photophysical and magnetic properties. *Dalton Trans.* **2020**, *49*, 16106–16124. [CrossRef]
108. Alexandru, M.G.; Visinescu, D.; Madalan, A.M.; Lloret, F.; Julve, M.; Andruh, M. $[W(bipy)(CN)_6]^-$: A Suitable Metalloligand in the Design of Heterotrimetallic Complexes. The First $Cu^{II}Ln^{III}W^{V}$ Trinuclear Complexes. *Inorg. Chem.* **2012**, *51*, 4906–4908. [CrossRef] [PubMed]
109. Díaz-Gallifa, P.; Fabelo, O.; Cañadillas-Delgado, L.; Pasán, J.; Labrador, A.; Lloret, F.; Julve, M.; Ruiz-Pérez, C. Synthesis, Crystal Structure and Magnetic Characterzation of a Series of $Cu^{II}$–$Ln^{III}$ Heterometallic (Ln = La, Ce, Pr, Nd and Sm) Metal-Organic compounds with a Unusual Single Crystal to Single Crystal Phase Transition. *Cryst. Growth Des.* **2013**, *13*, 4735–4745. [CrossRef]
110. Palion-Gazda, J.; Switlicka-Olszewska, A.; Machura, B.; Grancha, T.; Pardo, E.; Lloret, F.; Julve, M. High-Temperature Spin Crossover in a Mononuclear Six-Coordinate Cobalt(II) Complex. *Inorg. Chem.* **2014**, *53*, 10009–10011. [CrossRef] [PubMed]
111. Palion-Gazda, J.; Machura, B.; Kruszynski, R.; Grancha, T.; Moliner, N.; Lloret, F.; Julve, M. Spin Crossover in Double Salts Containing Six- and Four-Coordinate Cobalt(II) Ions. *Inorg. Chem.* **2017**, *56*, 6281–6296. [CrossRef]
112. Rabelo, R.; Toma, L.; Moliner, N.; Julve, M.; Lloret, F.; Pasán, J.; Ruiz-Pérez, C.; Ruiz-García, R.; Cano, J. Electroswitching of the single-molecule magnet behaviour in an octahedral spin crossover cobalt(II) complex with a redox-active pyrimidine ligand. *Chem. Commun.* **2020**, *56*, 12242–12245. [CrossRef]
113. Lloret Pastor, F. Diseño de Imanes Porosos Multifuncionales (Design of Multifunctional Porous Magnets). In Proceedings of the XVI National School of Molecular Materials, Santa Pola, Spain, 1–6 February 2015.
114. Batten, S.R.; Champness, N.R.; Chen, X.-M.; García-Martinez, J.; Kitagawa, S.; Ohrstrom, L.; O'Keeffe, M.; Paik Suh, M.; Reedijk, J. Coordination polymers, metal–organic frameworks and the need for terminology guidelines. *CrystEngComm* **2012**, *14*, 3001–3004. [CrossRef]
115. Reedijk, J. Terminology of metal–organic frameworks and coordination polymers (IUPAC Recommendations 2013). *Pure Appl. Chem.* **2013**, *85*, 1715–1724.
116. Ferrando-Soria, J.; Grancha, T.; Julve, M.; Cano, J.; Lloret, F.; Journaux, Y.; Pasán, J.; Ruiz-Pérez, C.; Pardo, E. Ligand effects on the dimensionality of oxamato-bridged mixed-metal open-framework magnets. *Chem. Commun.* **2012**, *48*, 3539–3541. [CrossRef]
117. Ferrando-Soria, J.; Serra-Crespo, P.; de Lange, M.; Gascón, J.; Kapteijn, F.; Julve, M.; Cano, J.; Lloret, F.; Pasán, J.; RuizPérez, C.; et al. Selective Gas and Vapor Sorption and Magnetic Sensing by an Isoreticular Mixed-Metal-Organic Framework. *J. Am. Chem. Soc.* **2012**, *134*, 15301–15304. [CrossRef]
118. Cangussu, D.; Pardo, E.; Dul, M.C.; Lescouezec, R.; Herson, P.; Journaux, Y.; Pedroso, E.F.; Pereira, C.L.M.; Stumpf, H.O.; Munoz, M.C.; et al. Rational Design of a New Class of Heterobimetallic Molecule-Based Magnets: Synthesis, Crystal Structures and Magnetic Properties of Oxamato-Bridged $M'_3M_2$ ($M' = Li^{I}$ and $Mn^{II}$; $M = Ni^{II}$ and $Co^{II}$) Open-Frameworks with a Three-Dimensional Honeycomb Architecture. *Inorg. Chim. Acta* **2008**, *361*, 3394–3402. [CrossRef]
119. Pardo, E.; Cangussu, D.; Dul, M.-C.; Lescouezec, R.; Herson, P.; Journaux, Y.; Pedroso, E.F.; Pereira, C.L.M.; Munoz, M.C.; Ruiz-García, R.; et al. A Metallacryptand-Based Manganese(II)–Cobalt(II) Ferrimagnet with a Three-Dimensional Honeycomb Open-Framework. *Angew. Chem. Int. Ed.* **2008**, *47*, 4211–4216. [CrossRef]
120. Ferrando-Soria, J.; Ruiz-García, R.; Cano, J.; Stiriba, S.-E.; Vallejo, J.; Castro, I.; Julve, M.; Lloret, F.; Amoros, P.; Pasán, J.; et al. Reversible Solvatomagnetic Switching in a Spongelike Manganese(II)–Copper(II) 3D Open Framework with a Pillared Square/Octogonal Layer Architecture. *Chem. Eur. J.* **2012**, *18*, 1608. [CrossRef]
121. Ferrando-Soria, J.; Pasán, J.; Ruiz-Pérez, C.; Journaux, Y.; Julve, M.; Lloret, F.; Cano, J.; Pardo, E. Spin Control in Oxamato-Based Manganese(II)–Copper(II) Coordination Polymers with Brick-Wall Layer Architecture. *Inorg. Chem.* **2011**, *50*, 8694–8696. [CrossRef]
122. Ferrando-Soria, J.; Khajavi, H.; Serra-Crespo, P.; Gascon, J.; Kapteijn, F.; Julve, M.; Lloret, F.; Pasán, J.; Ruiz-Pérez, C.; Journaux, Y.; et al. Highly Selective Chemical Sensing in a Luminiscent Nanoporous Magnet. *Adv. Mater.* **2012**, *24*, 5625–5629. [CrossRef]
123. Titos-Padilla, S.; Ruiz, J.; Herrera, J.M.; Brechin, E.K.; Wersndorfer, W.; Lloret, F.; Colacio, E. Dilution-Triggered SMM Behavior under Zero Field in a Luminescent $Zn_2Dy_2$ Tetranuclear Complex Incorporating Carbonato-Bridging Ligands Derived from Atmospheric $CO_2$ Fixation. *Inorg. Chem.* **2013**, *52*, 9620–9626. [CrossRef] [PubMed]
124. Mon, M.; Lloret, F.; Ferrando-Soria, J.; Marti-Gastaldo, C.; Armentano, D.; Pardo, E. Selective and Efficient Removal of Mercury from Aqueous Media with the Highly Flexible Arms of a BioMOF. *Angew. Chem. Int. Ed.* **2016**, *55*, 11167–11172. [CrossRef] [PubMed]
125. Hrimla, M.; Bahsis, L.; Boutouil, A.; Laamari, M.R.; Julve, M.; Stiriba, S.E. A combined computational and experimental study on the mild steel corrosion inhibition in hydrochloric acid by new multifunctional phosphonic acid containing 1,2,3-triazoles. *J. Adhesion Sci. Technol.* **2020**, *34*, 1741–1773. [CrossRef]

126. Cañadillas-Delgado, L.J.; Fabelo, O.; Ruiz-Pérez, C.; Delgado, F.S.; Julve, M.; Hernández-Molina, M.; Milagros Laz, M.; Lorenzo-Luis, P. Zeolite-like Nanoporous Gadolinium Complexes Incorporating Alkaline Cations. *Cryst. Growth Des.* **2006**, *6*, 87–93. [CrossRef]
127. Marinho, M.V.; Reis, D.O.; Oliveira, W.X.C.; Marques, L.F.; Stumpf, H.O.; Déniz, M.; Pasán, J.; Ruiz-Pérez, C.; Cano, J.; Lloret, F.; et al. Photoluminescent and Slow Magnetic Relaxation Studies on Lanthanide(III)-2,5-pyrazinedicarboxylate Frameworks. *Inorg. Chem.* **2017**, *56*, 2108–2123. [CrossRef]
128. Pardo, E.; Train, C.; Gontard, G.; Boubekeur, K.; Fabelo, O.; Liu, H.; Dkhil, B.; Lloret, F.; Nakagawa, K.; Tokoro, H.; et al. High Proton Conduction in a Chiral Ferromagnetic Metal-Organic Quartz-like Framework. *J. Am. Chem. Soc.* **2011**, *133*, 15328–15331. [CrossRef] [PubMed]
129. Canadillas-Delgado, L.; Fabelo, O.; Rodriguez-Velamazan, J.; Lemee-Cailleau, M.-H.; Mason, S.A.; Pardo, E.; Lloret, F.; Zhao, J.-P.; Bu, X.-H.; Simonet, V.; et al. The Role of Order-Disorder Transitions in the Quest for Molecular Multiferroics: Structural and Magnetic Neutron Studies of a Mixed Valence Iron(II)-Iron(III) Formate Framework. *J. Am. Chem. Soc.* **2012**, *134*, 19772–19781. [CrossRef] [PubMed]
130. Rabelo, R.; Stiriba, S.-E.; Cangussu, D.; Pereira, C.L.M.; Moliner, N.; Ruiz-García, R.; Cano, J.; Faus, J.; Journaux, Y.; Julve, M. When Molecular Magnetism Meets Supramolecular Chemistry: Multifunctional and Multiresponsive Dicopper(II) Metallacyclophanes as Proof-of-Concept for Single-Molecule Spintronics and Quantum Computing Technologies. *Magnetochemistry* **2020**, *6*, 69. [CrossRef]
131. Launay, J.P.; Verdaguer, M. *Electrons in Molecules*; Oxford University Press: Oxford, UK, 2018.
132. *Obituary*; Department of Inorganic Chemistry, University of Valencia: Valencia, Spain, 2024; Available online: https://www.uv.es/uvweb/quimica/es/novedades-del-departamento/defuncion-del-profesor-miquel-julve-olcina-1285923388171/Novetat.html?id=1286391974582 (accessed on 10 January 2025).

*magnetochemistry*

MDPI

*Article*

# Manganese (III) Compounds Derived from R-Salicylaldoxime and 9-Anthracenecarboxylate Ligands: A Study of Their Synthesis and Structural, Magnetic, and Luminescent Properties †

**Berta Casanovas [1], Ramon Vicente [1], Mercè Font-Bardía [2] and Mohamed Salah El Fallah [1,*]**

1 Departament de Química Inorgànica i Orgànica, Secció de Química Inorgànica and Institut de Nanociència i Nanotecnologia (IN2UB), Universitat de Barcelona, Martí i Franquès 1-11, 08028 Barcelona, Spain

2 Departament de Mineralogia, Cristal·Lografia i Dipòsits Minerals and Unitat de Difracció de RX. Centre Científic i Tecnològic de la Universitat de Barcelona (CCiTUB), Universitat de Barcelona, C/Solé i Sabarís, 1-3, 08028 Barcelona, Spain

* Correspondence: salah.elfallah@qi.ub.es; Tel.: +34-93-4039134

† During the preparation of this manuscript dedicated to Professors Miguel Julve and Francesco Lloret on the occasion of their retirements, and before submitting it to the Magnetochemistry journal, we received the sad news of the death of one of the honourees, Prof. Miguel Julve. We deeply regret the loss of one of the most distinguished professors in the field of coordination chemistry, especially in the area of molecular magnetism. We regret the loss of a great friend, a big communicator, and a giant scientist. Miguel, we will miss you, but you will live in our memory with the beautiful moments shared with you. Thank you for your friendship, humility and availability. "Descansa en paz Miguel".

**Abstract:** The reaction of Mn(II) salts in the air with different R-salicylaldehyde oximes and the sodium or cesium salts of 9-anthracenecarboxylato (9-AC) allows for the isolation of new six polynuclear compounds: $[Mn_3NaO(salox)_3(9\text{-}AC)_2(EtOH)_3H_2O]_n\cdot 2EtOH$ (**1**), $[Mn_3NaO(3\text{-}Me\text{-}salox)_3(9\text{-}AC)_2(EtOH)_3H_2O]_n\cdot EtOH$ (**2**), $[Mn_6O_2(salox)_6(9\text{-}AC)_2(EtOH)_2(H_2O)_2]\cdot 3EtOH$ (**3**), $[Mn_3O(3\text{-}Me\text{-}salox)_3(9\text{-}AC)(EtOH)_3(H_2O)]\cdot 1.8EtOH\cdot 3H_2O$ (**4**), $[Mn_6O_2(Me\text{-}salox)_6(9\text{-}AC)_2(EtOH)_4(H_2O)_2]\cdot 0.5H_2O$ (**5**), and $[Mn_6O_2(Et\text{-}salox)_6(9\text{-}AC)_2(EtOH)_4(H_2O)_2]\cdot 3EtOH$ (**6**). $H_2$salox is a salicylaldehyde oxime, $H_2$(3-Me-salox) is a 3-methyl-salicylaldehyde oxime, $H_2$Me-salox is a 1-(2-hydroxyphenyl)ethan-1-one oxime and a $H_2$-Et-salox is 1-(2-hydroxyphenyl)propan-1-one oxime. Structurally, compounds **1** and **2** consist of chains of trinuclear $\{Mn^{III}_3(\mu_3\text{-}O)(salox)_3\}^+$ units connected by $Na^+$ ions. Compounds **3**, **5**, and **6** are hexanuclear units formed by two parallel trinuclear units $\{Mn^{III}_3(\mu_3\text{-}O)(salox)_3\}^+$ or $\{Mn^{III}_3(\mu_3\text{-}O)(Me\text{-}salox)_3\}^+$ planes related through an inversion center. Compound **4** consists of two isolated $[Mn_3O(3\text{-}Me\text{-}salox)_3(9\text{-}AC)(EtOH)_3(H_2O)]$ trinuclear molecules in the unit cell showing crystallographic differences. Magnetic studies reveal a set of antiferromagnetic interactions in compounds **1** and **2** and a combination of antiferromagnetic and ferromagnetic interactions in compounds **3**, **5**, and **6**. In all cases, the magneto-structural correlation between the intramolecular $Mn^{III}$-N-O-$Mn^{III}$ torsion angle and the magnetic exchange within these units have been confirmed. For compounds **5** and **6**, *ac* magnetic measurements reveal the slow relaxation of magnetization with moderate energy barriers of 19.9 $cm^{-1}$ and 31.1 $cm^{-1}$, respectively. Absorbance and fluorescence measurements in solution show the transitions of the 9-anthracenecarboxylate chromophore for all the compounds.

**Keywords:** manganese coordination compounds; magnetic properties; single-molecule magnets; luminescence

## 1. Introduction

At the beginning of the year 2000, our research group described two new hexanuclear compounds with the formulas of $[Mn_6O_2(salox)_6(Me\text{-}COO)_2(EtOH)_4]$ and $[Mn_6O_2(salox)_6(Ph\text{-}COO)_2(EtOH)_4]$ [1]. These compounds are constructed by two triangular entities $\{Mn^{III}_3(\mu_3\text{-}O)\ (salox)_3\}^+$, where $salox^{2-}$ is the anion derived from the double deprotonation of the salicylaldoxime ligand represented in Scheme 1a. These kinds of ligands are commonly used for metal extraction or as anticorrosive agents in coating processes [2]. The structure of the trinuclear unit represented in Scheme 1b consists of three Mn(III) ions connected by a $\mu_3$-$O^{2-}$ anion, located at the center of the triangle, and by the nitrogen and oxygen atoms of the oxime groups of three R-$salox^{2-}$ ligands, which bridge each pair of Mn(III) ions. These ligands also coordinate terminally with each of the Mn(III) ions through the oxygen in the phenolate group. Numerous research groups call these compounds "inverse metallacrowns" because they feature a cyclic M-N-O repeat unit and contain an anion in the central cavity [3–6]. Due to the extensive synthetic possibilities arising from the use of different substituted salicylaldoximes, $H_2$-R-salox (R = H, $CH_3$ (Me), $CH_3CH_2$ (Et)...) [7–10], as well as the types of counterions or used solvents, many compounds based on the same trinuclear group $\{Mn^{III}{}_3(\mu_3\text{-}O)(R\text{-}salox)_3\}^+$ have been obtained. The most used counterions are carboxylates and azide anions. Generally, in these compounds, Mn(III) ions exhibit either an octahedral coordination geometry with Jahn–Teller distortions or a square pyramidal type. In both cases, the equatorial or basal positions lie in the same $[Mn_3]$ plane. Magnetic studies revealed that hexanuclear compounds exhibit a ground spin state of S = 4, derived from the ferromagnetic coupling between the two trinuclear units with S = 2, resulting from the intra-triangular antiferromagnetic coupling among the three high-spin Mn(III) ions ($S_{Mn(III)}$ = 2). Furthermore, it was observed that these compounds exhibit single-molecule magnet (SMM) behavior [1].

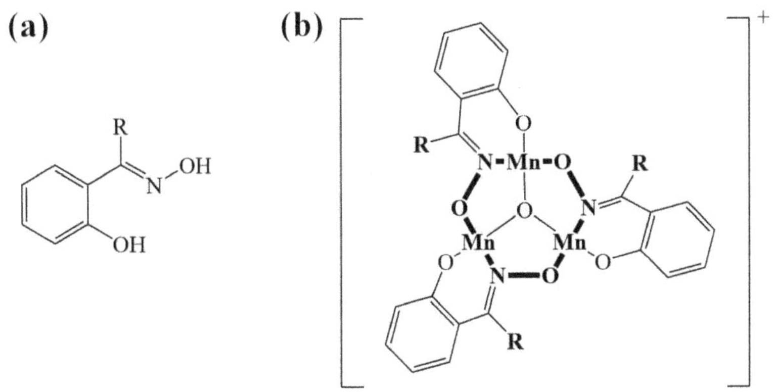


**Scheme 1.** (**a**) General representation of the $H_2$-R-salicylaldoxime ligands. (**b**) Trinuclear entity $\{Mn^{III}{}_3(\mu_3\text{-}O)(R\text{-}salox)_3\}^+$.

Several systematic studies on hexanuclear compounds with R-$salox^{2-}$ ligands having different alkyl groups and carboxylate ligands varying in size have corroborated that by increasing the volume of the R group in the substituted $salox^{2-}$ anions, the fundamental spin state can increase from S = 2 to S = 12, which is the maximum possible value for this kind of hexanuclear system as a consequence of the ferromagnetic coupling between all Mn(III) ions [11]. Subsequently, through a detailed observation of the structure and the magnetic properties of different hexanuclear systems, it was postulated that the generally weak coupling [12] between Mn(III) ions within the triangle is mainly dominated by the Mn–N–O–Mn torsion angle. A so-called "magic area" is present when this angle ranges between 30.4° and 31.3° [13]. According to the established magneto-structural correlation, if the angles are below 30.4°, antiferromagnetic coupling ($J < 0$) is expected, while torsions greater than 31.3° are expected to result in ferromagnetic exchange ($J > 0$). If the torsion angle falls within this range, the magnetic exchange between the ions can be either antiferromagnetic or ferromagnetic. The Mn–N–O–Mn torsion angle increases as the volume of R-substituent groups of the oxime group increases. The fundamental spin state of the triangle is determined by the contribution of each of the three couplings via the oxime group, and thus, taking the average value of the torsion angles is not valid. In all cases, it was observed that the coupling between triangles via the R-salox$^{2-}$ ligands is of a ferromagnetic nature. Moreover, when all the magnetic exchanges between Mn(III) ions are ferromagnetic, the larger the torsion angles, the higher the magnetization relaxation barrier. This experimental fact is attributed to an increase in the coupling constant, $J$, which results in the ground state, S, being more energetically isolated from the excited states, thus preventing tunneling relaxation [10]. In these triangular units, the anisotropy axes of Mn(III) are almost parallel and perpendicular to the $[Mn^{III}]_3$ triangular planes, resulting in the formation of compounds, with $d$ metals exhibiting the highest observed magnetization energy barrier [14]. The dominance of the Mn-N-O-Mn torsion angle value over the coupling between Mn(III) atoms observed in published hexanuclear compounds is valid for any compound containing triangular units $\{Mn^{III}_3(\mu_3\text{-O})(Rsalox)_3\}^+$ [15]. Within each triangular unit, depending on the torsion angles present in the compound, the total spin (S) can range from 0 to 6.

The compounds based on $\{Mn^{III}_3(\mu_3\text{-O})(R\text{-salox})_3\}^+$ units and carboxylate anions generally have the general formula $[Mn^{III}_xO_{1\text{-}2}\ (R\text{-salox})_x(R\text{-COO})_{1\text{-}2}(L)_{4\text{-}6}]$, where R = H, $CH_3$ (Me), $CH_2CH_3$ (Et), $C_6H_5$ (Ph)... [1,8–11,16], L = solvent (alcohols or water), and $x = 3$ or 6 [7]. In some instances, the use of poly-carboxylates has led to the formation of 1D or even 2D compounds based on the same unit core [17,18].

On the other hand, it is well known that the anthracene aromatic group of the carboxylate ligand shows luminescent properties [19] that can be maintained when the ligand is complexed, as has been observed in different manganese coordination compounds with different ligands that contain anthracene groups [20–23]. Therefore, the aim of this work is obtaining multifunctional compounds based on the $\{Mn^{III}_3(\mu_3\text{-O})(Rsalox)_3\}^+$ unit that exhibit interesting magnetic properties and anthracenecarboxylate ligands, where the presence of the $\pi$-conjugated system could offer luminescence properties to the Mn compounds. We also aimed to investigate how different cations ($Na^+$ or $Cs^+$) in alkaline carboxylate salt and the anion of the starting manganese salt ($NO_3^-$ or $ClO_4^-$) influence the structure and, consequently, the magnetic and luminescent properties of the new compounds. The incorporation of $Na^+$ ions in the synthesis of compounds featuring the $\{Mn^{III}_3(\mu_3\text{-O})(R\text{-salox})_3\}^+$ unit and carboxylate ligands has been previously explored to achieve new topologies. It has been observed that this cation can serve as a linkage point between $[Mn_3]$-based units, forming several structural architectures, such as dimers of trinuclear units [15,24], $[Mn_{18}]$ [25], as well as 1D [26,27] or 2D structures [28–30]. In all the compounds, $Na^+$ ions present a hexacoordinated environment, and it is observed that the alkaline cations keep the Mn(III) units well-isolated. As a result, their binding does not affect the intra-triangular magnetic properties. Additionally, we decided to investigate the effect of other alkaline ions with a larger ionic radius than $Na^+$, such as $Cs^+$, a variable that has not been explored to date. Also, we examine the effect of the substitution of hydrogen by methyl groups on the aromatic ring of the salicylaldoxime ligand on the Mn-N-O-Mn torsion angle, as this parameter has been poorly studied [8,24,27]. So, we present in this work six new Mn(III) compounds based on the $\{Mn^{III}_3(\mu_3\text{-O})(R\text{-salox})_3\}^+$ fragment: $[Mn_3NaO(salox)_3(9\text{-AC})_2(EtOH)_3H_2O]_n\cdot 2EtOH$ (**1**), $[Mn_3NaO(3\text{-Me-salox})_3(9\text{-AC})_2(EtOH)_3H_2O]_n\cdot EtOH$ (**2**), $[Mn_6O_2(salox)_6(9\text{-AC})_2(EtOH)_2(H_2O)_2]\cdot 3EtOH$ (**3**), $[Mn_3O(3\text{-Me-salox})_3(9\text{-AC})(EtOH)_3(H_2O)]\cdot 1.8EtOH\cdot 3H_2O$ (**4**), $[Mn_6O_2(Me\text{-salox})_6(9\text{-AC})_2(EtOH)_4(H_2O)_2]\cdot 0.5H_2O$ (**5**), and $[Mn_6O_2(Et\text{-salox})_6(9\text{-AC})_2(EtOH)_4(H_2O)_2]\cdot 3EtOH$ (**6**). The compounds have been obtained from different $H_2$-R-salox ligands and the sodium or cesium salt of 9-anthracenecarboxylate that are shown in Scheme 2.

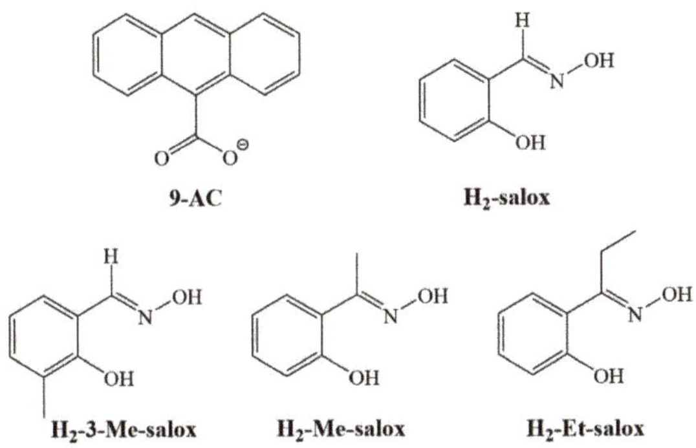


**Scheme 2.** Representation of the different organic ligands used in the synthesis of compounds **1–6**.

## 2. Results and Discussion

### *2.1. X-ray Crystal Structures*

Crystallographic data for compounds **1–6** are summarized in the Supplementary Materials in Table S1. Oxidation number +3 has been determined for all manganese ions present in compounds **1–6** [31] through bond distances, charge balance, and bond valence sum (BVS) analysis. The results of this analysis are presented in Table S2.

#### 2.1.1. $[Mn_3NaO(salox)_3(9\text{-}AC)_2(EtOH)_3H_2O]_n{\cdot}2EtOH$ (**1**) and $[Mn_3NaO(3\text{-}Me\text{-}salox)_3(9\text{-}AC)_2(EtOH)_3H_2O]_n{\cdot}EtOH$ (**2**)

Compound **1** crystallizes in the C2/c group of the monoclinic system, while compound **2** crystallizes in the P-1 space group of the triclinic system. Figure 1 shows the crystalline structures of compounds **1** and **2**. They consist of chains of trinuclear units $\{Mn^{III}{}_3(\mu_3\text{-}O)(salox)_3\}^+$ connected by $Na^+$ ions along the [001] direction of the crystal lattice for **1** and [010] for **2**. In compound **1**, zigzag-type chains with inverted parallel planes alternate with contiguous intersecting planes at a dihedral angle of 57.3°, whereas in compound **2**, the chains consist of parallel inverted $\{Mn^{III}{}_3(\mu_3\text{-}O)(salox)_3\}^+$ planes. Both compounds have the same asymmetric unit $[Mn_3O(R\text{-}salox)_3(9\text{-}AC)_2(EtOH)_3H_2O]$, R = H for **1** and $CH_3$ for **2** (Figure 1a), with an architecture not previously observed. Given the structural similarity between **1** and **2**, only the structure of compound **1** is described as an example.

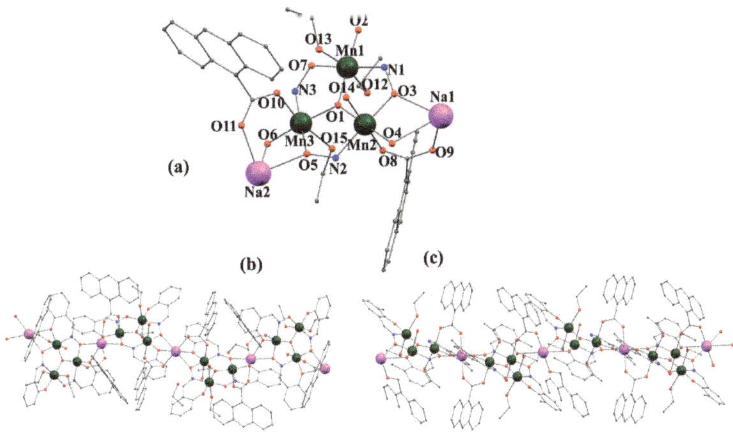


**Figure 1.** (**a**) Asymmetric units (partially labeled) for compounds **1** and **2**. (**b**,**c**) Chain structure representation of compounds **1** and **2**, respectively. Hydrogen atoms and solvent molecules have been omitted for clarity. Color code: dark green = Mn, pink = Na, red = O, blue = N, and gray = C.

The crystalline structure of **1** consists of chains of trinuclear units $\{Mn^{III}_3(\mu_3\text{-}O)(salox)_3\}^+$ connected through $Na^+$ ions. Table 1 lists the most important bond distances and angles for both compounds **1** and **2**. The Mn(III) ions form in **1** an almost equilateral triangle with distances between the vertices equal to 3.250 Å (Mn1···Mn2), 3.270 Å (Mn2···Mn3), and 3.232 Å (Mn1···Mn3), while in compound **2**, the observed distances are 3.262 Å (Mn1···Mn2), 3.279 Å (Mn2···Mn3), and 3.253 Å (Mn1···Mn3). Within each trinuclear unit, the three Mn(III) ions exhibit an octahedral geometry with Jahn–Teller distortions, with the equatorial planes of the octahedron lying within the same triangular plane. The four equatorial positions of the octahedron are occupied, in all three cases, by a bond with the $\mu_3\text{-}O^{2-}$ anion (O1), which also forms part of the plane formed by the three Mn(III) atoms with Mn-O1 distances ranging from 1.873(2) to 1.879 Å (1.8796(16) and 1.8916(18) Å for **2**), two bonds with nitrogen atoms (N1, N2, or N3) in the oxime group and oxygen atoms (O2, O4, or O6) in the phenoxo group of a $salox^{2-}$ ligand, and a bond with oxygen atoms (O3, O5, or O7) in the oxime group of another $salox^{2-}$ ligand. The Mn-N-O-Mn torsion angles are 28.6(3)° for Mn1-Mn2 (26.71(19)° for **2**), 5.4(3)° for Mn2-Mn3 (18.92(19)° for **2**), and 1.0(3)° for Mn1-Mn3 (21.6(2)° for **2**). The axial positions vary depending on the Mn(III) ion: for the Mn1 atom, these are occupied by two oxygen atoms, O12 and O13, from two ethanol molecules. On the other hand, the Mn2 and Mn3 ions each have an axial bond with the oxygen atoms O8 and O10, respectively, of two carboxylate ligands that act as bridges (in the *syn-syn* coordination mode) between these ions and the corresponding $Na^+$ cations (Na1 for Mn2 and Na2 for Mn3), forming the one-dimensional structure. The remaining axial positions are occupied by oxygen atom O14 of a water molecule in the case of the Mn2 ion and by oxygen atom O15 of an ethanol molecule in the case of Mn3.

**Table 1.** Selected bond distances (Å) and angles (°) for compounds **1** and **2**.

| | 1 | 2 | | 1 | 2 |
|---|---|---|---|---|---|
| Mn1-O1 | 1.879(2) | 1.8796(16) | Na1-O3 | 2.429(2) | 2.3117(17) |
| Mn1-O2 | 1.863(2) | 1.865(2) | Na1-O4 | 2.502(2) | 2.8331(18) |
| Mn1-O7 | 1.903(2) | 1.9043(19) | Na1-O9 | 2.332(2) | 2.3321(17) |
| Mn1-O12 | 2.268(2) | 2.3293(19) | Na1-O3″ | 2.429(2) | 2.3117(17) |
| Mn1-O13 | 2.348(3) | 2.3512(18) | Na1-O4″ | 2.502(2) | 2.8331(18) |
| Mn1-N1 | 2.000(2) | 2.002(2) | Na1-O9″ | 2.332(2) | 2.3321(17) |
| Mn2-O1 | 1.879(2) | 1.8831(16) | Na2-O5 | 2.454(2) | 2.3548(19) |
| Mn2-O3 | 1.925(2) | 1.9203(18) | Na2-O6 | 2.519(3) | 2.7059(16) |
| Mn2-O4 | 1.883(2) | 1.8794(17) | Na2-O11 | 2.353(3) | 2.3335(16) |
| Mn2-O8 | 2.249(2) | 2.2661(17) | Na2-O5′ | 2.454(2) | 2.3548(19) |
| Mn2-O14 | 2.266(2) | 2.2729(18) | Na2-O6′ | 2.519(3) | 2.7059(16) |
| Mn2-N2 | 2.012(2) | 2.003(2) | Na2-O11′ | 2.353(3) | 2.3335(16) |
| Mn3-O1 | 1.873(2) | 1.8916(18) | Mn1-O1-Mn3 | 118.95(11) | 120.22(9) |
| Mn3-O5 | 1.912(2) | 1.9188(16) | Mn1-O1-Mn2 | 119.77(11) | 119.19(8) |
| Mn3-O6 | 1.873(2) | 1.8838(19) | Mn2-O1-Mn3 | 121.25(11) | 120.59(8) |
| Mn3-O10 | 2.158(2) | 2.2053(17) | Mn2-O3-N1-Mn1 | 28.6(2) | 26.71(19) |
| Mn3-O15 | 2.539(2) | 2.3599(18) | Mn3-O5-N2-Mn2 | 5.4(3) | 18.92(19) |
| Mn3-N3 | 2.010(2) | 1.9979(19) | Mn1-O7-N3-Mn3 | 1.0(3) | 21.6(2) |

(′) = 1−x, y, 3/2−z (**1**); 1−x, −y, 1−z (**2**). (″) = 1−x, 1−y, 1−z (**1**); 1−x, 1−y, 1−z (**2**).

Each sodium cation, Na1 and Na2, exhibits an $[NaO_6]$ coordination environment constructed by O9 and O9′ (for Na1) or O11 and O11′ (for Na2) from two of the bridging carboxylate ligands, by O3 and O3′ (for Na1) or O5 and O5′ (for Na2) from the oxime groups of two different $salox^{2-}$ ligands, and by the phenoxo oxygen atoms O4 and O4′ (for Na1) or O6 and O6′ (for Na2) from two other $salox^{2-}$ ligands. The Na-O distances are in the range of 2.332(2)-2.502(2) Å for the Na1 atom (2.3117(17)-2.8331(18) Å in **2**) and in the range of 2.353(3)-2.519(3) Å for the Na2 atom (2.3335(16)-2.7059(16) Å in **2**). The three $salox^{2-}$ ligands in each trinuclear unit therefore exhibit three different coordination modes: one in the $\eta^2{:}\eta^1{:}\eta^1{:}\mu_3$ mode (Scheme 3a), another in the $\eta^2{:}\eta^2{:}\eta^1{:}\mu_4$ mode (Scheme 3b), and the remaining one in the $\eta^1{:}\eta^2{:}\eta^1{:}\mu_3$ mode (Scheme 3c). The Na1 ion adopts a distorted octahedral geometry (elongated axial position due to bonding with oxygen atoms O4 and O4′; Figure 2a), while the Na2 ion does not exhibit a regular geometry (Figure 2b). It is important to note that, in compound **2**, the Na2 ion exhibits an octahedral geometry, although it is axially distorted with the axial axis formed by O11 and O11′ (Figure 2c). The distortion of the coordination sphere of Na ions has been calculated using the SHAPE program based on continuous shape measures (CShMs) [32] yielding a deviation values of 3.95, 5.85, and 5.69 from the ideal octahedron (OC-6) for Na1 in **1**, Na2 in **1**, and Na2 in **2**, respectively (Table S4). This structural difference is due to the different dihedral angles between the contiguous $\{Mn^{III}_3(\mu_3\text{-}O)(salox)_3\}^+$ triangular planes in compounds **1** and **2**, as previously discussed.

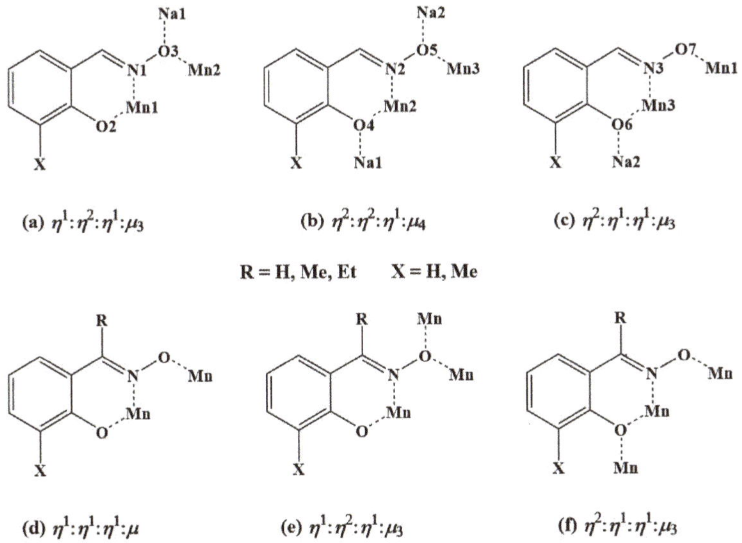


**Scheme 3.** Coordination modes of the R-salox$^{2-}$ ligands in compounds **1–6**.

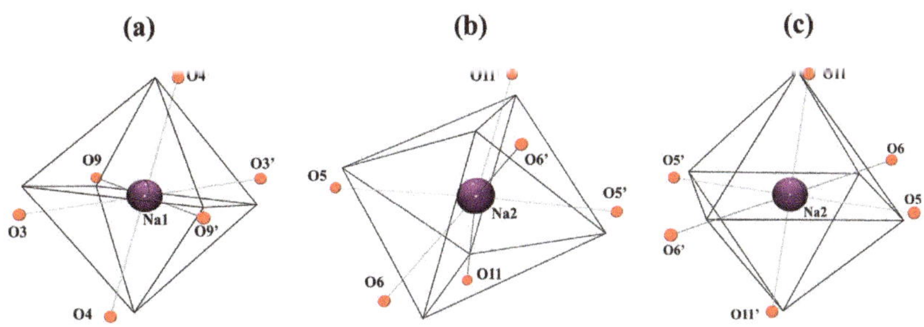


**Figure 2.** Coordination environments of cations (**a**) Na1, (**b**) Na2 for **1** and (**c**) Na2 for **2**.

2.1.2. Compound $[Mn_6O_2(salox)_6(9\text{-}AC)_2(EtOH)_2(H_2O)_2]\cdot 3EtOH$ (**3**·3EtOH)

The hexanuclear compound $[Mn_6O_2(salox)_6(9\text{-}AC)_2(EtOH)_2(H_2O)_2]\cdot 3EtOH$, shown in Figure 3, consists of two parallel trinuclear $\{Mn^{III}_3(\mu_3\text{-}O)(salox)_3\}^+$ planes, crystallographically related through an inversion center. The triangle formed by the three metal ions has sides with lengths of 3.267 Å (Mn1···Mn2), 3.245 Å (Mn2···Mn3), and 3.161 Å (Mn1···Mn3). The distances and angles of this compound are listed in Table 2

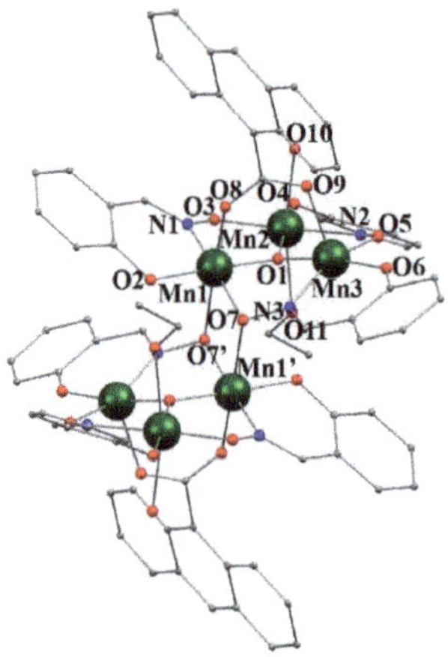


**Figure 3.** Structure of **3**, partially labeled. Hydrogen atoms have been omitted for clarity. Color code: dark green = Mn, red = O, blue = N, and gray = C.

**Table 2.** Most relevant bond distances (Å) and angles (°) of compounds **3**, **5**, and **6**.

| | 3 | 5 | 6 |
|---|---|---|---|
| Mn1-O1 | 1.8781(18) | 1.889(3) | 1.876(3) |
| Mn1-O2 | 1.8825(19) | 1.920(3) | 1.904(3) |
| Mn1-O7 | 1.960(2) | 1.955(3) | 1.951(3) |
| Mn1-O8 | 2.161(2) | 2.138(3) | 2.171(3) |
| Mn1-N1 | 1.993(3) | 2.007(5) | 2.011(4) |
| Mn1-O7′) | 2.332(2) | 2.434(3) | 2.415(3) |
| Mn2-O1 | 1.889(2) | 1.880(3) | 1.873(3) |
| Mn2-O3 | 1.931(2) | 1.924(3) | 1.921(3) |
| Mn2-O4 | 1.864(2) | 1.877(4) | 1.855(3) |
| Mn2-O10 | 2.272(2) | 2.262(3) | 2.373(5) |
| Mn2-O11 | 2.289(2) | 2.246(3) | 2.204(3) |
| Mn2-N2 | 2.023(2) | 2.025(4) | 2.001(4) |
| Mn3-O1 | 1.854(2) | 1.890(3) | 1.896(3) |
| Mn3-O5 | 1.927(2) | 1.919(3) | 1.916(3) |
| Mn3-O6 | 1.862(2) | 1.868(4) | 1.872(3) |
| Mn3-O9 | 2.132(2) | -- | -- |
| Mn3-O12 | -- | 2.246(3) | 2.231(3) |
| Mn3-O2′) | -- | 2.453(3) | 2.395(3) |
| Mn1-O1-Mn2 | 120.27(11) | 119.53(16) | 120.07(14) |
| Mn1-O1-Mn3 | 115.79(11) | 120.93(16) | 120.54(14) |
| Mn2-O1-Mn3 | 120.21(9) | 118.42(13) | 118.40(13) |
| Mn1-O2-Mn3′) | -- | 118.27(12) | 118.75(12) |
| Mn1-O7-Mn1′) | 98.05(8) | 94.23(12) | 96.48(11) |
| Mn2-O3-N1-Mn1 | 27.7(2) | 28.9(4) | 30.3(3) |
| Mn3-O5-N2-Mn2 | 21.8(3) | 34.7(4) | 37.3(3) |
| Mn1-O7-N3-Mn3 | 24.7(2) | 41.9(3) | 47.3(3) |

(′) = 1−x, 1−y, 1−z (**3**); 1−x, 1−y, −z (**5**); 1−x, 1−y, 1−z (**6**).

The two identical planes are connected via the two bonds between Mn1 and Mn1′ atoms with the bridging oxygen atoms O7′ and O7, respectively, from the oxime group of the corresponding salox$^{2-}$ ligand of the opposite triangular plane, which exhibit a $\eta^1$:$\eta^2$:$\eta^1$:$\mu_3$ coordination mode (Scheme 3e).

Within each triangular unit $\{Mn^{III}{}_3(\mu_3\text{-}O)(salox)_3\}^+$, the $\mu_3$-$O^{2-}$ (O1) anion, which is located at 0.210 Å above the plane formed by the three metal atoms, coordinates with the three Mn(III) ions. The three salox$^{2-}$ ligands act as bridges between two neighboring coplanar Mn(III) ions through the nitrogen (N1, N2, and N3) and oxygen (O3, O5, and O7) atoms of the oxime group, and also interact via the oxygen of the phenoxo group (O2, O4, and O6), exhibiting $\eta^1:\eta^1:\eta^1:\mu$ or $\eta^1:\eta^2:\eta^1:\mu_3$ coordination modes as shown in Scheme 3d,e, respectively. The torsion of the Mn–N–O–Mn angles is, in all cases, less than 31° (Mn1-Mn2 = 27.7(2)°, Mn2-Mn3 = 21.8(3)°, and Mn1-Mn3 = 24.7(2)°). Therefore, the Mn-Mn magnetic exchange via the salox$^{2-}$ ligands is expected to be antiferromagnetic. The Mn1 and Mn2 atoms have an octahedral coordination environment with Jahn–Teller distortions, while the Mn3 atom has a pyramidal geometry. The equatorial planes of the geometry of the three ions coincide with the triangular plane $\{Mn^{III}{}_3(\mu_3\text{-O})(salox)_3\}^+$. For Mn2, the axial positions are occupied by an oxygen atom from a water molecule (O10) and an oxygen atom from an EtOH molecule (O11). For Mn1, one of the axial positions is occupied, as mentioned before, by the oximate oxygen atom O7′ connecting the two trinuclear planes. The remaining axial position is occupied by one of the oxygen atoms (O8) from a 9-AC anion, which also binds, through a *syn-syn* coordination, to the Mn3 ion via O9 occupying the corresponding axial position. The axial axes of all manganese atoms are almost parallel to each other and perpendicular to the $\{Mn^{III}{}_3(\mu_3\text{-O})(salox)_3\}^+$ planes.

The $H_2O$ and EtOH molecules coordinated to Mn2 form intramolecular hydrogen bridges. The hydrogen atoms H11A of the coordinated EtOH molecules form a bond with the nearest phenolic oxygen atom O2, belonging to one of the salox$^{2-}$ ligands of a neighboring triangle. In the case of the $H_2O$ molecules, each hydrogen H10A forms a bridge with the oxygen atom O9 of the neighboring carboxylate, while the H10B atoms interact with the oxygen atom O1M of a free EtOH molecule. The hydrogen of this solvent molecule also interacts via a hydrogen bridge with the oximate oxygen atom O5 of a salox$^{2-}$ ligand belonging to a neighboring hexanuclear unit, forming chains of $[Mn_6O_2(salox)_6(9\text{-}AC)_2(EtOH)_2(H_2O)_2]$ molecules along the [010] direction of the cell (Figure S1). The structural parameters of these supramolecular contacts are listed in Table S3.

### 2.1.3. Compound $[Mn_3O(3\text{-}Me\text{-}salox)_3(9\text{-}AC)(EtOH)_3(H_2O)]\cdot 1.8EtOH\cdot 3H_2O$ ($\mathbf{4}\cdot 1.8EtOH\cdot 3H_2O$)

The asymmetric unit of coordination compound **4**, which crystallizes in the P-1 space group of the triclinic system, consists of two molecules $[Mn_3O(3\text{-}Me\text{-}salox)_3(9\text{-}AC)(EtOH)_3(H_2O)]$, referred to as A and B. These molecules exhibit differences in structural parameters but identical coordination. The most relevant distances and angles of molecules A and B in compound **4** are listed in Table 3. Figure 4 shows the A molecule.

**Table 3.** Selected bonds distances (Å) and angles (°) in fragments A and B of **4**.

| Compound 4 | | | | | |
|---|---|---|---|---|---|
| | A | B | | A | B |
| Mn1-O1 | 1.886(2) | 1.864(2) | Mn3-O1 | 1.867(2) | 1.890(2) |
| Mn1-O2 | 1.857(2) | 1.864(2) | Mn3-O5 | 1.921(2) | 1.897(2) |
| Mn1-O7 | 1.940(2) | 1.912(3) | Mn3-O6 | 1.853(3) | 1.889(2) |
| Mn1-O8 | 2.240(2) | 2.095(2) | Mn3-O9 | 2.220(2) | 2.224(2) |
| Mn1-O10 | 2.270(3) | 3.769(3) | Mn3-O13 | 2.349(3) | 2.282(3) |
| Mn1-N1 | 1.998(3) | 2.005(3) | Mn3-N3 | 2.012(3) | 2.008(3) |
| Mn2-O1 | 1.869(2) | 1.887(2) | Mn1-O1-Mn2 | 121.09(12) | 120.64(11) |
| Mn2-O3 | 1.916(2) | 1.928(2) | Mn1-O1-Mn3 | 116.73(12) | 114.59(11) |
| Mn2-O4 | 1.866(2) | 1.874(2) | Mn2-O1-Mn3 | 120.28(12) | 119.46(12) |
| Mn2-O11 | 2.283(3) | 2.231(3) | Mn2-O3-N1-Mn1 | 28.9(3) | 15.6(3) |
| Mn2-O12 | 2.353(3) | 2.233(3) | Mn3-O5-N2-Mn2 | 26.7(3) | 16.4(3) |
| Mn2-N2 | 2.021(3) | 2.006(3) | Mn1-O7-N3-Mn3 | 32.1(3) | 18.9(3) |

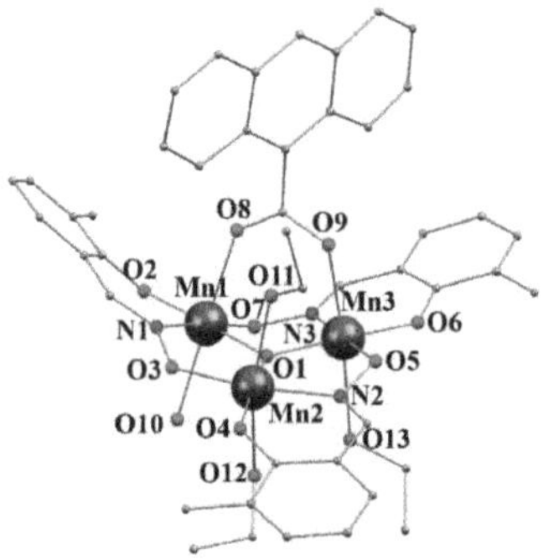


**Figure 4.** Partially labeled illustration of the molecule $[Mn_3O(3\text{-Me-salox})_3(9\text{-AC})(EtOH)_3(H_2O)](A)$ in compound **4**. Hydrogen atoms have been omitted for clarity.

Each $[Mn_3O(3\text{-Me-salox})_3(9\text{-AC})(EtOH)_3(H_2O)]$ A or B molecule shows the $\{Mn^{III}_3(\mu_3\text{-}O)(3\text{-Me-salox})_3\}^+$ plane coplanar with the equatorial bases of the axially distorted octahedral geometry of the three Mn(III) ions. The Mn1 and Mn3 ions are connected by a bridging 9-AC ligand in a *syn-syn* coordination mode by the oxygen atoms O8 and O9, which occupy one of the axial positions of each of the two Mn(III) ions. The remaining axial positions are occupied by O10 from a water molecule in the case of Mn1, and by O13 from an EtOH molecule in the case of the Mn3 ion. The Jahn–Teller axis of the Mn2 atom is established by the oxygen atoms O11 and O2 of two EtOH molecules.

The distances in the triangle between the three Mn(III) ions are 3.269 Å, 3.240 Å, and 3.196 Å in fragment A and 3.259 Å, 3.262 Å, and 3.159 Å in fragment B. The torsion angles of molecules A and B differ significantly. The Mn-N-O-Mn torsions are 28.9(3)°, 26.7(3)°, and 32.1(3)° in molecule A, and 15.6°, 16.4(3)°, and 18.9° in molecule B. Therefore, the $\{Mn^{III}_3(\mu_3\text{-}O)(3\text{-Me-salox})_3\}^+$ plane in molecule B is much less distorted than in molecule A. This causes the axial bonds Mn1B-O8B and Mn1B-10B to be much shorter and considerably longer, respectively, compared to the equivalent ones in molecule A.

The solvent molecules coordinated to the Mn(III) ions form hydrogen bonds. The structural parameters are collected in Table S3 (Supplementary Materials). For fragment A, the oxygen atom O10A of the water molecule forms a bond with one of the water molecules (O1W) that co-crystallizes with the compound and another bond with the oxygen atom O6B of the phenolate group of one of the 3-Me-salox$^{2-}$ ligands from the other neutral trinuclear fragment. The hydrogen atoms bonded to the oxygen atoms O11A, O12A, and O13A of the three coordinated ethanol molecules interact with oxygen atoms O9A, O10A, and O1W, which are from the bridging carboxylate ligand, the water molecule bonded to the Mn1 ion and the free water molecule, respectively. In fragment B, the water coordinated to the Mn1 ion (O10B) forms bridges with the oxygen atoms O5A and O9A of the other trinuclear fragment. One of the two ethanol atoms (O12B) coordinated to the Mn2 atom forms a hydrogen bond with the water molecule (O10B). The other solvent molecule interacts via O11B with the water molecule of crystallization O1W. The hydrogen bonded to the oxygen atom O13B of EtOH coordinated to the Mn3 atom also forms a bridge with the water molecule O10B. As a result of these supramolecular interactions, chains of $[Mn_3O(3\text{-Me-salox})_3(9\text{-AC})(EtOH)_3(H_2O)]$ units are constructed along the [100] direction of the crystal lattice (Figure S2).

### 2.1.4. Compounds $[Mn_6O_2(\text{Me-salox})_6(9\text{-AC})_2(EtOH)_4(H_2O)_2]\cdot 0.5H_2O$ (**5**·$0.5H_2O$) and $[Mn_6O_2(\text{Et-salox})_6(9\text{-AC})_2(EtOH)_4(H_2O)_2]\cdot 3EtOH$ (**6**·3EtOH)

Figure 5 shows the structure of compounds **5** and **6**. As previously explained, these two compounds are obtained starting from either the $Na^+$ carboxylate or $Cs^+$ carboxylate salts. Although the hexanuclear core of compound **6**·3EtOH, has been published previously [33], there are differences with respect to the crystallization solvent molecules in the unit cell; for compound **6**·3EtOH, three molecules of EtOH crystallize, whereas in the previously published results, compounds with the same hexanuclear core crystallize 0.66 molecules of EtOH and 0.33 molecules of $H_2O$.

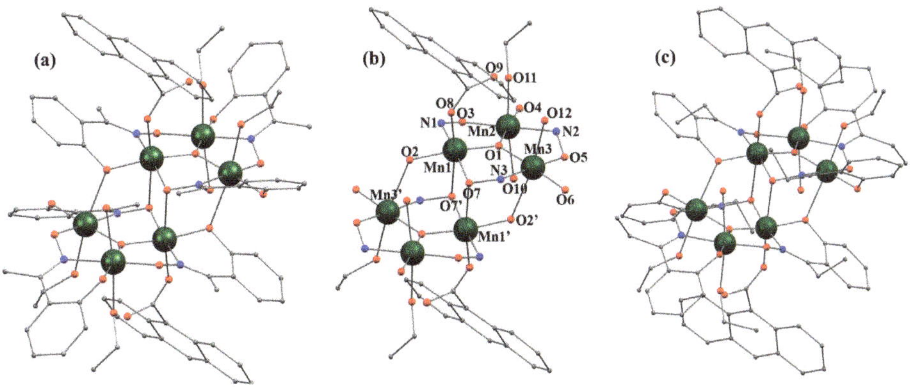


**Figure 5.** (**a**,**c**) Hexanuclear compounds **5** and **6**, respectively. Figure (**b**) represents the partially labeled hexanuclear core of these two structures. The hydrogen atoms and solvent molecules have been omitted for clarity. Color code: dark green = Mn, red = O, blue = N, and gray = C.

Given the high similarity between the structures of compounds **5** and **6**, the structural description of the hexanuclear cores are provided using compound **5** as the example.

The structure of the compound $[Mn_6O_2(Me\text{-}salox)_6(9\text{-}AC)_2(EtOH)_4(H_2O)_2]$ (**5**) consists of two identical parallel planes $\{Mn^{III}{}_3(\mu_3\text{-}O)(Me\text{-}salox)_3\}^+$ related by an inversion center. BVS calculations and the charge balance of the compound indicate that the six metals are in the +3-oxidation state (Table S2). Additionally, all Mn(III) ions exhibit an octahedral coordination geometry with axial Jahn–Teller distortions. The axial axes are almost completely parallel to each other and perpendicular to the two $\{Mn^{III}{}_3(\mu_3\text{-}O)(Me\text{-}salox)_3\}^+$ planes. In the $[Mn]_3$ triangles, the distances between the Mn(III) atoms are similar, being 3.256 Å (Mn1···Mn2), 3.238 Å (Mn2···Mn3), and 3.287 Å (Mn1···Mn3) for compound **5**, and 3.248 Å (Mn1···Mn2), 3.237 Å (Mn2···Mn3), and 3.275 Å (Mn1···Mn3) for compound **6**.

In each $\{Mn^{III}{}_3(\mu_3\text{-}O)(Me\text{-}salox)_3\}^+$ unit, the plane formed by the $[Mn_3]$ ions coincides with the equatorial plane of the coordination environment of these ions. One of the four equatorial positions in the coordination sphere of the Mn(III) ions is occupied by the $\mu_3\text{-}O^{2-}$ anion (O1), which is found slightly above the plane formed by the three metal atoms (0.116 Å for **5** and 0.109 Å for **6**). The remaining three equatorial positions are occupied by a phenoxo oxygen atom (O2, O4, or O6), a nitrogen atom (N1, N2, or N3) from the $Me\text{-}salox^{2-}$ ligand, and an oxygen atom from the oximate group (O3, O5, or O7) of another $Me\text{-}salox^{2-}$ ligand. The oximate group of each $Me\text{-}salox^{2-}$ ligand acts as a bridge between two coplanar Mn(III) ions, with Mn-N-O-Mn torsion angles of 28.9(4)° for Mn1-Mn2 (30.3(3)° for **6**), 34.7(4)° for Mn2-Mn3 (37.3(3)° for **6**), and 41.9(3)° for Mn1-Mn3 (47.3(3)° for **6**). The axial positions of the Mn1 ions are formed by O8 from a terminal anthracenecarboxylate and O7′ from the oximato group of an $Me\text{-}salox^{2-}$ anion placed at the opposite plane. For the Mn2 atom, the axial positions are occupied by two oxygen atoms, O10 and O11, which are part of a water molecule and an EtOH molecule, respectively. The axial bond of Mn3 with oxygen atom O2′ of the phenoxo group from one of the $Me\text{-}salox^{2-}$ entities also serves as a connection between the two $\{Mn^{III}{}_3(\mu_3\text{-}O)(Me\text{-}salox)_3\}^+$ planes. The remaining axial position is occupied by oxygen atom O12 from an EtOH molecule. Therefore, two of the six $Me\text{-}salox^{2-}$ ligands exhibit a $\eta^1{:}\eta^1{:}\eta^1{:}\mu$ coordination (Scheme 3d), acting as bridges between pairs of Mn(III) ions within the same triangular unit, and the remaining four also connect the two trinuclear planes via oximato group oxygens in the $\eta^1{:}\eta^2{:}\eta1{:}\mu_3$ coordination mode (Scheme 3e) or the hydroxo group ($\eta^2{:}\eta^1{:}\eta^1{:}\mu_3$ coordination mode; Scheme 3f).

In compound **5**, different intramolecular hydrogen bonds are observed. On the one hand, oxygen atom O9 of the carboxylate interacts with hydrogen atoms H11A and H12A of the two EtOH molecules coordinated to Mn2 and Mn3, respectively. On the other hand, the water coordinated to the Mn2 ion forms a hydrogen bond through hydrogen atom H10A with oxygen atom O2′ and through the H10B atom with the oxygen atom of a free water molecule (O1W). In the case of compound **6**, the H10B atom interacts with oxygen atom O7 of the oximate group of a nearby Me-salox$^{2-}$ ligand, while the hydrogen atom H10A forms a bond with a free EtOH molecule (O1M). Although it has not been possible to assign the hydrogen atoms of the free EtOH molecule (O1M), it is observed that this molecule forms a hydrogen bond with another EtOH molecule (O1M′), leading to a 1D arrangement of $[Mn_6O_2(Et\text{-}salox)_6(9\text{-}AC)_2(EtOH)_4(H_2O)_2]$ units along the [110] direction of the crystalline network. The distances and angles of these contacts are listed in Table S3.

It should be highlighted that the differences in the number and type of crystallization solvent molecules between compound **6**·3EtOH and the previously published compound **6**·0.66EtOH·0.33$H_2O$ [33] have been observed, and this affects the structure of the hexanuclear complex and, possibly, its magnetic properties. In the case of complex **6**·3EtOH, the Mn-N-O-Mn torsion angles within the $[Mn^{III}]_3$ plane are 30.3° (Mn1-Mn2), 37.3° (Mn2-Mn3), and 47.9° (Mn1-Mn3), whereas for the previously published compound **6**·0.66EtOH·0.33$H_2O$, the torsion angles are 42.30° (Mn1-Mn2), 25.6° (Mn2-Mn3), and 39.3° (Mn1-Mn3).

2.1.5. Structural Discussion

In compounds **1–6**, BVS calculations determined an oxidation state of +3 for the manganese ions present, even though Mn(II) was used initially. The oxidation of the Mn(II) cation is due to atmospheric oxygen. The deprotonation of the R-salicyladoxime species occurs via the anions of the starting Mn(II) salts and the bases present in the reaction medium. All attempts to obtain the manganese (II) anthracenecarboxylate salt failed. Therefore, it was decided to start with Mn(II) salts with nitrate and perchlorate anions, which are less susceptible to coordinate with this metal [34].

As previously mentioned, to obtain new topologies of compounds based on the triangular unit $\{Mn^{III}_3(\mu_3\text{-}O)(R\text{-}salox)_3\}^+$, new variables were introduced into the synthesis process. On one hand, the $H_2$-3-Me-salox ligand, with a $CH_3$ group at position 3 of the phenyl ring and the oxime group at position 1, was used. Another variable introduced was the alkali cation used in the starting salt of the anthracenecarboxylate anion. As summarized in Scheme 4, for the R-salox$^{2-}$ ligands with X = H and 3-Me, both having a hydrogen bonded to the carbon of the oxime group, it was observed that using sodium salt yielded chain-like structures $[Mn_3NaO(salox)_3(9\text{-}AC)_2(EtOH)_3H_2O]_n$·2nEtOH (**1**·2EtOH) and $[Mn_3NaO(3\text{-}Me\text{-}salox)_3(9\text{-}AC)_2(EtOH)_3H_2O]_n$·nEtOH (**2**·EtOH) consisting of triangles connected by $Na^+$ cations. On the other hand, the use of the cesium carboxylate salt did not produce chain-like compounds; instead, the hexanuclear compound $[Mn_6O_2(salox)_6(9\text{-}AC)_2(EtOH)_2(H_2O)_2]$·3EtOH (**3**·3EtOH) and the trinuclear compound $[Mn_3O(3\text{-}Me\text{-}salox)_3(9\text{-}AC)(EtOH)_3(H_2O)]$·1.8EtOH·3$H_2O$ (**4**·1.8EtOH·3$H_2O$) were obtained, due to the larger ionic radius of the $Cs^+$ cation compared to $Na^+$. For the Me-salox$^{2-}$ and Et-salox$^{2-}$ ligands, the hexanuclear compounds $[Mn_6O_2(Me\text{-}salox)_6(9\text{-}AC)_2(EtOH)_4(H_2O)_2]$·0.5$H_2O$ (**6**·0.5$H_2O$) and $[Mn_6O_2(Et\text{-}salox)_6(9\text{-}AC)_2(EtOH)_4(H_2O)_2]$·3EtOH (**6**·3EtOH), respectively, were obtained regardless of whether the cesium or sodium anthracenecarboxylate salt was used. This is due to the distortion caused by the alkyl substituent groups on the oxime group of the different R-salox$^{2-}$ ligands on the Mn-N-O-Mn torsion angles.

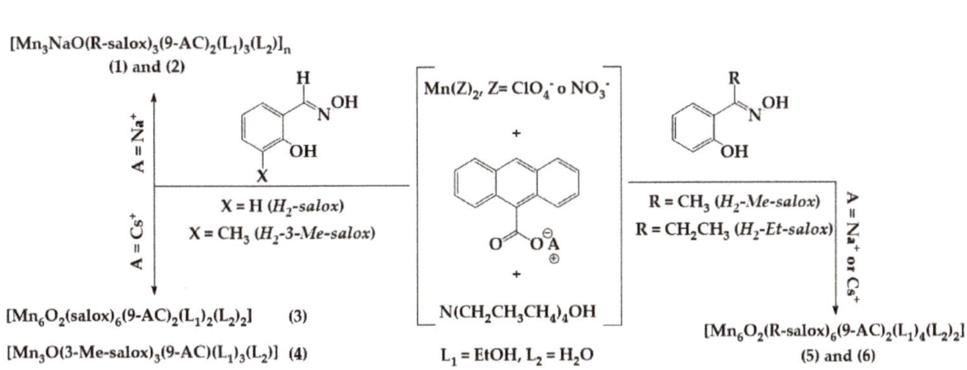


**Scheme 4.** Synthetic routes to obtain compounds **1–6**.

Generally, hexanuclear compounds with two $\{Mn^{III}_3(\mu_3\text{-}O)(R\text{-}salox)_3\}^+$ units present two types of structures. For small substituent group volumes (R = H), as in compound **3**, the two triangles tend to connect through two oxime bridges, with each Mn(III) ion exhibiting a square pyramidal coordination, while the other two Mn(III) ions have a distorted octahedral geometry [7,35]. The coordination of the carboxylate ligand in this type of architecture is syn-syn. On the other hand, when using R-salox$^{2-}$ with more voluminous groups, as in compounds **5** (R = Me) and **6** (R = Et), the inter-triangular Mn–O bridge distances with the oxime group oxygen atom tend to decrease due to the increased Mn–N–O–Mn distortion, and the connection between triangles also occurs via two phenolate bridges. As a result, all three Mn(III) ions are in a distorted octahedral environment. Another structural consequence of varying the R-salox$^{2-}$ ligand is that, as the volume of the R group increases, the triangles lose planarity, the carboxylate cannot coordinate in a syn-syn mode, and instead adopts a terminal mode, with the axial position of the Mn(III) occupied by a solvent molecule [6]. However, the different coordination modes of the carboxylate anion do not significantly influence the magnetic properties of the compounds.

## 2.2. Magnetic Properties

### 2.2.1. *dc* Magnetic Studies

The measurements of the magnetic susceptibility of compounds **1–6** were carried out in a range of temperatures, (2–300) K, under a magnetic field (dc) of 0.5 T for compound **1** and 0.3 T for compounds **2–6**. Figure 6a shows the thermal dependence of the susceptibility of these compounds in the form of the $\chi_M T$ product.

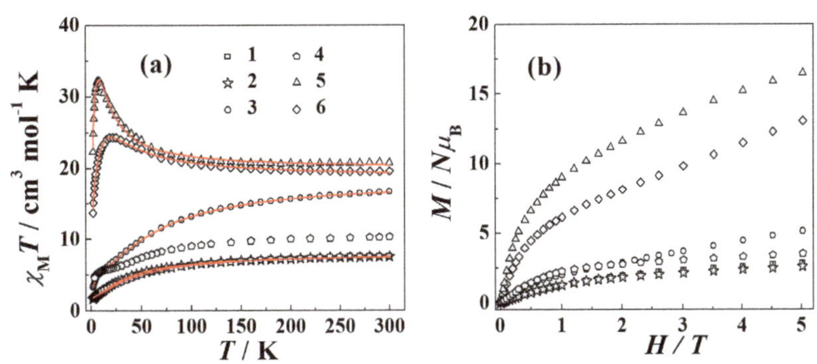


**Figure 6.** (**a**) $\chi_M T$ curves as a function of T for compounds **1–6**. The solid lines represent the theoretical fits according to the models discussed in the text. (**b**) Curves of magnetization ($M/N\mu_B$) as a function of the external magnetic field (H) for compounds **1–6**. The symbols have the same meanings as in (**a**).

Compounds **1** and **2** exhibit a $\chi_M T$ value at room temperatures of 7.51 and 7.49 $cm^3\ mol^{-1}$ K, respectively, which is significantly lower than the expected value for three non-interacting high-spin Mn(III) centers (9 $cm^3\ mol^{-1}$ K for S = 2 and g = 2.0). In both complexes, the $\chi_M T$ curve remains constant as the temperature decreases to approximately 90 K. From this temperature, a more pronounced drop in the curves is observed, reaching a $\chi_M T$ value of 1.60 $cm^3\ mol^{-1}$ K for **1** and 1.87 $cm^3\ mol^{-1}$K for **2** at 2 K. The experimental data could be fitted using a model assuming three types of magnetic interactions ($J_1$, $J_2$, and $J_3$, Equation (1), Scheme 5a) between each pair of neighboring Mn(III) ions within each triangle, through the oxo group (O1) and the corresponding oxime groups of the ligands derived from salicylaldoxime.

$$H = -J_1(S_1 \cdot S_2) - J_2(S_2 \cdot S_3) - J_3(S_1 \cdot S_3) \quad (1)$$

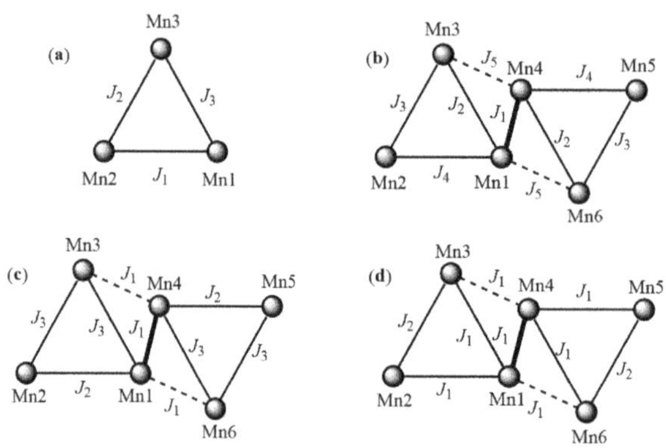


**Scheme 5.** Representation of the magnetic interaction models based on the Hamiltonians described in the text used to fit the experimental data of $\chi_M T$ as a function of T (**a**) for compounds **1** and **2**, (**b**) for **3**, (**c**) for **5** and **6**, and (**d**) for **6**·0.66EtOH·0.33$H_2O$ (**d**) [33].

The obtained values from the fittings indicate a set of antiferromagnetic interactions promoted by the torsion angles less than 30.4° within the triangles, being $J_1$ = −7.3 $cm^{-1}$, mboxemphJ$_2 \approx J_3$ = −2.5 $cm^{-1}$, g = 1.90 with an error R = 5.2 × $10^{-4}$, and S = 1 with an excited state S = 2 at 0.32 $cm^{-1}$ for **1**. For compound **2**, the values are $J_1$ = −6.02 $cm^{-1}$, $J_2 \approx J_3$ = −2.2 $cm^{-1}$, g = 1.88 with R = 2.1 × $10^{-4}$, and S = 1, with a very close energy excited state, S = 2 (at 0.43 $cm^{-1}$).

The susceptibility vs. temperature plot of hexanuclear compound **3** shows a $\chi_M T$ value of 16.62 $cm^3\ mol^{-1}$ K at 300 K, slightly lower than the expected 18 $cm^3\ mol^{-1}$ K for six fully uncoupled Mn(III) ions at this temperature, indicating possible antiferromagnetic interactions. As the temperature decreases, the $\chi_M T$ curve gradually decreases as well. At approximately 10 K, a change in the slope of the curve is observed, and for temperatures below this, $\chi_M T$ drops more abruptly, reaching a $\chi_M T$ value of 3.29 $cm^3\ mol^{-1}$ K at 2 K. The experimental measurements could be fitted based on a model that assumes five different types of magnetic interactions, depending on the different bridges between Mn(III) ions, as described in Equation (2) and Scheme 5b.

$$\begin{aligned} H = &-J_1(S_1 \cdot S_4) - J_2(S_1 \cdot S_3 + S_4 \cdot S_6) - J_3(S_2 \cdot S_3 + S_5 \cdot S_6) - J_4(S_1 \cdot S_2 + S_4 \cdot S_5) \\ &-J_5(S_3 \cdot S_4 + S_1 \cdot S_6) \end{aligned} \quad (2)$$

The best fitting parameters obtained yield values of $J_1$ = +3.8 $cm^{-1}$, $J_2$ = −6.0 $cm^{-1}$, $J_3$ = −9.9 $cm^{-1}$, $J_4$ = −5.2 $cm^{-1}$, $J_5$ = +1.2 $cm^{-1}$, g = 2.04 with R = 7.3 × $10^{-6}$, and S = 2 (with an excited state S = 3 located at 4.29 $cm^{-1}$). Therefore, the intra-triangular magnetic exchanges (with observed torsion angles less than 30.4°) are of the antiferromagnetic type, while the magnetic interaction between triangles is of the ferromagnetic type.

The $\chi_M T$ versus temperature curve of compound **4**, for only one of the triangular units, shows a value of 10.30 $cm^3\ mol^{-1}$ K at T = 300 K, slightly above the expected value for three fully uncoupled Mn(III) ions. As the temperature decreases, the curve declines, suggesting the presence of antiferromagnetic intramolecular couplings. $\chi_M T$ reaches a value of 3.73 $cm^3\ mol^{-1}$ K at T = 2 K. Due to the fact that **4** is composed by two [$Mn_3O$(3-Me-salox)$_3$(9-AC)(EtOH)3($H_2O$)] molecules with different structural parameters, an analysis of the susceptibility data has not been possible as it requires extensive parameterization.

For compounds **5** and **6**, the $\chi_M T$ values at room temperature are 20.75 and 19.55 $cm^3\ mol^{-1}$ K, respectively, above the expected value at this temperature (18 $cm^3\ mol^{-1}$ K), suggesting the existence of ferromagnetic coupling. As the temperature decreases, the curves gradually increase to 21.39 $cm^3\ mol^{-1}$ K for **5** and 20.80 $cm^3\ mol^{-1}$ K for **6** at T = 100 K. From this temperature, the $\chi_M T$ values rise rapidly, reaching maximum values of 32.16 $cm^3\ mol^{-1}$ K at T = 7 K (**5**) and 24.37 $cm^3\ mol^{-1}$ K at T = 19 K (**6**). Subsequently, the curves drop to $\chi_M T$ values of 22.30 and 13.67 $cm^3\ mol^{-1}$ K for **5** and **6**, respectively, at T = 2 K. The fitting of the experimental susceptibility measurements was performed for both compounds using Equation 3, considering three different types of magnetic exchange interactions (J), as represented in Scheme 5c. $J_1$ corresponds to the coupling between the two [$Mn_3$] triangles, $J_2$ to the coupling between Mn1 and Mn2 ions mediated by the oxime groups and the oxo bridge with torsion angles of 28.9° (**5**) and 30.3° (**6**), which is below the magic angle region, and $J_3$ corresponds to the intra-triangular coupling between Mn1-Mn3 and Mn2-Mn3 ions, with torsion angles within the ferromagnetic range for both compounds.

$$H = -J_1(S_1{\cdot}S_4 + S_3{\cdot}S_4 + S_1{\cdot}S_6) - J_2(S_1{\cdot}S_2 + S_4{\cdot}S_5) \\ -J_3(S_1{\cdot}S_3 + S_2{\cdot}S_3 + S_4{\cdot}S_6 + S_5{\cdot}S_6) \tag{3}$$

The fittings performed indicate values of $J_1$ = +2.3 $cm^{-1}$, $J_2$ = −2.0 $cm^{-1}$, $J_3$ = +1.7 $cm^{-1}$, and g = 2.09 with R = 1.2 × $10^{-4}$ and S = 5 with very close excited states, S = 6 (at 0.24 $cm^{-1}$) and S = 4 (at 0.49 $cm^{-1}$), for complex **5**. For compound **6**, the values obtained are $J_1$ = +3.18 $cm^{-1}$, $J_2$ = −5.2 $cm^{-1}$, $J_3$ = +3.5 $cm^{-1}$, g = 2.03 with R = 5.8 × $10^{-5}$, and S = 4 with an excited state of S = 5 at 0.22 $cm^{-1}$.

The results for compound **6**·3EtOH can be compared with those reported for **6**·0.66EtOH·0.33$H_2O$, where $\chi_M T$ versus temperature curves show similar $\chi_M T$ values at T = 300 K, being 19.55 $cm^3\ mol^{-1}$ K for **6**·3EtOH and 19.80 $cm^3\ mol^{-1}$ K for **6**·0.66EtOH·0.33$H_2O$. As the temperature decreases, the $\chi_M T$ curves of both compounds increase, but while the maximum $\chi_M T$ value for complex **6**·3EtOH is 24.37 $cm^3\ mol^{-1}$ K (at T = 100 K), compound **6**·0.66EtOH·0.33$H_2O$ shows a maximum $\chi_M T$ value of 50.53 $cm^3\ mol^{-1}$ K (at T ≈ 10 K). The fitting of the experimental $\chi_M T$ data of compound **6**·0.66EtOH·0.33$H_2O$ was performed considering a Hamiltonian with only two magnetic coupling constants (Scheme 5d), yielding parameters of $J_1$ = +3.50 $cm^{-1}$ and $J_2$ = −1.80 $cm^{-1}$, g = 2.00, and showing a ground state of S = 12 with an excited state of S = 11 at 0.79 $cm^{-1}$. Comparing these results with those obtained for compound **6**·3EtOH ($J_1$ = +3.18 $cm^{-1}$, $J_2$ = −5.2 $cm^{-1}$, $J_3$ = +3.5 $cm^{-1}$, g = 2.03, and S = 4) clearly shows that structural variations promote magnetic coupling between the Mn(III) ions with a greater contribution of antiferromagnetic exchanges in **6**·3EtOH, resulting in a lower-magnitude ground state, Table 4.

Magnetization measurements for compounds **1–6** were performed under magnetic fields in the range of [0–5] T at a temperature of 2 K. As shown in Figure 6b, no magnetic saturation was observed in any case, confirming the presence of spin-excited states very close in energy to the ground state. This is the reason why the experimental data could not be fitted.

**Table 4.** Torsion angles (°), expected and observed net coupling, and $J_i$ interactions.

| Compound | Mn-O-N-Mn (°) [(a)] | Expected Net Coupling | Observed Net Coupling | *Ji* (*i* = 1–5) ($cm^{-1}$) [(c)] |
|---|---|---|---|---|
| **1** | 28.6, 5.4, 1.0 | AF | AF | −7.3, −2.5, −2.5 |
| **2** | 26.7, 18.9, 21.6 | AF | AF | −6.02, −2.2, −2.2 |
| **3** | 27.7, 21.8, 24.7 | AF | AF | +3.8, −6.0, −9.9, −5.2, +1.2 |
| **4-A** [(b)] | 28.9, 26.7, 32.1 | AF | - | - |
| **4-B** [(b)] | 15.6, 16.4, 18.9 | AF | - | - |
| **5** | 28.9, 34.7, 41.9 | AF/F | F | +2.3, −2.0, +1.7 |
| **6·3EtOH** | 30.3, 37.3, 47.3 | AF/F | F | +3.18, −5.2, +3.5 |
| **6·0.66EtOH·0.33H$_2$O** [(d)] | 42.3, 25.6, 39.3 | AF/F | F | +3.5, −1.8 |

[(a)] Torsion angle within each triangular unit. [(b)] Compound **4** is formed by two [$Mn_3O$(3-Me-salox)$_3$(9-AC)(EtOH)3($H_2O$)] molecules with different structural parameters. The analysis of the susceptibility data has not been possible as it requires extensive parameterization. [(c)] The magnetic interaction, *Ji*, did not correspond to the same set in all the compounds; see the magnetic interaction representation in Scheme 5. [(d)] Compound reported in the reference [33].

### 2.2.2. Dynamic Magnetic Measurements (*ac*)

For compounds **5** and **6**, *ac* magnetization was performed at frequencies in the range of (5–1488) Hz for **5** and (1–1488) Hz for **6**, applying an *ac* field of $4 \times 10^{-4}$ T. Figure 7 shows the thermal dependencies of the in-phase ($\chi_M'$) and out-of-phase ($\chi_M''$) components of susceptibility for both compounds. Maxima were observed at approximately 3 K for **5** and 3.4 K for **6**. In the graphs of the two susceptibility components of compound **5**, a shoulder at higher temperatures can be discerned, as observed in other similar Mn(III) compounds [36], which may indicate the presence of a second relaxation process. However, the analysis of the dynamic magnetic properties of compound **5** was performed, only considering the maxima observed at low temperatures.

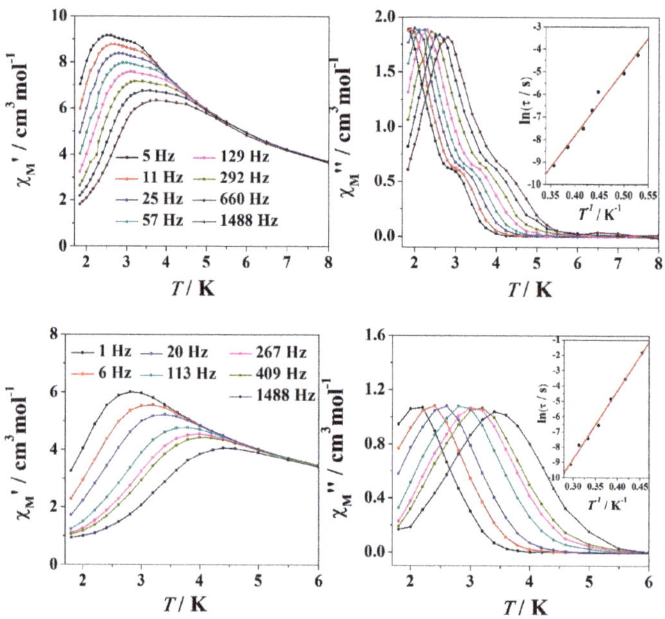


**Figure 7.** Curves $\chi_M'$ (**left**) and $\chi_M''$ (**right**) as a function of temperature for compound **5** (**top**) and for compound **6** (**bottom**). In the inset of each $\chi_M''$ graph is shown the dependence of ln (τ) versus $T^{-1}$. The red lines represent the fits of Arrhenius. The color lines have the same meanings as in (**left**).

The susceptibility components of both compounds also show frequency dependence, as the curves shift to lower temperatures with decreasing frequency, typical behavior of single-molecule magnets (SMMs). From the maxima of the $\chi_M''$ curves, the thermal dependence of the relaxation time for the process was obtained, which can be determined according to the Arrhenius law, yielding effective energy barrier ($U_{eff}$) and pre-exponential factor ($\tau_0$) values of 19.90 $cm^{-1}$ and $4.19 \times 10^{-9}$ s for **5**, and 31.1 $cm^{-1}$ and $2.27 \times 10^{-10}$ s for **6**, while those of the reported compound **6**·0.66EtOH·0.33$H_2O$ are 45.9 $cm^{-1}$ and $3.99 \times 10^{-10}$ s. The zero-field splitting parameter (D), calculated according to $U_{eff} = |D| \cdot S^2$, is 0.76 $cm^{-1}$ for **5** and 1.94 $cm^{-1}$ for **6**, are similar to the values found in other hexanuclear Mn(III) compounds [13,34].

### 2.3. Luminescence Studies

Due to the presence of the 9-anthracenecarboxylate ligand in compounds **1–6**, a UV-visible study was conducted for both the free ligand and the mentioned compounds (Figure 8). The spectra were measured in ethanolic solution at a concentration of $10^{-5}$ M. In all compounds, the absorbance peaks of the π–π* transitions of the carboxylate ligand were observed to be slightly shifted toward red compared to the spectrum of the free ligand. The decrease in the energy of the ligand transitions is due to its coordination with Mn(III) metals [33]. The quenching effect of Mn(III) and Na+ ions (in the cases of compounds **1** and **2**) decreases the signal intensity compared to that of the free ligand [37].

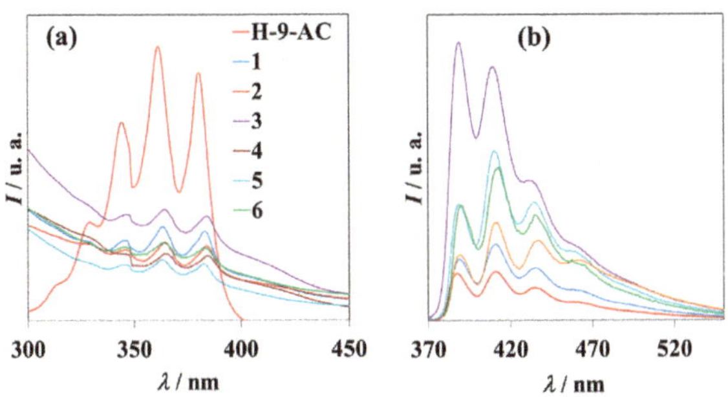


**Figure 8.** (**a**) Absorption of the Na(9-AC) species and of compounds **1–6**, and (**b**) emission spectra ($\lambda_{em}$ = 364 nm) of compounds **1–6** in ethanol solution ($10^{-5}$ M) measured at room temperature. Color lines have the same meanings as in (**a**).

In view of these results, the emission measurements (Figure 8), also in solution, reveal the ligand transitions under the excitation of 364 nm, centered approximately at 389 nm, 410 nm, 434 nm, and 460 nm (Figure 8b). The slight variations in the positions and relative intensities of the emission peaks are due to the different coordination modes of the 9-AC ligand in each compound [20].

## 3. Experimental Section

### 3.1. Starting Materials

$Mn(X)_2 \cdot 6H_2O$ salts (X = $ClO_4^-$, $NO_3^-$), 9-anthracenecarboxylic acid (H-9-AC)), sodium and cesium hydroxide, ethanol, tetrapropylammonium hydroxide, tetramethylammonium hydroxide, and tetrabutylammonium hydroxide were obtained from Sigma Aldrich and used without further purification.

The ligands $H_2$-3-Me-salox, $H_2$-Me-salox, and $H_2$-Et-salox were synthesized following the method previously described in the literature [38], employing aldehyde and EtOH as the solvents. Sodium and cesium 9-anthracenecarboxylato salts were prepared according to the synthesis previously described [39], starting from 9-anthracenecarboxylic acid and either sodium or cesium hydroxide. The syntheses of compounds **1–6** were conducted under aerobic conditions using the same experimental procedure, varying the anthracenecarboxylic salt and the ligand $H_2$-R-salox (A = $Na^+$ and R = H for **1**, A = $Na^+$ and R = 3-Me for **2**, A = $Cs^+$ and R = H for **3**, A = $Cs^+$ and R = 3-Me for **4**, A = $Na^+$ or $Cs^+$ and R = Me for **5** and A = $Na^+$ or $Cs^+$ and R = Et for **6**), as indicated in Scheme 4.

### 3.2. General Syntheses

CAUTION: Perchlorate salts of metal complexes with organic ligands are potentially explosive. Only a small amount of material should be prepared, and it should be handled with caution.

In the synthesis process, we started from an ethanolic solution of 20 mL containing 0.98 mmol of $Mn(ClO_4)_2{\cdot}6H_2O$, 0.98 mmol of the corresponding alkaline salt of anthracenecarboxylate, and the corresponding ligand $H_2$-R-salox. To the resulting mixture, which acquires a light brown hue, 0.5 mmols of tetrapropylammonium hydroxide, $(Pr_4N)OH$, were added. Immediately, a whitish solid precipitated and the solution darkened, turning dark green. The solution was left under agitation at room temperature for an hour and then filtered. To crystallize compounds **1–3**, **5**, and **6**, the slow layer diffusion method was followed, taking 10 mL of solution and 10 mL of a 1:1 mixture of diethyl ether and hexane. Two week later, dark-brown prismatic crystals were obtained. On the other hand, the crystallization of compound **4** was only possible after one month of slow evaporation of the solution. Compounds **1–6** can also been obtained starting from Mn(II) nitrate salt and using tetramethylammonium hydroxide or tetrabutylammonium hydroxide instead of $(Pr_4N)OH$.

The most relevant IR bands, Ref. [40], of compounds **1–6** are reported below:

For **1**: Selected IR bands (KBr pellet, $cm^{-1}$): 3435 (m, $\upsilon_{(O\text{–}H)}$), 2979-2884 (w, $\upsilon_{(\text{–}C\text{–}H)}$), 1635 (s, $\upsilon_{as(COO^-)}$), and 1601 (s, $\upsilon_{(C=N)}$).

For **2**: Selected IR bands (KBr pellet, $cm^{-1}$): 3444 (m, $\upsilon_{(O\text{–}H)}$), 2966-2842 (w, $\upsilon_{(\text{–}C\text{–}H)}$), 1625 (s, $\upsilon_{as(COO^-)}$), and 1589 (s, $\upsilon_{(C=N)}$).

For **3**: Selected IR bands (KBr pellet, $cm^{-1}$): 3416 (m, $\upsilon_{(O\text{–}H)}$), 1638 (s, $\upsilon_{as(COO^-)}$), and 1535 (m, $\upsilon_{(C=N)}$).

For **4**: Selected IR bands (KBr pellet, $cm^{-1}$): 3442 (m, $\upsilon_{(O\text{–}H)}$), 3028-2889 (w, $\upsilon_{(\text{–}C\text{–}H)}$), 1595 (s, $\upsilon_{as(COO^-)}$), and 1545 (s, $\upsilon_{(C=N)}$).

For **5**: Selected IR bands (KBr pellet, $cm^{-1}$): 3442 (m, $\upsilon_{(O\text{–}H)}$), 3050-2817 (w, $\upsilon_{(\text{–}C\text{–}H)}$), 1595 (s, $\upsilon_{as(COO^-)}$), and 1532 (s, $\upsilon_{(C=N)}$).

For **6**: Selected IR bands (KBr pellet, $cm^{-1}$) [a]: 3442 (m, $\upsilon_{(O\text{–}H)}$), 2986-2832 (w, $\upsilon_{(\text{–}C\text{–}H)}$), 1601 (s, $\upsilon_{as(COO^-)}$), and 1536 (s, $\upsilon_{(C=N)}$).

### 3.3. Physical Measurements and Magnetic Fits

Infrared spectra (4000–400 $cm^{-1}$) were recorded with a Perkin-Elmer 380-B spectrophotometer. The samples were prepared by mixing with KBr.

Fluorescence spectra were recorded with a Horiba Jobin Yvon SPEX Nanolog spectrophotometer at room temperature. For solid-state measurements, the samples were placed between two quartz plates.

Magnetic measurements were performed at the Mesures Magnètiques Unit from Scientific and Technological Centers (CCiTUB), Universitat de Barcelona, using a Quantum Design MPMS-XL SQUID magnetometer.

To fit the magnetic susceptibility data, the PHI program was used [41]. The error ($R$) associated with the divergence of the experimental values and calculations of the different fits was calculated according to the following equation:

$$R = \frac{\Sigma\left(\chi_M T_{exp}{}^2 - \chi_M T_{cal}{}^2\right)}{\Sigma\ \chi_M T_{exp}{}^2}$$

### 3.4. X-ray Crystallography

Single-crystal X-ray diffraction measurements were performed with a D8VENTURE diffractometer (Bruker) with a CMOS detector. The intensities were recorded with a graphite monochromator using Mo-K$\alpha$ radiation. Lorentz polarization and absorption corrections were performed on all the samples. The structures were solved by direct methods, using the computer program SHELXS-97 [42], and refined through full matrix least squares methods, using the computer program SHELXL-2014 [43]. The SHELXLE program was used as a graphical interface. Atoms, excluding hydrogen, were located from successive difference Fourier synthesis and refined with anisotropic thermal parameters ($F^2$). In most cases, the hydrogen atoms were calculated, assigning the isotropic temperature factors as 1.2 or 1.5 times, with respect to bonded atoms.

CCDC numbers: 2362267 (for **1**), 2362264 (for **2**) 2362265 (for **3**), 2362266 (for **4**), 2362268 (for **5**), and 2362269 (for **6**) contain the supplementary crystallographic data for this paper. These data can be obtained free of charge via https://www.ccdc.cam.ac.uk/structures/, accessed on 12 June 2024.

## 4. Conclusions

In this work, six new compounds obtained from different $H_2$-R-salox ligands (R = H for $H_2$-salox; 3-$CH_3$ for $H_2$-3-Me-salox; $CH_3$ for $H_2$-Me-salox; $CH_2CH_3$ for $H_2$-Et-salox) and 9-anthracenecarboxylate anions (9-AC) have been presented. When using R-salox$^{2-}$ ligands with hydrogen atoms coordinated to the carbon atom of the oxime group, it has been observed that the topology of the resulting compound changes depending on the presence of $Na^+$ or $Cs^+$ ions due to the different ionic radius values between these two alkali cations. Thus, from Na(9-AC), the complexes $[Mn_3NaO(salox)_3(9\text{-}AC)_2(EtOH)_3H_2O]n\cdot2EtOH$ (**1**·2EtOH) and $[Mn_3NaO(3\text{-}Me\text{-}salox)_3(9\text{-}AC)_2(EtOH)_3H_2O]n\cdot EtOH$ (**2**·EtOH) have been obtained, both with chain-like structures formed by $\{Mn^{III}{}_3(\mu_3\text{-}O)(R\text{-}salox)_3\}^+$ triangles connected by hexacoordinated $Na^+$ cations and carboxylate anions. In contrast, when changing the cation of the carboxylic salt to $Cs^+$, the low-nuclearity compounds $[Mn_6O_2(salox)_6(9\text{-}AC)_2(EtOH)_2(H_2O)_2]\cdot3EtOH$ (**3**·3EtOH) and $[Mn_3O(3\text{-}Me\text{-}salox)_3(9\text{-}AC)(EtOH)_3(H_2O)]\cdot1.8EtOH\cdot3H_2O$ (**4**·1.8EtOH·3H2O) were obtained.

Using $H_2$-R-salox ligands with alkyl groups coordinated to the carbon atom of the oxime group, the hexanuclear compounds $[Mn_6O_2(Me\text{-}salox)_6(9\text{-}AC)_2(EtOH)_4(H_2O)_2]\cdot0.5H_2O$ (**5**·0.5H2O) and $[Mn_6O_2(Et\text{-}salox)_6(9\text{-}AC)_2(EtOH)_4(H_2O)_2]\cdot3EtOH$ (**6**·3EtOH) were obtained, irrespective of whether the sodium or cesium salts of anthracenecarboxylate was used.

Compounds derived from 3-Me-salicylaldoxime, with the alkyl group bonded to the aromatic ring at an ortho position relative to the OH group, also follow the magneto-structural correlation of the compounds with $\{Mn^{III}{}_3(\mu_3\text{-}O)(R\text{-}salox)_3\}^+$ cores. Therefore, it is not only the volume of the substituent group that affects the torsion angle, but also its position relative to the oxime group bonded to Mn, as observed in other reported studies [35].

The magnetic study of compounds **1**–**6** indicates a set of antiferromagnetic interactions in compounds **1** and **2** and a combination of antiferromagnetic and ferromagnetic interactions in compounds **3**, **5**, and **6**. In all analyzed cases, the magneto-structural correlation between the intramolecular $Mn^{III}$-N-O-$Mn^{III}$ torsion angle and the magnetic exchange within these units has been confirmed. For compounds **5** and **6**, magnetic measurements have revealed the slow relaxation of the magnetization, with moderate energy barriers of 19.9 and 31.1 $cm^{-1}$, respectively.

Absorbance and fluorescence measurements in solution show the transitions of the 9-anthracenecarboxylate chromophore for compounds **1**–**6**. Consequently, compounds **5** and **6**, with single-molecule magnets and luminescent properties, can be considered bifunctional molecular materials.

**Supplementary Materials:** The following supporting information can be downloaded at: https://www.mdpi.com/article/10.3390/magnetochemistry10080055/s1, Table S1: Crystallographic data for compounds **1–6**; Table S2: BVS parameters for the Mn ions of complexes **1–6**; Table S3: Distances (d, where A = acceptor atom, D = donor atom) and angles of the different hydrogen bonds observed in compounds **1–6**. Table S4. Data of continuous shape measures calculations of Na ions in 1 and 2 deviations from the ideal octahedron structure (ML6). Figure S1. Packing configuration observed within the unit cell of 3 showing intra- and inter-molecular hydrogen bonds (blue discontinuous line). Figure S2. Representation of hydrogen bonds (blue dashed line) in compound **4**.

**Author Contributions:** The manuscript was prepared from the contributions of all authors. They specifically contributed as follows: B.C.: synthesis, characterization of obtained compounds, performance of the magnetic and photoluminescence measurements, and the analysis of the results of magnetic studies for all compounds. R.V.: design and development of the general idea of the research and contribution to the original draft preparation. M.F.-B.: performance of single-crystal X-ray diffraction experiments and refinement of the crystal structure. M.S.E.F.: supervision and preparation of the final version of the manuscript. All authors have read and agreed to the published version of the manuscript.

**Funding:** This work was supported by the Spanish Ministry of Science and Innovation (MICINN) Project PID2023-146166NB-I00.

**Data Availability Statement:** The data presented in this study are available in this article or the Supplementary Materials.

**Acknowledgments:** The authors would like to thank Sergio Caballero and Annia Tubau for their help with the shape calculations and the revision of structure **6**, respectively.

**Conflicts of Interest:** There are no conflicts of interest to declare.

## References

1. Milios, C.J.; Raptopoulou, C.P.; Terzis, A.; Lloret, F.; Vicente, R.; Perlepes, S.P.; Escuer, A. Hexanuclear Manganese(III) Single-Molecule Magnets. *Angew. Chem. Int. Ed.* **2004**, *43*, 210–212. [CrossRef] [PubMed]
2. Smith, A. The Structures of Phenolic Oximes and Their Complexes. *Coord. Chem. Rev.* **2003**, *241*, 61–85. [CrossRef]
3. Afrati, T.; Dendrinou-Samara, C.; Raptopoulou, C.; Terzis, A.; Tangoulis, V.; Tsipis, A.; Kessissoglou, D.P. Experimental and Theoretical Study of the Antisymmetric Magnetic Behavior of Copper *Inverse* -9-Metallacrown-3 Compounds. *Inorg. Chem.* **2008**, *47*, 7545–7555. [CrossRef] [PubMed]
4. Stamatatos, T.C.; Dionyssopoulou, S.; Efthymiou, G.; Kyritsis, P.; Raptopoulou, C.P.; Terzis, A.; Vicente, R.; Escuer, A.; Perlepes, S.P. The First Cobalt Metallacrowns: Preparation and Characterization of Mixed-Valence Cobalt(II/III), Inverse 12-Metallacrown-4 Complexes. *Inorg. Chem.* **2005**, *44*, 3374–3376. [CrossRef] [PubMed]
5. Stemmler, A.J.; Kampf, J.W.; Pecoraro, V.L. Synthesis and Crystal Structure of the First Inverse 12-Metallacrown-4. *Inorg. Chem.* **1995**, *34*, 2271–2272. [CrossRef]
6. Vlahopoulou, G.; Escuer, A.; Font-Bardia, M.; Calvet, T. Synthesis and Characterization of $Co^{III}_3$ Inverse Metallacrowns via Use of 6-Methyl-2-Pyridylaldoxime. *Inorg. Chem. Commun.* **2012**, *16*, 78–80. [CrossRef]
7. Flamourakis, A.G.; Kalofolias, D.A.; Siczek, M.; Lis, T.; Brechin, E.K.; Milios, C.J. New Members of the [$Mn_6$/Oxime] Family and Analogues with Converging [$Mn_3$] Planes. *J. Coord. Chem.* **2016**, *69*, 826–840. [CrossRef]
8. Perivolaris, A.; Fidelli, A.M.; Inglis, R.; Kessler, V.G.; Slawin, A.M.Z.; Brechin, E.K.; Papaefstathiou, G.S. A Family of Hexanuclear Mn(III) Single-Molecule Magnets. *J. Coord. Chem.* **2014**, *67*, 3972–3986. [CrossRef]
9. Hołyńska, M.; Dehnen, S. A Series of [$Mn_6$] Complexes with Terminal Functional Groups. *Z. Anorg. Allg. Chem.* **2012**, *638*, 763–769. [CrossRef]
10. Inglis, R.; Jones, L.F.; Milios, C.J.; Datta, S.; Collins, A.; Parsons, S.; Wernsdorfer, W.; Hill, S.; Perlepes, S.P.; Piligkos, S.; et al. Attempting to Understand (and Control) the Relationship between Structure and Magnetism in an Extended Family of $Mn_6$ Single-Molecule Magnets. *Dalton Trans.* **2009**, 3403–3412. [CrossRef]
11. Milios, C.J.; Vinslava, A.; Wernsdorfer, W.; Prescimone, A.; Wood, P.A.; Parsons, S.; Perlepes, S.P.; Christou, G.; Brechin, E.K. Spin Switching via Targeted Structural Distortion. *J. Am. Chem. Soc.* **2007**, *129*, 6547–6561. [CrossRef] [PubMed]
12. Chaudhuri, P. Homo- and Hetero-Polymetallic Exchange Coupled Metal-Oximates. *Coord. Chem. Rev.* **2003**, *243*, 143–190. [CrossRef]
13. Milios, C.J.; Inglis, R.; Vinslava, A.; Bagai, R.; Wernsdorfer, W.; Parsons, S.; Perlepes, S.P.; Christou, G.; Brechin, E.K. Toward a Magnetostructural Correlation for a Family of $Mn_6$ SMMs. *J. Am. Chem. Soc.* **2007**, *129*, 12505–12511. [CrossRef] [PubMed]
14. Milios, C.J.; Vinslava, A.; Wernsdorfer, W.; Moggach, S.; Parsons, S.; Perlepes, S.P.; Christou, G.; Brechin, E.K. A Record Anisotropy Barrier for a Single-Molecule Magnet. *J. Am. Chem. Soc.* **2007**, *129*, 2754–2755. [CrossRef]

15. Inglis, R.; Jones, L.F.; Mason, K.; Collins, A.; Moggach, S.A.; Parsons, S.; Perlepes, S.P.; Wernsdorfer, W.; Brechin, E.K. Ground Spin State Changes and 3 D Networks of Exchange Coupled $[Mn^{III}_3]$ Single-Molecule Magnets. *Chem.—A Eur. J.* **2008**, *14*, 9117–9121. [CrossRef] [PubMed]
16. Song, X.; Liu, R.; Zhang, S.; Li, L. A Rare Ferromagnetic Manganese(III) Hexanuclear Cluster. *Inorg. Chem. Commun.* **2010**, *13*, 828–830. [CrossRef]
17. An, G.-Y.; Cui, A.-L.; Kou, H.-Z. Assembly of Oximate-Bridged $Mn_6$ Cluster to a One-Dimensional Chain. *Inorg. Chem. Commun.* **2011**, *14*, 1475–1478. [CrossRef]
18. Haryono, M.; Kalisz, M.; Sibille, R.; Lescouëzec, R.; Fave, C.; Trippe-Allard, G.; Li, Y.; Seuleiman, M.; Rousselière, H.; Balkhy, A.M.; et al. One Dimensional Assembly of Mn6 Single Molecule Magnets Linked by Oligothiophene Bridges. *Dalton Trans.* **2010**, *39*, 4751–4756. [CrossRef] [PubMed]
19. Schwarz, F.P.; Wasik, S.P. Fluorescence Measurements of Benzene, Naphthalene, Anthracene, Pyrene, Fluoranthene, and Benzo[e]Pyrene in Water. *Anal. Chem.* **1976**, *48*, 524–528. [CrossRef]
20. Li, L.; Hu, T.-L.; Li, J.-R.; Wang, D.-Z.; Zeng, Y.-F.; Bu, X.-H. Metal–Organic Coordination Architectures of 9,10-Bis(N-Benzimidazolyl)Anthracene: Syntheses, Structures and Emission Properties. *CrystEngComm* **2007**, *9*, 412–420. [CrossRef]
21. Alaimo, A.A.; Takahashi, D.; Cunha-Silva, L.; Christou, G.; Stamatatos, T.C. Emissive $\{Mn_4^{III}Ca\}$ Clusters with Square Pyramidal Topologies: Syntheses and Structural, Spectroscopic, and Physicochemical Characterization. *Inorg. Chem.* **2015**, *54*, 2137–2151. [CrossRef] [PubMed]
22. Beedle, C.C.; Stephenson, C.J.; Heroux, K.J.; Wernsdorfer, W.; Hendrickson, D.N. Photoluminescent $Mn_4$ Single-Molecule Magnet. *Inorg. Chem.* **2008**, *47*, 10798–10800. [CrossRef] [PubMed]
23. Alexandropoulos, D.I.; Mowson, A.M.; Pilkington, M.; Bekiari, V.; Christou, G.; Stamatatos, T.C. Emissive Molecular Nanomagnets: Introducing Optical Properties in Triangular Oximato $\{Mn^{III}_3\}$ SMMs from the Deliberate Replacement of Simple Carboxylate Ligands with Their Fluorescent Analogues. *Dalton Trans.* **2014**, *43*, 1965–1969. [CrossRef]
24. Kalofolias, D.A.; Flamourakis, A.G.; Siczek, M.; Lis, T.; Milios, C.J. A Bulky Oxime for the Synthesis of Mn(III) Clusters. *J. Coord. Chem.* **2015**, *68*, 3472–3484. [CrossRef]
25. Manoli, M.; Inglis, R.; Piligkos, S.; Yanhua, L.; Wernsdorfer, W.; Brechin, E.K.; Tasiopoulos, A.J. A Hexameric $[Mn^{III}_{18}Na_6]$ Wheel Based on $[Mn^{III}_3O]^{7+}$ Sub-Units. *Chem. Commun.* **2016**, *52*, 12829–12832. [CrossRef] [PubMed]
26. Raptopoulou, C.P.; Boudalis, A.K.; Lazarou, K.N.; Psycharis, V.; Panopoulos, N.; Fardis, M.; Diamantopoulos, G.; Tuchagues, J.-P.; Mari, A.; Papavassiliou, G. Salicylaldoxime in Manganese(III) Carboxylate Chemistry: Synthesis, Structural Characterization and Physical Studies of Hexanuclear and Polymeric Complexes. *Polyhedron* **2008**, *27*, 3575–3586. [CrossRef]
27. Yang, C.-I.; Feng, P.-Y.; Chen, Y.-T.; Tsai, Y.-J.; Lee, G.-H.; Tsai, H.-L. Molecular Architecture Based on Manganese Triangles: Monomer, Dimer, and One-Dimensional Polymer. *Polyhedron* **2011**, *30*, 3265–3271. [CrossRef]
28. Geng, J.-P.; Wang, Z.-X.; He, X.; Xiao, H.-P.; Li, M.-X. A Novel 2D Coordination Polymer Based on Triangular-Shaped $[Mn_3O]$ Units Bridged by Sodium Ions and Benzene-1,2,4,5-Tetracarboxylate. *Inorg. Chem. Commun.* **2011**, *14*, 997–1000. [CrossRef]
29. Wu, L.-F.; Wang, Z.-X.; Geng, J.-P.; Xiao, H.-P.; Li, M.-X. A Two-Dimensional Coordination Polymer Constructed from Hexanuclear Manganese Units with Sodium Ions as Auxiliary Bridges. *Inorg. Chim. Acta* **2014**, *412*, 1–5. [CrossRef]
30. Mukherjee, S.; Mukherjee, P.S. Role of Dicarboxylate Linkers in $Mn^{III}$ -Salicylaldoximate Based Extended Structures: Synthesis, Structures, and Magnetic Behavior. *Chem.—A Eur. J.* **2013**, *19*, 17064–17074. [CrossRef]
31. Brown, I.D.; Altermatt, D. Bond-Valence Parameters Obtained from a Systematic Analysis of the Inorganic Crystal Structure Database. *Acta Crystallogr. B* **1985**, *41*, 244–247. [CrossRef]
32. ALVAREZ, S.; ALEMANY, P.; CASANOVA, D.; CIRERA, J.; LLUNELL, M.; AVNIR, D. Shape Maps and Polyhedral Interconversion Paths in Transition Metal Chemistry. *Coord. Chem. Rev.* **2005**, *249*, 1693–1708. [CrossRef]
33. Jones, L.F.; Cochrane, M.E.; Koivisto, B.D.; Leigh, D.A.; Perlepes, S.P.; Wernsdorfer, W.; Brechin, E.K. Tuning Magnetic Properties Using Targeted Structural Distortion: New Additions to a Family of $Mn_6$ Single-Molecule Magnets. *Inorg. Chim. Acta* **2008**, *361*, 3420–3426. [CrossRef]
34. Escuer, A.; Cordero, B.; Font-Bardia, M.; Teat, S.J.; Roubeau, O. Manganese–Salicyloximate Clusters Starting from $[Mn^{II}(Hfacac)_2]$: From $Mn_4$ to $Mn_{12}$. *Eur. J. Inorg. Chem.* **2016**, *2016*, 1232–1241. [CrossRef]
35. Milios, C.J.; Vinslava, A.; Whittaker, A.G.; Parsons, S.; Wernsdorfer, W.; Christou, G.; Perlepes, S.P.; Brechin, E.K. Microwave-Assisted Synthesis of a Hexanuclear $Mn^{III}$ Single-Molecule Magnet. *Inorg. Chem.* **2006**, *45*, 5272–5274. [CrossRef] [PubMed]
36. Tarushi, A.; Hatzidimitriou, A.G.; Estrader, M.; Kessissoglou, D.P.; Tangoulis, V.; Psomas, G. Toward Multifunctional Materials Incorporating Stepladder Manganese(III) Inverse-[9-MC-3]-Metallacrowns and Anti-Inflammatory Drugs. *Inorg. Chem.* **2017**, *56*, 7048–7057. [CrossRef] [PubMed]
37. Fleischauer, P.D.; Fleischauer, P. Photoluminescence of Transition Metal Coordination Compounds. *Chem. Rev.* **1970**, *70*, 199–230. [CrossRef]
38. Dunstan, W.R.; Henry, T.A. VIII.—Occurrence of Orthohydroxyacetophenone in the Volatile Oil of Chione Glabra. *J. Chem. Soc. Trans.* **1899**, *75*, 66–71. [CrossRef]
39. Ueno, A.; Moriwaki, F.; Osa, T.; Hamada, F.; Murai, K. Association, Photodimerization, and Induced-Fit Types of Host-Guest Complexation of Anthracene-Appended .Gamma.-Cyclodextrin Derivatives. *J. Am. Chem. Soc.* **1988**, *110*, 4323–4328. [CrossRef]
40. Nakamoto, K. *Infrared and Raman Spectra of Inorganic and Coordination Compounds*; Wiley: Hoboken, NJ, USA, 2008; ISBN 9780471743392.

41. Chilton, N.F.; Anderson, R.P.; Turner, L.D.; Soncini, A.; Murray, K.S. PHI: A Powerful New Program for the Analysis of Anisotropic Monomeric and Exchange-coupled Polynuclear *d* - and *f* -block Complexes. *J. Comput. Chem.* **2013**, *34*, 1164–1175. [CrossRef]
42. Sheldrick, G.M. A Short History of *SHELX*. *Acta Crystallogr. A* **2008**, *64*, 112–122. [CrossRef] [PubMed]
43. Sheldrick, G.M. Crystal Structure Refinement with *SHELXL*. *Acta Crystallogr. C Struct. Chem.* **2015**, *71*, 3–8. [CrossRef] [PubMed]

*magnetochemistry*

MDPI

*Article*

# Theoretical Hints to Optimize Energy Dissipation and Cell–Cell Response in Quantum Cellular Automata Based on Tetrameric and Bidimeric Cells †

**Andrew Palii [1,*], Shmuel Zilberg [2] and Boris Tsukerblat [2,3,*]**

1 Federal Research Center of Problems of Chemical Physics and Medical Chemistry, Chernogolovka 142432, Russia
2 Department of Chemical Sciences, Materials Research Center, Ariel University, Ariel 4076414, Israel; shmuel.zilberg@gmail.com
3 Department of Chemistry, Ben-Gurion University of the Negev, Beer-Sheva 84105, Israel
* Correspondence: andrew.palii@uv.es (A.P.); tsuker@bgu.ac.il (B.T.)

† This article was originally going to be dedicated to the distinguished scientists and our friends Miguel Julve and Francesco Lloret on their retirement. When this article was almost ready for submission to Magnetochemistry, we learned of the tragic news of the death of our dear friend Miguel, an outstanding scientist, internationally recognized for his invaluable contributions to coordination chemistry and molecular magnetism, with a great enthusiasm for science and a bright personality. We dedicate this article to his blessed memory. You will live on in our memory, dear Miguel.

**Abstract:** This article is largely oriented towards the theoretical foundations of the rational design of molecular cells for quantum cellular automata (QCA) devices with optimized properties. We apply the vibronic approach to the analysis of the two key properties of such molecular cells, namely the cell–cell response and energy dissipation in the course of the non-adiabatic switching of the electric field acting on the cell. We consider two kinds of square planar cells, namely cells represented by a two-electron tetrameric mixed valence (MV) cluster and bidimeric cells composed of two one-electron MV dimeric half-cells. The model includes vibronic coupling of the excess electrons with the breathing modes of the redox sites, electron transfer, intracell interelectronic Coulomb repulsion, and also the interaction of the cell with the electric field of polarized neighboring cells. For both kinds of cells, the heat release is shown to be minimal in the case of strong delocalization of excess electrons (weak vibronic coupling and/or strong electron transfer) exposed to a weak electric field. On the other hand, such a parametric regime proves to be incompatible with a strong nonlinear cell–cell response. To reach a compromise between low energy dissipation and a strong cell–cell response, we suggest using weakly interacting MV molecules with weak electron delocalization as cells. From this point of view, bidimeric cells are advantageous over tetrameric ones due to their smaller number of electron transfer pathways, resulting in a lower extent of electron delocalization. The distinct features of bidimeric cells, such as their two possible mutual arrangements ("side-by-side" and "head-to-tail"), are discussed as well. Finally, we briefly discuss some relevant results from a recent ab initio study on electron transfer and vibronic coupling from the perspective of the possibility of controlling the key parameters of molecular QCA cells.

**Keywords:** quantum cellular automata; molecular cells; cell–cell response; mixed valence molecules; electron transfer; vibronic coupling; electron delocalization; heat release

## 1. Introduction

Molecular quantum cellular automata (QCA) represent a new promising paradigm of transistorless electronics and computing in which digital (binary) information is encoded in charge configurations of mixed-valence (MV) molecules [1–3]. QCA technology holds promise as a new revolutionary technology in nanoelectronics and computing based on molecular units, having essential advantages over traditional electronic devices. The idea of molecular QCA has given rise to a new area of research combining chemistry, physics, molecular magnetism, and materials science, with an emphasis on 2D systems. Since the first proposal to use MV molecules as QCA cells or half-cells [3], numerous MV systems have been suggested for the molecular implementation of QCA (see examples in Refs. [4–18] and a recent exhaustive review [19]). Figure 1 shows two selected examples of square-planar molecular cells [3,8,20], which are tetraruthenium $Ru(III)_2Ru(II)_2$ complexes composed of a pair of famous Creutz–Taube ions [21] (Nobel Prize in Chemistry for discoveries on electron transfer). Each of these molecules comprises two excess holes, delocalized among the four sites occurring either via the bridges located on the edges of the square (Figure 1a) or through the central bridge (Figure 1b).

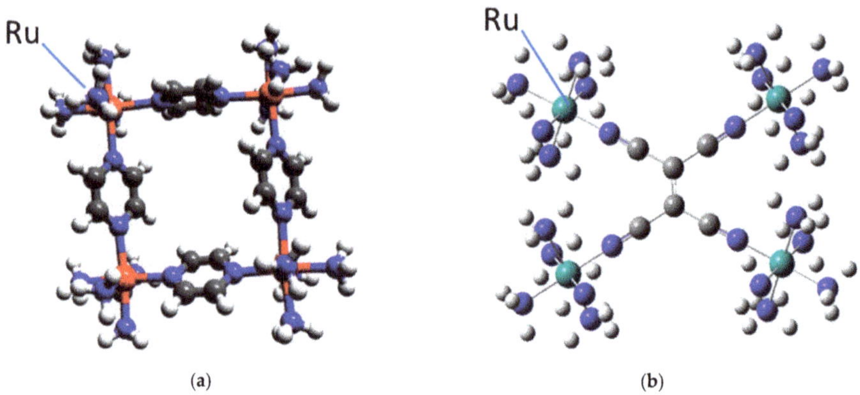


**Figure 1.** Two examples of the tetrameric square-planar molecular cell for QCA representing the tetraruthenium Creutz–Taube derivatives. (**a**) Illustration for the case of electron transfer along the sides and (**b**) the case of transfer along the diagonals.

These two topologies of the electron transfer pathways are closely related to the two kinds of spin arrangements (as will be discussed below), which in turn demonstrate the interrelation between the areas of the molecular magnetism and the QCA.

Theoretical work aimed at identifying and understanding the specific properties (sometimes conflicting with each other) of active molecules, which allows us to consider their suitability for use as QCA cells, has acquired a special role. Among the requirements that the MV molecule should meet in order to act as a cell or half-cell, two should be specifically mentioned. First, the molecule should exhibit low energy dissipation in the course of the switching cycle [22–27], and second, the molecule should be able to easily switch between the two charge configurations encoding binary information under the action of the electric field created by the polarized neighboring molecule, which manifests itself in the strongly nonlinear (stepwise) cell–cell response function. The latter requirement, first formulated for QCA based on quantum dots [28–32], is general for all kinds of QCA devices and all types of cells.

Concerning the first requirement, it has recently been shown that the higher the extent of delocalization of the excess charges in an MV dimer, the lower the specific heat release (heat release per one cell during one switching cycle) [26,27]. It follows from the vibronic theory of molecular QCA based on MV compounds [33–37] that the extent of delocalization is determined by the interplay of two competing interactions, namely electron transfer promoting delocalization and vibronic coupling with the breathing vibrations of the redox sites producing a self-trapping effect. Then, a high degree of delocalization means that electron transfer dominates over the vibronic coupling. In addition, heat release increases with an increase in the Coulomb interaction strength.

Regarding the second requirement, it has been demonstrated [33] that the extent of the electron delocalization should be low in order to obtain a strong nonlinear cell–cell response, and so the vibronic coupling should dominate over the electron transfer. This means that the two requirements are in some sense contradictory to each other. Indeed, if one uses weakly interacting MV dimers with strong electron delocalization to minimize the energy dissipation (first requirement), one faces an impossibility of fulfilling the second requirement, since in this case the cell–cell response proves to be weak and almost linear. By considering the simplest dimeric cells, we have demonstrated that the best way to reconcile low heat release and strong nonlinear cell–cell response is to use weakly delocalized MV dimers separated from each other by a long distance as cells in order to prevent strong interdimer Coulomb interaction [26,27]. This follows from the fact that such dimers can easily be polarized even by a rather weak Coulomb field, which, in turn, can significantly reduce the heat release in spite of the high extent of the electron localization.

In this work, based on the vibronic theory developed in Refs. [33–37], we generalize the approach proposed in Refs. [26,27] to the cases of more practically applicable molecular cells, which are more complex than the dimeric ones. These the square-planar cells composed of two one-electron MV dimers playing the role of half-cells ("bidimeric cells") and cells representing two-electron tetrameric square-planar MV molecules ("tetrameric cell"). A number of candidates for the molecular implementation of tetrameric and bidimeric cells have been considered [4–18], and also the attempt to compare the functional properties of these two kinds of cells has been addressed [34]. Still, this attempt has been restricted to a special limiting case of strong intracell Coulomb interaction, and, even more importantly, only the cell–cell response has been analyzed, while the energy dissipation remained outside of the scope of consideration in Ref. [34]. In this paper, based on the quantum-mechanical solution of the vibronic problems for such cells [36,37], we analyze both the dependences of the cell–cell response functions and the specific heat release in the nonadiabatic switching cycle on the electronic and vibronic parameters as well as on the strength of the cell–cell interaction. Then, we establish parametric regimes at which competing requirements of low heat release and strong nonlinear cell–cell response can be simultaneously fulfilled. Along the way, a complete comparative analysis of bidimeric and tetrameric cells is carried out in terms of their advantages for creating QCA devices. Finally, we analyze some important ab initio results for a number of specifically selected MV compounds, with a particular focus on the ability to control and optimize the key parameters responsible for functioning of the cells.

## 2. Description of the Molecular Cells within the Vibronic Approach

### *2.1. Hamiltonians of the Bidimeric and Tetrameric Working Cells*

We consider two interacting cells of the same type, with one of them playing the role of the driver-cell that acts as a source of the electric field (it is either bidimeric or tetrameric square cell $A'B'C'D'$, Figure 2) and the other one represents the working cell that can be polarized by this field (bidimeric or tetrameric cell $ABCD$). The square planar driver-cell is assumed to be polarized in such a way that the densities on the sites situated along one diagonal of the square are both equal to $\rho$, while the densities on the sites belonging to another diagonal are $1-\rho$ (Figure 2). Such a cell acts as an electric quadrupole.

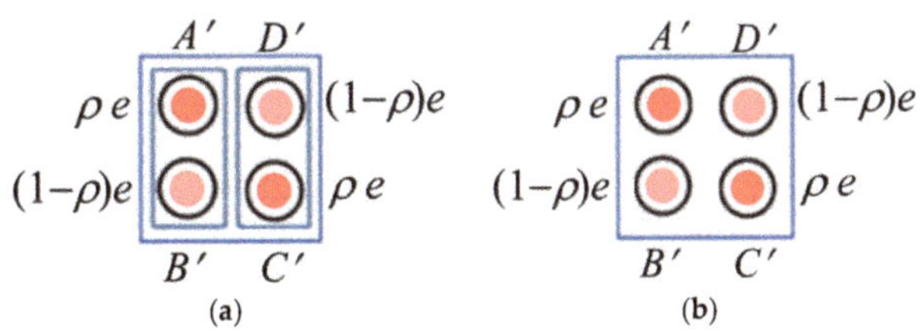


**Figure 2.** Schematic representation of the electronic populations of sites in the polarized bidimeric square-planar driver-cell (**a**) and in tetrameric driver-cell (**b**). Both such cells can be regarded as electric quadrupoles. Red balls denote the sites occupied by electrons.

Polarization of the square-planar driver-cell $A'B'C'D'$ (bidimeric and tetrameric) is defined as a normalized difference in populations ($\rho'_A$, etc.) of the two diagonals of the square, that is

$$P_{dc} = \frac{\rho'_A + \rho'_C - \rho'_B - \rho'_D}{\rho'_A + \rho'_C + \rho'_B + \rho'_D} = 2\rho - 1. \tag{1}$$

The polarized driver-cell creates a quadrupolar electric field inducing polarization of the neighboring working cell. The polarization $P_{wc}$ of the working cell is defined by similar expression. It is evident that the polarization of the working cell should depend on the polarization of the driver-cell. This dependence $P_{wc}(P_{dc})$ is the so-called "cell–cell response function" that is one of the key functional characteristics of QCA.

To calculate the cell–cell response function and the heat release, one needs a proper Hamiltonian of the working cell subjected to the action of the electric field created by the polarized driver-cell. The specific form of the Hamiltonian depends on the type of the cell under consideration, the employed approximations and also on the adopted vibronic model. In this work, we consider two types of square-planar *ABCD*-cells, namely bidimeric cells consisting of two dimeric one-electron half-cells *AB* and *CD* and tetrameric cells represented by a two-electron MV tetramer. As the vibronic model, we use the conventional Piepho–Krausz–Schatz (PKS) vibronic model, which takes into account the interaction of the excess electrons with the fully symmetric ("breathing") vibrations of the redox sites [38]. As to the approximations employed, for both considered types of cells, we consider the limiting case of strong Coulomb interaction inside the cell, which is defined by the inequality $t << U$, where $U$ is the difference between the intracell Coulomb energies of the two close (edges of the square cell) and two distant (diagonals of the cell) excess electrons, while $t$ is the one-electron transfer parameter. In the framework of this approximation, we also consider the general case of the arbitrary relationship between the parameters $t$ and $U$.

Let us start with the general case discussed previously in view of the problem of the evaluation of cell–cell response (see Refs. [35–37]). We briefly discuss the key issues related to this case with short comments. For the description of tetrameric and bidimeric cells, we use the following Hamiltonians:

$$\begin{aligned}\hat{H}^{ta}_{bidim} =& t\sum_{\sigma}(c^{+}_{A\sigma}c_{B\sigma} + c^{+}_{C\sigma}c_{D\sigma} + h.c.)\\ &+u_{AC}\hat{n}_A\hat{n}_C + u_{BD}\hat{n}_B\hat{n}_D + (U + u^{ta}_{AD})\hat{n}_A\hat{n}_D + (U + u^{ta}_{BC})\hat{n}_B\hat{n}_C\\ &+\hbar\omega\left(\hat{b}^{+}_1\hat{b}_1 + \hat{b}^{+}_2\hat{b}_2 + 1\right)(\hat{n}_A\hat{n}_C + \hat{n}_B\hat{n}_D + \hat{n}_A\hat{n}_D + \hat{n}_B\hat{n}_C)\\ &+v\left[\left(\hat{b}^{+}_1 + \hat{b}_1\right)(\hat{n}_A\hat{n}_C - \hat{n}_B\hat{n}_D) + \left(\hat{b}^{+}_2 + \hat{b}_2\right)(\hat{n}_A\hat{n}_D - \hat{n}_B\hat{n}_C)\right],\end{aligned} \tag{2}$$

$$\begin{aligned}\hat{H}_{tetr} =& t\sum_{\sigma}(c^{+}_{A\sigma}c_{B\sigma} + c^{+}_{B\sigma}c_{C\sigma} + c^{+}_{C\sigma}c_{D\sigma} + c^{+}_{A\sigma}c_{D\sigma} + h.c.)\\ &+u_{AC}\hat{n}_A\hat{n}_C + u_{BD}\hat{n}_B\hat{n}_D + (U + u_{AD})\hat{n}_A\hat{n}_D + (U + u_{BC})\hat{n}_B\hat{n}_C + (U + u_{AB})\hat{n}_A\hat{n}_B\\ &+(U + u_{CD})\hat{n}_C\hat{n}_D\\ &+\hbar\omega\left(\hat{b}^{+}_1\hat{b}_1 + \hat{b}^{+}_2\hat{b}_2 + \hat{b}^{+}_3\hat{b}_3 + \frac{3}{2}\right)(\hat{n}_A\hat{n}_C + \hat{n}_B\hat{n}_D + \hat{n}_A\hat{n}_D + \hat{n}_B\hat{n}_C + \hat{n}_A\hat{n}_B + \hat{n}_C\hat{n}_D)\\ &+v\left[\left(\hat{b}^{+}_1 + \hat{b}_1\right)(\hat{n}_A\hat{n}_C - \hat{n}_B\hat{n}_D) + \left(\hat{b}^{+}_2 + \hat{b}_2\right)(\hat{n}_A\hat{n}_D - \hat{n}_B\hat{n}_C) + \left(\hat{b}^{+}_3 + \hat{b}_3\right)(\hat{n}_A\hat{n}_B - \hat{n}_C\hat{n}_D)\right].\end{aligned} \tag{3}$$

The first terms in Equations (2) and (3) describe the electron transfer contributions, where $t$ is the electron transfer parameter, $c^{+}_{i\sigma}$ and $c_{i\sigma}$ are the fermionic creation and annihilation operators, and $\sigma$ is the projection of the spin of the electron. The form of the electron transfer Hamiltonian in Equation (2) assumes that the transfer in the bidimeric cell is allowed within each dimeric half-cell, and no transfer is possible between different half-cells. In contrast, in tetrameric cells, the transfer is allowed between each of the two nearest neighboring sites, while the transfer between the antipodal sites of the square (along the diagonal) is neglected. For this reason, the Hamiltonian in Equation (3) contains twice as many electron transfer connections than that in Equation (2). In this context, the tetramer shown in Figure 1a represents an example of a compound with predominant transfer between neighboring sites, while in the tetramer shown in Figure 1b, both the transfer processes occurring along the edges and the diagonal transfer processes can produce comparable contributions.

The second terms in Equations (2) and (3) describe the Stark interaction of bidimeric and tetrameric working cells with the quadrupolar electric field created by the polarized driver-cell, where $\hat{n}_i$ is the operator of the number of electrons on the site $i$, whose eigenvalues can either be 1 (site occupied by the electron) or 0 (empty site). For the description of the Stark interaction in the case of bidimeric cells, one needs four energies of the Coulomb interactions between the polarized driver-cell and the working cell with four possible different configurations of the excess electrons, which are denoted as $u_{AC}$, $u_{BD}$, $u^{ta}_{AD}$, and $u^{ta}_{BC}$ in Equation (2). These energies depend on the size of the cell, the distance between the cells, the driver-cell polarization $P_{dc}$ and the type of mutual arrangement of the interacting cells, which can be either "head-to-tail" or "side-by-side", as shown in Figure 3a,b, respectively. The index *ta* (type of arrangement) in Equation (2) is used to indicate the type of mutual arrangement of bidimeric cells. It should be noted that for diagonal-type electronic configurations *AC* and *BD*, the Coulomb energies are independent of the type of cells arrangement, and so the index *ta* is omitted in the notations of the corresponding Coulomb energies. As distinguished from the bidimeric cells, the Stark interaction in the case of tetrameric cells is described by six intercell Coulomb energies $u_{AC}$, $u_{BD}$, $u_{AD}$, $u_{BC}$, $u_{AB}$, and $u_{CD}$, as can be seen from Equation (3). The smaller number of intercell Coulomb energies for bidimeric cells is because for such cells, there are fewer electronic configurations as compared with the tetrameric cells, since configurations with two excess electrons per dimeric half-cell are forbidden. The dependences of the intercell Coulomb energies for tetrameric and bidimeric cells on the dimensions of the cell and the intercell distance are given in Refs. [35,37], respectively.

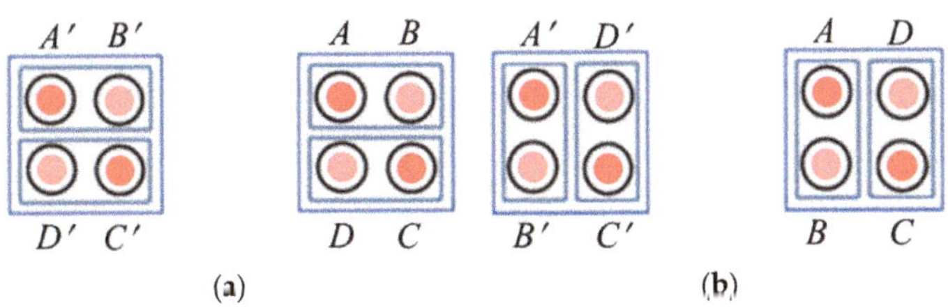


**Figure 3.** Head-to-tail (**a**) and side-by-side (**b**) mutual arrangements of the bidimeric square cells.

Besides the intercell interaction, the Hamiltonians, Equations (2) and (3), also include the contributions of the intracell Coulomb interaction. This interaction is described by the Coulomb energy gap $U$ between the diagonal-type electronic configurations *AC* and *BD* in which the two electrons are distributed far from each other, forming the ground Coulomb manifold of the cell and the excited configurations (*AD* and *BC* in the case of bidimeric cell and *AD*, *BC*, *AB* and *CD* in the case of tetrameric cell) with closely spaced electrons.

Finally, the last two terms in Equations (3) and (4) describe the contribution of active molecular vibrations. We apply the conventional vibronic Piepho–Krausz–Schatz (PKS) model [38], which duly takes into account all the key features of the localization–delocalization phenomenon responsible for the functional properties of the molecular cells. Within this model, the molecular vibrations are built as symmetry-adapted combinations of the full-symmetric ("breathing") vibrations of the constituent redox sites. It has been shown [37] that in the case of bidimeric cells, one should take into account two molecular vibrational out-of-phase modes. The first such mode is that describing compression of the two sites situated along one diagonal accompanied by expanding of the two sites on another diagonal of the square. The second mode describes compression (expansion) of the two closely spaced sites *A* and *D* belonging to different half-cells accompanied by expansion (compressions) of the other two sites $B$ and $C$. For tetrameric cells, in addition to these two active modes, one more active out-of-phase mode appears, which describes compression (expansion) of the two closely spaced sites $A$ and $B$ accompanied by expansion (compressions) of the other two sites $C$ and $D$ [35]. The contributions of the out-of-phase vibrations are expressed through the bosonic creation and annihilation operators $\hat{b}_j^+$ and $\hat{b}_j$, where the subindex j enumerates the active PKS vibrations ($j$ = 1, 2 in the case of bidimeric cell and $j$ = 1, 2, 3 for tetrameric cell). The terms, containing the products $\hat{b}_j^+\hat{b}_j$ in Equations (2) and (3), are the Hamiltonians of the free vibrations occurring with the frequency $\omega$, while the linear terms with respect to the operators $\hat{b}_j^+$ and $\hat{b}_j$ describe the interaction of the excess electron with the out-of-phase modes, i. e., the vibronic coupling, where $v$ is the vibronic coupling parameter. Note that all out-of-phase PKS vibrations occur with the same frequency $\omega$, which is equal to the frequency of the local breathing mode (it is assumed that the frequency of the breathing vibration is independent of the oxidation degree of the site), and also the vibronic coupling with all such vibrations is described by the only vibronic parameter $v$, since these modes are composed of the breathing vibrations of the equivalent redox sites.

Along with the general case, we also analyze the limit of strong intracell Coulomb interaction. This limit has been considered for bidimeric and tetrameric cells in [34] in the context of the analysis of the cell–cell response, while the problem of heat release has not been discussed. Considering the intracell Coulomb interaction as a zero-order Hamiltonian, and applying the perturbation theory to all other interactions, we reduce the initial Hamiltonians, Equations (2) and (3), to the following simpler ones:

$$\begin{aligned}\hat{H}_{tc} = k_{tc}t_{eff}\sum_{\sigma\sigma'}\left(c^+_{D\sigma}c^+_{B\sigma'}c_{C\sigma}c_{A\sigma'} + h.c.\right) + uP_{dc}(\hat{n}_A\hat{n}_C - \hat{n}_B\hat{n}_D) \\ +\hbar\omega\left(\hat{b}_1^+\hat{b}_1 + \frac{1}{2}\right)(\hat{n}_A\hat{n}_C + \hat{n}_B\hat{n}_D) + v\left(\hat{b}_1^+ + \hat{b}_1\right)(\hat{n}_A\hat{n}_C - \hat{n}_B\hat{n}_D),\end{aligned} \tag{4}$$

where $t_{eff} = t^2/U$ is the effective second-order electron transfer parameter, the index $tc$ specifies the kind of cell ($tc = tetr,\ bidim$), and $k_{tc}$ is a factor that is different for tetrameric and bidimeric cells, namely $k_{tetr} = 4$ and $k_{bidim} = 2$; finally, $u$ is the parameter that is equal to the half of the so-called kink energy, defined as the difference in Coulomb energies of pairs of cells with different and identical polarizations.

As distinguished from the general Hamiltonians, Equations (2) and (3), whose electronic contributions act within the full electronic basis containing six electronic configurations for tetrameric cells and four configurations for bidimeric cells, the electronic parts of the effective Hamiltonians, Equation (4), operate within the truncated electronic basis comprising only two ground electronic configurations, namely the diagonal-type configurations $AC$ and $BD$ possessing the minimal intracell Coulomb energy. The first term in Equation (4) describes the effective two-electron transfer (occurring via virtual one-electron transfers mixing the ground and excited electronic configurations), which transforms one diagonal configuration to another one. The second term in Equation (4) describes the intercell Stark interaction of the working cell with the polarized driver-cell with polarization $P_{dc}$. As distinguished from the general case, only one intercell Coulomb energy $uP_{dc}$ is required to describe such an interaction.

The last two terms in Equation (4) are the free harmonic oscillator Hamiltonian and the Hamiltonian of the linear vibronic coupling. Only the PKS out-of-phase vibration involving the two diagonals of the square is active within the truncated basis of diagonal configurations and so instead of two-mode and three-mode vibrations problems arising in the general case, we face a one-mode vibronic problem for both types of cells, provided that the limit of strong Coulomb intracell interaction is fulfilled.

### 2.2. *Quantum-Mechanical Vibronic Approach: Evaluation of Cell–Cell Response*

To calculate the eigenvalues and the eigenvectors of the Hamiltonians, Equations (2)–(4), one has to present these Hamiltonians as a matrix defined in the full electron-vibrational space. To achieve this, we use as the basis the product $|D, n_1, \ldots n_N\rangle \equiv |D\rangle\ |n_1, \ldots n_N\rangle$,. where $|D\rangle$ represents the electronic wave functions, each characterized by a certain electronic distribution $D$, and $|n_1, \ldots n_N\rangle$ represent the wave functions of free $N$-dimensional harmonic oscillator ($N$ = 1 for bidimeric and tetrameric cells, for which the limiting case of strong Coulomb interaction is satisfied, and $N$ = 2 and 3 for bidimeric and tetrameric cells with a violated limit of strong Coulomb interaction).

Diagonalization of the obtained matrices provides the sets of the vibronic eigenvalues and eigenvectors, which depend on the type of cell, its size, the distance between the cells, and the values of the parameters describing the intracell and intercell interactions. The knowledge of the vibronic wave functions evaluated with different values of the driver-cell polarization $P_{dc}$ allows us to calculate the dependence of the induced polarization $P_{wc}$ of the working cell on $P_{dc}$, that is, the cell–cell response function. Also, based on the solution of the vibronic problem, one can evaluate the specific heat release assuming fast nonadiabatic switching of the electric field acting on the working cell.

### 2.3. *Quantum–Mechanical Vibronic Approach: Evaluation of Specific Heat Release*

At this point, we introduce a simplifying assumption that restricts the generality of the results to some extent, but at the same time considerably facilitates the calculations and makes them more transparent. We assume that the process of repolarization of the working cell is caused by a sudden (nonadiabatic) switching of the field of the driver-cell, as schematically shown in Figure 4 for the case of bidimeric cells considered within the limit of strong intracell Coulomb interaction. The left side of Figure 4 shows a pair of cells, each of which is in the Boolean state **0**, which corresponds to the minimum energy of the Coulomb interaction between the cells. The middle part of Figure 4 shows an intermediate transition state arising upon sudden switching of the driver-cell from the state **0** to the state **1**, while the working cell still remains in the state **0**. This corresponds to the Coulomb excitation of the pair of cells (circled area in Figure 4), with the Coulomb excitation energy being just equal to the kink energy $2u$. At the last stage of the cycle, the excitation energy is converted into the heat through vibrational relaxation and transferred to the phonon bath, after which the pair of cells acquires a new stable charge configuration with both cells being in the state **1** (right part of Figure 4). It is seen that the kink energy can be considered as a measure of the specific heat release, or, more precisely, as will be shown below, as the maximum possible value of the specific heat release in the course of nonadiabatic switching.

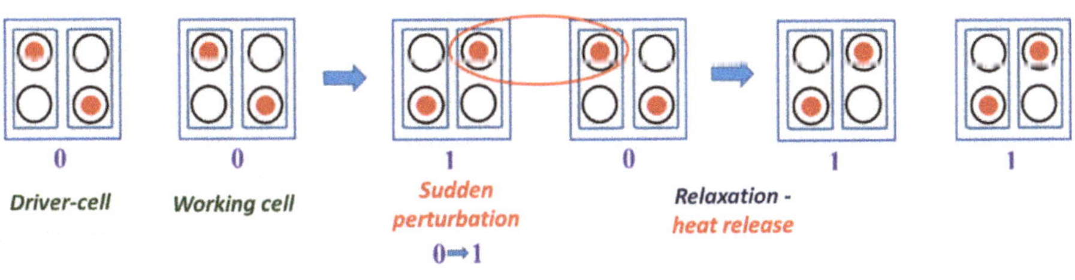


**Figure 4.** Schematic representation of the nonadiabatic switching cycle of a bidimeric working cell under the influence of a sudden change in the Coulomb field of the driver-cell. The image of the switching cycle for the tetrameric cell looks quite similar.

Note that in this consideration, switching is assumed to be so fast that there is no time for energy exchange between the cell and the phonon bath to occur in the course of the transition from the initial thermodynamically equilibrium state (left part of Figure 4) to the nonequilibrium state (middle part of Figure 4). Under such conditions, the bath does not influence the switching event and the energy exchange occurs only in the course of subsequent vibrational relaxation, resulting in the final equilibrium state (right part of Figure 4).

To calculate the heat release for all types of cells, one needs the eigenvalues and the eigenfunctions of the Hamiltonians, Equations (2)–(4), evaluated at two polarizations of the driver-cell, namely at $P_{dc} = -1$ (beginning of the switching cycle) and $P_{dc} = 1$ (end of the cycle). The energies and the wave functions of the working cell calculated at $P_{dc} = -1$ are denoted by $\widetilde{E}_{\widetilde{\nu}}$ and $|\widetilde{\nu}\rangle$, while $E_\nu$ and $|\nu\rangle$ are the energies and the wave functions calculated at $P_{dc} = 1$. The wave functions obtained for these values of $P_{dc}$ are the linear combinations of the basis vectors

$$\begin{aligned} |\widetilde{\nu}\rangle &= \sum_D |D\rangle \sum_{n_1 \ldots n_N} c^D\left(n_1 \ldots n_N, \widetilde{\nu}\right)|n_1 \ldots n_N\rangle, \\ |\nu\rangle &= \sum_D |D\rangle \sum_{n_1 \ldots n_N} c^D(n_1 \ldots n_N, \nu)|n_1 \ldots n_N\rangle, \end{aligned} \tag{5}$$

in which the coefficients $c^D(n_1 \ldots n_N, \widetilde{\nu})$ and $c^D(n_1 \ldots n_N, \nu)$ depend on the set of parameters describing the working cell, the size of the cell, the intercell distance, and the polarization of the driver-cell.

Upon the action of sudden (stepwise) perturbation of the form

$$\hat{V}(\tau) = \left[\hat{H}_{wc}(P_{dc} = 1) - \hat{H}_{wc}(P_{dc} = -1)\right]\theta(\tau) = 2u\theta(\tau), \tag{6}$$

expressed in terms of the Heaviside step function of time $\theta(\tau)$, the working cell being initially (at time $\tau < 0$) in a stationary state $\left|\widetilde{\nu}\right\rangle$ will occur after sudden switching (i.e., at $\tau > 0$) with a certain probability in a new stationary state $|\nu\rangle$, and, as follows from the nonstationary perturbation theory for the case of instant perturbation [39], this probability is defined as a square of the overlap integral

$$a_{\nu\widetilde{\nu}} = \left\langle \nu \middle| \widetilde{\nu} \right\rangle \tag{7}$$

of the vibronic wave functions calculated at $P_{dc} = -1$ and $P_{dc} = 1$.

Using the calculated probability amplitudes $a_{\nu\widetilde{\nu}}$ and the calculated eigenvalues $\widetilde{E}_{\widetilde{\nu}}$ and $E_\nu$, one can derive the following expression for the specific heat release $W$ occurring in course of the non-adiabatic switching:

$$W = \frac{\sum_\nu E_\nu \sum_{\widetilde{\nu}} (a_{\nu\widetilde{\nu}})^2 \exp(-\beta_T \widetilde{E}_{\widetilde{\nu}})}{\sum_\mu \sum_{\widetilde{\nu}} a_{\mu\widetilde{\nu}}{}^2 \exp(-\beta_T \widetilde{E}_{\widetilde{\nu}})} - \frac{\sum_\nu E_\nu \exp(-\beta_T E_\nu)}{\sum_\mu \exp(-\beta_T E_\mu)}, \tag{8}$$

where $\beta_T = 1/k_B T$. Although the derivation of Equation (8) in [27] has been carried out for the simplest case of a dimeric cell, this expression is valid for bidimeric and tetrameric cells as well, while all specifical features (both electronic and vibronic) of the cells and the approximations made so far are taken into account by using a specific form of the cell Hamiltonian.

## 3. Results and Discussion

### 3.1. Consideration of Bidimeric and Tetrameric Cells: Limit of Strong Coulomb Repulsion

Based on the general formalism described above, we study the energy dissipation for the two types of square-planar two-electron molecular QCA cells, namely for bidimeric cells composed of one-electron MV dimers, such as the $d^1$-$d^0$-type transition metal clusters or similar organic MV molecules, and for two-electron tetrameric cells represented by the $d^1$-$d^1$-$d^0$-$d^0$ type 3d-metal clusters and similar organic complexes.

Let us first consider the limiting case of strong Coulomb interaction when $U \gg t$. Before analyzing the results of calculation, it is appropriate to make some preliminary remarks concerning the spin-states of the system. Since we are dealing with the bi-electronic systems, their full spin ($S = 0$ and $S = 1$) is a good quantum number. At $U \gg t$, the energy pattern of a two-electron square-planar MV tetramer consists of two groups of levels separated by a large Coulomb gap $U$ [40]. The low-lying group corresponds to the localization of the mobile electrons at the antipodal sites lying on the diagonal of the square, while the electronic population of the adjacent sites lying on the sides of the square gives rise to the highly excited (having energies $\approx U$) group of levels. Figure 5a shows the pattern of the low-lying spin-singlets and spin-triplets evaluated for a square-planar MV tetramer of the $d^1$-$d^1$-$d^0$-$d^0$-type within the adopted model of the electron transfer pathways in which only the transfer occurring along the edge of the molecular square is assumed to be nonzero. One can see that mixing of different Coulomb configurations through the transfer splits the lower Coulomb manifold in such a way that the spin-singlet proves to be the ground state. The pattern of the potential curves evaluated within the PKS model (Figure 5b) shows that the vibronic ground state associated with the potential minima also remains the spin-singlet. Moreover, one can show that the spin-singlet proves to be the ground state even when the polarizing electric field created by a neighboring cell (driver-cell) is applied.

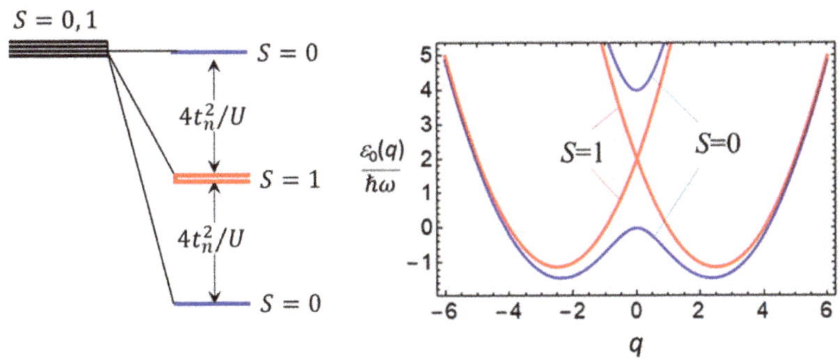


(a) (b)

**Figure 5.** Scheme of the low-lying spin levels of the two-electron square planar MV tetrameric system in the strong Coulomb repulsion limit (**a**) and the scheme of the adiabatic potentials in the PKS vibronic model with $\hbar\omega t_{eff}/\hbar\omega = 0.5$, $v/\hbar\omega = 2.5$, where $q$ is the vibrational coordinate of the out-of-phase PKS vibration, $\omega$ is the frequency of this vibration, and $v$ is the vibronic coupling parameter (**b**).

Here, we focus on the analysis of the cell–cell response and heat release in the low-temperature limit. In view of the above discussion and mentioned approximation, this means that in the case under consideration, the excited levels with $S = 1$ do not contribute to the low-temperature cell–cell response, meaning that the following consideration will be confined to the low-lying $S = 0$ states. Considering the power dissipation, one should note that the instant switching of the polarizing field does not induce the transitions with a spin change, so that in the low-temperature limit, the transitions from the ground spin-singlet occur only to the excited levels with $S = 0$. This means that in the model of the transfer pathways adopted here, the spin-vibronic levels of the tetrameric cells with $S = 1$ do not contribute to power dissipation in the low temperature limit. In the bidimeric cells, the two half-cells are not connected through the transfer processes and hence all their states are degenerated with respect to the total spin $S$ of the cell. Therefore, while considering the bidimeric cells, one can also select only the $S = 0$ pattern for the calculation of the cell's characteristics.

Based on the numerical solution of the one-mode vibronic problem, one can evaluate the vibronic energy levels and the corresponding wave functions of the cells, and also calculate the specific heat release and cell–cell response functions for different sets of cell parameters. These results should allow us to conduct a comparative analysis of the two types of cells in order to establish the best "architecture" of the molecular cells and the arrays of cells which would ensure the coexistence of a low energy dissipation with a strong substantially nonlinear cell–cell response.

Figure 6 shows the dependences of the specific heat release on the vibronic PKS coupling parameter, calculated for tetrameric cells at $\hbar\omega = 1605$ cm$^{-1}$, two values of $u$ shown in the plots and the following two values of $t_{eff}$ for each $u$: $t_{eff} = 100$ cm$^{-1}$ (dashed lines) and $t_{eff} = 50$ cm$^{-1}$ (solid lines). It is seen from Figure 6 that the heat release $W$ increases with increasing $\upsilon$, provided that $u$ and $t_{eff}$ are fixed, and also $W$ increases with decreasing $t_{eff}$ for fixed $u$ and $\upsilon$ values. In all cases, $W$ tends to the maximum value $W_{max} = 2u$, which is equal to the kink energy. It also follows from Figure 4 that for fixed $t_{eff}$ and $\upsilon$, the heat release increases with increasing $u$.

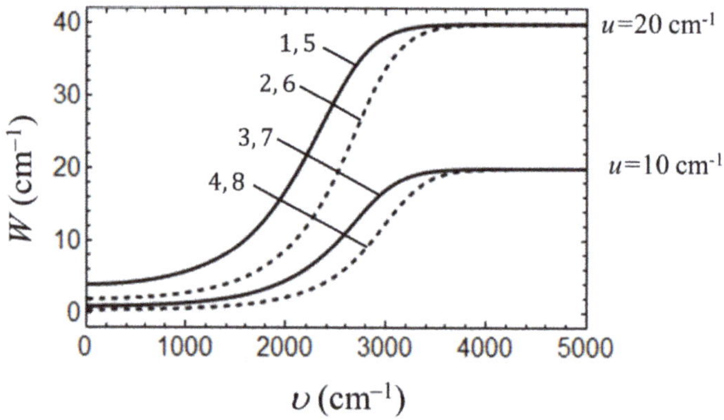


**Figure 6.** Dependences of low-temperature specific heat release on the vibronic PKS coupling parameter, calculated in the limit of strong Coulomb interaction inside the cell. Tetrameric cells: curve 1:u = 20 cm$^{-1}$, $t_{eff} = 50$ cm$^{-1}$; 2: u = 20 cm$^{-1}$, $t_{eff} = 100$ cm$^{-1}$; 3: u = 10 cm$^{-1}$, $t_{eff} = 50$ cm$^{-1}$; 4: u = 10 cm$^{-1}$, $t_{eff} = 100$ cm$^{-1}$. Bidimeric cells: 5: u = 20 cm$^{-1}$, $t_{eff} = 100$ cm$^{-1}$; 6: u = 20 cm$^{-1}$, $t_{eff} = 200$ cm$^{-1}$; 7: u = 10 cm$^{-1}$, $t_{eff} = 100$ cm$^{-1}$; 8: u = 10 cm$^{-1}$, $t_{eff} = 200$ cm$^{-1}$. In all calculations presented in this figure, the vibrational quantum is set to $\hbar\omega = 1605$ cm$^{-1}$.

It should be noted that exactly the same dependencies as in Figure 6 are obtained for bidimeric cells with two times larger $t_{eff}$ values, namely for $t_{eff} = 100$ cm$^{-1}$ (solid lines) and $t_{eff} = 200$ cm$^{-1}$ (dashed lines). This result is a consequence of the fact that in the limit of strong Coulomb interaction, the only difference between the Hamiltonians describing the two types of molecular cells is the factor in the matrix describing the effective second-order electron transfer, which is equal to 2 for a bidimeric cell and 4 for a tetrameric cell. Therefore, for a bidimeric cell, as in the case of a tetrameric cell, the specific heat release is a decreasing function of the parameters $t_{eff}$ and $\upsilon$ and an increasing function of the intercell Coulomb energy $u$.

Finally, Figure 6 allows us to compare the dependences of $W$ vs. $\upsilon$ evaluated in the low-temperature limit for the bidimeric and tetrameric cells with the same values $\hbar\omega = 1605$ cm$^{-1}$ and$t_{eff} = 100$ cm$^{-1}$. One can see that for all the same parameters, the heat release is lower for the tetrameric cell. This is explained by the fact that in a tetrameric cell, there are four electron transfer pathways (along the four sides of the square), while in a bidimeric cell, there are only two such pathways (along the two dimeric half-cells). For this reason, the contribution of effective transfer in the tetrameric cell turns out to be twice as large as in the dimer ($4t_{eff}$ against $2t_{eff}$), and, consequently, the tetrameric cell is characterized by a higher degree of electronic delocalization. It is seen that in the limit of strong vibronic coupling, the difference in the heat releases for the two kinds of cells disappears, since the effective transfer in both types of cells in this limit is almost completely suppressed. In contrast, in the case of moderate vibronic coupling, the heat release for a tetrameric cell can be significantly lower than that for a bidimeric cell.

The results obtained would seem to suggest that in order to reduce energy dissipation, one should use weakly interacting tetrameric or bidimeric cells characterized by weak vibronic coupling and or/strong electron transfer, and also that a tetrameric cell should exhibit better functional properties than a bidimeric cell. Below, we show that this statement is erroneous because it is based exclusively on the requirement of low energy dissipation and does not take into account the other key requirement for the molecular cell, that is the strong nonlinear response of the working cell.

When both of these requirements are considered simultaneously, one immediately arrives at the conclusion that the use of MV molecules with weak vibronic interaction and/or strong electron transfer as cells or half-cells is counterproductive. In fact, the MV molecules exhibiting such properties are strongly delocalized (systems belonging to the Robin and Day class III), and therefore they do not have their own electric dipole (or quadrupole) moment in a minimum of a single-well adiabatic potential. Therefore, a very high value of energy $u$ is required to polarize such a cell, which, in turn, leads to a dramatic increase in heat release. In contrast, using weakly delocalized (strong vibronic coupling and/or weak electron transfer) In this case, MV molecules acting as cells or half-cells ensures a strongly nonlinear cell–cell response, even providing for a weak intercell Coulomb interaction (small $u$). Therefore, one can expect that the heat release could be relatively low even if it is close to the upper limit $W = 2u$.

Also, a bidimeric cell should have an advantage over a tetrameric cell, since for a bi-dimeric cell, a strong nonlinear response can be achieved at lower values of $u$, which allows the heat release to be significant reduced.

The reasoning presented so far is essentially qualitative in nature and indicates only the range of parameters in which the required functional properties of the cells could be expected. The quantitative results confirming the advantages of weakly delocalized molecules as cells are shown in Figure 7a,b. These figures present the low-temperature cell–cell response functions $P_{wc}$ vs. $P_{dc}$ calculated for the tetrameric cells characterized by different sets of parameters and the corresponding specific heat release values $W$ evaluated with these sets of parameters.

Figure 7a shows the cell–cell response functions calculated for $\hbar\omega = 1605$ cm$^{-1}$ and $v = 3500$ cm$^{-1}$ and three different sets of $u$ and $t_{eff}$. Comparison of the three cases in Figure 6 shows that although the lowest heat release is achieved at $u = 10$ cm$^{-1}$ and $t_{eff} = 200$ cm$^{-1}$ ($W \approx 15.9$ cm$^{-1}$), that is, with a minimum value of $u$ and a maximum value of $t_{eff}$ (within the parameter sets used in Figure 7a), the cell–cell response in this case is rather weak and the function $P_{wc}(P_{dc})$ has a slight slope in a wide range of $P_{dc}$. This is definitely an unfavorable factor from the point of view of the functioning of QCA devices. To ensure a strong nonlinear cell–cell response at such large values of the effective transfer parameter, it is necessary to significantly enhance (by approximately five times) the intercell Coulomb interaction (in Figure 7a, this is the case when $u = 50$ cm$^{-1}$ and $t_{eff} = 200$ cm$^{-1}$), but such a strong increase in $u$ leads to a sharp (up to $W \approx 98.1$ cm$^{-1}$) increase in heat release. In contrast, for $u = 10$ cm$^{-1}$ and $t_{eff} = 50$ cm$^{-1}$ (weak intercell interaction and weak effective transfer), the cell–cell response function has a highly nonlinear stepwise shape (red curve in Figure 7a), and at the same time, the corresponding heat release value $W \approx 19.7$ cm$^{-1}$ turns out to be only slightly higher than that found at $u = 10$ cm$^{-1}$ and $t_{eff} = 200$ cm$^{-1}$. One can thus conclude that the optimal rational strategy to design highly efficient molecular QCA cells is using weakly interacting MV tetramers characterized by weak electron transfer.

Figure 7b shows the cell–cell response functions calculated at $\hbar\omega = 1605$ cm$^{-1}$ and $t_{eff} = 50$ cm$^{-1}$ and three different sets of $u$ and $v$. Although the lowest heat release occurs when the intercell interaction and the vibronic interaction are both relatively weak (when $u = 10$ cm$^{-1}$ and $v = 3000$ cm$^{-1}$, for which we obtain $W \approx 17$ cm$^{-1}$), the cell–cell response is weak and exhibits an almost linear slope, which makes such a parametric regime unsuitable for the proper functioning of QCA devices. To ensure a strong cell–cell response at such a weak vibronic coupling, it is necessary, as in the case considered above, to significantly enhance the intercell Coulomb interaction (case of $u = 40$ cm$^{-1}$ and $v = 3000$ cm$^{-1}$ in Figure 7b), but this leads to a sharp (up to $W \approx 79$ cm$^{-1}$) increase in the heat release. The favorable regime is that of weak intercell interaction and strong vibronic coupling (case of $u = 10$ cm$^{-1}$ and $v = 3500$ cm$^{-1}$ in Figure 7b), because in this case both the nonlinearity of the cell–cell response function and the relatively low heat release of $W \approx 19.7$ cm$^{-1}$ prove to be attainable (the red curve in Figure 7b, which coincides with red curve in Figure 7a, since both these plots are obtained with the same set of parameters). Therefore, it follows from Figure 7b that weakly interacting tetramers characterized by strong vibronic coupling are promising for use as cells.

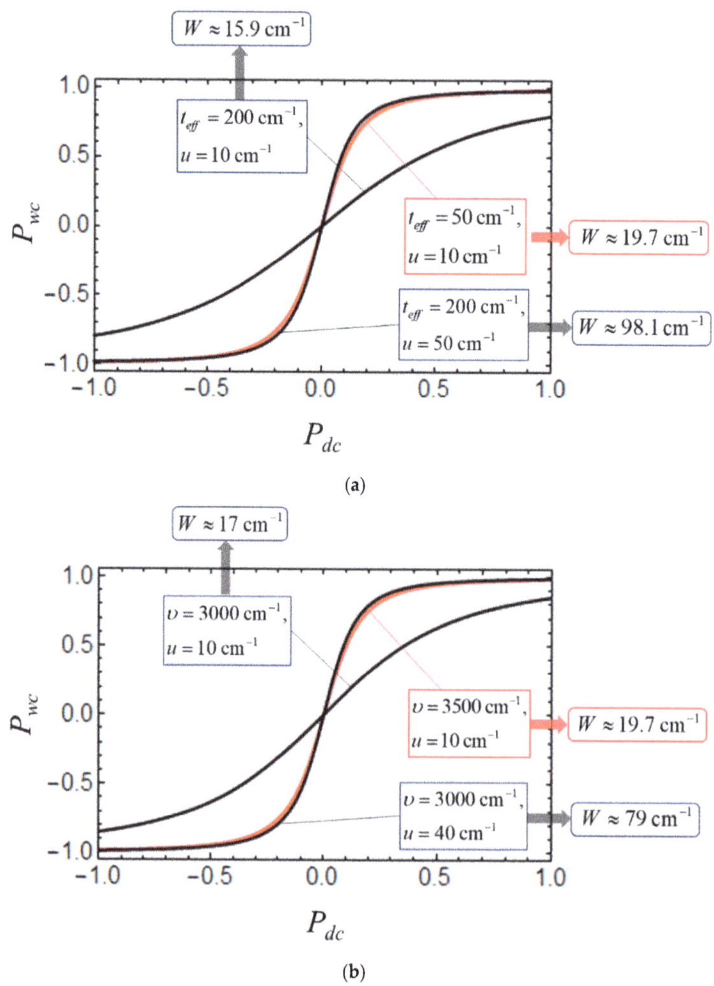


**Figure 7.** Low-temperature cell–cell response functions calculated in the limiting case of strong Coulomb interaction for tetrameric cells with $\hbar\omega = 1605$ cm$^{-1}$ and $v = 3500$ cm$^{-1}$ and the three sets of $t_{eff}$ and $u$ values shown in the plots (**a**) and for tetrameric cells with $\hbar\omega = 1605$ cm$^{-1}$ and $t_{eff} = 50$ cm$^{-1}$ and three sets of $v$ and $u$ values shown in the plots. (**b**) The corresponding calculated values of specific heat release are also shown in the plots.

Based on the observations following from Figure 7a,b, one can conclude that the high efficiency of tetrameric molecular QCA cells in which the limit of strong Coulomb interaction holds can be achieved for weakly delocalized MV molecules which are allowed to only weakly interact with each other. The first requirement can be satisfied provided that the transfer of the excess charge is weak and/or the vibronic coupling is strong, while in order to meet the second requirement, the tetramers should be located at long distances from each other and/or separated by groups of atoms capable to strongly shield the electric field.

The same is true for the bidimeric cells in the limiting case of strong Coulomb interaction, since they differ from the tetrameric cells only in the factor appearing in the effective transfer matrix. Moreover, the results obtained allow us to conclude that (ceteris paribus) the bidimeric molecular cells can have functional advantages over the tetrameric ones. This is explained by the weaker electron delocalization in the bidimeric cell, which means that strong nonlinear response for such a cell is achieved at lower values of $u$, which, in turn, helps to reduce the heat release.

### 3.2. Consideration of Bidimeric and Tetrameric Cells: Common Case of Arbitrary Coulomb Interaction

In addition to the limiting case of strong Coulomb interaction, an actual issue is the general case of an arbitrary interrelation between the Coulomb energy gap $U$ and the electron transfer parameter $t$ in tetrameric and bidimeric cells. The calculation of heat release and cell–cell response in this case is based on solving more complex multimode vibronic problems such as a three-mode problem for a tetrameric cell and a two-mode problem for a bidimeric cell. Along with the complications in the vibronic problem, an expansion of the electronic basis is also required. Thus, for a tetrameric cell, the electronic basis includes six states corresponding to six possible distributions of the two excess electrons in an MV tetramer. In the case of a bidimeric cell, the electronic distributions with two excess electrons per dimeric half-cell should be excluded, and so the electric basis is reduced to only four states. Note that in this general case, the intercell Coulomb interaction cannot be described by a single parameter $u$; instead, to describe such an interaction, we used the distance $c$ between the cells. Here, we use the parameters evaluated with the aid of ab initio quantum chemical calculations as representatives of the series of organic MV compounds, which includes oxidized norbornadiene and its polycyclic derivatives [41].

Figure 6 shows the dependence of the low-temperature specific heat release of a tetrameric cell on the vibronic interaction parameter evaluated at $t = 1573$ cm$^{-1}$, $\hbar\omega = 1605$ cm$^{-1}$, and $b = 6.973$ Å and two different intercell distances $c$. It is seen that the heat release increases with the increase in $v$ and saturates in the limit of strong vibronic coupling. It also follows from Figure 8 that the higher the heat release, the shorter the distance between the cells, and the stronger the Coulomb interaction between them. One can also conclude (the corresponding plots are not shown) that the heat release decreases with increasing $t$. Qualitatively, these results coincide with those described above for the limiting case of strong intracell Coulomb interaction.

By comparing the cell–cell response functions calculated for a tetrameric cell with three sets of $c$ and $v$ values (Figure 9), one can conclude that the most favorable case for the proper functioning of a cell is the case of strong vibronic interaction and a large intercell distance. This is the red curve in Figure 9, which corresponds to $c = 30$ Å and $v = 4932$ cm$^{-1}$. Only in this case does the cell demonstrate a strong nonlinear response in combination with a relatively low heat release ($W \approx 22$ cm$^{-1}$). Indeed, although in the case of weaker vibronic coupling $v = 4000$ cm$^{-1}$ and the same $c$ value, even lower heat release is obtained, the cell–cell response in this case turns out to be weak and has a slight slope (almost linear behavior). To improve the cell–cell response function while retaining this $v$ value, it is necessary to significantly bring the cells closer together, but such an approach of the cells leads to a significant (more than an order of magnitude) increase in the heat release. Calculations also show that cells separated from each other by long distances and characterized by small values of the parameter $t$ should demonstrate a strong nonlinear response in combination with low heat release. Qualitatively, these conclusions coincide with those made for the case of strong intracell Coulomb interaction. Note that the conclusions made for tetrameric cells based on the analysis of the dependencies presented in Figures 8 and 9 are also valid for the bidimeric cells with a violated limit of strong Coulomb interaction.

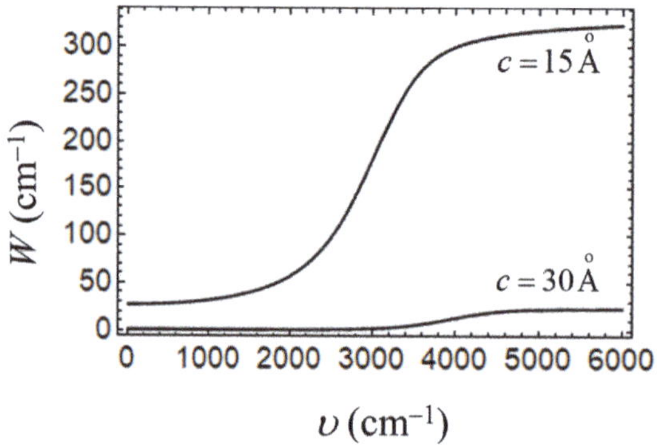


**Figure 8.** Dependences of specific heat release on the vibronic PKS coupling parameter, calculated in the low-temperature limit for a tetrameric cell with a violated limit of strong Coulomb interaction with $t = 1573\ \text{cm}^{-1}$, $\hbar\omega = 1605\ \text{cm}^{-1}$, and $b = 6.973$ Å and two intercell distances shown in the plots.

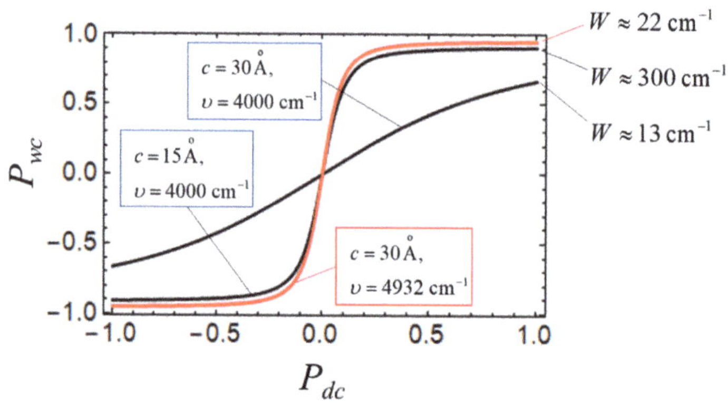


**Figure 9.** Low-temperature cell–cell response functions for a tetrameric cell evaluated beyond the limit of strong Coulomb interaction with $t = 1573\ \text{cm}^{-1}$, $\hbar\omega = 1605\ \text{cm}^{-1}$, and $b = 6.973$ Å and the three sets of $c$ and $\upsilon$ values shown in the plots. The corresponding calculated values of the specific heat release are also shown.

All of the regularities discussed so far are common to all types of cells and valid not only in the limiting case of strong Coulomb interaction but also in a general case of an arbitrary interrelation between the parameters $t$ and $U$. Besides these common features, bidimeric cells with a violated limit of strong Coulomb interaction also have important specific characteristics, arising from the fact that the intercell Coulomb interaction in this case proves to be dependent on the mutual arrangement of the interacting cells, which can be of either "side-by-side" or "head-to-tail" type, as discussed earlier (see Figure 3). The influence of the relative arrangement of bidimeric cells on their polarization characteristics was studied in detail in Ref. [34]. In this regard, it seems useful to analyze the influence of the relative arrangement of such cells on the energy dissipation under nonadiabatic switching conditions.

The results of calculations of the specific heat release for the two types of mutual arrangement of bidimeric cells are presented in Figure 10. One can see that the heat dissipation increases when increasing the vibronic coupling parameter, which is a common feature of all types of cells discussed so far. It also follows from Figure 10 that, all other things being equal, the heat release is lower for the head-to-tail-type arrangement of cells. This observation correlates rather well with the shapes of calculated cell–cell response functions shown in Figure 11, from which it follows that the saturation cell–cell response value is lower for the head-to-tail-type arrangement. The difference in specific heat release for the two arrangements decreases with the increase in the strength of the vibronic coupling, and fully disappears in the limit of strong vibronic coupling or, in other words, in the saturation region. Note that in this limit, the difference in response functions also disappears (see Ref. [37]), and at the same time, the cell–cell response function, which is the same for two mutual arrangements of cells, takes on a highly nonlinear stepwise shape. Thus, under the condition of strong vibronic coupling, both important functional characteristics of the cells (heat release and cell–cell response) become independent of their mutual arrangement.

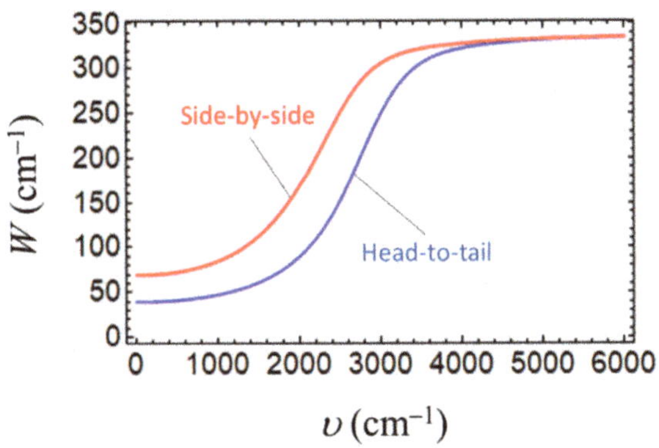


**Figure 10.** The dependences of low-temperature specific heat release on the vibronic parameter evaluated for the two mutual arrangements of the bidimeric cells with a violated limit of strong Coulomb interaction at $t = 1573\ \text{cm}^{-1}$, $\hbar\omega = 1605\ \text{cm}^{-1}$, $b = 6.973$ Å, and $c = 15$ Å.

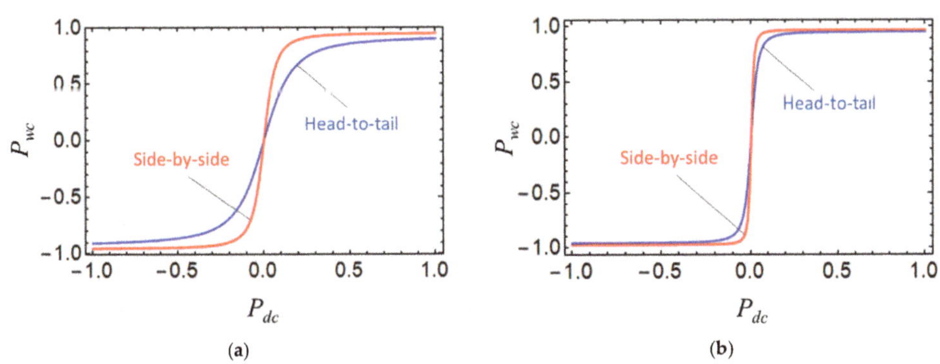


**Figure 11.** Cell–cell response functions evaluated for two mutual arrangements of bidimeric cells with a violated limit of strong Coulomb interaction at $t = 1573\ \text{cm}^{-1}$, $\hbar\omega = 1605\ \text{cm}^{-1}$, $b = 6.973$ Å, and $c = 15$ Å, and the following two values of $\upsilon$: $\upsilon = 3500\ \text{cm}^{-1}$ (**a**), $\upsilon = 4000\ \text{cm}^{-1}$ (**b**).

The latter result seems to be quite appealing from the point of view of the use of the bidimeric cells for building up QCA gates and circuits. In fact, almost all QCA devices are usually built in such a way that their architecture includes both types of mutual arrangements of the constituent cells, which are desired to be equivalent in terms of their functional characteristics (cell–cell response functions, heat release, etc.). Therefore, the presence of a strong vibronic interaction in a bidimeric cell makes it possible to simultaneously ensure the fulfillment of the two key conditions required for the proper functioning of the QCA devices, namely to obtain a strong nonlinear cell–cell response even for a weak intercell interaction, ensuring low energy dissipation, and also to eliminate undesired effects of the non-equivalence of the two types of cell arrangement.

## 4. Some Relevant Ab Initio Results and Possibility of Controlling the Key Parameters

The above theoretical consideration and conclusions are formulated within the framework of a semi-empirical approach dealing with adjustable parameters which are to be evaluated separately. In this section, we go beyond the parametric theory and briefly discuss the results of ab initio calculations of the key parameters such as electron transfer parameter $t$ and the vibronic coupling parameter $v$, while the intercell distance remains variable. The evaluation of these parameters and identification of their microscopic origin is necessary not only to identify the real areas of their changes that critically affect the heat release and cell–cell response, but also to reveal the possibilities for their rational control. The discussion below is inspired by the regularities recently found for the series of organic MV molecules in which either $t$ or $v$ has been shown to vary in a systematical way within each such series [41,42].

The first such series includes MV oxidized norbornadiene $[C_7H_8]^+$ and four of its polycyclic derivatives, $[C_{12}H_{12}]^+$, $[C_{17}H_{16}]^+$, $[C_{27}H_{24}]^+$, and $[C_{32}H_{28}]^+$ (compounds I–V in Ref. [41]). Each MV molecule in this series comprises two terminal C=C chromophores (redox sites) connected by a bridge, which mediates the intramolecular charge transfer [41]. The accuracy of the description of the vibronic and electron transfer parameters and consequently the degree of localization (and even possibility of trapping) in MV molecules crucially depend upon the level of the quantum-chemical approach employed. In particular, the single-configuration diabatic wave function could not provide a satisfactory description of the system. In particular, single-configuration methods are inadequate for the calculation of radical cations, because they lead to an adequate condition for charge-delocalized structures (see Ref. [41] references therein). The quantum-chemical study of norbornylogous compounds I–V in [41] included evaluation of the tunneling energy gap and determination of the fully optimized molecular structures. In this way, the barrier between the two with localized structures has been found along with Mulliken charges in symmetric delocalized and localized configurations. The CASSCF methodology [42,43] (GAMESS program suite [44,45]) has been used to study the localization the charge in I–V norbornylogous cation radicals. Numerical vibrational frequency calculations have been performed for the systems I, II and III using the GAMESS program. The cc-pVDZ basis set was used for calculations of I–III and the 6-311G* basis set was used for calculations of IV and V. High-level ab initio calculations combined with the parametric description within the vibronic model have revealed a quite weak dependence of the vibronic parameter and frequency of the vibrations associated with the redox sites on the length of the bridge, which justifies the applicability of PKS model for treating of these compounds. Unfortunately, we do not have the results of ab initio calculations for the tetrameric square-planar molecular cell for QCA representing the tetraruthenium Creutz–Taube derivatives mentioned so far. Due to the presence of a bridge connecting the Creutz–Taube dimers, the electronic properties and vibronic parameters of the tetraruthenium units are expected to differ significantly from those of the dimeric systems. That is why in this article, we limit ourselves to considering norbornylogous compounds, leaving the important case of tetraruthenium systems for detailed consideration in the future.

One can combine two identical MV dimers (e.g., two $[C_{12}H_{12}]^+$ (I) and $[C_{17}H_{16}]^+$ (II) molecules, as shown in Figure 12a,b) to obtain bidimeric square cells. One can adopt a reasonable assumption that the values of t and υ parameters are the same in the isolated dimeric entities and in the bidimeric cell combined of the dimeric units. In view of the results of the present study, one can conclude that the cells composed of dimers (in series I–V [41]) with longer bridges have greater polarizability than those consisting of dimers with shorter bridges even with smaller kink energy u, which, in turn, allows the lower power dissipation of such cells. This example shows the possibility of controlling the transfer parameter by modifying the bridge connecting the redox sites without changing the vibronic coupling.

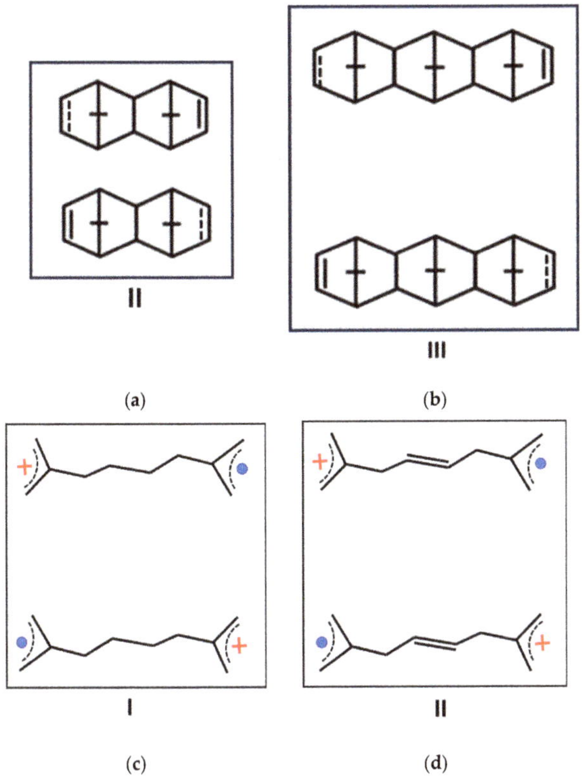


**Figure 12.** Schemes of the bidimeric cells composed of two MV dimers: $[C_{12}H_{12}]^{+}$ (**a**) and $[C_{17}H_{16}]^{+}$ (**b**) [41]; radical-cation forms of 1,4-diallyl-butane with a saturated bridge (**c**); radical-cation forms of 1,4-diallyl-butene-2 with an unsaturated bridge [46] (**d**).

Another series of organic MV dimers includes the radical cation forms of the parent 1,4-diallyl-butane (the system initially proposed by Lent et al. for molecular implementation of quantum cellular automata [1]) and its derivatives such as 1,4-diallyl-butene-2 (**II**), 1,4-diallyl-hexane (**III**), 1,4-diallyl-hexene-3(**IV**), 1,4-diallyl-octane (**V**) and 1,4-diallyl-octene-4 (**VI**) (numbered from **I** to **V** according to Ref. [46] containing a detailed quantum-chemical study of these compounds). In these MV compounds, the bridge of variable length connecting allyl groups playing the role of redox sites can be either saturated or unsaturated and modified by the $\pi$-spacer [46]. As distinguished from the series of oxidized norbornadiene derivatives in which one can control the electron transfer parameter $t$ through variation of the bridge length, in the series **I**–**V** based on 1,4-diallyl-butane, one can modify the vibronic coupling $v$ [46]. In fact, the parameter $v$ strongly depends on the electronic state of the bridge being essentially different in compounds with saturated and unsaturated bridges. This dependence has been explained by polarization of the $\pi$-chromophore on the bridge by the excess charge at the redox sites, which has been referred to as the polaronic mechanism of localization [46]. In view of the present study, this means that bidimeric cells (Figure 12a,b) composed of the dimers with saturated bridges (Figure 12b) are preferable in the design of a bidimeric cell, since they require lower kink energy $2u$ for their efficient polarization and so one can expect lower heat release for such cells.

Summarizing the results emanating from the ab initio consideration of the two families of MV organic compounds, one can see that the key parameters of MV systems used for the design of the QCA cells can be varied in a controllable manner, providing options to optimize the heat release and the shape of the cell–cell response. By changing either the length of the bridge or its electronic structure in constituent dimeric half-cells, one can tune polarization and dissipative properties of bidimeric cells.

## 5. Conclusions

In this work, we have generalized the vibronic approach to calculating the specific heat release during nonadiabatic switching, previously proposed for the simplest case of dimeric molecular cells, to the case of more complex and practically promising square-planar bidimeric and tetrameric cells. Using the general expression, Equation (8), we have evaluated the specific heat release based on previously found quantum-mechanical solutions of the vibronic problems, which are specific to different types of cells and also dependent on the approximations adopted in the course of their description. This expression is applicable to estimating temperature-variable specific heat release in bidimeric and tetrameric cells (or, more precisely, its upper limit) as a function of the electronic and vibronic parameters of the cell, as well as on the energy of the intercell Coulomb interaction.

First, based on the developed general approach, specific heat release and cell–cell response functions have been evaluated for molecular QCA based on bidimeric (composed of $d^1 - d^0$-type MV dimers or organic MV compounds playing the role of half-cells) and tetrameric (represented by $d^1 - d^1 - d^0 - d^0$-type clusters or organic MV compounds) two-electron square-planar cells, for which an important limiting case occurs, when the energy of the intracell interelectronic Coulomb repulsion significantly exceeds the electron transfer energy. The description of such cells has been performed with the aid of a quantum-mechanical solution of the vibronic problem arising in the framework of the vibronic PKS model, with this problem being significantly simplified in the limit of strong Coulomb interaction, reducing to a single-mode problem. Based on solving this problem, it has been shown that specific heat release in bidimeric and tetrameric cells increases with increasing vibronic coupling and/or weakening of electron transfer, reaching in the limit of strong vibronic coupling and/or weak transfer its maximum value equal to the Coulomb excitation energy of the pair of interacting cells. It has also been shown that, provided that the values of all parameters are the same, the heat release for a tetrameric cell is lower than for a bidimeric one. We have explained this difference by the fact that in a tetrameric cell, there are four electron transfer pathways (along the sides of the square), while in a bidimeric cell, there are only two such pathways (one within each dimeric half-cell). As a result, the tetrameric cell is characterized by a higher extent of the electronic delocalization, which leads to lower energy dissipation in the tetrameric cell. It follows from the results obtained that this difference disappears in the limit of strong vibronic coupling. This can be explained by the fact that in this limit, the effective electron transfer in both types of cells is almost fully suppressed.

Also, specific heat release in the course of nonadiabatic switching has been calculated for tetrameric and bidimeric cells in the general case of an arbitrary relationship between the energies of intracell Coulomb interaction and the electron transfer. These calculations are based on the numerical solutions of the three-mode vibronic problem for a tetrameric cell and two-mode problem for a bidimeric cell. Since, unlike the strong Coulomb interaction limit, in the general case the intercell Coulomb interaction cannot be described by using a single parameter, the intercell distance has been used to describe such interaction. It has been shown that as the vibronic coupling parameter increases, the heat release increases and reaches saturation in the limit of strong vibronic coupling, and the higher the heat release, the shorter the distance between the cells, i.e., the stronger the Coulomb interaction between them. Qualitatively, these results coincide with the results obtained in the limit of strong Coulomb interaction. In addition to these regularities, which are common to all types of cells, bidimeric cells with a violated limit of strong Coulomb interaction have been shown to possess some specific behavioral features arising from the dependence of the intercell Coulomb interaction on the mutual arrangement of the interacting cells, which can be either head-to-tail or side-by-side. We have demonstrated that, all other things being equal, heat release is lower in the case of a head-to-tail arrangement of cells, and the difference in specific heat release for the two arrangements decreases with increasing vibronic coupling, completely disappearing in the strong vibronic coupling limit. Therefore, bidimeric cells with a violated strong Coulomb interaction limit and strong vibronic coupling are equivalent to cells for which the strong Coulomb interaction limit is satisfied, since in both these cases, the mutual arrangement of the cells does not affect the heat release. This result seems to be important from the point of view of the rational design of QCA devices, as far as different types of cells' arrangement in such devices should be equivalent in terms of their functional characteristics, including energy dissipation.

Based on the developed vibronic approach aimed at the evaluation of the specific heat release in the course of nonadiabatic switching as well as the previously developed approach to calculating vibronic cell–cell response functions, a theoretical methodology for optimizing functional properties of cells of various types has been proposed. This methodology involves the procedure for establishing the parametric regimes, which provide low heat release in combination with a strong nonlinear response of the working cell to the electric field created by a polarized driver-cell. During the search for such optimal regimes, both the electronic and the vibronic parameters of the cell and also the intercell Coulomb interaction energies dependent on the distances between the cells have been varied. The use of the proposed methodology allows us to formulate criteria for the rational design of QCA based on the molecular cells. Thus, it has been shown that the optimal design strategy is to use weakly interacting molecular cells (cells located at long distances from each other or separated by the groups of atoms which can strongly screen the electric field) exhibiting weak delocalization of the excess charges. Suitable systems to satisfy the condition of weak delocalization are MV molecules characterized by weak electron transfer and/or strong vibronic coupling.

A comparative analysis of the efficiency of the two types of square-planar cells has been carried out from the point of view of the possibility of minimizing heat release, while maintaining strong nonlinear cell–cell response. We have thus concluded that bidimeric cells have an advantage over tetrameric cells. This advantage follows from the fact that, due to the weaker delocalization of the excess electrons in the bidimeric cells, a strong nonlinear response for such cells can be achieved with weaker intercell interactions as compared with tetrameric cells, which makes it possible to significantly reduce the heat release. Although the bidimeric cells are advantageous over tetrameric cells provided that the two kinds of mutual arrangements of the bidimeric cells (head-to-tail and side-by-side) are equivalent in terms of their functional properties (cell–cell response and heat release), such equivalence does not contradict the main criterion formulated above. Indeed, its fulfillment also requires the presence of weak electronic delocalization in the molecule.

Finally, we have discussed some relevant ab initio results of the study of electron transfer and vibronic coupling in view of the possibility of controlling the key parameters responsible for the functioning of molecular QCA cells. We elaborate upon the computer-aided methodology to control the electron transfer and the vibronic coupling, which are the interactions governing the heat release and the cell–cell response in molecular QCA.

The above-proposed methodology for optimizing the properties of molecular cells is based on the two key requirements that such cells should meet, namely low energy dissipation and a strong nonlinear cell–cell response. Along with these, there are other important conditions for the applicability of molecules as QCA cells, such as their ability to perform clocking [3,47–53], a special disposition of the counterion preventing violation of the high symmetry of the cell [15–17], and finally, the ability of the molecule to be attached to a substrate without breaking down its required properties (see, e.g., [18]). Elaboration of a more general approach to optimizing the properties of molecular cells, which takes into account all aforenamed criteria, as well as the synthesis of suitable molecular candidates, is still a challenge in the design of high-performance molecular QCA.

**Author Contributions:** Methodology, A.P. and B.T.; software, S.Z.; validation, S.Z.; formal analysis, A.P., S.Z. and B.T.; investigation, S.Z.; data curation, S.Z.; writing—review and editing, A.P. and B.T. All authors have read and agreed to the published version of the manuscript.

**Funding:** The work was in part (comparative analysis of bidimeric and tetrameric cells) performed with financial support from the Russian Science Foundation (A.P., project no. 20-13-00374-П) and in part (numerical solution of the two-mode and three-mode quantum-mechanical vibronic problems) with financial support from the Ministry of Science and Higher Education of the RF (A.P., state assignment No. 124013100858-3).

**Data Availability Statement:** The original contributions presented in the study are included in the article, further inquiries can be directed to the corresponding authors.

**Conflicts of Interest:** The authors declare no conflict of interest.

## References

1. Lent, C.S. Bypassing the transistor paradigm. *Science* **2000**, *288*, 1597–1599. [CrossRef]
2. Lieberman, M.; Chellamma, S.; Varughese, B.; Wang, Y.; Lent, C.; Bernstein, G.H.; Snider, G.; Peiris, F.C. Quantum-dot cellular automata at a molecular scale. *Ann. N. Y. Acad. Sci.* **2002**, *960*, 225–239. [CrossRef]
3. Lent, C.S.; Isaksen, B.; Lieberman, M. Molecular quantum-dot cellular automata. *J. Am. Chem. Soc.* **2003**, *125*, 1056–1063. [CrossRef] [PubMed]
4. Jiao, J.; Long, G.J.; Grandjean, F.; Beatty, A.M.; Fehlner, T.P. Building Blocks for the Molecular Expression of Quantum Cellular Automata. Isolation and Characterization of a Covalently Bonded Square Array of Two Ferrocenium and Two Ferrocene Complexes. *J. Am. Chem. Soc.* **2003**, *125*, 7522–7523. [CrossRef]
5. Li, Z.; Beatty, A.M.; Fehlner, T.P. Molecular QCA Cells. 1. Structure and Functionalization of an Unsymmetrical Dinuclear Mixed-Valence Complex for Surface Binding. *Inorg. Chem.* **2003**, *42*, 5707–5714. [CrossRef] [PubMed]
6. Li, Z.; Fehlner, T.P. Molecular QCA Cells. 2. Characterization of an Unsymmetrical Dinuclear Mixed-Valence Complex Bound to a Au Surface by an Organic Linker. *Inorg. Chem.* **2003**, *42*, 5715–5721. [CrossRef]
7. Qi, H.; Sharma, S.; Li, Z.; Snider, G.L.; Orlov, A.O.; Lent, C.S.; Fehlner, T.P. Molecular Quantum Cellular Automata Cells. Electric Field Driven Switching of a Silicon Surface Bound Array of Vertically Oriented Two-Dot Molecular Quantum Cellular Automata. *J. Am. Chem. Soc.* **2003**, *125*, 15250–15259. [CrossRef]
8. Braun-Sand, S.B.; Wiest, O. Theoretical Studies of Mixed Valence Transition Metal Complexes for Molecular Computing. *J. Phys. Chem. A* **2003**, *107*, 285–291. [CrossRef]
9. Braun-Sand, S.B.; Wiest, O. Biasing Mixed-Valence Transition Metal Complexes in Search of Bistable Complexes for Molecular Computing. *J. Phys. Chem. B* **2003**, *107*, 9624–9628. [CrossRef]
10. Qi, H.; Gupta, A.; Noll, B.C.; Snider, G.L.; Lu, Y.; Lent, C.; Fehlner, T.P. Dependence of Field Switched Ordered Arrays of Dinuclear Mixed-Valence Complexes on the Distance between the Redox Centers and the Size of the Counterions. *J. Am. Chem. Soc.* **2005**, *127*, 15218–15227. [CrossRef]
11. Jiao, J.; Long, G.J.; Rebbouh, L.; Grandjean, F.; Beatty, A.M.; Fehlner, T.P. Properties of a Mixed-Valence $(Fe^{II})_2(Fe^{III})_2$ Square Cell for Utilization in the Quantum Cellular Automata Paradigm for Molecular Electronics. *J. Am. Chem. Soc.* **2005**, *127*, 17819–17831. [CrossRef] [PubMed]
12. Lu, Y.; Lent, C.S. Theoretical Study of Molecular Quantum Dot Cellular Automata. *J. Comput. Electron.* **2005**, *4*, 115–118. [CrossRef]
13. Zhao, Y.; Guo, D.; Liu, Y.; He, C.; Duan, C. A Mixed-Valence $(Fe^{II})_2(Fe^{III})_2$ Square for Molecular Expression of Quantum Cellular Automata. *Chem. Commun.* **2008**, 5725–5727. [CrossRef] [PubMed]
14. Nemykin, V.N.; Rohde, G.T.; Barrett, C.D.; Hadt, R.G.; Bizzarri, C.; Galloni, P.; Floris, B.; Nowik, I.; Herber, R.H.; Marrani, A.G.; et al. Electron-Transfer Processes in Metal-Free Tetraferrocenylporphyrin. Understanding Internal Interactions to Access Mixed Valence States Potentially Useful for Quantum Cellular Automata. *J. Am. Chem. Soc.* **2009**, *131*, 14969–14978. [CrossRef] [PubMed]
15. Lu, Y.; Lent, C. Self-doping of molecular quantum-dot cellular automata: Mixed valence zwitterions. *Phys. Chem. Chem. Phys.* **2011**, *13*, 14928–14936. [CrossRef]
16. Lu, Y.; Lent, C.S. Counterion-free molecular quantum-dot cellular automata using mixed valence zwitterions—A double-dot derivative of the $[closo\text{-}1\text{-}CB_9H_{10}]^-$ cluster. *Chem. Phys. Lett.* **2013**, *582*, 86–89. [CrossRef]
17. Christie, A.; Forrest, R.P.; Corcelli, S.A.; Wasio, N.A.; Quardokus, R.C.; Brown, R.; Kandel, S.A.; Lu, Y.; Lent, C.S.; Henderson, K.W. Synthesis of a neutral mixed-valence diferrocenyl carborane for molecular quantum-dot cellular automata applications. *Angew. Chem.* **2015**, *127*, 15668–15671. [CrossRef]
18. Wang, X.; Yu, L.; Inakollu, V.S.S.; Pan, X.; Ma, J.; Yu, H. Molecular Quantum-Dot Cellular Automata Based on Diboryl Radical Anions. *J. Phys. Chem. C* **2018**, *122*, 2454–2460. [CrossRef]

19. Macrae, R.M. Mixed-valence realizations of quantum dot cellular automata. *J. Phys. Chem. Solids* **2023**, *177*, 111303. [CrossRef]
20. Aviram, A. Molecules for Memory, Logic, and Amplification. *J. Am. Chem. Soc.* **1988**, *125*, 5687–5692. [CrossRef]
21. Creutz, C.; Taube, H. Binuclear complexes of ruthenium ammines. *J. Am. Chem. Soc.* **1973**, *95*, 1086–1094. [CrossRef]
22. Timler, J.; Lent, C.S. Power gain and dissipation in quantum-dot cellular automata. *J. Appl. Phys.* **2002**, *91*, 823–831. [CrossRef]
23. Rahimi, E. Energy dissipation of quantum-dot cellular automata logic gates. *Micro Nano Lett.* **2016**, *11*, 369–371. [CrossRef]
24. Rahimi, E.; Reimers, J.R. Molecular quantum cellular automata cell design trade-offs: Latching vs. power dissipation. *Phys. Chem. Chem. Phys.* **2018**, *20*, 17881–17888. [CrossRef]
25. Pidaparthi, S.S.; Lent, C.S. Molecular reorganization energy in quantum-dot cellular automata switching. *J. Appl. Phys.* **2022**, *131*, 044502. [CrossRef]
26. Palii, A.; Belonovich, V.; Tsukerblat, B. In quest of optimal parametric regime of nonadiabatic switching ensuring low heat release in conjunction with high polarizability of mixed-valence molecular dimer. *Phys. Chem. Chem. Phys.* **2023**, *25*, 17526–17534. [CrossRef]
27. Palii, A.; Belonovich, V.; Aldoshin, S.; Zilberg, S.; Tsukerblat, B. Mixed-Valence Bridged Norbornylogous Compounds as Switchable Cells for Molecular Quantum Cellular Automata: A Compromise between High Polarizability and Low Power Dissipation. *J. Phys. Chem. A* **2023**, *127*, 9030–9039. [CrossRef]
28. Lent, C.S.; Tougaw, P.; Porod, W. Bistable saturation in coupled quantum dots for quantum cellular automata. *Appl. Phys. Lett.* **1993**, *62*, 714–716. [CrossRef]
29. Lent, C.S.; Tougaw, P.D. Lines of interacting quantum dot cells: A binary wire. *J. Appl. Phys.* **1993**, *74*, 6227–6233. [CrossRef]
30. Tougaw, P.; Lent, C. Logical Devices Implemented Using Quantum Cellular Automata. *J. Appl. Phys.* **1994**, *75*, 1818–1825. [CrossRef]
31. Lent, C.S.; Tougaw, P.D. A device architecture for computing with quantum dots. *Proc. IEEE* **1997**, *85*, 541–557. [CrossRef]
32. Porod, W.; Lent, C.S.; Bernstein, G.H.; Orlov, A.O.; Amlani, I.; Snider, G.L.; Merz, J.L. Quantum-dot cellular automata: Computing with coupled quantum dots. *Int. J. Electron.* **1999**, *86*, 549–590.
33. Tsukerblat, B.; Palii, A.; Clemente-Juan, J.M.; Coronado, E. Mixed-Valence Molecular Four-Dot Unit for Quantum Cellular Automata: Vibronic Self-Trapping and Cell-Cell Response. *J. Chem. Phys.* **2015**, *143*, 134307. [CrossRef] [PubMed]
34. Palii, A.; Zilberg, S.; Rybakov, A.; Tsukerblat, B. Double-Dimeric Versus Tetrameric Cells for Quantum Cellular Automata: A Semiempirical Approach to Evaluation of Cell−Cell Responses Combined with Quantum-Chemical Modeling of Molecular Structures. *J. Phys. Chem. C* **2019**, *123*, 22614–22623. [CrossRef]
35. Palii, A.; Aldoshin, S.; Tsukerblat, B. Towards the design of molecular cells for quantum cellular automata: Critical reconsideration of the parameter regime for achieving functionality. *Dalton Trans.* **2022**, *51*, 286–302. [CrossRef]
36. Palii, A.; Aldoshin, S.; Tsukerblat, B. Functional Properties of Tetrameric Molecular Cells for Quantum Cellular Automata: A Quantum-Mechanical Treatment Extended to the Range of Arbitrary Coulomb Repulsion. *Magnetochemistry* **2022**, *8*, 92. [CrossRef]
37. Tsukerblat, B.; Palii, A.; Zilberg, S.; Korchagin, D.; Aldoshin, S.; Clemente-Juan, J.M. Vibronic recovering of functionality of quantum cellular automata based on bi-dimeric square cells with violated condition of strong Coulomb repulsion. *J. Chem. Phys.* **2022**, *157*, 074308. [CrossRef]
38. Piepho, S.B.; Krausz, E.R.; Schatz, P.N. Vibronic coupling model for calculation of mixed valence absorption profiles. *J. Am. Chem. Soc.* **1978**, *100*, 2996–3005. [CrossRef]
39. Landau, L.D.; Lifshitz, E.M. *Quantum Mechanics (Non-Relativistic Theory)*, 3rd ed.; Pergamon Press: Oxford, UK, 1977.
40. Tsukerblat, B.; Palii, A.; Clemente-Juan, J.M. Self-trapping of charge polarized states in four-dot molecular quantum cellular automata: Bi-electronic tetrameric mixed-valence species. *Pure Appl. Chem.* **2015**, *87*, 271–282. [CrossRef]
41. Zilberg, S.; Stekolshik, Y.; Palii, A.; Tsukerblat, B. Controllable Electron Transfer in Mixed-Valence Bridged Norbornylogous Compounds: Ab Initio Calculation Combined with a Parametric Model and Through-Bond and Through-Space Interpretation. *J. Phys. Chem. A* **2022**, *126*, 2855–2878. [CrossRef]
42. Roos, B.O. The Complete Active Space Self-Consistent Field Method and its Applications in Electronic Structure Calculations. *Adv. Chem. Phys.* **1987**, *69*, 399–445.
43. Busch, T.; Degli Esposti, A.; Werner, H.-J. Analytical Energy Gradients for Multiconfiguration Self-Consistent Field Wave Functions with Frozen Core Orbitals. *J. Chem. Phys.* **1991**, *94*, 6708–6715. [CrossRef]
44. Schmidt, M.W.; Baldridge, K.K.; Boatz, J.A.; Elbert, S.T.; Gordon, M.S.; Jensen, J.H.; Koseki, S.; Matsunaga, N.; Nguyen, K.A.; Su, S.J.; et al. General Atomic and Molecular Electronic Structure System. *J. Comput. Chem.* **1993**, *14*, 1347–1363. [CrossRef]
45. Gordon, M.S.; Schmidt, M.W. Advances in Electronic Structure Theory: GAMESS a Decade Later. In *Theory and Applications of Computational Chemistry, The First Forty Years*; Dykstra, C.E., Frenking, G., Kim, K.S., Scuseria, G.E., Eds.; Elsevier: Amsterdam, The Netherlands, 2005; Chapter 41; p. 1167. Available online: http://www.msg.chem.iastate.edu/GAMESS/GAMESS.html (accessed on 25 September 2024).
46. Zilberg, S.; Tsukerblat, B.; Palii, A. Polaronic Mechanism of Vibronic Localization in Mixed-Valence Cation Radicals with a Non-Conjugated Chromophore on the Bridge. *J. Phys. Chem. A* **2023**, *127*, 3281−3292. [CrossRef]
47. Lent, C.S.; Isaksen, B. Clocked Molecular Quantum-Dot Cellular Automata. *IEEE Trans. Electron Devices* **2003**, *50*, 1890–1896. [CrossRef]
48. Blair, E.; Lent, C. Clock Topologies for Molecular Quantum-Dot Cellular Automata. *J. Low Power Electron. Appl.* **2018**, *8*, 31–44. [CrossRef]

49. Hennessy, K.; Lent, C.S. Clocking of molecular quantum-dot cellular automata. *J. Vac. Sci. Technol. B* **2001**, *19*, 1752–1755. [CrossRef]
50. Toth, G.; Lent, C. Quasiadiabatic switching for metal-island quantum-dot cellular automata. *J. Appl. Phys.* **1999**, *85*, 2977–2984. [CrossRef]
51. Lu, Y.; Liu, M.; Lent, C.S. Molecular quantum-dot cellular automata: From molecular structure to circuit dynamics. *J. Appl. Phys.* **2007**, *102*, 034311. [CrossRef]
52. Lent, C.S.; Liu, M.; Lu, Y. Bennett clocking of quantum-dot cellular automata and the limits to binary logic scaling. *Nanotechnology* **2006**, *17*, 4240–4251. [CrossRef]
53. Palii, A.; Aldoshin, S.; Tsukerblat, B. Theoretical insight into clocking in a molecular mixed valence cell of quantum cellular automata through the vibronic approach. *J. Chem. Phys.* **2024**, *160*, 014302. [CrossRef] [PubMed]

magnetochemistry

Article

# $[Mn^{III}_6Mn^{II}Na^I_2]$, $[Mn^{III}_3Mn^{II}Na^I]$, and $[Mn^{III}_3]$ Clusters Derived from Schiff Bases: Syntheses, Structures, and Magnetic Properties †

**Johannes Löhr [1,2], Mercè Font-Bardia [3], Júlia Mayans [1,4,*] and Albert Escuer [1,4,*]**

1 Departament de Química Inorgànica i Orgànica, Secció de Química Inorgànica, Universitat de Barcelona, Marti i Franques 1-11, 08028 Barcelona, Spain; loehrjohannes@web.de
2 Institute for Solar Energy Research Hamelin, 31860 Emmerthal, Germany
3 Departament de Mineralogia, Cristal·lografia i Dipòsits Minerals and Unitat de Difracció de R-X, Centre Científic i Tecnològic de la Universitat de Barcelona (CCiTUB), 08028 Barcelona, Spain
4 Institut de Nanociència i Nanotecnologia (IN2UB), Universitat de Barcelona, 08028 Barcelona, Spain
* Correspondence: julia.mayans@qi.ub.edu (J.M.); albert.escuer@qi.ub.edu (A.E.)

† Miquel and Paco, enormous scientists, incredible people and above all, extraordinary friends.

**Abstract:** The reaction of manganese halides with polydentate Schiff bases obtained by the condensation of 3-ethoxysalicylaldehyde and different amino alcohols, resulting in a $NO_3$ set of donors, yielded a series of manganese clusters with $\{Mn^{III}_6Mn^{II}Na_2\}$, $\{Mn^{III}_3Mn^{II}Na\}$, and $\{Mn^{III}_3\}$ metallic cores. The influence of the ligand substituents and the halide on the final nuclearity has been studied. Analysis of their static magnetic behaviour confirms the ground states of 19/2 for the $\{Mn^{III}_6Mn^{II}Na_2\}$ complexes, 7/2 for the $\{Mn^{III}_3Mn^{II}Na\}$ clusters, and 12/2 for the triangular $\{Mn^{III}_3\}$ systems, and a weak field induced a slow relaxation of the magnetization for the trinuclear complexes.

**Keywords:** manganese; crystal structure; mixed valence; chirality; magnetic properties

## 1. Introduction

The condensation of a primary amine with aldehydes or ketones yields Schiff bases, which are a class of highly tunable ligands as a function of the R–$NH_2$ and the R–C=O precursors. The design, taking into account shape, chirality, and the number and kind of donors, can induce the selective coordination of only one cation or to generate polynuclear systems with the desired properties after coordination to the adequate cations. The enormous variety of available amines and aldehydes that can be employed as precursors can generate ligands with specific topologies or donor atoms that can link either 3*d* or 4*f* cations that demonstrated to be of application in many fields like syntheses, [1] catalysis [2], luminescence [3], or pharmaceuticals [4]. Salicyl and o-vanillin are the most employed aldehyde precursors to reach tetradentate two-pocket ligands with a $NO_3$ set of donors.

The condensation of 3-ethoxysalicylaldehyde with substituted 2-aminoethanol, such 2-phenylglycinol or 2-Amino-1-phenylethanol, gives the polytopic ligands $H_2L1$ and $H_2L2$, respectively, see Scheme 1—top. Each ligand has a $NO_3$ set of donors able to link up to three cations, generating medium and large clusters.

Manganese chemistry with these ligands is dominated by the $\{Mn^{III}_3M'M''\}$ metallic cores that consist of a triangular bipyramid formed by one triangular arrangement of $Mn^{III}$ cations, linked through a $\mu_3$-O donor, and two apical M′ and M″ cations, see Scheme 1—bottom. The most common structure consists of pentanuclear $\{Mn^{III}_3Mn^{II}Na^I\}$ units [5–12] but some $\{Mn^{III}_3Mn^{II}Ca^{II}\}$ [5,7,10], $\{Mn^{III}_3Mn^{II}Mn^{II}\}$ [7,10], or $\{Mn^{III}_3Ln^{III}Na^I\}$ [13] cores have also been reported. This pentanuclear unit is very versatile and larger nuclearities have been reported when one of the apical cations is shared by two units, resulting in a nonanuclear $\{Mn^{III}_6Mn^{II}Na^I_2\}$ [8,10,12,14,15] or $\{Mn^{III}_6Ln^{III}Na^I_2\}$ [16,17] skeleton or decanuclear $\{Mn^{III}{}_6Mn^{II}{}_2Na^{I}{}_2\}$ systems when the $Mn^{II}$ cations from two or more units are linked by halides, pseudohalides, or methoxides [9,11,12,18–20].

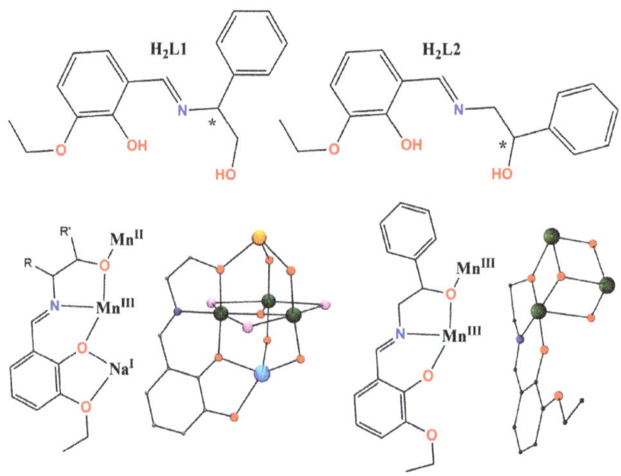


**Scheme 1.** (**Top**) Molecular structure of $H_2L1$ and $H_2L2$ ligands employed in this work. (**Bottom**) coordination modes found in complexes **1*R*, 2*R*a,b,** and **5** (**left**) or **3** and **4** (**right**) and the topologies of the corresponding clusters. Asterisks denote chiral C-atoms.

During the last years, we explored the reactivity of a family of related ligands derived from o-vanillin and 2-amino-1-ethanol in $Mn^{II,III}$, [8–10,12,17], $Ni^{II}$ [21,22], and $Fe^{III}$ chemistry [23], trying to elucidate the effect of the substituents on the C-atoms of amino alcohol. From this previous work, we realized that the size and position of the substituents (Me, Ph, isopropyl) is hardly insignificant and largely induces different shapes and nuclearities in the resulting clusters.

Following our previous work in this field, we have explored the influence of the ethoxo substituent on the salicylaldehyde ring and the position of the phenyl ring on the alcoxo arm of the Schiff base. The reaction of manganese(II) halides with ligands $H_2L1$ and $H_2L2$ yielded a series of clusters with different nuclearities ranging from tri- to nonanuclear with formulas $[Mn^{III}{}_6Mn^{II}Na_2(O)_2(L1)_6Cl_6]$ (**1*R***), $[Mn^{III}{}_6Mn^{II}Na_2(O)_2(L1)_6I_6]$ (**2*R*a**, **2*R*b**), $[Mn^{III}{}_3(L2)_3(O)(H_2O)_2(EtOH)]Cl$ (**3**), $[Mn^{III}{}_3(L2)_3(O)(H_2O)(MeOH)_2]Br$ (**4**), and $[Mn^{III}{}_3Mn^{II}Na(O)(L2)_3Cl_3(CH_3CN)]_2[Mn^{II}Br_4]$ (**5**). Structural characterization shows a variety of metallic cores as function of minor changes on the Schiff bases, such as the rare trinuclear topology found in complexes **3–4** and the unusual $Mn^{III}$-I-$Mn^{III}$ linkage found in complexes **2*R*a** and **2*R*b**. The study of their magnetic properties shows $Mn^{III}$–$Mn^{III}$ ferromagnetic interactions and $Mn^{II}$–$Mn^{III}$ antiferromagnetic coupling yielding $S$ ground states between 7/2 and 19/2.

## 2. Materials and Methods

### *2.1. Physical Measurements*

Magnetic susceptibility measurements were carried out on pressed polycrystalline samples at the Mesures Magnètiques Unit from Scientific and Technological Centers (CCi-TUB), Universitat de Barcelona, using a MPMS5 Quantum Design susceptometer (Quantum Design, Pfungstadt, Germany), working in the range 30–300 K under magnetic fields of 0.3 T and under a field of 0.03 T in the 30–2 K range to avoid saturation effects at low temperature. Diamagnetic corrections were estimated from Pascal tables. The fit of the magnetic experimental data were performed with the PHI programme [24]. CHN analyses were performed on vacuum dried samples. Infrared spectra (4000–400 $cm^{-1}$) were recorded from KBr pellets on a Bruker IFS-125 FT-IR spectrophotometer (Bruker-Spain, Madrid, Spain).

### 2.2. X-ray Crystallography

Red prism-like specimens of **1*R***, **2*R*a**, **2*Rb***, **3**, **4**, and **5** were employed for the single crystal X-ray crystallographic analysis. The X-ray intensity data were measured on a D8 Venture system equipped with a multilayer monochromator and a Mo microfocus. The frames were integrated with the Bruker SAINT software package using a narrow-frame algorithm. The structures were solved and refined using the Bruker SHELXTL-2014/7 Software [25]. Crystal data and refinement details for complexes **1** to **5** are summarized in Table S1. Complete crystallographic information can be found in the corresponding CIF files with the following CCDC numbers: 2382107 (for **1*R***), 2382108 (for **2*Ra***), 2382109 (for **2*Rb***), 2382110 (for **3**), 2382111 (for **4**), and 2382112 (for **5**), which contain the supplementary crystallographic data for this paper. These data can be obtained free of charge via https://www.ccdc.cam.ac.uk/structures accessed on 9 May 2024.

### 2.3. Synthesis

$H_2L1$ and $H_2L2$ Schiff bases. The ligands were synthesized by the reaction of 3-ethoxysalicylaldehyde with *R*-2-phenylglycinol or the *rac*-2-Amino-1-phenylethanol, respectively, following previously reported methods [12]. Methanolic or ethanolic solutions (20 mL) obtained mixing 0.605 g (3.65 mmol) of 3-ethoxysalicylaldehyde and 0.5 g (3.65 mmol) of 2-phenylglycinol or 2-amino-1-phenylethanol were heated at 80° for 30 min in a microwave Anton Parr Monowave-300 furnace and diluted to 100 mL of volume. Fractions of 10 mL (0.0365 mmol) were employed in the syntheses of the corresponding complexes.

$[Mn^{III}_6Mn^{II}Na_2(O)_2(L1)_6Cl_6]$ (**1*R***) was synthesized. A solution of $MnCl_2{\cdot}4H_2O$ (0.072 g, 0.365 mmol) and 0.024 g (0.365 mmol) of sodium azide in 15 mL of acetonitrile was mixed with 10 mL of the previously prepared methanolic solution of (*R*)-$H_2L1$ (0.365 mmol). The resulting yellow mixture was maintained at open air with vigorous stirring and after 10 min, the colour changed to dark brown. After 30 more minutes, the clear solution was layered with diethyl ether and well-formed dark crystals, adequate for X-ray diffraction, and these were collected after three days with a 50% yield (~0.12 g). Because the azide anion was not coordinated, the same product was obtained employing Na(OH) (0.015 g, 0.364 mmol) as the base. CHN for $C_{102}H_{102}Cl_6Mn_7N_6Na_2O_{20}$: calcd. (%): C, 51.58; H, 4.33; N, 3.54. Found: C, 51.0; H, 4.5; N, 3.4.

$[Mn^{III}_6Mn^{II}Na_2(O)_2(L1)_6I_6]$ (**2*R*a**) and $[Mn^{III}_6Mn^{II}Na_2(O)_2(L1)_6I_6]{\cdot}11\,CH_3CN$ (**2*Rb***) were synthesized. The synthesis was performed following the same procedure employed in **1*R*** but using 0.112 g (0.364 mmol) of manganese iodide instead the chloride salt. Crystallization of the complex, either from the layers with diethyl ether or slow evaporation, yielded a mixture of well-formed prismatic crystals and bars in a similar amount, and these were manually separated. Structural determination proved the identity of the prismatic crystals (**2*R*a**) and the bars (**2b**). Yield: 25% for both complexes (~0.10 g). CHN for $C_{102}H_{102}I_6Mn_7N_6Na_2O_{20}$: calcd. (%): C, 41.90; H, 3.52; N, 2.87. Found **2*R*a**: C, 42.1; H, 3.4; N, 2.9. Found **2*Rb***: C, 42.3; H, 3.6; N, 2.8.

$[Mn^{III}_3(L2)_3(O)(H_2O)_2(EtOH)]Cl{\cdot}0.5\,CH_3CN{\cdot}0.25EtOH$ (**3**) was synthesized. To a solution of $MnCl_2{\cdot}4H_2O$ (0.072 g, 0.365 mmol) in acetonitrile (10 mL) was added Cs(OH) (0.109 g, 0.728 mmol). To the resulting brown suspension, 10 mL of the previously prepared ethanolic solution of rac-$H_2L$ (0.364 mmol) was added. The solid was redissolved after some minutes of stirring, resulting a dark brown solution. Crystals adequate for X-ray diffraction were obtained after one week from the slow diffusion of diethyl ether. Yield: 50% (~0.08 g). CHN for $C_{53}H_{61}ClMn_3N_3O_{13}$: calcd. (%): C, 55.43; H, 5.35; N, 3.66. Found C, 55.8; H, 5.2; N, 3.5.

$[Mn^{III}_3(L2)_3(O)(H_2O)(MeOH)_2]Br{\cdot}CH_3CN{\cdot}MeOH$ (**4**) was synthesized. The complex was obtained following the same procedure employed for the synthesis of complex **4**, but starting from $MnBr_2{\cdot}H_2O$ (0.104 g, 0.364 mmol), $Et_3N$ (0.036 g, 0.364 mmol), and a methanolic solution of *rac*-$H_2L2$ (0.364 mmol). The slow diffusion of diethyl ether yields crystalline complex **4** in a 60% yield (~0.09 g). CHN for $C_{53}H_{61}BrMn_3N_3O_{13}$: calcd. (%): C, 53.37; H, 5.15; N, 3.52. Found C, 53.0; H, 5.3; N, 3.7.

$[Mn^{III}_3Mn^{II}Na(O)(L2)_3Cl_3(CH_3CN)]_2[Mn^{II}Br_4]·0.5\ H_2O$ (**5**) was synthesized. The synthesis of this complex was performed following the same procedure employed for compound **1** starting from $MnBr_2·4H_2O$ (0.104 g, (0.364 mmol) and 10 mL of the previously prepared ethanolic solution of *rac*-$H_2L2$. Layering the resulting dark solution with diethyl ether, hexagonal crystals of **5** were collected in a 70% yield (~0.09 g). CHN for $C_{106}H_{108}Br_{10}Mn_9N_8Na_2O_{20}$: calcd. (%): C, 40.37; H, 3.45; N, 3.55. Found C, 39.9; H, 3.6; N, 3.4.

Representative IR spectra (complexes **1*R***, **3,** and **5**) are shown in Figure S1.

## 3. Results and Discussion

### *3.1. Description of the Structures*

The oxidation states of the manganese cations have been assigned with respect to the shape of the coordination sphere, charge valence of the resulting formulas, and BSV calculations, as shown in Table S2.

#### 3.1.1. Crystal Structure of $[Mn^{III}_6Mn^{II}Na_2(O)_2(L1)_6Cl_6]$ (**1*R***), $[Mn^{III}_6Mn^{II}Na_2(O)_2(L1)_6I_6]$ (**2*Ra***), and $[Mn^{III}_6Mn^{II}Na_2(O)_2(L1)_6I_6]$ (**2*Rb***)

The three structures have a common core consisting in two $\{Mn_3^{III}Mn^{II}Na\}$ pentanuclear units sharing the $Mn^{II}$ cation. Selected bond parameters are reported in Table 1. To avoid repetitive plots, the structure of one of them (**2*R*a**) is shown in Figure 1 as a representative core. The pentanuclear units show a trigonal bipyramid arrangement with the $Mn^{III}$ cations in the equatorial plane with the sodium and the divalent manganese in the apical positions. One $\mu_3$-O donor is placed in the centre of the triangular arrangement of $Mn^{III}$ cations. Each pentanuclear fragment is assembled by three $L1^{2-}$ ligands linking the $Mn^{III}$ cations with the $Mn^{II}$ by means of alcoxo bridges and with the sodium cation by means of phenoxo bridges. The coordination sphere of the $Mn^{III}$ cations is formed by one tridentate $L1^{2-}$ ligand linking the alcoxo, phenoxo, and N-iminic donors, as well as the central oxo ligand and two bridging halide atoms. In all cases, the sodium cations are hexacoordinated and placed in the cavity formed by the three bidentate fragments composed by the phenoxo and the ethoxo O donors of the three $L1^{2-}$ ligands, resulting in an intermediate polyhedron between the octahedral and trigonal prism.

**Table 1.** Main bond distances (Å) and angles (deg.) for complexes **1*R*** and **2*R*a**.

| | 1*R* | 2*R*a |
|---|---|---|
| Mn1–O2 | 1.900(3) | 1.880(4) |
| Mn1–O3 | 1.899(3) | 1.903(3) |
| Mn1–O7 | 1.8810(8) | 1.8780(9) |
| Mn1–X1 | 2.595(1) | 2.942(1) |
| Mn1′–X1 | 2.701(1) | 3.021(1) |
| Mn2–O5 | 1.914(3) | 1.891(5) |
| Mn2–O6 | 1.886(3) | 1.900(4) |
| Mn2–O8 | 1.8761(9) | 1.872(1) |
| Mn2–X2 | 2.637(1) | 3.023(1) |
| Mn2′–X2 | 2.735(1) | 3.105(1) |
| Mn3–O3 | 2.201(3) | 2.202(3) |
| Mn3–O6 | 2.196(3) | 2.193(4) |
| Mn1–O7–Mn1′ | 119.2(5) | 119.17(5) |
| Mn2–O8–Mn2′ | 118.82(6) | 119.04(7) |
| Mn1–O3–Mn3 | 122.9(2) | 123.8(2) |
| Mn2–O6–Mn3 | 125.4(2) | 126.5(2) |
| Mn1–X1–Mn1′ | 75.54(4) | 65.79(3) |
| Mn2–X2–Mn2′ | 73.89(4) | 63.51(3) |

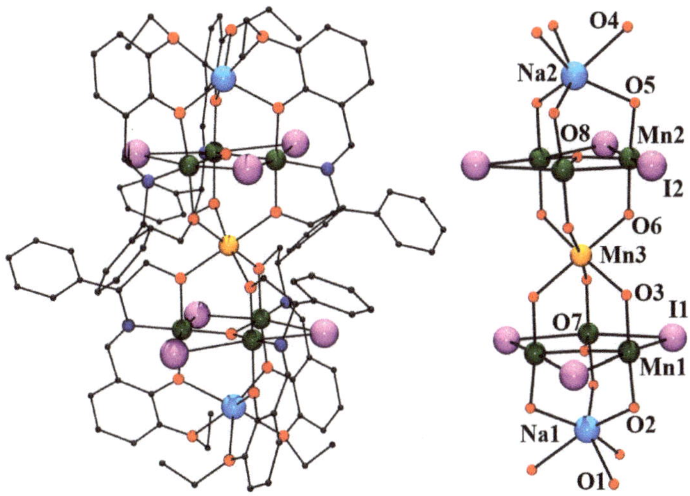


**Figure 1. Left**: view of the molecular structure of the nonanuclear $\{Mn^{III}_6Mn^{II}Na_2\}$ complex **2*R*a**. **Right**: labelled plot of the core of complex **2*R*a**. Atom numbering is the same for the isostructural chloro complex **1*R***.

Complexes **1*R*** and **2*R*a** possesses a $C_3$ axis that relates the $Mn^{III}$ and the $L1^{2-}$ ligands, but the lack of an inversion centre imposed by the chiral ligand (*R*3 space group) determines that the two pentanuclear subunits are not equivalent and there are small differences in the bond parameters around Mn(1) and Mn(2), as shown in Table 1. Complex **2*R*b** shows the same core as **1*R*** and **2*R*a,** but the loss of symmetry (from *R*3 to $P2_1$ space group) has the consequence that despite the bond parameters being very similar to those of **2*R*a,** all the cations are not equivalent, see Table S3.

### 3.1.2. Crystal Structure of $[Mn^{III}_3(L2)_3(O)(H_2O)_2(EtOH)]Cl$ (3) and $[Mn^{III}_3(L2)_3(O)(H_2O)(MeOH)_2]Br$ (4)

The two compounds have a common structure that consist of a triangular arrangement of $Mn^{III}$ cations linked by one $\mu_3$-O bridge and three $L2^{2-}$ ligands, three solvent molecules, and one halide counterion. The representative molecular structure of complex **3** is shown in Figure 2 and the main bond parameters for the two complexes are listed in Table 2. The $\mu_3$-O donor is placed 0.85 Å over the plane defined by the manganese cations and the Mn-O-Mn bond angles lie in the short interval of 101.69(6)°–102.36(6)°. Each $L2^{2-}$ ligand acts as tridentate by means of the deprotonated O-donors and the iminic nitrogen, whereas the O-ethoxy atoms remain uncoordinated. The coordination sphere of the $Mn^{III}$ cations is formed by the three donors from one $L2^{2-}$ ligand, two O-alcoxo donors, and one solvent molecule (water or alcohol). The $L2^{2-}$ ligand contains a chiral C-atom and the main plane of the (*R*) and (*S*) isomers are disordered and tilted around 50° with respect to the plane defined by the three cations, resulting a helical arrangement.

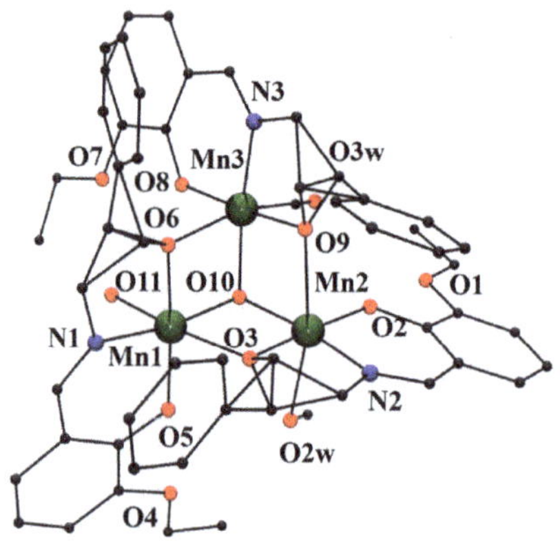


**Figure 2.** Partially labelled plot of the molecular structure of complex **3**. The C-atom in the vicinity of the O-alcoxo donors is disordered and corresponds to the (*R*) or (*S*) enantiomer of the ligand. Structure and atom numbering is the same for complex **4**.

**Table 2.** Main bond distances (Å) and angles (deg.) for the triangular complexes **3** and **4**.

| | 3 | 4 |
|---|---|---|
| Mn1–O3 | 2.255(2) | 2.227(1) |
| Mn1–O5 | 1.902(2) | 1.911(1) |
| Mn1–O6 | 1.882(1) | 1.877(1) |
| Mn1–O10 | 1.926(1) | 1.929(1) |
| Mn1–O11 | 2.293(3) | 2.278(2) |
| Mn1–N1 | 1.987(2) | 1.979(2) |
| Mn2–O2 | 1.910(2) | 1.908(1) |
| Mn2–O2w | 2.270(2) | 2.316(1) |
| Mn2–O3 | 1.889(1) | 1.868(1) |
| Mn2–O9 | 2.234(2) | 2.240(1) |
| Mn2–O10 | 1.921(1) | 1.915(1) |
| Mn2–N2 | 1.969(2) | 1.978(1) |
| Mn3–O3w | 2.266(2) | 2.279(2) |
| Mn3–O6 | 2.253(2) | 2.231(1) |
| Mn3–O8 | 1.902(2) | 1.890(1) |
| Mn3–O9 | 1.864(2) | 1.878(1) |
| Mn3–O10 | 1.929(1) | 1.921(1) |
| Mn3–N3 | 1.993(2) | 1.978(2) |
| Mn1–O3–Mn2 | 92.22(6) | 93.38(5) |
| Mn1-O6-Mn3 | 92.62(6) | 92.83(5) |
| Mn2–O9–Mn3 | 93.14(6) | 92.29(5) |
| Mn1–O10–Mn2 | 102.36(6) | 102.13(6) |
| Mn1–O10–Mn3 | 102.24(6) | 101.69(6) |
| Mn2–O10–Mn3 | 101.74(6) | 101.96(5) |

3.1.3. Crystal Structure of $[Mn^{III}_3Mn^{II}Na(O)(L2)_3Cl_3(CH_3CN)]_2[Mn^{II}Br_4]$ (**5**)

The structure of **5** consists in two non-equivalent $\{Mn^{III}Mn^{II}Na\}$ cationic pentanuclear units and one $[Mn^{II}Br_4]^{2-}$ counteranion. Bond parameters are similar in both molecules forming the global system (Table 3), which is shown in Figure 3.

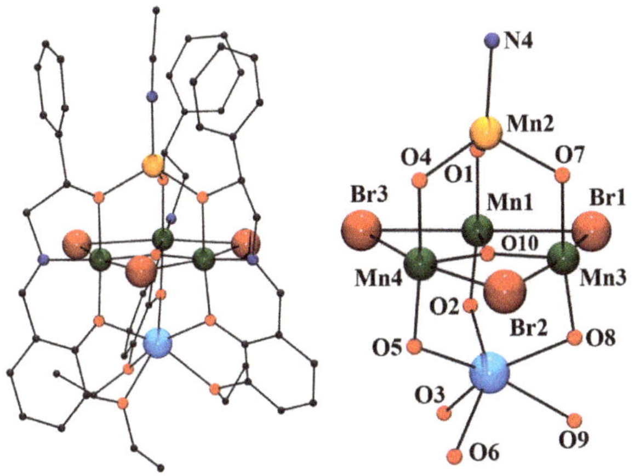


**Figure 3. Left**: view of the molecular structure of one pentanuclear $\{Mn^{III}Mn^{II}Na\}$ unit of complex **5**. **Right**: labelled plot of the core of the pentanuclear complex.

**Table 3.** Main bond distances (Å) and angles (deg.) for the two non-equivalent clusters present in the unit cell of compound **5**.

| | 5A | 5B |
|---|---|---|
| Mn1–O1 | 1.914(2) | 1.931(3) |
| Mn1–O2 | 1.867(3) | 1.883(3) |
| Mn1–O10 | 1.920(2) | 1.903(2) |
| Mn1–Br1 | 2.7608(7) | 2.7261(7) |
| Mn1–Br3 | 2.7725(7) | 2.7938(7) |
| Mn3–O7 | 1.921(2) | 1.920(3) |
| Mn3–O8 | 1.887(2) | 1.889(3) |
| Mn3–O10 | 1.895(2) | 1.890(2) |
| Mn3–Br1 | 2.8260(6) | 2.8403(7) |
| Mn3–Br2 | 2.7777(7) | 2.7730(7) |
| Mn4–O4 | 1.903(2) | 1.906(3) |
| Mn4–O5 | 1.890(2) | 1.889(3) |
| Mn4–O10 | 1.897(2) | 1.887(3) |
| Mn4–Br2 | 2.9163(6) | 2.7651(7) |
| Mn4–Br3 | 2.7785(6) | 2.8431(7) |
| Mn2–O1 | 2.061(2) | 2.036(3) |
| Mn2–O4 | 2.052(2) | 2.050(3) |
| Mn2–O7 | 2.054(2) | 2.064(3) |
| Mn2–N4 | 2.135(4) | 2.113(5) |
| Mn1–O10–Mn3 | 120.0(1) | 119.8(1) |
| Mn1–O10–Mn4 | 119.9(1) | 120.0(1) |
| Mn3–O10–Mn4 | 120.1(1) | 119.9(1) |
| Mn2–O1–Mn1 | 109.3(1) | 113.8(1) |
| Mn2–O7–Mn3 | 110.3(1) | 112.7(1) |
| Mn2–O4–Mn4 | 111.8(1) | 113.4(2) |
| Mn1–Br1–Mn3 | 72.50(2) | 72.22(3) |
| Mn1–Br3–Mn4 | 73.09(2) | 71.21(3) |
| Mn3–Br2–Mn4 | 70.41(2) | 72.36(3) |

The position of the cations and the linkage provided by the $L2^{2-}$ ligands in the pentanuclear units are similar to those described for complexes **1*R*** and **2*R*a/2*R*b**. However, in this case, the divalent manganese is tetracoordinated with a coordination sphere composed of three O-alcoxo donors from the corresponding Schiff bases and one acetonitrile ligand, resulting in a $NO_3$ environment. The tetrahedral arrangement around the $Mn^{II}$ imposes larger $Mn^{II}$-O-$Mn^{III}$ bond angles than the octahedral environment that takes values in the 109.3(1)°–113.8(1)° range.

*3.2. Structural and Synthetic Comments*

The chemistry of the manganese halides and the reported ligands is sensitive to the specific halide and the topology of the ligand. The substituents in the vicinity of the O-alkoxo donor induce steric hindrance that becomes the determinant for the final nuclearities of the clusters. In fact, the $\{Mn_6^{III}Mn^{II}\}$ core becomes not possible for the $L2^{2-}$ ligand and the products tend to lower nuclearities such as $\{Mn_3^{III}\}$ or $\{Mn_3^{III}Mn^{II}Na\}$ found in complexes **3**–**5**. It is remarkable that the characterization of clusters **3** and **4** had a triangular topology, which was unprecedented for the analogous ligand with a methoxy substituent. Ligand $H_2L2$ is able to give the usual $\{Mn_3^{III}Mn^{II}Na\}$ core, but probably due to small differences in solubility with respect to the methoxy-related ligand, can generate different cores.

Complexes **2*R*a** and **2*R*b** exhibit the unusual $Mn^{III}$-I linkage that a priori is poorly compatible due to the oxidant character of the trivalent manganese. Most of the reported complexes exhibiting this bond are mononuclear phthalocyanine derivatives [26–29], in which the $Mn^{III}$ cation is placed in a square pyramidal environment with the iodide donor in the apical elongated site. Also, some iodide–phosphine complexes [30,31] and some other mononuclear complexes with the Schiff base 2,2′-bis(3,5-di-t-butylsalicylaldimino)-6,6′-dimethylbiphenyl [32] show a square pyramidal environment or the macrocyclic cyclam [33] derivative with an elongated octahedral environment, which have been reported.

Due to the apparent instability of the $Mn^{III}/I^-$ system, this ligand has not been usually employed in manganese cluster chemistry. The only structurally characterized clusters combining $Mn^{III}$ and iodide are those reported by Brechin et al. in some $\{Mn^{II}_4Mn^{III}_6I_4\}$ derivatives of 2-Amino-2-methyl-1,3-propanediol and 2-Amino-2-ethyl-1,3-propanediol [34], as well as one hexanuclear complex derived from ethyl-salicyaldoxime that were the only clusters with $Mn^{III}$-I bonds, including one $\mu_3$-I bridge [35]. It is remarkable that complexes **2*R*a** and **2*R*b** become the second clusters characterized with this kind of bond.

The configuration of the complexes built from chiral ligands is determined by the ligand chirality, which can be transferred to the coordination sphere of the cations to the complex or to the crystal. This "predetermined chirality" is usually propagated in an opposite sense by the corresponding enantiomeric ligands, resulting in mirror image clusters or networks for each enantiomer [36,37]. Thus, starting from the racemic $H_2L2$ ligand, it could be expected that complexes **3**, **4**, or **5**, containing three ligands in their structures, must show the *RRR* and *SSS* arrangements related by inversion centres in the network. However, as was summarized in a previous work, this family of clusters often exhibits an unexpected mixing of enantiomers in the same molecule [12].

Complex **3** shows a complete disorder for the three ligands and complex **4** shows a helical arrangement of the three Schiff bases with respect to the plane defined by the manganese cations and a mixture of enantiomers, with one *R* enantiomer and two disordered *R*/*S* enantiomers, or one S enantiomer and two disordered *R*/*S* ligands, see Figure 4. Complex **5** exhibits an unusual mixture of enantiomers in the same molecule, with one of the molecules containing two *R* enantiomers and one *R*/*S* disordered ligand, and the molecule related by the inversion centre in the network contains two *S* enantiomers and one disordered *R*/*S* ligand, as shown in Figure 4.

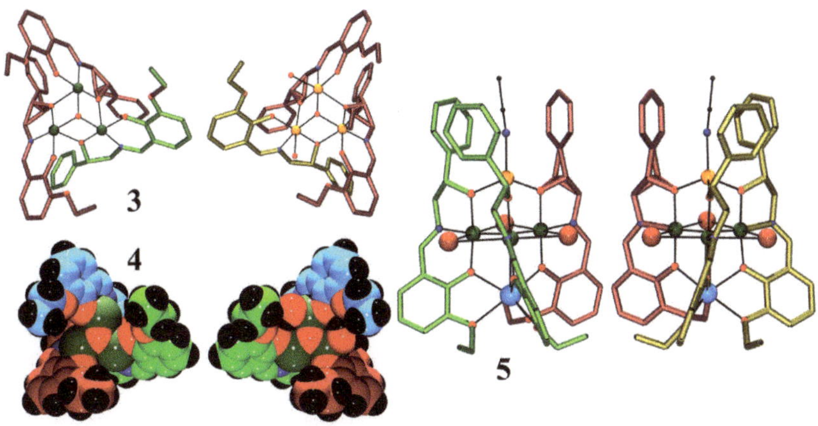


**Figure 4.** Arrangement of the enantiomers of the Schiff base for complexes **3** (**top**, **left**) and **5** (**right**). Colour key, Firebrick. Disordered *R*/*S* enantiomers: *S*-ligand, green; *R*-ligand, bronze. **Left–bottom**, space-fill plot of the two helicities of complex **4** (all ligands are disordered).

### 3.3. Magnetic Properties

Susceptibility measurements were performed in the 2–300 K temperature range on powdered pressed samples for complexes **1*R***, **2*R*a**, **4,** and **5**. Complexes **1*R*** and **2*R*a** show $\chi_M T$ values of 17.49 and 18.48 $cm^3 \cdot mol^{-1} \cdot K$ at room temperature, lower than the expected values for six $Mn^{III}$ and one $Mn^{II}$ isolated cations (22.375 $cm^3 \cdot mol^{-1} \cdot K$) that decrease slightly down to a minimum of 16.86 $cm^3 \cdot mol^{-1} \cdot K$ at 150 K and 17.68 $cm^3 \cdot mol^{-1} \cdot K$ at 110 K for **1*R*** and **2*R*a,** respectively. Below this minimum, the $\chi_M T$ values increase up to a maximum of 38.42 $cm^3 \cdot mol^{-1} \cdot K$ placed at 13 K for complex **1** and 37.24 $cm^3 \cdot mol^{-1} \cdot K$ at 10 K for complex **2*R*a** with a moderate decrease at lower temperatures, see Figure 5. The shape of the plot with a well-defined minimum is the signature of ferrimagnetic response for these two complexes.

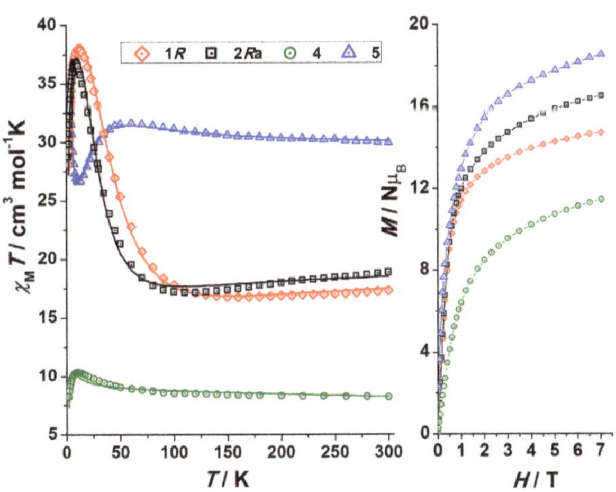


**Figure 5. Left**: $\chi_M T$ plots in terms of temperature for complexes **1*R***, **2*R*a**, **4,** and **5** (solid lines show the best fit of the data). **Right**: isothermal magnetization at 2 K for each compound.

Complex **4** shows the $\chi_M T$ value of 8.24 $cm^3 \cdot mol^{-1} \cdot K$ at room temperature, slightly lower than the expected value for three isolated $Mn^{III}$ cations of 9.00 $cm^3 \cdot mol^{-1} \cdot K$. On cooling, the $\chi_M T$ value increases slightly up to a maximum value of 10.37 $cm^3 \cdot mol^{-1} \cdot K$

at 5.5 K, followed by a weak decrease at low temperature, suggesting a very weak ferromagnetic interaction. Finally, the room temperature $\chi_M T$ value for complex **5** is 29.97 $cm^3 \cdot mol^{-1} \cdot K$, very close to the expected value for six $Mn^{III}$ and three $Mn^{II}$ isolated cations of 31.125 $cm^3 \cdot mol^{-1} \cdot K$. The $\chi_M T$ value increases up to 31.62 $cm^3 \cdot mol^{-1} \cdot K$ at 13 K and a further decrease below this temperature appears to be indicative of the presence of ferrimagnetic interactions, as shown in Figure 5.

The susceptibility plots for complexes **1*R*** and **2*R*a** were fitted according the two-$J$ Hamiltonian derived from Scheme 2.

$$H = -2J_1(S_1 \cdot S_2 + S_1 \cdot S_3 + S_2 \cdot S_3 + S_4 \cdot S_5 + S_4 \cdot S_6 + S_5 \cdot S_6) - 2J_2(S_1 \cdot S_7 + S_2 \cdot S_7 + S_3 \cdot S_7 + S_4 \cdot S_7 + S_5 \cdot S_7 + S_6 \cdot S_7)$$

This results in the best fit parameters $J_1$ = 1.73(8) $cm^{-1}$, $J_2$ = −3.81(6) $cm^{-1}$, and $g$ = 1.859 (2) for complex **1** and $J_1$ = 1.47(6) $cm^{-1}$, $J_2$ = −2.30(2) $cm^{-1}$, and $g$ = 1.867(3) for complex **2*R*a**. Magnetization measurements give unsaturated values of 14.70 and 16.50 $N\mu_\beta$ for complexes **1*R*** and **2*R*a,** respectively, slightly lower than the expected spin-only value of 17.7 $N\mu_\beta$, assuming a $g$ value of 1.86 obtained in the susceptibility fit. The deviation should be attributed to weak zero-field splitting of the $Mn^{III}$ cations that has not been included in the fits due to the large size of the matrix, out of the scope of usual fits.

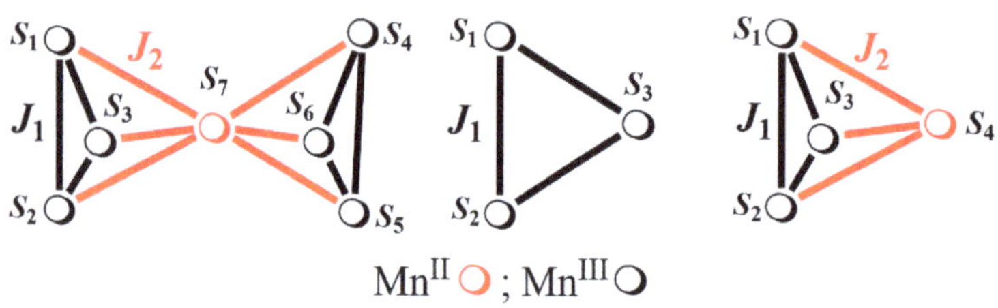


**Scheme 2.** Coupling scheme for the different topologies of complexes **1*R***, **2*R*a** (**left**), **4** (**centre**), and **5** (**right**). Red lines indicate the interactions involving the $Mn^{II}$ cation.

The fit of the experimental data for complex **4** according the spin-only Hamiltonian derived from Scheme 2 can be determined as follows:

$$H = -2J_1(S_1 \cdot S_2 + S_1 \cdot S_3 + S_2 \cdot S_3)$$

This does not reproduce adequately the weak susceptibility increase and thus, a $D_{ion}$ term was added. The best fit values were $J_1$ = 0.19(1) $cm^{-1}$, $D_{ion}$ = 2.04(8), and $g$ = 1.974(6). The magnetization measurement performed at 2 K reaches an unsaturated value of 11.45 $N\mu_\beta$, close to the expected value of 12 $N\mu_\beta$.

The susceptibility data for complex **5** were fitted, assuming the same set of interactions for the two non-equivalent units with the two-$J$ Hamiltonian derived from Scheme 2.

$$H = 2 \times (-2J_1(S_1 \cdot S_2 + S_1 \cdot S_3 + S_2 \cdot S_3) - 2J_2(S_1 \cdot S_4 + S_2 \cdot S_4 + S_3 \cdot S_4))$$

This includes the interactions inside the $Mn^{III}$ triangles ($J_1$) and the interaction between the $Mn^{II} \cdots Mn^{III}$ inside the two tetranuclear units. The contribution of one isolated $S = 5/2$ from the $[MnBr_4]^{2-}$ isolated counteranion, assumed as fully isotropic according to its high $S = 5/2$ spin ($^6A_1$ term), has been taken into account for the total $\chi_M T$ value. The best fit parameters were $J_1$ = 2.38(2) $cm^{-1}$, $J_2$ = −1.32(1) $cm^{-1}$, and $g$ = 1.939(1), implying a $S = 7/2$ spin ground state for each $\{NaMn^{II}Mn^{III}_3\}$ cluster. Isothermal magnetization measurement at 2 K reaches a value of 18.54 $N\mu_\beta$ under the maximum applied field, close to the expected 19 $N\mu_\beta$ value, confirming the $S = 7/2$ ground state for the two $\{Mn^{III}_3Mn^{II}Na^I\}$ units, see Figure 5—right.

The *ac* response of complexes **1*R***, **2*R*a**, **4,** and **5** was tested under different external fields (0–0.5 T) and a constant frequency of 1000 Hz and no out-of-phase signals were observed with the exception of weak tails in the $\chi''M(H)$ plots for triangular complex **3**, Figure S2, which was shown to be too weak for a further analysis. The lack of *ac* response despite the relatively high *S* ground states of 19/2 (**1*R***, **2*R*a**, **2*R*b**), 12/2 (**3**, **4**), and 7/2 (**5**) should be attributed to the unfavourable arrangement of the easy axis of the $Mn^{III}$ cations, which show a triangular arrangement that results in a low or negligible total anisotropy for the whole molecules, see Figure 6.

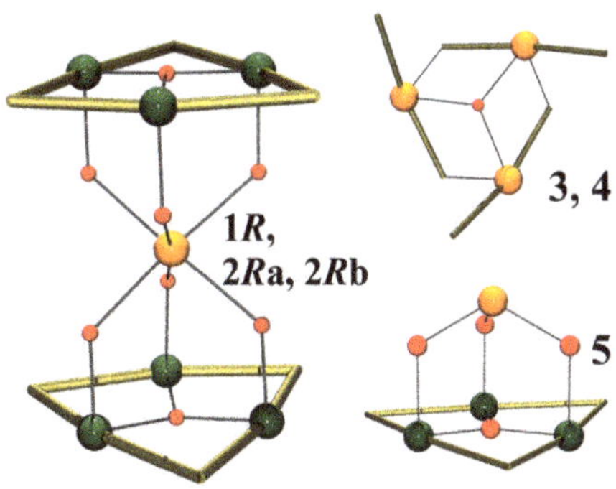


**Figure 6.** Core of the three topologies for complexes **1–5**, emphasizing the triangular arrangement of the elongated axis of the $Mn^{III}$ cations (bold bonds).

## 4. Conclusions

Depending on the position of the substituents on the amino alcohol fragment of the Schiff bases obtained by condensation with 3-ethoxysalicylaldehyde with the 1 or 2-phenyl derivatives of 2-amino-1-ethanol, a series of clusters with $\{Mn^{III}{}_6Mn^{II}Na_2\}$, $\{Mn^{III}{}_3Mn^{II}Na\}$, and $\{Mn^{III}{}_3\}$ nuclearity have been characterized. The magnetic response of these systems yields large *S* ground states comprised between 7/2 and 19/2 as a consequence of the $Mn^{III}$–$Mn^{III}$ ferromagnetic and $Mn^{II}$–$Mn^{III}$ antiferromagnetic coupling. The reported systems confirm that the mix of ligands with different (*R*)/(*S*) configurations in the same molecule instead of the expected separation of enantiomers is a common feature in this kind of cage with an odd number of chiral ligands.

**Supplementary Materials:** The following supporting information can be downloaded at: https://www.mdpi.com/article/10.3390/magnetochemistry10100076/s1, Table S1: Crystallographic data for compounds **1–5**; Table S2: BVS calculation for the manganese cations; Table S3: Selected bond distances (Å) and angles (deg.) for compound **2*R*b**; Figure S1: Infrared spectra for complexes **1*R***, **3** and **5**; Figure S2: AC susceptibility measurements vs. field (1000 Hz) for **4**.

**Author Contributions:** The manuscript was prepared with the contributions of all authors. They specifically contributed as follows: J.L.; synthesis, characterization, performance of the magnetic measurements, and the analysis of the results of magnetic studies. M.F.-B.; single-crystal X-ray diffraction collection and refinement of the crystal structures. J.M. and A.E.; conceptualization and methodology, writing—original draft preparation, and writing—review and editing. All authors have read and agreed to the published version of the manuscript.

**Funding:** This work was supported by the Spanish Ministry of Science and Innovation (MICINN) Project PID2023-146166NB-I00.

**Data Availability Statement:** The data presented in this study are available in this article or the Supplementary Materials.

**Conflicts of Interest:** The authors declare no conflicts of interest.

## References

1. Andruh, M. The exceptionally rich coordination chemistry generated by Schiff-base ligands derived from o-vanillin. *Dalton Trans.* **2015**, *44*, 16633–16653. [CrossRef] [PubMed]

2. Liu, X.; Manzur, C.; Novoa, N.; Celedón, S.; Hamon, J.R. Multidentate unsymmetrically-substituted Schiff bases and their metal complexes: Synthesis, functional materials properties, and applications to catalysis. *Coord. Chem. Rev.* **2018**, *357*, 144–172. [CrossRef]
3. Chen, S.; Fan, R.Q.; Gao, S.; Wang, X.; Yang, Y.L. Synthesis, crystal structure and effect of deuterated solvents and temperature on visible and near infrared luminescence of $N_4$-donor Schiff base lanthanide complexes. *J. Lumin.* **2014**, *149*, 75–85. [CrossRef]
4. Kaczmarek, M.T.; Zabiszak, M.; Nowak, M.; Jastrzab, R. Lanthanides: Schiff base complexes, applications in cancer diagnosis, therapy, and antibacterial activity. *Coord. Chem. Rev.* **2018**, *370*, 42–54. [CrossRef]
5. Hewitt, I.J.; Tang, J.-K.; Madhu, N.T.; Clerac, R.; Buth, G.; Anson, C.E.; Powell, A.K. A series of new structural models for the OEC in photosystem II. *Chem. Commun.* **2006**, 2650–2652. [CrossRef] [PubMed]
6. Fan, L.L.; Guo, F.S.; Yun, L.; Lin, Z.J.; Herchel, R.; Leng, J.D.; Ou, Y.C.; Tong, M.L. Chiral transition metal clusters from two enantiomeric schiff base ligands. Synthesis, structures, CD spectra and magnetic properties. *Dalton Trans.* **2010**, *39*, 1771–1780. [CrossRef]
7. Nayak, S.; Nayek, H.P.; Dehnen, S.; Powell, A.K.; Reedijk, J. Trigonal propeller-shaped $[Mn^{III}{}_3M^{II}Na]$ complexes (M = Mn, Ca): Structural and functional models for the dioxygen evolving centre of PSII. *Dalton Trans.* **2011**, *40*, 2699–2702. [CrossRef]
8. Escuer, A.; Mayans, J.; Font-Bardia, M.; Górecki, M.; Di Bari, L. Syntheses, structures, chiroptical and magnetic properties of chiral clusters built from Schiff bases: A novel $[Mn^{II}Mn^{III}{}_6Na^{I}{}_2]$ core. *Dalton Trans.* **2017**, *46*, 6514–6517. [CrossRef]
9. Mayans, J.; Font-Bardia, M.; Escuer, A. Triple halide bridges in chiral $Mn^{II}{}_2Mn^{III}{}_6Na^{I}{}_2$ cages: Structural and magnetic characterization. *Inorg. Chem.* **2018**, *57*, 926–929. [CrossRef]
10. Mayans, J.; Font-Bardia, M.; Di Bari, L.; Escuer, A. Chiral $[Mn^{II}Mn^{III}{}_3M']$ (M=$Na^I$, $Ca^{II}$, $Mn^{II}$] and $[Mn^{II}{}_2Mn^{III}{}_6Na^{I}{}_2]$ clusters built from an enantiomerically pure Schiff base: Synthetic, chiroptical and magnetic properties. *Chem. Eur. J.* **2018**, *24*, 18705–18717. [CrossRef]
11. Liu, J.; Suo, Q.; Song, J.; Zhang, L.; Gao, Y. A series of $[Na_2Mn^{II}{}_2Mn^{III}{}_6]$ clusters derived from trigonal bipyramidal $[NaMn^{II}Mn^{III}{}_3]$ subunits bridged by mono-, di- and tri-fold azide groups: Synthesis, crystal structures and magnetic properties. *Polyhedron* **2019**, *161*, 34–39. [CrossRef]
12. Pilichos, E.; Escuer, A.; Font-Bardia, M.; Mayans, J. Chiral versus non-chiral $[Mn^{III}{}_6Mn^{II}Na^{I}]$, $[Mn^{III}{}_6Mn^{II}{}_2Na^{I}{}_2]$ and $[Mn^{III}{}_3Mn^{II}Na^{I}]$ clusters derived from Schiff bases or the fight for symmetry. *Chem. Eur. J.* **2020**, *57*, 13053–13062. [CrossRef] [PubMed]
13. Chandrasekhar, V.; Dey, A.; Das, S.; Kundu, S. Syntheses and structures of a family of heterometallic pentanuclear $[Mn^{III}{}_3LnNa]$ (Ln = Dy, Tb, Gd and Nd) complexes: H-bonding reduces the nuclearity from nine to five. *CrystEngComm* **2014**, *16*, 1304–1310. [CrossRef]
14. Yang, L.G.; Wang, F.; Geng, C.H.; Yu, Z.C.; Wang, X.; Wang, K.; Zhang, Y.H.; Niu, Y.-S.; Li, D.-C. Heptanuclear manganese complexes with Schiff-base ligand: Syntheses, crystal structures and magnetic properties. *Chin. J. Inorg. Chem.* **2019**, *35*, 1651–1658.
15. Yang, L.G.; Wang, X.; Yu, Z.C.; Zhang, Y.H.; Li, S.L.; Wang, K.; Niu, Y.S.; Song, H.X. Synthesis, crystal structure and magnetic properties of heptanuclear manganese complexes with Schiff-base ligands. *Chin. J. Struct. Chem.* **2019**, *38*, 2135–2140.
16. Yang, P.P.; Wang, X.L.; Li, L.C.; Liao, D.Z. Synthesis, structure and magnetic properties of a novel family of heterometallic nonanuclear $Na^{I}{}_2Mn^{III}{}_6Ln^{III}$ (Ln = Eu, Gd, Tb, Dy) complexes. *Dalton Trans.* **2011**, *40*, 4155–4161. [CrossRef]
17. Mayans, J.; Font-Bardia, M.; Escuer, A. $Na_2Mn^{III}{}_6Ln^{III}$ clusters with a non-equivalent core: Chiral vs. *meso* isomerism. *Dalton Trans.* **2020**, *49*, 4216–4219. [CrossRef]
18. Liu, D.; Zhou, Q.; Chen, Y.; Yang, F.; Yu, Y.; Shi, Z.; Feng, S. Constructing octa- and hexadecanuclear manganese clusters from tetrahedral $Mn^{III}{}_3Mn^{II}$ cores bridged by quinquedentate Schiff base and versatile azide groups. *Dalton Trans.* **2010**, *39*, 5504–5508. [CrossRef]
19. Gole, B.; Mondal, K.C.; Mukherjee, P.S. Tuning nuclearity of clusters by positional change of functional group: Synthesis of polynuclear clusters, crystal structures and magnetic properties. *Inorg. Chim. Acta* **2014**, *415*, 151–164. [CrossRef]
20. Song, Y.; Zhang, G.; Qin, X.; Gao, Y.; Ding, S.; Wang, Y.; Du, C.; Liu, Z. Chiral $[NaMn^{II}Mn^{III}{}_3]$ and $[Na_2Mn^{II}{}_2Mn^{III}{}_6]$ clusters constructed by chiral multidentate Schiff-base ligands: Synthesis, structures, CD spectra and magnetic properties. *Dalton Trans.* **2014**, *43*, 3880–3887. [CrossRef]
21. Mayans, J.; Font-Bardia, M.; Di Bari, L.; Arrico, L.; Zinna, F.; Pescitelli, G.; Escuer, A. From mesocates to helicates: Structural, magnetic and chiro-optical studies on Nickel(II) supramolecular assemblies derived from tetradentate Schiff bases. *Chem. Eur. J.* **2018**, *24*, 7653–7663. [CrossRef] [PubMed]
22. Mayans, J.; Martin, A.; Font-Bardia, M.; Escuer, A. Chiral tetranuclear $Ni^{II}$ clusters derived from Schiff bases and azido co-ligands. *Polyhedron* **2018**, *150*, 10–14. [CrossRef]
23. Mayans, J.; Font-Bardia, M.; Escuer, A. Chiroptical and magnetic properties of star-shaped $Fe^{III}{}_4$ complexes from chiral Schiff bases. Structural and magnetic correlations based on continuous shape measures. *Dalton Trans.* **2018**, *47*, 8392–8401. [CrossRef] [PubMed]
24. Chilton, N.F.; Anderson, R.P.; Turner, L.D.; Soncini, A.; Murray, K.S. PHI: A powerful new program for the analysis of anisotropic monomeric and exchange-coupled polynuclear d- and f-block complexes. *J. Comput. Chem.* **2013**, *34*, 1164–1165. [CrossRef]
25. Sheldrick, G.M. *SHELXL-2014/7: Program for the Solution of Crystal Structures*; University of Göttingen: Göttingen, Germany, 2014.
26. Turner, P.; Gunter, M.J.; Skelton, B.W.; White, A.H. Crystal structures of the pentacoordinate bromo, isocyanato, iodo, acetato and isothiocyanato complexes of the meso-tetraphenylporphyrinatomanganese cation. *Aust. J. Chem.* **1998**, *51*, 835–852. [CrossRef]

27. Van Trieste, G.P.; Reid, K.A.; Hicks, M.H.; Das, A.; Figgins, M.T.; Bhuvanesh, N.; Ozarowski, A.; Telser, J.; Powers, D.C. Nitrene photochemistry of manganese N-haloamides. *Angew. Chem. Int. Ed.* **2021**, *60*, 26647–26655. [CrossRef]
28. McKearney, D.; Zhou, W.; Zellman, C.; Williams, V.E.; Leznoff, D.B. Facile tuning of strong near-IR absorption wavelengths in manganese(iii) phthalocyanines via axial ligand exchange. *Chem. Commun.* **2019**, *55*, 6696–6699. [CrossRef]
29. Bone, A.N.; Stavretis, S.E.; Krzystek, J.; Liu, Z.; Chen, Q.; Gai, Z.; Wang, X.; Steren, C.A.; Powers, X.B.; Podlesnyak, A.A.; et al. Manganese tetraphenylporphyrin bromide and iodide. Studies of structures and magnetic properties. *Polyhedron* **2020**, *184*, 114488. [CrossRef]
30. Beagley, B.; McAuliffe, C.A.; Minten, K.; Pritchard, R.G. The co-ordination of small molecules by manganese(II) phosphine complexes. Part 8. Irreversible oxidation of trimethylphosphinemanganese(II) dihalides, $[MnX_2(PMe_3)](X = Cl, Br, or I)$, to bis(trimethylphosphine)manganese(III) trihalides, $[MnX_3(PMe_3)_2]$. X-Ray crystal structure of trans-tri-iodobis-(trimethylphosphine)manganese(III), the first example of a trigonal-bipyramidal manganese(III) tertiary phosphine complex. *J. Chem. Soc. Dalton Trans.* **1987**, 1999–2003.
31. McAuliffe, C.A.; Godfrey, S.M.; Mackie, A.G.; Pritchard, R.G. The reaction of coarse-grain manganese powder with diiodotrimethylphosphorane to form $[Mn(PMe_3)I_2]_n$, which reacts with trace quantities of molecular oxygen to form the mixed (+2/+3) oxidation state complex $[Mn_2(PMe_3)_3I_5]PMe_3$. *Chem. Commun.* **1992**, 483–485. [CrossRef]
32. Sanders, C.J.; O'Shaughnessy, P.N.; Scott, P. Non-planar manganese Schiff-base complexes; synthesis and molecular structures. *Polyhedron* **2003**, *22*, 1617–1625. [CrossRef]
33. Mossin, S.; Sorensen, H.O.; Weihe, H.; Glerup, J.; Sotofte, I. Manganese(III) cyclam complexes with aqua, iodo, nitrito, perchlorato and acetic acid/acetato axial ligands. *Inorg. Chim. Acta* **2005**, *358*, 1096–1106. [CrossRef]
34. Manoli, M.; Collins, A.; Parsons, S.; Candini, A.; Evangelisti, M.; Brechin, E.K. Mixed-valent Mn supertetrahedra and planar discs as enhanced magnetic coolers. *J. Am. Chem. Soc.* **2008**, *130*, 11129–11139. [CrossRef] [PubMed]
35. Jones, L.F.; Inglis, R.; Cochrane, M.E.; Mason, K.; Collins, A.; Parsons, S.; Perlepes, S.P.; Brechin, E.K. New structural types and different oxidation levels in the family of $Mn_6$-oxime single-molecule magnets. *Dalton Trans.* **2008**, 6205–6210. [CrossRef]
36. Crassous, J. Chiral transfer in coordination complexes: Towards molecular materials. *Chem. Soc. Rev.* **2009**, *38*, 830–845. [CrossRef]
37. Miyake, H. Supramolecular chirality in dynamic coordination chemistry. *Symmetry* **2014**, *6*, 880–895. [CrossRef]

magnetochemistry

MDPI

Article

# Two New 2p–3d Metal Complexes with a Nitronyl-Nitroxide Ligand Derived from *o*-Vanillin: Synthesis, Crystals Structures and Magnetic Properties

**Cristian Andrei Spinu [1,2], Daniel O. T. A. Martins [3], Teodora Mocanu [4], Mihaela Hillebrand [5], Jean-Pascal Sutter [6,*], Floriana Tuna [3,*] and Marius Andruh [1,2,*]**

1 Inorganic Chemistry Laboratory, Faculty of Chemistry, University of Bucharest, Regina Elisabeta Blvd. 4-12, 030018 Bucharest, Romania; cristian.spinu@icoscdn.ro
2 C. D. Nenitzescu Institute of Organic Chemistry and Supramolecular Chemistry of the Romanian Academy, Splaiul Independenței, 202B, 060023 Bucharest, Romania
3 Department of Chemistry and Photon Science Institute, University of Manchester, Oxford Road, Manchester M13 9PL, UK; tavaresmartins.daniel@gmail.com
4 "Ilie Murgulescu" Institute of Physical Chemistry, Romanian Academy, Splaiul Independenței 202, 060021 Bucharest, Romania; teo_mocanu_sl@yahoo.com
5 Physical Chemistry Department, Faculty of Chemistry, University of Bucharest, 030018 Bucharest, Romania; mihaela.hillebrand@gmail.com
6 Laboratoire de Chimie de Coordination du CNRS (LCC), Université de Toulouse, CNRS, 31077 Toulouse, France
* Correspondence: jean-pascal.sutter@lcc-toulouse.fr (J.-P.S.); floriana.tuna@manchester.ac.uk (F.T.); marius.andruh@acad.ro (M.A.)

**Abstract:** Two new 2p–3d complexes, $(Et_3NH)[ML(hfac)_2]$, have been obtained using the nitronyl-nitroxide radical (HL) derived from 2-hydroxy-3-methoxy-5-nitrobenzaldehyde (M = Mn **1**; Co **2**). The two compounds are isomorphous and their structures consist of anionic mononuclear species, $[M(hfac)_2L]^-$, M = Mn **1**; Co **2**, and triethylammonium cations, $Et_3NH^+$. The metal ions adopt an octahedral geometry, being coordinated by phenoxido and aminoxyl oxygen atoms from the ligand and four oxygen atoms from the hexafluoroacetylacetonato ($hfac^-$) ligand. The cryomagnetic behaviors of the two compounds reveal relatively strong antiferromagnetic M(II)-Rad interactions ($J_{MnRad} = -191$ cm$^{-1}$, $J_{CoRad} = -166$ cm$^{-1}$ with $\boldsymbol{H} = -JS_MS_{Rad}$). The EPR spectra (X- and Q-band) of compound **1** below 70 K show the characteristical features of a $S = 2$ spin system with zero field splitting terms of $D = 0.26$ cm$^{-1}$ and $E = 0.031$ cm$^{-1}$.

**Keywords:** manganese(II) complexes; cobalt(II) complexes; nitronyl-nitroxide ligands; EPR spectra; cryomagnetic measurements

## 1. Introduction

Numerous molecular magnetic materials are obtained by combining organic radicals with either 3d or 4f metal ions [1–6]. Such paramagnetic ligands belong to different families: TEMPO, verdazyl, and nitronyl-nitroxide derivatives. The most popular radicals used as ligands are nitronyl-nitroxide (Nit) compounds, which generate a very rich coordination chemistry, consisting of discrete mono- and oligonuclear complexes up to coordination polymers (mainly 1D and 2D). We recall here that the first single chain magnet (SCM) [7], and SCMs with large coercive field [8,9] are 1D coordination polymers constructed from cobalt(II) ions and nitronyl-nitroxide ligands.

Let us briefly review the chemistry of manganese(II)-Nit molecular magnetic materials. The coordinating abilities of the Nit ligands can be tuned by ligand functionalization. A plethora of manganese(II) complexes have been synthesized with such ligands, varying from mononuclear [10–16] and oligonuclear species [10,17–25], to 1D coordination polymers [26,27]. Mononuclear complexes have been obtained by reacting 2-pyridyl [11,14], triphenylphosphine oxide [10], amide [12,13], and triazole [15,16] substituted nitronyl-nitroxide ligands, with manganese(II) chloride, perchlorate and, most frequently, hexafluoroacetylacetonate. With longer ligands displaying coordination sites well separated from one another, such as 4-pyridyl [17], 4-quinolyl [18], 4-(5-pyrimidyl)phenyl [19], 4-(1-imidazole)phenyl [20], 2-quinoxalinyl [21], 4-(carboxyl)phenyl [22], 3-(*m*-pyridyl)-2-thienyl [23], 4-(3-pyridinylmethoxy)phenyl [24], and 3-((1*H*-1,2,4-triazol-1-yl)methoxy)-phenyl [25], the resulted complexes are binuclear. A smaller ligand, 3-isobutyl-pyrazole substituted nitronyl-nitroxide, generates a 1D-chain [27]. 2D coordination polymers have been obtained from nitronyl-nitroxide ligands decorated with imidazole and benzimidazole fragments by using a molar ratio metal:ligand of 2:3. [28–30]. Low nuclearity manganese(II)-(nitronyl-nitroxide) complexes, particularly mononuclear ones, are useful models for magneto-structural correlations.

In this paper we report on the crystal structures and magnetic properties of two 3D metal complexes containing a nitronyl-nitroxide ligand, $(Et_3NH)[M(hfac)_2L]$, where M = Mn(II) **1**, Co(II) **2**, and HL = 2-(2-hydroxy-3-methoxy-5-nitrophenyl)-4,4,5,5-tetramethyl-4,5-dihydro-1H-imidazol-3-oxide-1-oxyl (Scheme 1).

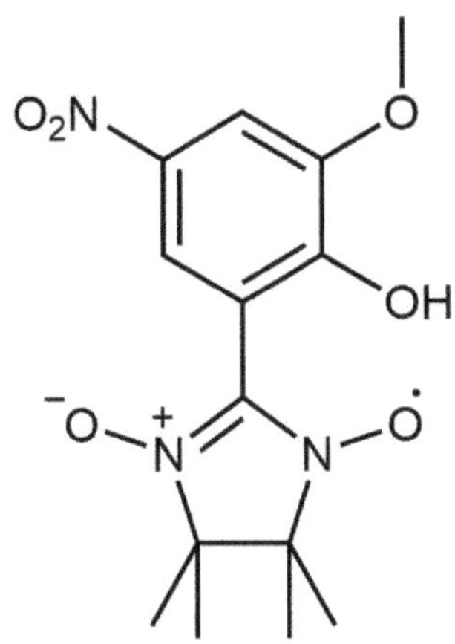


**Scheme 1.** Structure of HL ligand.

## 2. Experimental Part

### 2.1. Materials and Methods

The 2-(2-Hydroxy-3-methoxy-5-nitrophenyl)-4,4,5,5-tetramethyl-4,5-dihydro-1H-imidazol-3-oxide-1-oxyl (HL) ligand was synthesized as previously described [31]. All other reagents and solvents were commercially purchased and used without any further purification, if not stated otherwise. Elemental analyses (C, H, N) were performed on a EuroEA Elemental Analyzer.

### 2.2. Synthesis of the Complexes

#### 2.2.1. Synthesis of $(Et_3NH)[MnL(hfac)_2]$ **1**

$Mn(hfac)_2{\cdot}2H_2O$ (0.073 g, 0.1443 mmol) was dissolved in 10 mL of heptane and refluxed for 30 min. Then, after cooling down the solution, another 10 mL of $CH_2Cl_2$ solution containing HL (0.047 g, 0.1443 mmol) and triethylamine (0.0146 g, 0.1443 mmol, 20 μL) was added over. The solution was refluxed for 30 min, cooled down and filtered. After allowing the solvent to slowly evaporate for three days, violet needle shaped crystals of the product were obtained 0.070 g, yield 54%. Anal. Calcd for $C_{30}H_{35}O_{10}N_4F_{12}Mn$ (%): C, 40.28; H, 3.94; N, 6.26; Found: C, 40.01; H, 3.11; N, 5.91. Selected IR peaks ($cm^{-1}$): 3420 (w), 2992 (w), 2957 (w), 2837 (w), 2691 (w), 1649 (s), 1599 (w), 1555 (m), 1528 (m), 1503 (vs), 1455 (m), 1398 (w), 1366 (m), 1329 (s), 1308 (s), 1254 (vs), 1202 (s), 1144 (vs), 1103 (m), 1071 (w),

978 (w), 947 (w), 870 (w), 831 (w), 797 (w), 764 (w), 743 (w), 665 (m), 583 (w), 544 (w), 529 (w), 451 (w). UV-Vis (nm): 428, 547, 589.

2.2.2. Synthesis of $(Et_3NH)[CoL(hfac)_2]$ **2**

$Co(hfac)_2 \cdot 2H_2O$ (0.073 g, 0.1443 mmol) was dissolved in 10 mL of heptane and refluxed for 30 min. Then, after cooling down the solution, another 10 mL of $CH_2Cl_2$ solution containing HL (0.047 g, 0.1443 mmol) and triethylamine (0.0146 g, 0.1443 mmol, 20 μL) was added over. The solution was refluxed another 30 min, cooled down and filtered. After allowing the solvent to slowly evaporate for three days, violet needle shaped crystals of the product were obtained 0.085 g, yield 65%. Anal. Calcd for $C_{30}H_{35}O_{10}N_4F_{12}Co$ (%): C, 40.10; H, 3.93; N, 6.24; Found: C, 39.99; H, 3.84; N, 5.62. Selected IR peaks ($cm^{-1}$): 3092 (w), 2991 (w), 2951 (w), 2839 (w), 2685 (w), 1643 (s), 1601 (w), 1555 (m), 1526 (m), 1504 (s), 1452 (m), 1396 (w), 1364 (m), 1329 (m), 1310 (s), 1256 (vs), 1204 (vs), 1146 (vs), 1094 (m), 1070 (w), 980 (w), 947 (w), 870 (w), 849 (w), 795 (m), 764 (w), 744 (w), 671 (m), 584 (w), 544 (w), 528 (w), 451 (w). UV-Vis (nm): 426, 538, 610, 1230.

## 3. Physical Measurements

IR spectra were recorded on a FTIR Bruker Tensor V-37 spectrophotometer (KBr pellets) in the range of 4000–400 $cm^{-1}$ (Figure S1). UV-Vis diffuse reflectance spectra were recorded on a JASCO V-670 spectrophotometer on undiluted samples in the range 200–1600 nm. The X-Ray powder diffraction measurements (XRPD) were carried out on a Proto AXRD Benchtop using the Cu-Kα radiation with a wavelength of 1.54059 Å in the range 5–35° 2θ.

Continuous-wave (CW) electron paramagnetic resonance (EPR) spectra were recorded on a Bruker EMX 300 EPR spectrometer operating at X- (~9.4 GHz) or Q-band (~34 GHz) microwave frequencies, and at variable temperatures in the range 5 to 298 K. Spectra were simulated using the EasySpin 6.0.6 software [32].

Magnetic measurements were carried out with a Quantum Design MPMS 5S SQUID magnetometer in the temperature range 2−300 K. The measurements were performed on crystalline powders of the complexes mixed with grease (for Co) and hold in a gelatin capsule. The temperature dependences of the magnetization were measured in an applied field of 1 kOe and the isothermal field dependence of the magnetizations were collected up to 50 kOe. The molar susceptibility ($\chi_M$) was systematically corrected for sample holder, grease and for the diamagnetic contribution of all the atoms by using Pascal's tables [33]. AC susceptibility has been collected in zero field and with applied field with an AC frequency of 1000 Hz.

## 4. Crystal Structure Determination and Refinement

Suitable single crystals were mounted on glass fiber and X-Ray data were collected on a Rigaku XtaLAB Synergy, Single source at offset/far, HyPix (**1**) and STOE IPDS II (**2**) diffractometers using a graphite-monochromated Mo Kα radiation source (λ = 0.71073Å). The structures were solved by direct methods and refined by full-matrix least squares techniques based on $F^2$ using SHELXTL-2018 crystallographic software packages [34,35]. The non-H atoms were refined anisotropically. The H atoms attached to the C atoms were placed on calculated positions (riding model), while the H atoms on the N atoms of the triethylammonium cations were assigned using the residual peaks, and refined with isotropic thermal displacement parameters. The restraining DFIX and DANG commands were applied to rationalize the bond and angles parameters. A summary of the crystallographic data and the structure refinement for crystals **1** and **2** are given in Table 1. Selected bond distances and angle values are listed in Table 2. CCDC Reference numbers: 2,388,242 (**1**), 2,388,243 (**2**).

**Table 1.** Crystallographic data and structure refinement parameters for compounds **1** and **2**.

| Compound | 1-Mn | 2-Co |
|---|---|---|
| Formula | $C_{30}H_{35}N_4O_{10}F_{12}Mn$ | $C_{30}H_{35}N_4O_{10}F_{12}Co$ |
| Formula weight | 894.56 | 898.55 |
| Crystal system | monoclinic | monoclinic |
| Space group | $P2_1/c$ | $P2_1/c$ |
| $a$/Å | 20.4684(18) | 20.150(3) |
| $b$/Å | 9.5569(5) | 9.5624(8) |
| $c$/Å | 21.6281(17) | 21.501(3) |
| $\beta/°$ | 111.421(9) | 110.433(11) |
| $V$/Å$^3$ | 3928.5(6) | 3882.0(9) |
| Z | 4 | 4 |
| $D_c$/g cm$^{-3}$ | 1.509 | 1.537 |
| T/K | 293(2) | 293(2) |
| μ/mm$^{-1}$ | 0.448 | 0.556 |
| Reflections collected | 28,318 | 39,975 |
| Independent reflection | 6943 [$R_{int}$ = 0.0487] | 6840 [$R_{int}$ = 0.1921] |
| Final $R$ indices [$I > 2\sigma(I)$] | 0.0478; 0.1247 | 0.0773; 0.1353 |
| $R$ indices (all data) | 0.0772; 0.1404 | 0.1835; 0.1837 |
| Goodness-of-fit on $F^2$ | 1.031 | 1.088 |
| $\Delta\rho_{min}/\Delta\rho_{max}$ (e Å$^{-3}$) | −0.27/0.52 | −0.32/0.37 |

**Table 2.** Selected bond distances and angle values in compounds **1** and **2**.

| 1 | 2 |
|---|---|
| **Bonds length (Å)** | |
| Mn1–O1 = 2.191(2) | Co1–O1 = 2.097(4) |
| Mn1–O3 = 2.121(2) | Co1–O3 = 2.033(5) |
| Mn1–O7 = 2.172(2) | Co1–O7 = 2.080(5) |
| Mn1–O8 = 2.162(2) | Co1–O8 = 2.040(5) |
| Mn1–O9 = 2.129(2) | Co1–O9 = 2.036(5) |
| Mn1–O10 = 2.162(2) | Co1–O10 = 2.073(5) |
| **Angles (°)** | |
| O1 Mn1 O3 = 84.6(2) | O1 Co1 O3 = 88.8(3) |
| O1 Mn1 O7 = 90.9(2) | O1 Co1 O7 = 90.3(2) |
| O1 Mn1 O8 = 171.3(2) | O1 Co1 O8 = 172.6(2) |
| O1 Mn1 O9 = 81.1(2) | O1 Co1 O9 = 81.2(2) |
| O1 Mn1 O10 = 98.5(2) | O1 Co1 O10 = 96.4(2) |
| O3 Mn1 O7 = 94.42(1) | O3 Co1 O7 = 92.3(2) |
| O3 Mn1 O8 = 100.6(1) | O3 Co1 O8 = 98.1(2) |
| O3 Mn1 O9 = 163.2(1) | O3 Co1 O9 = 169.0(2) |
| O3 Mn1 O10 = 89.4(1) | O3 Co1 O10 = 87.5(2) |
| O7 Mn1 O8 = 81.8(2) | O7 Co1 O8 = 87.1(2) |
| O7 Mn1 O9 = 94.7(1) | O7 Co1 O9 = 92.4(2) |
| O7 Mn1 O10 = 169.9(2) | O7 Co1 O10 = 173.2(3) |
| O8 Mn1 O9 = 94.5(1) | O8 Co1 O9 = 91.9(2) |
| O8 Mn1 O10 = 88.3(1) | O8 Co1 O10 = 86.2(2) |
| O9 Mn1 O10 = 83.8(1) | O9 Co1 O10 = 88.9(2) |

## 5. Results and Discussion

The Nit ligand, HL, has been synthesized following Ullman's protocol starting from 2-hydroxy-3-methoxy-5-nitrobenzaldehyde (*o*-vanillin with the 5 position previously protected with a nitro group) [31]. We have tested the ability of HL molecules to form complexes by reacting it with $M(hfac)_2$, M = Ni(II), Zn(II), in the presence of $Et_3N$, to deprotonate the phenolic group [31]. The manganese and cobalt complexes, $(Et_3NH)[Mn(hfac)_2L]$ **1**, $(Et_3NH)[Co(hfac)_2L]$ **2**, are isomorphous with the nickel and zinc derivatives mentioned above. Therefore, we describe briefly the crystal structure of the manganese derivative. Compound $(Et_3NH)[Mn(hfac)_2L]$ **1** consists of an anionic mononuclear species, $[Mn(hfac)_2L]^-$ (Figure 1), and triethylammonium cations, $Et_3NH^+$. The metal ion adopts an octahedral coordination sphere, coordinated by phenoxido and aminoxyl oxygen atoms from the ligand and four oxygen atoms from the hexafluoroacetylacetonato ($hfac^-$) ligand. The Mn(II)–O bonds vary from 2.121(2) to 2.191(2) Å. The N1–O1 bond length from the coordinated aminoxyl group is 1.292(3) Å, slightly longer than the uncoordinated N2–O2 group (1.274(3) Å).

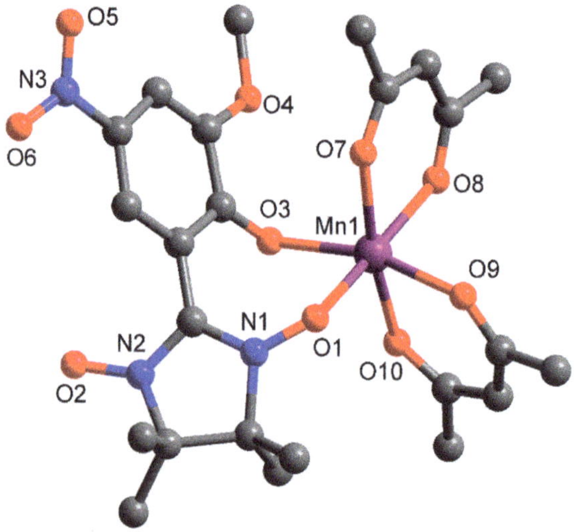


**Figure 1.** The X-ray structure of the complex anion in **1**; the hydrogen and fluorine atoms have been omitted for clarity.

In compound **1** the $CF_3$ groups are disordered over two crystallographic positions assigned as A (0.43) and B (0.57). For both crystallographic models, the packing diagram reveals a network of intermolecular F···F interactions (2.548–3.087 Å) connecting the mononuclear units at supramolecular level (Figures S2 and S3). The crystal structure of compound **2** is represented in Figure S4. The purity of the crystalline compounds, **1** and **2**, has been proved by powder X-Ray diffraction: the experimental diffractograms correspond well to the simulated ones (Figures S5 and S6).

The diffuse reflectance UV-Vis spectra of the ligand and the two complexes are displayed in Figure S7. Compound **2** shows, apart from the bands arising from the organic ligands, one d-d band at $^4T_1 \rightarrow {}^4T_2$ (1230 nm), assuming the *O* point group [36].

### *5.1. Magnetic Properties of* **1** *and* **2**

Complexes **1** and **2** were studied by magnetometry, results are plotted as $\chi_M T = f(T)$ (where $\chi_M$ stands for the molar magnetic susceptibility) and $M = f(H)$ in Figure 2 and Figure 4, respectively. The $\chi_M T$ product for **1** at 300 K is 3.27 $cm^3\ mol^{-1}$ K, which is smaller than the expected value of 4.75 $cm^3 mol^{-1} K$ for uncoupled $S = 5/2$ ($Mn^{II}$) and $S = 1/2$ (radical) spins, suggesting significant antiferromagnetic coupling between Mn(II) and the paramagnetic ligand. The $\chi_M T$ value decreases as the temperature is reduced reaching a plateau below 150 K at about 3.1 $cm^3\ mol^{-1}$ K, and drops to a lower value below 50 K (Figure 2). The recorded $\chi_M T$ plateau corresponds to a $S = 2$ spin ground state resulted from the antiferromagnetic $Mn^{II}$—radical coupling. The data have been analyzed considering a $S = 5/2$ and a $S = 1/2$ in exchange interaction ($J$), possible weak inter-molecular interactions have been considered within the mean-field approximation ($zJ'$). Modeling was done using the PHI program [37]. The best fit to the data gives the following values (with respect to $\boldsymbol{H}$ =

$-JS_{Mn}S_{Rad}$): $J = -191$ cm$^{-1}$; $g_{Mn} = 2.032$; $g_{Rad} = 2.0$ (fixed), with very weak intermolecular interactions, $zJ' = -0.019$ cm$^{-1}$. The magnetization vs. field curve obtained at 2 K, reaches saturation to high field with a value of 3.92 $\mu_B$, a slightly lower than the value anticipated for an $S = 2$ ground state.

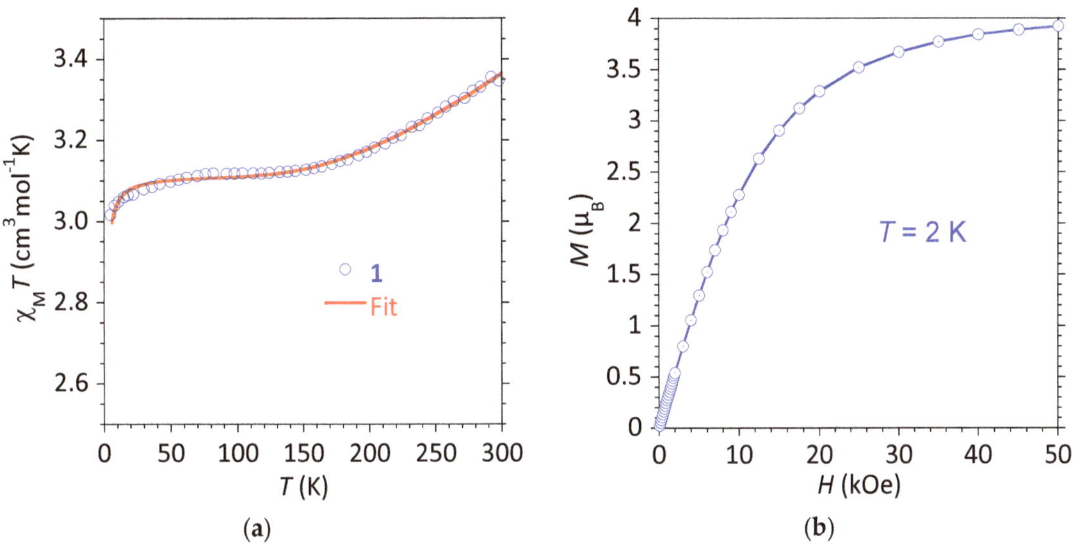


**Figure 2.** (**a**) Experimental (○) and calculated (—) temperature dependence of $\chi_M T$ and (**b**) field dependence of the magnetization at 2 K for compound **1**.

In the literature, the Mn(II)-Nit coupling is antiferromagnetic, with the exchange parameter $J$ ranging from −0.74 to −213 cm$^{-1}$ [10–16], which is strongly influenced by the values of the Mn–O–N and Mn–O–N–C (dihedral) angles, as well as the Mn–O(Nit) distance. Mn–O–N angles of ca. 125° lead to quite strong antiferromagnetic coupling between manganese (II) and the radical. Representative examples are presented in Table 3.

**Table 3.** Geometrical parameters and $J$ values (with respect to $H = -JS_{Mn}S_{Rad}$) for selected mononuclearMn(II)-Nit complexes.

| Compound | $J_{Mn\text{-}NIT}$ (cm$^{-1}$) | Mn-O (Å) | Mn-O-N (°) | Mn-O-N-C (°) | Ref. |
|---|---|---|---|---|---|
| [Mn(hfac)$_2$(oPONit)] [a] | −213 | 2.150(4) | 124.7 | 83.1 | [10] |
| [Mn(hfac)$_2$L$^1$] [b] | −114.4 | 2.181(2) | 124.5 | 28.7 | [12] |
| [Mn(NIT2Py)(NTB)](ClO$_4$)$_2$ [c] | −0.74 | 2.133(3) | 132.9 | 20.4 | [14] |
| [Mn(hfac)$_2$L$^3$] [d] | −193.4 | 2.129(3) | 123.6 | 47.8 | [15] |
| [Mn(hfac)$_2$(4-Me-3-Nit-trz)] [e] | −99.2 | 2.158(2) | 117.2 | 51.9 | [16] |
| (Et$_3$NH)[MnL(hfac)$_2$] **1** | −191 | 2.191(2) | 128.1 | 57.9 | this work |

[a] oPONit = (o-nitronyl nitroxide-phenyl)diphenylphosphine oxide. [b] L$^1$ = 4,4,5,5-tetramethyl-4,5-dihydro-1H-imidazole-3-oxide-1-oxyl-2-carboxylic acid amide. [c] NIT2Py = 2-(2′-pyridyl)-4,4,5,5-tetramethylimidazoline-1-oxyl-3-oxide. [d] L$^3$ = 4,4,5,5-tetramethyl-2-(2-phenyl-1,2,3-triazole-4-yl)imidazoline-1–oxyl-3-oxide. [e] 4-Me-3-Nit-trz = 2-[3-(4-methyl-1,2,4-triazolyl)]-4,4,5,5-tetramethylimidazoline-1-oxyl-3-oxide.

The strong antiferromagnetic Mn(II)–Rad interaction is confirmed by DFT calculations, which were performed using the Gaussian09 program [38], maintaining the crystal geometry without optimization. Starting with the spins of the Mn(II) ion and of the radical, 5/2 and 1/2 respectively, two states were considered: a high spin (HS), $S = 3$, $2S + 1 = 7$ and a low spin (LS), in which the radical spin is antiparallel to that of the Mn ion, $S = 2$, $2S + 1 = 5$. The energy of the low spin state, hereafter labeled as the Broken Symmetry (BS) state, was calculated by the fragment procedure as implemented in the Gaussian09 program and further checked for its stability. Two functionals, uB3LYP [39] and uM06 [40] were used with the TZVP basis set [41]. The exchange coupling constant, $J$, was calculated using the literature Equations (1)–(3) [42,43]:

$$J = (E_{BS} - E_{HS})/(2S_{Mn}S_{Rad} + S_{Rad}) \quad (1)$$

$$J = 2(E_{BS} - E_{HS})/S^2{}_{max}\ (S_{max} = S_1 + S_2) \quad (2)$$

$$J = 2(E_{BS} - E_{HS})/(<S^2>_{HS} - <S^2>_{BS}) \quad (3)$$

The results presented in Table 4 reflect the antiferromagnetic behavior of the complex, the broken spin state (BS) being the ground state; as far as the $J$ value is concerned, Equation (2) led to the closest agreement with the experimental value, $J_{exp} = -191\ cm^{-1}$. It can also be remarked that both functionals lead to quite similar results.

**Table 4.** Energies of the low (BS) and high spin (HS) states and $J$ values calculated using Equations (1)–(3).

| Functional | $E_{BS}$ (Ha) | $E_{HS}$ (Ha) | $J$ ($cm^{-1}$) | | |
|---|---|---|---|---|---|
| | | | (Equation (1)) | (Equation (2)) | (Equation (3)) |
| uB3Lyp | −4192.3621525 | −4192.3581071 | −295.95 | **−197.30** | −297.90 |
| uM06 | −4190.8464697 | −4190.8426809 | −277.18 | **−184.79** | −279.09 |

The spin isodensity surfaces corresponding to the two spin states are displayed in Figure 3a (BS-state) and Figure 3b (HS-state). For both states the spin density is mainly localized on the Mn ion and to a lower extent on the nitronyl groups.

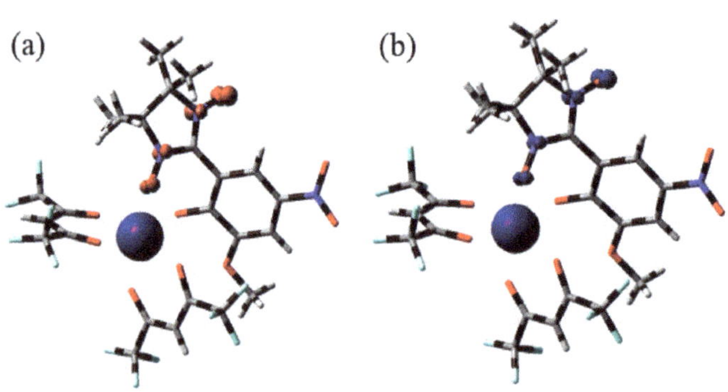


**Figure 3.** The spin density surfaces (isodensity = 0.04) for the BS (**a**) and HS (**b**) spin states at the uB3LYP/TZVP level.

The magnetic data of compound **2**, in the form of $\chi_M T$ vs. $T$ and $M$ vs. $H/T$, are shown in Figure 4. Due to the significant orbital contribution in octahedral geometry, the Co(II) ion is characterized by a $\chi_M T$ value that overcome the calculated spin-only value of 1.875 cm$^3$ mol$^{-1}$ K and usually approaches 3 cm$^3$ mol$^{-1}$ at 298 K [33,44,45]. Therefore, in the absence of any interaction between Co and ligand, an approximate $\chi_M T$ value of 3.375 cm$^3$ mol$^{-1}$ K would be expected for compound **2**. However, the measured $\chi_M T$ for **2** is 2.26 cm$^3$ mol$^{-1}$ K at room temperature, and it decreases as the temperature is lowered. This suggests that strong antiferromagnetic coupling occurs between the Co(II) and the radical. The $\chi_M T(T)$ curve shows a steady decrease below 300 K, then an inflection is noticed around 130 K. Below this temperature, a faster decrease occurs, because of the thermal depopulation of the excited Kramers doublets of the Co(II) ions. Fitting of the data has assumed contributions from $J_{CoRad}$, $D$ for Co(II), and possible intermolecular interactions considered within a mean field approximation ($zJ'$). The analysis was performed simultaneously on $\chi_M T$ and $M$ = f($H/T$). The following values have been obtained: $g_{Co}$ = 2.417, $J = -166$ cm$^{-1}$, $D$ = 65 cm$^{-1}$. The antiferromagnetic Co(II)-Rad interaction is quite strong (formalism: $\boldsymbol{H} = -JS_{Co}S_{Rad}$). However, some uncertainty applies because $g_{Co}$ is not known precisely (that for the radical was fixed to 2 for the calculations). The field dependence of the magnetization is also well reproduced, as it is plotted in the form of $M$ = f($H/T$) in Figure 4b.

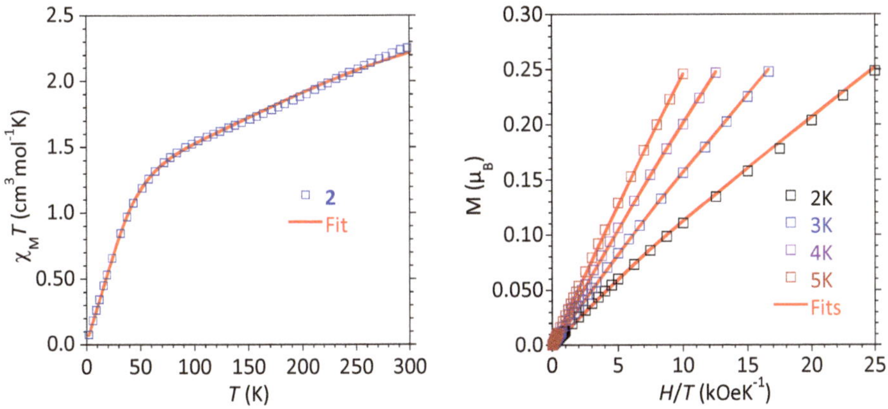


(**a**) (**b**)

**Figure 4.** Experimental (□) and best fits (full red lines) for (**a**) $\chi_M T$ = f($T$) and (**b**) M = f($H/T$) behaviors for compound **2**.

If the magnetic data are represented as $\chi_M$ vs. $T$ plot (Figure S8), after a plateau observed at 30 K, a Curie tail (i.e., an increase of the magnetic susceptibility) is clearly observed for lower $T$ indicating the presence of a paramagnetic impurity. This $\chi_M$ vs. $T$ curve was perfectly reproduced when a fraction (0.01) of a $S = 3/2$ impurity was considered. The contribution of this 0.01 fraction of S = 3/2 was taken into account for the analyses of $\chi_M T$ and M = f($H$) discussed above.

AC magnetic susceptibility measurements between 2 and 15 K did not reveal any evidence for slow relaxation of the magnetization for this compound.

### 5.2. Electron Paramagnetic Resonance (EPR) Data for 1

Continuous-wave (CW) EPR spectra for powder samples of **1** were recorded at X- (ca. 9.4 GHz) and Q-band (ca. 34 GHz) and several temperatures (Figures 5 and 6 respectively). Spectra recorded below 70 K show the clear characteristics of a $S = 2$ spin state that was stabilized by antiferromagnetic coupling between the spins of Mn(II) and Nit radical (see above). Here the zero-field splitting (ZFS) interaction $D[S_z^2 - (1/3)S(S+1)] - E(S_x^2 - S_y^2)$ splits the levels of an $S = 2$ spin state into two non-Kramer doublets, $|\pm2\rangle$ and $|\pm1\rangle$, and a singlet corresponding to the $m_S = 0$ [46]. Under the application of a magnetic field EPR transitions can occur within and between these splitting components leading to a multi-line structure as observed in Figures 5 and 6. Simulation of the Q-band EPR spectrum at 9 K (Figure 7) using EasySpin [32], and the spin Hamiltonian (4):

$$\mathcal{H} = gBS\mu_B + D\left[S_z{}^2 - (1/3)S(S+1)\right] + E\left(S_x{}^2 - S_y{}^2\right) \quad (4)$$

with $S = 2$, and $D$ and $E$ as the axial and rhombic zero-field splitting terms, respectively, provided $D = 0.26\ \text{cm}^{-1}$ and $E = 0.031\ \text{cm}^{-1}$. ZFS is present in systems which $S > 1/2$ and is a result from both SOC and spin-spin coupling (SSC) interactions, the former being more prominent in transition metals. Although perfectly symmetric half-occupied *d*-orbitals should not present any SOC, some may arise if the compound deviates from a perfect symmetry [47].

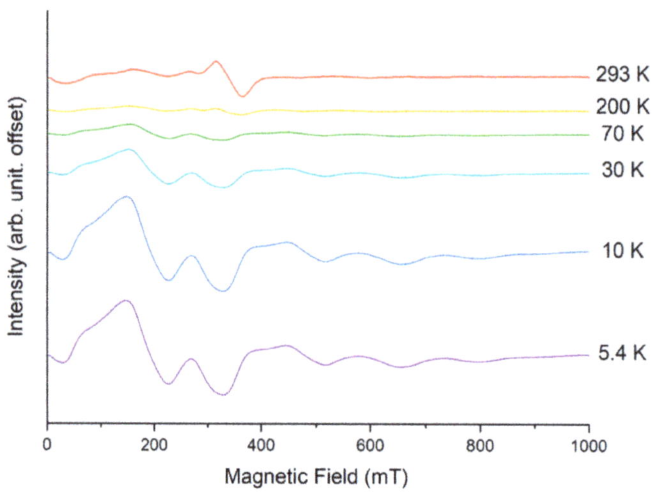


**Figure 5.** X-band (9.4 GHz) CW-EPR spectra of powder samples of **1** at temperatures between 5.4 and 293 K.

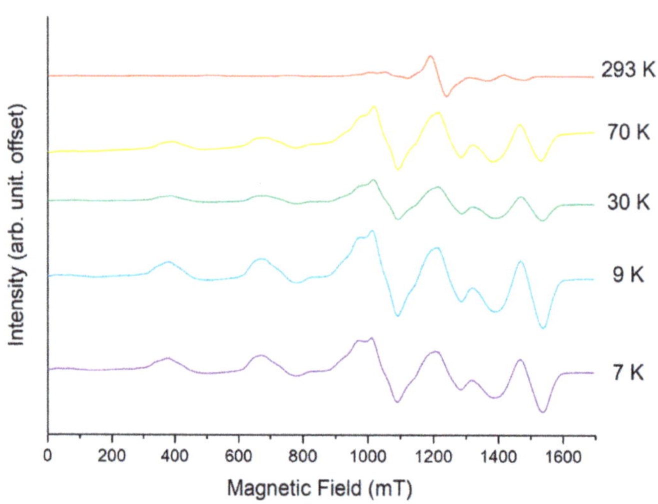


**Figure 6.** Q-band (34 GHz) CW-EPR spectra of powdered **1** at temperatures between 7 and 293 K.

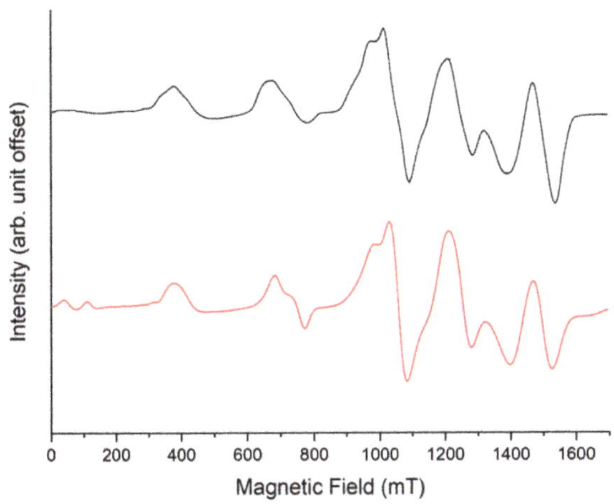


**Figure 7.** Experimental (black) and simulated (red) Q-band (34 GHz) CW-EPR spectra of powdered compound **1** at 9 K (g = 2, $D = 0.26\ cm^{-1}$, $E = 0.031\ cm^{-1}$, and H-Strain $(x, y, z)$ = 1300, 2100, 2000 MHz).

## 6. Conclusions

Two 2p–3d mononuclear complexes have been synthesized and structurally characterized. For both Mn(II) and Co(II) compounds strong antiferromagnetic metal-radical interaction was found, with a significant Zero Field Splitting effect ($D$). The EPR investigation of the manganese derivative reveals the signature of an $S = 2$ ground state, in agreement with the magnetic measurements.

**Supplementary Materials:** The following supporting information can be downloaded at: https://www.mdpi.com/article/10.3390/magnetochemistry10110086/s1, Figure S1. Infrared spectra for HL and complexes **1** and **2**; Figure S2. The packing diagram of crystal **1** with the representation of F···F interactions taking into account the disorder of $CF_3$ group in position assigned A (0.43 occupancy); Figure S3. The packing diagram of crystal **1** with the representation of F···F interactions taking into account the disorder of $CF_3$ group in position assigned B (0.57 occupancy); Figure S4. The X-ray structure of the complex anion in **2**; the hydrogen and fluorine atoms have been omitted for clarity; Figure S5. Measured (red) and calculated (black) powder X-ray diffraction patterns of **1**; Figure S6. Measured (red) and calculated (black) powder X-ray diffraction patterns of **2**; Figure S7. Diffuse reflectance spectra for HL and complexes **1** and **2**; Figure S8. Experimental (□) and calculated (—) $\chi_M = f(T)$ for compound **2**. The best fit parameters are discussed in the main text.

**Author Contributions:** The manuscript was prepared with the contributions of all authors. They specifically contributed as follows: C.A.S.—Investigation; D.O.T.A.M.—Investigation. T.M.—Data curation; M.H.—Investigation; J.-P.S.—Investigation, Formal analysis; F.T.—Formal analysis; M.A., F.T. and J.-P.S.—Writing original draft, supervision. All authors have read and agreed to the published version of the manuscript.

**Funding:** This research was funded by UEFISCDI (Project PN-III-P4-ID-PCE-2016-0308I) and EPSRC UK (grants EP/W014521/1, EP/V035231/1 and EP/X034523/1).

**Institutional Review Board Statement:** Not applicable.

**Informed Consent Statement:** Not applicable.

**Data Availability Statement:** The original contributions presented in the study are included in the article and Supplementary Materials, further inquiries can be directed to the corresponding authors.

**Acknowledgments:** Authors are grateful to M. J.-F. Menier (LCC, Toulouse) for technical assistance in magnetic data collections and M. Răducă (ICOS, Bucharest) for X-ray powder diffraction data collections. We further acknowledge the UK National EPR Facility at the University of Manchester for EPR data collection.

**Conflicts of Interest:** The authors declare no conflict of interest.

## References

1. Vaz, M.G.F.; Andruh, M. Molecule-based magnetic materials constructed from paramagnetic organic ligands and two different metal ions. *Coord. Chem. Rev.* **2021**, *427*, 213611. [CrossRef]
2. Meng, X.; Shi, W.; Cheng, P. Magnetism in one-dimensional metal–nitronyl nitroxide radical system. *Coord. Chem. Rev.* **2019**, *378*, 134–150. [CrossRef]
3. Demir, S.; Jeon, I.-R.; Long, J.R.; Harris, T.D. Radical ligand-containing single-molecule magnets. *Coord. Chem. Rev.* **2015**, *289–290*, 149. [CrossRef]
4. Brook, D.J.R. Coordination Chemistry of Verdazyl Radicals. *Comm. Inorg. Chem.* **2015**, *35*, 1–17. [CrossRef]
5. Miller, J.S. Magnetically ordered molecule-based materials. *Chem. Soc. Rev.* **2011**, *40*, 3266–3296. [CrossRef]
6. Luneau, D.; Rey, P. Magnetism of metal-nitroxide compounds involving bis-chelating imidazole and benzimidazole substituted nitronyl nitroxide free radicals. *Coord. Chem. Rev.* **2005**, *249*, 2591–2611. [CrossRef]
7. Caneschi, A.; Gatteschi, D.; Lalioti, N.; Sangregorio, C.; Sessoli, R.; Venturi, G.; Vindigni, A.; Rettori, A.; Pini, M.G.; Novak, M.A. Cobalt(II)-Nitronyl Nitroxide Chains as Molecular Magnetic Nanowires. *Angew. Chem. Int. Ed.* **2001**, *40*, 1760–1763. [CrossRef]
8. Liu, X.; Feng, X.; Meihaus, K.R.; Meng, X.; Zhang, Y.; Li, L.; Liu, J.-L.; Pedersen, K.S.; Keller, L.; Shi, W.; et al. Coercive Fields Above 6 T in Two Cobalt(II)–Radical Chain Compounds. *Angew. Chem.* **2020**, *132*, 10697–10705. [CrossRef]
9. Vaz, M.G.F.; Cassaro, R.A.A.; Akpinar, H.; Schlueter, J.A.; Lahti, P.M.; Novak, M. A Cobalt Pyrenylnitronylnitroxide Single-Chain Magnet with High Coercivity and Record Blocking Temperature. *Chem.-Eur. J.* **2014**, *20*, 5460–5467. [CrossRef]
10. Rancurel, C.; Leznoff, D.B.; Sutter, J.-P.; Golhen, S.; Ouahab, L.; Kliava, J.; Kahn, O. Synthesis, structure, and magnetism of mono- and binuclear manganese(II) compounds of nitronyl nitroxide substituted phosphine oxides. *Inorg. Chem.* **1999**, *38*, 4753–4758. [CrossRef]
11. Luneau, D.; Risoan, G.; Rey, P.; Grand, A.; Caneschi, A.; Gatteschi, D.; Laugier, J. Transition metal derivatives of a chelating nitronyl nitroxide ligand. Nickel(II) and manganese(II) complexes. *Inorg. Chem.* **1993**, *32*, 5616–5622. [CrossRef]
12. Tretyakov, E.V.; Koreneva, O.V.; Romanenko, G.V.; Shvedenkov, Y.G.; Ovcharenko, V.I. Synthesis, structure and magnetism of $M(hfac)_2$ complexes with spin labelled amides. *Polyhedron* **2004**, *23*, 763–772. [CrossRef]
13. Yu, Y.-X.; Hu, H.-M.; Zhang, D.-Q.; Wang, Z.-Y.; Zhu, D.-B. Two New Nitronyl Nitroxide Radicals and Their Complexes with $M(hfac)_2$ [M=Co(II), Ni(II), Mn(II)]: Syntheses, Crystal Structures, and Magnetic Characterizations. *Chin. J. Chem.* **2007**, *25*, 1259–1266. [CrossRef]
14. Zhang, C.-X.; Zhang, Y.-Y.; Zhang, H.-L.; Zhao, Y.; Sun, Y.-Q. Synthesis, structures and magnetic properties of nickel(II), manganic(II) and zinc(II) complexes containing pyridyl-substituted nitronyl nitroxide and tris(2-benzimidazolymethyl)amine. *Inorg. Chim. Acta* **2009**, *362*, 5231–5236. [CrossRef]
15. Zhang, Y.-J.; Wang, J.-J.; Huang, Q.; Chen, J. Crystal structures and magnetic properties of nitronyl nitroxide NIT2-phtrz and its M(II) hexafluoroacetylacetonate (M = manganese, copper) complexes $[M(hfac)_2(NIT2\text{-}phtrz)]$. *Transit. Met. Chem.* **2012**, *37*, 743–749. [CrossRef]
16. Li, T.; Shi, X.J.; Chen, P.Y.; Yu, S.J.; Tian, L. A series bi-spin transition metal(II) complexes based on triazole nitronyl nitroxide radical. *Inorg. Chim. Acta* **2017**, *461*, 206–212. [CrossRef]
17. Caneschi, A.; Gatteschi, D.; Sessoli, R.; Rey, P. Structure and magnetic properties of a ring of four spins formed by manganese(II) and a pyridine substituted nitronyl nitroxide. *Inorg. Chim. Acta* **1991**, *184*, 67–71. [CrossRef]
18. Wang, H.; Liu, Z.; Liu, C.; Zhang, D.; Lü, Z.; Geng, H.; Shuai, Z.; Zhu, D. Coordination Complexes of 2-(4-Quinolyl)nitronyl Nitroxide with $M(hfac)_2$ [M = Mn(II), Co(II), and Cu(II)]: Syntheses, Crystal Structures, and Magnetic Characterization. *Inorg. Chem.* **2004**, *43*, 4091–4098. [CrossRef]
19. Zhang, J.-Y.; Liu, C.-M.; Zhang, D.-Q.; Gao, S.; Zhu, D.-B. Syntheses, crystal structures, and magnetic properties of two cyclic dimer $M_2L_2$ complexes constructed from a new nitronyl nitroxide ligand and $M(hfac)_2$ ($M = Cu^{2+}, Mn^{2+}$). *Inorg. Chim. Acta* **2007**, *360*, 3553–3559. [CrossRef]
20. Liu, R.-N.; Li, L.-C.; Xing, X.-Y.; Liao, D.-Z. Cyclic metal–radical complexes based on 2-[4-(1-imidazole)phenyl]-4,4,5,5-tetramethylimidazoline-1-oxyl-3-oxide: Syntheses, crystal structures and magnetic properties. *Inorg. Chim. Acta* **2009**, *362*, 2253–2258. [CrossRef]
21. Wang, C.; Ma, Y.; Wang, Y.; Wang, Q.; Li, L.; Cheng, P.; Liao, D. A New Quinoxalinyl-Substituted Nitronyl Nitroxide Radical and its Five-Spin $Cu^{II}$ and Four-Spin $Mn^{II}$ Complexes: Syntheses, Crystal Structures, and Magnetic Properties. *Aust. J. Chem.* **2012**, *65*, 672–679. [CrossRef]
22. Tian, H.; Li, Y.; Zhang, C.; Mei, X.; Hu, P.; Li, L. Two novel cyclic dimer $M_2L_2$ complexes constructed from 2-[(4-carboxyl)phenyl]-4,4,5,5-tetramethylimidazoline-1-oxyl-3 oxide: Crystal structure and magnetic properties. *Polyhedron* **2013**, *52*, 1053–1058. [CrossRef]
23. Guo, J.-N.; Wang, J.-J.; Sun, G.-F.; Li, H.-D.; Li, L.-C. A novel nitronyl nitroxide radical containing thiophene and pyridine rings and its manganese(II) complex: Synthesis, structure, and magnetic properties. *J. Coord. Chem.* **2017**, *70*, 1926–1935. [CrossRef]

24. Zhu, M.; Lou, D.; Deng, X.; Zhang, L.; Zhang, W.; Lü, Y. A functional nitroxide ligand builds up two 2p–3d complexes with different spin ground states and a 2p–3d–4f chain of rings. *CrystEngComm* **2018**, *20*, 2583–2592. [CrossRef]
25. Yang, M.; Xie, S.; Liang, X.; Zhang, Y.; Dong, W. A novel functional nitronyl nitroxide and its manganese and cobalt complexes: Synthesis, structures and magnetic properties. *Polyhedron* **2019**, *161*, 132–136. [CrossRef]
26. Lin, H.-H.; Mohanta, S.; Lee, C.-J.; Wei, H.-H. Syntheses, Crystal Engineering, and Magnetic Property of a Dicyanamide Bridged Three-Dimensional Manganese(II)–Nitronyl Nitroxide Coordination Polymer Derived from a New Radical. *Inorg. Chem.* **2003**, *42*, 1584–1589. [CrossRef] [PubMed]
27. Gao, Y.-L.; Tian, C.; Gao, P.; Jiao, Y.; Gong, Y.; Ma, X. Synthesis, crystal structures, and magnetic properties of nitroxide radical-coordinated manganese(II) and cobalt(II) complexes. *J. Chem. Res.* **2021**, *45*, 788–794. [CrossRef]
28. Fegy, K.; Luneau, D.; Ohm, T.; Paulsen, C.; Rey, P. Two-Dimensional Nitroxide-Based Molecular Magnetic Materials. *Angew. Chem. Int. Ed.* **1998**, *37*, 1270–1273. [CrossRef]
29. Luneau, D.; Borta, A.; Chumakov, Y.; Jacquot, J.-F.; Jeanneau, E.; Lescop, C.; Rey, P. Molecular magnets based on two-dimensional Mn(II)–nitronyl nitroxide frameworks in layered structures. *Inorg. Chim. Acta* **2008**, *361*, 3669–3676. [CrossRef]
30. Lecourt, C.; Izumi, Y.; Khrouz, L.; Toche, F.; Chiriac, R.; Bélanger-Desmarais, N.; Reber, C.; Fabelo, O.; Inoue, K.; Desroches, C.; et al. Thermally-induced hysteretic valence tautomeric conversions in the solid state via two-step labile electron transfers in manganese-nitronyl nitroxide 2D-frameworks. *Dalton Trans.* **2020**, *49*, 15646–15662. [CrossRef]
31. Spinu, C.A.; Pichon, C.; Ionita, G.; Mocanu, T.; Calancea, S.; Raduca, M.; Sutter, J.-P.; Hillebrand, M.; Andruh, M. Synthesis, crystal structure, magnetic, spectroscopic, and theoretical investigations of two new nitronyl-nitroxide complexes. *J. Coord. Chem.* **2021**, *74*, 279–293. [CrossRef]
32. Stoll, S.; Schweiger, A. EasySpin, a comprehensive software package for spectral simulation and analysis in EPR. *J. Magn. Reson.* **2006**, *178*, 42–55. [CrossRef]
33. Kahn, O. *Molecular Magnetism*, 2nd ed.; Dover Publications: New York, NY, USA, 2021.
34. Sheldrick, G.M. Crystal structure refinement with SHELXL. *Acta Cryst.* **2015**, *C71*, 3–8.
35. Sheldrick, G.M. SHELXT—Integrated space-group and crystal structure determination. *Acta Cryst.* **2015**, *A71*, 3–8. [CrossRef]
36. Lever, A.B.P. *Inorganic Electronic Spectroscopy*, 2nd ed.; Elsevier: Amsterdam, The Netherlands, 1984; pp. 479–489.
37. Chilton, N.F.; Anderson, R.P.; Turner, L.D.; Soncini, A.; Murray, K.S. PHI: A powerful new program for the analysis of anisotropic monomeric and exchange-coupled polynuclear d- and f-block complexes. *J. Comput. Chem* **2013**, *34*, 1164–1175. [CrossRef]
38. Frisch, M.J.; Trucks, G.W.; Schlegel, H.B.; Scuseria, G.E.; Robb, M.A.; Cheeseman, J.R.; Scalmani, G.; Barone, V.; Mennucci, B.; Petersson, G.A.; et al. *GAUSSIAN 09*; Revision, C.01; Gaussian, Inc.: Wallingford, CT, USA, 2009.
39. Becke, A.D. Density-functional thermochemistry. III. The role of exact exchange. *J. Chem. Phys.* **1993**, *98*, 5648–5652. [CrossRef]
40. Valero, R.; Costa, R.; Moreira, I.d.P.R.; Truhlar, D.G.; Illas, F. Performance of the M06 family of exchange-correlation functionals for predicting magnetic coupling in organic and inorganic molecules. *J. Chem. Phys.* **2008**, *128*, 114103. [CrossRef]
41. Weigend, F.; Ahlrichs, R. Balanced basis sets of split valence, triple zeta valence and quadruple zeta valence quality for H to Rn: Design and assessment of accuracy. *Phys. Chem. Chem. Phys.* **2005**, *7*, 3297–3305. [CrossRef]
42. Ruiz, E.; Cano, J.; Alvarez, S.; Alemany, P. Broken symmetry approach to calculation of exchange coupling constants for homobinuclear and heterobinuclear transition metal complexes. *J. Comput. Chem.* **1999**, *20*, 1391–1400. [CrossRef]
43. Nishino, M.; Yamanaka, S.; Yoshioka, Y.; Yamaguchi, K. Theoretical Approaches to Direct Exchange Couplings between Divalent Chromium Ions in Naked Dimers, Tetramers, and Clusters. *J. Phys. Chem. A* **1997**, *101*, 705–712. [CrossRef]
44. Palii, A.V.; Korchagin, D.V.; Yureva, E.A.; Akimov, A.V.; Misochko, E.Y.; Shilov, G.V.; Talantsev, A.D.; Morgunov, R.B.; Aldoshin, S.M.; Tsukerblat, B.S. Single-Ion Magnet $Et_4N[Co^{II}(hfac)_3]$ with Nonuniaxial Anisotropy: Synthesis, Experimental Characterization, and Theoretical Modeling. *Inorg. Chem.* **2016**, *55*, 9696–9706. [CrossRef] [PubMed]
45. Lloret, F.; Julve, M.; Cano, J.; Ruiz-García, R.; Pardo, E. Magnetic properties of six-coordinated high-spin cobalt(II) complexes: Theoretical background and its application. *Inorg. Chim. Acta* **2008**, *361*, 3432–3445. [CrossRef]
46. Abragam, A.; Bleaney, B. *Electron Paramagnetic Resonance of Transition Metal Ions*; Clarendon Press: Oxford, UK, 1970; pp. 209–211.
47. Telser, J. EPR Interactions—Zero-field Splittings. In *eMagRes*; John Wiley & Sons, Ltd.: Chichester, UK, 2017; Volume 6, pp. 207–234.

*magnetochemistry*

*Review*

# Recent Insights into Magneto-Structural Properties of Co(II) Dicyanamide Coordination Compounds †

**Anna Świtlicka**

Institute of Chemistry, University of Silesia, Szkolna 9, 40-006 Katowice, Poland; anna.switlicka@us.edu.pl

† This article was originally going to be dedicated to the distinguished scientists and my friends Miguel Julve and Francesco Lloret on their retirement. When this article was almost ready for submission to Magnetochemistry, I learned of the tragic news of the death of my dear friend Miguel, an outstanding scientist, internationally recognized for his invaluable contributions to coordination chemistry and molecular magnetism, with a great enthusiasm for science and a bright personality. I dedicate this article to his blessed memory. You will live on in our memory, dear Miguel.

**Abstract:** In recent decades, the chemistry of transition metal coordination compounds has undergone continuous development at both scientific and application levels. The diversity of metal(II) complexes, along with their structural features and physicochemical properties, makes them attractive for a wide range of applications. The dicyanamide ion ($N(CN)_2$) has the ability to form various transition metal compounds characterized by different architectures and topologies. The interaction of $\pi$-electrons from the nitrile groups with the $\pi$-system of the central nitrogen may enable electron delocalization, potentially facilitating electron transfer between the metal centers through the bridging dicyanamide (dca) ligand. This review focuses on dca–Co(II) compounds and, after a brief introduction, the structural aspects and magnetic properties are analyzed in detail.

**Keywords:** cobalt(II) complexes; dicyanamide ion; coordination polymers; structural studies; magnetic properties

## 1. Introduction

Before dicyanamide (cyanoguanidine, $N(CN)_2^-$, dca) gained popularity in supramolecular architecture and crystal engineering, it was primarily employed in a completely different branch of chemistry. This popular, versatile ligand was used as an effective nitrification inhibitor, helping to reduce leaching losses by retaining applied nitrogen in the ammoniacal form [1]. It has been proven that the addition of *dca* to ammonium or urea salt increased the yields of crops including rice, wheat and citrus trees [1]. Dicyanamide was also successfully tested in hydrophobic ionic liquids (ILs) for efficient extraction of Cu(II), Ni(II), Pb(II) and Cd(II) from water [2]. Regardless of its initial applications, the boomerang-shaped dicyanamide anion is widely known as an effective linker between 3d and 4d metal ions, making it perfect for the design and construction of coordination polymers (CPs) with different frameworks and topologies. Importantly, dicyanamide–metal coordination compounds have demonstrated significant diversity in the application horizon in the fields of molecular sensors, electrical conductivity, non-linear optical activity (NLO), catalysis, molecular magnetism, microporous materials for gas adsorption and storage [3–12].

Preliminary studies of the first compounds of the formula $[M(dca)_2]$ (M = Cu(II), Hg(II), Pb(II) were performed by Madelung and Kern in 1922, and later were continued by Köhler and Hvastijová [13–15]. Shortly thereafter, in 1940, E. W. Hughes made a successful attempt to determine the crystal structure of dicyanamide by confronting other proposed structures of this compound (Deposition Number: 1498379) [16]. Many of these early complexes have been re-investigated using modern X-ray analysis methods. The $dca^-$ geometry favors the action of the anion as a terminal or bridging ligand, and prevents the formation of chelated coordination species. In homo- or heteronuclear complexes, the dicyanamide anion has been shown to be a versatile ligand and can bind two to five metal ions in a variety of ways, either through its terminal nitrile nitrogen group(s), the central amide nitrogen, or both. These possible coordination modes have been illustrated in Scheme 1. In the literature, various naming conventions have been used for dca coordination modes. Following the most common notification, the form $\mu_{x,y,z}$ will be used, where x, y and z are represent the nitrogen atoms which are coordinating with Co(II) ion(s). The dca ligand can have the values of 1, 3 or 5, where 1 or 5 are assigned to a nitrile nitrogen atom, while 3 is attributed to an amide nitrogen atom. However, for the sake of thoroughness, an alternative notation will also be discussed below.

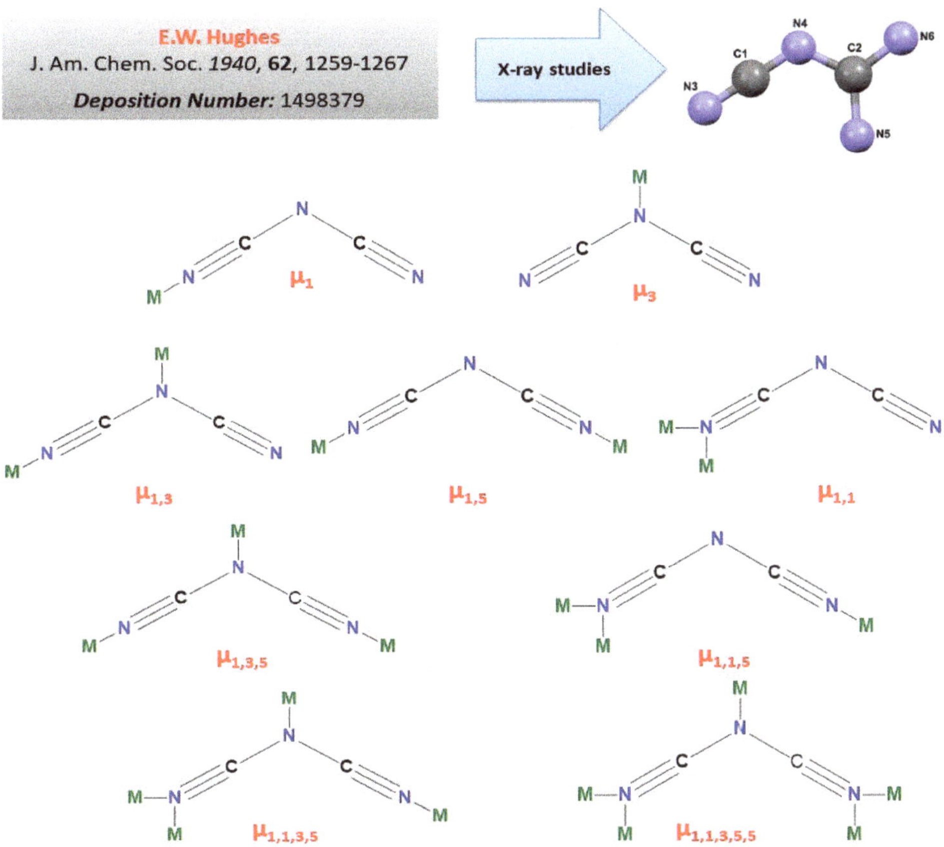


**Scheme 1.** Coordination modes of dicyanamide ligand [16].

The coordination bonding modes of $dca^-$ with divalent metal(II) ions in most compounds are primarily established through its two terminal N-nitrile groups ($\mu_1$) or through less common central amide nitrogen ($\mu_3$), or through both. The most common coordination modes of the dicyanamide ion in coordination polymers are $\mu_{1,5}$ (*end-to-end, EE*) and $\mu_{1,1}$ (*end-to-on, EO*). Other bonding modes such as $\mu$-$_{1,1,5}$, $\mu$-$_{1,3,5}$, $\mu$-$_{1,1,3,5}$ or $\mu$-$_{1,1,3,5,5}$ have also been isolated and structurally studied [17,18]. Other possibilities of impossible notations for the bridging dca ion include bidentate $\mu_2$-bonding ($\mu_{1,5}$, $\mu_{1,3}$ and $\mu_{1,1}$), $\mu_3$-bonding for $\mu$-$_{1,1,5}$ and $\mu$-$_{1,3,5}$, $\mu_4$-bonding for $\mu$-$_{1,1,3,5}$ and finally $\mu_5$-bonding for $\mu$-$_{1,1,3,5,5}$. In the literature, the letter "$\eta$" is sometimes used instead of "$\mu$". Typically, the "$\eta$" denotes coordination involving more than one bridging atom; therefore, notations such as $\mu$-$_{1,3,5}$, $\mu$-$_{1,1,3,5}$ or $\mu$-$_{1,1,3,5,5}$ can be alternatively written as $\eta^3$-1,3,5, $\eta^4$-1,1,3,5, or $\eta^5$-1,1,3,5,5, respectively.

A search of the Cambridge Structural Database (CSD, version 2024.2.0) reveals 1466 structures of transition metal complexes containing dicyanamide as a ligand (either terminal or bridging) [19]. Further refinement of this initial hitlist shows that the cobalt(II) complexes constitute 14% of all structures, ranking third after copper(II) and manganese(II) compounds. Although new structures of metal(II) complexes based on dicyanamide bring great excitement to scientists, these compounds are studied not only because of their structural attractiveness, but also because of their intriguing and fascinating magnetic properties. Since the discovery of long-range magnetic ordering in the isomorphous series of α-[M(dca)$_2$] compounds [M = Cr, Mn, Fe, Co, Ni, Cu], these kinds of compounds have received a lot of attention. This review provides an overview of the chemistry of dicyanamide-Co(II) complexes, in which magneto-structural relationships have been extensively explored. Emphasis is placed on recent results for monomers, dimers and coordination polymers.

## 2. Synthetic Methods of Dicyanamide Co(II) Complexes

The major challenge in synthesizing dicyanamide compounds is obtaining single crystals suitable for X-ray measurements, particularly due to their tendency to form poorly soluble crystalline precipitates. Improving synthesis methods is necessary, because the reactions with the same starting materials, assuming the synthesis method changes, can sometimes lead to different coordination compounds. Another common factor is the dependence of the resulting complex compound on the cobalt(II) salt used and/or the stoichiometry of the reaction. Representative examples of such Co(II) compounds will be presented and discussed below.

Based on the analyzed dca–Co(II) compounds, two main synthetic strategies were found to be used.

- *Saturation Method*
  The synthesis is carried out by mixing cobalt(II) salts with the appropriate amount of the ligand, and then a dicyanamide solution is added. The resulting solution is kept for evaporation at room temperature and X-ray-quality crystals grow in saturated solutions, usually few days.
- *Diffusion Method*
  This method is especially dedicated to those compounds that precipitate during the reaction in the form of poorly soluble or non-soluble precipitates. Generally, two approaches can be used. The first technique involves dissolving the reactants in two different solvents. Then, the layers of these two solutions should be dropped on top of each other and separated by a layer of pure solvent. The second one assumes that the compound was obtained using the traditional method. Then, the impurities from the reaction solution have to be filtered off and the obtained filtrate should be placed under a layer of precipitating solvent. This type of crystallization is most often carried out in a narrow vial, i.e., a test tube or an H-tube, and requires some skill because the boundary between the two solutions must be clear and distinct. Crystals most often form at the phase boundary after a few weeks. The following examples of compounds can serve as representatives of these two approaches: [Co(L1)(dca)$_2$] [20], [Co(phen)(dca)$_2$] [21], [Co(bpp)(dca)$_2$]$_n$ [22], [Co(4-bpmp)$_2$(dca)$_2$(H$_2$O)$_2$]$_n$ [23], {[Co(dca)$_2$(H$_2$O)$_2$]·(hmt)}$_n$ [24], or [Co(HFlu)$_2$(dca)$_2$]$_n$ [25].

Interestingly, L. Váhovská et al. obtained two complexes with the same formula [Co(phen)(dca)$_2$] using the saturation method [21]. The first compound appeared as an orange precipitate during synthesis, while the second crystallized after two weeks of slow evaporation in the dark. Both structures are compared in the section "*Description of Mononuclear and Dinuclear Complexes*". The list of abbreviations is presented at the end of the article.

## 3. Description of Mononuclear and Dinuclear Complexes

*Summarized structural data can be found in Table S1 in Supplementary Materials*

Based on the results from the CSD, it can be concluded that mononuclear dca-Co(II) complexes are relatively uncommon, with there being only 12 structures of cobalt(II) complexes with a terminal dca ligand. On the other hand, examples of dimeric compounds are even rarer, with only two structurally characterized examples. There may be various reasons for this, but it seems that the divergent nature of the coordinating nitrile groups in dca is a key factor [26]. Of course, it does not imply that these complexes are difficult to synthesize, but their lower prevalence could be due to their reduced attractiveness compared to coordination polymers. Among the mononuclear compounds, only one Co(II) complex is known in which terminal dicyanamide coordinates with a metal atom via amide nitrogen. In complex [Co($H_2$biim)$_2$(dca)$_2$], two bidendate N-donor ligands are coordinated to the octahedral Co(II) ion forming an equatorial plane, while the axial positions are occupied by amide nitrogen from two *dca* anions [23]. Typically, coordination through the nitrile nitrogen is preferred due to its higher electronegativity and more delocalized electrons. The formation of this unique complex is thought to be influenced by the electron-donating properties of $H_2$biim, which increase the Co(II) cation's affinity for the softer amide nitrogen over the nitrile group of the dca anion. (Figure 1). Analysis of the structural data shows that the Co–$N_{H2biim}$ lengths fall in the typical range of 2.168(12)–2.1435(12) Å, consistent with bond lengths observed in other mononuclear cobalt(II) compounds (see Tables 1 and S1). This is the only similarity of this compound in comparison to the remaining eleven [Co(L)$_{1–4}$(dca)$_2$] complexes, where L is a mono–, bi–, tri– or tetradentate organic ligand and dicyanamide coordinates with the Co(II) ion through nitrile nitrogen (see Tables 1 and S1). In [Co($H_2$biim)$_2$(dca)$_2$], the Co–$N_{dca}$ bond lengths [2.1912(13) Å] are longer compared to the bond lengths in other cobalt(II) complexes, which these range from 2.003(3) Å to 2.1546(15) Å. This difference results from the diversity of the coordination modes of the dicyanamide anion.

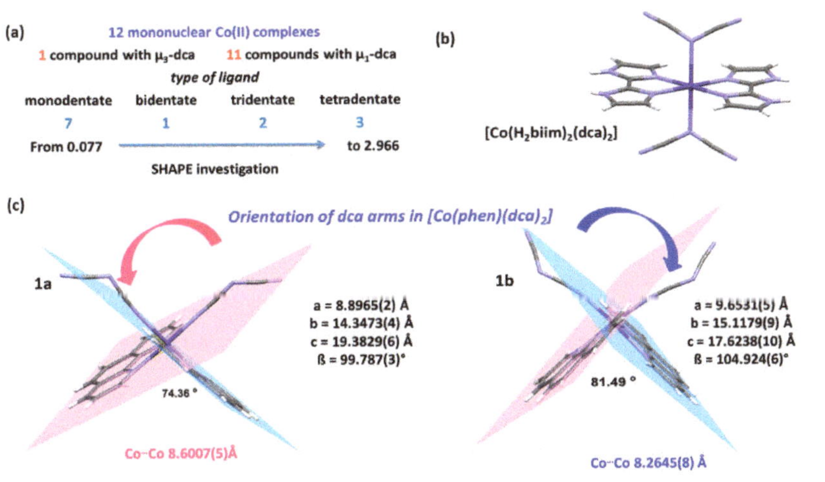


**Figure 1.** Summary of data for mononuclear Co(II) compounds (**a**); molecular structure of [Co($H_2$biim)$_2$(dca)$_2$] [27] (**b**); structural differences between two structures with the same formula [Co(phen)(dca)$_2$] [21] (**c**).

In [Co($H_2$biim)$_2$(dca)$_2$], the Co–N–C angles are 118.72° and 120.47(10)°, while the C–N–C angle has one of the lowest values of 119.27(13)°, probably due to the fact that the nitrogen in the C–N–C group participates in bonding to the metal atom. Interestingly, the metal···metal separation of 7.162(1) Å is one of the lowest values. In other mononuclear Co(II) compounds, metal atoms are generally better separated, with distances ranging from 7.187(1) to 9.686(2) Å.

**Table 1.** Structural parameters for the selected cobalt(II) mononuclear complexes (for the full version, see Table S1).

| Compound | Co–N$_{ligand}$ [Å] | Co–N$_{DCA}$ [Å] | Angles [°] Co–NC$_{(dca)}$ | Angles [°] C–N–C$_{(dca)}$ | Co···Co [Å] | Ref. |
|---|---|---|---|---|---|---|
| **Monodentate ligands** | | | | | | |
| [Co(bim)$_4$(dca)$_2$] | 2.1546(15)<br>2.1489(15) | 2.1575(18) | 155.32(18) | 122.8(3) | 9.686(2) | [28] |
| **Bidentate ligands** | | | | | | |
| [Co(H$_2$biim)$_2$(dca)$_2$] | 2.1435(12)<br>2.1268(12)<br>2.1435(12)<br>2.1258(12) | 2.1912(13)<br>2.1912(13) | 118.82(10)<br>120.47(10) | 119.27(13) | 7.162(1) | [27] |
| [Co(phen)$_2$(dca)$_2$]<br>Crystal 1a | **193 K** / **293 K**<br>2.149(1) / 2.142(2)<br>2.147(1) / 2.142(2)<br>2.145(1) / 2.141(2)<br>2.173(2) / 2.166(2) | **193 K** / **293 K**<br>2.093(2) / 2.092(2)<br>2.070(2) / 2.066(2) | 173.8(1)<br>169.4(2) | 121.4(2)<br>122.4(2) | 8.6007(5) | [21] |
| [Co(phen)$_2$(dca)$_2$]<br>Crystal 1b | **293 K**<br>2.139(3)<br>2.175(3)<br>2.132(3)<br>2.157(3) | **293 K**<br>2.071(3)<br>2.069(3) | 171.3(3)<br>168.3(3) | 124.6(4)<br>125.8(4) | 8.2645(8) | [21] |
| **Tridentate organic ligand** | | | | | | |
| [Co(tppz)$_2$](dca)$_2$] | **125 K** / **330 K**<br>1.855(5) / 1.870(3)<br>1.917(5) / 1.912(3)<br>1.977(4) / 2.003(2)<br>2.155(4) / 2.128(2) | –<br>uncoordinated dca | – | **125 K**<br>120.5(6)<br>**330 K**<br>123.2(7) | **125 K**<br>8.9361(7)<br>**330 K**<br>9.0081(4) Å | [29] |
| **Tetradentate organic ligand** | | | | | | |
| [Co(L$^1$)(dca)$_2$] | **100 K** / **280 K**<br>2.340(2) / 2.002(2)<br>1.920(2) / 2.351(2)<br>2.315(2) / 1.999(3)<br>1.909(2) / 2.324(2) | **100 K** / **280 K**<br>1.927(2) / 2.003(3)<br>1.942(2) / 2.024(3) | **100 K** / **280 K**<br>173.7(2) / 170.4(3)<br>166.5(2) / 164.3(3) | **100 K** / **280 K**<br>117.8(2) / 121.9(4)<br>116.2(1) / 118.0(3) | **100 K** / **280 K**<br>8.9101(6) / 9.0233(8) | [20] |

**H$_2$BiIm** = 2,2A-biimidazole; **L$^1$** = N,N′-di-tertbutyl-2,11-diaza[3,3](2,6)pyridinophane; **bim** = 1 benzylimidazole; **tppz** = 2,3,5,6-tetrakis(2-pyridyl)pyrazine; **phen** = 1,10-phenanthroline.

Generally, the *cis* configuration of dicyanamide in Co(II) complexes is favored over the *trans* conformation. *Trans*-coordinated dca has been observed only in compounds based on monodentate derivatives, with four N-donor ligands of the same type in [Co(bim)$_4$(dca)$_2$] [28] or with two different types of ligands in [Co(4-bpo)$_2$(H$_2$O)$_2$(dca)$_2$](H$_2$O)$_2$ [30] and [Co(nitppy)$_2$(H$_2$O)$_2$(dca)$_2$] [31]. In the cases of complexes based on tetradentate and bidentate ligands (with the exception of the previously discussed [Co(H$_2$biim)$_2$(dca)$_2$]), the dca anions are coordinated in a *cis* arrangement. Introducing tridentate N-donor derivatives into the coordination environment of the Co(II) ion results in the lack of coordination of the dca ion. Such a situation occurred only in one complex [Co(tppz)$_2$](dca)$_2$, where two N–donor molecules coordinate to the cobalt(II) ion in a *mer* fashion, equatorially through pyridyl and axially via pyrazine nitrogen atoms, while dicyanamide plays a role of a counteranion [29].

Among the mononuclear Co(II) complexes, an intriguing example concerns two compounds of the same formula, [Co(phen)(dca)$_2$], which were obtained by two different synthetic methods [21] (see Figure 1c). The first noticeable difference between the structures is the orientation of the *dca* arms, which also results in differences in structural parameters [$N_{phen}$–Co–$N_{phen}$ = 92.15°/96.02°; Co–N–C = 173.83° and 169.36°/171.31° and 168.33°]. This small change involves another change in the arrangement of *phen* ligands, which are best described by dihedral angles 72.71(2)° and 81.56(7)°. As a result, differences in Co···Co separations and the supramolecular arrangement have been observed. Moreover, these two compounds exhibit different packing structures. In the first one (1a), the weak C–H···N interactions link molecules in layers, which are further bound into a 3D supramolecular network via face-to-face $\pi\cdots\pi$ interactions. In the second complex, due to weaker $\pi\cdots\pi$ stacking, chains instead of layers are created. Due to the different orientation of the *dca* and phen ligands, C–H···N interactions link the [Co(phen)(dca)$_2$] molecules into a 3D net, not layers. These structural differences will prove to be important in describing the axial *zfs* parameters (see *Magnetic properties in Terms of Structural Relationship*).

The coordination geometry of the cobalt(II) atom in mononuclear dicyanamide complexes has also been analyzed using the Continuous Shape Measures methodology [32]. The lowest Shape value with respect to the octahedral geometry is observed in Co(II) systems based on monodentate ligands (0.077–0.122), while the most distorted octahedron has been confirmed in the case of Co(II) with tridentate and tetradentate derivatives (1.316–2.966) (see Figure 1a and Table S2). These findings highlight the impact of ligand denticity on the distortion of the coordination environment in Co(II) complexes.

As mentioned above, dimeric structures of dicyananamide cobalt(II) complexes are extremely rare, with only two examples reported: $[Co(bpm)_2(dca)]_2(ClO_4)_2$ [33] and $[Co(tptz)(H_2O)(dca)]_2(ClO_4)_2$ [34]. In both complexes, metal atoms are bridged by two dca anions in a $\mu_{1,5}$ *(end–to–end)* fashion. Each cobalt(II) atom is octahedrally coordinated by two nitrogen atoms of dca ligands and four donor atoms from one molecule of tridentate *tptz* and one molecule of water in $[Co(tptz)(H_2O)(dca)]_2(ClO_4)_2$ or two molecules of bidentate *bpm* derivatives in $[Co(bpm)_2(dca)]_2(ClO_4)_2$. Bond lengths of Co–$N_{Ligand}$/$O_{H_2O}$ are longer than those of Co–$N_{dca}$ by ~0.06 Å (see Table S3). The Co···Co distance spanned by the bridging ligands varies in the range of 7.377(3)–7.592(2) Å, while the shortest intermolecular metal–metal distance is within 9.266(3)–10.577(2) Å.

## 4. Structural Diversity in the Family of Coordination Polymers

*Summarized structural data can be found in Tables S3–S5 in Supplementary Materials*

The dca-Co(II) coordination polymers are much more diverse than the previous one. Only in the group of one-dimensional cobalt(II) coordination polymers can we distinguish four subgroups among which the dicyanamide ion plays different roles.

The first two groups include 13 and 22 Co(II) structures with single and double $\mu_{1,5}$-dca bridges, respectively. These chains generally adopt either a linear or *zig-zag* geometry. Searching structural relationships between both groups, it can be concluded that the tri– or tetradentate ligands are favored during the formation of Co(II) chains with a single $\mu_{1,5}$-$dca^-$. On the other hand, in the case of double-bridged 1D dca-Co(II) systems, definitely monodentate ligands predominate, leading to the formation of linear chains. Among the first group, the most common motif is the linear chain in complexes $[Co(etpybzam)(dca)_2]_n$ [35], $\{[Co(imph)_4(dca)](ClO_4)\cdot 4(EtOH)(H_2O)\}_n]$ [36], $[Co(L1)(dca)(ClO_4)_2(MeOH)]_n$ [37], $\{[Co(enbzpy)(dca)](PF_6)\}_n$ [38], $[Co(3\text{-}pyo)(3\text{-}pyOH)(dca)(MeOH)]_n$ [39], $\{[Co(enbzpy(dca)](ClO_4)\}_n$ [40], and $\{[Co(pypz)(dca)(H_2O)]\cdot dca\}_n$ [41] followed by the five examples of *zig-zag* chains in $[Co(azpyph)(dca)]BF_4\ MeOH]_n$ [42], $\{[Co(bpm)_2(dca)](ClO_4)\}_n$ [30], $\{[Co(bdpb)(dca)](ClO_4)\}_n$ [43], $[Co(pybiu)(dca)]_n$ [44] and two exceptions: the helical arrangements in $\{[Co(H_2BiIm)_2(dca)_2]Cl\}_n$ [27] and $\{[Co(H_3daps)(dca)](MeOH)_2(MeCN)\}_n$ [45]. The type of ligand affects the binding mode of the dca linker, which in turn influences the intra- and interchain Co···Co distances—structural parameters that are important for further studies of magnetic properties (Tables 2 and S3). Multidentate ligands used for a single $\mu_{1,5}$-Co(II) CP provide longer distances by about ~0.6Å between metal atoms in comparison to compounds, where cobalt(II) atoms are connected through a double *dca* linker.

**Table 2.** Structural parameters for the selected cobalt(II) 1D coordination polymers (for the full version, see Table S3).

| Compound | Co–$N_{ligand}$ [Å] | Co–$N_{DCA}$ [Å] | Angles [°] | | Co···Co in Chain or (Between Chains) | Ref. |
|---|---|---|---|---|---|---|
| | | | Co–$NC_{(dca)}$ | C–N–$C_{(dca)}$ | | |
| Single $\mu_{1,5}$-dca bridge | | | | | | |
| $\{[Co(H_2BiIm)_2(dca)_2]Cl\}_n$ | 2.1386(18)<br>2.1593(18)<br>2.1372(18)<br>2.1325(18) | 2.1900(19) 2.1133(19) | 124.4(2)<br>172.3(2) | 120.4(2) | 7.2860(5)<br>7.9690(6) | [27] |

**Table 2.** *Cont.*

| Compound | Co–N$_{ligand}$ [Å] | Co–N$_{DCA}$ [Å] | Angles [°] | | Co···Co in Chain or (Between Chains) | Ref. |
|---|---|---|---|---|---|---|
| | | | Co–NC$_{(dca)}$ | C–N–C$_{(dca)}$ | | |
| {[Co(pypz)(dca)(H$_2$O)]·dca}$_n$ | 2.1082(18)<br>2.1891(19)<br>2.1685(18) | 2.057(2)<br>2.096(2) | 155.6(2) 165.03(19) | 121.7(2) | 7.447(6)<br>8.4287(7) | [41] |
| [Co(pybiu)(dca)]$_n$ | 1.986(3)<br>1.952(4)<br>1.966(3) | 2.330(4)<br>1.956(5) | 121.3(4)<br>158.1(4) | 120.6(5) | | [44] |
| **Double μ$_{1,5}$-dca bridge** | | | | | | |
| {[Co(azpyph)(dca)](BF$_4$)·MeOH}$_n$ | 2.309(2)<br>2.041(2)<br>2.346(3)<br>2.045(2) | 2.070(3)<br>2.100(3) | 156.3(2) 177.2(2) | 125.3(3) | 8.3790(9)<br>9.276(1) | [42] |
| {[Co(dca)$_2$(H$_2$O)$_2$]·(hmt)}$_n$ | 2.082<br>2.082 | 2.123<br>2.118<br>2.123<br>2.118 | 162.0<br>162.6 | 118.2(2)<br>118.2(2) | 7.3617<br>7.1243 | [24] |
| **Double bridges (ligand + dca)** | | | | | | |
| [Co$_2$(tppz)(dca)$_4$]$_n$ | 2.156(2)<br>2.156(2)<br>2.114(2)<br>2.114(2)<br>2.140(2)<br>2.140(2) | 2.044(2)<br>2.044(2)<br>2.141(2)<br>2.141(2) | 168.9(2) 152.3(2) | 119.1(2) | 7.377 A<br>5.9262(5) | [46] |
| **dca as a co-ligand** | | | | | | |
| [Co(bimb)$_2$(dca)$_2$]$_n$ | 2.168<br>2.154<br>2.168<br>2.154 | 2.139<br>2.139 | 169.9<br>169.9 | 124.1(2) 124.1(2) | 10.146<br>7.583 | [23] |
| [Co(bte)$_2$(dca)$_2$]$_n$ | 2.165<br>2.165<br>2.124<br>2.124 | 2.132<br>2.132 | 157.0 | 119.4(3) | 8.345<br>9.597 | [47] |

**H$_2$BiIm** = 2,2′-biimidazole; **azpyph** = N,N′-di-tert-butyl-2,11 diaza[3,3](2,6)pyridinophane; **bimb** = 1,4-bis(im idazol-1-yl)-butane; **hmt** = hexamethylenetetramine; **pybiu** = N-(picolinoyl)-biurate; **tppz**= tetra-2-pyridylpyrazine (2,3,5,6-tetrakis(2-pyridyl)pyrazine); **bte** = 1,2-bis(1,2,4-triazol-1-yl)ethane; **pypz** = 2,6-bis(pyrazol-1-yl)pyridine.

The second-to-last group of 1D structures includes five Co(II) compounds, in which two types of linkers have been identified: the dicyanamide anion and an organic ligand with appropriately arranged donor atoms. Finally, the last group consists of three compounds where the dca ligand does not serve a linking function. Both groups are relatively small, containing only four examples each. Nevertheless, an intriguing question arises: what factor determines whether the dca$^-$ is eliminated from, or retained in, its role as a bridge? This, of course, can be partially addressed during the design and synthesis stage through the use of the following:

(1) *Ligands with appropriately arranged donor atoms*, which usually increase the chance of a double bridge. This has been observed for multidentate tetra-2-pyridylpyrazine (*tppz*) in [Co$_2$(tppz)(dca)$_4$]$_n$ [46] or anionic heterodonor nicotinate anion (*nic*) in {[Co$_3$(nic)$_4$(dca)$_2$(H$_2$O)$_8$]·(H$_2$O)$_2$}$_n$ [48].
(2) *Flexible, easily bent organic ligands*—if the ligand contains an additional spacer between the aromatic rings containing donor atoms, then there is a high probability that the *dca* anion will not participate in the formation of bridges. Such a situation was observed in [Co(bimb)$_2$(dca)$_2$]$_n$ [23] and [Co(4-bpmp)$_2$(dca)$_2$(H$_2$O)$_2$]$_n$ [49], where butane in *bimb* and piperazine in *4-bpmp* had separation functions, respectively. Also noteworthy is the compound [Co(bte)$_2$(dca)$_2$]$_n$ [47], where the CH$_2$=CH$_2$ group acts as an effective spacer, reducing *dca* to the role of a co-ligand. Interestingly, replacing this group with a disulfide in [Co(bpds)(dca)$_2$]$_n$ [49] causes the *dca* ion to return to its role as a linker. The reason seems to lie in the bond lengths: the distance between carbon atoms in CH$_2$=CH$_2$ in [Co(bte)$_2$(dca)$_2$]$_n$ is 1.520(1) Å, which makes it more rigid in comparison to the disulfide group [1.782(5) Å] in [Co(bpds)(dca)$_2$]$_n$.

(3) Of course, utilizing the geometrical properties of organic ligands is an important and powerful tool in the design of coordination compounds. Therefore, the donor-acceptor strength of organic derivatives is a crucial factor in coordination chemistry and should be taken into consideration.

A very interesting case is that of the pair of complexes $[Co(pydz)_2(dca)_2]_n$ and $[Co(pydz)(dca)_2]_n$ [50]. The first one is a typical 1D coordination polymer, where each of the cobalt(II) atoms are bound via double *end–to–end* ($\mu_{1,5}$) dicyanamide. What is interesting is that the second compound has been prepared through the thermal decomposition of ligand-rich precursor $[Co(pydz)_2(dca)_2]_n$. The $[Co(pydz)(dca)_2]_n$ is derived from a 1D compound; however, dicyanamide's coordination of $\mu$-$_{1,3,5}$ and $\mu$-$_{1,5}$ manners provides it with a layer structure, which definitively ends the topic of 1D polymers and begins the discussion of 2D polymers. In this unique structure, the slightly distorted octahedral cobalt(II) cations are coordinated by five *dca* ions and one pyridazine (*pydz*) ligand. In structure two, a type of dicyanamide bridges exists to form linear Co–$(dca)_2$–Co chains. The central nitrogen atom of $\mu$-$_{1,3,5}$–dca binds chains into layers, in which each cation is coordinated by two $\mu$-$_{1,5}$ and three $\mu$-$_{1,3,5}$ dca anions (Figure 2a).

The same unique type of dicyanamide coordination mode has been confirmed in $[Co(H_2O)(dca)_2{\cdot}phz]_n$ [51]. There are two crystallographically independent octahedral metal(II) atoms and two types of $dca^-$ ligands (one bidentate $\mu_{1,5}$ and one tridentate $\mu_{1,3,5}$). The resulting sheets can be described as being composed of two alternating chains, which are perpendicular to each other, but propagate in the same direction. *Phz* molecules are stacked between these layers, but do not coordinate directly to the metal atoms. Apart from the above-described Co(II) compounds, the family of two-dimensional coordination polymers is rich in structural gems. As before, here, five subgroups can be distinguished. The first one concerns unique coordination modes of the *dca*$^-$ anion and three compounds of the formulas $\{[Co(dca)_2]_2{\cdot}(pzdo)\}_n$ [52], $\{[Co(dca)_2]_2{\cdot}(mpdo)\}_n$ [52], and $[Co_2(dmdpy)(dca)_2]_n$ [53] should be mentioned. In the first two structures, each of the crystallographically independent metal(II) ions (Co1 and Co2) are connected by double $\mu_{1,5}$ and $\mu_3$–dca bridges, leading to a triangular 2D lattice (Figure 2b). On the other hand, in the $[Co_2(dmdpy)(dca)_2]_n$ [53], two types of bridges, $\mu_{1,3}$ and $\mu_{1,5}$ coordination modes, bind metal atoms, forming 2D sheets (Table 3, Figure 2c).

As can be expected, the most numerous are two subgroups of 2D Co(II) compounds, where the central atoms are connected by a single $\mu_{1,5}$-dca bridge (16 structures) and by a pair of bridges: double + single $\mu_{1,5}$-dca (9 structures). Comparing both groups, two interesting differences can be noticed:

- *Role of the accompanying ligand*

In general, the use of a bidentate ligand leads to the formation of compounds with two types of single + double $\mu_{1,5}$-dca bridges. On the other hand, 2D Co(II) compounds based on a single $\mu_{1,5}$ linker are formed in the presence of a monodentate ligand.

- *Motifs*

The most diverse in terms of motifs are cobalt(II) compounds based on two types of bridges. The topology of the herringbone-waved grid has been confirmed for $[Co(dca)_2(phen)]_n$ and $[Co(4,7\text{-}dmphen)(dca)_2]_n$ [54], a brick wall was observed in $[Co(dmpzm)(dca)_2]_n$ [55]. Structures of $[Co(ptzda)_2(dca)]_n$ [56] and $\{(Ph_4As)[Co(dca)_3]\}_n$ [57] consist of layers of ladder-like chains joined into a 2D net by an *end-to-end* dca anion. However, in most complexes of the second group, single $\mu_{1,5}$ dicyanamide connects Co(II) atoms into a rhombus grid. The only structurally different exceptions are the compounds $[Co(bzim)_2(dca)_2]_n$ [58] and $[Co(DAT)_2(dca)_2]_n$ [59]. In both dca–Co(II) systems, metal cations are link into a 2D network, forming an hourglass-shaped 24-membered $Co_4(dca)_4$ metallacycle (Figure 3a–e).

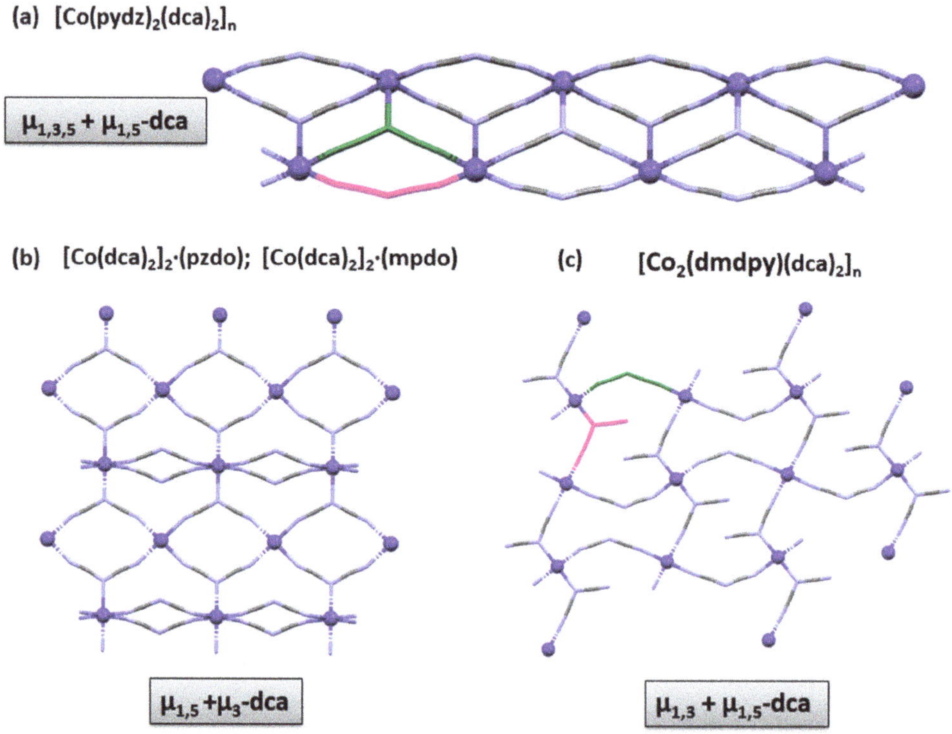


**Figure 2.** Summary of data for 2D Co(II) CP with two types of *dca* bridges [10,52,60]. The molecules of co-ligands have been omitted for clarity.

Despite these differences, the geometry of the dicyanoamide ions does not differ in any significant way. This, in turn, does not affect the metal···metal distances, which for both groups oscillate in the ranges of 7.274(4)–8.968(2) and 7.201(1)–8.748(1), respectively.

The second to last family gather five Co(II) complexes with mixed linkers—a $dca^-$ anion together with a molecule of an organic ligand both play roles as bridges. The common feature for all these compounds with formulas $\{[Co(btrm)_2(dca)]ClO_4\}_n$ [61], $\{[Co(pypypz)_2(dca)]\cdot(OH)(gly)_2\}_n$ [62], $[Co(DPNDI)(dca)]_n\cdot 2nDMF\cdot nH_2O$ [63], $[Co(bnzd)(dca)_2]_n$ [30], $[Co(3\text{-}bpo)(dca)_2]_n$ [30] and $\{[Co(bipy)(dca)_2\cdot 0.5H_2O\cdot 0.5MeOH\}_n$ [64] is the presence of an organic ligand with appropriately arranged donor centers, whose structure suggests it was designed as an organic linker. This is best evidenced by the Co···Co distances measured through an organic bridge, which range from 8.910(1) in $\{[Co(btrm)_2(dca)]ClO_4\}_n$ to 14.527(1) Å in $\{[Co(pypypz)_2(dca)]\cdot(OH)(gly)_2\}_n$. Interestingly, changing the synthesis of complex $\{[Co(pypypz)_2(dca)]\cdot(OH)(gly)_2\}_n$ [61] through changes in the selection of the solvent led to surprising changes. In two obtained complexes, $[Co(pypypz)_2(dca)]_n$ and $\{[Co(pypypz)_2(dca)_2]\cdot(H_2O)\}_n$ [61], the N–donor ligand acts as a linker eliminating the $dca^-$ of the bridging role. In those structures, the distance between the nearest cobalt(II) atoms is even greater ([16.465(1) Å in $[Co(pypypz)_2(dca)]_n$; 17.535(1) Å in $\{[Co(pypypz)_2(dca)_2](H_2O)\}_n$). In comparing these three compounds, the reason seems to be quite simple: in $\{[Co(pypypz)_2(dca)]\cdot(OH)(gly)_2\}_n$, rectangular interlayer channels are occupied by the intercalated glycol guest molecules and each glycol forms a pair of strong O–H···N bonds with two pyrazole nitrogen atoms from one adjacent layer. (Tables 3 and S5)

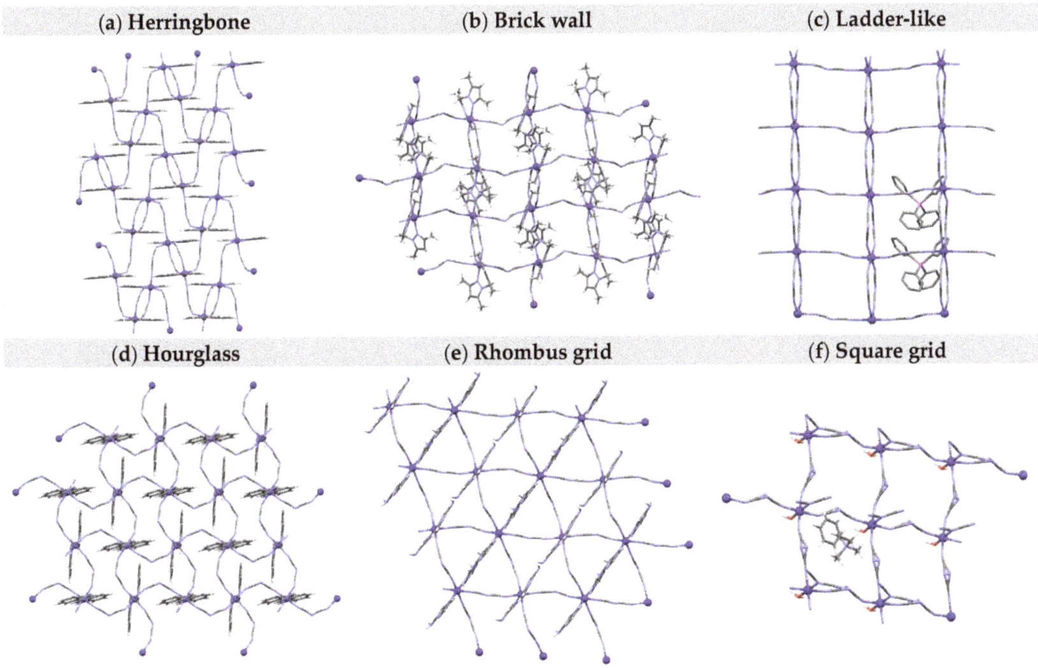


**Figure 3.** Representative motifs in 2D Co(II) CP based on dca$^-$ linker: (**a**) $[Co(phen)(dca)_2]_n$ [54]; (**b**) $[Co(dmpzm)(dca)_2]_n$ [56]; (**c**) $\{Ph_4As)[Co(dca)_3]\}_n$ [57]; (**d**) $[Co(bzim)_2(dca)_2]_n$ [58]; (**e**) $\{[Co(dca)_2(NH_2pyz)_2]{\cdot}H_2O]\}_n$ [65]; (**f**) $BeTriMe[Co(dca)_3(H_2O)]_n$ [66].

The final compound, $\{BeTriMe[Co(dca)_3(H_2O)]\}_n$, is difficult to assign to a distinct group [66]. In this square-grid network, the octahedral metal cations are bound by single $\mu_{1,5}$-dca units, while the roles of the accompanying ligands are fulfilled by water molecules and non-bridging dca anions, coordinating in a $\mu_3$ manner. The Co–N$_{\mu1,5dca}$ bond lengths fall in the range of 2.082(3)–2.118(3) Å, while the bond distance Co–N$_{\mu3dca}$ is slightly longer [2.225(4) Å]. The non-bridging *dca* ligands together with the hydrogen-bonded coordinated water form a stable honeycomb-like assembly, and the presence of O–H···N interactions causes this elongation of this bond (Figure 3f).

**Table 3.** Structural parameters for the selected cobalt(II) 2D coordination polymers (for the full version, see Table S4).

| Compound | Co–N$_{ligand}$ [Å] | Co–N$_{DCA}$ [Å] | | Angles [°] | | Co···Co | Ref. |
|---|---|---|---|---|---|---|---|
| | | | | Co–NC$_{(dca)}$ | C-N-C$_{(dca)}$ | | |
| | | $\mu_3$ + $\mu_{1,5}$ | | | | | |
| $[Co(pzdo)(dca)_2]_n$ | Co1 | Co1 | Co(2) | $\mu_{1,5}$ | 119.9 | Co1···Co1 | [52] |
| | 2.138 | $\mu_{1,5}$ | $\mu_{1,5}$ | 149.5 | 120.8 | 7.4313 | |
| | 2.138 | 2.099 | 2.089 | 149.5 | | Co2···Co2 | |
| | | 2.099 | 2.089 | $\mu_3$ | | 7.4313 | |
| | | 2.099 | 2.089 | 160.5 | | Co1···Co2 | |
| | | 2.099 | 2.089 | 160.5 | | 6.0188 | |
| | | | $\mu_3$ | | | | |
| | | | 2.179 | | | | |
| | | | 2.179 | | | | |

**Table 3.** *Cont.*

| Compound | Co–N$_{ligand}$ [Å] | Co–N$_{DCA}$ [Å] | Angles [°] Co–NC$_{(dca)}$ | C–N–C$_{(dca)}$ | Co···Co | Ref. |
|---|---|---|---|---|---|---|
| | | $\mu_3$ + $\mu_{1,3,5}$ | | | | |
| [Co(H$_2$O)(dca)$_2$·phz]$_n$ | 2.049 | $\mu_{1,5}$<br>2.080<br>2.086<br>$\mu_{1,3,5}$<br>2.213<br>2.121<br>2.163 | $\mu_{1,5}$<br>159.4<br>160.4<br>$\mu_{1,3,5}$<br>164.4<br>163.0<br>120.9<br>122.5 | 119.5(2)<br>116.4(2) | 7.4110 | [51] |
| | | **Single $\mu_{1,5}$ + double $\mu_{1,5}$** | | | | |
| {[Ph$_4$P]{Co[dca]$_3$} | - | Single<br>2.155<br>2.155<br>Double<br>2.118<br>2.118<br>2.114<br>2.114 | Double<br>161.8<br>159.6<br>Single<br>161.8<br>149.5 | 118.8(2)<br>126.5(6) | Double<br>7.4426<br>Single<br>8.7948 | [57] |
| | | **Single $\mu_{1,5}$** | | | | |
| [Co(atz)$_2$(dca)$_2$]$_n$ | 2.168<br>2.168 | 2.097<br>2.097<br>2.089<br>2.089 | | | 8.0415 | [67] |
| [Co(bim)$_2$(dca)$_2$]$_n$ | 2.128(3)<br>2.134(3) | 2.190(3)<br>2.220(3)<br>2.209(3)<br>2.219(3) | 169.5(3)<br>158.3(3) | 125.1(4)<br>123.8(4) | 8.927(2)<br>8.968(2) | [28] |
| | | **Single $\mu_{1,5}$ + ligand** | | | | |
| {[Co(btrm)$_2$(dca)]ClO$_4$}$_n$ | Co1<br>2.140(6)<br>2.146(6)<br>2.142(6)<br>2.150(6)<br>Co2<br>2.136<br>2.157 | Co1<br>single<br>2.072(5)<br>2.082(5)<br>Co2<br>single<br>2.085<br>2.085 | 160.3(6)<br>164.4 | 122.6(9) | Co1···Co1<br>8.894(4)<br>Co2···Co2<br>8.526 | [68] |
| | | **Three types of bridges** | | | | |
| [Co(3-bpo)(dca)$_2$]$_n$ | 2.175(5)<br>2.160(5) | 2.091(5)<br>2.136(5)<br>2.103(4)<br>2.114(4) | 154.3(4)<br>164.1(4)<br>161.6(4)<br>156.4(4) | 121.8(5)<br>119.1(4) | 7.383(3)<br>7.587(2) | [30] |
| | | **dca as a co-ligand** | | | | |
| {[Co(pypypz)$_2$(dca)](OH)(gly)$_2$}$_n$ | 2.182<br>2.163 | 2.163<br>2.182<br>2.175<br>2.175 | 166.3<br>171.1<br>173.6<br>171.1 | 129.0(5)<br>129.0(5) | 14.537 | [61] |

**atz** = 2-amino-1,3,5-triazine; **bim** = 1-benzylimidazole; **btrm** = 1,2-bis(1,2,4-triazole-1-yl)methane; **3-bpo** = 2,5-bis(3-pyridyl)-1,3,4-oxadiazole; **pzdo** = pyrazinedioxide; **phz** = phenazine; **pypypz** = bis[3,5- dimethyl-4-(4′-pyridyl)pyrazol-1-yl]methane.

Paradoxically, the last, but historically the most important, is the class of 3D dca-Co(II) coordination polymers. The observation of long-range magnetic order in dicyaniamide transition metal complexes of the formula [M(dca)$_2$]$_n$ (M = Co, Cu, Ni, Fe, Mn etc.) caused great interest and excitement and overall it resulted in a large number of articles that appeared in short time intervals. This research ranges from studies focused on structural aspects related to the employment of the dicyanamide ion in the design and synthesis of d-block coordination compounds, up to issues related to the expansion of horizons related to new examples of molecular network materials. Thanks to these first works, it is possible to distinguish four main groups in the 3D polymer family. The first one includes the well-known rutile ($TiO_2$) networks [M(dca)$_2$]$_n$ with six connecting Co(II) centers and three connecting centers (*dca* ligands) in the ratio of 1:2 [69]. The cobalt(II) atoms are six-coordinated with four nitrile nitrogens from *dca*$^-$ in the equatorial plane and two amide nitrogens from two axial dicyanamide ligands. In this group, two types of bridges have been observed: $\mu_{1,3,5}$ [70–73] and $\mu_{1,5}$ [74–76] (see Table S6). In Co(II) structures based on $\mu_{1,3,5}$-dca, the bond lengths between the metal cations and nitrile nitrogens [2.085(1)–2.097(2) Å] are shorter than those of the Co–$N_{amide}$ [2.150(1)–2.153(1) Å]. This is a typical feature for these compounds because in all isostructural metal(II) structures, the M(II) cation–amide nitrogen bonds are longer than the M(II)–nitrile bonds [70]. The dicyanamide bridges show their tendency to bend in angles of Co–$N_{nitrile}$C and C–N–C which vary from 159.9–164.0(2)° to 117.1–118.6°, respectively. In those structures, where Co(II) centers are bound via $\mu_{1,5}$–dca, the Co–N–C angles tend to be linear [160.5(4)–179.6°] with the exceptions of {$(MePh_3P)[Co(dca)_3]$}$_n$ and {$(EtPh_3P)[Co(dca)_3]$}$_n$ [76]. In these two structures, the network consists of sheets, which are cross linked into a 3D net with hexagonal channels, and what is rare in this structure is that the Schläfi symbol of this net is $4^6.6^4$. The second subgroup is represented by $[Co(Phpyk)(dca)_2]_n$ [77], $[Co(tmeda)(dca)_2]_n$ [78] and $[Co(deen)(dca)_2]_n$ [79], which are compounds, where organic ligands coordinate with the cobalt(II) ion and each metal atom displays a distorted octahedral coordination environment, and is linked by single $\mu_{1,5}$-dca bridges (in an *end–to–end* fashion). In $[Co(Phpyk)(dca)_2]_n$, *dca* linkers gives rise to a 3D net of a $6^6$ topology, which in most cases refers to diamondoid network. No less interesting is $[Co(tmeda)(dca)_2]_n$, where 3D helical CP is built from helical chains, which are further linked by a single dicyanamide anion. In $[Co(deen)(dca)_2]_n$, the *dca* ligand behaves as an interconnecting grid to give rise to a scaffolding-like 3D structure with distorted rectangular cavities [size: 8.1 × 6.9 Å]. Following the trend of the previous two groups of coordination polymers, the co-ligands can also act as linkers and such a situation can be observed in {$[Co(bpe)_2(dca)](dca)_2{\cdot}4H_2O$}$_n$ [80], {$[Co(bte)(dca)_2]{\cdot}H_2O$}$_n$ [47], $[Co(bpeado)(dca)_2]_n$ [81], $[Co(bpy)(dca)_2]_n$ [82], $[Co(dbtp)(dca)_2]_n$ [83], $[Co(bipy)(dca)_2]_n$ [64], {[Co(bipim(dca)](dca)}$_n$ [84] and {$[Co(bipim)(dca)]_2{\cdot}(ClO_4)_2{\cdot}(CH_3OH)_2{\cdot}H_2O$}$_n$ [84]. In the first of the six Co(II) examples, the authors employed organic derivatives with well-arranged nitrogen as a donor atom that is already known from the discussion concerning 2D dca-Co(II) CP. This type of derivatives like, i.e., bpe (1,2-bis(4-pyridyl)ethane) or *bte* (1,2-bis(1,2,4-triazol-1-yl)ethane) were widely used in the 2000s, guarantees the separation of metal(II) atoms over large distances (>9 Å), which in turn ensures synthetic success in the synthesis of 2D or even 3D polymers. In analyzed compounds, the longest Co···Co distances [14.0137(1) Å] was observed in $[Co(bpeado)(dca)_2]_n$ [81]. What is interesting, the shortest metal···metal separation [6.809(1) Å] is in {$[Co(bte)(dca)_2]{\cdot}H_2O$}$_n$ [47], and hydrogen bonding interaction between the solvent water molecule and the uncoordinated nitrogen atom of the triazole ring of *bte* is responsible for this state of affairs. In two last complexes {[Co(bipym(dca)](dca)}$_n$ and {$[Co(bipym)(dca)]_2{\cdot}(ClO_4)_2{\cdot}(CH_3OH)_2{\cdot}H_2O$}$_n$ [84] the bis-bidentate 2,2′-bipyrimidine (*bipym*) with multidentate coordination sites plays a role of organic, bridging ligand (see Table S6). The last group to be discussed contain only one example of formula {$[Co_4(CH_3O)_4(CH_3OH)_4(dca)_4]$}$_n$ [85]. These compound is built up from cubane $Co_4O_4$ core, where Co(II) cations display octahedral coordination sphere. Finally, the cubane building blocks are connected to other four cores by eight $\mu_{1,5}$-dca bridging mode. An interesting fact is the synthesis of this complex, because it was obtain during solvatothermal synthesis between Nadca and Co(II) salt in methanol at 75 °C and this is only one examples of such synthesis in the group of 3D network. Usually described Co(II) compounds were prepared by direct reaction of metal(II) salt, *dca* and organic ligands in different solvent at room temperature.

## 5. Magnetic Properties in Terms of Structural Relationships

*Summarized magnetic data can be found in Tables S1, S3–S5 in Supplementary Materials*

As previously mentioned, the phenomenon of long-range magnetic ordering in the isomorphous series of α-$[M(dca)_2]_n$ has driven extensive research into this group of compounds. It has been demonstrated that the nature of the ordering is closely dependent on the specific d-block metal involved. The rutile-like 3D networks formed by six coordinated metal ions and three connecting connectors ($\mu_{1,3,5}$-dca) in the 1:2 metal to dca molar ratio exhibit a broad diversity of long-range magnetic ordering at very low temperatures. This magnetic behavior depends on the nature of the metal ion, such as the weakly ferromagnetic Co(II) (9 K) Cu(II) (<1.7 K) and Ni(II) (21 K)], whereas Cr (47 K), Mn (16 K) and Fe (19 K) complexes are canted antiferromagnets [26]. The modification of the M-dca architecture by employing an N-donor organic ligand into the structures resulted in the formation of numerous architectures of the general formula $[M(dca)_2(L)_n]_n$ (n = 1 or 2), including chains, 2D (4,4) nets, triangular lattices with $\mu_{1,3,5}$- and $\mu_{1,5}$- dca linkers, and even 3D networks. In contrast to the rutile-like α-$[M(dca)_2]_n$ compounds, many magneto-structurally studied heteroleptic compounds display very weak antiferromagnetic couplings ($J < 1\ cm^{-1}$) due to the poor ability of the extended $\mu_{1,5}$-dca bridges to mediate electronic interaction. The family of high-spin cobalt(II) complexes with $\mu_{1,5}$-dca bridges, although they are somewhat hard due to the presence of first-order spin–orbit coupling in the ground state, can be very interesting because they can exhibit a slow relaxation of magnetization, behaving as single-ion magnets (SIMs). The origin of this magnetic behavior arises from the large magnetic anisotropy of the Co(II) ion, combined with a flexible zero-field that depends on its coordination geometry and the degree of distortion in its environment, as well as the sufficient magnetic isolation of the metal centers (as mentioned above, due to their poor ability for transmitting magnetic interactions).

This chapter will present the results of magnetic studies in the context of the previously discussed structural relationships. All the studied dicyanamide Co(II) coordination compounds have been extensively investigated using a wide variety of techniques ranging from DC magnetic susceptibilities and magnetization, AC magnetic susceptibilities, Mössbauer spectroscopy and EPR spectroscopy. However, upon reviewing all of these materials, it is noteworthy that the greatest interest appears to be focused on Co(II) systems, often referred to as "single-ion magnet" (SIM) or spin crossover (SCO) Co(II) materials.

In the relatively small group of mononuclear dca-Co(II)complexes, two examples showing spin crossover behavior can be distinguished. In 2017, J. Palion-Gazda, M. Julve and their co-workers studied the thermally induced spin crossover (SCO) behavior of the $[Co(tppz)_2](dca)_2$ from a high-spin state (HS; S = 3/2) at a high temperature region to a low-spin state (LS; S =1/2) at lower temperatures, with the low-spin phase being reached at $T \leq 200$ K. Additionally, X-band EPR measurements were performed to characterize the low-spin state, revealing a quasi-isotropic spin-doublet rather than hyperfine splitting, which appeared as eight lines [29]. The SCO phenomenon was also observed in the *cis*-dicyanamido complex of formula $[Co(L1)(dca)_2]$, which displays a reversible thermoinduced HS to LS with a $T_{1/2}$ value around 238 K [20]. The measured $\chi T$ was 2.55 $cm^3\ mol^{-1}$ K at 400 K and it was a consistent value for one Co(II) ion in the HS state (S = 3/2, g = 2.0 and $\chi T$ = 1.875 $cm^3\ mol^{-1}$ K). Upon decreasing the temperature, the value of $\chi T$ decreased gradually to reach a plateau with the values of 0.52–0.49 $cm^3\ mol^{-1}$ K in the temperature range of 80–20 K, which in turn is expected for the LS Co(II) ion (S = 1/2, g = 2.0, $\chi T$ = 0.375 $cm^3\ mol^{-1}$ K) [86]. As an alternative to the EPR technique, an electrochemical method including cyclic voltammetry, square wave voltammetry and differential pulse voltammetry was used to study the nature of HS/LS. Voltammograms of $[Co(L1)(dca)_2]$ displayed a reversible oxidation for HS Co(II) to LS Co(III) at 0.05 V vs, $Fc^+/Fc$ ($E_{pc}$ = 0.13 V and $E_{pa}$ = −0.03 V) as well as a reduction process at −1.67 V vs. $Fcc^+/Fc$. These two compounds represent a rare example of the successful use of a pseudohalide in the design and synthesis of SCO cobalt(II) systems.

Considerations regarding the dynamic mechanisms for magnetic relaxation were carried out for two complexes, $[Co(bim)_4(dca)_2]$ [28] and $[Co(12\text{-}TMC)(dca)_2]$ [87]. In the first complex, frequency-dependent alternating current magnetic susceptibility measurements were carried out below 8 K down to 2 K and this showed no out-of-phase ($\chi''$) signal under the 0 G static field, suggesting a fast tunneling of the magnetization. However, the frequency depended on the in-phase ($\chi'$) and out-of-phase ($\chi''$) under the external dc fields (of 500, 1000 and 2500 G). The value of the relaxation time followed the Arrhenius law characteristic of a thermally activated mechanism ($\tau = \tau_0 \exp(U_{eff}/kT)$). At 1000 G, the relaxation time $\tau_0$ and activation energy $U_{eff}$ ($E_a$) are equal to $0.87 \cdot 10^{-6}$ s and 7.74 $cm^{-1}$, respectively. What is interesting, for this complex, is that an almost perfect semicircle in the Cole–Cole plot was observed in temperature ranges of 2.5–3.5 K, which could be fitted using the Debye model. On the other hand, Y-Q. Zhang, Z. Wang et al. showed that in [Co(12-TMC)(dca)$_2$], QTM and direct processes do not dominate in the whole studied temperature range [87]. The Raman process dominates in the low-temperature region, while the Orbach mechanism dominates in the high-temperature range. The fitting of the data gave the following parameters of $\tau_0$ = 1.14 × $10^{-8}$ s and $U_{eff}$ = 27.31 cm$^{-1}$. Furthermore, the axial (D) and/or rhombic (E) zero-field splitting parameter were calculated and analyzed [Table S1]. An interesting situation has been found in two isomers of formula [Co(phen)$_2$(dca)$_2$] [21]. The temperature dependences of magnetic susceptibility for both compounds are quite similar, but the magnetization data show saturation at a higher value in the second isomer. Furthermore, the fitting procedure applied gave D = 99.1 cm$^{-1}$ for the first isomer, and D = 84.8 cm$^{-1}$ for the second one. The differences in the packing of these two compounds, discussed in detail in the structural part of this paper, are the reason for such results. The $\mu_3$ coordination mode found in [Co(H$_2$biim)$_2$(dca)$_2$] does not seem to significantly affect the result of parameter D which is 23.3 cm$^{-1}$ [27]. Similar values have been found in other octahedral pseudohalide Co(II) complexes. In the family of binuclear Co(II) compounds, the magnetic properties of [Co(tptz)(H$_2$O)(dca)]$_2$(ClO$_4$)$_2$ have been studied in detail. Variable temperature (300–2 K) magnetic measurements of this complex reveal that dicyanamide ligands transmit a weak antiferromagnetic interaction between the Co(II) centers. Typically for octahedral Co(II), the $\chi$T value continuously decreased from room temperature to 20 K [$\chi_M$T = 6.019 cm$^3$ mol$^{-1}$ K at 300 K and $\chi_M$T = 4.41 cm$^3$ mol$^{-1}$ K at 20 K]. Below 20 K, the presence of a small number of ferromagnetic or canted antiferromagnetic ordered impurities, i.e., $\alpha$-or $\beta$-[Co(dca)$_2$], has been detected [34].

In the family of 1D Co(II) coordination polymers, two examples of Co(II) SCO systems have been carefully studied. In {[Co(enbzpy)(dca)](PF$_6$)}$_n$ [38] and {[Co(enbzpy)(dca)](ClO$_4$)}$_n$ [40], the $\chi$T value [2.67 and 2.43 cm$^3$ mol$^{-1}$ K] decreases gradually until it reaches a temperature at which it becomes almost constant, which has been attributed to a low-spin Co(II) atom. Surprisingly, for {[Co(enbzpy)(dca)](ClO$_4$)}$_n$, the HS$\leftrightarrow$LS transition occurs with a small hysteresis loop of 3 K in width. The presence of a hysteresis loop was confirmed by calorimetric measurements and repeated over four times in the 227–160 K temperature range to demonstrate the robust SCO behavior in this Co(II) coordination polymer. Moreover, in both cases, the significant role of innocent counteranions in the SCO behavior of the polycationic coordination polymers has been carefully examined. Interestingly, a typical feature in almost all octahedral cobalt(II) complexes is that the value of $\chi_M$T or $\mu_{eff}$ at room temperature is usually greater than that expected for one isolated spin-only ion [$\chi_M$T = 1.87 cm$^3$ mol$^{-1}$ K for a S = 3/2 ion], which can be attributed to the significant orbital contribution. Another common observation is the differences in the $\chi_M$T value at lower temperatures (T < 15 K) when small magnetic fields are applied. This feature indicates the presence of a small amount of ferromagnetic or canted antiferromagnetic ordered impurities, likely $\alpha$- or $\beta$-[Co(dca)$_2$]$_n$, which have $T_c$ at 9–10 K [69,86,88]. Unfortunately, dca-Co(II) derivatives tend to decompose after some time and the presence of these magnetic ordered impurities is very frequent. At room temperature, typical values of $\chi_M$T vary in the range of 2.25–3.36 cm$^3$ mol$^{-1}$ K. With decreasing temperature, the $\chi_M$T usually remains almost constant until ~50 K, after which it rapidly decreases until reaching 2 K [see Table S4]. The magnetic behavior of these Co(II) complexes corresponds to a weak antiferromagnetically coupled system. The obtained data were analyzed and fitted in a variety of ways in order to calculate Curie (C) and Weiss ($\theta$) constants and coupling constant J values. To evaluate these magnetic parameters, different approaches have been used. Generally, some parts of the compounds were fitted using a classical spin Heisenberg model for a chain or by the fitting of the low-temperature region to an Ising chain model [89]. Also, an analytical expression for the infinite chain of classical spins was derived by Fisher through the Hamiltonian method [90]. However, reasonable results were obtained using the phenomenological approach for some low-dimensional Co(II) systems that allows us to estimate the strength of the antiferromagnetic exchange interactions [91]. Rueff and coworkers postulated the equation $\chi_M T = A \exp(E_1/kT) + B \exp(E_2/kT)$, where A + B represents the Curie constant, while $E_1$ and $E_2$

equal the activation energies corresponding to the spin–orbit coupling, and the antiferromagnetic exchange interaction, respectively. The C, θ and J values vary in the ranges of 1.394–3.64 $cm^3$ $mol^{-1}$ K, −42 to +0.1 K and −24.8 to −1.1 $cm^{-1}$, respectively (Table S4), in five complexes, namely $\{[Co(H_2BiIm)_2(dca)_2]Cl\}_n$ [27], $\{[Co(azpyph)(dca)](BF_4)\cdot MeOH\}_n$ [42], $[Co(5,6\text{-}(Me)_2\text{-}bzim)_2(dca)_2]_n$ [92], $[Co(5\text{-}Mebzim)_2(dca)_2]_n$ [92] and $\{[Co(H_3daps)(dca)](MeOH)_2(MeCN)\}_n$ [45]. Also, an analysis of the axial zero-field splitting parameter (D) has been made. It is well known that the D parameter is sensitive to many factors, which makes it difficult to find specific relationships. However, for octahedral cobalt(II) complexes, a positive D parameter is rather expected, but a negative D value has also recently been observed [93,94]. In $\{[Co(azpyph)(dca)](BF_4)\cdot MeOH\}_n$ [42], magnetic data were treated using the spin Hamiltonian $\hat{H} = D(\hat{S}x^2 - \hat{S}^2/3) + E(\hat{S}x^2 - \hat{S}y^2) + \mu_B B_g \hat{S}a - zj< \hat{S}a > \hat{S}a$. An analysis of the experimental data provided the D, E/D and molecular field correction parameter zj equal to −72.3 $cm^{-1}$, 0.0 and −0.273 $cm^{-1}$, respectively. On the other hand, in $\{[Co(H_2BiIm)_2(dca)_2]Cl\}_n$, the axial *zfs* parameter was +40.3 $cm^{-1}$. The fitting DC magnetic data for $[Co(5,6\text{-}(Me)_2\text{-}bzim)_2(dca)_2]_n$ and $[Co(5\text{-}Mebzim)_2(dca)_2]_n$ [92] also gave a positive value of D, which was 63(19) and 60(8) $cm^{-1}$. Interestingly, FIRMS measurements were performed for these two samples. For the first compound, there was a zero-field transition at 134.0 $cm^{-1}$, which represents the splitting between the $m_s = \pm 1/2$ and $m_s = \pm 3/2$ Kramers doublets. The D parameter (62.5 $cm^{-1}$) calculated from equation $\Delta = 2\sqrt{D^2 + 3E^2}$ has good agreement with magnetic data fitting. The same procedures were performed for $[Co(5\text{-}Mebzim)_2(dca)_2]_n$, and the splitting between the Kramers doublets seen in the FIRMS spectra was 124 $cm^{-1}$. Further analysis demonstrates that the FIRMS spectra show good agreement with the experimental data using the same parameters, giving D = 60 $cm^{-1}$ and E = 9 $cm^{-1}$. The same Hamiltonian equation was used to determine the *zfs* parameters in $\{Co(H_3daps)(dca)](MeOH)_2(MeCN)\}_n$ [45]. The best fit presented a positive value of D = 41.3(5) $cm^{-1}$, E = 0.81(3) $cm^{-1}$ and a very weak antiferromagnetic exchange interaction, J = 0.004 $cm^{-1}$. Furthermore, to analyze the electronic structure and magnetic anisotropy of the Co(II) complex, quantum chemical calculations with ORCA4.0 and MOLCAS8.2 were performed. However, the last program showed anisotropic parameters that were a little bit higher in comparison with ORCA 4.0. Finally, the *ac* measurements were performed for $\{[Co(H_3daps)(dca)](MeOH)_2(MeCN)\}_n$ [45], $\{[Co(pypz)(dca)(H_2O)]\cdot dca\}_n$ [41], $[Co(5,6\text{-}(Me)_2\text{-}bzim)_2(dca)_2]_n$ [92] and $[Co(5\text{-}Mebzim)_2(dca)_2]_n$ [92]. Unfortunately, no out-of-phase signal on AC susceptibility was observed in the absence of an external magnetic field; therefore, those complexes are not natural single-ion magnets. The reason for this lies in the strong QTM (quantum tunneling of magnetization), which leads to fast relaxation as commonly observed in transition metal-based compounds [95]. With the increasing external field, the $\chi_M''$ component rises and passes through a maximum, which confirms that dca–Co(II) 1D complexes exhibit a field-supported slow magnetic relaxation. The nature of the semicircle of the Cole–Cole plot showed the single relaxation process in $\{[Co(H_3daps)(dca)](MeOH)_2(MeCN)\}_n$ [45]. The generalized Debye model was used to extract the magnetization relaxation time (τ) and, later, the fitting of the Arrhenius using well-known equation $\ln(\tau) = \ln(\tau_0) + U_{eff}/kT$, where the energy barrier is 9.9 K (6.88 $cm^{-1}$), while $\tau_0 = 5.3 \times 10^{-6}$ s. In this compound, S. Konar and co-workers demonstrate that the relaxation process does not follow the Orbach process. The magnetization relaxation process is mainly controlled by QTM together with some shortcut path through virtual states (the Raman process). A similar situation was observed in $\{[Co(pypz)(dca)(H_2O)]\cdot dca\}_n$ [41], where both QTM and Orbach processes are involved in relaxation processes. The energy barrier and relation time are equal to 103 K and $\tau_0 = 1.2 \times 10^{-11}$ s. However, the authors note that the spin reversal $U_{eff}$ might be underestimated because the last three data do not fit perfectly on the straight line. For this reason, further analysis is carried out and the involvement of the Raman process or even the phonon bottleneck process is considered. Interestingly, for $[Co(5,6\text{-}(Me)_2\text{-}bzim)_2(dca)_2]_n$ [92] and $[Co(5\text{-}Mebzim)_2(dca)_2]_n$ [92], no indication of the quantum tunneling of magnetization was found based on the Arrhenius-like plot. It has been proven that a single Raman-like term is insufficient to reproduce the whole data set. However, when two of them were considered, combined with phonon bottleneck relaxation mechanisms, the relaxation parameters could be accurately determined.

Interesting structural-magnetic relationships can be observed in three 2D coordination polymers with the formulas $[Co(pzdo)(dca)_2]_n$ [52], $[Co(mpdo)(dca)_2]_n$ [52] and $[Co(2,5\text{-}dmpdo)(dca)_2]_n$ [96]. The ferromagnetic ordering confirmed in the first two complexes results from the existence of $\mu_3$- and $\mu_{1,5}$-dca coordination modes. On the other hand, bridging ligand 2,5-dmpdo in the third compound was found to mediate antiferromagnetic coupling [60,97,98]. In the first two complexes, the magnetic susceptibility above 40–50 K obeys the Curie–Weiss law, with the Weiss constant C equal to −9.4 K and −13.7 K and the Curie constant C equal to 3.17 and 3.34 $cm^3\ mol^{-1}$ K, respectively. The divergence of $\lambda_M T$ below 20 K is attributed to the ferromagnetic couplings between the effective spins of Co(II). For both complexes, FCM and ZFCM measurements in a low field show the occurrence of two long-range ferromagnetic transitions. The temperature dependence of ac magnetic susceptibility confirms phase transitions below 9 and 3 K in $[Co(pzdo)(dca)_2]_n$ and 1.9 K and 8.6 K in $[Co(mpdo)(dca)_2]_n$. Additionally, the field dependence of the magnetization measurements shows the rapid saturation of magnetization, which is a typical feature of long-range ferromagnetic ordering. Furthermore, hysteresis measured at 1.74 K in $[Co(pzdo)(dca)_2]_n$ and 1.78 K in $[Co(mpdo)(dca)_2]_n$ reveals that both complexes are soft magnets. The magnetic behavior of $[Co(2,5\text{-}dmpdo)(dca)_2]_n$ [60] is quite different from the two previous one compounds. The Curie and Weiss constants are 6.61 $cm^3\ mol^{-1}$ K and −33.5 K, respectively. The antiferromagnetic ordering was confirmed using the field dependence of the magnetization, which was measured below the Neel temperature ($T_N$ = 10.8 K). $[Co(H_2O)(dca)_2{\cdot}phz]_n$ has attracted considerable attention due to the presence of $\mu_3$− and $\mu_{1,3,5}$-dca coordination modes in its structure, along with the topological relationship between 2D sheets and 3D rutile-like structures [51]. Unfortunately, no bifurcation in the FCM and ZFCM plots was observed due to the presence of weak intra-network ferromagnetic coupling. In other Co(II) complexes based on single, double or mixed (single and double) dicyanamide bridges, temperature-dependent susceptibilities obey the Curie–Weiss law. At room temperature, the values of $\lambda_M T$ are in the range of 2.30–4.21 $cm^3\ mol^{-1}$ K and when the temperature is lowered, $\lambda_M T$ slowly decreased and reaches 1.21–2.85 $cm^3\ mol^{-1}$ K at ~2 K. The observed decrease typically can be the result of antiferromagnetic coupling between Co(II) ions or the depopulation of the higher-energy Kramers doublets of the cobalt(II) centers [99,100] (see Table S5, ESI). The magnetic data were fitted to the Curie–Weiss law, leading to C and θ constants in the ranges of 2.29–4.47 $cm^3\ mol^{-1}$ and −4.62–20.09 K, respectively. Only in one example, $[Co(bzim)_2(dca)_2]_n$ [58], does the θ constant have a positive sign (17.14 K), and together with $zJ$ = 0.648 $cm^{-1}$ indicates weak ferromagnetic coupling between the Co(II) ions. Moreover, for eight complexes: $[Co(biq)(dca)_2]_n$ [101], $\{[Co(NH_2pyz)_2(dca)_2]{\cdot}H_2O\}_n$ [65], $[Co_3(HOpyz)_5(dca)_6(H_2O)_2]_n$ [65], $[Co(atz)_2(dca)_2]_n$ [67], $[Co(bim)_2(dca)_2]_n$ [28], $[Co(bmim)_2(dca)_2]_n$ [28], $[Co(dmdpy)(dca)_2]_n$ [53] and $[Co(2\text{-}Mebzim)(dca)_2]_n$ [92], estimations of the magnitude of anisotropy and exchange interactions were conducted. In all of them, the axial *zfs* parameter was positive and ranged from +66.4 to +95.4 $cm^{-1}$. Bearing in mind that the increase in EPR signal intensity with decreasing temperature supports and confirms a positive value of D, EPR spectra were recorded in low temperatures in the X-band spectrometers. In the case of $[Co(biq)(dca)_2]_n$ [101], only a very broad featureless line was observed, caused by intermolecular interactions. However, the typical EPR spectrum for Co(II) compounds with a positive D was observed in $\{[Co(NH_2pyz)_2(dca)_2]{\cdot}H_2O\}_n$ [65] and $[Co_3(HOpyz)_5(dca)_6(H_2O)_2]_n$ [65]. Interestingly, the FIRMS spectrum of $[Co(2\text{-}Mebzim)(dca)_2]_n$ [92] showed multiple zero-field transitions at around 138 $cm^{-1}$. To estimate the value of D, a simulation using a spin Hamiltonian was conducted, resulting in D = 67 $cm^{-1}$ and E = 10 $cm^{-1}$, which were in agreement with the experimental data.

In 2015, B. Machura, M. Julve and co-workers confirmed the single-ion magnet (SIM) behavior for a 2D coordination polymer of formula $[Co(atz)_2(dca)_2]_n$ with the cobalt(II) ions bridged by single $\mu_{1,5}$-dicyanamide bridges. One year later, the same research team published magneto-structural studies on $[Co(bim)_2(dca)_2]_n$ [28] and $[Co(bmim)_2(dca)_2]_n$ [28] in comparison to $[Co(atz)_2(dca)_2]_n$. In the following years, numerous papers were published describing the phenomenon of slow magnetic relaxation in cobalt(II) coordination polymers. These studies highlighted the potential of cobalt(II) complexes as single-ion magnets (SIMs) due to their ability to exhibit a slow relaxation of magnetization under certain conditions. The coordination environment, the nature of the bridging ligands, and the geometry of the cobalt(II) centers were identified as key factors influencing this magnetic behavior, expanding the understanding of SIM phenomena in 3d metal-based systems.

Of course, it is impossible to ignore the characteristics of the magnetic properties of α-$[Co(dca)_2]_n$ and β-$[Co(dca)_2]_n$ described in detail by two research teams in 1998 [70,102,103]. The long-range ferromagnetic ordering with a Curie temperature $T_C$ of 9 K was confirmed for $[Co(dca)_2]$. The authors anticipated ferromagnetic properties after calculating a positive Weiss constant, θ = +6.1 K. In a small field, the $\chi T$ value increase sharply, reaching a much higher value than expected for a short-range order. The ferromagnetic behavior of $[Co(dca)_2]_n$ was supported by the low-field field-cooled (FC) and zero-field-cooled (ZFC) temperature dependence of the magnetization and by the 5.5 T saturation magnetization. To confirm the magnetic ordering, the temperature dependencies of the in-phase (real, $\lambda'(T)$) and out-of-phase (imaginary, $\lambda''(T)$) components' ac susceptibilities were measured. The studies demonstrated clear ferromagnetic coupling due to the presence of a $\lambda''(T) \neq 0$ component [104,105]. Since then, many articles have been published exploring transition metal complexes exhibiting long-range ferromagnetic order, similar to the rutile-like compound $[Co(dca)_2]_n$, though with varying degrees of success. The 3D complexes $\{(MePh_3P)[Co(dca)_3]\}_n$ and $\{(EtPh_3P)[Co(dca)_3]\}_n$ exhibit a decreasing magnetic moment from 4.9 μB at room temperature to 3.5 μB at 4 K, which is typical for antiferromagnetic coupling. A sharp increase at 9 K in low fields was observed during measurements for $\{(EtPh_3P)[Co(dca)_3]\}_n$, but this was only indicated in the presence of traces of α-$[Co(dca)_2]_n$ [76]. A similar situation was observed for $[Co(dbtp)(dca)_2]_n$ and $\{[Co(bte)(dca)_2]\cdot H_2O\}_n$, where the $\mu_{eff}$ values remained independent from the applied field, varying between 0.01 and 1 T in the 4–100 K region, ruling out any long-range magnetic ordering [47,83]. In the remaining three-dimensional Co(II) compounds, the negative value of the Weiss constant clearly indicates the transmission of antiferromagnetic interactions through the dca bridges (see Table S6).

## 6. Conclusions

The chemistry of coordination compounds of d-block metals, particularly those involving cobalt(II) ions, has consistently amazed researchers with the subtle beauty of their crystal structures and their diverse physicochemical properties. Given the remarkable simplicity of the dicyanamide ligand, this anionic pseudohalide ligand has proven to be one of the most excellent inorganic spacers to use to construct magnetic coordination compounds, displaying multiple bridging coordination modes. The versatility of $dcac^-$ has enabled the design and synthesis of mononuclear and dinuclear species, as well as numerous one-, two- and three-dimensional coordination polymers with diverse magnetic couplings, ranging from antiferromagnetism to ferromagnetism, including slow magnetic relaxation phenomena in single-ion magnet (SIM) systems.

Reflecting on past research, it is evident that transition metal compounds based on the dca anion have contributed significantly to the fundamental knowledge across various fields. This article focuses on the magnetic–structural relationships in cobalt(II) compounds, representing only a small part of the deeper amount of research on dicyanamide complex compounds of copper(II), manganese(II, III), nickel(II), iron(II), zinc(II) and other d-block metal cations, which have been thoroughly investigated from diverse physicochemical properties.

What about future perspectives? It might seem that the chemistry of the dicyanamide anion has been fully explored, but recent articles in top scientific journals prove otherwise. Over the past five years, researchers have shown that there is still much to discover—not only in the field of coordination chemistry, but also in the development of dicyanamide-based ionic liquids [106,107] or in electrochemistry, where β-[Zn(dca)$_2$]$_2$ was efficiently used as a negative electrode material for lithium-ion batteries [108].

**Supplementary Materials:** The following supporting information can be downloaded at: https://www.mdpi.com/article/10.3390/magnetochemistry10110090/s1, Table S1: Magneto-structural data for mononuclear dca-Co(II) compounds; Table S2: Shape values for mononuclear Co(II) complexes; Table S3: Magneto-structural data for binuclear dca-Co(II) compounds; Table S4: Magneto-structural data for one-dimensional dca-Co(II) coordination polymers; Table S5: Magneto-structural data for two-dimensional dca-Co(II) coordination polymer; Table S6. Magneto-structural data for three-dimensional dca-Co(II) coordination polymer.

**Funding:** The research activities were co-financed by the funds granted under the Research Excellence Initiative of the University of Silesia in Katowice (POB-2).

**Data Availability Statement:** Not applicable.

**Conflicts of Interest:** The author declares no conflicts of interest.

## Abbreviations

| | | | |
|---|---|---|---|
| **atz** | 2-amino-1,3,5-triazine | **H$_3$daps** | 2,6-bis(1-salicyloylhydrazonoethyl) pyridine |
| **azpyph** | N,N′-di-tert-butyl-2,11 diaza[3,3](2,6)pyridinophane | **L$^1$** | N,N′-di-tertbutyl-2,11-diaza[3,3](2,6)pyridinophane |
| **bim** | 1-benzylimidazole | **nitppy** | 2-(4′-pyridyl)-4,4,5,5–tetramethylimidazoline-1-oxyl-3-oxide |
| **biq** | 2,2′-biquinoline | **nic** | isonicotinamide |
| **bimb** | 1,4-bis(im idazol-1-yl)-butane | **NH$_2$pyz** | 2-aminopyrazine |
| **bipy** | 4,4′-bipyridine | **MeCN** | acetonitrile |
| **bipim** | 2,2-bipyrimidine | **mpdo** | 2-methyl pyrazine dioxide |
| **bmim** | 1-benzyl-2-methylimidazole | **phz** | phenazine |
| **bpp** | 1,3-bis(4′-pyridyl)propane | **phen** | 1,10-phenanthroline |
| **bpm** | bis[(3,5-di methyl)pyrazolyl]methane | **pybiu** | N-(picolinoyl)-biurate |
| **bte** | 1,2-bis(1,2,4-triazol-1-yl)ethane | **pydz** | pyridazine |
| **bpds** | 4,4′-bipyridyl disulfide | **pzdo** | pyrazinedioxide |
| **bpeado** | 1,2-bis(4-pyridyl)ethane-N,N′-dioxide | | |
| **BeTriMe** | benzyltrimethylammonium | **ptzda** | 2,4-diamino-6-pyridyl-1,3,5-triazine |
| **bdpb** | N,N-bis(3,5-dimethyl-1H-pyrazol-1-yl) methyl-N2-phenylethane-1,2-diamine | **pypypz** | bis[3,5- dimethyl-4-(49-pyridyl)pyrazol-1-yl]methane |
| **bpe** | 1,2-bis(4-pyridyl)ethane | **Phpyk** | 2-benzoylpyridine |
| **bte** | 1,2-bis(1,2,4-triazol-1-yl)ethane | **pypz** | 2,6- bis(pyrazol-1-yl)pyridine |
| **btrm** | 1,2-bis(1,2,4-triazole-1-yl)methane | **pzda** | 2,4-diamino-6-pyridyl-1,3,5-triazine |
| **bzim** | benzimidazole | **tmeda** | tetra methylethylenediamine |
| **DAT** | 1,5-diaminotetrazole | **tptz** | 2,4,6-tris(2-pyridyl)-1,3,5-triazine |
| **dmdpy** | 5,5′-dimethyl-2,2′-dipyridine | **tppz** | 2,3,5,6-tetrakis(2-pyridyl)pyrazine |
| **dmpzm** | bis(3,5-dimethylpyrazolyl)methane | **3-bpo** | 2,5-bis(3-pyridyl)-1,3,4-oxadiazole |
| **deen** | N,N-diethyl-ethylenediamine | **3-pyo** | 3-hydroxypyridine |
| **dbtp** | 1,3-di(benzotriazol-1 yl)propane | **4-bpo** | 2,5-bis(4-pyridyl)-1,3,4-oxadiazole |
| **etpybzam** | (N,N-diethyl,N′-(pyridin-2-yl)benzylidene)ethane-1,2-diamine | **4-bpmp** | bis(4-pyridylmethyl)piperazine |
| **imph** | 4-(imidazol 1-yl)phenol | **4,7-dmphen** | 4,7-dimethylphenanthroline |
| **HOpyz** | 2-hydroxypyrazine | **4-bpmp** | bis(4-pyridylmethyl)piperazine |
| **hmt** | hexamethylenetetramine | **5,6-(Me)$_2$-bzim** | 5,6-dimethylbenzimidazole |
| **HFlu** | fluconazole | **5-Mebzim** | 5-methyl benzimidazole |
| **H$_2$biim** | 2,2′-biimidazole | | |

## References


1. Serna, M.D.; Legaz, E.; Primo-Millo, E. Improvement of the N fertilizer efficiency with dicyandiamide (*dcd*) in citrus trees. *Fertil. Res.* **1996**, *43*, 137–142. [CrossRef]
2. Boudesocque, S.; Mohamadou, A.; Dupont, L.; Martinez, A.; Déchamps, I. Use of dicyanamide ionic liquids for extraction of metal ions. *RSC Adv.* **2016**, *6*, 107894–107904. [CrossRef]
3. Halder, G.J.; Kepert, C.J.; Moubaraki, B.; Murray, K.S.; Cashion, J.D. Guest-dependent spin crossover in a nanoporous molecular framework material. *Science* **2002**, *298*, 1762–1765. [CrossRef] [PubMed]
4. Gong, Y.N.; Huang, Y.L.; Jiang, L.; Lu, T.-B. A luminescent microporous metal–organic framework with highly selective $CO_2$ adsorption and sensing of nitro explosives. *Inorg. Chem.* **2014**, *53*, 9457–9459. [CrossRef] [PubMed]
5. Getman, R.B.; Bae, Y.-S.; Wilmer, C.E.; Snurr, R.Q. Review and analysis of molecular simulations of methane, hydrogen, and acetylene storage in metal–organic frameworks. *Chem. Rev.* **2012**, *112*, 703–723. [CrossRef]
6. Rao, C.N.R.; Ranganathan, A.; Pedireddi, V.R.; Raju, A.R. A novel hybrid layer compound containing silver sheets and an organic spacer. *Chem. Commun.* **2000**, *1*, 39–40. [CrossRef]
7. Ruben, M.; Lehn, J.-M.; Vaughan, G. Synthesis of ionisable [2,2] grid-type metallo-arrays and reversible protonic modulation of the optical properties of the $[Co^{II}_4L_4]^{8+}$ species. *Chem. Commun.* **2003**, *12*, 1338–1339. [CrossRef]
8. Ivanova, B.; Spiteller, M. $Ag^I$ and $Zn^{II}$ complexes with possible application as NLO materials—Crystal structures and properties. *Polyhedron* **2011**, *30*, 241–245. [CrossRef]
9. Ma, L.-F.; Wang, L.-Y.; Wang, Y.-Y.; Batten, S.R.; Wang, J.-G. Self-assembly of a series of cobalt(II) coordination polymers constructed from $H_2$tbip and dipyridyl-based ligands. *Inorg. Chem.* **2009**, *48*, 915–924. [CrossRef]
10. Wriedt, M.; Sellmer, S.; Näther, C. Coordination polymer changing its magnetic properties and colour by thermal decomposition: Synthesis, structure and properties of new thiocyanato iron(II) coordination polymers based on 4,4′-bipyridine as ligand. *Dalton Trans.* **2009**, 7975–7984. [CrossRef]
11. Journaux, Y.; Ferrando-Soria, J.; Pardo, E.; Ruiz-Garcia, R.; Julve, M.; Lloret, F.; Cano, J.; Li, Y.; Lisnard, L.; Yu, P.H.; et al. Design of Magnetic Coordination Polymers Built from Polyoxalamide Ligands: A Thirty Year Story. *Eur. J. Inorg. Chem.* **2018**, 228–247. [CrossRef]
12. Batten, S.R.; Neville, S.M.; Turner, D.R. *Coordination Polymers. Design, Analysis and Application*; RSC Publishing: Cambridge, UK, 2008. [CrossRef]
13. Madelung, W.; Kern, E. Über Dicyanamid. *Ann. Chem.* **1922**, *427*, 1–26. [CrossRef]
14. Köhler, H. Beitrage zur Chemie des Dicyanamid- und des Tricyanmethanidions. I. Die Bildung von Übergangsmetall-Pyridin-Komplexen. *Z. Anorg. Allg. Chem.* **1964**, *331*, 237–248. [CrossRef]
15. Hvastijová, M.; Kohout, J.; Wusterhausen, H.; Köhler, H. Strukturelle Eigenschaften der Kupfer(II)-dicyanamid- und-tricyanmethanid-Komplexe substituierter Pyridine. *Z. Anorg. Allg. Chem.* **1984**, *510*, 37–45. [CrossRef]
16. Hughes, E.W. The crystal structure of dicyanamide. *J. Am. Chem. Soc.* **1940**, *62*, 1259–1267. [CrossRef]
17. Zhang, S.L.; Shen, B.W.; Song, X.D.; Wang, Y.L.; Li, S.S. Ni-Na Coordination Polymer Bridged by Dicyanamide: Synthesis, Structures, Spectra and Thermal Stability. *J. Chem. Crystallogr.* **2023**, *53*, 138–144. [CrossRef]
18. Merabet, L.; Vologzhanina, A.V.; Setifi, Z.; Kabouba, L.; Setifi, F. Topological motifs in dicyanamides of transition metals. *CrystEngComm* **2022**, *24*, 4740–4747. [CrossRef]
19. Groom, C.R.; Bruno, I.J.; Lightfoot, M.P.; Ward, S.C. The Cambridge Structural Database. *Acta Crystallogr. Sect. B Struct. Sci. Cryst. Eng. Mater.* **2016**, *72*, 171–179. [CrossRef]
20. Ghosh, S.; Selvamani, S.; Mehta, S.; Mondal, A. Reversible thermo-induced spin crossover in a mononuclear *cis*-dicyanamido-cobalt(II) complex containing a macrocyclic tetradentate ligand. *Dalton Trans.* **2020**, *49*, 9208–9212. [CrossRef]
21. Váhovská, L.; Potocnáka, I.; Vitushkina, S.; Dušek, M.; Titiš, J.; Boca, R. Low-dimensional compounds containing cyanido groups. XXVI. Crystal structure, spectroscopic and magnetic properties of Co(II) complexes with non-linear pseudohalide ligands. *Polyhedron* **2014**, *81*, 396–408. [CrossRef]
22. Gao, E.-Q.; Bai, S.-Q.; Wang, Z.-M.; Yan, C.H. One- and two-dimensional metal–dicyanamido complexes with a f lexible bridging co-ligand: Structural and magnetic properties. *Dalton Trans.* **2003**, 1759–1764. [CrossRef]
23. Wang, Y.; Qi, Y. Six New Coordination Polymers: From Dimer to 2D Network based on the Combination of *cis*-Bis(imidazole) Spacers with Cobalt(II) Cations. *Z. Anorg. Allg. Chem.* **2014**, *640*, 2609–2615. [CrossRef]
24. Manna, S.C.; Ghosh, A.K.; Ribas, J.; Drew, M.G.B.; Lin, C.-N.; Zangrando, E.; Chaudhuri, N.R. Synthesis, crystal structure, magnetic behavior and thermal property of three polynuclear complexes: $[M(dca)_2(H_2O)_2]_n(hmt)_n$ [M=Mn(II), Co(II)] and $[Co(dca)_2(bpds)]_n$ [dca, dicyanamide; hmt, hexamethylenetetramine; bpds, 4,4′-bipyridyl disulfide]. *Inorg. Chim. Acta* **2006**, *359*, 1395–1403. [CrossRef]
25. Zhao, X.J.; Wang, Q.; Du, M. Metal-directed assembly of 1-D and 2-D coordination polymers with fluconazole and dicyanamide co-ligand. *Inorg. Chim. Acta* **2007**, *360*, 1970–1976. [CrossRef]
26. Turner, D.R.; Chesman, A.S.R.; Murray, K.S.; Deacon, G.B.; Batten, S.R. The chemistry and complexes of small cyano anions, *Chem. Commun.* **2011**, *47*, 10189–10210.
27. Marshall, S.R.; Incarvito, C.D.; Shum, W.W.; Rheingold, A.L.; Miller, J.S. Novel coordination of dicyanamide, $[N(CN)_2]_2$: Preferential binding of the amide nitrogen. *Chem. Commun.* **2002**, 3006–3007. [CrossRef]

28. Świtlicka-Olszewska, A.; Palion-Gazda, J.; Klemens, T.; Machura, B.; Vallejo, J.; Cano, J.; Lloret, F.; Julve, M. Single-ion magnet behaviour in mononuclear and two-dimensional dicyanamide-containing cobalt(II) complexes. *Dalton Trans.* **2016**, *45*, 10181–10193. [CrossRef]
29. Palion-Gazda, J.; Machura, B.; Kruszynski, R.; Grancha, T.; Moliner, N.; Lloret, F.; Julve, M. Spin Crossover in Double Salts Containing Six-and Four-Coordinate Cobalt(II) Ions. *Inorg. Chem.* **2017**, *56*, 6281–6296. [CrossRef]
30. Du, M.; Wang, Q.; Li, C.-P.; Zhao, X.J.; Ribas, J. Coordination Assemblies of $Co^{II}/Cu^{II}/Zn^{II}/Cd^{II}$ with 2,5-Bipyridyl-1,3,4-Oxadiazole and Dicyanamide Anion: Structural Diversification and Properties. *Cryst. Growth Des.* **2010**, *10*, 3285–3296. [CrossRef]
31. Zhou, H.-B.; Wang, S.-P.; Liu, Z.-Q.; Liao, D.-Z.; Jiang, Z.-H.; Yan, S.-P.; Cheng, P. Supramolecular assemblies of new 2-D complexes: Syntheses, crystal structures, spectroscopic and magnetic properties of $\{[MnAu_2(CN)_4(NITpPy)_2(H_2O)_2]\}_n$ and $\{[Co(N(CN)_2)_2(NITpPy)_2(H_2O)_2]\}_n$. *Inorg. Chim. Acta* **2006**, *359*, 533–540. [CrossRef]
32. Lunell, M.; Casanova, D.; Cirera, J.; Alemany, P.; Alvarez, S. *SHAPE*, version 2.0. Universidad de Barcelona: Barcelona, Spain, 2010.
33. Shen, W.Z.; Chen, X.Y.; Cheng, P.; Liao, D.Z.; Yan, S.P.; Jiang, Z.H. Cobalt(II) Complexes with Dicyanamide—From Binuclear Entities to Chains. *Z. Anorg. Allg. Chem.* **2003**, *629*, 2591–2595. [CrossRef]
34. Das, A.; Marschner, C.; Cano, J.; Baumgartner, J.; Ribas, J.; El Fallah, M.S.; Mitra, S. Synthesis, crystal structures and magnetic behaviors of two dicyanmide bridged di– and polynuclear complexes of cobalt(II) derived from 2,4,6-tris(2-pyridyl)1,3,5-trizine and imidazole. *Polyhedron* **2009**, *28*, 2436–2442. [CrossRef]
35. Kundu, S.; Roy, S.; Bhar, K.; Ghosh, R.; Lin, C.-H.; Ribas, J.; Ghosh, B.K. Syntheses, structures and magnetic properties of two one-dimensional coordination polymers of cobalt(II) and nickel(II) dicyanamide containing a tridentate N-donor Schiff base. *J. Mol. Struct.* **2013**, *1038*, 78–85. [CrossRef]
36. Zheng, L.; Hu, S.; Zhou, A. Two three-dimensional supramolecular networks built from metallophenols through hydrogen bonds and coordinate bonds. *Trans. Met. Chem.* **2015**, *40*, 637–642. [CrossRef]
37. Banerjee, S.; Halder, S.; Brandão, P.; Gómez García, C.J.; Benmansour, S.; Saha, A. Synthesis and characterization of a novel dicyanamide-bridged Co(II) 1-D coordination polymer with a $N_4$-donor Schiff base ligand. *Inorg. Chim. Acta* **2017**, *464*, 65–73. [CrossRef]
38. Roy, S.; Choubey, S.; Bhar, K.; Sikdar, N.; Costa, J.S.; Mitrad, P.; Ghosh, B.K. Counter anion dependent gradual spin transition in a 1D cobalt(II) coordination polimer. *Dalton Trans.* **2015**, *44*, 7774–7776. [CrossRef]
39. Mautner, F.A.; Traber, M.; Fischer, R.C.; Massoud, S.S.; Vicente, R. Synthesis, crystal structures, spectral and magnetic properties of 1-D polymeric dicyanamido-metal(II) complexes. *Polyhedron* **2017**, *138*, 13–20. [CrossRef]
40. Bhar, K.; Khan, S.; Costa, J.S.; Ribas, J.; Roubeau, O.; Mitra, P.; Ghosh, B.K. Crystallographic Evidence for Reversible Symmetry Breaking in a Spin-Crossover d7 Cobalt(II) Coordination Polymer. *Angew. Chem. Int. Ed.* **2012**, *51*, 2142–2145. [CrossRef]
41. Świtlicka, A.; Bieńko, D.C.; Titiš, J.; Machura, B.; Penkala, M.; Bieńko, A.; Rajnák, C.; Boča, R.; Ozarowski, A. Slow magnetic relaxation in hexacoordinated cobalt(II) field-induced single-ion magnets. *Inorg. Chem. Front.* **2020**, *7*, 2637–2650. [CrossRef]
42. Ghosh, S.; Kamily, S.; Rouzieres, M.; Herchel, R.; Mehta, S.; Mondal, A. Reversible Spin-State Switching and Tuning of Nuclearity and Dimensionality via Nonlinear Pseudohalides in Cobalt(II) Complexes. *Inorg. Chem.* **2020**, *59*, 17638–17649. [CrossRef]
43. Munshi, S.J.; Sadhu, M.H.; Kundu, S.; Savani, C.; Kumar, S.B. Synthesis, characterization and structures of binuclear copper(II) and polynuclear cobalt(II), nickel(II) and cadmium(II) complexes involving $N_4$-donor pyrazolyl based ligand and dicyanamide as bridging ligand. *J. Mol. Struct.* **2020**, *1209*, 127984. [CrossRef]
44. Biswas, M.; Pilet, G.; El Fallah, M.S.; Ribas, J.; Mitra, S. Design of a flexible ligand for the construction of a one-dimensional metal-organic coordination polymer. *Inorg. Chim. Acta* **2008**, *361*, 387–392. [CrossRef]
45. Mondal, A.K.; Mondal, A.; Konar, S. Slow Magnetic Relaxation in a One-Dimensional Coordination Polymer Constructed from Hepta-Coordinate Cobalt(II) Nodes. *Magnetochemistry* **2020**, *6*, 45. [CrossRef]
46. Hsu, G.-Y.; Chen, C.-W.; Cheng, S.C.; Lin, S.-H.; Wei, H.-H.; Lee, C.-J. Structure and magnetic properties of one-dimensional metal complexes constructed from alternating dicyanamide linked through binuclear metal tetra-2-pyridylpyrazine subunits. *Polyhedron* **2005**, *24*, 487–494. [CrossRef]
47. Li, B.L.; Wan, X.-Y.; Zhu, X.; Gao, S.; Zhang, Y. Synthesis, crystal structure and magnetic behavior of two cobalt coordination polymers with 1,2-bis(1,2,4-triazol-1-yl)ethane and dicyanamide. *Polyhedron* **2007**, *26*, 5219–5224. [CrossRef]
48. Kutasi, A.M.; Batten, S.R.; Harris, A.R.; Moubaraki, B.; Murray, K.S. Structure and magnetism of the ladder-like coordination polymer $Co_3(dca)_2(nic)_4(H_2O)_8 \cdot 2H_2O$[dca = dicyanamide anion, $N(CN)_2^-$; nic = nicotinate anion]. *CrystEngComm* **2002**, *4*, 202–204. [CrossRef]
49. Hardy, A.M.; LaDuca, R.L. Synthesis and structure of a cobalt dicyanamide chain coordination polymer incorporating a long-spanning hydrogen-bonding capable diimine with a novel binodal (4,6)-connected supramolecular topology. *Inorg. Chem. Commun.* **2009**, *12*, 308–311. [CrossRef]
50. Wriedt, M.; Näther, C. Directed synthesis of μ-1,3,5 bridged dicyanamides by thermal decomposition of μ-1,5 bridged precursor compounds. *Dalton Trans.* **2011**, *40*, 886–898. [CrossRef]
51. Kutasi, A.M.; Batten, S.R.; Moubaraki, B.; Murray, K.S. New 2D coordination polymers containing both bi- and tri-dentate dicyanamide bridges and intercalated phenazine. *J. Chem. Soc. Dalton Trans.* **2002**, 819–821. [CrossRef]

52. Sun, H.L.; Gao, S.; Ma, B.Q.; Su, G. Long-Range Ferromagnetic Ordering in Two-Dimensional Coordination Polymers $Co[N(CN)_2]_2(L)$ [L = Pyrazine Dioxide (pzdo) and 2-Methyl Pyrazine Dioxide (mpdo)] with Dual μ– and $\mu_3$-$[N(CN)_2]$ Bridges. *Inorg. Chem.* **2003**, *42*, 5399–5404. [CrossRef]
53. Lopes, L.B.; Correa, C.C.; Guedes, G.P.; Vaz, M.G.F.; Diniz, R.; Machado, F.C. Two new coordination polymers involving Mn(II), Co(II), dicyanamide anion and the nitrogen ligand 5,5′-dimethyl-2,2′-dipyridine: Crystal structures and magnetic properties. *Polyhedron* **2013**, *50*, 16–21. [CrossRef]
54. Armentano, D.; De Munno, G.; Guerra, F.; Julve, M.; Lloret, F. Ligand Effects on the Structures of Extended Networks of Dicyanamide-Containing Transition-Metal Ions. *Inorg. Chem.* **2006**, *45*, 4626–4636. [CrossRef] [PubMed]
55. Li, Q.Y.; Zhang, W.H.; Li, H.X.; Tang, X.Y.; Lang, J.P.; Zhang, Y.; Wang, X.Y.; Gao, S. Construction of a Novel 2D Polymer $[Co(dmpzm)(dca)_2]_\infty$ from Reaction of a Mononuclear Complex $[Co(dmpzm)Cl_2]$ with Sodium Dicyanamide (dca) [dmpzm=bis(3,5-dimethylpyrazolyl)methane]. *Chin. J. Chem.* **2006**, *24*, 1716–1720. [CrossRef]
56. Li, J.Y.; Xie, M.J.; Jiang, J.; Chang, Q.W.; Ye, Q.S.; Liu, W.P.; Chen, J.L.; Ning, P. Synthesis and Crystal Structure of Two Polydimensional Molecular Architectures from Cobalt(II), Copper(II) Complexes of 2,4-Diamino-6-pyridyl-1,3,5-triazine. *Asian J. Chem.* **2014**, *26*, 419–422. [CrossRef]
57. van der Werff, P.M.; Batten, S.R.; Jensen, P.; Moubaraki, B.; Murray, K.S.; Tan, E.H.-K. Structure and magnetism of anionic dicyanamidometallate extended networks of types $(Ph_4As)[M^{II}(dca)_3]$ and $(Ph_4As)_2[M_2{}^{II}(dca)_6(H_2O)]\cdot H_2O\cdot xCH_3OH$, where $dca{=}N(CN)_2{}^-$ and $M^{II}$=Co, Ni). *Polyhedron* **2001**, *20*, 1129–1136. [CrossRef]
58. Wang, S.; Wang, L.; Li, B.; Zhang, Y. Synthesis, Structure and Magnetic Properties of a New Two-dimensional (4,4) Network Cobalt Coordination Polymer with Dicyanamide and Benzimidazole Ligands. *J. Chem. Crystallogr.* **2009**, *39*, 221–224. [CrossRef]
59. Xu, Y.; Wang, Y.; Zhong, Y.; Lei, G.; Li, Z.; Zhang, J.; Zhang, T. High-Energy Metal−Organic Frameworks with a Dicyanamide Linker for Hypergolic Fuels. *Inorg. Chem.* **2021**, *60*, 5100–5106. [CrossRef]
60. Sun, H.L.; Ma, B.Q.; Gao, S.; Su, G. Pyrazine dioxide bridged two-dimensional antiferromagnets $[M(NCS)_2(pzdo)_2]$ (M = Mn, Co; pzdo = pyrazine dioxide). *Chem. Commun.* **2001**, 2586–2587. [CrossRef]
61. Du, M.; Wang, X.-G.; Zhang, Z.-H.; Tang, L.-F.; Zhao, X.-J. Solvent-directed layered Co(II) coordination polymers with unusual solid-state properties: From a nanoporous framework to the dense polythreading 3-D aggregation. *CrystEngComm* **2006**, *8*, 788–793. [CrossRef]
62. Xu, W.-J.; Chen, K.-P.; Zhang, Y.; Ma, Y.; Li, Q.-W.; Wang, Q.-L. Two new Co(II) coordination polymers based on redox-active ligands: Structure, Chromism and Magnetism. *J. Mol. Struct.* **2021**, *1231*, 129948. [CrossRef]
63. Khan, S.; Roy, S.; Bhar, K.; Ghosh, R.; Lin, C.-H.; Ribas, J.; Ghosh, B.K. Syntheses, structures and magnetic properties of two neutral coordination polymers of cobalt(II) containing a tailored aromatic diamine and pseudohalides as bridging units: Control of dimensionality by varying pseudohalide. *Inorg. Chim. Acta* **2013**, *398*, 40–45. [CrossRef]
64. Jensen, P.; Batten, S.R.; Moubaraki, B.; Murray, K.S. Syntheses, crystal structures, and magnetic properties of first row transition metal coordination polymers containing dicyanamide and 4,4-bipyridine. *J. Chem. Soc. Dalton Trans.* **2002**, 3712–3722. [CrossRef]
65. Palion-Gazda, J.; Świtlicka, A.; Choroba, K.; Machura, B.; Kruszynski, R.; Julve, M. Influence of the pyrazine substituent on the structure and magnetic properties of dicyanamide-bridged cobalt(II) complexes. *Dalton Trans.* **2019**, *48*, 17266–17280. [CrossRef] [PubMed]
66. Mączka, M.; Gągor, A.; Stroppa, A.; Gonçalves, J.N.; Zaręba, J.K.; Stefańska, D.; Pikul, A.; Drozd, M.; Sieradzki, A. Two-dimensional metal dicyanamide frameworks of $BeTriMe[M(dca)_3(H_2O)]$ (BeTriMe = benzyltrimethylammonium; dca = dicyanamide; M = $Mn^{2+}$, $Co^{2+}$, $Ni^{2+}$): Coexistence of polar and magnetic orders and nonlinear optical threshold temperature sensing. *J. Mat. Chem. C* **2020**, *8*, 11735–11747. [CrossRef]
67. Palion-Gazda, J.; Klemens, T.; Machura, B.; Vallejo, J.; Lloret, F.; Julve, M. Single ion magnet behaviour in a two-dimensional network of dicyanamide-bridged cobalt(II) ions. *Dalton Trans.* **2015**, *44*, 2989–2992. [CrossRef]
68. Chen, X.Y.; Cheng, P.; Yan, S.-P.; Liao, D.-Z.; Jiang, Z.-H. Grid-like Cobalt(II) Layered Complex with an Unprecedented Double 1,2-Bis(1,2,4-triazole-1-yl)methane Bridges. *Z. Anorg. Allg. Chem.* **2005**, *631*, 3104–3107. [CrossRef]
69. Batten, S.R.; Murray, K.S. Structure and magnetism of coordination polymers containing dicyanamide and tricyanomethanide ions. *Coord. Chem. Rev.* **2003**, *246*, 103–130. [CrossRef]
70. Jensen, P.; Batten, S.R.; Fallon, G.D.; Moubaraki, B.; Murray, K.S.; Price, D.J. Structural isomers of $M(dca)_2$ molecule-based magnets. Crystal structure of tetrahedrally coordinated sheet-like β-$Zn(dca)_2$ and β-$Co/Zn(dca)_2$, and the octahedrally coordinated rutile-like α-$Co(dca)_2$, where $dca^-$ = dicyanamide, $N(CN)_2{}^-$, and magnetism of β-$Co(dca)_2$. *Chem. Commun.* **1999**, 177–178. [CrossRef]
71. Kurmoo, M.; Kepert, C.J. Hard magnets based on transition metal complexes with the dicyanamide anion, $\{N(CN)_2\}^-$. *N. J. Chem.* **1998**, 1515–1524. [CrossRef]
72. Wu, L.N.; He, M.Q.; Li, M.X.; Nfor, E.N.; Wang, Z.X. Inorganic-organic hybrid materials with methylviologen dication confined in magnetic hosts: Synthesis, crystal structures and magnetic properties. *Inorg. Chem. Commun.* **2020**, *116*, 107908. [CrossRef]
73. van der Werff, P.M.; Martinez-Ferrero, E.; Batten, S.R.; Jensen, P.; Ruiz-Perez, C.; Almeida, M.; Waerenborgh, J.C.; Cashion, J.D.; Moubaraki, B.; Gal'an-Mascaros, J.R.; et al. Hybrid materials containing organometallic cations and 3-D anionic metal dicyanamide networks of type $[Cp^*{}_2M][M'(dca)_3]$. *Dalton Trans.* **2005**, 285–290. [CrossRef] [PubMed]
74. Bermudez-Garcia, J.M.; Sanchez-Andujar, M.; Yanez-Vilar, S.; Castro-Garcia, S.; Artiaga, R.; Lopez-Beceiro, J.; Botana, L.; Alegriade, A.; Senaris-Rodriguez, M.A. Multiple phase and dielectric transitions on a novel multi-sensitive $[TPrA][M(dca)_3]$ (M: $Fe^{2+}$,$Co^{2+}$ and $Ni^{2+}$) hybrid inorganic–organic perovskite family. *J. Mater. Chem. C* **2016**, *4*, 4889–4898. [CrossRef]

75. Tong, M.L.; Ru, J.; Wu, Y.M.; Chen, X.-M.; Chang, H.-C.; Mochizuki, K.; Kitagawa, S. Cation-templated construction of three-dimensional α-Po cubic-type [M(dca)$_3$] networks. Syntheses, structures and magnetic properties of A[M(dca)$_3$] (dca = dicyanamide; for A = benzyltributylammonium, M = $Mn^{2+}$,$Co^{2+}$; for A = benzyltriethylammonium, M = $Mn^{2+}$,$Fe^{2+}$). *N. J. Chem.* **2003**, *27*, 779–782.
76. van der Werff, P.M.; Batten, S.R.; Jensen, P.; Moubaraki, B.; Murray, K.S.; Cashion, J.D. Structure and Magnetism of 3D Anionic Metal Dicyanamide (MePh$_3$P)[M(dca)$_3$](M = Fe, Co, Ni) and (EtPh$_3$P)[M(dca)$_3$](M = Mn, Co, Ni) Networks. *Cryst. Growth Des.* **2004**, *4*, 503–508. [CrossRef]
77. Ghosh, T.; Chattopadhyay, T.; Das, S.; Mondal, S.; Suresh, E.; Zangrando, E.; Das, D. Thiocyanate and Dicyanamide Anion Controlled Nuclearity in Mn, Co, Ni, Cu, and Zn Metal Complexes with Hemilabile Ligand 2-Benzoylpyridine. *Cryst. Growth Des.* **2011**, *11*, 3198–3205. [CrossRef]
78. Luo, J.; Zhou, X.G.; Gao, S.; Weng, L.H.; Shao, Z.H.; Zhang, C.M.; Li, Y.R.; Zhang, J.; Cai, R.F. Syntheses, structures and magnetic properties of two novel 3D helical dicyanamide complexes containing polyamine ligand. *Inorg. Chem. Commun.* **2004**, *7*, 669–672. [CrossRef]
79. Mal, D.; Sen, R.; Rentschler, E.; Okamoto, K.; Miyashita, Y.; Koner, S. Fully interlocked three-dimensional molecular scaffolding: Synthesis, X-ray structure, magnetic and nitrogen sorption study. *Inorg. Chim. Acta* **2012**, *385*, 27–30. [CrossRef]
80. Haldar, R.; Maji, T.K. Selective carbon dioxide uptake and crystal-to-crystal transformation: Porous 3D framework to 1D chain triggered by conformational change of the spacer. *CrystEngComm* **2012**, *14*, 684–690. [CrossRef]
81. Sun, H.-L.; Gao, S.; Ma, B.-Q.; Batten, S.R. 3D Self-penetrating coordination network constructed by dicyanamide and 1,2-bis(4-pyridyl)ethane-N,N′-dioxide (bpeado). *CrystEngComm* **2004**, *6*, 579–583. [CrossRef]
82. Sun, B.-W.; Gao, S.; Ma, B.-Q.; Wang, Z.-M.N. A two-fold interpenetrated three-dimensional cobalt(II) complex with dual dicyanamide and 4,4′-bipyridine bridges. *N. J. Chem.* **2000**, *24*, 953–954. [CrossRef]
83. Jones, L.F.; O'Dea, L.; Offermann, D.A.; Jensen, P.; Moubaraki, B.; Murray, K.S. Benzotriazole based 1-D, 2-D and 3-D metal dicyanamide and tricyanomethanide coordination networks. *Polyhedron* **2006**, *25*, 360–372. [CrossRef]
84. Chen, C.W.; Wei, H.H.; Lee, C.J. Two New 3D Networks Co(II) Complexes Constructed via the Bridging Dicyanamide and 2,2-Bipyrimidine Ligands: Structures and Magnetic Properties. *J. Chin. Chem. Soc.* **2006**, *53*, 1291–1296. [CrossRef]
85. Lin, Z.; Li, Z.; Zhang, H. Syntheses, Structures, and Magnetic Properties of Two Novel Three-Dimensional Frameworks Built from $M_4O_4$ Cubanes and Dicyanamide Bridges. *Cryst. Growth. Des.* **2007**, *7*, 589–591. [CrossRef]
86. Lloret, F.; Julve, M.; Cano, J.; Ruiz-García, R.; Pardo, E. Magnetic properties of six-coordinated high-spin cobalt(II) complexes: Theoretical background and its application. *Inorg. Chim. Acta* **2008**, *361*, 3432–3445. [CrossRef]
87. Cui, H.-H.; Zhang, Y.-Q.; Xue, Z.-L. Magnetic anisotropy and slow magnetic relaxation processes of cobalt(II)-pseudohalide complexes. *Dalton Trans.* **2019**, *48*, 10743–10752. [CrossRef]
88. Colacio, E.; Lloret, F.; Maimoun, I.B.; Kiveka, R.; Sillanpa, R.; Suarez-Varela, J. A New Type of Anionic Metal Dicyanamide Extended Networks through [Cu(N$_4$-macrocycle)]$^{2+}$ Cation Templation. Structure and Magnetic Propertie. *Inorg. Chem.* **2003**, *42*, 2720–2724. [CrossRef]
89. Fisher, M.E. Perpendicular Susceptibility of the Ising Model. *J. Math. Phys.* **1963**, *4*, 124–135. [CrossRef]
90. Fisher, M.E. Magnetism in One-Dimensional Systems—The Heisenberg Model for Infinite Spin. *Am. J. Phys.* **1964**, *32*, 343–346. [CrossRef]
91. Rueff, J.-M.; Masciocchi, N.; Rabu, P.; Sironi, A.; Skoulios, A. Structure and Magnetism of a Polycrystalline Transition Metal Soap – $Co^{II}$[OOC(CH$_2$)$_{10}$COO](H$_2$O)$_2$. *Eur. J. Inorg. Chem.* **2001**, 2843–2848. [CrossRef]
92. Świtlicka, A.; Bieńko, D.C.; Machura, B.; Rajnák, C.; Bieńko, A.; Kozieł, S.; Boča, R.; Ozarowski, A.; Ozerov, M. Non-traditional thermal behavior of Co(II) coordination networks showing slow magnetic relaxation. *Inorg. Chem. Front.* **2021**, *8*, 4356–4366. [CrossRef]
93. Váhovská, L.; Bukrynov, O.; Potočňák, I.; Čižmár, E.; Kliuikov, A.; Vitushkina, S.; Dušek, M.; Herchel, R. New Cobalt(II) Field-Induced Single-Molecule Magnet and the First Example of a Cobalt(III) Complex with Tridentate Binding of a Deprotonated 4-Amino-3,5-bis(pyridin-2-yl)-1,2,4-Triazole Ligand. *Eur. J. Inorg. Chem.* **2019**, *2019*, 250–261. [CrossRef]
94. Titiš, J.; Boča, R. Magnetostructural D Correlations in Hexacoordinated Cobalt(II)Complexes. *Inorg. Chem.* **2011**, *50*, 11838–11845. [CrossRef] [PubMed]
95. Craig, G.A.; Murrie, M. 3d single-ion magnets. *Chem. Soc. Rev.* **2015**, *44*, 2135–2147. [CrossRef] [PubMed]
96. Sun, H.-L.; Wang, Z.-M.; Gao, S. Synthesis, Crystal Structures, and Magnetism of Cobalt Coordination Polymers Based on Dicyanamide and Pyrazine-dioxide Derivatives. *Inorg. Chem.* **2005**, *44*, 2169–2176. [CrossRef]
97. Ma, B.Q.; Sun, H.L.; Gao, S.; Su, G. A Novel Azido and Pyrazine-Dioxide Bridged Three-Dimensional Manganese(II) Network with Antiferromagnetic Ordering ($T_N$ = 62 K) and a Spin Flop State. *Chem. Mater.* **2001**, *13*, 1946–1948. [CrossRef]
98. Sun, H.L.; Gao, S.; Ma, B.Q.; Su, G.; Batten, S.R. Structures and Magnetism of a Series Mn(II) Coordination Polymers Containing Pyrazine-Dioxide Derivatives and Different Anions. *Cryst. Growth Des.* **2005**, *5*, 269–277. [CrossRef]
99. Herrera, J.M.; Bleuzen, A.; Dromzée, Y.; Julve, M.; Lloret, F.; Verdaguer, M. Crystal Structures and Magnetic Properties of Two Octacyanotungstate(IV) and (V)-Cobalt(II) Three-Dimensional Bimetallic Frameworks. *Inorg. Chem.* **2003**, *42*, 7052–7059. [CrossRef]
100. Banci, L.; Bencini, A.; Gatteschi, D.; Zanchini, C. Spectral Structural Correlations in High Spin Cobalt(II) Complexes *Struct. Bond.* **1982**, *52*, 37.

101. Kliuikov, A.; Bukrynov, O.; Čižmár, E.; Váhovská, L.; Vitushkina, S.; Samoľová, E.; Potočňák, I. Syntheses, structures and magnetic properties of two isostructural dicyanamide-bridged 2D polymers. *New J. Chem.* **2021**, *45*, 7117–7128. [CrossRef]
102. Batten, S.R.; Jensen, P.; Moubaraki, B.; Murray, K.S.; Robson, R. Structure and molecular magnetism of the rutile-related compounds $M(dca)_2$, M = $Co^{II}$, $Ni^{II}$, $Cu^{II}$, dca = dicyanamide, $N(CN)_2^-$. *Chem. Commun.* **1998**, 439–440. [CrossRef]
103. Manson, J.L.; Kmety, C.R.; Huang, Q.-Z.; Lynn, J.W.; Bendele, G.M.; Pagola, S.; Stephens, P.W.; Liable-Sands, L.M.; Rheingold, A.L.; Epstein, A.J.; et al. Structure and Magnetic Ordering of $M^{II}[N(CN)_2]_2$ (M = Co, Ni), *Chem. Mater.* **1998**, *10*, 2552–2560.
104. Mukherjee, S.; Ranganathan, R.; Roy, S.B. Linear and nonlinear ac susceptibility of the canted-spin system: $Ce(Fe_{0.96}Al_{0.04})_2$. *Phys. Rev. B* **1994**, *50*, 1084. [CrossRef] [PubMed]
105. Mydosh, J.A. *Spin Glasses: An Experimental Introduction*; Taylor &Francis: London, UK; Washington, DC, USA, 1993.
106. Tang, S.F.; Smetana, V.; Mishra, M.K.; Kelley, S.P.; Renier, O.; Rogers, R.D.; Mudring, A.V. Forcing Dicyanamide Coordination to f-Elements by Dissolution in Dicyanamide-Based Ionic Liquids. *Inorg. Chem.* **2020**, *59*, 7227–7237. [CrossRef] [PubMed]
107. Hunger, J.; Roy, S.; Grechko, M.; Bonn, M. Dynamics of Dicyanamide in Ionic Liquids is Dominated by Local Interactions. *J. Phys. Chem. B* **2019**, *123*, 1831–1839. [CrossRef]
108. Qiao, X.; Corkett, A.J.; Muller, P.C.; Wu, X.; Zhang, L.; Wu, D.; Wang, Y.; Cai, G.; Wang, C.; Yin, Y.; et al. Zinc Dicyanamide: A Potential High-Capacity Negative Electrode for Li-Ion Batteries. *ACS Appl. Mater. Interfaces* **2024**, *16*, 43574–43581. [CrossRef]

magnetochemistry

MDPI

Article

# Mesoxalate-Bridged Heptanuclear Copper(II) Complexes: Structure and Magnetic Properties †

**Beatriz Gil-Hernández [1,2], Simon Millan [3], Irina Gruber [3], Christoph Janiak [3], Carlos J. Gómez-García [4] and Joaquín Sanchiz [1,2,*]**

1 Departamento de Química, Facultad de Ciencias, Universidad de La Laguna, 38206 La Laguna, Santa Cruz de Tenerife, Spain; beagher@ull.edu.es
2 Instituto Universitario de Materiales y Nanotecnología (IMN), Universidad de La Laguna, 38206 La Laguna, Santa Cruz de Tenerife, Spain
3 Institut für Anorganische Chemie und Strukturchemie, Heinrich-Heine-Universität, 40204 Düsseldorf, Germany; janiak@uni-duesseldorf.de (C.J.)
4 Departmento de Química Inorgánica, Universidad de Valencia, Dr Moliner 50, 46100 Burjasot, Valencia, Spain; carlos.gomez@uv.es
* Correspondence: jsanchiz@ull.edu.es; Tel.: +34-922845425

† This work is dedicated to Prof. Miguel Julve and Prof. Francesc Lloret. With this work, we wish to contribute to the tribute and international recognition of our colleagues Paco and Miguel. We are deeply saddened by Miguel's recent passing, and we cannot let this occasion pass without paying our small tribute to him and his work. We have been fortunate to know both Paco and Miguel for a long time, and we have witnessed their active participation in the development of the field known as molecular magnetism. Thanks to the brilliance and creativity of their work, they have advanced knowledge in this field and inspired many researchers who have joined their cause. Now, all of us express our gratitude by offering them this recognition. Thank you both for sharing your wisdom and motivation with us.

**Abstract:** Two new compounds, $(NH_4)_4[Cu_7(Hmesox)_6(H_2O)_8]\cdot 10H_2O$ (**1**) and $[Ru(bpy)_3]_4[Cu_7(Hmesox)_6Cl_2]Cl_2\cdot 2CH_3CN\cdot 12H_2O$ (**2**), were prepared and their structures were solved by single-crystal X-ray diffraction (mesoxalic acid = dihydroxypropanedioic acid, $H_4$mesox). The triply deprotonated mesoxalate anion acts as a chelating and bridging ligand with its carboxylate and alkoxide groups, forming the anionic heptanuclear copper(II) clusters $[Cu_7(Hmesox)_6(H_2O)_8]^{4-}$ and $[Cu_7(Hmesox)_6Cl_2]^{6-}$ in **1** and **2**, respectively. Variable temperature magnetic studies revealed antiferromagnetic interactions in **1** and the coexistence of ferro and antiferromagnetic interactions in **2**. The $[Ru(bpy)_3]^{2+}$ cations provided luminescent properties to compound **2**.

**Keywords:** heptanuclear Cu(II) complexes; mesoxalate; magnetic properties; crystal structure; luminescence

## 1. Introduction

The field formerly known as molecular magnetism underwent significant development, both experimentally and theoretically, in the 1980s, and since then, there has been a continuous stream of research in this area, yielding a plethora of compounds with intriguing magnetic properties and behaviours. Two key figures in the advancement of this field have undoubtedly been Professors Miguel Julve and Francesc Lloret from the University of Valencia. Sadly, Professor Julve passed away this year, and Professor Lloret will retire (although not completely) during 2024. Thanks to their pioneering work and the efforts of an entire community of researchers, numerous advancements have occurred in this field. The University of La Laguna has also contributed to the growth of this field, and we take this opportunity to pay tribute to our colleagues.

Many of the studies conducted by these researchers, especially in the early stages, focused on the investigation of magneto-structural correlations with ligands containing carboxylate and hydroxyl groups [1–5]. To a large extent, these studies established the factors influencing the intensity and nature of the magnetic interaction between paramagnetic ions, particularly $Cu^{II}$ ions, bridged by those ligands. At the University of La Laguna, some researchers have collaborated over the years with Professors Julve and Lloret, studying the chemistry and properties of malonate as a ligand [6], and more recently utilizing the mesoxalate ligand (the conjugated base of 2,2-dihydroxypropanedioic acid, Scheme 1), which contains not only two carboxylate groups but also two alcohol groups, mediating magnetic interaction not only through the carboxylate but also through at least one alkoxide group. Combined with $Cu^{II}$, this ligand yields trinuclear $[Cu_3(Hmesox)_3]^{3-}$ species that can be used as a ferromagnetically or antiferromagnetically coupled secondary building units [7,8]. Combining this trinuclear unit with $Cu^{II}$ or other 3d metallic ions such as $Mn^{II}$, $Co^{II}$, $Ni^{II}$ or some $Ln^{III}$ ions, we have prepared chiral three-dimensional MOFs, some of them exhibiting long-range magnetic ordering or proton conduction [9,10].

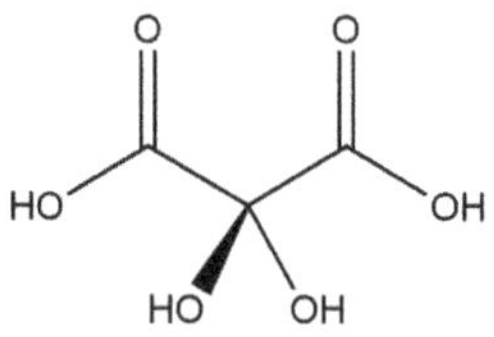

**Scheme 1.** Mesoxalic acid, $H_4$mesox.

However, up to now, we have not studied the formation of species of lower dimensionality, such as discrete high-nuclearity molecules, and we scarcely explored the preparation of compounds that combine other properties apart from magnetism and proton conduction. In this work, we describe two new compounds which are the first examples of mesoxalate-bridged heptanuclear copper(II) clusters $(NH_4)_4[Cu_7(Hmesox)_6(H_2O)_8]\cdot 10H_2O$ (**1**) and $[Ru(bpy)_3]_4[Cu_7(Hmesox)_6Cl_2]Cl_2\cdot 2\,CH_3CN\cdot 12H_2O$ (**2**). Heptanuclear copper(II) clusters is a topic of current interest in molecular magnetism, and compounds exhibiting ferromagnetic and antiferromagnetic interactions and less frequent magnetic phenomena, such as antisymmetric exchange interactions, have been recently prepared and studied [11,12]. Compounds **1** and **2** are both built from the $[Cu_3(Hmesox)_3]^{3-}$ secondary building unit and exhibit different magnetic properties due to small structural differences. Moreover, compound **2** incorporates a luminescent counterion and displays additional luminescent properties.

## 2. Materials and Methods

### 2.1. Materials

Copper (II) acetate hydrate, basic copper (II) carbonate, mesoxalic acid disodium salt, cation exchange resin Amberlite IR-120, 20% aqueous ammonia, acetonitrile, methanol and diethyl ether were purchased from commercial sources and used as received. Tris(bipyridine)ruthenium(II) chloride hexahydrate, $[Ru(bpy)_3]Cl_2\cdot 6H_2O$, was obtained by the method reported by Bromhead and Young [13].

### 2.2. Characterization

Elemental CHN analysis was performed with a Perkin-Elmer CHN 2400 (Perkin-Elmer, Waltham, MA, USA). IR spectra (400–4000 $cm^{-1}$) were recorded on a Thermo-Nicolet avatar 360 FT-IR spectrometer with the sample prepared as KBr disks. X-ray powder diffraction patterns on the polycrystalline samples of 1–2 (Supplementary Material: Figures S1 and S2, respectively) were collected with a PANalytical X'pert X-ray diffractometer (Cu $K_{\alpha,1}$ radiation = 1.54184 Å) at room temperature.

Magnetic measurements were performed in a Quantum Design MPMS XL-5 magnetometer with an applied magnetic field of 1 T (in the temperature range 15–300 K) and with an applied magnetic field of 100 mT (in the range 2–15 K). The susceptibility data were corrected for the sample holder, measured under the same conditions and for the diamagnetic contribution, evaluated using Pascal's constants [14].

Emission spectra were obtained by exciting the samples with light from a 350 W Xe arc lamp passed through a 0.25 m double monochromator (Spex 1680, Spex Industries, Metuchen, NJ, USA). Fluorescence was detected using a 0.25 m monochromator (Spex 1681) with a photomultiplier (Hamamatsu R928, Hamamatsu Photonics, Hamamatsu, Japan). The luminescence decay was measured using an optical parametric oscillator laser (EKSPLA, Vilnius, Lithuania), with excitation at 480 nm and detection at 600 nm. The signal acquired with the photomultiplier was registered in an oscilloscope (Tektronix 2430, Tectronix Inc., Beaverton, OR, USA) controlled by a computer.

### 2.3. Synthesis of $(NH_4)_4[Cu_7(Hmesox)_6(H_2O)_8]\cdot 10H_2O$ (**1**)

Solid disodium mesoxalate (180 mg, 1 mmol) was added to a suspension of the cation exchange resin Amberlite IR-120 (1.5 g in 5 mL of water) in order to obtain a mesoxalic acid solution ($H_4$mesox). This suspension was filtered through a Büchner funnel, and the filtrate was collected in a beaker. To this solution, 20% aqueous ammonia (115 µL, 1.5 mmol) was added, followed by the addition of 8 mL of a copper(II) acetate aqueous solution (200 mg, 1.1 mmol) under continuous stirring. Prismatic blue single-crystals of 1, obtained after one week, were filtered and washed with water and air-dried at room-temperature. Yield: 148 mg (54%). IR spectra main peaks ($cm^{-1}$): 3447, 3207, 1654 (s,b), 1400 (s), 1346 (s), 1084 (s), 865 (w), 819 (w), 668 (w), 485 (w). Anal. Calc. for air-dried $C_{18}H_{50}Cu_7N_4O_{50}$ (four crystallization molecules were lost respect to the single crystal): C, 13.79; H, 3.22; N, 3.57%. Found: C, 13.83; H, 3.22; N, 3.45%.

### 2.4. Synthesis of $[Ru(bpy)_3]_4[Cu_7(Hmesox)_6Cl_2]Cl_2\cdot 2CH_3CN\cdot 12H_2O$ (**2**)

A mesoxalic acid solution (0.5 mmol, 5 mL) was obtained as in the synthesis of 1. To this solution, solid basic copper(II) carbonate (55 mg, 0.25 mmol) was added, and the suspension was stirred and heated at 40 °C for 20 min. The resultant solution was filtered, diluted to 10 mL, and solid copper(II) acetate (50 mg, 0.25 mmol) was added under stirring. Finally, 4 mL of a 50:50 water/acetonitrile solution of [Ru(bpy)3]Cl2 (250 mg, 0.33 mmol) was added. The solution was allowed to stand at room-temperature, and prismatic dark-red single-crystals of 2 were obtained within 5 days. Yield 95 mg, 29%. IR ($cm^{-1}$): 3428 (b), 1647 (s,b), 1420 (s), 1336 (s), 1102 (s), 865 (w), 822 (w), 774 (w), 659 (w). Anal. Calc. for air-dried $C_{142}H_{132}Cl_4Cu_7N_{26}O_{48}Ru_4$: C, 43.02; H, 3.36; N, 9.19%. Found: C, 42.95; H, 3.33; N, 9.13%.

### 2.5. X-Ray Data Collection and Structure Refinement

Suitable single-crystals of **1** and **2** were selected in the microscope, covered with protective oil and mounted on a 50 µm loop. Data collection for compound **1** was performed on a Bruker Kappa APEX2 CCD X-ray diffractometer (Bruker AXS Inc., Madison, WI, USA) with a microfocus tube, Cu-K$\alpha$ radiation ($\lambda$ = 1.5418 Å) and a multilayer mirror system. The data were collected using $\omega$ scans, with the collection performed using APEX2 [15], cell refinement carried out with SMART, data reduction performed using SAINT [15] and experimental absorption correction performed with SADABS [16]. The structure was solved by direct methods (SHELXS-2016), refinement was achieved by full-matrix least squares on F2 using the SHELXL-2016 program suite [17], and the graphical user interface (GUI) ShelXle was used [18]. Crystal data and details on the structure refinement are given in Table 1 and Appendix A. Hydrogen atoms on the mesoxalate hydroxyl groups were positioned geometrically (O-H = 0.83 Å) and refined using a riding model (AFIX 83) with Uiso(H) = 1.5 Ueq(O). H atoms on the oxygen atoms of both coordination and crystallization water molecules could not be located. H atoms of the ammonium cations could not be located either. Nevertheless, all of them were included in the UNIT card for the calculation of the sum formula. The data collection for compound **2** was performed on a Bruker–Nonius Kappa CCD diffractometer equipped with graphite-monochromated

Mo-K$\alpha$ ($\lambda$ = 0.71073 Å) using COLLECT [19]. Data reduction was carried out with HKL DENZO and SCALEPACK [20], and experimental absorption correction was applied using SADABS [16]. The structure was solved by direct methods using SHELXS97 [21] and refinement was achieved by full-matrix least squares on F2 using the SHELXL97 program suite [21]. Crystal data and details on the structure refinement are given in Table 1 and Appendix A. The $\theta$ range is relatively low because the crystals obtained were very small and did not diffract well beyond a certain threshold. Despite several attempts to obtain higher-quality crystals, these efforts were unsuccessful. The hydrogen atoms for aromatic C–H and methyl groups were positioned geometrically (C–H = 0.94 Å for aromatic C–H and C–H = 0.97 Å for $CH_3$) and refined using a riding model (AFIX 43 for aromatic C–H, rotating group refinement AFIX 137 for $CH_3$), with Uiso(H) = 1.2Ueq (C–H) and Uiso(H) = 1.5 Ueq ($CH_3$), respectively. The H atoms of the hydroxyl central mesoxalate groups could not be found or refined. The H atoms of the oxygen atoms of crystallization water could not be located, but they were included in the UNIT card for the calculation of the sum formula. The carbon atoms of the bipyridine ligands of Ru2 were refined only isotropically in view of the low number of reflections and the reflection/parameter ratio. The pyridyl groups of Ru1 were refined as a regular hexagon (AFIX 66/65) to lower the number of independent parameters for a still reasonable reflection/parameter ratio.

**Table 1.** Summary of crystallographic data for complexes **1** and **2**.

| Compound | 1 | 2 |
|---|---|---|
| Empirical formula | $C_{18}H_{58}Cu_7N_4O_{54}$ | $C_{142}H_{132}Cl_4Cu_7N_{26}O_{48}Ru_4$ |
| M (g mol$^{-1}$) | 1639.46 | 3955.55 |
| Temperature (K) | 296 | 296 |
| $\lambda$ (Å) | 1.54178 | 0.71073 |
| Crystal system | Monoclinic | Monoclinic |
| Space group | $P2_1/c$ | $P2_1/c$ |
| a (Å) | 10.1046(5) | 13.6168(2) |
| b (Å) | 32.7505(16) | 22.0785(3) |
| c (Å) | 7.8273(4) | 24.6442(5) |
| $\beta$ (°) | 104.742(2) | 90.395(1) |
| V (Å$^3$) | 2505.0(2) | 7408.8(2) |
| Z | 2 | 2 |
| $D_{calc}$ (g cm$^{-3}$) | 2.174 | 1.762 |
| $\mu$ (mm$^{-1}$) | 4.56 | 1.55 |
| Theta range (°) | 4.5–67.2 | 4.4–21.2 |
| Unique reflections | 26,616 | 27,089 |
| $R_{int}$ | 0.043 | 0.029 |
| GOF on $F^2$ | 1.07 | 1.05 |
| $R_1$ [I > 2σ(I)] [a] | 0.0663 | 0.044 |
| $wR_2$ [I > 2σ(I)] [b] | 0.193 | 0.110 |
| CCDC number | 1909005 | 1909004 |

[a] $R_1 = [\Sigma(||F_o| - |F_c||)/\Sigma|F_o|]$; [b] $wR_2 = [\Sigma[w(F_o^2 - F_c^2)^2]/\Sigma[w(F_o^2)^2]]^{1/2}$.

## 3. Results and Discussion

### 3.1. Crystal Structures of Compounds 1 and 2

#### 3.1.1. Crystal Structure of $(NH_4)_4[Cu_7(Hmesox)_6(H_2O)_8]\cdot 10\,H_2O$ (1)

Compound **1** crystallizes in the monoclinic centrosymmetric space group $P2_1/c$. The asymmetric unit contains three mesoxalate ligands, four copper atoms, two counterbalancing ammonium cations, four coordination water molecules and five crystallization water molecules. The structure of **1** consists of ammonium cations, crystallization water molecules and an anionic centrosymmetric heptanuclear copper(II) cluster, which can be seen as two trinuclear copper(II)-mesoxalate entities, $[Cu_3(Hmesox)_3(H_2O)_3]^{3-}$, linked by a central copper(II) ion (Cu1) (Figure 1).

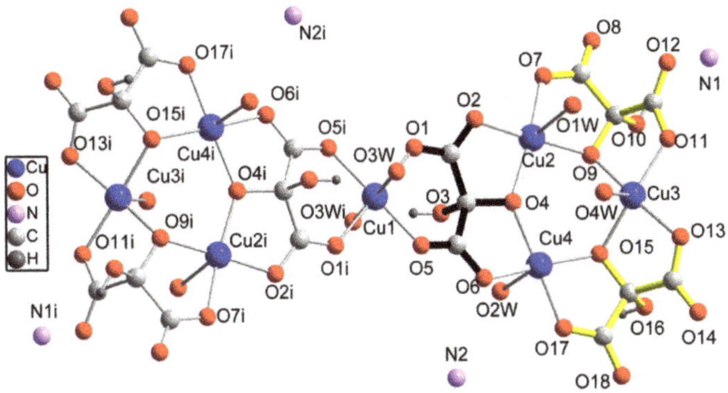


**Figure 1.** Structure of the centrosymmetric $[Cu_7(Hmesox)_6(H_2O)_8]^{4-}$ heptanuclear cluster in **1** with the nitrogen atoms of the ammonium counter ions. Central mesoxalate highlighted in black, peripheral mesoxalates highlighted in yellow. C and H atom labels are omitted for clarity. Symmetry code: (i) 1 − x, 1 − y, −z.

All the mesoxalate ligands are triply deprotonated, $HC_3O_6^{3-}$ $(Hmesox)^{3-}$ keeping only one of their four protons and acting as coordinating and bridging ligand among the copper atoms. The two peripheral mesoxalates (shown in yellow in Figure 1) act as blocking, but also as bridging ligands between Cu4–Cu3 and Cu3–Cu2 through μ-alkoxido oxygens (O15 and O9, respectively). The central mesoxalate (shown in black in Figure 1) acts as $\mu_3$-bridging ligand connecting Cu4–Cu2–Cu1 centres. Cu2 and Cu4 are connected through a μ-alkoxido-oxygen, whereas Cu2–Cu1 and Cu4–Cu1 connections are performed with two *anti-anti* carboxylate groups. One of the central hydroxyl groups of all the mesoxalate ligands remains protonated and uncoordinated. The mesoxalate-alkoxido and carboxylato oxygens occupy the basal positions of the Cu2, Cu3 and Cu4 atoms and form five-membered chelate rings with the skeleton of the mesoxalate ligand. These three copper ions display a square-pyramidal coordination geometry with τ parameters of 0.15 for Cu2, 0.12 for Cu3 and 0.03 for Cu4 [22]. The basal Cu–O distances for all these $Cu^{II}$ ions lie in the range 1.9298(1)–1.9525(1) Å (Table 2). The apical positions are occupied by a water molecule in the three Cu ions with longer distances ranging from 2.4198(1) to 2.4717(1) Å (Table 2). All these bond distances are in agreement with other copper(II)–mesoxalate complexes [23]. Cu2, Cu3 and Cu4 atoms assemble with the three mesoxalate ligands into trinuclear $[Cu_3(Hmesox)_3(H_2O)_3]^{3-}$ entities, which appear in other Cu(II)-mesoxalate compounds [23,24]. The trinuclear unit exhibits a distorted isosceles triangular structure with two small [Cu4–O15–Cu3 = 119.218(2)° and Cu3–O9–Cu2 = 118.779(2)°] and one long [Cu4–O4–Cu2 = 139.387(2)°] Cu–O–Cu bridging angles (Table 2).

The Cu1 atom sits on an inversion centre and bridges the two trinuclear entities forming the heptanuclear copper(II)-mesoxalate cluster. The mesoxalate ligands around this copper(II) ion form two six-membered chelate rings. The 4 + 2 Jahn-Teller distorted-octahedral coordination environment of Cu1 is completed with two water molecules filling the axial positions at distances of 2.4275(1) Å. The heptanuclear cluster has a net charge of −4, which is balanced by 4 ammonium cations.

The supramolecular packing (Figure 2) of these heptanuclear copper (II) clusters takes place through hydrogen bonding among the ammonium cations and the coordinated water molecules that stabilize the three-dimensional structure (Table 3).

**Table 2.** Selected bond lengths (Å) and angles (°) for compound **1**.

| Atoms | Distance (Å) | Atoms | Distance (Å) | Atoms | Angle (°) |
|---|---|---|---|---|---|
| Cu1–O5 | 1.940(5) | Cu3–O9 | 1.935(4) | Cu4–O4–Cu2 | 139.4(2) |
| Cu1–O1 | 1.952(5) | Cu3–O15 | 1.948(4) | Cu2–O9–Cu3 | 118.8(2) |
| Cu1–O3W | 2.427(5) | Cu3–O4W | 2.437(5) | Cu3–O15–Cu4 | 119.2(2) |
| Cu2–O9 | 1.926(4) | Cu4–O4 | 1.953(4) | | |
| Cu2–O2 | 1.933(5) | Cu4–O17 | 1.935(5) | | |
| Cu2–O7 | 1.939(5) | Cu4–O6 | 1.945(4) | | |
| Cu2–O4 | 1.957(4) | Cu4–O15 | 1.949(4) | | |
| Cu3–O11 | 1.930(4) | Cu1–Cu2 | 5.514(6) | | |
| Cu3–O13 | 1.930(4) | Cu1–Cu4 | 5.528(6) | | |

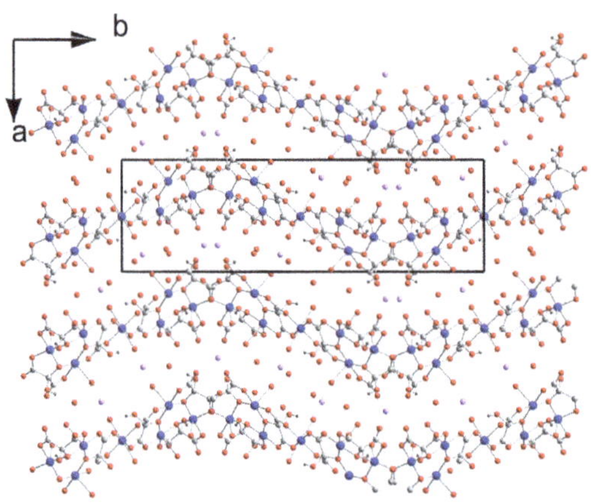


**Figure 2.** Supramolecular packing of the heptanuclear copper(II) clusters in compound **1** (same colour code as in Figure 1).

**Table 3.** Short intermolecular contacts involved in hydrogen bonding (Å) for compound **1** (the H atoms of the water and ammonium molecules could not be located or refined).

| D–H···A | D···A | D–H···A | D···A | D–H···A | D···A |
|---|---|---|---|---|---|
| O1W···O9W | 2.835 | O2W···N2 $^{iii}$ | 2.779 | O12···N1 $^{vi}$ | 2.866 |
| O2···N2 | 2.803 | O7···O3W $^{iv}$ | 2.723 | O18···N1 $^{iii}$ | 2.836 |
| O2W···O5W $^{ii}$ | 2.554 | O12···N1 $^{v}$ | 2.818 | O18···N1 $^{vii}$ | 2.865 |

ii = −1 + x, y, −1 + z; iii = −1 + x, y, z; iv = 1 − x, −y, −z; v = x, y, −1 + z; vi = x, 0.5 − y, −0.5 + z; vii = −1 + x, 0.5 − y, −0.5 + z.

3.1.2. Crystal Structure of $[Ru(bpy)_3]_4[Cu_7(Hmesox)_6Cl_2]Cl_2 \cdot 2CH_3CN \cdot 12H_2O$ (**2**)

Compound **2** also crystallizes in the monoclinic centrosymmetric space group $P2_1/c$. The asymmetric unit contains three mesoxalate ligands, four copper atoms, two chloride anions, two $[Ru(bpy)_3]^{2+}$ cations, one $CH_3CN$ and five crystallization water molecules (Figure 3). The centrosymmetric heptanuclear copper(II) cluster is very similar to that of **1**: it consists of two $[Cu_3(Hmesox)_3Cl]^{4-}$ trinuclear units bridged by a central copper(II) ion. In the asymmetric unit, two of the mesoxalate ligands (depicted in yellow in Figure 3) act as chelating and bridging agents between two copper(II) ions (Cu2–Cu3 and Cu3–Cu4), whereas the third one (shown in black in Figure 3) acts as a $\mu_3$-bridge between Cu1–Cu2–Cu4.

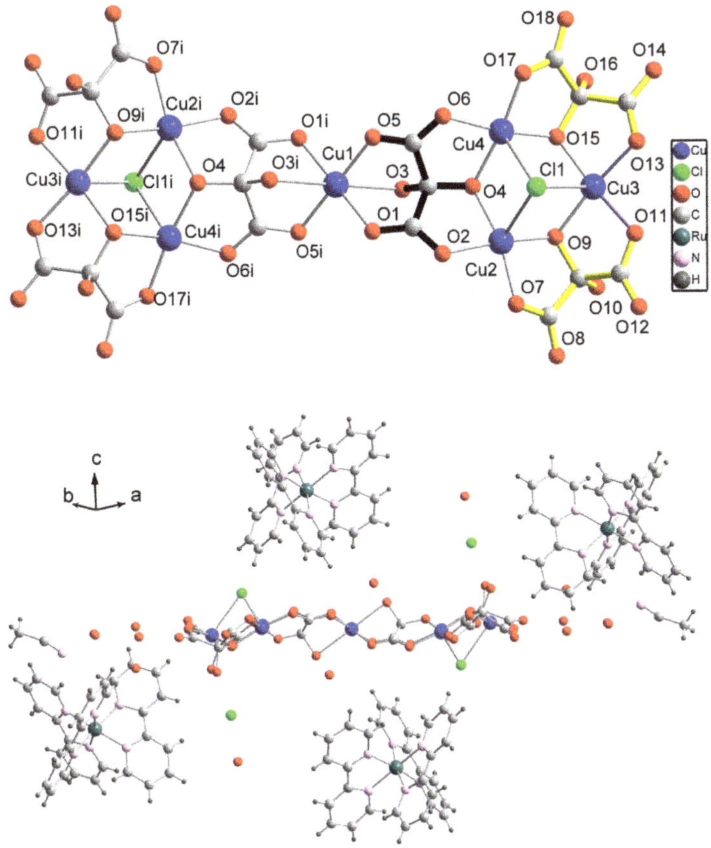


**Figure 3.** (**top**) Molecular structure of the heptanuclear cluster in **2**, with the labelling scheme of the main atoms. Symmetry code: (i) 1 − x, 1 − y, −z (**bottom**). View of one heptanuclear cluster surrounded by four $[Ru(bpy)_3]^{2+}$ cations, crystallization water and $CH_3CN$ molecules.

The bridging modes and the protonation degree of the mesoxalate ligands are analogous to those observed in **1**. The copper(II) ions in the triangular units, Cu2, Cu3 and Cu4, display very slightly distorted square-pyramidal environments with $\tau$ parameters of 0.15, 0.02 and 0.06, respectively [22]. They all present four mesoxalate oxygens in the basal positions (two alkoxido and two carboxylate-oxygens from two different mesoxalate ligands) and a $\mu_3$-$Cl^-$ in the apical position. The basal Cu–O and Cu–Cl bond distances are in the range 1.903(4)–1.981(4) Å and 2.643(2)–2.821(2) Å, respectively (Table 4). These values are similar to those observed in other related compounds [23]. The $\mu_3$-$Cl^-$ bridge at the apical position is one of the main differences between the heptanuclear clusters of **1** and **2** and results in much smaller Cu–O–Cu bridging angles. which have capital importance in the magnetic behaviour. The other main difference is the elongated and trigonally-distorted octahedral environment of Cu1, resulting from the coordination of four mesoxalate carboxylate-oxygens (from two mesoxalate ligands, as in **1**) and two mesoxalate hydroxyl groups (in contrast to the two water molecules observed in **1**). The basal

Cu1–O1 and Cu1–O5 distances are 1.954(4) Å and 1.984(4) Å, respectively, and the long axial one is Cu1–O3 2.459(4) Å.

**Table 4.** Selected bond lengths (Å) and angles (°) for compound **2**.

| Atoms | Distance (Å) | Atoms | Distance (Å) | Atoms | Angle (°) |
|---|---|---|---|---|---|
| Cu1–O5 | 1.954(4) | Cu4–Cl1 | 2.728(2) | Cu2–Cl1–Cu4 | 74.46(5) |
| Cu1–O1 | 1.984(4) | Ru1–N4 | 2.056(3) | Cu2–O9–Cu3 | 113.18(19) |
| Cu2–O7 | 1.925(5) | Ru1–N6 | 2.059(3) | Cu4–O4–Cu2 | 111.39(18) |
| Cu2–O9 | 1.932(4) | Ru1–N1 | 2.063(3) | Cu4–O4–Cu2 | 111.39(18) |
| Cu1–O3 | 2.459(4) | Ru1–N5 | 2.067(2) | Cu4–O15–Cu3 | 112.34(19) |
| Cu2–O2 | 1.966(4) | Ru1–N2 | 2.070(3) | Cu4–O15–Cu3 | 112.34(19) |
| Cu2–O4 | 1.981(4) | Ru1–N3 | 2.070(3) | | |
| Cu2–Cl1 | 2.643(2) | Ru2–N12 | 2.053(5) | | |
| Cu3–O13 | 1.914(5) | Ru2–N11 | 2.055(5) | | |
| Cu3–O11 | 1.914(5) | Ru2–N8 | 2.055(5) | | |
| Cu3–O15 | 1.946(4) | Ru2–N9 | 2.056(5) | | |
| Cu3–O9 | 1.947(4) | Ru2–N7 | 2.062(5) | | |
| Cu3–Cl1 | 2.821(2) | Ru2–N10 | 2.067(5) | | |
| Cu4–O17 | 1.903(4) | Cu1–Cu2 | 5.3078(7) | | |
| Cu4–O15 | 1.946(4) | Cu1–Cu4 | 5.2820(7) | | |

The three-dimensional structure is made up of the alternation of cationic and anionic layers that interact electrostatically and stack along the *c* direction (Figure 4 left). The anionic layers contain the $[Cu_7(Hmesox)_6Cl_2]^{6-}$ heptanuclear clusters, whereas the $[Ru(bpy)_3]^{2+}$ cations and $Cl^-$ anions give rise to $\{[Ru(bpy)_3]_2Cl\}_n^{3n+}$ cationic layers.

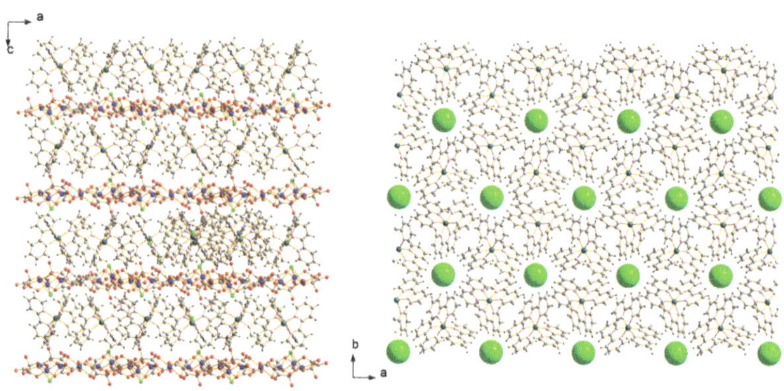


**Figure 4. (left)** View of the anionic layers containing the heptanuclear $[Cu_7(Hmesox)_6Cl_2]^{6-}$ units alternating along the *c* direction with cationic layers containing $\{[Ru(bpy)_3]_2Cl\}_n^{3n+}$ units in compound **2** (**right**). Arrangement of the chloride (space-filling green spheres) and $[Ru(bpy)_3]^{2+}$ ions in the cationic $\{[Ru(bpy)_3]_2Cl\}_n^{3n+}$ layers. (Same colour code as in Figure 3).

In the cationic $\{[Ru(bpy)_3]_2Cl\}_n^{3+}$ layers, each chloride ion is surrounded by six $[Ru(bpy)_3]^{2+}$ cations which exist in both enantiomeric forms, Λ and Δ, related by an inversion centre, resulting in a nonchiral structure (Figure 4 right). The $[Ru(bpy)_3]^{2+}$ cations display bonding distances and angles in the range of previously reported compounds [25,26]. The three-dimensional structure is stabilized by $CH_3CN$ and $H_2O$ crystallization solvent molecules.

### 3.2. *Magnetic Properties of Compounds* **1** *and* **2**

The thermal variation of the product of the molar magnetic susceptibility per $Cu_7$ cluster times the temperature ($\chi_m T$) for compound **1** shows a room-temperature value of

2.15 $cm^3$ K $mol^{-1}$ (Figure 5), which is appreciably lower than that expected value for seven non-interacting $Cu^{II}$ ions (2.8 $cm^3$ K $mol^{-1}$ for $g$ = 2.1) [27]. The $\chi_m T$ values continuously decrease on cooling the sample and reach a plateau at around 1.14 $cm^3$ K $mol^{-1}$ between 70 and 20 K and a progressive decrease below 20 K (Figure 5). The low value at room-temperature and the shape of the plot denote a dominant moderate-strong antiferromagnetic interaction among the copper(II) ions through the alkoxide and carboxylate bridges.

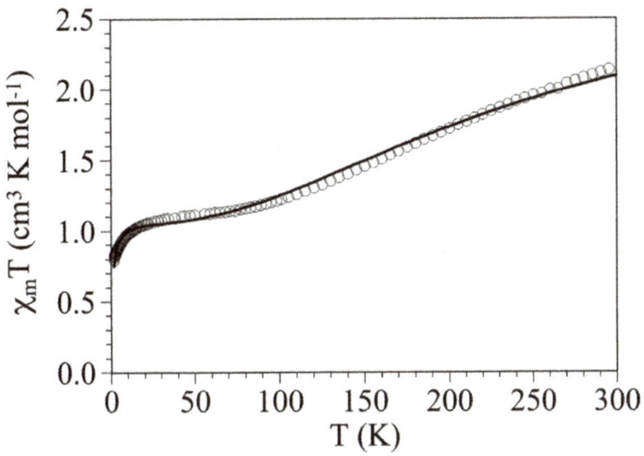


**Figure 5.** Temperature dependence of $\chi_m T$ for **1**. The solid line corresponds to the best fit to the model (see text).

In order to analyse the different exchange interactions existing in **1**, we can divide the plot into three regions attending to the different slopes observed. In the high-temperature region, with a fast decay of the $\chi_m T$ values, a strong antiferromagnetic coupling is operative, which more likely corresponds to antiferromagnetic interactions among the copper(II) ions of the trinuclear units through the mesoxalate–alkoxido linkers. This region concludes with two triangles, each with a $S = ½$ spin ground state, yet which do not interact strongly with the central copper(II) ion (also $S = ½$),leading to a plateau with a $\chi_m T$ value of around 1.14 $cm^3$ K $mol^{-1}$, which is the value expected for three non-interacting $S = 1/2$ ions. These three ions interact at very low temperatures through the carboxylate group and produce the additional progressive decrease in the $\chi_m T$ values. If we look in detail at the structure of compound **1**, we can distinguish three different coupling pathways (Scheme 2): two of them couple the $Cu^{II}$ ions through alkoxide bridges and one through carboxylate ones. The magnetic-exchange coupling constants $J_1$ and $J_2$ correspond to the coupling through the alkoxide bridges, while $J_3$ corresponds to the interaction through the carboxylate bridges.

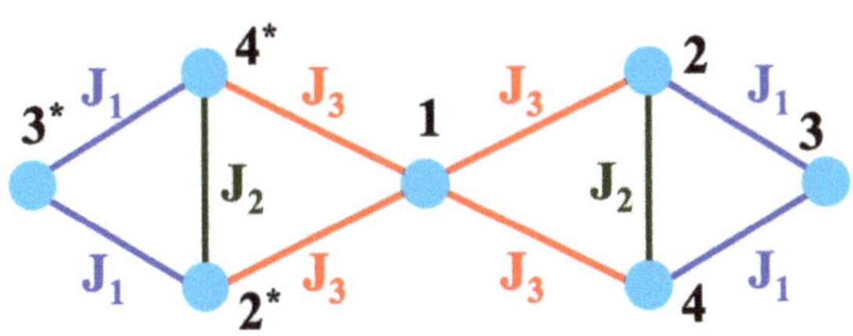


**Scheme 2.** Magnetic coupling scheme in compound **1**. Magnetic coupling through alkoxide bridges is shown in blue and green ($J_1$ and $J_2$), that through carboxylate bridges is shown in red ($J_3$). Symmetry code: (*) 1 − x, 1 − y, −z.

The copper(II) ions in an octahedral or square pyramidal environment have their unpaired electron mainly in the $d_{x^2-y^2}$ orbital; thus, the carboxylate and the alkoxido oxygens filling the basal positions of the square pyramidal environment function as mediators of the magnetic coupling in these compounds. In order to simplify the number of adjustable parameters, we assume $J_{23} = J_{34} = J_1$, since both bridges have very similar Cu–O–Cu structural parameters (Cu4–O15–Cu3 = 119.218(1)° and Cu2–O9–Cu3 = 118.779(1)°) but different from those of the third bridge in the triangle (Cu4–O4–Cu2 = 139.387(2)°), for which $J_2$ is defined. On the other hand, we observe that the two exchange pathways connecting Cu1 with Cu2 and Cu4 though *anti-anti* carboxylate bridges are very similar and assume that they are equivalent. Of course, given the centro-symmetry of the $Cu_7$ cluster, we can write the following: $J_{12} = J_{14} = J_{12*} = J_{14*} = J_3$. Based on this magnetic exchange scheme, the phenomenological Hamiltonian can be written as follows (Scheme 2):

$$H_1 = -J_1\,(S_2S_3 + S_3S_4 + S_{2*}S_{3*} + S_{3*}S_{4*}) - J_2\,(S_2S_4 + S_{2*}S_{4*}) - J_3\,(S_1S_2 + S_1S_4 + S_1S_{2*} + S_1S_{4*})$$

The magnetic susceptibility data were analysed with PHI software (version 3.1.5) [28], and a good fit was obtained with the following parameters: $g = 2.225(2)$, $J_1 = 120.5(3)$ cm$^{-1}$, $J_2 = -203.1(5)$ cm$^{-1}$, $J_3 = -86.0(4)$ cm$^{-1}$ and $zj = -0.18(2)$ cm$^{-1}$ (the solid line in Figure 5 represents $zj$, the intercluster interaction, modeled using the mean field approximation). The values obtained indicate the occurrence of antiferromagnetic interactions through the alkoxido and carboxylate bridges, confirming the starting hypothesis.

Previous studies conducted by our research group showed that the most important parameter in the interaction between Cu(II) ions through the alkoxide bridge of the mesoxalate ligand is the Cu–O–Cu bond angle. For angles below 114°, a ferromagnetic interaction can occur, whereas for angles above 114°, the interaction becomes antiferromagnetic. Moreover, the strength of the antiferromagnetic interaction increases as the Cu–O–Cu angle increases (Figure 6) [23].

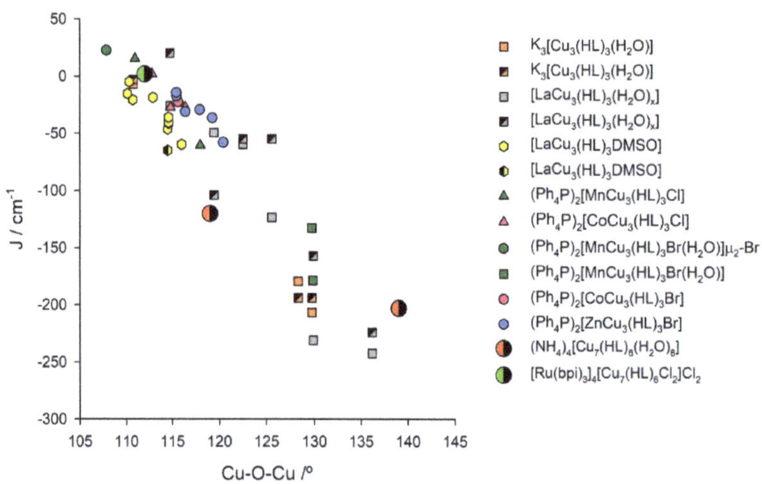


**Figure 6.** Dependence of the magnetic coupling constant, $J$, with the Cu–O–Cu bond angle in copper(II) mesoxalate compounds. Solid symbols partially filled in black correspond to experimental values, while the remaining symbols correspond to calculated values. Compounds **1** and **2** are shown with bigger size symbols to enhance their visibility. $H_4L$ = mesoxalic acid.

In compound **1**, we find a moderate antiferromagnetic coupling between Cu2 and Cu3 and between Cu3 and Cu4, with bridging Cu–O–Cu angles around 119° ($J_1 = -120.5(3)$ cm$^{-1}$). As expected, a stronger antiferromagnetic interaction occurs between Cu2 and Cu4, with a Cu–O–Cu bridging angle of 139.387(2)° ($J_2 = -203.1(5)$ cm$^{-1}$). Therefore, the values found in **1** show the expected trend and are within the expected range observed in previous studies [23]. Additionally, as observed experimentally, $J_3$ ($-86.0(4)$ cm$^{-1}$) is also expected to be antiferromagnetic since it corresponds to the coupling of two $Cu^{II}$ ions via a single *anti-anti* carboxylate bridge (Scheme 3) [1,29–31]. However, this magnetic coupling constant is more difficult to compare because most studies were conducted with oxalate ligands connecting two $Cu^{II}$ ions through a double *anti-anti* carboxylate bridge [2,32,33]. Furthermore, the bridging mode of the $\mu_3$-mesoxalate in compound **1** is unique, since it couples antiferromagnetically two copper(II) ions via an alkoxide bridge, and these two atoms are, at the same time, antiferromagnetically coupled with a third copper(II) via two single *anti-anti* carboxylate groups, leading to a situation that may exhibit spin frustration [34,35].

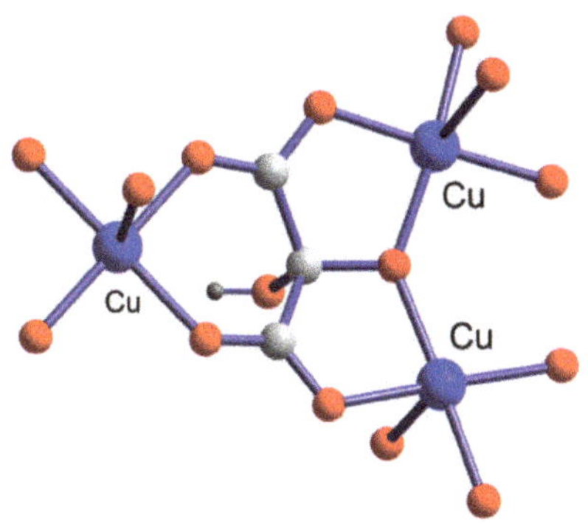


**Scheme 3.** $\mu_3$-bridging mode of the Hmesoxalate ligand in compound **1**.

The thermal dependence of the $\chi_m T$ product for compound **2** per Cu7 cluster shows, at room temperature, a value of 2.6 $cm^3$ K $mol^{-1}$ (Figure 7). This value is somewhat lower than the expected one for seven isolated spin doublets (2.8 $cm^3$ K $mol^{-1}$ for $g$ = 2.1). When the sample is cooled, $\chi_m T$ decreases to reach a minimum value of 2.4 $cm^3$ K $mol^{-1}$ at around 100 K. Below this temperature, $\chi_m T$ increases to reach a maximum value of 3.0 $cm^3$ K $mol^{-1}$ at 22 K, followed by a sharp decrease below this temperature (Figure 7). This behaviour is indicative of the occurrence of both antiferromagnetic and ferromagnetic interactions in the heptanuclear unit.

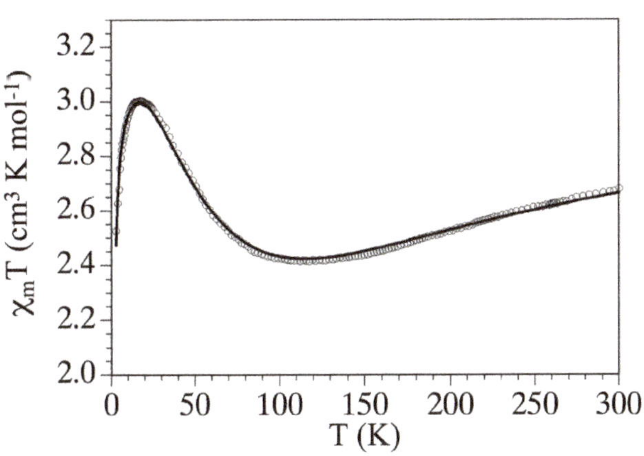


**Figure 7.** Temperature dependence of $\chi_m T$ for **2**. The solid line corresponds to the best fit to the model (see text).

Here again, we have a very complex molecule with seven spin carriers and many magnetic exchange pathways. In order to avoid the over parametrization of the model, looking at the symmetry and at the nature of the bridging groups, we tried to simplify the system. In previous studies, we found that a ferromagnetic coupling can be found in copper(II)/mesoxalate trinuclear units with Cu–O–Cu bridging angles lower than 114° [23]. Since the three Cu–O–Cu bond angles are similar along the three sides of the triangle (Cu3–O15–Cu4 = 112.34(2)°, Cu4–O4–Cu2 = 111.39(18)° and Cu2–O9–Cu3 = 113.18(19)°, Table 4), we use the same coupling constant for the three pathways ($J_1$ in Scheme 4). As we did in compound **1**, we assume that the coupling between the central Cu1 centre and Cu2 and Cu4 are also antiferromagnetic since they correspond to the *anti-anti* carboxylate bridges ($J_2$ in Scheme 4).

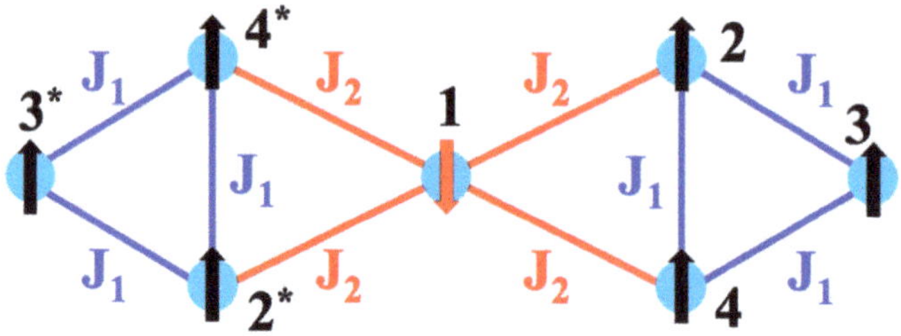


**Scheme 4.** Magnetic coupling scheme in compound **2**. The magnetic coupling through alkoxide and carboxylate bridges are shown in blue ($J_1$) and red ($J_2$), respectively. ). Symmetry code: (*) 1 − x, 1 − y, −z.

Using this exchange scheme, the phenomenological Hamiltonian used to fit the magnetic data can be written as follows:

$$H_2 = -J_1 (S_2S_3 + S_3S_4 + S_2S_4 + S_{2^*}S_{3^*} + S_{3^*}S_{4^*} + S_{2^*}S_{4^*}) - J_2 (S_1S_2 + S_1S_4 + S_1S_{2^*} + S_1S_{4^*})$$

The magnetic susceptibility data were fitted with the PHI software (version 3.1.5) [28], and a very good fit was obtained in the whole temperature range with the following parameters: $g$ = 2.1688(3), $J_1$ = 1.96(4) cm$^{-1}$, $J_2$ = −114.9(2) cm$^{-1}$ and zj = −0.129(1) cm$^{-1}$ (solid line in Figure 7). On one hand, in contrast to compound **1** in **2,** we find a ferromagnetic coupling within both mesoxalate–alkoxido-bridged Cu2–Cu3–Cu4 triangles. On the other side, as observed in compound **1**, the central Cu1 is antiferromagnetically coupled with the Cu2 and Cu4 centres of the triangles. This behaviour agrees with the presence of similar *anti-anti* carboxylate bridges connecting the copper(II) centres, as in **1**. The value obtained for $J_1$ is weak and ferromagnetic, in agreement with our previous results, showing that for Cu–O–Cu bond angles below 114°, the coupling is ferromagnetic (see above). In compound **2**, these angles are Cu3–O15–Cu4 = 112.34(2)°, Cu4–O4–Cu2 = 111.39(18)° and Cu2–O9–Cu3 = 113.18(19)°; therefore, a weak ferromagnetic coupling is expected, in agreement with the experimental result ($J_1$ = 1.96(4) cm$^{-1}$). On the other hand, the coupling through the two single *anti-anti* carboxylate bridges for this compound is somewhat stronger to that found in **1**. This behaviour can be understood considering the shorter Cu1–Cu2 and Cu1–Cu4 distances compared to those found in **1** (Tables 2 and 4).

The ferromagnetic $J_1$ and antiferromagnetic $J_2$ coupling constants lead to a spin distribution, as shown in Scheme 4, where the six spins on the copper(II) triangles are aligned in a parallel way to each other and antiparallel to the central Cu1 spin, leading to a $S$ = 5/2 spin ground state. Accordingly, the isothermal magnetizations at 2 and 5 K show saturation values very close to 5.0 $\mu_B$, which is the expected value for a $S$ = 5/2 ground spin sate (Figure 8). Furthermore, they can be satisfactorily reproduced with a Brillouin function for an $S$ = 5/2 ground spin state (solid lines in Figure 8) with $g$ values close to 2.

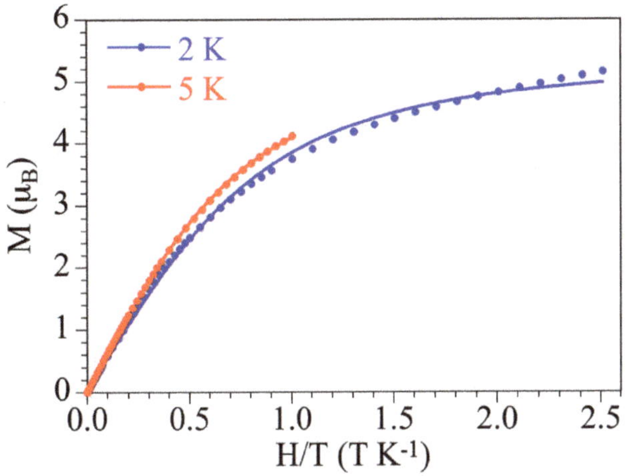


**Figure 8.** Isothermal magnetization of compound **2** at 2 and 5 K. The solid lines are the best fit to the Brillouin function for an $S = 5/2$ ground spin state.

*3.3. Spectroscopic Measurements*

The excitation spectrum of a solid-state sample of compound **2**, monitored at 600 nm, shows two prominent bands around 325 and 480 nm (Figure 9). These bands can be assigned to charge–transfer d→π* transitions, similar to those observed in $[Ru(bpy)_3]Cl_2$ [36]. Moreover, when the compound is excited at 480 nm, an intense emission band is detected around 600 nm. These results are similar to those found in the $[Ru(bpy)_3]Cl_2$ [36], showing an emission band at 600 nm with a lifetime around 0.7–0.9 μs at room temperature. This emission band is associated with a triplet–singlet transition, but with a short lifetime because the selection rules are partially removed.

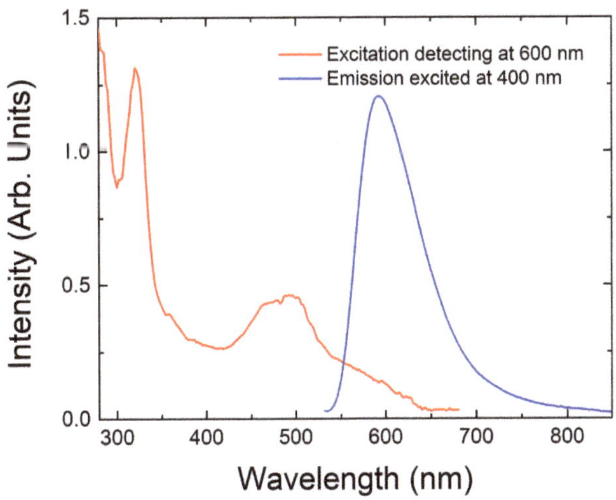


**Figure 9.** Emission and excitation spectra of compound **2** obtained, respectively, under excitation at 480 nm and detection at 600 nm.

## 4. Conclusions

We prepared two new compounds: $(NH_4)_4[Cu_7(Hmesox)_6(H_2O)_8]\cdot 10H_2O$ (**1**) and $[Ru(bpy)_3]_4[Cu_7(Hmesox)_6Cl_2]Cl_2\cdot 2CH_3CN\cdot 12H_2O$ (**2**) containing very similar centrosymmetric heptanuclear mesoxalate-bridged copper(II) clusters formed by two $Cu_3$ triangles connected to a central Cu atom. The structural analysis of both heptanuclear $Cu^{II}$ clusters show that the copper(II) ions in the triangles are connected via alkoxido bridges, whereas the central Cu atom is connected to two of the three Cu centres of each triangle via *anti-anti* carboxylate bridges. In compound **1**, the structural parameters indicate the presence of two different coupling constants inside the triangle ($J_1$ and $J_2$), whereas in compound **2**, both constants are the same ($J_1$). The magnetic properties show that in compound **1**, $J_1$ and $J_2$ are antiferromagnetic, in agreement with the Cu–O–Cu bond angles, leading to a spin frustration inside the triangle. In contrast, in compound **2**, $J_1$ is weak and ferromagnetic. In both compounds, the coupling through the anti-anti carboxylate bridges between the central Cu centre and the triangles ($J_3$ in 1 and $J_2$ in 2) are moderate and antiferromagnetic, as expected given the presence of *anti-anti* carboxylate bridges. The use of these 3$J$ and 2$J$ models for compounds **1** and **2**, respectively, allows us to reproduce, in a very satisfactory way, the magnetic data of both compounds. In all cases, the J values found agree with previous magneto-structural correlations. Comparing **1** and **2**, we observe that small differences in the Cu–O–Cu bridging angles are responsible for the different magnetic behaviour. Finally, the anionic character of these heptanuclear clusters allows the incorporation of counterions that can introduce other properties in addition to the magnetic ones. Thus, compound **2** contains a luminescent $[Ru(bpy)_3]^{2+}$ counterion that displays its characteristic luminescent properties.

**Supplementary Materials:** The following supporting information can be downloaded at: https://www.mdpi.com/article/10.3390/magnetochemistry10120093/s1, Figure S1: Powder X-Ray Diffraction pattern of compound **1** compared to the pattern simulated from the single crystal structure; Figure S2: Powder X-Ray Diffraction pattern of compound **2** compared to the pattern simulated from the single crystal structure.

**Author Contributions:** Conceptualization, B.G.-H. and J.S.; methodology, B.G.-H. and J.S.; validation, B.G.-H. and J.S.; Investigation, B.G.-H., S.M., I.G., J.S. and C.J.G.-G.; data curation, I.G., S.M., B.G.-H. and C.J.G.-G.; writing—original draft preparation, J.S., C.J.G.-G. and B.G.-H.; writing—review and editing, B.G.-H., J.S., C.J.G.-G. and C.J.; funding acquisition, C.J.G.-G. and C.J. All authors have read and agreed to the published version of the manuscript.

**Funding:** C. J. G. G. acknowledges the support of the Advanced Materials program, supported by the Spanish MCIN with funding from the European Union NextGeneration EU (PRTR-C17.I1) and the Generalitat Valenciana (project MFA-2022-057). We also acknowledge the Grant PID2021-125907NB-I00, funded by MCIN/AEI/10.13039/501100011033 and by "ERDF A way of making Europe", as well as project CIPROM-2022-060 from the Generalitat Valenciana, for financial support.

**Data Availability Statement:** The data will be provided upon request to the corresponding author.

**Acknowledgments:** JS and BGH thank the Departamento de Química de la Universidad de La Laguna for financial support and SEGAI-ULL for X-ray measurements.

**Conflicts of Interest:** The authors declare no conflicts of interest.

## Appendix A

CCDC 1909005 and 1909004 contain the supplementary crystallographic data for compounds **1** and **2**, respectively. These data can be obtained free of charge via http://www.ccdc.cam.ac.uk/conts/retrieving.html or from Cambridge Crystallographic Data Centre, 12 Union Road, Cambridge CB2 1EZ, UK; fax: (+44) 1223-336-033; or e-mail: deposit@ccdc.cam.ac.uk.

## References

1. Julve, M.; Verdaguer, M.; Kahn, O.; Gleizes, A.; Philoche-Levisalles, M. Tunable exchange interaction in. mu.-oxalato copper(II) dinuclear complexes. *Inorg. Chem.* **1983**, *22*, 368–370. [CrossRef]
2. Julve, M.; Gleizes, A.; Chamoreau, L.M.; Ruiz, E.; Verdaguer, M. Antiferromagnetic Interactions in Copper(II) μ-Oxalato Dinuclear Complexes: The Role of the Counterion. *Eur. J. Inorg. Chem.* **2017**, *2018*, 509–516. [CrossRef]

3. Julve, M.; Faus, J.; Verdaguer, M.; Gleizes, A. Copper(II), a chemical Janus: Two different (oxalato)(bipyridyl)copper(II) complexes in one single crystal. Structure and magnetic properties. *J. Am. Chem. Soc.* **2002**, *106*, 8306–8308. [CrossRef]
4. Castro, I.; Calatayud, M.L.; Orts-Arroyo, M.; Marino, N.; De Munno, G.; Lloret, F.; Ruiz-García, R.; Julve, M. Oxalato as polyatomic coordination center and magnetic coupler in copper(II)-polypyrazole inverse polynuclear complexes and coordination polymers. *Coord. Chem. Rev.* **2022**, *471*, 214730. [CrossRef]
5. Sletten, J.; Soerensen, A.; Julve, M.; Journaux, Y. A tetranuclear hydroxo-bridged copper(II) cluster of the cubane type. Preparation and structural and magnetic characterization of tetrakis [(2,2′-bipyridyl)hydroxocopper(II)] hexafluorophosphate. *Inorg. Chem.* **1990**, *29*, 5054–5058. [CrossRef]
6. Ruiz-Perez, C.; Sanchiz, J.; Molina, M.H.; Lloret, F.; Julve, M. Ferromagnetism in malonato-bridged copper(II) complexes. synthesis, crystal structures, and magnetic properties of $\{[Cu(H_2O)_3][Cu(mal)_2(H_2O)]\}_n$ and $\{[Cu(H_2O)_4]_2[Cu(mal)_2(H_2O)]\}[Cu(mal)_2(H_2O)_2]$ $\{[Cu(H_2O)_4][Cu(mal)_2(H_2O)_2]\}$ ($H_2$mal = malonic Acid). *Inorg. Chem.* **2000**, *39*, 1363–1370. [CrossRef]
7. Sanchiz, J.; Pasán, J.; Fabelo, O.; Lloret, F.; Julve, M.; Ruiz-Pérez, C. $[Cu_3(Hmesox)_3]^{3-}$: A Precursor for the Rational Design of Chiral Molecule-Based Magnets ($H_4$mesox = 2-dihydroxymalonic acid). *Inorg. Chem.* **2010**, *49*, 7880–7889. [CrossRef]
8. Gil-Hernández, B.; Gili, P.; Sanchiz, J. A ferromagnetically coupled copper(II) trinuclear secondary building unit as precursor for the preparation of molecule-based magnets. *Inorg. Chim. Acta* **2011**, *371*, 47–52. [CrossRef]
9. Gil-Hernández, B.; Gili, P.; Vieth, J.K.; Janiak, C.; Sanchiz, J. Magnetic ordering in two molecule-based (10,3)-a nets prepared from a copper(II) trinuclear secondary building unit. *Inorg. Chem.* **2010**, *49*, 7478–7490. [CrossRef]
10. Gil-Hernández, B.; Savvin, S.; Makhloufi, G.; Núñez, P.; Janiak, C.; Sanchiz, J. Proton conduction and long-range ferrimagnetic ordering in two isostructural copper(II) mesoxalate metal-organic frameworks. *Inorg. Chem.* **2015**, *54*, 1597–1605. [CrossRef]
11. Mistry, S.; Sutter, J.P.; Natarajan, S. Stabilization of Cu7 clusters in azide networks: Syntheses, structures and magnetic behaviour. *Dalton Trans.* **2016**, *45*, 5140–5150. [CrossRef]
12. Ferrer, S.; Aznar, E.; Lloret, F.; Castineiras, A.; Liu-Gonzalez, M.; Borras, J. One-dimensional metal-organic framework with unprecedented heptanuclear copper units. *Inorg. Chem.* **2007**, *46*, 372–374. [CrossRef] [PubMed]
13. Bromhead, J.A.; Young, C.G. Tris(2,2′-Bipyridine)Ruthenium(II) Dichloride Hexahydrate. *Inorg. Synth.* **1982**, *21*, 127. [CrossRef]
14. Bain, G.A.; Berry, J.F. Diamagnetic Corrections and Pascal's Constants. *J. Chem. Educ.* **2008**, *85*, 532–536. [CrossRef]
15. Bruker AXS. *APEX2*; Bruker: Billerica, MA, USA, 2012.
16. Sheldrick, G.M. *Program SADABS: Area-Detector Absorption Correction*; University of Göttingen: Göttingen, Germany, 1996.
17. Sheldrick, G.M. Crystal structure refinement with SHELXL. *Acta Crystallogr. C Struct. Chem.* **2015**, *71*, 3–8. [CrossRef] [PubMed]
18. Hubschle, C.B.; Sheldrick, G.M.; Dittrich, B. ShelXle: A Qt graphical user interface for SHELXL. *J. Appl. Crystallogr.* **2011**, *44*, 1281–1284. [CrossRef]
19. Hooft, R.W.W. *Collect*; Nonius BV: Delft, The Netherlands, 1998.
20. Sheldrick, G.M. *SADABS, Program for Area Detector Adsorption Correction*; Institute for Inorganic Chemistry, University of Göttingen: Göttingen, Germany, 1996.
21. Sheldrick, G.M. *SHELXL-97, Program for the Refinement of Crystal Structure from Diffraction Data*; University of Göttingen: Göttingen, Germany, 1997.
22. Addison, A.W.; Rao, T.N.; Reedijk, J.; van Rijn, J.; Verschoor, G.C. Synthesis, structure, and spectroscopic properties of copper(II) compounds containing nitrogen–sulphur donor ligands; the crystal and molecular structure of aqua [1,7-bis(N-methylbenzimidazol-2′-yl)-2,6-dithiaheptane] copper(II) perchlorate. *J. Chem. Soc. Dalton Trans.* **1984**, 1349–1356. [CrossRef]
23. Gil-Hernández, B.; Calahorro, A.J.; Gili, P.; Sanchiz, J. Effect of the apical ligand on the geometry and magnetic properties of copper(II)/mesoxalate trinuclear units. *Dalton Trans.* **2017**, *46*, 5260–5268. [CrossRef]
24. Gil-Hernández, B.; Gili, P.; Pasán, J.; Sanchiz, J.; Ruiz-Pérez, C. Two-dimensional (6,3) networks obtained with the $\{Cu_3(Hmesox)_3\}^{3-}$ secondary building unit ($H_4$mesox = mesoxalic acid). *CrystEngComm* **2012**, *14*, 4289–4297. [CrossRef]
25. Harrowfield, J.M.; Sobolev, A.N. The Crystal Structure of Tris(2,2′-bipyridine)ruthenium(II) Perchlorate. *Aust. J. Chem.* **1994**, *47*, 763. [CrossRef]
26. Rillema, D.P.; Jones, D.S. Structure of tris(2,2′-bipyridyl)ruthenium(II) hexafluorophosphate, $[Ru(bipy)_3][PF_6]_2$; X-ray crystallographic determination. *J. Chem. Soc. Chem. Commun.* **1979**, 849–851. [CrossRef]
27. Kahn, O. *Molecular Magnetism*; VCH: New York, NY, USA, 1993.
28. Chilton, N.F.; Anderson, R.P.; Turner, L.D.; Soncini, A.; Murray, K.S. PHI: A powerful new program for the analysis of anisotropic monomeric and exchange-coupled polynuclear d- and f-block complexes. *J. Comput. Chem.* **2013**, *34*, 1164–1175. [CrossRef] [PubMed]
29. Pasán, J.; Delgado, F.S.; Rodríguez-Martín, Y.; Hernández-Molina, M.; Ruiz-Pérez, C.; Sanchiz, J.; Lloret, F.; Julve, M. Malonate-based copper(II) coordination compounds: Ferromagnetic coupling controlled by dicarboxylates. *Polyhedron* **2003**, *22*, 2143–2153. [CrossRef]
30. Rodríguez-Martín, Y.; Hernández-Molina, M.; Delgado, F.S.; Pasán, J.; Ruiz-Pérez, C.; Sanchiz, J.; Lloret, F.; Julve, M. Structural versatility of the malonate ligand as a tool for crystal engineering in the design of molecular magnets. *CrystEngComm* **2002**, *4*, 522–535. [CrossRef]
31. Delgado, F.S.; Ruiz-Pérez, C.; Sanchiz, J.; Lloret, F.; Julve, M. Versatile supramolecular self-assembly Part II. Network formation and magnetic behaviour of copper(II) malonate anions in ammonium derivatives. *CrystEngComm* **2006**, *8*, 530–544. [CrossRef]

32. Julve, M.; Verdaguer, M.; Gleizes, A.; Philoche-Levisalles, M.; Kahn, O. Design of .mu.-oxalato copper(II) binuclear complexes exhibiting expected magnetic properties. *Inorg. Chem.* **1984**, *23*, 3808–3818. [CrossRef]
33. Calatayud, M.L.; Orts-Arroyo, M.; Julve, M.; Lloret, F.; Marino, N.; De Munno, G.; Ruiz-García, R.; Castro, I. Magneto-structural correlations in asymmetric oxalato-bridged dicopper(II) complexes with polymethyl-substituted pyrazole ligands. *J. Coord. Chem.* **2018**, *71*, 657–674. [CrossRef]
34. Ferrer, S.; Lloret, F.; Bertomeu, I.; Alzuet, G.; Borrás, J.; García-Granda, S.; Liu-González, M.; Haasnoot, J.G. Cyclic Trinuclear and Chain of Cyclic Trinuclear Copper(II) Complexes Containing a Pyramidal $Cu_3O(H)$ Core. Crystal Structures and Magnetic Properties of $[Cu_3(\mu_3\text{-OH})(aaat)_3(H_2O)_3](NO_3)_2 \cdot H_2O$ [aaat = 3-Acetylamino-5-amino-1,2,4-triazolate] and $\{[Cu_3(\mu_3\text{-OH})(aat)_3(\mu_3\text{-}SO_4)] \cdot 6H_2O\}_n$ [aat = 3-Acetylamino-1,2,4-triazolate]: New Cases of Spin-Frustrated Systems. *Inorg. Chem.* **2002**, *41*, 5821–5830. [CrossRef]
35. Ferrer, S.; Lloret, F.; Pardo, E.; Clemente-Juan, J.M.; Liu-Gonzalez, M.; Garcia-Granda, S. Antisymmetric exchange in triangular tricopper(II) complexes: Correlation among structural, magnetic, and electron paramagnetic resonance parameters. *Inorg. Chem.* **2012**, *51*, 985–1001. [CrossRef]
36. Lytle, F.E.; Hercules, D.M. Luminescence of tris (2, 2′-bipyridine) ruthenium (II) dichloride. *J. Am. Chem. Soc.* **1969**, *91*, 253–257. [CrossRef]

magnetochemistry

MDPI

*Article*

# Building Up a Hexacopper(II)-Pyrazolate/Oxamate Magnetic Complex with Rare Ethane-1,2-Dioxide (–$OCH_2CH_2O$–) as a Bridge Between Copper(II) Units

**Willian X. C. Oliveira [1,*], Victor G. Araújo [1], Carlos B. Pinheiro [2], Miguel Julve [3,†] and Cynthia L. M. Pereira [1,*]**

1 Departamento de Química, Instituto de Ciências Exatas, UFMG—Universidade Federal de Minas Gerais, Av. Pres. Antônio Carlos, 6627, Pampulha, Belo Horizonte 31270-901, MG, Brazil
2 Departamento de Física, Instituto de Ciências Exatas, UFMG—Universidade Federal de Minas Gerais, Av. Pres. Antônio Carlos, 6627, Pampulha, Belo Horizonte 31270-901, MG, Brazil
3 Instituto de Ciencia Molecular (ICMol), Universitat de València, C/Catedrático José Beltrán 2, 46980 Paterna, València, Spain
* Correspondence: wxcoliveira@ufmg.br (W.X.C.O.); cynthialopes@ufmg.br (C.L.M.P.)
† Dedicated to the memory of Prof. Miguel Julve, who kindly accepted to supervise Willian X. C. Oliveira during part of his PhD in Spain (2014–2015), where the compound presented in this work was synthesized for the first time. He also participated in the discussion of the results before his untimely passing.

**Abstract:** The synthesis, structural, and magnetic characterization of a novel neutral copper(II) hexanuclear complex $[Cu_6(en)_4(OCH_2CH_2O)_2(pyox)_4]\cdot 3eg\cdot en\cdot 12H_2O$ (**1**) was investigated [en = ethylenediamine, eg = ethylene glycol, and $H_2pyox$ = 4-(1H-pyrazole-4-yl)phenylene-N-oxamic acid]. The crystal structure of **1**, obtained by the single-crystal X-ray diffraction technique, revealed that the hexacopper(II) complex is built from two linear tricopper(II) complex subunits. Each subunit contains two $[Cu(en)]^{2+}$ moieties connected to a $[Cu(OCH_2CH_2O)]$ unit by two $pyox^{2-}$ ligands acting as $\mu$-$\kappa$N:$\kappa$N′ bridges, as well as a $[OCH_2CH_2O]^{2-}$ ligand, which is ultimately found in the $\mu_3$-$\kappa$O,O′:$\kappa$O:$\kappa$O′ coordination form. The subunits are connected via the amide portion of the $pyox^{2-}$ ligand, linked to copper atoms in the other subunit. They occupy the apical coordination positions, leading the trinuclear copper(II) segments to be almost perpendicular. The structural, chemical, and spectroscopic characterizations evidenced that ethylene glycol acted both as a solvent and a reactant upon deprotonation, forming the –$OCH_2CH_2O$– ligand due to the basic crystallization environment. DC magnetic studies revealed a strong antiferromagnetic interaction between the copper atoms within the trinuclear subunits, influenced by alkoxide and pyrazolate bridging ligands. Our findings offer new insights into the structural and magnetic properties of copper(II) complexes, enhancing the understanding of metal–ligand interactions in supramolecular chemistry.

**Keywords:** molecular magnetism; hexacopper(II); oxamate; pyrazolate; ethylene glycol; ethane-1,2-dioxide

## 1. Introduction

Polynuclear complexes have attracted significant attention in coordination chemistry, particularly their magnetic properties and potential applications in molecular materials with tunable magnetic behavior [1,2]. To achieve magnetic interactions between metal ions, orbital overlapping through the ligands is essential. In this context, bridging ligands with $sp^2$ or $sp$-hybridized atoms effectively facilitate spin density communication between metal centers [1,3,4]. Notably, oxamate and pyrazolate ligands have proven highly effective in bridging magnetic metal ions.

When oxamate acts as a bis-bidentate ligand ($\mu$-$\kappa^2$N,O:$\kappa^2$O′,O″ coordination mode), it can achieve coupling constants exceeding $-400\ cm^{-1}$ [5,6]. Pyrazolate, on the other hand, in its $\mu$-$\kappa$N:$\kappa$N′ coordination form, exhibits coupling capabilities that are strongly influenced by the angle between the metal–nitrogen bonds, though it is not far behind oxamate in terms of performance [7]. Thus, incorporating both oxamate and pyrazolate moieties within the same ligand holds great promise for creating strongly coupled polymetallic systems.

On the other hand, synthesizing polynuclear complexes can be very challenging, and self-assembly processes can significantly facilitate this, especially when faced with poorly soluble products. Allowing the components to adopt the most stable conformation spontaneously has already led to the discovery of remarkable magnetic polynuclear compounds, such as the square-like $[Cr^{III}_2(dmso)_8\{\mu\text{-}cis\text{-}Nb^{V}O_2(C_2O_4)_2\}_2]$ [8]; the temperature influenced the formation of the 1D ladder-like coordination polymer $[Mn_4(H_2mpba)_4(H_2O)_{12}]_n$ $\{[Mn_8Cu_8(mpba)_8(H_2O)_{24}]_n\}\cdot 29.5nH_2O$ and 2D coordination polymer $[Mn_4Cu_4(mpba)_4(H_2O)_9]_n\cdot 14nH_2O$ [9]; the discrete $[Cu(bipy)(H_2mpba)]_2\cdot 2H_2O$ and 1D coordination polymer $[Cu(bipy)(H_2mpba)]_n\cdot n$dmso systems, which can interchange based on solvent soaking [10]; the 2D coordination polymer with magnetization hysteresis $[Co_2Cu_2(mpba)_2(H_2O)_6]_n\cdot 6nH_2O$ [11]; and so on ($H_4$mpba = *N,N′*-1,3-phenilenebis(oxamic acid), bipy = 2,2′-bipyridine). This synthetic strategy often reveals unusual phenomena, including ligand displacement [8], solvent coordination, ligand transformation [12], and, in the case of this work, the formation of the rare ligand ethane-1,2-dioxide from ethylene glycol (eg) solvent doubly deprotonated in the reaction–crystallization process.

The ethane-1,2-dioxide ($OCH_2CH_2O$) ligand is only found as a bridging ligand when coordinated to transition metal ions [13–21] and group 13–15 metal complexes from the periodic table [21–31]. It has been observed that using the diol itself leads to natural deprotonation, forming the polynuclear complex upon moderate heating (100–200 °C) [13,15,18,20,21,23,24,27,31] or via proton exchange in alkoxide complexes [14,16,24,28]. Manganese examples with ethane-1,2-dioxide are the only ones where the simple addition of a weak base is enough to deprotonate the ethylene glycol at room temperature [9,17]. While several polynuclear complexes containing ethane-1,2-dioxide have been reported, only one other crystal structure presenting it with copper(II) ions has been reported [20].

This contribution explores a novel hexanuclear copper(II) complex coordinated by a ligand combining the oxamate and pyrazolate groups—pyox$^{2-}$ ($H_2$pyox = 4-(1H-pyrazole-4-yl)phenylene-N-oxamic acid), the ethane-1,2-dioxide ligand and ethylenediamine. To the best of our knowledge, this is a rare example from the literature that uses the ethane-1,2-dioxide ligand to connect copper atoms (Scheme 1). The synergistic interaction between these three distinct ligands allows for an in-depth investigation of the correlations between structure and magnetic behavior, with potential applications in developing new functional magnetic materials [32].

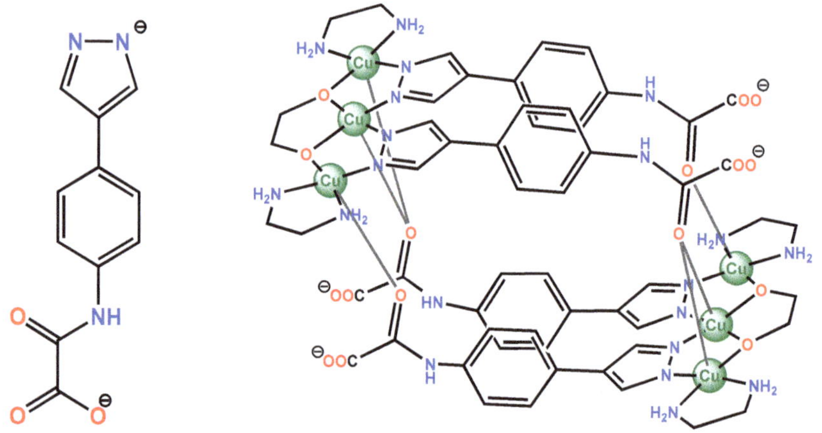

(a) (b)

**Scheme 1.** Molecular structures of (**a**) the pyox$^{2-}$ ligand and (**b**) the hexanuclear complex **1**. The grey bonds in (**b**) highlight the connection of the trimetallic units to form the hexanuclear.

## 2. Experimental Method

All chemicals were used without further purification. The compound ($n$-$Bu_4N$)Hpyox·0.5$H_2O$ was prepared as previously described in the literature [33,34]. [Cu(en)$_2$]($NO_3$)$_2$ (en = ethylenediamine) preparation followed the synthetic procedure described by Rafat et al. [35].

Synthesis of [$Cu_6$(en)$_4$($OCH_2CH_2O$)$_2$(pyox)$_4$]·3eg·en·12$H_2O$ (**1**). In the bottom of a small test tube (~5 mL capacity) was placed 0.5 mL of a solution containing 10.0 mg (32.5 μmol) of [Cu(en)$_2$]($NO_3$)$_2$ and 1.00 mg (4.14 μmol) of Cu($NO_3$)$_2$·3$H_2O$ dissolved in a mixture of ethylene glycol (eg) and methanol 1:1 $v/v$. Then, 2.0 mL of a mixture of methanol and ethylene glycol 4:1 $v/v$ containing 1% in volume of $Bu_4NOH$ (40% weight in water) was carefully placed over the bottom solution to form a separate layer. Finally, a third layer comprising a methanolic solution (0.250 mL) of ($n$-$Bu_4N$)Hpyox·0.5 $H_2O$ (30.0 mg, 62.5 μmol) was placed on the top. The tube was sealed and the materials were left to diffuse at room temperature (20 °C) for two weeks. After this time, dark blue crystals were observed in the tube wall. The crystals were collected, washed with methanol, and left to dry in an open atmosphere for 48 h. This synthesis was replicated using several test tubes, diffusing simultaneously to obtain enough material for all studies. Yield: 71.0% (6.25 mg; 2.94 μmol). Elemental Analysis [Exp.(calcd.)] for $C_{64}H_{116}N_{20}O_{36}Cu_6$ (2123.01 g mol$^{-1}$): %C 37.28 (36.21), %H 4.63 (5.51), %N 14.17 (13.20), %Cu 17.89 (17.96). IR (cm$^{-1}$): 3250 and 3129 (ν N–H), 1632 (ν C=O), 1044 (ν C–O).

### 2.1. Chemical Characterizations

The thermogravimetric (TG) and differential thermal analysis (DTA) of **1** was carried out in a Shimadzu DTG-60H thermobalance using a *ca.* 5.0 mg sample in an alumina crucible, using synthetic air flow at 50.0 mL min$^{-1}$, from room temperature to 800 °C (Figure S1). The IR spectrum was obtained in a Perkin Elmer FTIR GX spectrometer in ATR mode, from 400 to 4000 cm$^{-1}$, at 4 cm$^{-1}$ spectral resolution (Figure S2). Elemental analysis was carried out in a CHN 2400 Perkin-Elmer analyzer (PerkinElmer, Waltham, MA, USA) for carbon, hydrogen, and nitrogen content analysis, and a VARIAN AA240FS spectrophotometer (VARIAN, Palo Alto, CA, USA) with flame atomization was used for copper content analysis.

### 2.2. X-Ray Diffraction

X-ray diffraction data collection on single crystals **1** was performed with a Rigaku Gemini diffractometer using Mo-Kα radiations (0.71073 Å). Measurements were performed at 120 K. Data integration and the reflections for all compounds were scaled with the CRYSALIS suite [36]. Final unit cell parameters were based on the fitting of all reflection positions. Analytical absorption corrections and the space group identification were performed using the CRYSALIS suite [36]. The structures of all compounds were solved by direct methods using the SUPERFLIP program [37]. For each compound, the positions of all atoms could be unambiguously assigned on consecutive difference Fourier maps. Refinements were performed using SHELXL-2019 [38] based on $F^2$ through the full-matrix least-squares routine. All non-hydrogen atoms and non-water molecules were refined with anisotropic atomic displacement parameters. The *SQUEEZE* technique [39] was applied to remove the scattered electronic density over the tubular voids occupied by disordered solvent molecules. The process removed 483 electrons over 1361 Å$^3$ per unit cell. All hydrogen atoms were located in Fourier difference maps and included as fixed contributions according to the riding model [40]. The following restrain values were used in the structural refinements: C–H = 0.93 Å and $U_{iso}$(H) = 1.2 $U_{eq}$(C) for the aromatic carbon atoms, C–H = 0.97 Å and $U_{iso}$(H) = 1.2 $U_{eq}$(C) for methylene carbon atoms, N–H = 0.89 Å and $U_{iso}$(H) = 1.2 $U_{eq}$(N) for aromatic amides, and N–H = 0.91 Å and $U_{iso}$(H) = 1.2 $U_{eq}$(N) for coordinated amines. Bulk and single crystal identities were checked by comparing the calculated (from single crystals) and experimental X-ray powder diffraction patterns (PXRD). The PXRD of **1** was obtained in a PanAnalytical Empyrean diffractometer, using Cu-Kα radiation in $\theta$-$\theta$ mode (0.02° s$^{-1}$) with crushed single crystals mounted in a gyratory zero-background silicon sample holder. Figures of the crystalline structure of **1** were prepared using Mercury® [41] and Vesta [42] software. The crystal structure data were deposited under CCDC code 238974 and can be obtained free of charge from The Cambridge Crystallographic Data Centre via www.ccdc.cam.ac.uk/data_request/cif (accessed on 30 October 2024). Crystal data, data collection, temperature, and structure refinement details are summarized in Table 1.

**Table 1.** Summary of the crystallographic parameters of **1**.

| Compound | 1 |
|---|---|
| Chemical Formula | $C_{56}H_{68}N_{20}O_{16}Cu_6$ |
| $M$/g mol$^{-1}$ | 1658.54 * |
| $\lambda$/Å | 0.71073 |
| Crystal System | Triclinic |
| Space Group | $P\bar{1}$ |
| $a$/Å | 13.9356(4) |
| $b$/Å | 15.5338(5) |
| $c$/Å | 22.5074(7) |
| $\alpha$/° | 72.610(3) |
| $\beta$/° | 76.845(3) |
| $\gamma$/° | 67.435(3) |
| Volume/Å$^3$ | 4258.2(3) |
| $T$/K | 120(2) |
| Z | 2 |
| $\rho_{calc}$/g cm$^{-3}$ | 1.294 |
| Crystal size/mm$^3$ | 0.131 × 0.279 × 0.435 |
| μ/mm$^{-1}$ | 1.535 |
| Reflections collected ($R_{int}$) | 51,218 (0.0525) |
| Independent reflections | 17,441 |
| Reflections with $I \geq 2\sigma$ ($I$) | 11,551 |
| $R$ ($R$ total) [a] | 0.0566 (0.1427) |
| $wR$ (wR total) [b] | 0.0849 (0.1568) |
| $S$ [c] | 1.048 |
| $\rho_{max}$, $\rho_{min}$/$e$ Å$^{-3}$ | +0.988; −0.520 |
| CCDC number | 2,389,741 |

[a] $R = \Sigma||F_o| - |F_c||/\Sigma|F_o|$. [b] $wR = [\Sigma w(|F_o|^2 - |F_c|^2)^2/\Sigma w|F_o|^2]^{1/2}$. [c] $S = [\Sigma w(|F_o|^2 - |F_c|^2)^2/(n_o - n_p)]^{1/2}$, where $w \propto 1/\sigma$, $n_o$ = observed parameters, and $n_p$ = fitted parameters. * The *SQUEEZE* tool was applied to remove disordered solvent molecules inside the crystal structure pores.

### 2.3. Magnetic Measurements

The magnetic susceptibility measurements were carried out on single crystals **1** in the temperature range of 3.0–265 K with a Quantum Design SQUID magnetometer and using an applied magnetic field of 0.1 T. The sample was inserted in the SQUID at 265 K to quickly freeze it to minimize crystallization solvent evaporation during the air removal process. Upon reaching 3.0 K, the temperature was fixated, and the magnetization was evaluated by varying the magnetic field in the 0.1–7.0 T range. Diamagnetic corrections of the constituent atoms were estimated from Pascal's constants [43]. Experimental susceptibilities were also corrected for the magnetization of the sample holder (a plastic bag).

## 3. Results

### 3.1. Synthesis and Product Formation Comments

Compound **1** was obtained in a basic medium. Neutral or acidic environments did not lead to the formation of **1**. The medium was tested by basifying it with a strong base ($n$-$Bu_4NOH$) or a weak base ($Et_3N$), but only the strong one led to the formation of **1**. The strong base is necessary to deprotonate the pyrazole group—which could be deprotonated with weaker bases—and to deprotonate the coordinated ethylene glycol and promote the formation of the ethane-1,2-dioxide ligand. It is the second time that ethylene glycol has been observed in a doubly deprotonated form, acting as a bridging ligand for three copper(II) ions. It is the first time it has led to the formation of a discrete polynuclear compound. Its deprotonation is highly supported due to the lack of vibrations in the IR spectrum characteristic of nitrate anions, which, if present, would indicate that the hexanuclear complex is cationic due to the coordination of ethylene glycol in its neutral form. Its deprotonation is also supported by the fact that a crystallization ethylene glycol molecule (before the *SQUEEZE* tool was applied) is guiding its O–H group towards the coordinated oxygen atoms, indicating a lack of hydrogen atoms in the coordinated ethane-1,2-dioxide, acting as a hydrogen bond acceptor.

The thermal analysis (Figure S1) was used to characterize the solvent content in the voids. It revealed an initial mass loss of 10.1% between room temperature and 130 °C, as an endothermic event. It is associated with the evaporation of crystallization solvent molecules. Between 130 °C and 400 °C, four exothermic events were observed related to organic matter oxidation. After that, up to 800 °C, a plateau is observed at 22.5% of the residual mass, indicating complete oxidation of the organic matter, with CuO most likely as the residue. To investigate the solvent loss, we must consider that the methanol and ethylene glycol used in the synthesis of **1** were used without a drying process, meaning that they had a considerable water content, which must be occupying the voids. Additionally, after letting the sample dry in the open atmosphere and at room temperature, the methanol molecules inside the voids can evaporate and be replaced by the moisture in the air. The first weight loss corresponds to 12 water molecules per hexanuclear complex (calculated = 10.2%). The organic matter oxidation masks the release of crystallization ethylene glycol and ethylenediamine molecules, but the weight loss of 68.1% is associated with four pyox ligands, five ethylene glycol molecules (two in the form of ethane-1,2-dioxide ligand and three of crystallization), and five ethylenediamines (four as ligands and one of crystallization), which, when combined, correspond to 70.3% of **1′**s mass. The residual mass corresponds to six moles of CuO per complex, as expected (calculated = 22.5%).

To support the characterization of the void content, the *SQUEEZE* tool of *PLATON* was used to remove all electron density associated with the crystallization solvent molecules. Per unit cell, a total of 483 electrons were removed, spread across 1360 Å$^3$. Since Z = 2 for **1**, a total of 242 electrons are associated with the crystallization molecules per hexanuclear unit. Combining the 12 water molecules, the three ethylene glycol molecules, and one ethylenediamine group ($C_8H_{50}N_2O_{18}$) resulted in 220 electrons. Considering the atomic volume (approximately. 14.0 Å$^3$ for non-hydrogen atoms, i.e., atomic radii of about 1.5 Å, and 5.6 Å$^3$ for hydrogen atoms, with atomic radii of about 1.1 Å), these molecules would occupy a volume 667 Å$^3$, almost half of the volume occupied by the electron density eliminated using the *SQUEEZE* tool. The combination of the electron count and squeezed volume is very close to the proposed void content, in agreement with the observed thermal analysis, and also aligns with the results of both elemental and thermogravimetric analysis.

### 3.2. Crystal Structure Description

Compound **1** is a neutral copper(II) hexanuclear complex crystallized in the centrosymmetric space group $P\overline{1}$, as seen in Figure 1a. It crystallizes with several solvent molecules that were omitted due to their high disorder, occupying a large tubular void. This polynuclear complex is composed of six metal atoms, four pyox$^{2-}$ ligands, four ethylenediamine ligands, and two doubly deprotonated ethylene glycol ligands. It consists of two subunits, each containing half of the complex atoms, resulting in two trimetallic subunits, as shown in Figure 1b,c. The three copper atoms in each subunit are in a linear arrangement, linked by pyrazolate groups in a $\mu$-$\kappa$N:$\kappa$N′ coordination fashion. The central metal ion is coordinated to ethane-1,2-dioxide, while the terminal copper atoms are each coordinated to one ethylenediamine molecule, as well as one oxygen atom of the ethane-1,2-dioxide ligand, which is found in the $\mu_3$-$\kappa^2$OO′:$\kappa$O:$\kappa$O′ double-bridging mode. Thus, the metal atoms are bridged by pyrazolate and ethane-1,2-dioxide ligands, forming two five-membered rings fused on the central copper atom. The metal–ligand distances corroborate the presence of ethane-1,2-dioxide, as they are in the range of 1.918–1.964 Å, shorter than those of the ethylenediamine and pyrazolate ligands, whose bond lengths range from 1.981 to 2.018 Å (see Table 2), and also shorter than the other observed examples of ethylene glycol coordinated to divalent copper ions [44–46].

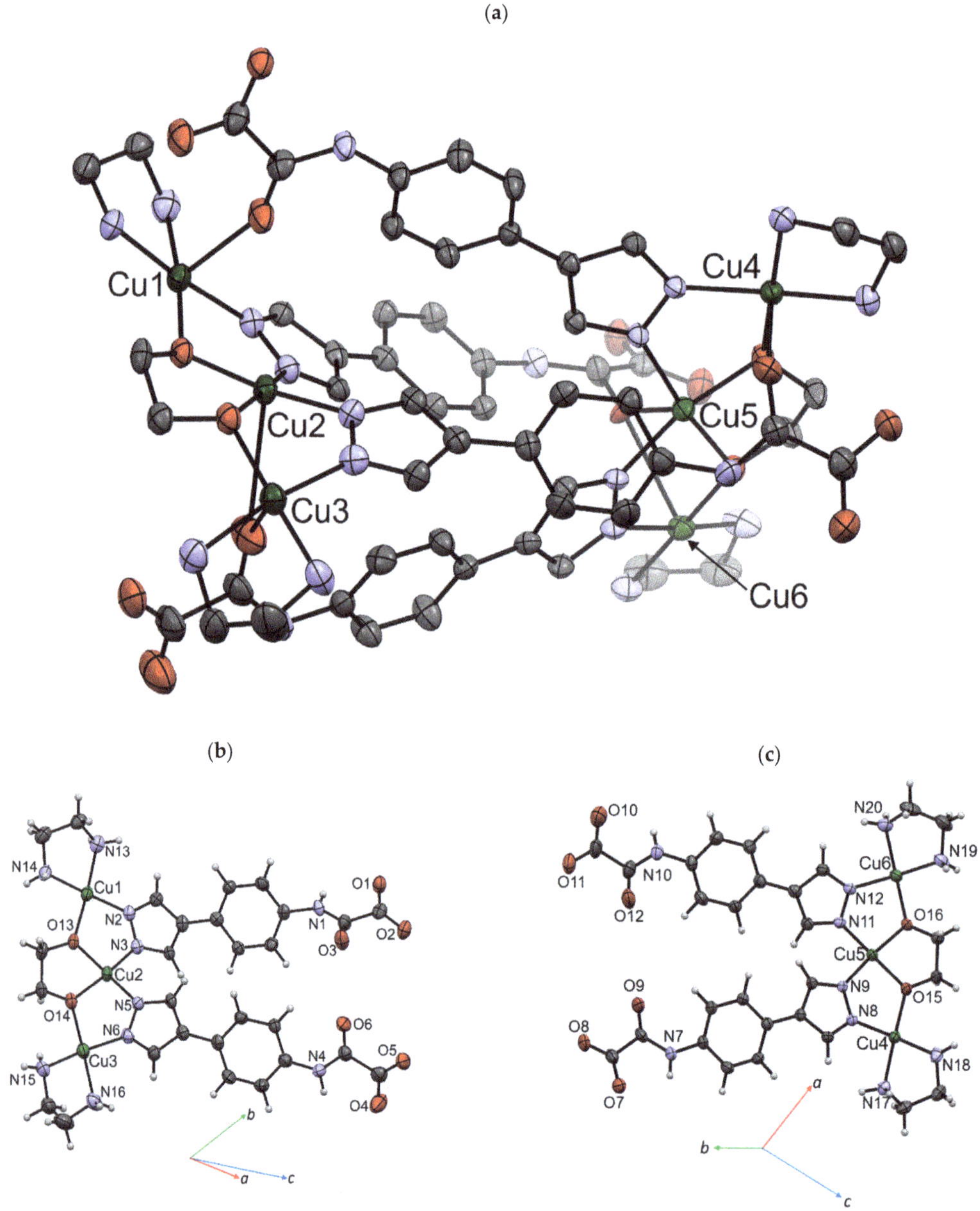


**Figure 1.** (**a**) Crystal structure of the hexanuclear complex **1**. To understand the tridimensional conformation of this complex, lighter hues indicate atoms positioned deeper within the structure. (**b**,**c**) The two subunits showcase pyrazolate and ethylene glycolate bridges. Atom color scheme: carbon—gray; hydrogen—white spheres, oxygen—red, nitrogen—light blue, and copper—dark green. Ellipsoids represent 50% of probability levels.

**Table 2.** Selected bond lengths and angles for metal atoms in **1**.

| Cu1 | | | | Cu2 | | | | Cu3 | | | |
|---|---|---|---|---|---|---|---|---|---|---|---|
| Cu1–L/Å | | L–Cu1–L/° | | Cu2–L/Å | | L–Cu2–L/° | | Cu3–L/Å | | L–Cu3–L/° | |
| Cu1–N2 | 1.981(3) | N2–Cu1–N13 | 94.94(14) | Cu2–N3 | 1.952(3) | N3–Cu2–N5 | 100.11(13) | Cu3–N6 | 1.964(3) | N6–Cu3–N15 | 173.16(16) |
| Cu1–N13 | 2.029(4) | N2–Cu1–N14 | 176.47(14) | Cu2–N5 | 1.961(3) | N3–Cu2–O9 | 88.8(3) | Cu3–N15 | 1.992(4) | N6–Cu3–N16 | 93.04(15) |
| Cu1–N14 | 1.993(3) | N2–Cu1–O12 | 92.24(10) | Cu2–O9 | 2.71(1) | N3–Cu2–O13 | 87.13(12) | Cu3–N16 | 2.019(4) | N6–Cu3–O9 | 78.0(7) |
| Cu1–O12 | 2.527(4) | N2–Cu1–O13 | 85.56(12) | Cu2–O13 | 1.926(3) | N3–Cu2–O14 | 169.07(13) | Cu3–O9 | 2.64(2) | N6–Cu3–O14 | 87.95(13) |
| Cu1–O13 | 1.918(3) | N13–Cu1–N14 | 83.87(15) | Cu2–O14 | 1.921(3) | N5–Cu2–O9 | 86.2(3) | Cu3–O14 | 1.940(3) | N15–Cu3–N16 | 84.25(16) |
| | | N13–Cu1–O12 | 84.61(10) | | | N5–Cu2–O13 | 163.58(13) | | | N15–Cu3–O9 | 96.3(7) |
| | | N13–Cu1–O13 | 176.10(13) | | | N5–Cu2–O14 | 88.53(13) | | | N15–Cu3–O14 | 94.58(14) |
| | | N14–Cu1–O12 | 85.35(10) | | | O9–Cu2–O13 | 108.9(3) | | | N16–Cu3–O9 | 100.8(7) |
| | | N14–Cu1–O13 | 95.86(13) | | | O9–Cu2–O14 | 85.1(3) | | | N16–Cu3–O14 | 178.08(14) |
| | | O12–Cu1–O13 | 99.24(10) | | | O13–Cu2–O14 | 86.30(12) | | | O9–Cu3–O14 | 77.8(7) |
| Cu4 | | | | Cu5 | | | | Cu6 | | | |
| Cu4–L/Å | | L–Cu4–L/° | | Cu5–L/Å | | L–Cu5–L/° | | Cu6–L/Å | | L–Cu6–L/° | |
| Cu4–N8 | 1.983(3) | N8–Cu4–N17 | 95.70(15) | Cu5–N9 | 1.952(3) | N9–Cu5–N12 | 99.68(14) | Cu6–N11 | 1.963(4) | N11–Cu6–N19 | 163.27(16) |
| Cu4–N17 | 2.016(4) | N8–Cu4–N18 | 173.99(16) | Cu5–N12 | 1.956(3) | N9–Cu5–O6 | 97.18(10) | Cu6–N19 | 2.018(4) | N11–Cu6–N20 | 93.10(15) |
| Cu4–N18 | 2.021(3) | N8–Cu4–O3 | 87.02(10) | Cu6–O6 | 2.934(4) | N9–Cu5–O15 | 86.11(13) | Cu6–N20 | 2.018(4) | N11–Cu6–O6 | 85.25(10) |
| Cu4–O3 | 2.442(3) | N8–Cu4–O15 | 85.74(13) | Cu5–O15 | 1.930(3) | N9–Cu5–O16 | 168.82(14) | Cu6–O6 | 2.846(4) | N11–Cu6–O16 | 87.19(13) |
| Cu4–O15 | 1.921(3) | N17–Cu4–N18 | 83.63(15) | Cu5–O16 | 1.951(3) | N12–Cu5–O6 | 84.66(10) | Cu6–O16 | 1.964(3) | N19–Cu6–N20 | 84.26(16) |
| | | N17–Cu4–O3 | 95.48(19) | | | N12–Cu5–O15 | 168.58(14) | | | N19–Cu6–O6 | 79.74(19) |
| | | N17–Cu4–O15 | 171.64(15) | | | N12–Cu5–O16 | 88.78(13) | | | N19–Cu6–O16 | 96.89(14) |
| | | N18–Cu4–O3 | 87.11(19) | | | O6–Cu5–O15 | 104.49(10) | | | N20–Cu6–O16 | 174.90(16) |
| | | N18–Cu4–O15 | 95.79(13) | | | O6–Cu5–O16 | 76.22(10) | | | N20–Cu6–O6 | 106.82(19) |
| | | O3–Cu4–O15 | 92.81(10) | | | O15–Cu5–O16 | 86.86(12) | | | O6–Cu6–O16 | 78.29(10) |

The sum of the bridging ligands culminates in the copper atoms being very close to each other inside each trinuclear subunit. The intramolecular metal–metal distances are 3.3317(6) Å for Cu1···Cu2, 3.2572(7) Å for Cu2···Cu3 and 6.3156(7) Å for Cu1···Cu3, with an angle of 146.88(2)° for Cu1···Cu2···Cu3. For the second trinuclear subunit, the distances are equal to 3.2787(7) Å for Cu4···Cu5, 3.2264(9) Å for Cu5···Cu6, and 6.2565(8) Å Cu4···Cu6, with an angle of 148.22(2)° for Cu4···Cu5···Cu6. The angle between the coordination planes of the connected metal atoms is equal to 30.9(4)° for Cu1N2N13N14 and Cu2N3N5O13O14 mean planes, 35.8(4)° for the angle between the latter and the Cu3N6N15N16O14 mean plane, 37.8(5)° for Cu4N8N17N18O15 and Cu5N9N11O15O16 mean planes, and finally 47.5(4)° between the latter and the Cu6N12N19N20O16 mean plane. The torsion angle around the pyrazolate N–N bond is equal to −4.1(1.4)° for Cu1–N2–N3–Cu2, −5.5(1.2)° for Cu2–N5–N6–Cu3, 0.3(1.4)° for Cu4–N8–N9–Cu5, and −0.43(1.3)° for Cu5–N11–N12–Cu6. It is also important to note the angle concerning the alkoxide bridging ligand of the ethane-1,2-dioxide ligand: 120.23(14)° for Cu1–O13–Cu2, 115.20(14)° for Cu2–O14–Cu3, 116.76(14)° for Cu4–O15–Cu5, and 110.96(14)° for Cu5–O16–Cu6.

The oxamate portion of each subunit plays a role in connecting the subunits. The amide oxygen atoms coordinate with the copper atoms of the other subunit. In each subunit, one amide group is monodentate, while the other bridges the central and a terminal copper atom. Consequently, all copper atoms are found in five coordination spheres, as shown in Figure 2a,b. The outer copper atoms (Cu1, Cu3, Cu4, and Cu6) are in a $CuN_3O_2$ environment, while the central ones (Cu2 and Cu5) are in a $CuN_2O_3$ coordination sphere. The trigonality parameter $\tau$ varies from 0.01 to 0.19, indicating a coordination geometry closer to square-based pyramidal ($\tau = 0$ represents perfect square-based geometry, and $\tau = 1$ represents trigonal bipyramidal geometry) [47]. The major distortion in the five-coordinate sphere for the copper atoms arises from the apical amide oxygen atom, which is shared by the Cu2, Cu3, Cu5, and Cu6 ligands. This sharing results in several geometric restrictions that cause both bond length elongation and bond angle deviation from the ideal (see Table 2). Due to this significant distortion in the coordination geometry and to corroborate the results obtained from the trigonality parameter analysis, the SHAPE software [48] was employed, using the metal and ligand atomic positions extracted from the crystal structure. Although deviations from perfect coordination geometries were confirmed for all copper atoms, they are closest to square-based geometry (See Table S2 in Supplementary Information). Finally, it is important to highlight that in the hexanuclear complex, the trimetallic units are entirely separated from each other. The closest intramolecular metal–metal distance is 9.5055(3) Å for Cu2···Cu5.

Although the carboxylate group has a higher electron density than the carbonyl portion of the amide, its inability to coordinate with copper atoms is attributed to molecular recognition involving intramolecular hydrogen bonds. All carboxylate groups are found to interact with the N–H groups of ethylenediamine (see Figure 2). The combination of intramolecular hydrogen bonds and the coordination of the metal by the amide carbonyl stabilizes the complex more than if the carboxylate were only coordinated with the copper, as the apical Cu–O bond lengths are quite large and might not provide sufficient stabilization if the carboxylate were solely coordinated.

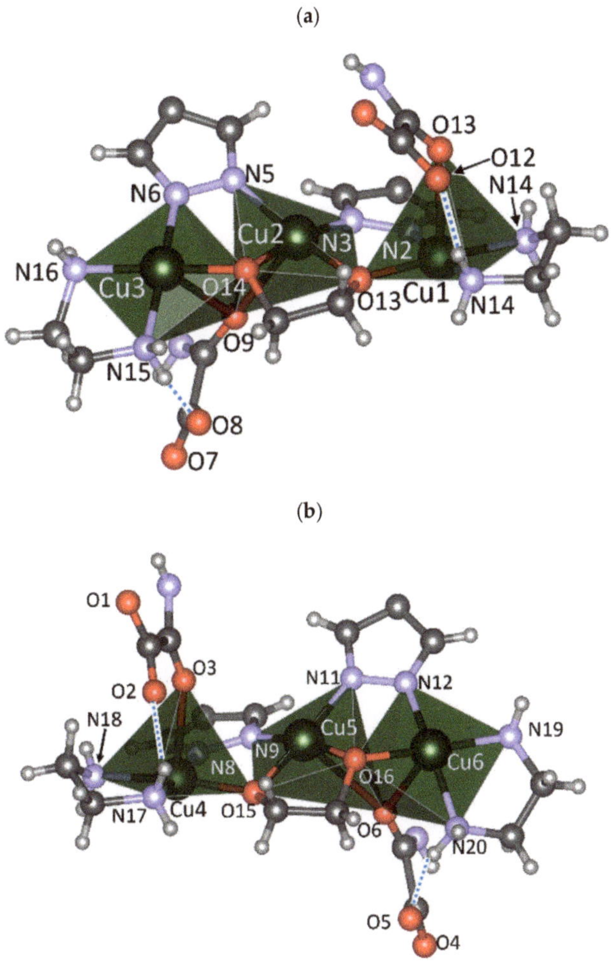


**Figure 2.** Coordination polyhedra of (**a**) Cu1, Cu2, Cu3 and (**b**) Cu4, Cu5, Cu6 in **1**. Dashed lines represent intramolecular ethylenediamine-to-carboxylate hydrogen bonds. Color codes: carbon—gray; hydrogen—white spheres, oxygen—red, nitrogen—light blue, and copper—dark green.

From a supramolecular perspective, crystal packing is dictated by the hydrogen bonds between the hexanuclear units (see all hydrogen bond geometry in Table S1 in the Supporting Information). The N–H groups of ethylenediamine and oxamate amide portion act as hydrogen bond donors, recognizing carboxylate groups around them. The hexanuclear complexes interact via amide-to-carboxylate bonds in dimeric forms along with the crystallographic ***a***-axis This means that while one oxamate amide recognizes the neighboring oxamate carboxylate portion [N1–H1···O7$^{i}$, i = $x + 1, y, z$], the latter interacts back, with its amide engaging with the carboxylate of the first oxamate [N7$^{i}$–H7$^{i}$···O1]. This interaction is defined by Etter et al. [49] as $R_2^2(10)$, indicating that it forms a ring with two hydrogen bond donors (both amide N–H) and two acceptors (one carboxylate of each oxamate) composed of ten atoms (two ···H–N–C–C–O··· units). This interaction forms a supramolecular 1D structure (See Figure 3a). In addition, also along the crystallographic ***a***-axis direction, ethylenediamine groups are connecting to the neighboring supramolecular chain's carboxylate, and once again, it interacts back, forming a motif denominated as $R_4^2(8)$. This means that two ethylenediamine $NH_2$ groups interact with two terminal oxygen atoms in a ring configuration, with a total of eight interacting atoms [N19–H19A···O15 and N19–H19B···O15$^{iv}$, iv = $-x + 1$, $-y + 2$, $-z + 1$]. Thus, the combination of both interactions leads to a supramolecular 1D double chain along the ***a***-axis, as shown in Figure 3a. These double chains interact along the crystallographic ***b***- and ***c***-axis via ethylenediamine–carboxylate discrete hydrogen bonds [N16—H16B···O10$^{iii}$, iii = $-x$, $-y + 2$, $-z$; N20–H20–O11$^{v}$, v = $-x + 1$, $-y + 2$, $-z$] to reach a three-dimensional supramolecular structure. The sum of all interactions forms a net, leaving a large zigzag tubular void, which hosts disordered crystallization solvent molecules removed from the crystal structure using the *SQUEEZE* technique, as depicted in Figure 3b.

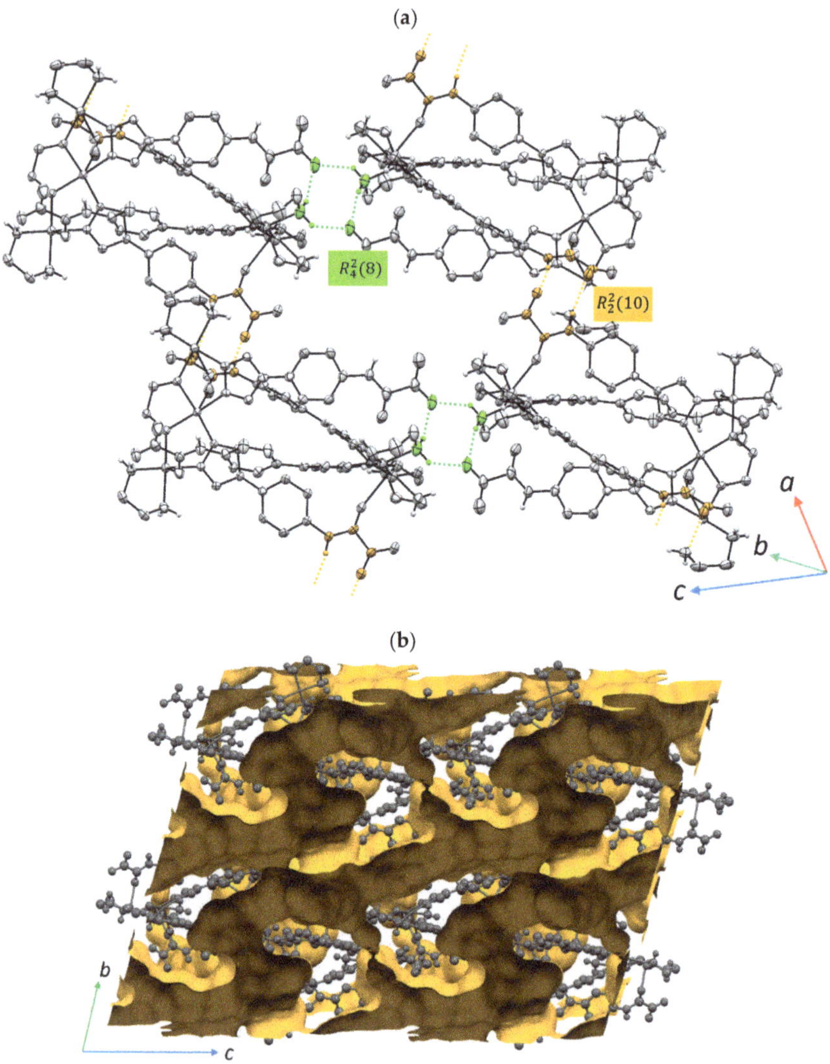


**Figure 3.** (**a**) Supramolecular double chain formed by the combination of amide-to-carboxylate hydrogen bond interactions (in orange) and ethylenediamine-to-carboxylate interactions (in green) along the crystallographic *a*-axis. For clarity, all atoms but those involved in the formation of $R_2^2(10)$ and $R_4^2(8)$ are colored in grey. (**b**) Zigzag void formed by the crystal packing of **1** (in orange). Once again, the atoms in the hexanuclear complexes are colored in gray for the sake of clarity.

### 3.3. Magnetic Studies

The direct current (DC) magnetic susceptibility measurement shows that at 265 K, the $\chi_M T$ is equal to 1.44 cm$^3$ mol$^{-1}$ K and is in a decreasing regime as the sample is cooled down (Figure 4). The $\chi_M T$ expected value for six copper(II) atoms, with respect to the spin-only approximation, is 2.48 cm$^3$ mol$^{-1}$ K ($S = 1/2$ and $g = 2.1$). The lower experimental value is associated with an active antiferromagnetic regime at high temperatures, indicating a somewhat strong coupling between the copper atoms. Upon cooling, a plateau is reached at 50 K, and the $\chi_M T$ is equal to 0.84 cm$^3$ mol$^{-1}$ K, which is close to the expected value for two isolated spin doublets (0.83 cm$^3$ mol$^{-1}$ K). The observed $\chi_M T$ and the magnetic behavior can be related to an antiferromagnetic coupling between the copper(II) atoms in the trinuclear subunits and a very weak to negligible magnetic interaction between the subunits, although they are connected via pyox ligands. Below 15 K, a second antiferromagnetic regime is observed, evidenced by a decrease in the $\chi_M T$ value. This probably occurs due to dipolar interactions throughout the space of the neighboring hexanuclear complexes. This coupling mechanism is evidenced by the low temperature at which the second antiferromagnetic regime occurs and the slight decrease observed at the lowest temperature (3.0 K) that accounts for a $\chi_M T$ equal to 0.78 cm$^3$ mol$^{-1}$ K, which is also a signature of a very weak antiferromagnetic coupling.

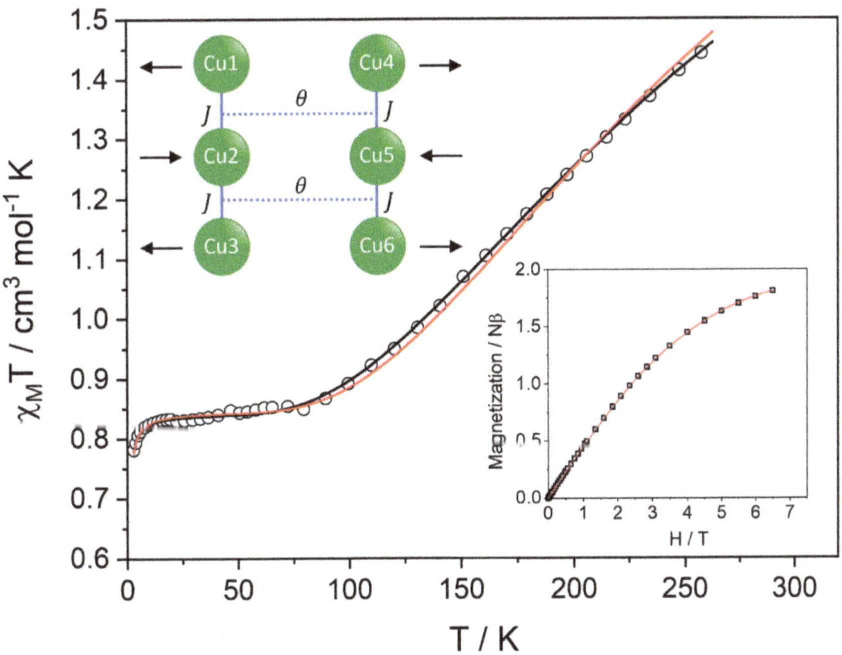


**Figure 4.** $\chi_M T$ vs. $T$ curve at 0.1 T. Circles represent the experimental data; the black line is the fitting performed using Equation (2), and the red lines represent the fitting using PHI software. Insets show the spin diagram for compound **1** and the field-dependent magnetization curve at 3.0 K.

The magnetization curve as a function of the applied field ($M$ vs. $H$) at 3.0 K revealed incomplete saturation at 7.0 T ($M_{sat}$ = 1.87 Nβ). This corroborates the presence of two spin doublets due to antiferromagnetic coupling between the metal atoms in each trinuclear subunit ($S_{total}$ = 1/2), which is expected to exhibit a saturation of the magnetization of 2.1 Nβ. However, as observed, at 3.0 K, in the $\chi_M T$ vs. $T$ curve, the active antiferromagnetic dipolar interactions may cause the magnetization to not reach the expected saturation value.

To investigate the magnitude of the magnetic coupling, the susceptibility data were fitted considering the coupling between the copper atoms in the subunits with the central copper atom coupled to the outer ones, as indicated in Equation (1). $J'$ and $J''$ are the magnetic coupling constants for each central atom and the outer ones in the different subunits. The terms $j'$ and $j''$ are the magnetic coupling between the outer metal atoms in each trinuclear subunit

$$\begin{aligned} H = -J'(S_{Cu1}\cdot S_{Cu2} + S_{Cu2}\cdot S_{Cu3}) + j'(S_{Cu1}\cdot S_{Cu3}) \\ -J''(S_{Cu4}\cdot S_{Cu5} + S_{Cu5}\cdot S_{Cu6}) + j''(S_{Cu4}\cdot S_{Cu6}) \end{aligned} \tag{1}$$

It is important to remark that the magnetic coupling between the outer copper atoms can be neglected, since they are not directly bonded. One can also assume that the subunits can be considered magnetically equivalent because the coordination parameters around the metal atoms are very similar, although they are not symmetry related. It will simplify the Hamiltonian for the presented molecule, as described in Equation (2). Therefore, the fitted coupling constant $J$ will be the average of the coupling constants $J'$ and $J''$ for each trinuclear subunit.

$$H = -2J(S_{Cu1}\cdot S_{Cu2} + S_{Cu2}\cdot S_{Cu3}) \tag{2}$$

Following the Hamiltonian, the susceptibility must adapt to Equation (3) [50]

$$\begin{aligned} \chi_M T(T) &= \frac{Ng^2\beta^2}{2k(T-\theta)}\frac{A}{B} \\ A &= 1 + e^{-\frac{J}{kT}} + 10e^{\frac{J}{2kT}} \\ B &= 1 + e^{-\frac{J}{kT}} + 2e^{\frac{J}{2kT}} \end{aligned} \tag{3}$$

$N$, $g$, $\beta$, and $k$ are the Avogadro number, the average Landé factor for all copper(II) atoms, the Bohr magneton, and the Boltzmann constant, respectively. $\theta$ is the deviation from the Curie law indicating dipolar interactions, which only occur at very low temperatures. An excellent fitting between the experimental and calculated data was obtained with $J = -106.3$ cm$^{-1}$, $g$ = 2.12, and $\theta = -0.23$ K with $R = 2.9 \times 10^{-5}$. The $\chi_M T$ curve fitting was also performed with PHI software (version 3.1.6) [51], which provided very close results, obtaining $J = -116.5(3)$ cm$^{-1}$, $g$ = 2.12(1), and $\theta = -0.08(1)$ K with $R = 1.4 \times 10^{-3}$ ($R$ is the agreement factor defined as $R = \Sigma(P_{obs} - P_{calc})^2/\Sigma(P_{obs})^2$, with $P$ being the physical parameter). This convergence in both analytical and numerical methods is indicative that the trinuclear subunits are almost magnetically isolated and that the provided simplified Hamiltonian in Equation (2) describes the magnetic behavior of this copper(II) hexanuclear compound very well. A comparison between the curve adjustments is shown in Figure 4. The obtained Landé factor is consistent with that of copper(II) ions, and the found coupling constants fall within the range typically found for complexes containing three metal ions in a linear arrangement, as can be concluded from the content of Table 3.

It is important to note that when only oxamate bridges are present in the trinuclear units, blocking ligands are important to orient the overlapping magnetic orbitals. In the presence of meridional orienting blocking ligands (Table 3), the bidentate oxamate can only bite the metal in apical–basal positions, causing a large angle between the basal planes and thus weakening antiferromagnetic coupling, as observed in [{Cu(bpca)}$_2$Cu(opba)(H$_2$O)]·H$_2$O [bpca = bis(2-pyridylcarbonyl)amidate; opba = *N,N′*-phenylene-1,2-bis(oxamate)] [50]; and [{Cubpca)}$_2${Cu(dmopba)(H$_2$O)}]$_2$·4H$_2$O [dmopba = *N,N′*-4,5-dimethylphenylene-1,2-bis(oxamate)] [52]. When bidentate-blocking ligands are employed, the oxamate usually occupies two basal positions, leading to a smaller angle between the basal planes and drastically increasing magnetic orbital overlapping. For instance, in [{Cu($Me_4$en)($H_2O$)}$_2$Cu(pba)($ClO_4$)] [53] and [{Cu($Me_4$en)($H_2O$)}$_2$Cu(pba)($H_2O$)]($PF_6$)$_2$ ($Me_4$en = *N,N,N′N′*-tetramethylethylenediamine; pba = *N,N′*-propylbis(oxamate)] [54], the angles between the basal planes are lower than the previous examples, exhibiting antiferromagnetic coupling that is ten times more intense than observed for [{Cubpca)}$_2${Cu(dmopba)($H_2O$)}]$_2$·4$H_2O$ (*J* values of −334.4, −342.1, and −31.96 $cm^{-1}$, respectively). In addition, for [{Cu($Me_4$en)($H_2O$)}$_2$Cu(pba)($ClO_4$)] and [{Cu($Me_4$en)($H_2O$)}$_2$Cu(pba)($H_2O$)]($PF_6$)$_2$, it is observed that the one with the lower angle between the basal planes presented a slightly stronger antiferromagnetic coupling constant due to more optimal orbital overlapping. Ferromagnetic couplings in the oxamate derivatives bridging examples arise from the bridging of metal ions with orthogonal magnetic orbitals [55,56] or different bridging forms other than $\mu$-$\kappa^2$N,O:$\kappa^2$O′,O″ [52,57,58].

On the other hand, both alkoxide and pyrazolate ligands lead to antiferromagnetic coupling between copper(II) ions. For the pyrazolate group, it is observed that the torsion angle around the N–N bond is a determinant for orbital overlapping. A smaller torsion angle means that the basal planes of the linked metal atoms are closer to coplanarity, and the spin density can be delocalized over the plane of the pyrazolate aromatic ring. To exemplify, both [$Cu_3$(L1)$_2$($NO_3$)$_2$($H_2O$)$_2$]·3$H_2O$ and [$Cu_3$(L1)$_2$$Cl_2$($CH_3OH$)$_2$]·2$H_2O$ [$H_2$*L*1 = 6,6-((1E,10E)-hydrazine-1,2-diylidenebis(methanylylidene))bis(2-methoxyphenol)] [59] have almost the same alkoxide bridging angles, but the small variation of the torsion angle from 28.86° to 30.54° causes the antiferromagnetic coupling to weaken from −113.1 to −98.3 $cm^{-1}$. For alkoxide ligands, the direct bridging angle is crucial to magnetic overlapping. The bridging angle closest to 120° indicates better fitting to $sp^2$ hybridization for the oxygen atom; thus, better magnetic overlapping can arise from the ligand. As seen in Table 3, the strongly antiferromagnetically coupled atoms have a Cu–O–Cu angle close to 120°, suppressing even the downside of having a somewhat greater torsion angle along the pyrazolate N–N bond. However, it can be concluded that this bridging angle has the drawback that even small variations can drastically impact the magnetic coupling between the bridged metal atoms.

Finally, the strength of copper(II) magnetic coupling depends not solely on the nature of the ligands. Copper(II) ions can have $d_{z^2}$ and $d_{x^2-y^2}$ as their magnetic orbitals, and thus small variations in ion coordination geometry, as well as the angle and/or bond length between the bridging atoms/ligands, cause variations in the crystal field and consequently in the metal's magnetic orbitals. Since each orbital resides in a different atomic axis, swapping them will significantly affect the quality of orbital overlap, finally resulting in the wide range of magnetic couplings constants observed in the literature.

**Table 3.** Magneto-structural properties of some oxamate-bridged or pyrazolate/oxo-bridged linear arranged trinuclear copper(II) compounds.

| Compound | Cu–NN–Cu [a]/° | Cu–O–Cu [b]/° | Mean Planes Angle [c]/° | $J/cm^{-1}$ | $g$ | $\theta$ or $zJ'/K$ | Ref. |
|---|---|---|---|---|---|---|---|
| 1 | −4.1 (1.4)/−5.5 (1.2)/ 0.3 (1.4)/−0.43 (1.3) | 120.23(14)/115.20(14)/ 116.76(14)/110.96(14) | 30.9(4)/35.8(4)/37.8(5)/47.5(4) | −106.3 [d] | 2.12 [d] | −0.23 [d] | This work |
| | | | | −116.5 [d] | 2.12 [d] | −0.08 [d] | |
| $[\{Cu(bpca)\}_2Cu(opba)(H_2O)]\cdot H_2O$ | --- [e] | --- [e] | 78.45/84.62 | −65.8 | 2.08 | --- | [50] |
| $[\{Cu(bpca)\}_2\{Cu(dmopba)(H_2O)\}]_2\cdot 4H_2O$ | --- [e] | --- [e] | 78.09/81.67 | −31.96/+1.34 [f] | 2.12 | --- | [52] |
| $[\{Cu(Me_4en)(H_2O)\}_2Cu(pba)(ClO_4)]$ | --- [e] | --- [e] | 11.29/18.50 | −334.4 | 2.08 | −0.36 | [53] |
| $[\{Cu(Me_4en)(H_2O)\}_2Cu(pba)(H_2O)](PF_6)_2$ | --- [e] | --- [e] | 7.94/10.75 | −342.1 | 2.14 | −0.61 | [54] |
| $[Cu_3(L1)_2(BF_4)(H_2O)_2]BF_4$ | 40.5(2)/33.0(1) | 116.23(7)/118.40(7) | 17.04 | −170 | 2.16 | --- | [59] |
| $[Cu_3(L1)_2(NO_3)_2(H_2O)_2]\cdot 3H_2O$ | 28.86 | 116.95 | 5.01 | −113.1 | 2.08 | --- | |
| $[Cu_3(L1)_2Cl_2(CH_3OH)_2]\cdot 2H_2O$ | 30.54 | 116.89 | 4.64 | −98.3 | 2.07 | --- | |
| $[Cu_3(L1)_2(ClO_4)_2(H_2O)_2]$ | 40.19/28.92 | 117.67/117.91 | 16.20 | −249.7 | 2.14 | --- | |
| $\{[Cu_3(L2)_2(H_2O)_2](ClO_4)_4\}\cdot 4H_2O$ | 17.2(8)/13.7(8) | 120.5(3)/116.6(3) | 8.19/18.36 | −179 | 2.0/2.2 | --- | [60] |
| $[Cu_3(HYDRAV)_2Cl_2]$ | 35.9 | 115 | 0.96/11.7 | −7.21 | 2.03 | --- | [61] |
| $\{[Cu_3(L3)(L4)_2(NO_3)_2(H_2O)_2]NO_3\cdot 1.5H_2O\}n$ | −6.9(8)/8.0(8) | 119.3(3)/124.5(3) | 22.9/33.6 | −149/−175 | 2.20 | −0.7 | [62] |

[a] Torsion angle along N–N bond. [b] Oxo bridge angle. [c] For metal atoms with NC = 5 (square-based pyramid geometry), basal planes were used. For those with NC = 6 (octahedral geometry), planes perpendicular to the elongated bonds' direction were considered. [d] $J$, $g$, and $zJ/\theta$ were calculated using Equation (2) (top line) and using PHI software (bottom line). [e] Not present for oxamate-bridged metal atoms. [f] Ferromagnetic coupling due to *syn-anti* $\mu^3$-carboxylate from oxamate group. Ligand abbreviations: bpca = bis(2-pyridylcarbonyl)amidate; opba = *N,N′*-phenylene-1,2-bis(oxamate); dmopba = *N,N′*-4,5-dimethylphnylene-1,2-bis(oxamate); $Me_4en$ = *N,N,N′,N′*-tetramethylethylenediamine; pba = *N,N′*-propilbis(oxamate); $H_2L1$ = 6,6-((1E,10E)-hydrazine-1,2-diylidenebis(methanylylidene))bis(2-methoxyphenol); HL2 = [(5-amino-1H-1,2,4-triazol-3-yl)amino](1H-imidazol-5-yl)methanolate. HYDRAV2 = Schiff base made with 1-(hydrazineylidenemethyl)naphthalen2-ol, o-vanillin and (2-hydroxy-3-methoxybenzaldehyde); $H_2L3$ = 4-[bis(pyridin-2-yl-methanol)]amino1,2,4-triazole; L4 = 4-amino-4H-1,2,4-triazole.

## 4. Conclusions

In this study, we report the synthesis, crystal structure, and magnetic properties of a novel hexacopper(II) complex, in which ethylene glycol serves both as a solvent for its crystallization and as a ligand in its doubly deprotonated form, ethane-1,2-dioxide. The crystal structure reveals that the hexanuclear complex consists of two linear trinuclear copper(II) units, with the central copper atom partially coordinated by ethane-1,2-dioxide, while each terminal copper atom coordinates one ethylenediamine molecule. Both pyrazolate (from the pyox ligand) and alkoxide groups (from ethane-1,2-dioxide) serve as bridging ligands between the external and central copper ions. The two trinuclear units are connected by the coordination of a carbonyl oxygen atom (from the oxamate portion of the pyox ligand) to the copper centers of the opposing trinuclear unit. However, the distance between them is relatively large and involves a pathway consisting of non-resonant $sp^2$ atoms. DC magnetic measurements indicate that the copper ions exhibit strong antiferromagnetic interactions at high temperatures, indicating an active antiferromagnetic regime. The plateau observed at 50 K corresponds to two isolated spin doublets, suggesting that the trinuclear units are not magnetically connected. This is consistent with the assumption that they are only linked to the extremities of the pyox ligand. The magnetic data were fitted using analytical and numerical methods, providing the best fits of the exchange interaction parameters of $J = -106.3$ cm$^{-1}$ and $g = 2.12$ and $J = -116.5$ cm$^{-1}$ and $g = 2.12$, respectively. The found values align with previously reported values for pyrazolate- and alkoxide-bridged copper(II) complexes.

**Supplementary Materials:** The following supporting information can be downloaded at: https://www.mdpi.com/article/10.3390/magnetochemistry10120094/s1, TG and DTA curves for **1** (Figure S1), Infrared spectrum of **1** (Figure S2), Comparison between experimental and calculated polycrystalline X-ray diffraction for **1** (Figure S3), a table containing the Hydrogen bond geometry for **1** (Table S1) and the table for Continuous SHAPE calculations for copper atoms in 1 (Table S2).

**Author Contributions:** W.X.C.O. and V.G.A. performed the synthesis and experimental characterization. W.X.C.O. and C.B.P. collected the single crystal data, solved, and refined the crystal structure. W.X.C.O. and M.J. carried out the magnetic measurements and interpretation. C.L.M.P., W.X.C.O. and M.J. guided the students, supervised all the experiments, and was responsible for funding acquisition. W.X.C.O. and C.L.M.P. wrote this paper. All authors have read and agreed to the published version of the manuscript, except for M.J., who sadly passed away a few months before the writing conclusion.

**Funding:** Authors are thankful to CNPQ (420443/2018-5, 309874/2021-1), FAPEMIG (APQ-01948-22 and APQ-05311-23) and PRPq-UFMG for financial support, to CAPES, CNPQ and PRPq-UFMG for the scholarships. The authors are also thankful to LabCri (http://www.labcri.ufmg.br/) for providing the single crystal and polycrystalline diffraction facilities and the Bioanalytical Facility NEPS-DQ (https://ne.qui.ufmg.br) for all chemical characterizations.

**Data Availability Statement:** Data are available upon request to the corresponding authors.

**Conflicts of Interest:** The authors declare no conflicts of interest.

## References

1. Castellano, M.; Ruiz-García, R.; Cano, J.; Ferrando-Soria, J.; Pardo, E.; Fortea-Pérez, F.R.; Stiriba, S.-E.; Barros, W.P.; Stumpf, H.O.; Cañadillas-Delgado, L.; et al. Metallosupramolecular Approach toward Multifunctional Magnetic Devices for Molecular Spintronics. *Coord. Chem. Rev.* **2015**, *303*, 110–138. [CrossRef]
2. Journaux, Y.; Ferrando-Soria, J.; Pardo, E.; Ruiz-Garcia, R.; Julve, M.; Lloret, F.; Cano, J.; Li, Y.; Lisnard, L.; Yu, P.; et al. Design of Magnetic Coordination Polymers Built from Polyoxalamide Ligands: A Thirty Year Story. *Eur. J. Inorg. Chem.* **2018**, *2018*, 228–247. [CrossRef]
3. Ferrando-Soria, J.; Vallejo, J.; Castellano, M.; Martínez-Lillo, J.; Pardo, E.; Cano, J.; Castro, I.; Lloret, F.; Ruiz-García, R.; Julve, M. Molecular Magnetism, Quo Vadis? A Historical Perspective from a Coordination Chemist Viewpoint. *Coord. Chem. Rev.* **2017**, *339*, 17–103. [CrossRef]
4. Pereira, C.L.M.; Doriguetto, A.C.; Konzen, C.; Meira-Belo, L.C.; Leitão, U.A.; Fernandes, N.G.; Mascarenhas, Y.P.; Ellena, J.; Brandl, A.L.; Knobel, M.; et al. A Crystalline Phase Transition and Optical Properties in a $Co^{II}Cu^{II}$ Oxamato-Bridged Ferrimagnetic Chain. *Eur. J. Inorg. Chem.* **2005**, *2005*, 5018–5025. [CrossRef]

5. Journaux, Y.; Sletten, J.; Kahn, O. Tunable Interaction in (μ-Oxamido)Copper(II) Binuclear Complexes. *Inorg. Chem.* **1985**, *24*, 4063–4069. [CrossRef]
6. Oliveira, W.X.C.; Ribeiro, M.A.; Pinheiro, C.B.; Nunes, W.C.; Julve, M.; Journaux, Y.; Stumpf, H.O.; Pereira, C.L.M. Magneto-Structural Study of an Oxamato-Bridged $Pd^{II}Co^{II}$ Chain: X-ray Crystallographic Evidence of a Single-Crystal-to-Single-Crystal Phase Transition. *Eur. J. Inorg. Chem.* **2012**, *2012*, 5685–5693. [CrossRef]
7. Das, A.K.; De, A.; Yadav, P.; Lloret, F.; Mukherjee, R. $Cu^{II}_2$, $Cu^{II}_4$ and $Cu^{II}_6$ Complexes with 3-(2-Pyridyl)Pyrazolate. Structure, Magnetism and Core Interconversion. *Polyhedron* **2019**, *171*, 365–373. [CrossRef]
8. Oliveira, W.X.C.; Pereira, C.L.M.; Pinheiro, C.B.; Cano, J.; Lloret, F.; Julve, M. Relatively Strong Intramolecular Antiferromagnetic Coupling in a Neutral $Cr^{III}_2Nb^{V}_2$ Heterobimetallic Molecular Square. *Chem. Commun.* **2015**, *51*, 11806–11809. [CrossRef] [PubMed]
9. Ferrando-Soria, J.; Grancha, T.; Pasán, J.; Ruiz-Pérez, C.; Cañadillas-Delgado, L.; Journaux, Y.; Julve, M.; Cano, J.; Lloret, F.; Pardo, E. Solid-State Aggregation of Metallacyclophane-Based $Mn^{II}Cu^{II}$ One-Dimensional Ladders. *Inorg. Chem.* **2012**, *51*, 7019–7021. [CrossRef] [PubMed]
10. Simões, T.R.G.; do Pim, W.D.; Silva, I.F.; Oliveira, W.X.C.; Pinheiro, C.B.; Pereira, C.L.M.; Lloret, F.; Julve, M.; Stumpf, H.O. Solvent-Driven Dimensionality Control in Molecular Systems Containing $Cu^{II}$, 2,2′-Bipyridine and an Oxamato-Based Ligand. *CrystEngComm* **2013**, *15*, 10165. [CrossRef]
11. Pereira, C.L.M.; Pedroso, E.F.; Stumpf, H.O.; Novak, M.A.; Ricard, L.; Ruiz-García, R.; Rivière, E.; Journaux, Y. A $Cu^{II}Co^{II}$ Metallacyclophane-Based Metamagnet with a Corrugated Brick-Wall Sheet Architecture. *Angew. Chem. Int. Ed.* **2004**, *43*, 956–958. [CrossRef] [PubMed]
12. Mon, M.; Ferrando-Soria, J.; Grancha, T.; Fortea-Pérez, F.R.; Gascon, J.; Leyva-Pérez, A.; Armentano, D.; Pardo, E. Selective Gold Recovery and Catalysis in a Highly Flexible Methionine-Decorated Metal–Organic Framework. *J. Am. Chem. Soc.* **2016**, *138*, 7864–7867. [CrossRef] [PubMed]
13. Lysova, A.A.; Samsonenko, D.G.; Kovalenko, K.A.; Nizovtsev, A.S.; Dybtsev, D.N.; Fedin, V.P. A Series of Mesoporous Metal-Organic Frameworks with Tunable Windows Sizes and Exceptionally High Ethane over Ethylene Adsorption Selectivity. *Angew. Chem. Int. Ed.* **2020**, *59*, 20561–20567. [CrossRef] [PubMed]
14. Jones, A.C.; Williams, P.A.; Bickley, J.F.; Steiner, A.; Davies, H.O.; Leedham, T.J.; Awaluddin, A.; Pemble, M.E.; Critchlow, G.W. Synthesis and Crystal Structures of Two New Titanium Alkoxy–Diolate Complexes. Potential Precursors for Oxide Ceramics. *J. Mater. Chem.* **2001**, *11*, 1428–1433. [CrossRef]
15. Lysova, A.A.; Samsonenko, D.G.; Dorovatovskii, P.V.; Lazarenko, V.A.; Khrustalev, V.N.; Kovalenko, K.A.; Dybtsev, D.N.; Fedin, V.P. Tuning the Molecular and Cationic Affinity in a Series of Multifunctional Metal–Organic Frameworks Based on Dodecanuclear Zn(II) Carboxylate Wheels. *J. Am. Chem. Soc.* **2019**, *141*, 17260–17269. [CrossRef]
16. Pajot, N.; Papiernik, R.; Hubert-Pfalzgraf, L.G.; Vaissermann, J.; Parraud, S. Metal-Assisted Activation of the C–O Bond of 2-Hydroxyethylmethacrylate. Synthesis and Molecular Structure of $Ti_5(OPr^i)_9(\mu\text{-}OPr^i)(\mu,\eta^2\text{-}OC_2H_4O)(\mu_3,\eta^2\text{-}OC_2H_4O)_3(\mu_4,\eta^2\text{-}OC_2H_4O)$. *J. Chem. Soc. Chem. Commun.* **1995**, *17*, 1817–1819. [CrossRef]
17. Hänninen, M.M.; Välivaara, J.; Mota, A.J.; Colacio, E.; Lloret, F.; Sillanpää, R. Ferromagnetic Dinuclear Mixed-Valence Mn(II)/Mn(III) Complexes: Building Blocks for the Higher Nuclearity Complexes. Structure, Magnetic Properties, and Density Functional Theory Calculations. *Inorg. Chem.* **2013**, *52*, 2228–2241. [CrossRef] [PubMed]
18. Chan, S.L.; Gao, S.; Chui, S.S.; Shek, L.; Huang, J.; Che, C. Supramolecular Self-Assembly of 1D and 3D Heterometallic Coordination Polymers with Triruthenium Building Blocks. *Chem. Eur. J.* **2012**, *18*, 11228–11237. [CrossRef]
19. Audhya, A.; Maity, M.; Abtab, S.M.T.; Mathonière, C.; Kalisz, M.; Clérac, R. Polyalcohols as Ancillary Ligands in Manganese–Oxime Chemistry: Syntheses, Structures and Magnetic Properties of a Series of Trinuclear Complexes Involving a Linear $Mn^{II}$–$Mn^{IV}$–$Mn^{II}$ Core. *Polyhedron* **2012**, *33*, 353–359. [CrossRef]
20. Teichert, J.; Doert, T.; Ruck, M. Mechanisms of the Polyol Reduction of Copper(II) Salts Depending on the Anion Type and Diol Chain Length. *Dalton Trans.* **2018**, *47*, 14085–14093. [CrossRef] [PubMed]
21. George, T.; Brosseau, C.L.; Masuda, J.D. Electrochemical and X-Ray Structural Evidence of Multiple Molybdenum Precursor Candidates from a Reported Non-Aqueous Electrodeposition of Molybdenum Disulfide. *RSC Adv.* **2023**, *13*, 32199–32216. [CrossRef] [PubMed]
22. Xiao, H.; Zhou, J.; Hu, F. Two New Oxyiodoplumbates: The Unique 3-D Hybrid Oxyiodoplumbate Based on Neutral 2-D$[Pb_2I_4]_n$ Layers. *Dalton Trans.* **2018**, *47*, 8442–8447. [CrossRef] [PubMed]
23. Liu, F.-F.; Wang, D.; Chen, G.-H.; Zhang, J.; Zhang, L. Ligands Hydrophobicity Dependent Electrocatalytic $CO_2$ Reduction Activities of $Sn_5$-Oxo Clusters. *J. Solid State Chem.* **2023**, *321*, 123918. [CrossRef]
24. Bates, P.A.; Hursthouse, M.B.; Davies, A.G.; Slater, S.D. The Structure of 2,2-Di-t-Butyl-1,3,2-Dioxa-, -Oxathia-, and -Dithia-Stannolanes: A Study by Solution and Solid State NMR and Single Crystal X-Ray Diffraction. *J. Organomet. Chem.* **1989**, *363*, 45–60. [CrossRef]
25. Biros, S.M.; Bridgewater, B.M.; Villeges-Estrada, A.; Tanski, J.M.; Parkin, G. Antimony Ethylene Glycolate and Catecholate Compounds: Structural Characterization of Polyesterification Catalysts. *Inorg. Chem.* **2002**, *41*, 4051–4057. [CrossRef] [PubMed]
26. Uhl, W.; Gerding, R.; Vester, A. The Decomposition Reaction of the Dialene Radical Anion $[R_2Al-AlR_2]^-$ in DME(II): Crystal Strctures of the Aluminum Glycolates $[R_2Al(OCH_2)_2]K(DME)$ and $R_2Al(OCH)_2)_2Al(R)(OCH_2)_2AlR_2$ (R = $CH(SiMe_3)_2$). *J. Organomet. Chem.* **1996**, *513*, 163–172. [CrossRef]

27. Xiao, H.; Zhou, J.; Liu, X. Two New 3-D Cadmium Bromoplumbates: The Only Example of Heterometallic Bromoplumbate Based on Crown [$Cd(Pb_4O_4)Br_2$] Clusters. *Dalton Trans.* **2018**, *47*, 4833–4839. [CrossRef] [PubMed]
28. McMahon, C.N.; Alemany, L.; Callender, R.L.; Bott, S.G.; Barron, A.R. Reaction of $Al(^tBu)_3$ with Ethylene Glycol: Intermediates to Aluminum Alkoxide (Alucone) Preceramic Polymers. *Chem. Mat.* **1999**, *11*, 3181–3188. [CrossRef]
29. Aguirre-Díaz, L.M.; Gándara, F.; Iglesias, M.; Snejko, N.; Gutiérrez-Puebla, E.; Monge, M.Á. Tunable Catalytic Activity of Solid Solution Metal–Organic Frameworks in One-Pot Multicomponent Reactions. *J. Am. Chem. Soc.* **2015**, *137*, 6132–6135. [CrossRef] [PubMed]
30. Cruickshank, M.C.; Glasser, L.S.D. A Penta-Co-Ordinated Aluminate Dimer; X-Ray Crystal Structure. *J. Chem. Soc. Chem. Commun.* **1985**, *2*, 84. [CrossRef]
31. Teichert, J.; Block, T.; Pöttgen, R.; Doert, T.; Ruck, M. Tin and Lead Alkoxides of Ethylene Glycol and Glycerol and Their Decomposition to Oxide Materials. *Eur. J. Inorg. Chem.* **2019**, *2019*, 3820–3831. [CrossRef]
32. Rabelo, R.; Stiriba, S.-E.; Cangussu, D.; Pereira, C.; Moliner, N.; Ruiz-García, R.; Cano, J.; Faus, J.; Journaux, Y.; Julve, M. When Molecular Magnetism Meets Supramolecular Chemistry: Multifunctional and Multiresponsive Dicopper(II) Metallacyclophanes as Proof-of-Concept for Single-Molecule Spintronics and Quantum Computing Technologies? *Magnetochemistry* **2020**, *6*, 69. [CrossRef]
33. Oliveira, W.X.C.; Pinheiro, C.B.; Journaux, Y.; Julve, M.; Pereira, C.L.M. Effects on the Magnetic Interaction Caused by Molecular Recognition in Complexes of 1,2-Azole-Based Oxamate and $[Cu(Bpca)]^+$ Units. *CrystEngComm* **2024**, *26*, 647–665. [CrossRef]
34. Oliveira, W.X.C.; do Pim, W.D.; Pinheiro, C.B.; Journaux, Y.; Julve, M.; Pereira, C.L.M. Monitoring the Hydrogen Bond Net Configuration and the Dimensionality of Aniline and Phenyloxamate by Adding 1*H*-Pyrazole and Isoxazole as Substituents for Molecular Self-Recognition. *CrystEngComm* **2019**, *21*, 2818–2833. [CrossRef]
35. Rafat, F.; Siddiqi, K.S. Lutfullah Synthesis and Characterization of Trinuclear Complexes Containing Cu(II) and Ti(IV). *Synth. React. Inorg. Met. Org. Chem.* **2004**, *34*, 763–774. [CrossRef]
36. Rigaku, O.D. *CrysAlis PRO*; Rigaku Corporation: Oxford, UK, 2022.
37. Palatinus, L.; Chapuis, G. *SUPERFLIP*—A Computer Program for the Solution of Crystal Structures by Charge Flipping in Arbitrary Dimensions. *J. Appl. Crystallogr.* **2007**, *40*, 786–790. [CrossRef]
38. Sheldrick, G.M. Crystal Structure Refinement with *SHELXL*. *Acta Crystallogr. C Struct. Chem.* **2015**, *71*, 3–8. [CrossRef]
39. Spek, A.L. *PLATON* SQUEEZE: A Tool for the Calculation of the Disordered Solvent Contribution to the Calculated Structure Factors. *Acta Crystallogr. C Struct. Chem.* **2015**, *71*, 9–18. [CrossRef]
40. Farrugia, L.J. *WinGX* and *ORTEP for Windows*: An Update. *J. Appl. Crystallogr.* **2012**, *45*, 849–854. [CrossRef]
41. Macrae, C.F.; Sovago, I.; Cottrell, S.J.; Galek, P.T.A.; McCabe, P.; Pidcock, E.; Platings, M.; Shields, G.P.; Stevens, J.S.; Towler, M.; et al. *Mercury 4.0*: From Visualization to Analysis, Design and Prediction. *J. Appl. Crystallogr.* **2020**, *53*, 226–235. [CrossRef] [PubMed]
42. Momma, K.; Izumi, F. *VESTA*: A Three-Dimensional Visualization System for Electronic and Structural Analysis. *J. Appl. Crystallogr.* **2008**, *41*, 653–658. [CrossRef]
43. Bain, G.A.; Berry, J.F. Diamagnetic Corrections and Pascal's Constants. *J. Chem. Educ.* **2008**, *85*, 532. [CrossRef]
44. Li, Y.; Zhao, B.; Xue, J.-P.; Xie, J.; Yao, Z.-S.; Tao, J. Giant Single-Crystal-to-Single-Crystal Transformations Associated with Chiral Interconversion Induced by Elimination of Chelating Ligands. *Nat. Commun.* **2021**, *12*, 6908. [CrossRef] [PubMed]
45. Labadi, I.; Parkanyi, L.; Grobelny, R.; Mrozinski, J. Redetermination of Crystal Structure, EPR and Magnetic Data of Tris(1,2-Ethanediol) Copper(II) Sulphate. *Polyhedron* **1994**, *13*, 2767–2774. [CrossRef]
46. Antti, B.-M.; Røst, E.; Valde, G.; Andresen, A.F.; Sandström, M. The Molecular and Crystal Structures of [$CuCl_2(C_2H_6O_2)$] and [$CuCl_2(C_2H_6O_2)$].½$H_2O$; Two Compounds Containing Neutral Dichloro(1,2-Ethanediol)Copper(II) Molecules. *Acta Chem. Scand.* **1976**, *30a*, 405–410. [CrossRef]
47. Addison, A.W.; Rao, T.N.; Reedijk, J.; van Rijn, J.; Verschoor, G.C. Synthesis, Structure, and Spectroscopic Properties of Copper(II) Compounds Containing Nitrogen–Sulphur Donor Ligands; the Crystal and Molecular Structure of Aqua[1,7-Bis(N-Methylbenzimidazol-2′-Yl)-2,6-Dithiaheptane]Copper(II) Perchlorate. *J. Chem. Soc. Dalton Trans.* **1984**, 1349–1356. [CrossRef]
48. Llunell, P.M.; Alemany, S. *Alvarez SHAPE 2013, Version 2.1*; The Electronic Structure Groups: Barcelona, Spain, 2013.
49. Etter, M.C.; MacDonald, J.C.; Bernstein, J. Graph-Set Analysis of Hydrogen-Bond Patterns in Organic Crystals. *Acta Crystallogr. B* **1990**, *46*, 256–262. [CrossRef] [PubMed]
50. Simões, T.R.G.; Mambrini, R.V.; Reis, D.O.; Marinho, M.V.; Ribeiro, M.A.; Pinheiro, C.B.; Ferrando-Soria, J.; Déniz, M.; Ruiz-Pérez, C.; Cangussu, D.; et al. Copper(II) Assembling with Bis(2-Pyridylcarbonyl)Amidate and N,N′-2,2-Phenylenebis(Oxamate). *Dalton Trans.* **2013**, *42*, 5778. [CrossRef]
51. Chilton, N.F.; Anderson, R.P.; Turner, L.D.; Soncini, A.; Murray, K.S. PHI: A Powerful New Program for the Analysis of Anisotropic Monomeric and Exchange-coupled Polynuclear *d*- and *f*-block Complexes. *J. Comput. Chem.* **2013**, *34*, 1164–1175. [CrossRef]
52. Lage, A.L.A.; Ribeiro, L.A.; Doriguetto, A.C.; Pinheiro, C.B.; Nunes, W.C.; Pedroso, E.F.; Pereira, C.L.M. Solid-State Self-Assembly of a Linear Hexanuclear Copper(II) Oxamate Complex with Alternating Antiferro- and Ferromagnetic Coupling. *Magnetochemistry* **2022**, *8*, 116. [CrossRef]
53. Costa, R.; Garcia, A.; Ribas, J.; Mallah, T.; Journaux, Y.; Sletten, J.; Solans, X.; Rodriguez, V. Tailored Magnetic Properties in Trinuclear Copper(II) Complexes: Synthesis, Structure, and Magnetic Properties of Complexes Derived from [1,3-Propanediylbis(Oxamato)]Cuprate(II) ($[Cu(Pba)]^{2-}$). *Inorg. Chem.* **1993**, *32*, 3733–3742. [CrossRef]

54. Tercero, J.; Diaz, C.; Ribas, J.; Mahía, J.; Maestro, M.; Solans, X. Synthesis, Characterization, and Magnetic Properties of New Complexes Based on Self-Assembled Homotrinuclear Units $Cu^{II}$–$Cu^{II}$–$Cu^{II}$. *J. Chem. Soc. Dalton Trans.* **2002**, *9*, 2040–2046. [CrossRef]
55. Li, Y.-T.; Yan, C.-W.; Guan, H.-S. Synthesis and Magnetic Studies of Oxalato-Bridged Copper(II)–Chromium(III)–Copper(II) and Copper(II)–Iron(III)–Copper(II) Heterotrinuclear Complexes. *Polyhedron* **2003**, *22*, 3223–3230. [CrossRef]
56. Cortes, R.; Urtiaga, M.K.; Lezama, L.; Arriortua, M.I.; Rojo, T. A Ferromagnetic Copper(II)-Vanadium(IV) Oxide .Mu.-Oxalato Complex: Crystallographic Structure and Spectroscopic and Magnetic Properties. *Inorg. Chem.* **1994**, *33*, 829–832. [CrossRef]
57. Oliveira, W.X.C.; Pinheiro, C.B.; da Costa, M.M.; Fontes, A.P.S.; Nunes, W.C.; Lloret, F.; Julve, M.; Pereira, C.L.M. Crystal Engineering Applied to Modulate the Structure and Magnetic Properties of Oxamate Complexes Containing the $[Cu(bpca)]^+$ Cation. *Cryst. Growth Des.* **2016**, *16*, 4094–4107. [CrossRef]
58. da Cunha, T.T.; da Silveira, C.O.C.; Barbosa, V.M.M.; Oliveira, W.X.C.; da Silva Júnior, E.N.; Ferreira, F.F.; Pedroso, E.F.; Pereira, C.L.M. Ferromagnetic Coupling in a Dicopper(II) Oxamate Complex Bridged by Carboxylate Groups. *CrystEngComm* **2021**, *23*, 1885–1897. [CrossRef]
59. Adhikary, A.; Ghosh, K. Fine Tuning of Coordination Environments by Anions on a Series of Cu(II) Dihydrazide Complexes: Syntheses, Structures, Magnetic Properties and Solution Phase Anion Exchange. *Polyhedron* **2019**, *168*, 37–47. [CrossRef]
60. Cruz, C.; Audebrand, N.; Páez-Hernández, D.; Paredes-García, V. Novel Linear Trinuclear $Cu^{II}$ Compound with Trapped Chiral Hemiaminal Ligand: Magnetostructural Study. *Magnetochemistry* **2023**, *9*, 175. [CrossRef]
61. Gholizadeh Dogaheh, S.; Khanmohammadi, H.; Sañudo, E.C. A New Trinuclear N–N Bridged Cu(II) Complex with an Asymmetric Schiff Base Ligand Derived from Hydrazine. *Polyhedron* **2017**, *133*, 48–53. [CrossRef]
62. Bazhina, E.S.; Bovkunova, A.A.; Medved'ko, A.V.; Efimov, N.N.; Kiskin, M.A.; Eremenko, I.L. Unusual Polynuclear Copper(II) Complexes with a Schiff-Base Ligand Containing Pyridyl and 1,2,4-Triazolyl Rings. *J. Clust. Sci.* **2019**, *30*, 1267–1275. [CrossRef]

*magnetochemistry* MDPI

*Article*

# Metallacrown of $Ce^{III}Cu^{II}_5$: Synthesis, Structural Characterization and Insights for Nanoparticles †

**Jésio D. Tempesta [1], Fábio Faria Paiva [1], Leonildo A. Ferreira [2], Rafaela M. R. da Silva [1], Luckerman D. G. Botelho [1], Iara M. L. Rosa [1], Caio Cesar Candido [3], Angelo Marcio Gomes [4], Wallace C. Nunes [5], Guilherme P. Guedes [6] and Maria Vanda Marinho [1,*,‡]**

1 LaPI, Instituto de Química, Universidade Federal de Alfenas, Campus Santa Clara, Alfenas 37133-840, MG, Brazil; jesio.tempesta@sou.unifal-mg.edu.br (J.D.T.); fabio.faria.paiva@gmail.com (F.F.P.); rafaela.silva@sou.unifal-mg.edu.br (R.M.R.d.S.); luckerman.botelho@sou.unifal-mg.edu.br (L.D.G.B.); cristalografia.landre@gmail.com (I.M.L.R.)

2 Institute de Química, Universidade Federal de Goiás, Campus Samambaia, Goiânia 74690-900, GO, Brazil; leonildoferreira@ufg.br

3 LabcrystAL, Institute de Química, Universidade Federal de Alfenas, Campus Santa Clara, Alfenas 37133-840, MG, Brazil; caio.candido@sou.unifal-mg.edu.br

4 Instituto de Física, Universidade Federal do Rio de Janeiro, Cidade Universitária, Rio de Janeiro 21941-972, RJ, Brazil; amgomes@if.ufrj.br

5 Instituto de Física, Universidade Federal Fluminense, Niterói 24210-346, RJ, Brazil; wcnunes@id.uff.br

6 Instituto de Química, Universidade Federal Fluminense, Niterói 24020-141, RJ, Brazil; guilherme_guedes@id.uff.br

* Correspondence: maria.marinho@unifal-mg.edu.br

† M. V. Marinho dedicates this article to the beloved memory of the great friend and Professor Miguel Julve.

‡ Dedicated to the brilliant careers of M. Julve and F. Lloret.

**Abstract:** The heterobimetallic 15-MC-5 metallacrown of formula $[CeCu_5(5mpzHA)_5(NO_3)(H_2O)_7]\cdot 2NO_3\cdot 7H_2O$, designated **MC-Ce**, was synthesized using 5-methyl-2-pyrazinehydroxamic acid (5mpzHA) as a linker, reacting with $Ce^{III}$ and $Cu^{II}$ salts under mild conditions. Single-crystal X-ray diffraction analysis reveals a crown-like $[Cu_5Ce(5mpzHA)_5]$ core, characteristic of a 15-MC-5 system, with five $Cu^{II}$ atoms at the rim of the crown and the $Ce^{III}$ ion occupying the dome of the crown, with water molecules, oxygen atoms and one nitrate anion filling the nine-coordination sphere around the $Ce^{III}$ ion, which exhibits a distorted spherical tricapped trigonal prism geometry. The thermogravimetric analysis evidences successive mass losses due to the removal of water molecules and decomposition of the structure after 217 °C, whereas the PXRD analysis of the thermal decomposition residue reveals the presence of copper and copper/cerium oxide particles. These nanocomposite materials were also synthesized using the metallacrown **MC-Ce** under a hydrothermal method in the presence of multi-walled carbon nanotubes (MWCNTs), affording insights that this metallacrown can act as a source precursor for the synthesis of these mixed cerium/copper oxide nanomaterials. The experimental $\chi MT$ value in **MC-Ce** at room temperature is 3.175 $cm^3$ $mol^{-1}$ K, which is higher than the calculated one for one magnetically isolated $Ce^{III}$ plus five $Cu^{II}$ ions, probably due to the antiferromagnetic interactions among $Cu^{II}$ ions within the metallacrown hoop plus the thermal depopulation of $JZ$ sublevels of $Ce^{III}$ ground state (5/2), which exhibit a small splitting under the anisotropic ligand field effects. The $\chi MT$ decreases continuously until it reaches the value of 0.80 $cm^3$ $mol^{-1}$ K at 10 K, reinforcing the presence of intramolecular antiferromagnetic interactions.

**Keywords:** 3d–4f metallacrown; cerium and copper oxide nanoparticles; magnetic properties

## 1. Introduction

Metallacrowns (MCs) are structures containing a cyclic core formed by repeating –[M–N–O]$_n$– units with an X-MC-Y topological representation [1] that have been developed very rapidly within the metallamacrocycle field [2]. The scope of the size of MCs in the solid state built by the $Cu^{II}$ center, for example, has been found to be 9-MC-3, 12-MC-4, $Ln^{III}$ [15-MC-5], and $Ln^{III}$ [18-MC-6], as summarized by M. V. Marinho et al. [3]. We have focused on MC chemistry with one particular 15-metallacrown-5 [3], assembled by the reaction between 5-methyl-2-pyrazinehydroxamic acid (5mpzHA) with copper(II) and europium(III) salts.

Examining the formation of MCs based on the $Ce^{III}$ ion into the MC ring, there are only two examples [4,5], and even though it is well known that the $Ce^{III}$ ion has only one unpaired electron, $4f^1$, the expected ground state (L = 3, and S = ½) is $^2F_{5/2}$, [6] which contributes to its striking feature of magnetic anisotropy [7]. Interestingly $Ce^{III}$-based SMMs are found [7–10], as well as templates for the synthesis of copper and cerium oxide nanoparticles [4,5], which opened new avenues for applications in sensors, fuel cells, coatings, and catalysts. [5]

Thus, in this contribution, we present the synthesis, results of thermal analysis, and spectroscopic, X-ray diffraction, and magnetic properties of a 15-MC-5 metallacrown of formula $[CeCu_5(5mpzHA)_5(NO_3)(H_2O)_7]\cdot 2NO_3\cdot 7H_2O$ (**MC-Ce**). In addition, we prepared nanocomposites of cerium and copper oxides supported on multi-walled carbon nanotubes (MWCNTs) under hydrothermal conditions, using water as the sole solvent in a stainless-steel autoclave at 190 °C. This synthetic approach was motivated by the nature of the metallacrown precursor **MC-Ce**, which contains metal–oxygen bonds with required metal atoms and undergoes thermal decomposition at relatively low temperatures. These characteristics enabled it to function as a single-source precursor, offering strong support for forming $Cu_2O/CeO_2$/MWCNTs nanocomposites.

## 2. Experimental

### 2.1. Materials

Cerium(III) nitrate hexahydrate (99.9%), 5-methyl-2-pyrazinecarboxylic acid (5mpca) (98%), and multi-walled carbon nanotube (MWCNT, 50–90 mm diameter, >95% carbon) were purchased from Sigma-Aldrich. Copper(II) acetate hydrate (98%) was purchased from Synth (Brazil). All chemicals were used as purchased, without further purification. 5-methyl-2-pyrazinehydroxamic acid (5mpzHA) was synthesized as described in the literature [3]. **MC-Ce** was synthesized following a reported method for the $Eu^{III}$ analog [3], reacting $Ce(NO_3)_3\cdot 6H_2O$ salt with $Cu(OAC)_2\cdot H_2O$ salt and 5mpzHA acid, as described above:

$[CeCu_5(5mpzHA)_5(NO_3)(H_2O)_7]\cdot 2NO_3\cdot 7H_2O$ (**MC-Ce**): 5mpzHA (0.017 g, 0.110 mmol), and $Cu(AcO)_2\cdot H_2O$ (0.023 g, 0.110 mmol) were dissolved in methanol (10 mL) and the solution was kept under stirring at 60 °C for 1 h. A methanolic solution (10 mL) of $Ce(NO_3)_3\cdot 6H_2O$ (0.009 g, 0.021 mmol) was added and the reaction was stirred at 60 °C for an additional 1 h. The solution color changed from brown to emerald green 15 min after adding cerium(III) salt. Thus, the solution was filtered and undisturbed at 18 °C, yielding dark green plate-like crystals within 2–4 weeks. Yield: 45%. ATR-FTIR ($\nu$/cm$^{-1}$): $3355_{(\nu O\text{-}H)}$; $3239_{(\nu O\text{-}H)}$; $3033_{(\nu C\text{-}H)}$; $1604_{(\nu C=O)}$; $1580_{(\nu CC/CN)}$; $1544_{(\nu C\text{-}Car)}$; $1481_{(\nu N\text{-}O(nitrate))}$; $1453_{(\delta C\text{-}H)}$; $1382_{(\nu N\text{-}O(nitrate))}$; 1299, $1250_{(\nu N\text{-}O(nitrate))}$; $1199_{(\delta C\text{-}N\text{-}O)}$; $1070_{(\delta C\text{-}H)}$; $1047_{(\nu N\text{-}O)}$; $991_{(\nu N\text{-}O(nitrate))}$.

Copper and cerium oxides on MWCNTs were synthesized following a procedure analogous to that described in the literature [4,5] with only minor adjustments to the quantities of precursors. A quantity of MWCNTs (0.015 g) was added to aqueous metallacrown **MC-Ce** (0.015 g) in 5 mL of water, and the mixture was stirred and sonicated for 1.5 h at room temperature to obtain a suspension. The mixture was transferred to a 25 mL Teflon-lined stainless steel autoclave and heated at 190 °C for one day. The resulting precipitate was separated by centrifugation, washed with distilled water and acetone, and dried in an oven at 110 °C.

### 2.2. Characterization

The FT-IR spectrum was collected using a Shimadzu model FT-IR-Prestige 21 ATR and spectral scan between 4000 and 400 cm$^{-1}$. Thermal analyses (TGA) (Netzsch, Germany, STA 449 F3 Jupiter) were performed with the simultaneous thermal analyzer in the temperature range 30–1100 °C by using alumina crucibles and ca. 10 mg of the sample. A dinitrogen flow of 100 $cm^3$ $min^{-1}$ with a heating rate of 10 °C $min^{-1}$ was used. The Powder X-ray diffraction data were collected at room temperature using a Rigaku Ultima IV (Rigaku, Japan) with Cu-K$\alpha$ radiation (1.5418 Å, 40 kV, 30 mA) from 3° to 50° and 10 to 100° (2$\theta$) using a step size of 0.02° (Figure S1, in Supporting Information). The magnetic measurements were performed in a SQUID magnetometer (model MPMS3), Quantum Design Inc., San Diego, CA, USA) for dc magnetic measurements and a PPMS for ac magnetic susceptibility measurements, both from the Quantum Design company. Data were collected for the polycrystalline sample crushed in a VSM powder sample holder.

### 2.3. X-Ray Crystallography

X-ray diffraction data collection for **MC-Ce** was performed with a GEMINI-A Ultra diffractometer (Rigaku-Oxford Diffraction, USA) diffractometer equipped with Mo-K$\alpha$ radiation ($\lambda$ = 0.71073 Å). Data integration and scaling of the reflections were performed with CrysAlisPro 1.171.40.42a [11]. Final unit cell parameters were based on the fitting of all reflection positions. Analytical absorption corrections were performed using the CrysAlisPro 1.171.40.42a. [12] Compound **MC-Ce** displayed large accessible voids, which are filled with disordered crystallization solvent molecules or counterions. Efforts to unambiguously identify all of them were unsuccessful and the solvent mask routine implemented on the Olex2 program was used [13]. One void presenting a volume of 552 Å$^3$ was found per unit cell, accounting for 182 electrons per unit cell. This number of electrons is consistent with the presence of one nitrate and six crystallization water molecules per formula unit. This assumption is consistent with elemental analysis and thermal results. The MERCURY [14] program was used to prepare artwork representations. Crystal data and details of the data collection and refinement for **MC-Ce** are listed in Table 1. All crystallographic data are available in the supporting information section as CIF files. CCDC number is 2388656.

**Table 1.** Summary of the crystal data and refinement details for **MC-Ce**.

| | |
|---|---|
| Formula | $C_{30}H_{53}Cu_5CeN_{18}O_{33}$ |
| Formula weight (g·mol$^{-1}$) | 1651.69 |
| Crystal system | Triclinic |
| Space group | $P\overline{1}$ |
| T/K | 293 |
| Wavelength (Å) | 0.71073 |
| $a$/Å | 11.6894(2) |
| $b$/Å | 15.4536(1) |
| $c$/Å | 16.1745(5) |
| $\alpha$ (°) | 79.469(2) |
| $\beta$ (°) | 87.062(2) |
| $\gamma$ (°) | 81.301(2) |
| V (Å$^3$) | 2838.72(13) |
| Z | 2 |
| $\rho$/mg cm$^{-3}$ | 1.733 |
| $\mu$ (mm$^{-1}$) | 2.71 |
| $F$(000) | 1470 |
| Crystal size (mm) | 0.35 × 0.25 × 0.03 |
| $\theta$ range/(°) | 2.2–29.6 |
| Reflns. with $I > 2\sigma(I)$ | 9434 |
| Independent reflns. | 13,295 |
| $R_{int}$ | 0.047 |
| N°. of parameters/GOF on F$^2$ | 736/1.11 |
| $R$, w$R$ [$I > 2\sigma(I)$] | 0.045, 0.105 |
| Largest diff. peak/hole (e Å$^{-3}$) | 0.97/−0.79 |

w = $1/[\sigma^2(Fo^2) + (0.0405P)^2 + 0.0994P]$ where $P = (Fo^2 + 2Fc^2)/3$.

## 3. Results and Discussion

The main characteristic observed in the IR spectrum is the vibrational stretching frequency at 1604 $cm^{-1}$ of the $\nu_{C=O}$ band, which indicates the double deprotonation of the 5mpzHA ligand and the coordination of the amide-type nitrogen and oxygen atoms to metal ions. The coordination of the ligand can also be noted by the presence of the νCC/CN at 1580 $cm^{-1}$ as well as bands at 1480, 1398, 1337, and 1297 $cm^{-1}$ due to the presence of nitrate anions [3].

The thermogravimetric analysis (Figure 1) shows an initial weight loss between 39 and 217 °C, attributed to removing coordinated and uncoordinated water molecules, totaling 14 water molecules (obsd. 14.6; calcd. 15.3%). Subsequent weight loss steps are observed up to 900 °C, likely corresponding to the decomposition of **MC-Ce**, involving the loss of nitrate anions (3 molecules) and the 5-methyl-2-pyrazinehydroximate ($5mpzA^{2-}$, 3 molecules), resulting in 41.4% of the compound (obsd. 38.7%). The significant residue can be associated with the presence of copper, copper oxide, and cerium oxide phases in the sample [Cu + 2 $Cu_2O$ + $CeO_2$] [4,5], as evidenced by the PXRD analysis (Figure 2).

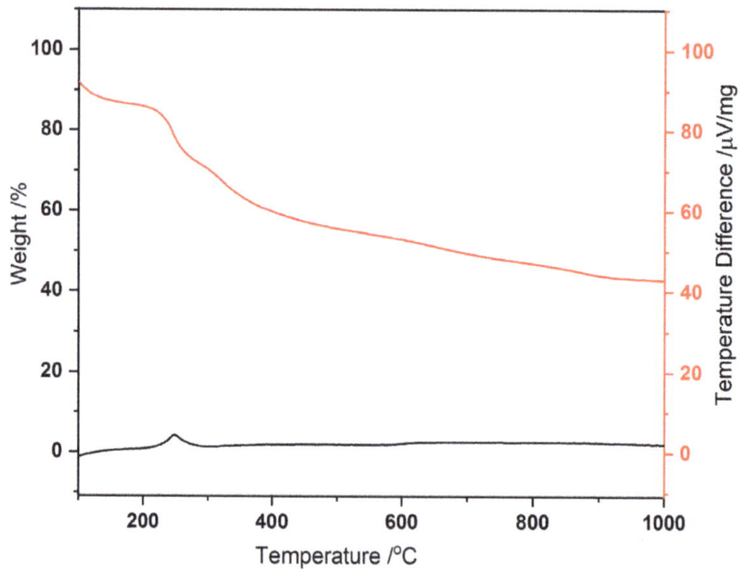


**Figure 1.** Thermal decomposition of MC-Ce under $N_2$, showing TG (red) and DSC (black) curves.

In addition, our powder X-ray investigation indicates that the solid obtained from the reaction of the precursor **MC-Ce** with MWNTs contains copper and cerium oxides, as previously reported for other lanthanide-copper 15-MC-5 compounds [4,5] (see Figure 3).

Thus, our findings based on an X-ray investigation of the residue from the **MC-Ce** thermal decomposition in an inert atmosphere show the obtaining of cerium and copper oxide mixture, whose nanocomposite materials were also evidenced by reacting with the **MC-Ce** with MWCNT under hydrothermal conditions. To our knowledge, this is the third report involving metallacrown systems for preparing cerium–copper mixed oxides. Earlier, 15-MC-5 complexes and MWCNTs were used as unique precursors via a simple hydrothermal method [4,5] by S. Y. Ketkov et al. using a polynuclear metallamacrocyclic complex and $Ce(H_2O)_4$[15MCCuGlyha-5]$Cl_3$ [4] with glycinehydroxamic acid (Glyha), as well as by I. L. Eremenko et al. [5] using a macrocycle [$Ce(H_2O)_4$[15-MCCuRha-5] with R = Glyha and α-phenylalaninehydroxamic acid (Phalaha). In this case, cerium and copper oxide nanocrystals of various sizes and morphologies were obtained [5]. In our study, we employ the metallacrown [$CeCu_5(5mpzHA)_5(NO_3)(H_2O)_7$]$\cdot 2NO_3 \cdot 7H_2O$ (**MC-Ce**) leveraging its high solubility in mixtures of water and polar solvents, such as ethanol and methanol, as well as its decomposition properties at a relatively low temperature (around 217 °C) and under hydrothermal method also utilizing MWCNTs as templates, to achieve similar objectives.

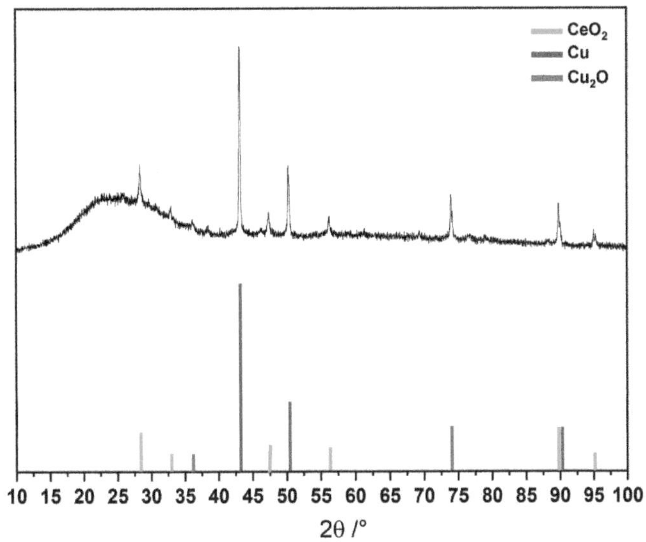


**Figure 2.** PXRD analysis of final product after thermal decomposition of **MC-Ce**.

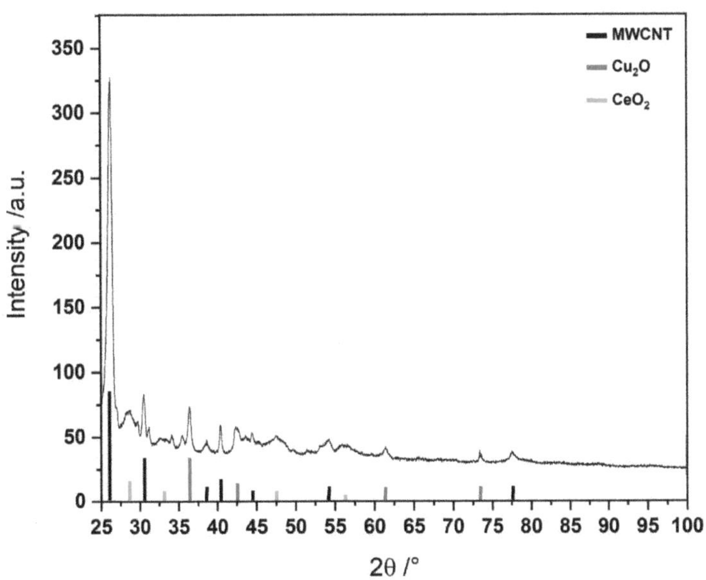


**Figure 3.** PXRD pattern in the presence of MWCNTs at 190 °C for 24 h using **MC-Ce** as a single route for nanoparticles.

In addition, PXRD experiments were also carried out with the crushed crystals of **MC-Ce**, showing a good coincidence between the experimental and calculated patterns (Figure S1, Supporting Information), confirming that the single-crystal X-ray structure is the same as the as-synthesized bulk material.

### 3.1. X-Ray Diffraction

The crystal structure of **MC-Ce** was solved in the centrosymmetric triclinic $P\bar{1}$ space group, probing the formation of a typical 15-MC-5 metallacrown with the general formula $[CeCu_5(5mpzHA)_5(NO_3)(H_2O)_7]\cdot 2NO_3\cdot 7H_2O$. The asymmetric unit is composed of a crown-like $[Cu_5Ce(5mpzHA)_5]$ core, containing five deprotonated $5mpzHA^{2-}$ ligands, five $Cu^{II}$ ions, one $Ce^{III}$ ion through a $\mu_3$-$\kappa^2N,N'$:$\kappa^2O,O'$:$\kappa O'$ chelating–bridging mode (Figure 4). Thermal ellipsoids are shown in Figure S2 (ESI).

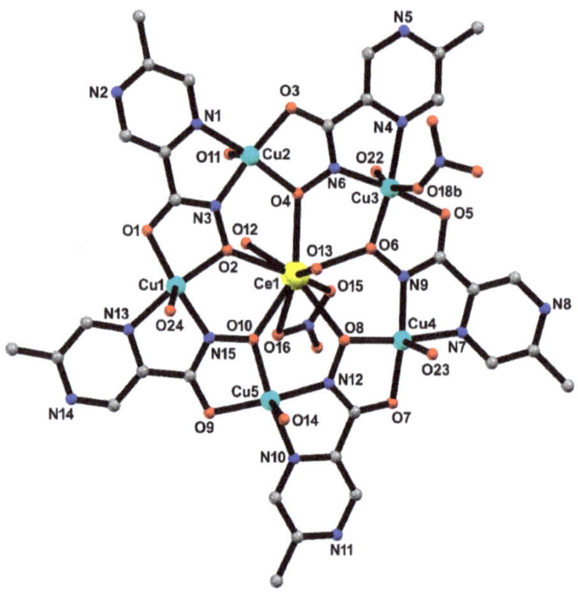


**Figure 4.** The asymmetric unit of MC-Ce with selected atom labeling. Hydrogen atoms and uncoordinated water molecules, and part of the disorder on the coordinated nitrate ion to Cu3, were omitted for clarity. Color codes: carbon (grey), copper (cyan), oxygen (red), nitrogen (blue), and cerium (yellow).

The $Ce^{III}$ ion is nine-coordinated to five 5mpzHA oxygen atoms, two water molecules oxygen atoms, and two nitrate group oxygen atoms, exhibiting a coordination environment of a spherical tricapped trigonal prism geometry. The SHAPE software (Version 2.1) [15] was used to determine the degree of distortion of the coordination polyhedral of the central $Ce^{III}$ ion to **MC-Ce** concerning the ideal nine-vertex coordination polyhedral $CeO_9$. The lowest continuous shape measurement (CShM) values correspond to the spherical tricapped trigonal prism ($D_{3h}$) (CShM = 5.874), spherical capped square antiprism ($C_{4v}$) (CShM = 6.649), and Muffin ($C_s$) (CShM = 6.549) (values in Table S1). The Ce–O bond distances in **MC-Ce** are in the range from 2.489(3) to 2.653(3) Å, where the distortion observed around the $Ce^{III}$ geometry can be associated with the chelating nitrate ion that fills the coordination sphere, showing mostly longer bond distances [2.617(3) to 2.653(3) Å] [16].

Four copper (II) ions are pentacoordinate, exhibiting a distorted square pyramidal geometry, where each one is coordinated to the 5mpzHA ligands (basal plane) and one water molecule (apical position). The values for the trigonality parameter $\tau$ [17] are 0.11, 0.05, 0.011, and 0.042 for Cu1, Cu2, Cu4, and Cu5, respectively. The copper(II) ions are shifted from the mean basal plane by 0.043 (Cu1), 0.125 (Cu2), 0.052 (Cu3), 0.108 (Cu4), and 0.168 (Cu5) Å toward the apical position.

One copper(II) ion is coordinated to the 5mpzHA ligands in the basal plane, one water molecule, and one nitrate anion in the apical position, resulting in a distorted octahedral environment. The aqua ligands completing the copper(II) coordination sphere exhibit relatively longer bond distances [Cu1-O(24) = 2.784(4); Cu2-O(11) = 2.388(3); Cu3-O(22) = 2.531(3); Cu4-O(23) = 2.443(4); and Cu5-O(14) = 2.370(3) Å] compared to the bond distances in 5mpzHA (see Table 2).

The average shortest intramolecular Ce···Cu and Cu···Cu distances are 3.9292(6) Å and 4.5945(8) Å, respectively (Table 2). The shortest intermolecular distances between metal ions in the crystal packing are 5.8957(7) Å (Cu4···Cu5i), 6.5831(6) Å (Ce1···$Cu5^{i}$), and 7.9041(5) Å (Ce1···$Ce1^{i}$), where $I = 1 - x, 1 - y, 1 - z$. In addition, the O-Ce-O angles of the $5mpzHA^{2-}$ ligands range from 68.71(9)° (O2-Ce1-O10) to 71.68(9)° (O6-Ce-O8).

**Table 2.** Selected distances (in Å) for **MC-Ce**.

| Atoms Label | Distance | Atoms Label | Distance |
|---|---|---|---|
| Cu1—N13 | 2.007(4) | Cu5—O9 | 1.944(3) |
| Cu1—N15 | 1.908(3) | Cu5—O10 | 1.928(4) |
| Cu1—O1 | 1.941(3) | Cu5—O14 | 2.370(3) |
| Cu1—O2 | 1.922(3) | Ce1—O2 | 2.522(3) |
| Cu1—O24 | 2.784(4) | Ce1—O4 | 2.489(3) |
| Cu2—N1 | 2.015(3) | Ce1—O6 | 2.496(3) |
| Cu2—N3 | 1.918(4) | Ce1—O8 | 2.502(3) |
| Cu2—O3 | 1.931(3) | Ce1—O10 | 2.505(3) |
| Cu2—O4 | 1.927(3) | Ce1—O12 | 2.530(3) |
| Cu2—O11 | 2.388(3) | Ce1—O13 | 2.593(3) |
| Cu3—N4 | 2.029(4) | Ce1—O15 | 2.617(3) |
| Cu3—N6 | 1.919(3) | Ce1—O16 | 2.653(3) |
| Cu3—O5 | 1.963(3) | Ce1···Cu1 | 3.9426(7) |
| Cu3—O6 | 1.930(3) | Ce1···Cu2 | 3.9567(6) |
| Cu3—O18b | 2.69(1) | Ce1···Cu3 | 3.9704(8) |
| Cu3—O22 | 2.531(3) | Ce1···Cu4 | 3.8821(6) |
| Cu4—N7 | 1.992(3) | Ce1···Cu5 | 3.8944(7) |
| Cu4—N9 | 1.927(3) | Cu1···Cu2 | 4.5727(9) |
| Cu4—O7 | 1.946(3) | Cu2···Cu3 | 4.5987(8) |
| Cu4—O8 | 1.928(3) | Cu3···Cu4 | 4.6205(8) |
| Cu4—O23 | 2.443(3) | Cu4···Cu5 | 4.5929(8) |
| Cu5—N10 | 1.996(4) | Cu5···Cu1 | 4.5880(7) |
| Cu5—N12 | 1.919(3) | | |

In **MC-Ce**, supramolecular dimers with a Ce···Ce separation of 7.9041(5) Å are formed via a series of O–H···O hydrogen bonds [O13–H13a···$O14^{i}$ = 170(6)°, O13···$O14^{i}$ = 2.857(5) Å; O23–H23b···$O24^{i}$ = 172(4)°, O23···$O24^{i}$ = 2.855(5) Å; and O14-H14a···$O23^{i}$ = 156(7), O14···O23i = 2.835(5); O14-H14b···O18bi = 143(5), O14···O18bi = 2.711(13) (i) = $1 - x$, $1 - y$, $1 - z$)], as shown in Figure 5 (top). Moreover, a network of hydrogen bonds involving the crystallization water molecule (O25), aqua ligands (O11, O22, and O24), and the pyrazine nitrogen atoms (N5 and N14) contributed to the stabilization of the crystal packing (Figure 5, bottom). Geometric parameters for these intermolecular interactions are provided in Table S3-ESI.

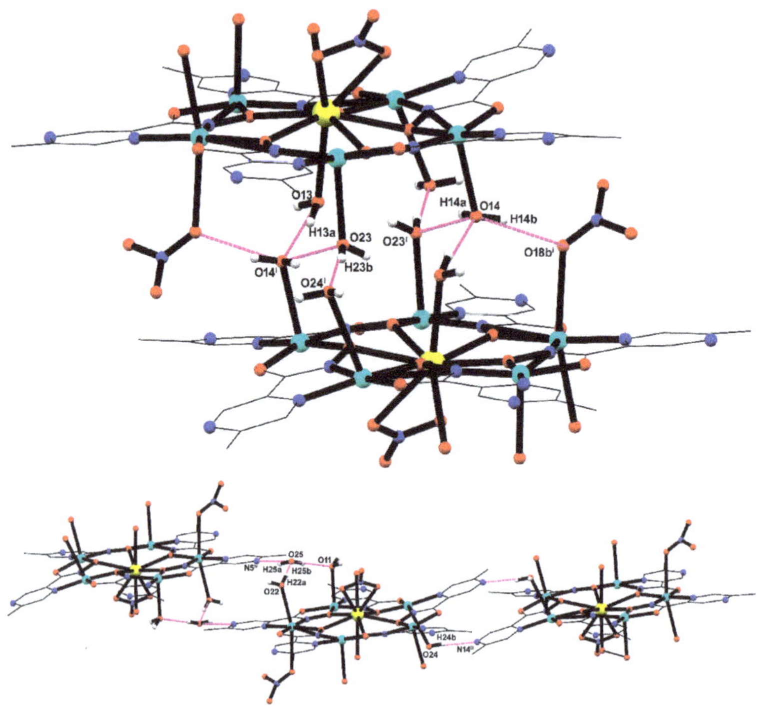


**Figure 5.** Supramolecular dimers are formed by hydrogen bonds involving coordinated water molecules and nitrate oxygen atom (**top**), and intermolecular interactions between crystallization water molecules, water ligands, and pyrazine nitrogen atom (**bottom**).

## 3.2. Magnetic Properties

*dc* Magnetic susceptibility was recorded in the 2–300 K temperature range for **MC-Ce** under a dc magnetic field of 1000 Oe. The thermal dependence of the $\chi_M$T product is shown in Figure 6, and it has a temperature behavior similar to that observed for metallacrown with other lanthanide metallacrown complexes [3,18]. The experimental $\chi_M$T value at room temperature is 3.175 cm$^3$ mol$^{-1}$ K, which is higher than the calculated one for one magnetically isolated Ce$^{III}$ ($J$ = 5/2, $g$ = 6/7, $\chi_M$T = 0.80 cm$^3$ mol$^{-1}$) plus five Cu$^{II}$ ions ($S$ = 1/2, $g$ = 2, $\chi_M$T = 1.875 cm$^3$ mol$^{-1}$ K). Upon cooling, $\chi_M$T decreases continuously until it reaches the value 0.80 cm$^3$ mol$^{-1}$ K at 10 K; this magnetic behavior is attributed to the antiferromagnetic interactions among Cu$^{II}$ ions within the metallacrown hoop plus the thermal depopulation of $J_Z$ sublevels of Ce$^{III}$ ground state (5/2), which exhibit a small splitting under the anisotropic ligand field effects. At around 10 K, it exhibits a further slight decrease, which also suggests the presence of intramolecular antiferromagnetic interactions. The experimentally obtained $\chi_M$T data for the **MC-Ce** were fitted using the previously reported additive model by means of the PHI program with a Hamiltonian (Equation (1)) that considers the exchange interactions between adjacent Cu$^{II}$ ions ($\hat{H}_{Ex(Cu\text{-}Cu)}$) to be dominant. The magnetic contribution of Ce$^{III}$ was considered by adding to the fitting model a free ion approximation to account for the magnetic anisotropy by using the anisotropic Hamiltonian ($\hat{H}_{CF(Ce)}$). The exchange interaction between Cu$^{II}$ and Ce$^{III}$ is expected to be very weak, and it was neglected to avoid overparameterization. To account for weak magnetic interactions in the system, including intermolecular magnetic interactions and exchange interactions between $Cu^{II}$ and $Ce^{III}$, a molecular field model was applied (Equation (4)):

$$\hat{H} = \hat{H}_{Ex(Cu\text{-}Cu)} + \hat{H}_{CF(Ce)} + \text{Zeeman terms} \quad (1)$$

$$\hat{H}_{Ex(Cu\text{-}Cu)} = 2J\,(\hat{S}_1\cdot\hat{S}_2 + \hat{S}_2\cdot\hat{S}_3 + \hat{S}_3\cdot\hat{S}_4 + \hat{S}_4\cdot\hat{S}_5 + \hat{S}_5\cdot\hat{S}_1) \quad (2)$$

$$\hat{H}_{CF(Ce)} = B_2{}^0\hat{O}_2{}^0 + B_4{}^0\hat{O}_4{}^0 \quad (3)$$

$$\chi_M T = \chi_M T/[1 - (2zJ'\chi_M T/N_A g^2\,\mu_B{}^2)] \quad (4)$$

in which $J$ is the isotropic exchange interaction parameter between adjacent $Cu^{II}$ ions in the metallacrown unit, and $\hat{S}_i$ is the spin operator for the corresponding $Cu^{II}$ ions. The anisotropic Hamiltonian considers the second and fourth-order axial anisotropy terms ($B_2{}^0$ and $B_4{}^0$) with the corresponding Stevens' operator equivalents ($\hat{O}_2{}^0$ and $\hat{O}_4{}^0$). As shown in Figure 6, the calculated curve (solid line) matches the experimental data in the investigated temperature range quite well. The best fit was achieved with $J = -51.8(1)$ cm$^{-1}$, $zJ' = -0.90(5)$ cm$^{-1}$, TIP = 0.003 cm$^3$ mol$^{-1}$ (TIP is the temperature-independent paramagnetism), $g_{Cu}$= 2.148(2), and $g_{Ce}$= 0.857(3). These values are close to previously reported data for similar metallacrown core $Ln^{III}$-$Cu^{II}$5 [9] obtained for the anisotropy coefficients ($B_2{}^0 = -13.0(2)$, $B_4{}^0 = -0.468(10)$ cm$^{-1}$), and are in the same order of values observed for other complexes based on $Ce^{III}$ ions [18].

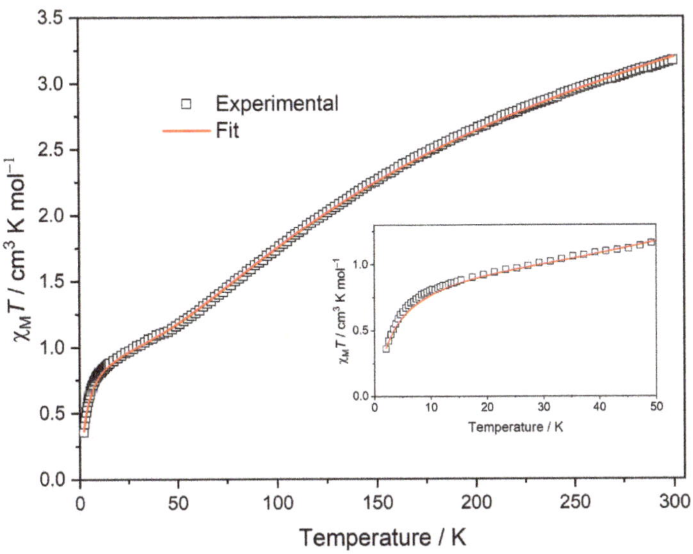


**Figure 6.** Temperature dependence of the $\chi_M$T product of **MC-Ce** was measured by applying the 1 kOe dc field. The solid line is the best fit (see the main text for details). Inset: a magnified view in the temperature range of 2–50 K.

The experimental magnetic field dependence of the magnetization at three different temperatures (2, 4, and 8 K) is shown in Figure 7. When the external magnetic field reaches 7 T, the magnetization attains a maximum value of 1.45 NμB. Despite the gradual increases at a high field, no saturation of the magnetization is observed, suggesting significant anisotropy. At low temperatures, the contribution of the adjacent uncompensated $Cu^{II}$ ion (S = 1/2) is expected alongside the contribution of the $Ce^{III}$ ion (S = 1/2), which would result in a saturation magnetization of 2 NμB. In addition to the anisotropy effect, as highlighted in the inset of Figure 7, the large discrepancy between the experimental value and the expected saturation magnetization value may be attributed to a frustration effect on the uncompensated $Cu^{II}$ ions [18]. This discrepancy is evident when the simulated magnetization curves, calculated using PHI with parameters derived from the susceptibility fit, are compared to the experimental magnetization data, shown in Figure S3.

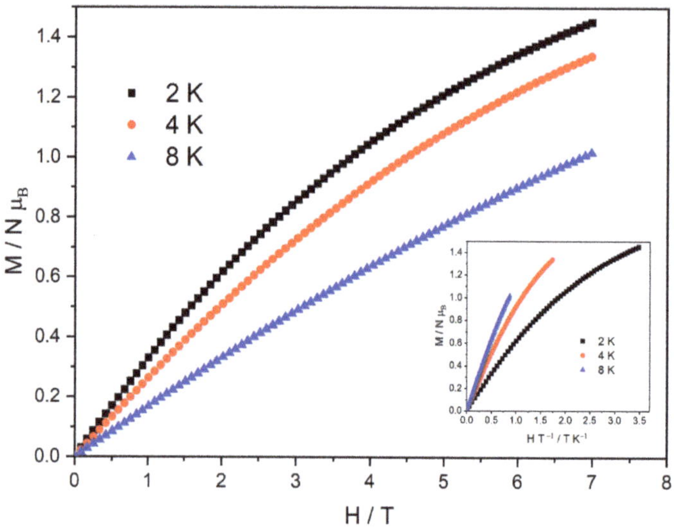


**Figure 7.** Isothermal M vs. H plot for complex **MC-Ce** measured at 2, 4, and 8~K. Inset: reduced magnetization plot (M vs. H/T) measured at 2, 4, and 8 K.

To explore the possible dynamics of the magnetization of the Ce–Cu5 complex, alternating current (ac) magnetic susceptibility measurements were conducted in the temperature range of 2.0–10.0 K under an applied field up to 10 kOe. This study showed the lack of any significant out-of-phase signal for **MC-Ce** under zero and non-zero applied dc fields, evidencing that the **MC-Ce** does not exhibit slow magnetic relaxation.

## 4. Conclusions

This work describes a new 15-MC-5 metallacrown containing $Ce^{III}$ and $Cu^{II}$ centers of the formula $[CeCu_5(5mpzHA)_5(NO_3)(H_2O)_7]\cdot 2NO_3\cdot 7H_2O$ (**MC-Ce**). The $Ce^{III}$ ion features a nine-vertex $CeO_9$ spherical tricapped trigonal prism coordination environment. The thermal analysis revealed the collapse of the structure after 217 °C, and the phase composition of the residue was determined using PXRD, showing the presence of copper and copper/cerium oxide particles. These nanocomposite materials were synthesized using the metallacrown **MC-Ce** via a hydrothermal method at 190 °C in the presence of MWCNTs, providing evidence that this metallacrown can serve as a template for the synthesis of nanoparticles. Inspired by this result, further investigations into the functionality of these nanocomposites could reveal their potential applications. The experimental $\chi_M T$ value at room temperature was 3.175 cm$^3$ mol$^{-1}$ K, higher than the calculated one for one magnetically isolated $Ce^{III}$ plus five $Cu^{II}$ ions ($\chi_M T$ = 1.875 cm$^3$ mol$^{-1}$ K). The $\chi_M T$ decreases continuously until it reaches the value of 0.80 cm$^3$ mol$^{-1}$ K at 10 K, where this magnetic behavior was attributed to the antiferromagnetic interactions among $Cu^{II}$ ions within the metallacrown hoop plus the thermal depopulation of $J_Z$ sublevels of $Ce^{III}$ ground state (5/2), which exhibit a small splitting under the anisotropic ligand field effects.

**Supplementary Materials:** The following supporting information can be downloaded at: https://www.mdpi.com/article/10.3390/magnetochemistry10120096/s1, Figure S1. Experimental (red) and calculated (black) PXRD patterns for **MC-Ce**; Figure S2. Thermal ellipsoids of the asymmetric unit of **MC-Ce** drawn at 50% of probability. Color codes: carbon (grey), copper (cyan), oxygen (red), nitrogen (blue), and cerium (yellow); Figure S3. Isothermal magnetization (M) versus magnetic field (H) plot for **MC-Ce** at 2, 4, and 8~K. Table S1. Geometric analysis of the coordination environment of the cerium(III) ion in **MC-Ce**, showing the site symmetry approximation derived from continuous shape measures; Table S2. Selected interatomic bond angles (deg) for the **MC-Ce**; Table S3. Hydrogen bond observed in the supramolecular crystal packing (values in Å and deg).

**Author Contributions:** J.D.T. and F.F.P. contributed equally. J.D.T., F.F.P. and L.A.F. synthesized and characterized the system; R.M.R.d.S., L.D.G.B. and C.C.C. participated in the characterization by thermal analysis and Powder X-ray diffraction; I.M.L.R. and G.P.G. collected and treated single-crystal X-ray diffraction data; A.M.G. and W.C.N. collected and performed the magnetic data acquisition and processing; M.V.M. participated in supervising the synthesis, analyses, and interpretation besides writing and reviewing all the data of the paper. All authors have read and agreed to the published version of the manuscript.

**Funding:** This work was carried out with support from FAPEMIG and CNPq Foundations, Brazil (process number: BPQ-APQ-06579-24); FAPEMIG (CEX-APQ-01597-17, APQ-00544-23, APQ-05218-23); FAPERJ (E-26/010.000978/2019, E-26/010.1553/2019, E-26/201.314/2022); CNPq (307135/2023-3); CAPES (001); PIBIC/PIBITI_CNPq, PIBICT_Fapemig, PRPPG-UNIFAL-MG.

**Data Availability Statement:** The data presented in this study are available in this article or the Supplementary Materials.

**Acknowledgments:** The Fundação de Amparo à Pesquisa do Estado de Minas Gerais supported this work (FAPEMIG, CEX-APQ-01597-17, APQ-00544-23, APQ-05218-23). M.V.M. extends their sincere appreciation for the support provided for this project, which was carried out with support from FAPEMIG and CNPq Foundations, Brazil (process number: BPQ-APQ06579-24). C.C.C., F.F.P. and R.M.R.d.S. thank CAPES Foundation (Brazil, CAPES, finance code: 001). W.C.N. thanks FAPERJ (projects E-26/010.000978/2019 and E-26/010.1553/2019). G.P.G. thanks FAPERJ (project E-26/201.314/2022) and CNPq (project 307135/2023-3). J.D.T. and L.D.G.B. thank PIBIC/PIBITI_CNPq, PIBICT_Fapemig, and UNIFAL-MG for their undergraduate research fellowships. M.V.M. wishes to thank PROAP and PRPPG (UNIFAL-MG) and the laboratories of UNIFAL-MG: (IR spectroscopy, TG/DSC analysis (FINEP 01/10/0798/00), powder XRD Multiuser Laboratory (LabcrystAL), and support of LabCRI (UFMG) for the use of SXRD diffraction facilities. M.V.M. dedicates this article to the beloved memory of great friend and Miguel Julve, whose scientific contributions and unwavering dedication to the field are incalculable. His passion for discovery and generosity in sharing knowledge continue to inspire all who had the privilege to work alongside him. He will be greatly missed both personally and professionally.

**Conflicts of Interest:** The authors declare no competing financial interests.

## References


1. Mezei, G.; Zaleski, C.M.; Pecoraro, V.L. Structural and Functional Evolution of Metallacrowns. *Chem. Rev.* **2007**, *107*, 4933–5003. [CrossRef] [PubMed]
2. Lutter, J.C.; Zaleski, C.M.; Pecoraro, V.L. Metallacrowns: Supramolecular Constructs with Potential in Extended Solids, Solution-State Dynamics, Molecular Magnetism, and Imaging. In *Advances in Inorganic Chemistry*; Academic Press Inc.: Cambridge, MA, USA, 2018; Volume 71, pp. 177–246.
3. Paiva, F.F.; Ferreira, L.A.; Rosa, I.M.L.; da Silva, R.M.R.; Sigoli, F.; Cambraia Alves, O.; Garcia, F.; Guedes, G.P.; Vanda Marinho, M. Heterobimetallic Metallacrown of $Eu^{III}Cu^{II}_5$ with 5-Methyl-2-Pyrazinehydroxamic Acid: Synthesis, Crystal Structure, Magnetism, and the Influence of $Cu^{II}$ Ions on the Photoluminescent Properties. *Polyhedron* **2021**, *209*, 115–466. [CrossRef]
4. Kremlev, K.V.; Samsonov, M.A.; Zabrodina, G.S.; Arapova, A.V.; Yunin, P.A.; Tatarsky, D.A.; Plyusnin, P.E.; Katkova, M.A.; Ketkov, S.Y. Copper(II)-Cerium(III) 15-Metallacrown-5 Based on Glycinehydroxamic Acid as a New Precursor for Heterobimetallic Composite Materials on Carbon Nanotubes. *Polyhedron* **2016**, *114*, 96–100. [CrossRef]
5. Katkova, M.A.; Kremlev, K.V.; Zabrodina, G.S.; Rumyantcev, R.V.; Gazhulina, A.P.; Gusev, S.A.; Ketkov, S.Y.; Fomina, I.G.; Eremenko, I.L. Polynuclear Aminohydroximate Metallamacrocyclic Cu(II)-Ce(III) Complexes: A Facile Route to Intricate Nanostructures of Copper and Cerium Oxides. *Eur. J. Inorg. Chem.* **2019**, *2019*, 1002–1010. [CrossRef]
6. Wada, H.; Ooka, S.; Yamamura, T.; Kajiwara, T. Light Lanthanide Complexes with Crown Ether and Its Aza Derivative Which Show Slow Magnetic Relaxation Behaviors. *Inorg. Chem.* **2017**, *56*, 147–155. [CrossRef] [PubMed]

7. Pointillart, F.; Bernot, K.; Sessoli, R.; Gatteschi, D. Effects of 3d-4f Magnetic Exchange Interactions on the Dynamics of the Magnetization of $Dy^{III}$-$M^{II}$-$Dy^{III}$ Trinuclear Clusters. *Chem. Eur. J.* **2007**, *13*, 1602–1609. [CrossRef] [PubMed]
8. Mautner, F.A.; Bierbaumer, F.; Fischer, R.C.; Tubau, À.; Speed, S.; Ruiz, E.; Massoud, S.S.; Vicente, R.; Gómez-Coca, S. Insights into the Spin Dynamics of Mononuclear Cerium(III) Single-Molecule Magnets. *Inorg. Chem.* **2022**, *61*, 11124–11136. [CrossRef] [PubMed]
9. Pavlishchuk, A.V.; Kolotilov, S.V.; Zeller, M.; Lofland, S.E.; Addison, A.W. Magnetic Properties of $Ln^{III}$–$Cu^{II}$ 15-Metallacrown-5 Dimers with Terephthalate ($Ln^{III}$ = Pr, Nd, Sm, Eu). *Eur. J. Inorg. Chem.* **2018**, *2018*, 3504–3511. [CrossRef]
10. Le Roy, J.J.; Korobkov, I.; Kim, J.E.; Schelter, E.J.; Murugesu, M. Structural and Magnetic Conformation of a Cerocene $[Ce(COT'')_2]^-$ Exhibiting a Uniconfigurational f1 Ground State and Slow-Magnetic Relaxation. *Dalton Trans.* **2014**, *43*, 2737–2740. [CrossRef] [PubMed]
11. Rigaku Oxford Diffraction. *CrysAlisPro*; Agilent: Santa Clara, CA, USA, 2014.
12. Clark, R.C.; Clark, J.S.R. The analytical calculation of absorption in multifaceted crystal. *Acta Cryst.* **1995**, *A51*, 887–897. [CrossRef]
13. Dolomanov, O.V.; Bourhis, L.J.; Gildea, R.J.; Howard, J.A.K.; Puschmann, H. OLEX2: A Complete Structure Solution, Refinement and Analysis Program. *J. Appl. Crystallogr.* **2009**, *42*, 339–341. [CrossRef]
14. MacRae, C.F.; Sovago, I.; Cottrell, S.J.; Galek, P.T.A.; McCabe, P.; Pidcock, E.; Platings, M.; Shields, G.P.; Stevens, J.S.; Towler, M.; et al. Mercury 4.0: From Visualization to Analysis, Design and Prediction. *J. Appl. Crystallogr.* **2020**, *53*, 226–235. [CrossRef] [PubMed]
15. Llunell, M.; Casanova, D.; Cirera, J.; Alemany, P.; Alvarez, S. *SHAPE: Program for the Stereochemical Analysis of Molecular Fragments by Means of Continuous Shape Measures and Associated Tools*; Version 2.1; Universitat de Barcelona: Barcelona, Spain, 2013.
16. Costa, A.I.; da Silva, R.M.R.; Botelho, L.D.G.; Coelho, S.F.N.; Sigoli, F.A.; Honorato, J.; Ellena, J.; Martins, F.T.; Gomes, A.M.; Nunes, W.C.; et al. Intensity and Lifetime Ratiometric Luminescent Thermometer Based on a Tb(III) Coordination Polymer. *Dalton Trans.* **2024**, *53*, 3994–4004. [CrossRef] [PubMed]
17. Addison, A.W.; Rao, T.N.; Reedijk, J.; Van Rijn, J.; Verschoor, G.C. Synthesis, Structure, and Spectroscopic Properties of Copper(II) Compounds Containing Nitrogen-Sulphur Donor Ligands; the Crystal and Molecular Structure of Aqua[1,7-Bis(N-Methylbenzimidazol-2′;-Yl)-2,6-Dithiaheptane]Copper(II) Perchlorate. *J. Chem. Soc. Dalton Trans.* **1984**, 1349–1356. [CrossRef]
18. Hino, S.; Maeda, M.; Yamashita, K.; Kataoka, Y.; Nakano, M.; Yamamura, T.; Nojiri, H.; Kofu, M.; Yamamuro, O.; Kajiwara, T. Linear Trinuclear Zn(II)-Ce(III)-Zn(II) Complex Which Behaves as a Single-Molecule Magnet. *Dalton Trans.* **2013**, *42*, 2683–2686. [CrossRef] [PubMed]

magnetochemistry

MDPI

*Article*

# Controlled Zn(II) to Co(II) Transmetalation in a Metal–Organic Framework Inducing Single-Ion Magnet Behavior †

**Paula Escamilla [1], Nicolás Moliner [1], Donatella Armentano [2], Emilio Pardo [1], Jesús Ferrando-Soria [1] and Thais Grancha [1,*]**

1 Instituto de Ciencia Molecular (ICMol), Universidad de Valencia, 46980 Valencia, Spain; paula.escamilla@uv.es (P.E.); fernando.moliner@uv.es (N.M.); emilio.pardo@uv.es (E.P.); jesus.ferrando@uv.es (J.F.-S.)

2 Dipartimento di Chimica e Tecnologie Chimiche (CTC), Università della Calabria, 87036 Rende, Italy; donatella.armentano@unicat.it

* Correspondence: thais.grancha@uv.es

† This contribution serves as a modest acknowledgement to two of the greatest coordination chemists the world has known. Yet this is not solely about their undeniably exceptional scientific achievements; it is a heartfelt acknowledgment of their greatness as individuals. Paco and Miguel are more than PhD supervisors and world-renowned scientists—they are family to us. We are who we are because of them. Although Miguel is no longer with us, each time we enter the lab, where he loved to be, he is with us in spirit. Paco remains our guide and inspiration on this side of life.

**Abstract:** The intrinsic characteristic features of metal–organic frameworks offer unique, great opportunities to develop novel materials with applications in very diverse fields. Aiming to take advantage of these, the application of post-synthetic methodologies has revealed itself to be a powerful approach to the isolation and structuration of metal ions, molecules, or more complex species, either within MOF channels or reticulated at their network, rendering novel and exciting MOFs with new or improved functionalities. Herein, we report the partial post-synthetic metal exchange of Zn(II) metal ions by Co(II) ones in water-stable three-dimensional $CaZn_6$-MOF **1**, derived from the amino acid *S*-methyl-*L*-cysteine, allowing us to obtain two novel MOFs with increasing contents of the Co(II) ions $\mathbf{Co_{4\%}@1}$ and $\mathbf{Co_{8\%}@1}$. Remarkably, the presented post-synthetic metal exchange methodology has two relevant implications for us: (*i*) it allowed us to obtain two novel MOFs, which were not accessible by direct synthesis, and (*ii*) enabled us to transform physical properties within this family of isoreticular MOFs from the diamagnetic pristine MOF **1** to MOFs $\mathbf{Co_{4\%}@1}$ and $\mathbf{Co_{8\%}@1}$, exhibiting field-induced, frequency-dependent, alternating current magnetic susceptibility signals, which are characteristic features of single-molecule magnets.

**Keywords:** metal–organic framework; molecular magnetism; single-ion magnet; cobalt(II)

## 1. Introduction

Single-ion magnets (SIMs) represent the ultimate limit of the miniaturization of single-molecule magnets (SMMs) [1–10]. Their potential applications as high-density magnetic memories and quantum computing devices in the emerging field of molecular spintronics [11–15] have attracted the attention of many groups working in the fields of molecular magnetism [16–20] and multifunctional magnetic materials [14,21–24]. In fact, during the last decade, great advances have been achieved in efforts to increase phase memory times and the temperature need to observe magnetic bistability [25–28]. SIMs have been reported with very distinct metal ions, including lanthanides [8,29] and actinides [30,31], but also with first- [32–34] and third-row [35,36] transition metal ions. An interesting family of SIMs are Co(II) complexes. Their inherently large magnetic anisotropy, due to first-order spin–orbit coupling (SOC), allow for the obtention of Co(II)-based SIMs with different coordination geometries and signs of magnetic anisotropy [32,37,38]. However, despite the commendable advances achieved in the field, several daunting challenges remain to be solved, with the controlled spatial organization of magnetic molecules to build solid ordered arrays while retaining their intrinsic individual magnetic properties being one of the greatest ones.

Metal–organic frameworks (MOFs) [39] have emerged as a novel generation of porous materials [40,41] due to their unique differentiating features, such as their modulability and rich host–guest chemistry [42] and high porosity and crystallinity [43,44]. Since their advent, MOFs have shown noteworthy applications in very diverse fields, from gas adsorption and separation [45,46] and catalysis [47,48] to drug delivery [49] and magnetism [50,51], recently being used, for example, in water remediation [52] and chemical nanoreactors as well [53]. This, in part, is a direct consequence of the commendable synthetic efforts to achieve great control of the MOF structure obtained. In fact, even though total control is not always possible to achieve, the different pre- and post-synthetic methodologies [54–57] used have made it possible to precisely tune the ratio of distinct components and to gain control of MOF dimensionality and topology and, consequently, of the MOF's final properties.

Thus, MOFs, a priori, represent outstanding frameworks for taking serious steps forward toward gaining control of the spatial organization of magnetic molecules. Indeed, seminal works have been devoted to this end, where the main synthetic strategies can mainly be classified into the following: (i) the direct self-assembly of metal ions with selected polytopic ligands, including mixtures of two distinct metal ions (doping) [37,58,59], (ii) the encapsulation of preformed SMMs [60,61], and (iii) post-synthetic metal exchange [62]. Regarding this last approach, we have already successfully performed it. We used a diamagnetic $Ca^{II}Co^{III}$ 3D MOF as the initial framework, and, post-synthetically, we replaced all $Ca^{II}$ ions with different lanthanide metal ions, where the final MOFs with Kramers ions ($Dy^{III}$ and $Er^{III}$) exhibited SIM behavior [24]. Also, recently, partial post-synthetic metal exchange was reported in a Zn-based MOF, replacing Zn(II) with Co(II) in different, very low atomic percentages [62]. Here, underpinned in our treasured knowledge of MOFs and SIMs, we further explore the latter route by performing partial post-synthetic metal exchange in a diamagnetic heterobimetallic $CaZn_6$-MOF (Figure 1), replacing Zn(II) with Co(II) metal ions in different controlled atomic percentages (4 and 8%), and we study how this different degree of metal heterogeneity influences the magnetic properties of the obtained $Ca^{II}Zn^{II}Co^{II}$-MOF. The resulting novel mixed-metal MOFs exhibit field-induced, frequency-dependent, alternating relaxation typical of single-molecule magnets.

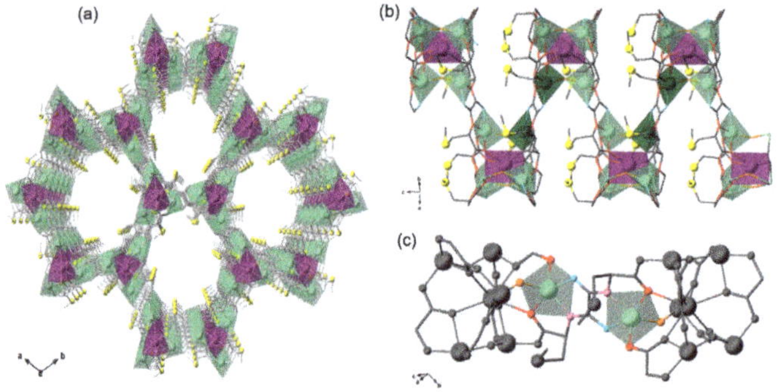


**Figure 1.** Crystal structure of **1**. (**a**) Perspective view of 3D porous network of **1** along *c* crystallographic axis. (**b**) Fragment of **1** highlighting the connection between dizinc(II) units. Dizinc units are represented by pale and dark polyhedra, whilst Ca and S are depicted as magenta polyhedra and yellow spheres; N, O, and C atoms are depicted as blue, red, and grey sticks. (**c**) View in detail of the dizinc(II) moiety highlighting the pentacoordinated Zn(II) ions and those atoms involved in their coordination geometry. Color code: Zn, pale green; amidate-N, cyan; carbonyl-O, pale pink; carboxylate-O, red; water/hydroxide-O, orange; rest of the structure is depicted as grey balls and sticks.

## 2. Materials and Methods

### 2.1. Materials

All reagents were obtained from commercial sources (Merck-Aldrich, Darmstadt, Germany) and used without further purification unless otherwise indicated. Reactions were performed in conventional round-bottomed flasks or sealed vials equipped with a magnetic stirrer. $\{CaZn_6[(S,S)\text{-Mecysmox}]_3(OH)_2(H_2O)\}\cdot 12\,H_2O$ (**1**) was prepared following a previously reported procedure [63].

### 2.2. Preparation of $\{CaCo_{0.24}Zn_{5.76}[(S,S)\text{-Mecysmox}]_3(OH)_2(H_2O)\}\cdot 14\,H_2O$ (**Co$_{4\%}$@1**) and $\{CaCo_{0.48}Zn_{5.52}[(S,S)\text{-Mecysmox}]_3(OH)_2(H_2O)\}\cdot 16\,H_2O$ (**Co$_{8\%}$@1**)

Polycrystalline powders of 1 (0.2 g, 0.12 mmol) were suspended in water solutions in two independent vials with different amounts of $Co(NO_3)_2\cdot 6\,H_2O$ (0.01 and 0.10 g for **Co$_{4\%}$@1** and **Co$_{8\%}$@1**, respectively) for 1 h under mild stirring. Then, the resulting solids were isolated by filtration, thoroughly washed with water and methanol, and air dried. **Co$_{4\%}$@1**: Anal.: calcd for $C_{30}H_{68}N_6O_{35}S_6CaCo_{0.24}Zn_{5.76}$ (1689.5): C, 21.32; H, 4.05; S, 11.08; N, 4.84%. Found: C, 21.37; H, 3.99; S, 11.12; N, 5.01%. **Co$_{8\%}$@1**: Anal.: calcd for $C_{30}H_{72}N_6O_{37}S_6CaCo_{0.48}Zn_{5.52}$ (1735.9): C, 20.75; H, 4.18; S, 11.52; N, 5.03%. Found: C, 20.83; H, 4.09; S, 11.56; N, 5.11%.

### 2.3. Physical Techniques

Elemental (C, H, S, and N) and ICP-MS analyses were performed at the Microanalytical Service of the Universitat de València. The thermogravimetric analysis was performed on crystalline samples under a dry $N_2$ atmosphere with a Mettler Toledo TGA/STDA 851e thermobalance, manufactured by Mettler Toledo, Greifensee, Switzerland, operating at a heating rate of 10 °C min$^{-1}$. A scanning electron microscopy (SEM) elemental analysis was carried out for **1**, **Co$_{4\%}$@1**, and **Co$_{8\%}$@1**, using a HITACHI S-4800 electron microscope, sourced from HITACHI High-Technologies Corporation, Tokyo, Japan, coupled with an energy dispersive X-ray (EDX) detector. Data were analyzed with the QUANTAX 400 system manufactured by Bruker AXS, Karlsruhe, Germany.

### 2.4. Gas Adsorption

The $N_2$ adsorption–desorption isotherms at 77 K were carried out on polycrystalline samples of **1**, **Co$_{4\%}$@1**, and **Co$_{8\%}$@1** with a BELSORP-mini-X instrument manufactured by BEL Japan, Inc., Osaka, Japan. The samples were first activated with methanol and then evacuated at 343 K during 16 h under $10^{-6}$ Torr prior to their analysis.

### 2.5. X-Ray Powder Diffraction Measurements

Polycrystalline samples of **1**, **Co$_{4\%}$@1**, and **Co$_{8\%}$@1** were introduced into a 0.5 mm borosilicate capillary prior to being mounted and aligned on an Empyrean PANalytical powder diffractometer, manufactured by PANalytical, Almelo, The Netherlands, using Cu K$\alpha$ radiation ($\lambda$ = 1.54056 Å). Five repeated measurements were collected at room temperature (2$\theta$ = 2–45°) and merged in a single diffractogram.

### 2.6. Magnetic Measurements

Direct current (dc) magnetic susceptibility measurements (2–300 K) under applied dc magnetic fields of 5000 G ($T \geq 20$ K) and 2500 G ($2 \leq T \leq 20$ K) and variable-field (0–5 T) magnetization measurements (2.0–10 K) on finely crushed crystals of **Co$_{4\%}$@1** and **Co$_{8\%}$@1** (mixed with grease to avoid the crystallite orientation) were carried out with a Quantum Design MPMS-XL7 SQUID magnetometer, manufactured by Quantum Design, San Diego, CA, USA. Variable-temperature (2–8 K) alternating current (ac) magnetic susceptibility measurements for **Co$_{4\%}$@1** and **Co$_{8\%}$@1** were performed under an applied dc field of 1 and 2.5 T, a ±5.0 G oscillating field, and at frequencies between 0.1 and 10.0 kHz with a Quantum Design Physical Property Measurement System (PPMS) manufactured by

Quantum Design, San Diego, CA, USA. The susceptibility data were corrected for the diamagnetism of both the sample holder and the constituent atoms.

The magnetic properties of octahedral cobalt(II) complexes are usually dominated by a first-order spin–orbit coupling (SOC). The coupling between the spin ($S$ = 3/2) and orbital momentum ($L$ = 1)—using the *T-P* isomorphism [64]—in a distorted octahedral environment generates six Kramers doublets. Among these, the two lowest-energy doublets are usually the only states populated below 300 K. Consequently, a phenomenological approach based on a zero-field splitting (ZFS) of a state $S$ = 3/2, considering only these doublets, was employed. In the spin Hamiltonian (Equation (1)), the parameters $D$ and $E$ represent the axial and rhombic magnetic anisotropy, respectively.

$$\hat{H}_{GS} = \beta H\left(g_x S_x + g_y S_y + g_z S_z\right) + D\left(\hat{S}_z^2 - \frac{S(S+1)}{3}\right) + E\left(\hat{S}_x^2 - \hat{S}_y^2\right) \quad (1)$$

The activation energy ($Ea$) in **Co$_{4\%}$@1** and **Co$_{8\%}$@1**, based on the phenomenological approach employed, was evaluated using the equation $E_a = 2\sqrt{D^2 + 3E}$, where the value of $E$ value is always positive.

## 3. Results and Discussion

### 3.1. Synthesis and Characterization

In this work, we present the solid-state post-synthetic partial metal exchange of Zn(II) metal ions by Co(II) ones, in a previously reported water-stable three-dimensional MOF, derived from the amino acid *S*-methyl-*L*-cysteine, with the formula {$CaZn_6$[(*S*,*S*)-Mecysmox]$_3(OH)_2(H_2O)$}·12 $H_2O$ (**1**) (Mecysmox = bis[*S*-methylcysteine]oxalyl diamide), allowing us to obtain two novel MOFs with distinct contents of Co(II) ions of formulas {$CaCo_{0.24}Zn_{5.76}$[(*S*,*S*)-Mecysmox]$_3(OH)_2(H_2O)$}·14 $H_2O$ (**Co$_{4\%}$@1**) and {$CaCo_{0.48}Zn_{5.52}$[(*S*,*S*)-Mecysmox]$_3(OH)_2(H_2O)$}·16 $H_2O$ (**Co$_{8\%}$@1**). Interestingly, direct synthesis attempts—adapting the previously reported procedure for **1** [63]—using the oxamidate-based ligand $H_2Me_2$-(*S*,*S*)-Mecysmox in a basic methanolic solution with mixtures of aqueous solutions of $Co(NO_3)_2$·6 $H_2O$, $Zn(NO_3)_2$·6 $H_2O$ and $CaCl_2$ were unsuccessful in synthesizing **Co$_{4\%}$@1** and **Co$_{8\%}$@1**.

**Co$_{4\%}$@1** and **Co$_{8\%}$@1** were obtained by immersion of the pristine polycrystalline powder of **1** in aqueous solutions containing different amounts of $Co(NO_3)_2$·6$H_2O$ (0.01 and 0.10 g for **Co$_{4\%}$@1** and **Co$_{8\%}$@1**, respectively) for 1 h under mild stirring. This process was carefully monitored visually, confirming the absence of powder dissolution and reprecipitation. The nature, pureness. and identity of **Co$_{4\%}$@1** and **Co$_{8\%}$@1** were conclusively established through a combination of physical techniques, including elemental analyses (C, H, S, and N), inductively coupled plasma mass spectrometry (ICP-MS), scanning electron microscopy with energy dispersive X-ray spectroscopy (SEM-EDX), powder X-ray diffraction (PXRD), nitrogen adsorption isotherms at 77 K, and thermogravimetric analyses (TGAs) (see Supporting Information). ICP-MS and SEM-EDX enabled us to determine the atomic percentages of metal ions in each MOF (Tables S1 and S2), and together with elemental analyses and TGAs allowed us to establish their chemical formulas (see Section 2). A comparison of ICP-MS and SEM-EDX data for **1**, **Co$_{4\%}$@1**, and **Co$_{8\%}$@1** confirmed that only Zn(II) ions in **1** have been replaced with Co(II) ions, since the atomic ratios between M(II) and Ca(II)—where M = Zn for **1** and Zn + Co for **Co$_{4\%}$@1** and **Co$_{8\%}$@1**—remained unchanged. These results also allowed the quantification of the percentage of substituting Co(II) ions in each compound. We determined Co(II) contents of 4 and 8%, relative to Zn(II)ions, in **Co$_{4\%}$@1** and **Co$_{8\%}$@1**, respectively. SEM micrographs and EDX elemental mapping confirmed a homogeneous distribution of Zn(II), Ca(II), and Co(II) ions in **1**, **Co$_{4\%}$@1**, and **Co$_{8\%}$@1** (Figures S1–S3). The TGAs of **Co$_{4\%}$@1** and **Co$_{8\%}$@1** (Figure S4) revealed a rapid mass loss from room temperature to ca. 100 K, followed by a plateau until decomposition started at ca. 300 K. The calculated percentage weight loss values at 150 K were 14.5% for **Co$_{4\%}$@1** and 16.8% for **Co$_{8\%}$@1**, corresponding to the release of 14 and 16 water molecules, respectively. The PXRD analyses evidenced a match between the theoretical pattern of **1**, derived from previously reported single-crystal X-ray diffraction data, and the experimental patterns of each sample (Figure 2). This confirmed the purity of **Co$_{4\%}$@1** and **Co$_{8\%}$@1**, as well as their isoreticular nature with respect to **1**. Additionally, the preservation of permanent porosity in **Co$_{4\%}$@1** and **Co$_{8\%}$@1** after the post-synthetic partial metal exchange was demonstrated by means of nitrogen adsorption isotherms at 77 K (Figure S5). All samples exhibited fully reversible type I isotherms, characteristic of microporous materials with permanent microporosity (Figure S5). The estimated Brunauer–Emmett–Teller (BET) surface areas [706 (**1**), 694 (**Co$_{4\%}$@1**), and 686 m$^2$ g$^{-1}$ (**Co$_{8\%}$@1**)] as well as the calculated pore widths [0.6622 (**1**), 0.6387 (**Co$_{4\%}$@1**), and 0.6589 nm (**Co$_{8\%}$@1**)] (Figure S6) were nearly identical, confirming the preservation of permanent porosity. Interestingly, the presented post-synthetic partial metal exchange leads to a change in the magnetic properties of the pristine MOF **1**, resulting in new MOFs of the oxamidate family, which exhibit SIM behavior (see magnetic properties below).

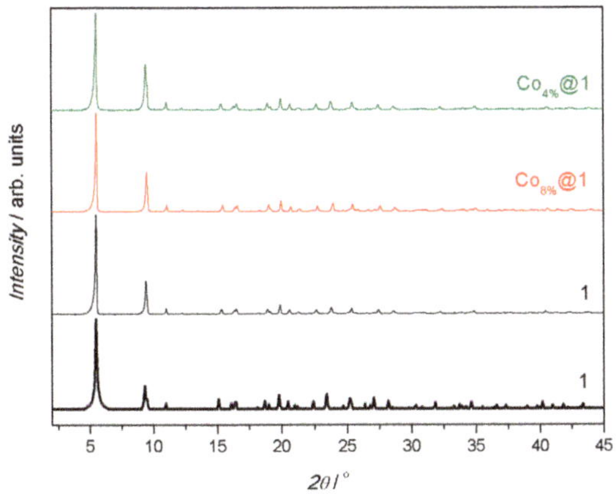


**Figure 2.** Calculated (black bold lines) and experimental (solid lines) PXRD patterns of **1** (black), **Co$_{8\%}$@1** (red), and **Co$_{4\%}$@1** (green) in the 2θ range of 2.0–45.0° at room temperature.

### *3.2. Description of the Structure*

Although the crystal structure of the parent MOF **1** has been previously reported, it is worth describing it here, particularly to understand the possible coordination environments and geometries of the Co(II) ions substituting those of Zn(II) in compounds **Co$_{4\%}$@1** and **Co$_{8\%}$@1**, which, in turn, are responsible for the SIM behavior. MOF **1** crystallizes in the chiral *P*6$_3$ space group and consists of a chiral honeycomb three-dimensional (3D) calcium(II)–zinc(II) network (Figure 1) featuring hexagonal channels. The network is constructed from trans-oxamidato-bridged dizinc(II) units {Zn$^{II}_2$[(*S*,*S*)-Mecysmox]} (Figure 1a), which act as linkers between Ca(II) cations through the carboxylate groups and are further interconnected to neighboring dizinc(II) moieties by water–hydroxide groups (Figure 1a). The Zn(II) ions are five-coordinate, exhibiting a trigonal bipyramidal geometry. They coordinate one amidate nitrogen atom, one carbonyl oxygen atom, one carboxylate oxygen atom, one carboxylate oxygen atom from a neighboring dizinc(II) unit, and one water–hydroxide bridging group (in a 1:2 statistical distribution). Thus, a priori, the potential coordination environment of substituting Co(II) ions should be the same as that of the replaced Zn(II) ones. However, with a lack of crystal structure, it is difficult to establish their precise coordination environment, being possible other environments, such as octahedral. Each calcium(II) ion is nine-coordinate in a distorted monocapped square antiprismatic geometry (CaO$_9$), built by six carboxylate oxygen atoms from six oxamidato groups and three oxygen atoms from water–hydroxide groups (Figure 1).

### 3.3. *Magnetic Properties*

#### 3.3.1. Static (dc) Magnetic Properties

The dc magnetic properties of compounds **Co$_{4\%}$@1** and **Co$_{8\%}$@1**, as the $\chi_M T$ against the $T$ plot, are shown in Figure 3, with $\chi_M$ being the normalized dc molar susceptibility per one cobalt(II) per formula unit (the raw $\chi_M T$ against the $T$ plot is shown in Figure S7). The $\chi_M T(T)$ data for **Co$_{4\%}$@1** and **Co$_{8\%}$@1** are typical for a high-spin d$^7$ Co(II) site. The $\chi_M T$ value at room temperature is 2.52 cm$^3$ K mol$^{-1}$ for **Co$_{4\%}$@1** and **Co$_{8\%}$@1**, which is consistent with previously reported mononuclear cobalt(II) complexes with unquenched orbital momentum contributions [37,65]. Upon cooling, $\chi_M T$ decreases smoothly down to 1.42 (**Co$_{4\%}$@1**) and 1.56 (**Co$_{8\%}$@1**) cm$^3$ K mol$^{-1}$ at 2 K, revealing the existence of a significant ZFS arising from the second-order spin–orbit coupling (SOC) of the Co(II) ion (Figure 3). This was further supported by the $M$ vs. $H$ plots, with magnetization values of 1.71 (**Co$_{4\%}$@1**) and 1.95 $N\beta$ (**Co$_{8\%}$@1**) at 3 K and 5 T (Figure 4). These values are well below the saturation magnetization for a high-spin d$^7$ Co(II)ion ($M_{sat} = gSN\beta = 3\ N\beta$), which suggests the presence of significant axial magnetic anisotropy. Here, where the ground $\pm3/2$ and excited $\pm1/2$ Kramers doublets for D < 0, or the opposite for D > 0, are well separated due to sizeable ZFS effects operating on the quartet ground state, the magnetization for the ground Kramers doublet is the only one to be considered. Additionally, as observed in other highly anisotropic complexes reported [65], the isofield magnetization curves at low temperatures (2–10 K) are nearly superimposable for **Co$_{4\%}$@1** and **Co$_{8\%}$@1**. Thus, $M(H/T)$ curves were not useful in accurately determining the ZFS parameters but allowed us to conclude that the ZFS is quite large.

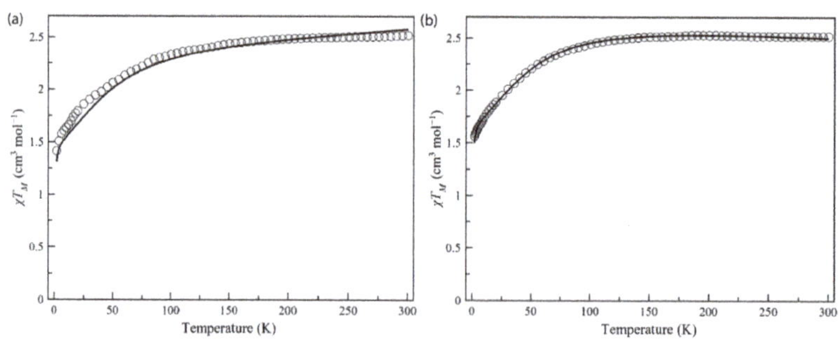


**Figure 3.** Temperature dependence of the product of the direct current (dc) molar magnetic susceptibility by the temperature ($\chi_M T$) for **Co$_{4\%}$@1** (**a**) and **Co$_{8\%}$@1** (**b**). The solid lines correspond to the best-fit curves considering ZFS effects (see text and Tables S3 and S4).

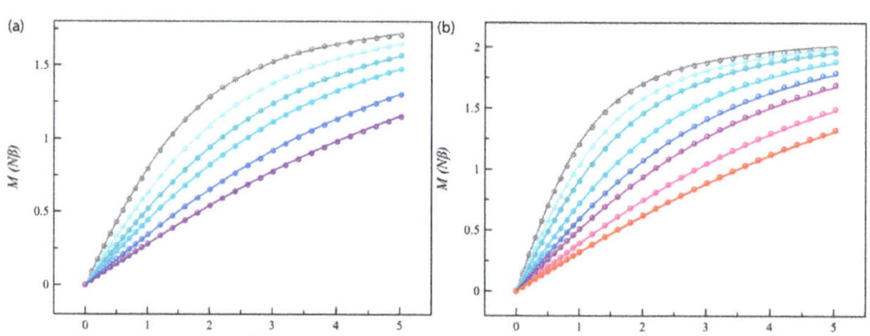


**Figure 4.** Field dependence of the magnetization for **Co$_{4\%}$@1** (**a**) and **Co$_{8\%}$@1** (**b**) at low temperatures ($T$ = 2.0–10 K, from grey to red). The solid lines correspond to the best-fit curves considering the ZFS effects (see text and Tables S3 and S4).

In order to quantify the relevant parameters, the experimental data for $\chi_M T$ vs. $T$ and $M$ vs. $H$ of **Co$_{4\%}$@1** and **Co$_{8\%}$@1** were simultaneously simulated using the PHI program [66] through the spin Hamiltonian considering ZFS parameters (Equation (1)) for an isolated $S = 3/2$ state (solid line in Figures 3, 4, S7 and S8 and Tables S3 and S4), with axial ($D$) and rhombic ($E$) ZFS parameters. As the experimental measurements were performed on finely crushed polycrystalline samples, $g_x$ and $g_y$ components of the g-factor were considered indistinguishable. Thus, only the parallel and perpendicular components were retained ($g_x = g_y = g_\perp$ and $g_z = g_\parallel$). The best-fit values of the parameters $g_\parallel, g_\perp$, $D$, $E$, $E/D$ were 2.24, 2.28, $-55.25$ cm$^{-1}$, 3.06 cm$^{-1}$, and 0.055 (**Co$_{4\%}$@1**) and 2.41, 2.35, $-50.80$ cm$^{-1}$, 11.94 cm$^{-1}$, and 0.235 (**Co$_{8\%}$@1**) cm$^{-1}$, which are in agreement with the literature values [37,65].

### 3.3.2. Dynamic (ac) Magnetic Properties

Alternating current (ac) magnetic measurements of **Co$_{4\%}$@1** and **Co$_{8\%}$@1** were carried out in the temperature range from 2 to 8 K to investigate their magnetization dynamics. No frequency dependence was observed for the ac components in the absence of a magnetic field ($H_{dc}$), a feature typically attributed to fast relaxation via quantum tunneling of magnetization. However, when different external static fields (1 and 2.5 kG) were applied, frequency-dependent in-phase ($\chi_M{}'$) and out-of-phase ($\chi_M{}''$) signals were observed (Figures 5 and S9–S11), which are consistent with a slow magnetic relaxation process typical of high-spin cobalt(II) SIMs.

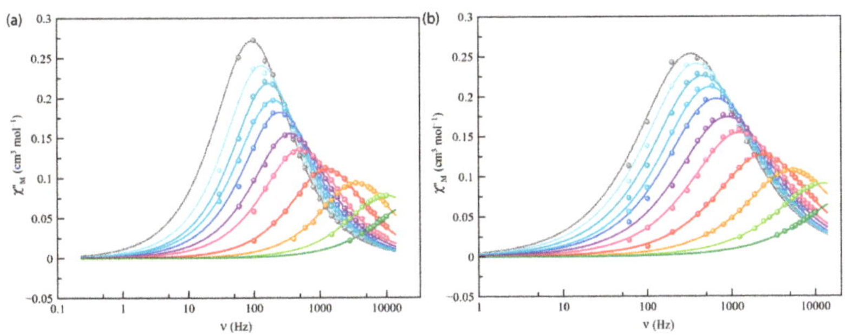


**Figure 5.** Frequency dependence of $\chi_M{}''$ for **Co$_{4\%}$@1** (**a**) and **Co$_{8\%}$@1** (**b**) under applied static field of 2500 G with a +5.0 G oscillating field in the temperature range of 2–8 K (from grey to green). The solid lines represent the best-fit curves simulated using the generalized Debye model.

Cole–Cole plots for **Co$_{4\%}$@1** and **Co$_{8\%}$@1** at 2–8 K under different applied static fields (1.0–2.5 kG) (Figures 6 and S12) showed almost perfect semicircles, which were fitted using the Debye model (solid lines in Figures 6 and S12). The obtained low values of the $\alpha$ parameters at different temperatures and dc magnetic fields [$\alpha$ = 0.099 (8.0 K)–0.121 (2.0 K) and 0.112 (8.0 K)–0.187 (2.0 K) for **Co$_{4\%}$@1** and **Co$_{8\%}$@1** under applied static field of 1000 G, respectively] (see Tables S3 and S4) support a single relaxation process. Moreover, increasing values of $\alpha$ with decreasing temperature reveal the appearance of intermolecular interactions within both frameworks of **Co$_{4\%}$@1** and **Co$_{8\%}$@1**. The semicircles in the Cole–Cole plots are slightly more flattened in **Co$_{8\%}$@1** than in **Co$_{4\%}$@1** (Figures 6 and S12). Thus, slightly larger $\alpha$ parameters were obtained for **Co$_{8\%}$@1** (Table S4). However, the $\alpha$ parameters are well below the values for spin-glass behavior ($\alpha \geq 0.4$).

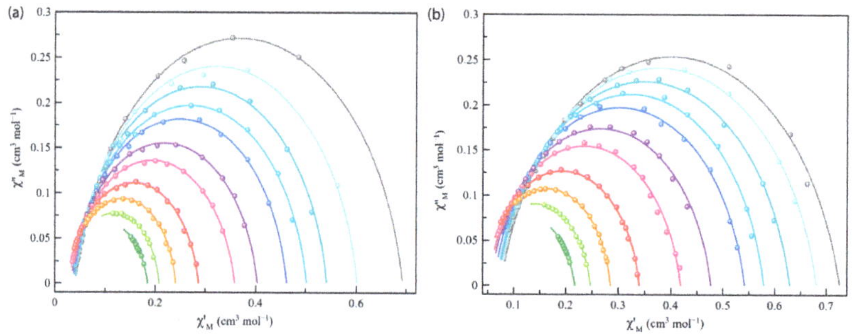


**Figure 6.** Cole–Cole plots of **Co4%@1** (**a**) and **Co8%@1** (**b**) at 2–8 K (from grey to green) under applied static field of 2500 G. The solid lines represent the best-fit curves.

Aiming to quantify the magnetic relaxation times, a joint analysis of the $\chi_M'$ and $\chi_M''$ vs. frequency ($\upsilon$) data (Figures 5 and S9) was performed using the generalized Debye model. Two different approaches are recommended to represent these times depending on the nature of the dominant relaxation mechanism involved. After qualitative scrutiny, we established that ln$\tau$ vs. ln$T$ plots (Figures 7 and S13) were the most appropriate for analyzing the predominant relaxation mechanisms, commonly associated with Raman or direct-like mechanisms ($\tau^{-1} \propto T^n$). Ideally, such plots should show a linear dependency with slope $n$. When $n$ approaches a value close to 2 or 8, the optical or acoustic phonon-assisted Raman mechanism dominates the fastest spin reversal, while a value close to 1 indicates that the direct mechanism plays the main role. Here, we would like to note that the ideal behavior is often far away from reality in reported examples in the literature. Consequently, the confluence of various relaxation mechanisms operating at different temperatures is a common outcome. However, we aimed to keep this analysis simple and avoid overparameterization, which would eventually lead to better fitting but lack physical significance. In our case, ln$\tau$ vs. ln$T$ plots for **Co4%@1** and **Co8%@1** at the two studied fields evidenced a deviation from linearity. The least-squares fit of these data yielded $n$ values, which indicate that Raman mechanisms assisted by optical and acoustic phonons rule the relaxation of the magnetization at low and high temperatures, respectively (Table 1).

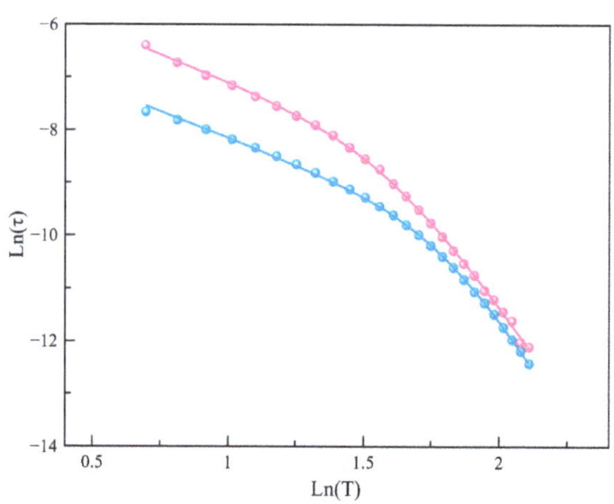


**Figure 7.** The ln$\tau$ versus ln$T$ plot for the calculated magnetic relaxation times of **Co4%@1** (pink) and **Co8%@1** (blue) under an applied static field of 2500 G. The solid lines represent the best-fit curves.

**Table 1.** Selected parameters from the least-squares fit of the ac magnetic data of **$Co_{4\%}$@1** and **$Co_{8\%}$@1** [a].

| | $H_{dc}$ (kOe) | $C_1$ [b] ($s^{-1}$ $K^{-n}$) | $n_1$ [b] | $C_2$ [b] ($s^{-1}$ $K^{-n}$) | $n_2$ [b] |
|---|---|---|---|---|---|
| **$Co_{4\%}$@1** | 1.0 | 210 ± 11 | 2.52 ± 0.05 | 0.0018 ± 0.0007 | 8.84 ± 0.18 |
| | 2.5 | 153 ± 13 | 2.04 ± 0.09 | 0.024 ± 0.007 | 7.49 ± 0.15 |
| **$Co_{8\%}$@1** | 1.0 | 961 ± 30 | 1.94 ± 0.03 | 0.0037 ± 0.0013 | 8.58 ± 0.17 |
| | 2.5 | 575 ± 16 | 1.80 ± 0.03 | 0.017 ± 0.003 | 7.77 ± 0.08 |

[a] The fits correspond to the double Raman relaxation model $\tau^{-1} = C_1T^{n_1} + C_2T^{n_2}$. [b] Coefficient and polynomial factor for the Raman process.

## 4. Conclusions

In this work, we applied a partial transmetalation post-synthetic strategy to achieve two new MOFs that could not be obtained by direct synthesis. Through the controlled exchange of Zn(II) with Co(II) ions in the pristine diamagnetic Zn-MOF, two MOFs with 4% and 8% Co(II) content were obtained, exhibiting SIM behavior. This underexplored strategy holds great potential to enrich the elemental composition of an MOF in a controlled manner, thereby achieving a higher degree of complexity that leads to new physicochemical properties. In particular, it serves as an excellent tool for the preparation of molecular-based magnetic materials.

**Supplementary Materials:** The following supporting information can be downloaded at https://www.mdpi.com/article/10.3390/magnetochemistry10120099/s1, Figures S1–S3: SEM-EDX micrographs; Figure S4: Thermogravimetric analysis (TGA); Figures S5 and S6: Textural properties; Figures S7–S13: Magnetic dc and ac data. Table S1: Selected data for SEM-EDX; Table S2: Selected data for ICP-MS; Table S3: Selected magnetic data for **$Co_{4\%}$@1** at different dc applied fields; Table S4: Selected magnetic data for **$Co_{8\%}$@1** at different dc applied fields.

**Author Contributions:** Conceptualization, E.P., J.F.-S. and T.G.; methodology, P.E., N.M. and D.A.; formal analysis, N.M., J.F.-S. and T.G.; investigation, E.P., J.F.-S. and T.G.; data curation, J.F.-S. and N.M.; writing—original draft preparation, P.E., J.F.-S., N.M. and T.G.; writing—review and editing, J.F.-S., E.P. and T.G.; supervision, J.F.-S. and T.G.; funding acquisition, E.P. and J.F.-S. All authors have read and agreed to the published version of the manuscript.

**Funding:** This research was funded by the Ministero dell'Istruzione, dell'Università e della Ricerca (Italy) and the MICINN (Spain) (Project PID2022-136349OB-I00 and Excellence Unit "Maria de Maeztu" CEX2019-000919-M). D.A. also acknowledges the financial support of the European Union-NextGenerationEU under the National Recovery and Resilience Plan (NRRP) of Ministero dell'Università e della Ricerca (MUR) (Project code PE0000021, Network 4 Energy Sustainable Transition—NEST). Thanks are also extended to the "Generalitat Valenciana" (Project PROMETEO/2021/054). E.P. acknowledges the financial support of the European Research Council under the European Union's Horizon 2020 research and innovation programme/ERC Grant Agreement No 814804, MOF-reactors. Thanks are also extended to the Ramón y Cajal Program, RYC2019-027940-I (J.F.-S.). This study forms part of the Advanced Materials programme (MFA/2022/048) and was supported by MCIN with funding from European Union NextGenerationEU (PRTR-C17.I1) and by Generalitat Valenciana.

**Institutional Review Board Statement:** Not applicable.

**Informed Consent Statement:** Not applicable.

**Data Availability Statement:** The datasets generated during and/or analyzed during the current study are available from the corresponding author on reasonable request.

**Conflicts of Interest:** The authors declare no conflicts of interest.

## References

1. Gatteschi, D.; Sessoli, R.; Villain, J. *Molecular Nanomagnets;* Oxford University Press: Oxford, UK, 2006.
2. Aromi, G.; Brechin, E.K. Synthesis of 3d Metallic Single-Molecule Magnets. In *Structure and Bonding;* Winpenny, R., Ed.; Springer: Berlin, Germany, 2006; Volume 122, pp. 1–67.

3. AlDamen, M.A.; Clemente-Juan, J.M.; Coronado, E.; Martí-Gastaldo, C.; Gaita-Ariño, A. Mononuclear Lanthanide Single-Molecule Magnets Based on Polyoxometalates. *J. Am. Chem. Soc.* **2008**, *130*, 8874–8875. [CrossRef] [PubMed]
4. Woodruff, D.N.; Winpenny, R.E.P.; Layfield, R.A. Lanthanide Single-Molecule Magnets. *Chem. Rev.* **2013**, *113*, 5110–5148. [CrossRef] [PubMed]
5. Craig, G.A.; Murrie, M. 3d single-ion magnets. *Chem. Soc. Rev.* **2015**, *44*, 2135–2147. [CrossRef]
6. Meng, Y.-S.; Jiang, S.-D.; Wang, B.-W.; Gao, S. Understanding the Magnetic Anisotropy toward Single-Ion Magnets. *Acc. Chem. Res.* **2016**, *49*, 2381–2389. [CrossRef] [PubMed]
7. Liu, J.-L.; Chen, Y.-C.; Tong, M.-L. Symmetry strategies for high performance lanthanide-based single-molecule magnets. *Chem. Soc. Rev.* **2018**, *47*, 2431–2453. [CrossRef] [PubMed]
8. Dey, A.; Kalita, P.; Chandrasekhar, V. Lanthanide(III)-Based Single-Ion Magnets. *ACS Omega* **2018**, *3*, 9462–9475. [CrossRef] [PubMed]
9. Coronado, E. Molecular magnetism: From chemical design to spin control in molecules, materials and devices. *Nat. Rev. Mater.* **2020**, *5*, 87–104. [CrossRef]
10. Zabala-Lekuona, A.; Seco, J.M.; Colacio, E. Single-Molecule Magnets: From Mn12-ac to dysprosium metallocenes, a travel in time. *Coord. Chem. Rev.* **2021**, *441*, 213984. [CrossRef]
11. Castellano, M.; Ruiz-García, R.; Cano, J.; Ferrando-Soria, J.; Pardo, E.; Fortea-Pérez, F.R.; Stiriba, S.-E.; Barros, W.P.; Stumpf, H.O.; Cañadillas-Delgado, L.; et al. Metallosupramolecular approach toward multifunctional magnetic devices for molecular spintronics. *Coord. Chem. Rev.* **2015**, *303*, 110–138. [CrossRef]
12. Cornia, A.; Seneor, P. Spintronics the Molecular Way. *Nat. Mater.* **2017**, *16*, 505–506. [CrossRef]
13. Gobbi, M.; Novak, M.A.; Del Barco, E. Molecular spintronics. *J. Appl. Phys.* **2019**, *125*, 240401. [CrossRef]
14. Forment-Aliaga, A.; Gaita-Ariño, A. Chiral, magnetic, molecule-based materials: A chemical path toward spintronics and quantum nanodevices. *J. Appl. Phys.* **2022**, *132*, 180901. [CrossRef]
15. Wang, K.; Yang, M.; Luo, J. *Spintronics: Materials, Devices, and Applications*; Wiley: Weinheim, Germany, 2022.
16. Ferrando-Soria, J.; Vallejo, J.; Castellano, M.; Martínez-Lillo, J.; Pardo, E.; Cano, J.; Castro, I.; Lloret, F.; Ruiz-García, R.; Julve, M. Molecular magnetism, quo vadis? A historical perspective from a coordination chemist viewpoint. *Coord. Chem. Rev.* **2017**, *339*, 17–103. [CrossRef]
17. Yuste, C.; Castellano, M.; Ferrando-Soria, J.; Stiriba, S.-E.; Marino, N.; Julve, M.; Lloret, F.; Ruiz-García, R.; Cano, J. Review: From computational design to the synthesis of molecular magnetic wires for single-molecule spintronics and quantum computing nanotechnologies. *J. Coord. Chem.* **2022**, *75*, 2359–2383. [CrossRef]
18. Kumar, P.; Swain, A.; Acharya, J.; Li, Y.; Kumar, V.; Rajaraman, G.; Colacio, E.; Chandrasekhar, V. Synthesis, Structure, and Zero-Field SMM Behavior of Homometallic Dy2, Dy4, and Dy6 Complexes. *Inorg. Chem.* **2022**, *61*, 11600–11621. [CrossRef]
19. Benmansour, S.; Pintado-Zaldo, C.; Martínez-Ponce, J.; Hernández-Paredes, A.; Valero-Martínez, A.; Gómez-Benmansour, M.; Gómez-García, C.J. The Versatility of Ethylene Glycol to Tune the Dimensionality and Magnetic Properties in DyIII-Anilato-Based Single-Ion Magnets. *Cryst. Growth Des.* **2023**, *23*, 1269–1280. [CrossRef]
20. Little, E.J.; Mrozek, J.; Rogers, C.J.; Liu, J.; McInnes, E.J.L.; Bowen, A.M.; Ardavan, A.; Winpenny, R.E.P. Experimental realisation of multi-qubit gates using electron paramagnetic resonance. *Nat. Commun.* **2023**, *14*, 7029. [CrossRef]
21. Grancha, T.; Ferrando-Soria, J.; Castellano, M.; Julve, M.; Pasán, J.; Armentano, D.; Pardo, E. Oxamato-based coordination polymers: Recent advances in multifunctional magnetic materials. *Chem. Commun.* **2014**, *50*, 7569–7585. [CrossRef]
22. Estrader, M.; Salinas-Uber, J.; Barrios, L.A.; Garcia, J.; Lloyd-Williams, P.; Roubeau, O.; Teat, S.J.; Aromí, G. A Magneto-optical Molecular Device: Interplay of Spin Crossover, Luminescence, Photomagnetism, and Photochromism. *Angew. Chem. Int. Ed.* **2017**, *56*, 15622–15627. [CrossRef]
23. Castellano, M.; Barros, W.P.; Ferrando-Soria, J.; Julve, M.; Lloret, F.; Pasán, J.; Ruiz-Pérez, C.; Cañadillas-Delgado, L.; Ruiz-García, R.; Cano, J. Dicopper(II) metallacyclophanes with photoswitchable oligoacene spacers: A joint experimental and computational study on molecular magnetic photoswitches. *J. Coord. Chem.* **2018**, *71*, 675–692. [CrossRef]
24. Kalinke, L.H.G.; Cangussu, D.; Mon, M.; Bruno, R.; Tiburcio, E.; Lloret, F.; Armentano, D.; Pardo, E.; Ferrando-Soria, J. Metal-Organic Frameworks as Playgrounds for Reticulate Single-Molecule Magnets. *Inorg. Chem.* **2019**, *58*, 14498–14506. [CrossRef] [PubMed]
25. Clérac, R.; Miyasaka, H.; Yamashita, M.; Coulon, C. Evidence for Single-Chain Magnet Behavior in a $Mn^{III}$–$Ni^{II}$ Chain Designed with High Spin Magnetic Units: A Route to High Temperature Metastable Magnets. *J. Am. Chem. Soc.* **2002**, *124*, 12837–12844. [CrossRef]
26. Liu, J.; Chen, Y.-C.; Liu, J.-L.; Vieru, V.; Ungur, L.; Jia, J.-H.; Chibotaru, L.F.; Lan, Y.; Wernsdorfer, W.; Gao, S.; et al. A Stable Pentagonal Bipyramidal Dy(III) Single-Ion Magnet with a Record Magnetization Reversal Barrier over 1000K. *J. Am. Chem. Soc.* **2016**, *138*, 5441–5450. [CrossRef] [PubMed]
27. Escalera-Moreno, L.; Baldoví, J.J.; Gaita-Ariño, A.; Coronado, E. Exploring the High-Temperature Frontier in Molecular Nanomagnets: From Lanthanides to Actinides. *Inorg. Chem.* **2019**, *58*, 11883–11892. [CrossRef] [PubMed]
28. Feng, M.; Ruan, Z.-Y.; Chen, Y.-C.; Tong, M.-L. Physical stimulus and chemical modulations of bistable molecular magnetic materials. *Chem. Commun.* **2020**, *56*, 13702–13718. [CrossRef]

29. McAdams, S.G.; Ariciu, A.-M.; Kostopoulos, A.K.; Walsh, J.P.S.; Tuna, F. Molecular single-ion magnets based on lanthanides and actinides: Design considerations and new advances in the context of quantum technologies. *Coord. Chem. Rev.* **2017**, *346*, 216–239. [CrossRef]
30. Meihaus, K.R.; Long, J.R. Actinide-based single-molecule magnets. *Dalton Trans.* **2015**, *44*, 2517–2528. [CrossRef]
31. Kindra, D.R.; Evans, W.J. Magnetic Susceptibility of Uranium Complexes. *Chem. Rev.* **2014**, *114*, 8865–8882. [CrossRef]
32. Sahu, P.K.; Kharel, R.; Shome, S.; Goswami, S.; Konar, S. Understanding the unceasing evolution of Co(II) based single-ion magnets. *Coord. Chem. Rev.* **2023**, *475*, 214871. [CrossRef]
33. Bamberger, H.; Albold, U.; Dubnická-Midlíková, J.; Su, C.-Y.; Deibel, N.; Hunger, D.; Hallmen, P.P.; Neugebauer, P.; Beerhues, J.; Demeshko, S.; et al. Iron(II), Cobalt(II), and Nickel(II) Complexes of Bis(sulfonamido)benzenes: Redox Properties, Large Zero-Field Splittings, and Single-Ion Magnets. *Inorg. Chem.* **2021**, *60*, 2953–2963. [CrossRef]
34. Alexandru, M.-G.; Visinescu, D.; Shova, S.; Cano, J.; Moliner, N.; Lloret, F.; Julve, M. A Chain of Vertex-Sharing $\{Co^{III}{}_2Co^{II}{}_2\}n$ Squares with Single-Ion Magnet Behavior. *Magnetochemistry* **2023**, *9*, 130. [CrossRef]
35. Woodall, C.; Craig, G.; Prescimone, A.; Misek, M.; Cano, J.; Faus, J.; Probert, M.R.; Parsons, S.; Moggach, S.; Martínez-Lillo, J.; et al. Pressure induced enhancement of the magnetic ordering temperature in rhenium(IV) monomers. *Nat. Commun.* **2016**, *7*, 13870. [CrossRef] [PubMed]
36. Orts-Arroyo, M.; Moliner, N.; Lloret, F.; Martínez-Lillo, J. Ferromagnetic Coupling and Single-Ion Magnet Phenomenon in Mononuclear Ruthenium(III) Complexes Based on Guanine Nucleobase. *Magnetochemistry* **2022**, *8*, 93. [CrossRef]
37. Vallejo, J.; Fortea-Pérez, F.R.; Pardo, E.; Benmansour, S.; Castro, I.; Krzystek, J.; Armentano, D.; Cano, J. Guest-dependent single-ion magnet behaviour in a cobalt(ii) metal–organic framework. *Chem. Sci.* **2016**, *7*, 2286–2293. [CrossRef]
38. Gomez-Coca, S.; Cremades, E.; Aliaga-Alcalde, N.; Ruiz, E. Mononuclear Single-Molecule Magnets: Tailoring the Magnetic Anisotropy of First-Row Transition-Metal Complexes. *J. Am. Chem. Soc.* **2013**, *135*, 7010–7018. [CrossRef]
39. Zhou, H.-C.; Long, J.R.; Yaghi, O.M. Introduction to Metal–Organic Frameworks. *Chem. Rev.* **2012**, *112*, 673–674. [CrossRef]
40. Maurin, G.; Serre, C.; Cooper, A.; Férey, G. The New Age of MOFs and of Their Porous-Related Solids. *Chem. Soc. Rev.* **2017**, *46*, 3104–3107. [CrossRef]
41. Freund, R.; Zaremba, O.; Arnauts, G.; Ameloot, R.; Skorupskii, G.; Dincă, M.; Bavykina, A.; Gascon, J.; Ejsmont, A.; Goscianska, J.; et al. The Current Status of MOF and COF Applications. *Angew. Chem. Int. Ed.* **2021**, *60*, 23975–24001. [CrossRef]
42. Chae, H.K.; Siberio-Pérez, D.Y.; Kim, J.; Go, Y.; Eddaoudi, M.; Matzger, A.J.; O'Keeffe, M.; Yaghi, O.M. A Route to High Surface Area, Porosity and Inclusion of Large Molecules in Crystals. *Nature* **2004**, *427*, 523–527. [CrossRef]
43. Mon, M.; Bruno, R.; Ferrando-Soria, J.; Bartella, L.; Di Donna, L.; Talia, M.; Lappano, R.; Maggiolini, M.; Armentano, D.; Pardo, E. Crystallographic Snapshots of Host-Guest Interactions in Drugs@metal-Organic Frameworks: Towards Mimicking Molecular Recognition Processes. *Mater. Horiz.* **2018**, *5*, 683–690. [CrossRef]
44. Fortea-Pérez, F.R.; Mon, M.; Ferrando-Soria, J.; Boronat, M.; Leyva-Pérez, A.; Corma, A.; Herrera, J.M.; Osadchii, D.; Gascon, J.; Armentano, D.; et al. The MOF-Driven Synthesis of Supported Palladium Clusters with Catalytic Activity for Carbene-Mediated Chemistry. *Nat. Mater.* **2017**, *16*, 760–766. [CrossRef] [PubMed]
45. Furukawa, H.; Cordova, K.E.; O'Keefee, M.; Yaghi, O.M. The Chemistry and Applications of Metal-Organic Frameworks. *Science* **2013**, *341*, 1230444. [CrossRef] [PubMed]
46. Grancha, T.; Mon, M.; Ferrando-Soria, J.; Gascon, J.; Seoane, B.; Ramos-Fernandez, E.V.; Armentano, D.; Pardo, E. Tuning the selectivity of light hydrocarbons in natural gas in a family of isoreticular MOFs. *J. Mater. Chem. A* **2017**, *5*, 11032–11039. [CrossRef]
47. Wang, Q.; Astruc, D. State of the Art and Prospects in Metal–Organic Framework (MOF)-Based and MOF-Derived Nanocatalysis. *Chem. Rev.* **2020**, *120*, 1438–1511. [CrossRef]
48. Mon, M.; Ferrando-Soria, J.; Grancha, T.; Fortea-Pérez, F.R.; Gascon, J.; Leyva-Pérez, A.; Armentano, D.; Pardo, E. Selective Gold Recovery and Catalysis in a Highly Flexible Methionine-Decorated Metal–Organic Framework. *J. Am. Chem. Soc.* **2016**, *138*, 7864–7867. [CrossRef]
49. Horcajada, P.; Serre, C.; Maurin, G.; Ramsahye, N.A.; Balas, F.; Vallet-Regí, M.; Sebban, M.; Taulelle, F.; Férey, G. Flexible Porous Metal-Organic Frameworks for a Controlled Drug Delivery. *J. Am. Chem. Soc.* **2008**, *130*, 6774–6780. [CrossRef]
50. Mínguez-Espallargas, G.; Coronado, E. Magnetic functionalities in MOFs: From the framework to the pore. *Chem. Soc. Rev.* **2018**, *47*, 533–557. [CrossRef]
51. Thorarinsdottir, A.E.; Harris, T.D. Metal–Organic Framework Magnets. *Chem. Rev.* **2020**, *120*, 8716–8789. [CrossRef]
52. Mon, M.; Bruno, R.; Ferrando-Soria, J.; Armentano, D.; Pardo, E. Metal–organic framework technologies for water remediation: Towards a sustainable ecosystem. *J. Mater. Chem. A* **2018**, *6*, 4912–4947. [CrossRef]
53. Escamilla, P.; Guerra, W.D.; Leyva-Pérez, A.; Armentado, D.; Ferrando-Soria, J.; Pardo, E. Metal–organic frameworks as chemical nanoreactors for the preparation of catalytically active metal compounds. *Chem. Commun.* **2023**, *59*, 836–851. [CrossRef]
54. Yuan, S.; Zou, L.; Qin, J.-S.; Li, J.; Huang, L.; Feng, L.; Wang, X.; Bosch, M.; Alsalme, A.; Cagin, T.; et al. Construction of hierarchically porous metal–organic frameworks through linker labilization. *Nat. Commun.* **2017**, *8*, 15356. [CrossRef] [PubMed]
55. Ji, Z.; Wang, H.; Canossa, S.; Wuttke, S.; Yaghi, O.M. Pore Chemistry of Metal–Organic Frameworks. *Adv. Funct. Mater.* **2020**, *30*, 2000238. [CrossRef]
56. Grancha, T.; Ferrando-Soria, J.; Zhou, H.-C.; Gascon, J.; Seoane, B.; Pasán, J.; Fabelo, O.; Julve, M.; Pardo, E. Postsynthetic Improvement of the Physical Properties in a Metal–Organic Framework through a Single Crystal to Single Crystal Transmetallation. *Angew. Chem. Int. Ed.* **2015**, *54*, 6521–6525. [CrossRef]

57. Mandal, S.; Natarajan, S.; Mani, P.; Pankajakshan, A. Post-Synthetic Modification of Metal–Organic Frameworks Toward Applications. *Adv. Funct. Mater.* **2021**, *31*, 2006291. [CrossRef]
58. Miyasaka, H.; Nakata, K.; Sugiura, K.; Yamashita, M.; Clérac, R. A Three-Dimensional Ferrimagnet Composed of Mixed-Valence $Mn_4$ Clusters Linked by an $\{Mn[N(CN)_2]_6\}^{4-}$ Unit. *Angew. Chem. Int. Ed.* **2004**, *43*, 707–711. [CrossRef]
59. Liu, K.; Li, K.; Zhang, X.; Shi, W.; Cheng, P. Constraining and Tuning the Coordination Geometry of a Lanthanide Ion in Metal–Organic Frameworks: Approach toward a Single-Molecule Magnet. *Inorg. Chem.* **2015**, *54*, 10224–10231. [CrossRef]
60. Aulakh, D.; Pyser, J.B.; Zhang, X.; Yakovenko, A.A.; Dunbar, K.R.; Wriedt, M. Metal–Organic Frameworks as Platforms for the Controlled Nanostructuring of Single-Molecule Magnets. *J. Am. Chem. Soc.* **2015**, *137*, 9254–9257. [CrossRef]
61. Mon, M.; Pascual-Álvarez, A.; Grancha, T.; Cano, J.; Ferrando-Soria, J.; Lloret, F.; Gascon, J.; Pasán, J.; Armentano, D.; Pardo, E. Solid-State Molecular Nanomagnet Inclusion into a Magnetic Metal–Organic Framework: Interplay of the Magnetic Properties. *Chem. Eur. J.* **2016**, *22*, 539–545. [CrossRef]
62. Wakizaka, M.; Ishikawa, R.; Tanaka, H.; Gupta, S.; Takaishi, S.; Yamashita, M. Creation of a Fiel-Induced Co(II) Single-Ion Magnet by doping into a Zn(II) Diamagnetic Metal-Organic Framework. *Small* **2023**, *19*, 2301966. [CrossRef]
63. Escamilla, P.; Bartella, L.; Sanz-Navarro, S.; Percoco, R.M.; Di Donna, L.; Prejanò, M.; Marino, T.; Ferrando-Soria, J.; Armentano, D.; Leyva-Pérez, A.; et al. Degradation of Penicillinic Antibiotics and β-Lactamase Enzymatic Catalysis in a Biomimetic Zn-Based Metal–Organic Framework. *Chem. Eur. J.* **2023**, *29*, e202301325. [CrossRef]
64. Lloret, F.; Julve, M.; Cano, J.; Ruiz-García, R.; Pardo, E. Magnetic properties of six-coordinated high-spin cobalt(II) complexes: Theoretical background and its application. *Inorg. Chim. Acta* **2008**, *361*, 3432–3445. [CrossRef]
65. Rabelo, R.; Toma, L.; Julve, M.; Lloret, F.; Pasán, J.; Cangussu, D.; Ruiz-García, R.; Cano, J. How the spin state tunes the slow magnetic relaxation field dependence in spin crossover cobalt(II) complexes. *Dalton Trans.* **2024**, *53*, 5507–5520. [CrossRef] [PubMed]
66. Chilton, N.F.; Anderson, R.P.; Turner, L.D.; Soncini, A.; Murray, K.S. PHI: A powerful new program for the analysis of anisotropic monomeric and exchange-coupled polynuclear d- and *f*-block complexes. *J. Comput. Chem.* **2013**, *34*, 1164–1175. [CrossRef]

magnetochemistry

MDPI

*Article*

# A Trinuclear Co(II) Complex Based on the Tris-Dioxolene Triphenylene Non-Innocent Bridge: Complementary Redox, Magnetic Behavior and Theoretical Calculations †

**Aristide Colin [1], Yiting Wang [1], François Lambert [1], Nathalie Bridonneau [1,*], Nicolas Suaud [2,*], Régis Guillot [1], Eric Rivière [1], Zakaria Halime [1], Nathalie Guihéry [2], Shin-ichi Ohkoshi [3] and Talal Mallah [1,*]**

[1] Institut de Chimie Moléculaire et des Matériaux d'Orsay, CNRS, Université Paris-Saclay, 17, Avenue des Sciences, 91400 Orsay, France; aristide.colin@universite-paris-saclay.fr (A.C.); yitingwang527829@outlook.com (Y.W.); francois.lambert.saclay@gmail.com (F.L.); regis.guillot@universite-paris-saclay.fr (R.G.); eric.riviere@universite-paris-saclay.fr (E.R.); zakaria.halime@universite-paris-saclay.fr (Z.H.)

[2] Laboratoire de Chimie et Physique Quantiques (LCPQ), Université de Toulouse, CNRS, 118 Route de Narbonne, 31062 Toulouse, France; nathalie.guihery@irsamc.ups-tlse.fr

[3] Department of Chemistry, School of Science, The University of Tokyo, 7-3-1 Hongo, Bunkyo-ku, Tokyo 113-0033, Japan; ohkoshi@chem.s.u-tokyo.ac.jp

* Correspondence: nathalie.bridonneau@universite-paris-saclay.fr (N.B.); suaud@irsamc.ups-tlse.fr (N.S.); talal.mallah@universite-paris-saclay.fr (T.M.)

† This article is dedicated to the long and outstanding scientific careers of Miguel Julve and Paco Lloret, two pioneers in the fields of coordination chemistry and magnetochemistry, two friends whose collaboration over more than 40 years impacted the domain of molecular magnetism and is an example for the young generation. Sadly, Miguel left us on 9 July 2024. We have not only lost a brilliant chemist and colleague, an outstanding teacher, highly esteemed by his students and his collaborators all over the world and particularly in France, but, even more, a close friend with a fine sense of humor, a fast mind and a generosity that we will never forget.

**Abstract:** A trinuclear Co(II)-containing complex was assembled using the non-innocent hexahydroxytriphenylene bridging ligand. Cyclovoltammetry and spectroelectrochemistry studies revealed that the central ligand sustained four reversible redox events, leading to different species with diverse optical behavior. Complementary analysis of the molecular structure confirmed by ab initio theoretical calculations were consistent with the bridge in the tris-semiquinone (sq) state for the trinuclear complex. The exchange coupling among the electrons of the bridge resulted in a spin doublet ($s = \frac{1}{2}$) localized close to one of the three $Co^{2+}$ ions, as suggested by the experimental magnetic data. The central doublet underwent one large antiferromagnetic exchange coupling with one Co(II) and almost no coupling with the two other metal ions.

**Keywords:** hexahydroxytriphenylene; non-innocent bridge; trinuclear Co complex; spectroelectrochemistry; Co magnetism; ab initio calculations

## 1. Introduction

Hexahydroxytriphenylene (noted $H_6HHTP$; $C_{18}O_6H_{12}$, Figure 1a) may, when deprotonated, present seven oxidation states from the fully reduced tris catecholate ($HHTP^{6-}$; [cat-cat-cat]) to the fully oxidized tris quinone (HHTP; [q-q-q]), as has been shown by Ward et al., when the ligand bridges Ru(II)-containing complexes [1–3]. However, this non-innocent redox active ligand was used only recently for the design of polynuclear complexes with magnetic ions, therefore adding a new functionality and providing original multifunctional polynuclear complexes. The trinuclear complexes were prepared with first row transition paramagnetic metal ions ($Ni^{2+}$ and $Cu^{2+}$)) [4–7]. For all these molecules, the most stable state of the bridging HHTP is [sq-sq-sq] (sq = semiquinone). In our effort to understand the electronic and magnetic behavior of these complexes, we have shown that, because the bridging ligand adopts the mesomeric form $HHTP^{3-}$-1 (Figure 1b) instead of a three-fold symmetry axis ($HHTP^{3-}$-2, Figure 1c), the interaction between the three unpaired sq electrons stabilizes a doublet ground state ($s = \frac{1}{2}$) that is mainly localized on one of the OCCO dioxolene moieties, leading to a large antiferromagnetic exchange coupling with one of the $Ni^{2+}$ metal ions, and one ferro- and one antiferromagnetic coupling with the other two $Ni^{2+}$, of weaker magnitudes [7]. Both experimental data (structure and magnetic) and ab initio theoretical calculations supported this view of the electronic structure, independently from the nature of the metal ions and from their geometry that can be a distorted octahedron, a distorted square pyramid or a distorted trigonal bipyramid [7].

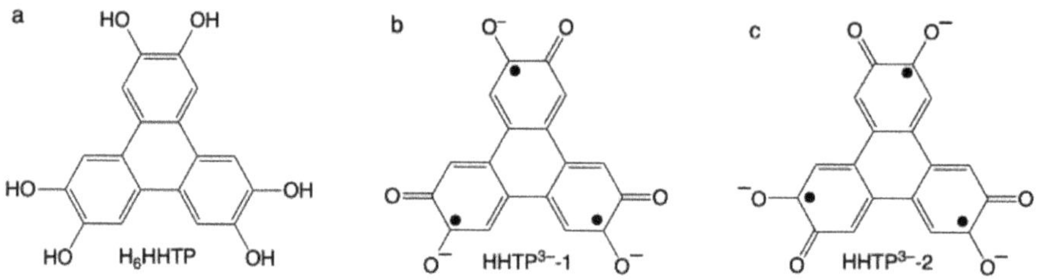


**Figure 1.** Chemical formula for $H_6HHTP$ (**a**), the mesomeric form of the tris-semiquinone state $HHTP^{3-}$ that lacks three-fold symmetry axis (**b**), and the mesomeric form of the tris-semiquinone with a three-fold symmetry axis (**c**).

The aim of this article was (i) to extend our study to a trinuclear complex with the paramagnetic $Co^{2+}$ metal ions ($s = 3/2$) where each metal center possesses three unpaired electrons, (ii) to verify whether the tris-semiquinone state remains the most stable and (iii) to assess to which extent the presence of $Co^{2+}$ instead of $Ni^{2+}$ metal ions affects the electronic structure of the HHTP, the exchange coupling within the complex, the redox and the optical behaviors. A trinuclear complex containing Co was reported but the metal ions were diamagnetic in the oxidation state III, probably because the capping tetradentate ligand imposes a strong ligand field that stabilizes Co(III) [8]. We therefore used $[Co(BH(Tp^{Ph,Ph})_3)]^+$, (where $BH(Tp^{Ph,Ph})_3$ is the tris(3,5-diphenyl-1H-pyrazol-1-yl)(hydridoborate)), as a building block with a tridentate capping ligand that induces a relatively weak ligand field and is expected to stabilize Co(II) in the high spin state [9].

We report the preparation (details of the synthesis and full characterization are given in SI), the crystal structure, the redox, the optical and the magnetic behavior of the trinuclear complex of formula $[HHTP\{Co(BH(Tp^{Ph,Ph})_3)\}]$ noted $Co_3HHTP$ in the following. Ab initio calculations were also carried out in order to determine the electronic structure of the complex and to facilitate the analysis of its magnetic properties.

## 2. Materials and Methods

### 2.1. Cyclic Voltammetry

The cyclic voltammetry experiments were carried out on solutions of 1 mM of the complex dissolved in dichloromethane with $(NBu_4)PF_6$ at 0.1 M as supporting electrolyte. The potentiostat was an AUTOLAB PGSTAT320 from Metrohm Autolab, and the three-electrode setup comprised a glassy carbon working electrode (Ø = 1mm), a platinum wire as counter electrode and a saturated calomel electrode as reference. Ferrocene was used as an internal reference and the presented voltammogram was acquired with a voltage sweeping speed of 100 mV/s.

### 2.2. Spectroelectrochemistry

UV–vis-NIR spectra were recorded using a Cary 5000 UV–vis-NIR spectrometer from Varian (Palo Alto, CA, USA) in double beam mode. Quartz cuvettes of 1 cm optical path length were used. For the spectroelectrochemical tandem measurements, the Varian Cary 5000 UV–vis-NIR spectrometer (Palo Alto, CA, USA) was used in single beam mode. The cuvette from Pine Research was made out of quartz and had a 1.7 mm optical path length. A combined platinum electrode was used as a counter and working electrode as well as reference electrode. The working electrode was of mesh-type. For reduction experiments, an argon flux was adapted on the cuvette in order to perform the reduction under a controlled atmosphere. The applied voltage was delivered by an Autolab PGSTAT204 potentiostat from Metrohm Autolab; Villebon Courtaboeuf, France.

### 2.3. Magnetic Measurements

The magnetic data were acquired using a MPMS XL7 Quantum Design SQUID magnetometer (San Diego, CA, USA) operating in the 2–300 K temperature range and in the 0–6 T field range. Crystalline powder of $Co_3$HHTP was ground and a pellet was made in order to prevent torquing effects at low temperatures and high magnetic fields. Diamagnetic correction of the sample holder, and of the compound using a Pascal scheme, were performed.

### 2.4. Preparation of the Different Compounds

The synthesis and full characterization of the precursors and of $Co_3$HHTP are given in Supplementary Materials.

### 2.5. Single Crystal X-Ray Diffraction

X-ray diffraction data for $Co_3$HHTP were collected by using a VENTURE PHOTON 100 Bruker diffractometer (Billerica, MA, USA) with Micro-focus IuS source Mo-K$\alpha$ radiation. The crystal of the compounds was selected under an optical microscope and glued in paratone oil. The crystal was mounted on a CryoLoop (Hampton Research; Aliso Viejo, CA, USA) with Paratone-N (Hampton Research) as cryoprotectant and then placed in a nitrogen-gas stream at low temperature. The temperature of the crystal was maintained at the selected value by means of a cooling device by a N-Helix Cryostream cooling device (Oxford Cryosystems, Long Hanborough, UK) with an accuracy of $\pm 1$ K. Data reduction was accomplished using SAINT V7.53a. The substantial redundancy in data allowed a semi-empirical absorption correction (SADABS V2.10) to be applied on the basis of multiple measurements of equivalent reflections. The structures were solved by direct methods using SHELXS-97 [10], and refined against $F^2$ by full-matrix least-squares techniques using SHELXL-2018 [11], with anisotropic displacement parameters for all non-hydrogen atoms. Hydrogen atoms were introduced into the calculations as a riding model with isotropic thermal parameters. All calculations were performed by using the crystal structure crystallographic software package WINGX [12]. CCDC 2403865 contains the supplementary crystallographic data for this paper. These data can be obtained free of charge from the Cambridge Crystallographic Data Centre and Fachinformationszentrum Karlsruhe via "http://www.ccdc.cam.ac.uk/structures/ accessed on 9 December 2024".

### 2.6. Ab Initio Calculations

Ab initio wave function-based calculations were performed using the ORCA5 code [13]. The treatment of the spin-orbit couplings (SOC) was achieved using the spin-orbit state interaction (SO-SI) method [14]. Calculations were carried out on a model complex where Co was replaced by Zn (noted $Zn_3$HHTP) in order to examine the electronic structure of HHTP. For this model, Def2-TZVP atomic basis sets were used for all atoms of the HHTP ligand (C and O: 5s3p2d1f) and Def2-SVP atomic basis sets for other atoms (Zn: 5s3p2d1f, B, C, N and O: 3s2p1d, H:2s1p). For $Co_3$HHTP, Def2-TZVP atomic basis sets were used for all atoms (Co: 6s4p4d1f, B, C, N and O: 5s3p2d1f) except H, for which Def2-SVP atomic basis sets were used (2s1p) and for the $Zn_2$Co1, $Zn_2$Co2 and $Zn_2$Co3 models (see Section 3.2 for the definition of these model complexes), Def2-TZVP atomic basis sets were used for the Co ions and the atoms of its coordination sphere (Co: 6s4p4d1f, N and O: 5s3p2d1f) while Def2-SVP atomic basis sets were used for other atoms (Zn: 5s3p2d1f, B, C, N and O: 3s2p1d, H: 2s1p).

## 3. Results and Discussion

### *3.1. Crystal Structure*

The crystal data collection and refinement parameters for $Co_3$HHTP are given in Table S1. $Co_3$HHTP is a trinuclear complex made from the assembly of a central HHTP ligand bridging three [Co(BH($Tp^{Ph,Ph}$)$_3$)] complexes through the three dioxolene groups (Figures 2 and S1).

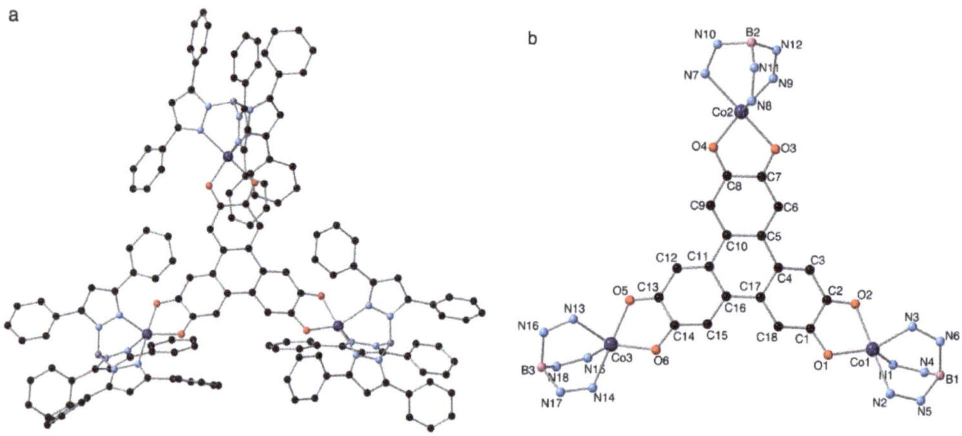


**Figure 2.** View of the molecular structure of $Co_3$HHTP (**a**) and with atom numbering, and with part of the peripheral tridentate ligand removed (**b**).

The Co atoms were pentacoordinate with a geometry close to trigonal bipyramidal (tbp; Addison parameter [15] $\tau$ close to 0.65). The pseudo three-fold symmetry axes of the three Co(II) complexes lay within the plane of the molecule and corresponded to the N2-O2, N7-O3 and N14-O5 directions for Co1, Co2 and Co3, respectively. The bond distances in the coordination sphere of the Co atoms were shorter in the tbp plane than along the pseudo three-fold axes (Table 1). The difference in the C-O bond lengths for the OCCO dioxolene moieties (Table 2) indicates that the unpaired sq electrons close to Co1 and Co3 were located on C1 and C14 (Figure 2b). For the OCCO moiety close to Co2, the difference in the C-O bond lengths was much smaller, indicating a relative delocalization of the sq unpaired electron over the OCCO moiety. The Co-O bond lengths were correlated with the C-O distances. The short Co-O bond distances were those with the oxygen atoms that had the longer C-O bonds (i.e., O1, O4 and O6 for Co1, Co2 and Co3, respectively), those that were formally negatively charged.

**Table 1.** Bond distances (in Å) in the coordination sphere of Co atoms for $Co_3$HHTP.

| Co-Ligand Distances Along the Pseudo Three-Fold Axes | | Co-Ligand Distances in the Equatorial Plane | | |
|---|---|---|---|---|
| Co1-O2 | Co1-N2 | Co1-O1 | Co1-N1 | Co1-N3 |
| 2.1113(17) | 2.143(2) | 1.9606(18) | 2.047(2) | 2.042(2) |
| Co2-O3 | Co2-N7 | Co2-O4 | Co2-N8 | Co2-N9 |
| 2.0788(19) | 2.152(2) | 1.9602(18) | 2.039(2) | 2.032(2) |
| Co3-O5 | Co3-N14 | Co3-O6 | Co3-N15 | Co3-N13 |
| 2.1113(17) | 2.114(2) | 1.9612(19) | 2.055(2) | 2.039(2) |

**Table 2.** Bond distances and their difference (in Å) for the OCCO moieties for $Co_3$HHTP.

| C1-O1 | C2-O2 | Δ(C-O)_Co1 |
|---|---|---|
| 1.313(3) | 1.258(3) | 0.055(6) |
| C8-O4 | C7-O3 | Δ(C-O)_Co2 |
| 1.291(3) | 1.280(3) | 0.011(6) |
| C14-O6 | C13-O5 | Δ(C-O)_Co3 |
| 1.303(3) | 1.257(3) | 0.046(6) |

These structural characteristics are consistent with the electronic structure of the central HHTP ligand depicted in Figure 1b, where the sq unpaired electrons of the OCCO moieties close to Co1 and Co3 are separated by 5 C-C bonds. The interaction between the three $s = \frac{1}{2}$ spins of the three sq unpaired electrons led to two doublet and one quadruplet states. The electronic structure that emerged from the crystallographic analysis should, therefore, have led to a ground doublet spin state ($s = \frac{1}{2}$) well separated in energy from the other doublet and the quadruplet states, because of a large antiferromagnetic exchange coupling between the two sq electrons separated by 5 C-C bonds. Theoretical calculations were, thus, carried out in order to confirm the electronic structure that emerged from the crystallographic structural analysis and to evaluate the exchange coupling between the central spin and those of the $Co^{2+}$ metal ions.

### 3.2. Theoretical Calculations

For the ab initio wave function-based calculations on the model complex $Zn_3$HHTP, the orbitals were optimized at the complete active space self-consistent field CAS(3,3)SCF level with an active space containing 3 electrons in 3 HHTP orbitals. Then, calculations on a CAS(12,12) active space (12 electrons, 3 from the HHTP and 9 from the three Co(II) in 12 orbitals) were performed on the whole complex. Finally, we carried out calculations on one-electron oxidized model complexes noted as $Zn_2$Co1, $Zn_2$Co2 and $Zn_2$Co3, where one Zn was replaced by one Co for the three sites Co1, Co2 and Co3, respectively, with the CAS(7,5) active space involving the 7 valence electrons of each Co(II) in their 5 3d orbitals in order to determine the axial ($D$) and rhombic ($E$) zero-field splitting ($ZFS$) parameters of the three Co fragments corresponding to the spin Hamiltonian:

$$\hat{H}_{ZFS} = D\left[\hat{s}_z^2 - \frac{s(s+1)}{3}\right] + E\left(\hat{s}_x^2 - \hat{s}_y^2\right) \quad (1)$$

The molecular orbitals of the model complex $Zn_3$HHTP optimized for the doublet ground state at the CAS(3,3)SCF level showed one bonding (HOMO), one non-bonding (SOMO) and one antibonding (LUMO) orbital, as depicted in Figure 3. The SOMO (Figure 3b) was mainly localized on the OCCO moiety close to Co2 and had almost no contribution to the two other OCCO groups.

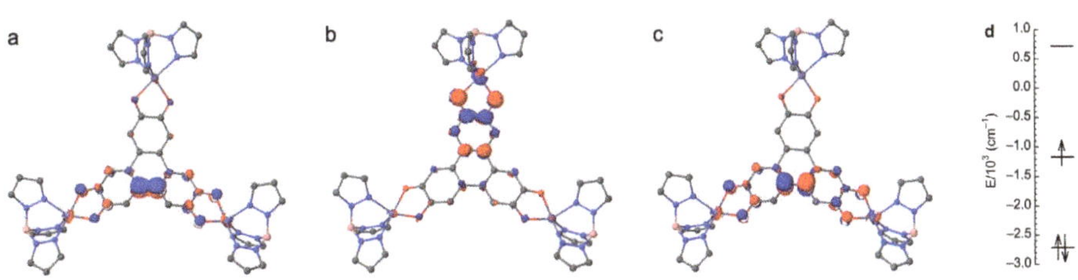


**Figure 3.** Frontier $\pi$ orbitals HOMO (**a**), SOMO (**b**) and LUMO (**c**) for $Zn_3$HHTP computed at the CAS(3,3)SCF level with electron occupation numbers 1.91, 1.00 and 0.09, respectively, and MO energy diagram (**d**).

The low energy spectrum was calculated using state average CAS(3/3)SCF + NEVPT2 calculations for the 2 lowest $s = \frac{1}{2}$ and the lowest $s = 3/2$ spin states. It showed a doublet ground state ($s = \frac{1}{2}$), an excited doublet ($s = \frac{1}{2}$) at 5144 cm$^{-1}$ and a quadruplet ($s = 3/2$) at 5468 cm$^{-1}$. Enlarging the active space did not change the spectrum qualitatively: the energy of excited $s = \frac{1}{2}$ and the $s = 3/2$ states with respect to the ground state were 4705 cm$^{-1}$ and 6566 cm$^{-1}$, respectively, for CAS(7,7)SCF + NEVPT2 calculations. The central ligand can thus be described as having one unpaired electron ($s_L = \frac{1}{2}$) expected to interact mainly with the unpaired electrons of Co2.

We then studied $Co_3$HHTP using a state average CAS(12,12)SCF calculation for all states generated by the interaction of three $s = 3/2$ on the $Co^{2+}$ ions and one $s_L = 1/2$ on the bridging ligand (i.e., one $S = 5$, three $S = 4$, five $S = 3$, seven $S = 2$, six S = 1 and two $S = 0$ states). The obtained molecular orbital energy diagram for $Co_3$HHTP (Figure 4) showed three sets of orbitals: one set where the orbitals were localized only on Co(II) with no contribution on HHTP (Figure 4, blue lines and Figure S2 for orbitals representation); one set with one bonding and one antibonding MOs localized on HHTP (Figure 4, green lines) that was almost the same as those computed for $Zn_3$HHTP (Figure 3a,c); and one set with a bonding and an antibonding MOs involving the SOMO of HHTP and one d orbital of Co2 (Figure 4, red lines). This orbital energy diagram indicates that the electronic structure of $Co_3$HHTP was mainly governed by the interaction between the SOMO of HHTP and one d orbital of one of the three $Co^{2+}$ metal ions, namely Co2. We, therefore, expected that the magnetic behavior of the trinuclear complex would be mainly governed by the exchange coupling between the spins of these two moieties, and by the local anisotropy of $Co^{2+}$ ions that also needed to be evaluated.

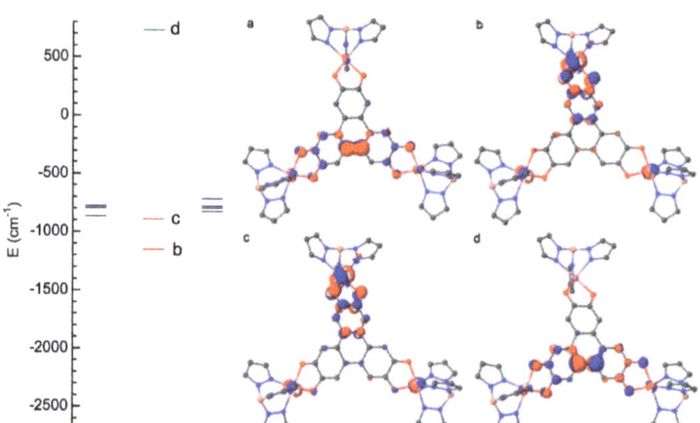


**Figure 4.** Energy levels of the MOs calculated for $Co_3$HHTP at the CAS(12,12)SCF level and representation of the MOs localized on HHTP (**a**,**d**) and of the MOs resulting from the interaction between the SOMO of HHTP and one *d* orbital of one Co(II) (**b**,**c**). A representation of the MOs localized on the Co atoms is given in Supplementary Materials (Figure S2).

The corresponding model Hamiltonian involves the exchange coupling parameters $J_2$ between the ligand doublet spin momentum $\hat{s}_L$ and the Co2 spin momentum $\hat{s}_2$ and the local anisotropies of the three Co(II) species:

$$\hat{H} = -J_2 \hat{s}_L.\hat{s}_2 + \sum_k \left( \hat{s}_k.\bar{\bar{D}}_k.\hat{s}_k \right) \quad (2)$$

The CAS(12,12)SCF + NEVPT2 energy spectrum (Table S2) confirmed a strong antiferromagnetic coupling $J_2$ between the ligand and Co2 spins of the order of 65 cm$^{-1}$ (see SI). It resulted in an effective local metal-ligand ground triplet state ($\hat{s}_{Co2-L} = 1$) and an excited quintet state ($\hat{s}'_{Co2-L} = 2$). Indeed, the analysis of the spectrum showed a low energy first set of 10 quasi-degenerate states (spectrum width ~7cm$^{-1}$) corresponding to small couplings (few cm$^{-1}$) between $\hat{s}_{Co2-L}$ and $\hat{s}_1$ and $\hat{s}_3$, and a second set (14 states) higher in energy ($-2J_2$ = 130 cm$^{-1}$, bandwidth ~20 cm$^{-1}$) corresponding as well to small couplings (few cm$^{-1}$) between $\hat{s}'_{Co2-L}$ and $\hat{s}_1$ and $\hat{s}_3$.

The ZFS parameters (Table 3) were extracted using an already reported method [16]. We found relatively large negative $D$ values for the three Co(II) fragments with a large rhombicity (we note that the maximum rhombicity corresponded to $E/|D| = 0.33$). As ZFS values were particularly large, we checked that the first-order SOC contributions were negligible. The easy axis of magnetization for the three Co(II) fragments (Figure 5) was aligned along the Co-O short bonds and not along the pseudo three-fold axis of the tbp. In order to analyze the origin of the negative $D$ values, we focused on one Co(II) species (Co2), since the interpretation of the results was the same for the three metal ions (see Table S3 and Table S4 for the composition of the wave function of the ground and excited $s = 3/2$ states obtained from calculations on the Co1 and Co2 sites). The calculation also provided the contribution at the second order of perturbation of each excited state to the overall $D$ value. The first excited state was mainly responsible of the large negative $D$ values for the Co(II) species (Tables S3 and S4). If we consider the Co2 fragment, the contribution of the first excited state was found equal to –53 cm$^{-1}$ (Table S3) and corresponded to an excitation between $d_{xy}$ and $d_{x^2-y^2}$ orbitals that are linear combinations of the $l = 2$, $m_l = \pm 2$ spherical harmonics, and therefore it was the $\hat{l}_z\hat{s}_z$ part of the spin-orbit operator that coupled this state to the ground state. When SO couples through $\hat{l}_z\hat{s}_z$ states of the same spin, it stabilizes the largest $|m_s|$ values ($\pm 3/2$) of the ground state and therefore contributes negatively to $D$ [17,18]. The third excited state was obtained by an excitation between the $d_{yz}$ and $d_{xz}$ orbitals (linear combinations of $m_l = \pm 1$) and had, for the same reasons, a negative contribution to $D$ but weaker in magnitude ($-15.6$ cm$^{-1}$ instead of $-53.1$ cm$^{-1}$) because of the weaker $|m_l|$ values and to the larger energy separation with the ground state (5451 cm$^{-1}$ instead of 1138 cm$^{-1}$). The second and the fourth excited states had positive contributions (Table S3), but of much weaker magnitude so that the overall $D$ value remained negative and large. For Co1, the same analysis justified its negative $D$ value (Table S4).

**Table 3.** Axial $D$ and rhombic $E$ ZFS parameters (in cm$^{-1}$) calculated using the SO-SI method with the CAS(7,5)NEVPT2 energies for the three Co(II) fragments.

| ZFS Parameters | Co1 | Co2 | Co3 |
|---|---|---|---|
| $D$ | −67.1 | −58.9 | −63.7 |
| $E$ | 16.4 | 15.2 | 15.5 |
| $E/\|D\|$ | 0.24 | 0.26 | 0.24 |

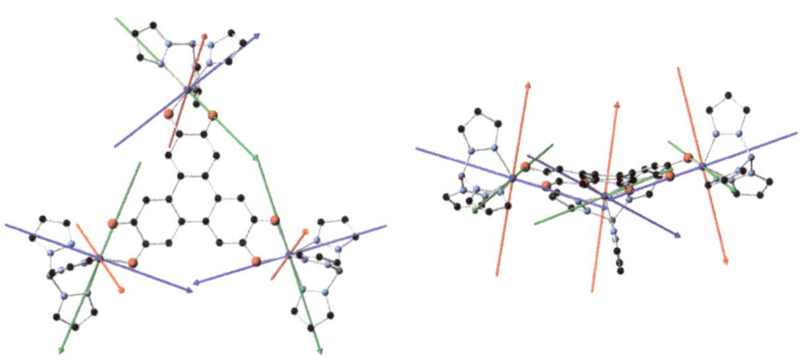

**Figure 5.** Relative orientation of the $\overline{\overline{D}}$ tensor axes (blue, easy axis: green, intermediate axis and red, hard axis) for the three Co moieties in $Co_3$HHTP.

### 3.3. Redox and Optical Properties

In order to investigate the redox and optical behavior of $Co_3HHTP$, we performed cyclovoltammetry (CV) and spectroelectrochemistry studies to evaluate the stability of the different species. Those data were compared with the previously reported Ni(II)-containing compound of general formula [HHTP{Ni(BH(Tp$^{Ph,Ph}$)$_3$)}] that possesses a similar structure with the same capping ligand (noted $Ni_3HHTP$ in the following) and where HHTP is also in the [sq-sq-sq] state [4]. The CV of $Co_3HHTP$ showed four reversible redox waves similar to those observed for the Ni(II) complex; three one-electron reduction waves and one one-electron oxidation wave (Figure 6, Table S5). The first reduction wave had the same redox potential as that of $Ni_3HHTP$, while the other waves were slightly shifted anodically, indicating a higher energy requirement for the oxidation process and more energetically accessible second and third reduction processes for $Co_3HHTP$ compared to $Ni_3HHTP$. In the latter, the redox processes were mainly located on HHTP because Ni(II) is redox-inactive in the observed potential window. The very close redox potential values for $Co_3HHTP$ indicate that the redox processes were also not metal-centered as for $Ni_3HHTP$ and that they, therefore, mainly involved the bridging ligand HHTP.

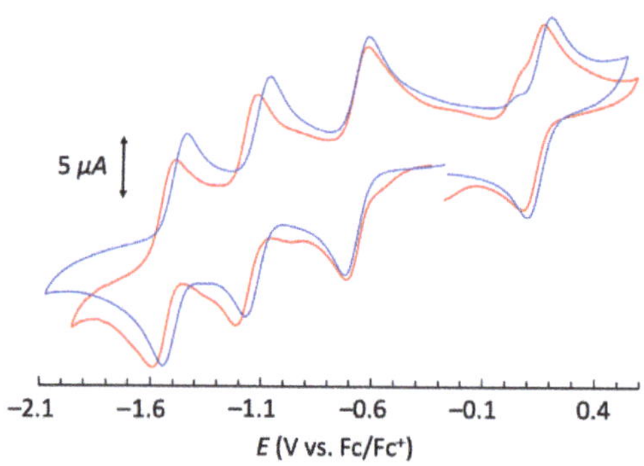


**Figure 6.** Cyclic voltammogram of 1 mM of $Co_3HHTP$ (blue trace) and of $Ni_3HHTP$ (red trace) in argon-degassed dichloromethane containing 0.1 M $[Bu_4N]PF_6$ at 25 °C and recorded using a glassy carbon working electrode, a platinum mesh as the counter electrode and a saturated calomel electrode (SCE) as the reference electrode.

The electronic spectra of $Co_3HHTP$ as well as its one-electron reduced and one-electron oxidized analogues were recorded using UV–vis-NIR spectroelectrochemistry. The three species showed two sets of absorption bands: in the near infrared (NIR) and in the visible region (Figure 7). We focused the following on the NIR region. $Co_3HHTP$ displayed an absorption band at 1140 nm that was red-shifted to 1315 nm (this band was probably due to two transitions at 1292 and 1370 nm) upon a one-electron reduction and blue-shifted to 1183 nm for the one-electron oxidized species. In order to propose a qualitative assignment for these bands, we compared the electronic spectra to those of $Ni_3HHTP$ (Figure 8).

The absorption bands in the NIR region of $Co_3HHTP$ and $Ni_3HHTP$ (Figure 8a) were almost identical, leading to the conclusion that the nature of the metal ions had almost no effect on the electronic transitions and that, therefore, these bands were due to transition(s) within the HHTP when it was in the [sq-sq-sq] state. In fact, with this hypothesis one could expect three absorption bands for the [sq-sq-sq] HHTP based on the MOs energy diagram of the model complex $Zn_3HHTP$ (Figure 3d), corresponding to the promotion of an electron between the three couple of MOs: HOMO → SOMO, SOMO → LUMO and HOMO → LUMO. The first two transitions were expected at lower energy than the latter. Figure 8a displayed two absorption bands at 1139 and 952 nm that may have been due to the transition expected at higher energy.

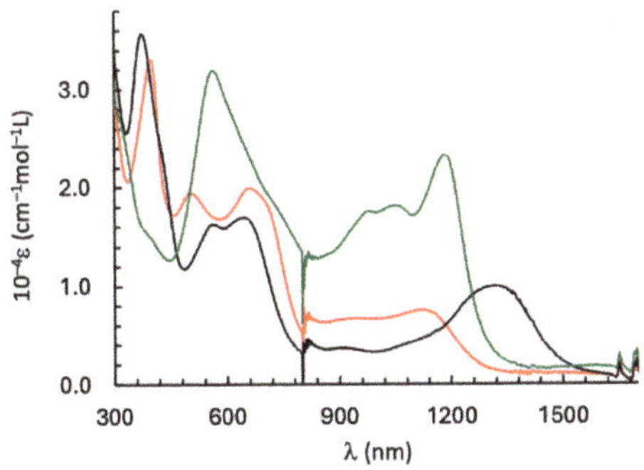


**Figure 7.** Electronic absorption spectra of $Co_3$HHTP (red trace), the one-electron reduced species (black trace) and the one-electron oxidized species (green trace) in dichloromethane.

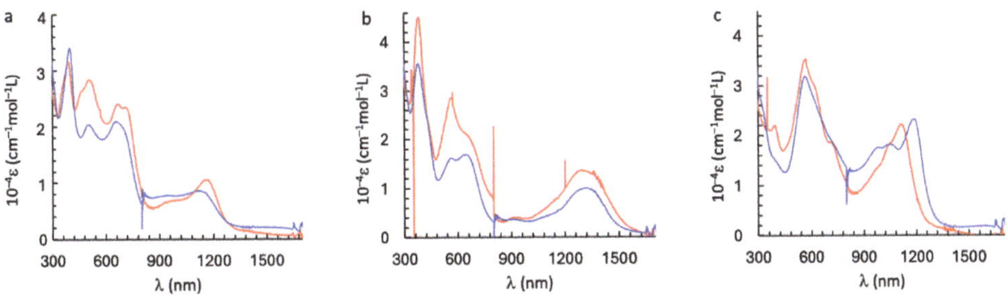


**Figure 8.** Electronic spectra recorded in UV–vis-NIR spectroelectrochemistry experiments for (**a**) **1** and $Ni_3$HHTP, (**b**) the one-electron reduced species and (**c**) the one-electron oxidized species, blue and red traces for the Co- and Ni-containing species, respectively, in dichloromethane containing 0.1 M $[Bu_4N]PF_6$ at 25 °C.

The spectra of the one-electron reduced species (Figure 8b) were also very similar in the NIR region, confirming that these transitions were mainly centered on the bridging ligand that was in the [sq-sq-cat] state and that the nature of the metal ions did not have a significant effect on the electronic structure of the bridging ligand. The situation was different for the one-electron oxidized species (Figure 8c), where a blue-shift of 60 nm (468 cm$^{-1}$) was observed for the most intense band when going from the Co to the Ni-containing complexes. The analysis performed here was only qualitative. A better understanding of the impact of the metal ions on the optical properties of the trinuclear species and their redox behavior requires an assignment of the different absorption bands that can be eventually performed with the help of time-dependent density functional calculations that are underway.

### 3.4. Magnetic Properties

The magnetic properties of $Co_3$HHTP were investigated by measuring the thermal dependence of the $\chi T$ product (Figure 9a) and the magnetization vs. the magnetic field at T = 2, 4, 6 K (Figure 9b). The value of the $\chi T$ product at T = 300 K (8.51 cm$^3$mol$^{-1}$K) may have corresponded to one $s = \frac{1}{2}$ with $g = 2$ and three $s = 3/2$ with $g = 2.4$ non-interacting species. Such a $g$ value for Co(II) is not unreasonable. Upon cooling, $\chi T$ decreased monotonically until a value of 7.5 cm$^3$mol$^{-1}$K at T = 75 K and then more abruptly. It reached a value of 3 cm$^3$mol$^{-1}$K at 2 K. The relatively large $\chi T$ value at low temperature indicated a magnetic ground state and the decrease an antiferromagnetic coupling with certainly magnetic anisotropy, as indicated from theoretical calculations. The

M = f(B/T) curves were not superimposable, which confirmed the presence of magnetic anisotropy within the complex.

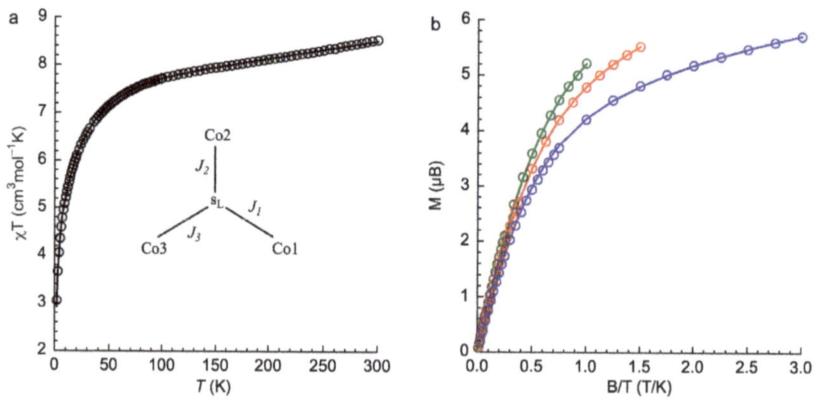


**Figure 9.** χT = f(T) plot and exchange coupling scheme (**a**) and M = f(B/T) plots at $T$ = 6 (green), 4 (red) and 2 (blue) K (**b**) for $Co_3$HHTP; experimental data (○) and fit (—) with the parameters in Table 4.

**Table 4.** Magnetic data fit parameters for $Co_3$HHTP. The ZFS and the exchange parameters in $cm^{-1}$.

| $g_1 = g_2 = g_3$ | $D_1 = D_3$ | $D_2$ | $E_1 = E_3$ | $E_2$ | $J_1 = J_3$ | $J_2$ | TIP |
|---|---|---|---|---|---|---|---|
| 2.55(2) | −22(2) | −60(3) | 3.3(3) | 6.7(5) | −0.1 | −129(4) | $1.1 \times 10^{-3}$ |

The data were fitted using PHI [19], assuming a model with a central $s = \frac{1}{2}$ and three peripheral $s = 3/2$ spins introducing TIP in the fit procedure. The number of parameters was large (three $g$ values, six ZFS parameters and three exchange coupling parameters). We therefore proceeded in several steps in order to delimit the values of the different parameters and relied on theoretical calculation by considering one large antiferromagnetic exchange coupling and two very weak, on one hand, and on the other hand, we assumed first the same or very close ZFS parameters for the three Co(II) species. Considering this last assumption, it was not possible to reach a satisfactory solution. We therefore used different ZFS parameters but still negative ones because distorted pentacoordinate Co(II) complexes were found to have negative $D$ values in many cases [17,20–22]. The best fit parameters are given in Table 4.

Overall, the parameters that best fit the magnetic data were qualitatively in agreement with those extracted from calculations, even though the figures may have differed when we examined those corresponding to each Co(II) species. There was a large antiferromagnetic exchange coupling between the unpaired electron of HHTP and *one* of the Co(II) species and two very weak couplings with the two other Co(II), as calculations predicted. The axial ZFS $D$ values were all negative, but we found weaker values (in absolute value) for two Co(II), while for the third Co(II) the $D$ value was almost the same as that extracted from calculations. The rhombic ZFS parameters of $E$ found from the fit were weaker than those evaluated from calculations.

## 4. Conclusions

We report the preparation, under aerobic conditions, of a trinuclear Co(II)-containing complex using the tris-dioxolene triphenylene bridging ligand (HHTP). The crystal structure analysis and particularly the C-O bond distances of the OCCO moieties of the ligand suggested that HHTP was in the [sq-sq-sq] state even when the synthesis was carried out starting from the tris-catecholate species. The cyclovoltammetry study confirmed that, in solution, the complex remained in the [sq-sq-sq] state with four accessible species, namely [cat-cat-cat], [cat-cat-sq], [cat-sq-sq] and [sq-sq-q]. The spectroelectrochemical studies of $Co_3$HHTP and $Ni_3$HHTP indicated that, qualitatively, the electronic structure of HHTP was independent of the nature of the metal ions for the [cat-sq-sq], and the [sq-sq-sq] states. Ab initio calculations were consistent with a strong antiferromagnetic exchange coupling among the three unpaired electrons of the HHTP and led to a ground doublet state localized mainly on one OCCO moiety close to one of the three peripheral Co(II) complexes, as already found for $Ni_3$HHTP. Consequently, the magnetic properties were the result of a large antiferromagnetic exchange coupling with one Co(II) and much weaker couplings with the other two. The main difference in the magnetic properties between **1** and $Ni_3$HHTP was that the two weak exchange coupling parameters were much weaker for $Co_3$HHTP than for $Ni_3$HHTP (less than 1 $cm^{-1}$ for $Co_3$HHTP and larger than 20 $cm^{-1}$ for $Ni_3$HHTP [7]), as found from fitting the magnetic data. This was due to the larger localization of the central doublet of HHTP for $Co_3$HHTP than for $Ni_3$HHTP (Figure 10a,b)). This larger localization stemmed from a larger distortion of the HHTP skeleton in $Co_3$HHTP than in $Ni_3$HHTP (Figure 10c,d).

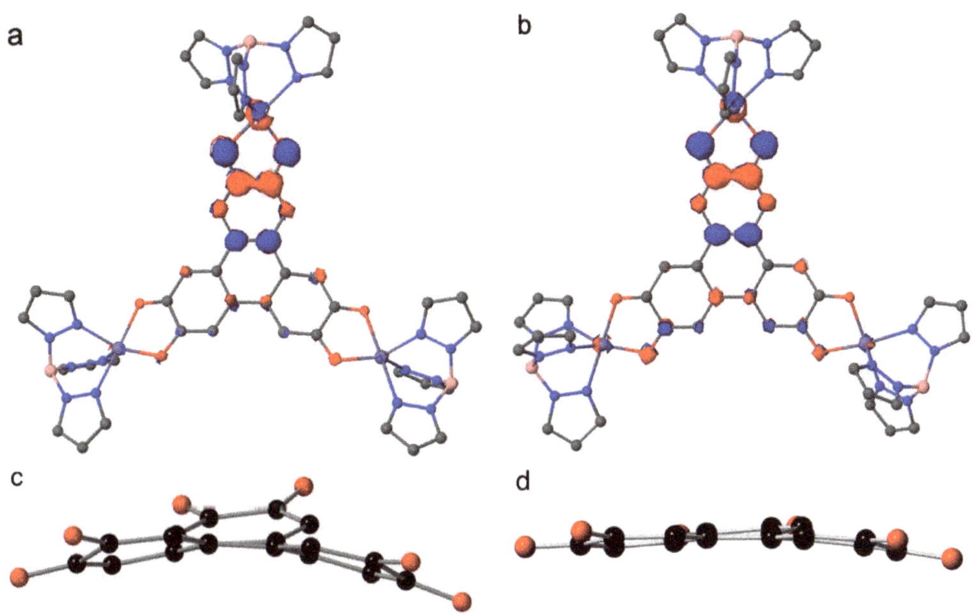


**Figure 10.** Representation of the SOMOs of the model complexes $Zn_3$HHTP for $Co_3$HHTP (**a**) and for $Ni_3$HHTP (**b**), where a larger localization of the orbital can be observed for $Co_3$HHTP; and view of the geometry of HHTP for the two complexes with the central cycle of HHTP perpendicular to the plane, highlighting its larger distortion for $Co_3$HHTP (**c**) than for $Ni_3$HHTP (**d**).

**Supplementary Materials:** The following supporting information can be downloaded at: https://www.mdpi.com/article/10.3390/magnetochemistry10120102/s1, Figure S1: An ORTEP drawing of compound $Co_3$HHTP. Thermal ellipsoids are shown at the 30% level. Solvent molecules ($C_3H_6O$ & $H_2O$) were omitted for clarity. Hydrogen atoms are omitted for clarity; Figure S2: Molecular orbitals localized on the Co calculated for $Co_3$HHTP at the CAS(12,12)SCF level; Table S1: Crystallographic data and structure refinement details for $Co_3$HHTP; Table S2: Energy spectrum of $Co_3$HHTP obtained from CAS(12/12)SCF + NEVPT2 calculations; Table S3: Contribution to $D$ of the first four excited states for the Co2 magnetic center. The occupation of the orbitals in the dominant determinants of each state are given for the following orbital sequence $|z^2\ xz\ yz\ x^2-y^2\ xy|$ followed by their weight in the CAS(7,5) wave function. Determinants contributing to less than 10% are not reported. GS = ground state, ESi = excited states numbering i = 1 to 4; Table S4: Contribution to *D* of the first four excited states for the Co1 magnetic center. The occupation of the orbitals in the dominant determinants of each state are given for the following orbital sequence | $z^2$ xz yz $x^2 - y^2$ xy | followed by their weight in the CAS(7,5) wave function. Determinants contributing to less than 10% are not reported. GS = ground state, ESi = excited states numbering i = 1 to 4; Table S5: Anodic ($E_a$) and cathodic ($E_c$) peak potentials, half-wave potentials and differences of the half-wave potentials from the CV of $Ni_3$HHTP and for $Co_3$HHTP.

**Author Contributions:** Conceptualization, T.M.; synthesis, A.C., Y.W. and F.L.; theory, N.S. and N.G.; crystallography, R.G.; magnetic study and analysis, E.R. and A.C., electrochemistry, and optical measurements, A.C. and Z.H.; data analysis, A.C., Y.W. and N.B.; supervision, N.B., F.L. and S.-i.O.; writing—original draft, T.M; writing—reviewing and editing, N.B., Z.H., N.G., T.M. and S.-i.O. All authors have read and agreed to the published version of the manuscript.

**Funding:** This work is supported by the ADI 2021 project funded by the IDEX Paris-Saclay, ANR-11-IDEX-0003-02 and JST SPRING (JPMJSP2108). This work was also supported by the ANR (project SMOLMAGIQ ANR-20-CE29-0010) and Grant-in-Aid for Scientific Research (A) from the Japan Society for the Promotion of Science (JSPS) (20H00369).

**Data Availability Statement:** The original contributions presented in the study are included in the article/Supplementary Materials. Further inquiries can be directed to the corresponding author.

**Acknowledgments:** A.C. acknowledge the University Paris-Saclay and the University of Tokyo for financial support within the framework of the double diploma agreement. The authors thank the CNRS (Centre de la Recherche Scientifique) for financial support.

**Conflicts of Interest:** The authors declare no conflicts of interest.

## References

1. Barthram, A.M.; Cleary, L.; Kowallick, R.; Ward, M.D. A new redox-tunable near-IR dye based on a trinuclear ruthenium(II) complex of hexahydroxytriphenylene. *Chem. Commun.* **1998**, *126*, 2695–2696. [CrossRef]
2. Ward, M.D.; McCleverty, J.A. Non-innocent behaviour in mononuclear and polynuclear complexes: Consequences for redox and electronic spectroscopic properties. *J. Chem. Soc. Dalton Trans.* **2002**, *3*, 275–288. [CrossRef]
3. Grange, C.S.; Meijer, A.J.H.M.; Ward, M.D. Trinuclear ruthenium dioxolene complexes based on the bridging ligand hexahydroxytriphenylene: Electrochemistry, spectroscopy, and near-infrared electrochromic behaviour associated with a reversible seven-membered redox chain. *Dalton Trans.* **2010**, *39*, 200–211. [CrossRef] [PubMed]
4. Wang, Y.T.; Lambert, F.; Rivière, E.; Guillot, R.; Herrero, C.; Tissot, A.; Halime, Z.; Mallah, T. Electronic and spin delocalization in a switchable trinuclear triphenylene trisemiquinone bridged Ni complex. *Chem. Commun.* **2019**, *55*, 12336–12339. [CrossRef] [PubMed]
5. Yang, L.M.; He, X.; Dinca, M. Triphenylene-Bridged Trinuclear Complexes of Cu: Models for Spin Interactions in Two-Dimensional Electrically Conductive Metal-Organic Frameworks. *J. Am. Chem. Soc.* **2019**, *141*, 10475–10480. [CrossRef] [PubMed]
6. Yang, L.M.; Dinca, M. Redox Ladder of Ni Complexes with Closed Shell, Mono-, and Diradical Triphenylene Units: Molecular Models for Conductive 2D MOFs. *Angew. Chem. Int. Ed.* **2021**, *60*, 23784–23789. [CrossRef]
7. Suaud, N.; Colin, A.; Bouammali, M.; Mallah, T.; Guihéry, N. Understanding the Electronic Structure of Magnetic Trinuclear Complexes Based on the Tris-dioxolene Triphenylene Non-innocent Bridging Ligand, a Theoretical Study. *Chem. Eur. J.* **2024**, *30*, e202302256. [CrossRef]
8. Suenaga, Y.; Inada, H.; Inomata, M.; Yamaguchi, R.; Okubo, T.; Maekawa, M.; Kuroda-Sowa, T. Crystal Structure and Characterization of Trinuclear Cobalt(III) Complex with 2,3,6,7,10,11-Hexahydroxytriphenylene. *Chem. Lett.* **2014**, *43*, 562–564. [CrossRef]
9. Uehara, K.; Hikichi, S.; Akita, M. Highly labile cationic tris-acetonitrile complexes, [TpM(NCMe)]OTf (M = Ni, Co; Tp: Hydrotrispyrazolylborato, R = Ph, Me and iPr): Versatile precursors for Tp-containing nickel and cobalt complexes. *J. Chem. Soc. Dalton Trans.* **2002**, 3529–3538. [CrossRef]
10. Sheldrick, G.M. *SHELXS-97, Program for Crystal Structure Solution*; University of Göttingen: Göttingen, Germany, 1997.
11. Sheldrick, G. A short history of SHELX. *Acta Christ. Sec. A* **2008**, *64*, 112–122.
12. Farrugia, L.J. WinGX suite for small-molecule single-crystal crystallography. *J. Appl. Cryst.* **1999**, *32*, 837. [CrossRef]
13. Neese, F.; Wennmohs, F.; Becker, U.; Riplinger, C. The ORCA quantum chemistry program package. *J. Chem. Phys.* **2020**, *152*, 224108. [CrossRef]
14. Malmqvist, P.Å.; Roos, B.O.; Schimmelpfennig, B. The restricted active space (RAS) state interaction approach with spin-orbit coupling. *Chem. Phys. Lett.* **2002**, *357*, 230–240. [CrossRef]

15. Addison, A.W.; Rao, T.N.; Reedijk, J.; Vanrijn, J.; Verschoor, G.C. Synthesis, Structure, and Spectroscopic Properties of Copper(II) Compounds Containing Nitrogen Sulfur Donor Ligands—The Crystal and Molecular-Structure of Aqua[1,7-Bis(N-Methylbenzimidazol-2′-Yl)-2,6-Dithiaheptane]Copper(Ii) Perchlorate. *J. Chem. Soc. Dalton Trans.* **1984**, *7*, 1349–1356. [CrossRef]
16. Maurice, R.; Bastardis, R.; de Graaf, C.; Suaud, N.; Mallah, T.; Guihéry, N. Universal Theoretical Approach to Extract Anisotropic Spin Hamiltonians. *J. Chem. Th. Comput.* **2009**, *5*, 2977–2984. [CrossRef]
17. Gomez-Coca, S.; Cremades, E.; Aliaga-Alcalde, N.; Ruiz, E. Mononuclear Single-Molecule Magnets: Tailoring the Magnetic Anisotropy of First-Row Transition-Metal Complexes. *J. Am. Chem. Soc.* **2013**, *135*, 7010–7018. [CrossRef]
18. Ruamps, R.; Batchelor, L.J.; Maurice, R.; Gogoi, N.; Jiménez-Lozano, P.; Guihéry, N.; de Graaf, C.; Barra, A.L.; Sutter, J.P.; Mallah, T. Origin of the Magnetic Anisotropy in Heptacoordinate Ni and Co Complexes. *Chem. Eur. J.* **2013**, *19*, 950–956. [CrossRef] [PubMed]
19. Chilton, N.F.; Anderson, R.P.; Turner, L.D.; Soncini, A.; Murray, K.S. PHI: A powerful new program for the analysis of anisotropic monomeric and exchange-coupled polynuclear d- and f-block complexes. *J. Comput. Chem.* **2013**, *34*, 1164–1175. [CrossRef]
20. Ruamps, R.; Batchelor, L.J.; Guillot, R.; Zakhia, G.; Barra, A.L.; Wernsdorfer, W.; Guihéry, N.; Mallah, T. Ising-type magnetic anisotropy and single molecule magnet behaviour in mononuclear trigonal bipyramidal Co(II) complexes. *Chem. Sci.* **2014**, *5*, 3418–3424. [CrossRef]
21. Cahier, B.; Perfetti, M.; Zakhia, G.; Naoufal, D.; El-Khatib, F.; Guillot, R.; Rivière, E.; Sessoli, R.; Barra, A.L.; Guihéry, N.; et al. Magnetic Anisotropy in Pentacoordinate Ni and Co Complexes: Unraveling Electronic and Geometrical Contributions. *Chem. Eur. J.* **2017**, *23*, 3648–3657. [CrossRef]
22. Devkota, L.; SantaLucia, D.J.; Wheaton, A.M.; Pienkos, A.J.; Lindeman, S.V.; Krzystek, J.; Ozerov, M.; Berry, J.F.; Telser, J.; Fiedler, A.T. Spectroscopic and Magnetic Studies of Co(II) Scorpionate Complexes: Is There a Halide Effect on Magnetic Anisotropy? *Inorg. Chem.* **2023**, *62*, 5984–6002. [CrossRef] [PubMed]

magnetochemistry

MDPI

*Article*

# Calixarene-like Lanthanide Single-Ion Magnets Based on $Nd^{III}$, $Gd^{III}$, $Tb^{III}$ and $Dy^{III}$ Oxamato Complexes †

**Tamyris T. da Cunha [1,2], João Honorato de Araujo-Neto [3], Meiry E. Alvarenga [4], Felipe Terra Martins [4], Emerson F. Pedroso [5], Davor L. Mariano [6], Wallace C. Nunes [7], Nicolás Moliner [8], Francesc Lloret [8], Miguel Julve [8,‡] and Cynthia L. M. Pereira [1,*]**

1 Departamento de Química, Instituto de Ciências Exatas, Universidade Federal de Minas Gerais, Av. Antônio Carlos 6627, Pampulha, Belo Horizonte 31270-901, MG, Brazil
2 Campus Itabirito, Instituto Federal de Educação, Ciência e Tecnologia de Minas Gerais, R. José Benedito, 139, Santa Efigênia, Itabirito 35450-000, MG, Brazil
3 Instituto de Química, Universidade de São Paulo, Av. Prof. Lineu Prestes, 748-Butantã, São Paulo 05508-900, SP, Brazil
4 Instituto de Química, Universidade Federal de Goiás, Campus Samambaia, Setor Itatiaia, Caixa Postal 131, Goiânia 74001-970, GO, Brazil
5 Laboratório de Tecnologias Integradas (Intechlab), Centro Federal de Educação Tecnológica de Minas Gerais, Av. Amazonas 5523, Belo Horizonte 30421-169, MG, Brazil
6 Instituto de Física, Universidade Federal do Rio de Janeiro (IF-UFRJ), Av. Athos da Silveira Ramos, 149, Cidade Universitária, Rio de Janeiro 24210-340, RJ, Brazil
7 Instituto de Física, Universidade Federal Fluminense, Av. Gal. Milton Tavares de Souza, s/n°, Campus da Praia Vermelha, Niterói 24210-346, RJ, Brazil
8 Departament de Química Inorgànica, Instituto de Ciencia Molecular (ICMol), Universitat de València, C/Catedrático José Beltrán 2, 46980 Paterna, València, Spain
* Correspondence: cynthialopes@ufmg.br

† We dedicate this manuscript to Prof. Francesc Lloret and Prof. Miguel Julve, brilliant scientists, to honor their careers on the occasion of their retirement.

‡ Prof. Miguel Julve passed away while preparing this contribution. We do not have words to express our feelings. Still, we are very grateful for the luck of knowing Prof. Miguel Julve, a beautiful person, beloved friend, and tireless scientific collaborator. Thank you, Miguel.

**Abstract:** In this work, we describe the synthesis, crystal structures and magnetic properties of four air-stable mononuclear lanthanide(III) complexes with the *N*-(2,4,6-trimethylphenyl)oxamate (Htmpa) of formula: $n$-$Bu_4N[Nd(Htmpa)_4(H_2O)]\cdot 4H_2O$ (**1**), $n$-$Bu_4N[Gd(Htmpa)_4(H_2O)]\cdot 3DMSO\cdot 2H_2O$ (**2**), $n$-$Bu_4N[Tb(Htmpa)_4(H_2O)]\cdot 3DMSO\cdot 1H_2O$ (**3**) and $n$-$Bu_4N[Dy(Htmpa)_4(H_2O)]\cdot 3DMSO\cdot 2H_2O$ (**4**) ($n$-$Bu_4N^+$ = $n$-tetrabutylammonium; DMSO = dimethylsulfoxide). Their crystal structures reveal the occurrence of calixarene-type monoanionic species containing all-*cis*-disposed Htmpa ligands and one water molecule coordinated with the respective $Ln^{III}$ ion (Ln = Nd, Gd, Tb and Dy), featuring a nine-coordinated environment with muffin (MFF-9) (**1**) or spherical-capped square antiprism (CSAPR-9) (**2–4**) geometry. The major difference between their crystal structures is related to the nature of crystallization solvent molecules, either water (**1**) or both DMSO and water (**2–4**). The intermolecular hydrogen bonds among the self-complementary Htmpa ligands in all four compounds mediated a 2 D supramolecular network in the solid state. Direct-current (dc) magnetic properties for **1–4** show typical behavior for the ground state terms of the $Ln^{III}$ ions [$^4I_{9/2}$ (Nd); $^8S_{7/2}$(Gd), $^7F_6$ (Tb), $^6H^{15/2}$ (Dy)]. Alternating-current (ac) magnetic measurements reveal the presence of slow magnetic relaxation without the presence of a dc field only for **4**. In contrast, field-induced slow magnetic relaxation behavior was found in complexes **1**, **2** and **3**.

**Keywords:** oxamate; lanthanide; single-ion magnet; supramolecular chemistry; neodymium(III); gadolinium(III); terbium(III); dysprosium(III)

## 1. Introduction

The search for new magnetic materials that exhibit slow magnetic relaxation (SMR) phenomena has received significant attention from research groups worldwide. These materials hold the potential for advancing next-generation technologies, including spin qubits and spin qudits for quantum computing, quantum information processing (QIP), spintronics and more [1].

Two synthetic strategies have been employed to obtain these materials: the "top-down" and "bottom-up" approaches. In the "top-down" approach, high-density data storage and material candidates for quantum technologies are produced by reducing the size of nanoparticles through physical methods. However, this process often results in inhomogeneous particle sizes, which is undesirable due to the lack of control over the final properties. The alternative "bottom-up" approach offers a promising route for synthesizing homogeneous molecules under mild, reproducible conditions. This method relies on self-organizing molecules exhibiting the desired properties [1]. This molecular approach employs organic linkers, enabling the incorporation of diverse paramagnetic ions to give slow-magnetic relaxing molecules, molecular assemblies or supramolecular coordination polymers [2,3].

In 2003, the discovery of terbium(III) and dysprosium(III) double-decker phthalocyanine (Pc) complexes, $(Bu_4N)[LnPc_2]$ (Ln = $Tb^{3+}$ or $Dy^{3+}$), identified as single-ion magnets (SIMs), marked a significant shift in focus from polynuclear d-block complexes to mononuclear f-block systems as promising candidates for technological applications [4]. Ln SIMs, as well as Ln SMMs (SMM being single-molecule magnet), exhibited SMR and stepwise open-loop magnetic hysteresis of pure molecular origin due to the presence of a single highly anisotropic lanthanide(III) ion due to the combined effect of first-order spin-orbit coupling (SOC) and ligand-field (LF) effects that cause the splitting of the $2J + 1$ components of the ground state characterized by a total angular momentum $J = |S - L|$ or $J = S + L$ for early (Ln = Ce − Eu) or late $Ln^{III}$ ions (Ln = Tb − Yb). In the absence of additional interactions, Kramers $Ln^{III}$ ions (Ln = $Ce^{3+}$, $Nd^{3+}$, $Sm^{3+}$, $Gd^{3+}$, $Dy^{3+}$, $Er^{3+}$, $Yb^{3+}$, $Lu^{3+}$) are, in principle, more suitable for the development of SIMs as single-ion memory devices compared with non-Kramers ions (Ln = $Pr^{3+}$, $Pm^{3+}$, $Eu^{3+}$, $Tb^{3+}$, $Ho^{3+}$, $Tm^{3+}$). However, dipolar or hyperfine interactions can induce transverse fields that break the degeneracy of Kramers doublets ($\pm m_J$), facilitating undesired fast relaxation processes. This phenomenon is associated with quantum tunneling of magnetization (QTM) phenomena. Notably, the suppression of QTM can be achieved by diluting paramagnetic complexes within an isostructural diamagnetic matrix, thereby reducing dipolar interactions or by applying an external static magnetic field [1,5].

The magnetic properties of Ln SIMs can be modulated through ligand design. Due to the large radii of $Ln^{III}$ ions and their electrostatic bonding, they usually form high coordination number (CN) complexes, lacking geometrical control. Several results indicate that $U_{eff}$ increases by reducing the CN of lanthanide(III) complexes, indicating that a high axial crystal field could lead to SMR via the second and third excited states [6]. Despite $Dy^{III}$ being a very promising ion to investigate its magnetic properties, the research diversity involving other $Ln^{III}$ ions is crucial for scientific advancement and the development of materials for various technological applications, such as molecular magnetic refrigeration, magnetic sensors, data storage materials or light emission conversion diodes [7–10].

Our work involving molecular self-assembly of aromatic polyoxalamide ligands and lanthanide(III) ions as Ln SIMs started some years ago through the study of mononuclear dysprosium(III) complex $Me_4N[Dy(Hdmpa)_4]\cdot 2CH_3CN$ (Hdmpa = *N*-2,6-(dimethylphenyl) oxamato; $Me_4N^+$ is the tetramethylammonium) [11]. This complex is eight-coordinate through four oxamate ligands, arising from a hydrogen-bonded supramolecular 2D network that exhibited field-induced SIM behavior below 9.0 K with $U_{eff}$ of ~69.1 $cm^{-1}$. Using the *N*-4-(hydroxyphenyl)oxamate (4-HOpa) ligand, a series of isostructural nonacoordinate complexes of formula $Na[Ln(4\text{-}HOpa)_4(H_2O)]\cdot 2H_2O$ [Ln = $Dy^{3+}$, $Gd^{3+}$, $Ho^{3+}$ and $Eu^{3+}$] with a hydrogen-bonded 2D supramolecular was described [12]. The dysprosium(III) and gadolinium(III) complexes exhibited SMR phenomena below 10 K. In principle, SMR behavior was not expected for the magnetically isotropic $Gd^{III}$ ion because it does not present SOC. Its ground state is $^8S_{7/2}$ with no angular momentum [13], so the energy barrier for spin reversal is very small because of the negligible value of the zero-field splitting component ($D$~0.1 $cm^{-1}$). Hence, to observe SMR in $Gd^{3+}$ ions, the metal environments should be distorted, leading to a specific anisotropy by mixing the low-lying states. This behavior can also occur with 3d-block ions like high-spin $Mn^{2+}$ ions in octahedral environments ($^6A_{1g}$), which shows a negligible $D$ value ($D$~0.1 $cm^{-1}$) [14]. Therefore, field-induced SMR behavior could appear in $Gd^{3+}$ or $Mn^{2+}$ complexes and is unrelated to a significant energy barrier. More recently, we demonstrated that by increasing the rigidity of the crystal structure, as in the 1D coordination polymer $\{Gd(Hphpa)_3(DMSO)_2\}_n \cdot nH_2O$ [Hphpa = *N*-(phenyl)oxamate (DMSO = dimethylsulfoxide)], field-induced SMR appeared below 33 K, the highest temperature range found for gadolinium(III) complex up to date [15].

The first examples of Ln SIMs with calixarene-like oxamate motifs were reported a few years ago. These compounds were synthesized using the *N*-(4-Xphenyl)oxamate ligands (H-4-Xphpa with X = Cl or F), affording isostructural compounds of formula $n$-$Bu_4N[Ln(4\text{-}Xphpa)_4(H_2O)] \cdot nH_2O$ ($n$-$Bu_4N^+$ = $n$-tetrabutylammonium with Ln = Eu, Gd, Dy, Tb, Nd, Ho, Y and La) [16,17]. In this series, the crystal structures remember a calixarene structure, but there is no connection between the phenyl rings, as seen in a classical calixarene. The connection is achieved by the $Ln^{III}$ ion through the coordination to the oxygen atoms from the all-*cis*-disposed ligands so that each aromatic ring is almost facing one another, remembering a calyx, with a *trans* apical solvent molecule completing the nine-coordinate sphere of the $Ln^{III}$ ion. In such complexes, only $Gd^{3+}$, $Dy^{3+}$, $Tb^{3+}$ and $Ho^{3+}$ compounds present field-induced SMR behavior, in which an energy barrier of 66 $cm^{-1}$ was observed for the $Dy^{3+}$ complex below 7.0 K.

Recently, the gadolinium complex $Na[Gd(4\text{-}HOpa)_4(H_2O)] \cdot 2H_2O$ was successfully tested as a spin qubit, so that is a suitable candidate for quantum technologies below 10 K [18]. Its diamagnetically diluted complex (1Gd0.12%) with yttrium shows relatively high spin-lattice and spin-relaxation times ($T_1$ = 1.66 ms at 4 K and $T_m$ = 4.25 μs at 8 K).

In this contribution, we report the synthesis, crystal structures and magnetic properties of the novel series of mononuclear lanthanide(III) complexes with the *N*-(2,4,6-trimethylphenyl)oxamate ligand (Htmpa) of general formula $n$-$Bu_4N[Ln(Htmpa)_4(H_2O)] \cdot xDMSO \cdot yH_2O$ [Ln = $Nd^{3+}$ (**1**), $Gd^{3+}$ (**2**), $Tb^{3+}$ (**3**) and $Dy^{3+}$ (**4**), with x = 0 and y = 4 (**1**), x = 2 and y = 2 (**2** and **4**), and x = 2 and y = 1 (**3**)] (Scheme 1). Complexes **1**–**3** behave as field-induced SIMs, while **4** exhibits SIM behavior without applying an external magnetic field.

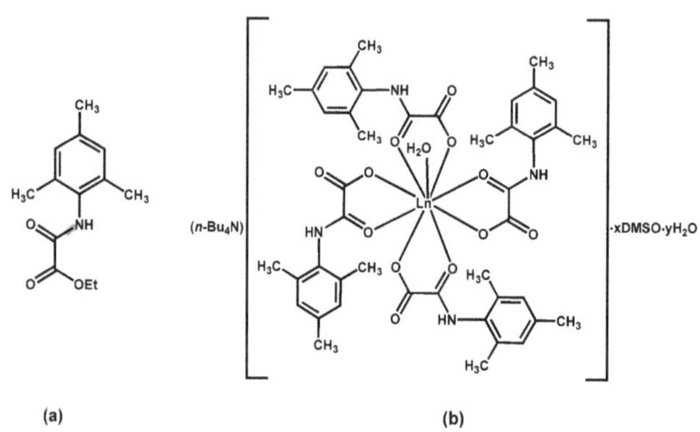


**Scheme 1.** (**a**) Chemical structure of the proligand *N*-(2,4,6-trimethylphenyl)oxamic acid ethyl ester (EtHtmpa) and (**b**) their mononuclear lanthanide(III) oxamate complexes.

## 2. Experimental

### 2.1. Materials and Methods

All chemicals and solvents used in this work were of analytical grade and used as received. The proligand EtHtmpa was obtained as previously reported [19,20]. Elemental analyses (CHN) were carried out with a Perkin-Elmer 2400 analyzer. IR spectra were recorded from KBr pellets on a PerkinElmer 882 spectrophotometer in the 4000 and 400 $cm^{-1}$ range. The thermal analysis (TG/DTA, thermogravimetric/differential thermal analysis) was conducted with a DTG-60H Shimadzu instrument using ~3.0 mg of each complex placed individually in an alumina crucible. The solid samples were heated at 10 °C $min^{-1}$ from room temperature to 700 °C under a synthetic air atmosphere (flow rate of 100 mL $min^{-1}$). X-ray powder diffraction patterns (PXRD) were taken on the Shimadzu XRD-7000 X-ray equipment, using $CuK_{\alpha}$ radiation (λ = 1.54056 Å), in the 4–40° range, with the goniometer speed equal to 0.5 $°min^{-1}$ and 1 s of sampling time. The simulated PXRD patterns were generated with Mercury software (version 2022.3.0), using the crystal data from the single-crystal X-ray diffraction cif files [21].

### 2.2. Magnetic Measurements

Variable-temperature (2–300 K) direct current (dc) magnetic susceptibility under applied fields of 0.20 kOe or 5.0 kOe were carried out using a Quantum Design SQUID magnetometer model MPMS-XL-5. Variable-field magnetization measurements were carried out using a Quantum Design SQUID magnetometer model MPMS-XL-5 at 2, 3, 4, 5, 7 and 10 K. Variable-temperature alternate current (ac) magnetic susceptibility measurements were carried out using a Model Quantum Design PPMS-9 operating with a small amplitude ac field (5 Oe) and scanning within the frequency range of 0.1–10 kHz at different applied dc magnetic fields for complexes **1–4**. The dc and ac measurements were carried out in polycrystalline samples of complexes **1–4**, previously placed in a gelatin capsule and with eicosane to prevent crystal movement. Corrections of diamagnetism of the samples and for the sample holder were also applied.

### 2.3. Crystal Data Collection and Refining

X-ray diffraction data collections on single crystals of compounds **1–4** were performed using a Rigaku Synergy diffractometer with a Mo-Kα (0.71073 Å) source at 293 K. The structures were solved using the SHELXS program (Version 2014/6) in Olex2 suit [22,23]. Data integration and scaling of the reflections for all compounds were performed with the CRYSALISPRO 43.90 suite [24]. Final unit cell parameters were based on the fitting of all reflection positions. Analytical absorption corrections and space group identification were performed using the CRYSALISPRO 43.90 suite [24]. Their structures were solved by direct methods using the SUPERFLIP 1.0 program [25]. The solvent mask tool from the Olex2 program was utilized to mask disordered solvents during the final refinement [23]. The corresponding electron counts removed were equivalent to 2.4 water molecules (24 electrons) for complex **1**, 1.1 DMSO molecules (46 electrons) for complex **2**, 0.6 water molecules (6 electrons) and 0.9 DMSO molecules (38 electrons) for complex **3** and 1.1 DMSO molecules (46 electrons) for complex **4**. For each compound, the positions of all atoms could be unambiguously assigned on consecutive difference Fourier maps. Refinements were performed using SHELXL based on F2 through the full-matrix least-square routine [26]. All non-hydrogen atoms were refined with anisotropic atomic displacement parameters. All hydrogen atoms were located in different maps and included as fixed contributions according to the riding model [27]. For the aromatic carbon atoms and amine groups C–H = 0.97 Å, N–H = 0.86 Å and $U_{iso}(H)$ = 1.2 $U_{eq}(C/N)$, for aliphatic carbon atoms C–H and N–H = 0.96 Å and $U_{iso}(H)$ = 1.5 Ueq(C/N) and finally for water molecules O–H = 1.00 Å and $U_{iso}(H)$ = 1.5 $U_{eq}(O)$. Molecular graphics were generated with MERCURY software (version 2022.3.0) [21]. Crystal data, experimental details and refinement results are summarized in Table 1.

**Table 1.** X-ray diffraction data collection and refinement parameters for complexes **1–4**.

| Compound | 1 | 2 | 3 | 4 |
|---|---|---|---|---|
| Formula | $C_{60}H_{90.8}N_5O_{15.4}Nd$ | $C_{66.2}H_{106.6}N_5O_{17}S_{3.1}Gd$ | $C_{65.8}H_{104.6}N_5O_{16.5}S_{2.9}Tb$ | $C_{66.2}H_{106.6}N_5O_{17.1}S_{3.1}Dy$ |
| MW/g mol$^{-1}$ | 1272.85 | 1502.79 | 1481.63 | 1508.04 |
| T/K | 293 | 293 | 293 | 293 |
| λ/Å | 0.71073 | 0.71073 | 0.71073 | 0.71073 |
| Crystal system | Monoclinic | Monoclinic | Monoclinic | Monoclinic |
| Space group | $P2/n$ | $P2_1/n$ | $P2_1/n$ | $P2_1/n$ |
| a/Å | 21.5741(6) | 15.4140(3) | 15.4180(5) | 15.4368(4) |
| b/Å | 11.9763(3) | 30.8190(8) | 30.8696(11) | 30.8542(12) |
| c/Å | 25.0931(7) | 16.1547(4) | 16.0716(5) | 16.1669(4) |
| β/° | 90.861(3) | 90.248(2) | 90.250(3) | 90.165(2) |
| V/Å$^3$ | 6482.8(3) | 7674.1(3) | 7649.2(4) | 7700.1(4) |
| Z | 4 | 4 | 4 | 4 |
| $\rho$/Mg m$^{-3}$ | 1.260 | 1.226 | 1.216 | 1.301 |
| μ/mm$^{-1}$ | 0.862 | 0.979 | 1.038 | 1.118 |
| F(000) | 2580.0 | 2972.0 | 2936.0 | 3165.0 |
| Reflections collected | 12,320 | 14,582 | 14,533 | 88,831 |
| R [a], wR [b] [I > 2σ(I)] | 0.0522;0.1223 | 0.0572, 0.1518 | 0.0626, 0.1616 | 0.0529, 0.1119 |
| R [a], wR [b] (all data) | 0.0570 | 0.0839 | 0.0874 | 0.0786 |
| S [c] | 1.053 | 1.164 | 1.088 | 1.124 |
| $\rho_{max}$ and $\rho_{min}$/e Å$^-$ | 0.9759/−1.3070 | 3.598/−1.417 | 3.153/−1.619 | 2.87/−1.08 |
| CCDC | 2,380,352 | 2,380,353 | 2,380,351 | 2,380,354 |

[a] $R = \Sigma \| |F_o| - |F_c| \| / \Sigma |F_o|$. [b] $wR = [\Sigma w(|F_o|^2 - |F_c|^2)^2 / \Sigma w |F_o|^2]^{1/2}$. [c] $S = [\Sigma w(|F_o|^2 - |F_c|^2)^2/(n_o - np)]^{1/2}$, where $w \propto 1/\sigma$, $n_o$ = observed and $n_p$ = fitted parameters.

### 2.4. General Procedure for Synthesis of Compounds n-$Bu_4N[Ln(Htmpa)_4(H_2O)]\cdot x$DMSO$\cdot yH_2O$ [Ln = $Nd^{3+}$ (1), $Gd^{3+}$ (2), $Tb^{3+}$ (3) and $Dy^{3+}$ (4)]

A mixture was prepared by combining deionized water (30 mL), the proligand EtHtmpa (480 mg, 2.04 mmol) and tetrabutylammonium hydroxide solution 40% w/v in water (1.33 mL, 2.04 mmol). This resultant mixture was maintained at 60 °C for 30 min under continuous stirring. After the hydrolysis, the mixture was cooled to room temperature, and dimethylsulfoxide (42 mL) was added. Subsequently, a solution containing the correspondent metal salt (0.51 mmol) previously dissolved in deionized water (6.0 mL) was added dropwise with constant agitation. The resulting mixture was heated to 60 °C for 15 min and then filtered, leaving the filtrate to rest at room temperature. After 5 days, suitable crystals for X-ray diffraction experiments were obtained. These crystals were filtered and air-dried at room temperature for 24 h. The correspondent salts employed in the synthesis were $Ln(CH_3COO)_3\cdot nH_2O$ [Ln = $Nd^{3+}$ (n = 1); $Gd^{3+}$ (n = 4), $Dy^{3+}$ (n = 3)] or $TbCl_3\cdot 6H_2O$.

*n*-$Bu_4N[Nd(Htmpa)_4(H_2O)]\cdot 4H_2O$ (**1**): yield 54% (362 mg; 0.28 mmol). Elemental analysis (calcd.) for $C_{60}H_{94}N_5O_{17}Nd$ (1301.67 g mol$^{-1}$): C 54.89 (55.36), H 6.95 (7.28), N 4.99 (5.38)%. IR (KBr, cm$^{-1}$): 3424 ($\nu_{O\text{-}H}$), 3342, 3250 ($\nu_{N\text{-}H}$), 2962, 2924 and 2874 ($\nu_{C\text{-}H}$), 1654 ($\nu_{C=O}$), 1528, 1484 ($\nu_{C=C}$), 1374 ($\delta_{N\text{-}H}$), 1232 ($\nu_{C\text{-}O}$), 850, 766 ($\gamma_{C\text{-}H}$) cm$^{-1}$.

*n*-$Bu_4N[Gd(Htmpa)_4(H_2O)]\cdot 3DMSO\cdot 2H_2O$ (**2**): yield 31% (222 mg; 0.15 mmol). Elemental analysis (calcd.) for $C_{66}H_{108}N_5O_{18}S_3Gd$ (1513.04 g mol$^{-1}$): C 52.32 (52.39), H 7.09 (7.20), N 4.82 (4.63)%. IR (KBr, cm$^{-1}$): 3422 ($\nu_{O\text{-}H}$), 3328, 3224 ($\nu_{N\text{-}H}$), 2966, 2918, 2876 ($\nu_{C\text{-}H}$), 1644 ($\nu_{C=O}$), 1534, 1486 ($\nu_{C=C}$), 1374 ($\delta_{N\text{-}H}$), 1240 ($\nu_{C\text{-}O}$), 1030 ($\nu_{S=O}$), 852, 766 ($\gamma_{C\text{-}H}$) cm$^{-1}$.

*n*-$Bu_4N[Tb(Htmpa)_4(H_2O)]\cdot 3DMSO\cdot 1H_2O$ (**3**): yield 64% (490 mg; 0.33 mmol). Elemental analysis (calcd.) for $C_{66}H_{106}N_5O_{17}S_3Tb$ (1496.70 g mol$^{-1}$): C 52.88 (52.97), H 6.94 (7.14), N 4.78 (4.68)%. IR (KBr, cm$^{-1}$): 3422 ($\nu_{O\text{-}H}$), 3328, 3224 ($\nu_{N\text{-}H}$), 2966, 2920, 2876 ($\nu_{C\text{-}H}$), 1644 ($\nu_{C=O}$), 1534, 1486 ($\nu_{C=C}$), 1376 ($\delta_{N\text{-}H}$), 1240 ($\nu_{C\text{-}O}$), 1028 ($\nu_{S=O}$), 852, 768 ($\gamma_{C\text{-}H}$) cm$^{-1}$.

*n*-$Bu_4N[Dy(Htmpa)_4(H_2O)]\cdot 3DMSO\cdot 2H_2O$ (**4**): yield 67% (490 mg; 0.32 mmol). Elemental analysis (calcd.) for $C_{66}H_{108}N_5O_{17}S_3Dy$ (1518.29 g mol$^{-1}$): C 52.93 (52.21), H 6.78 (7.17), N 4.78 (4.61)%. IR (KBr, cm$^{-1}$): 3422 ($\nu_{O\text{-}H}$), 3328, 3224 ($\nu_{N\text{-}H}$), 2966, 2918, 2878 ($\nu_{C\text{-}H}$), 1644 ($\nu_{C=O}$), 1534, 1486 ($\nu_{C=C}$), 1376 ($\delta_{N\text{-}H}$), 1240 ($\nu_{S=O}$), 1026 ($\nu_{S=O}$), 852, 768 ($\gamma_{C\text{-}H}$) cm$^{-1}$.

## 3. Results and Discussion

Complexes **1–4** were synthesized through the in situ hydrolysis of the ethyl ester with tetrabutylammonium hydroxide, as described for several oxamate complexes containing transition metal ions of the first series [28,29], followed by a stoichiometric reaction with the correspondent lanthanide(III) acetate salt in DMSO/water solution (Scheme 2). In this contribution, we demonstrate again that it is possible to obtain the lanthanide(III) oxamate compounds through one-pot synthesis without isolation of the *n*-$Bu_4N$(EtHmpa) salt [20]. Single crystals of the appropriate quality to solve the crystal structure using an X-ray diffraction experiment were obtained.

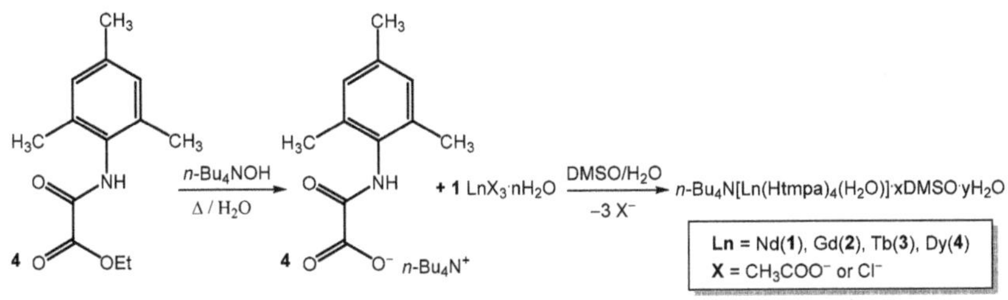


**Scheme 2.** Synthetic procedure for complexes **1–4**.

Elemental analysis, IR spectroscopy and thermogravimetric analysis confirm the chemical structure of all compounds. The IR spectra of complexes **1–4** (Figure S1) are very similar and exhibit the characteristic stretching vibrations corresponding to the presence of OH groups (3424–3422 cm$^{-1}$) and NH groups (3328 and 3224 cm$^{-1}$) [20]. Additionally, the downward shifting of the stretching vibrations of C=O groups (1654–1644 cm$^{-1}$) relative to the proligand EtHtmpa (1732 and 1680 cm$^{-1}$) is strong evidence of the ester hydrolysis and further coordination of these groups to metal ions.

Thermal analysis (TG/DTA) for complexes **1–4** (Figures S2–S5) agrees with the results proposed by elemental analysis and SCXRD data. For **1**, the first weight loss event (4.9%) occurs in the 38–140 °C range, corresponding to the release of approximately 3.6 water molecules (calcd.: 4.8%). Then, upon continuous heating until 250 °C, a small weight loss occurs (0.6%), which can be associated with 0.4 water molecules (calcd.: 0.4%). Complex **2** exhibits two weight losses from room temperature to 135 °C, running at two steps. The first one, in the range of 38–100 °C (6.7%), involves releasing one DMSO and one water molecule of crystallization (calc.: 6.3%). Then, upon heating, a second weight loss occurs until 149 °C (12.5%), associated with the loss of two DMSO molecules of crystallization and one water of coordination (calcd.: 12.6%). The resulting unsolvate species *n*-$Bu_4N[Gd(Htmpa)_4]$ is stable until 229 °C, when further heating provides the decomposition of the complex. The TG curves of **3** and **4** have a similar profile, with two events occurring from room temperature until 142 °C for **3** and 147 °C for **4**. From room temperature until 90 °C, a loss of 6.44% (calcd.: 6.38%) and 6.36% (calcd.: 6.40%) of the initial mass occurs for **3** and **4**, respectively. It can be associated with releasing one DMSO and one water crystallization molecule. Upon continuous heating, a second event occurs at 142 °C (**3**) and 146.5 °C (**4**), releasing 10.56% (**3**) and 11.69% (**4**) of the mass, which corroborates the removal of two DMSO crystallization molecules for **3** (calcd.: 10.37%) and two DMSO and one water crystallization molecule for **4** (calcd.: 11.60%).

The X-ray diffraction patterns of **2–4** are similar, differing from **1** (Figures S6–S9), confirming their isostructurality. Yet, all of them agree with the simulated X-ray diffraction patterns extracted from the correspondent retrieved from the cif files, suggesting that the bulks have the same structure as the single crystals, supporting the purity of the samples.

### 3.1. Crystal Structures

Complex **1** crystallizes in the $P2/n$ space group of the monoclinic system, with different unit cell dimensions for **1** compared to **2–4**, which in turn are then stated to be isostructural and crystallize in the $P2_1/n$ space group. In fact, **1** differs from **2–4** regarding the spatial organization of mononuclear units and the nature and number of crystallization solvent molecules. Hence, **1** presents four crystallization water molecules, while **2** and **4** present two water and three DMSO molecules. The thermal analysis of **3** revealed one water and three DMSO molecules of crystallization. There is some difference in water content found from crystal structure refinement and TG/elemental analysis, which can be assigned either to non-structural hydration or to smeared electron density corresponding to disordered structural water, which cannot be distinguished from residual electronic density in crystal structures bearing heavy metal atoms.

Due to the lanthanide contraction caused by the progressive electron occupation of the inner 4f orbitals, the radii of the lanthanides decrease slowly along the lanthanide series. Therefore, the ability to accommodate ligands around the $Ln^{3+}$ cation is similar for the series, consistent with the structural variations observed among the monomers in this series. Thus, due to the similarities between all three structures of **2–4**, we selected the structure of **3** for a detailed discussion when compared to that of **1**.

The structures of **1** and **3** comprise four all-*cis*-disposed Htmpa ligand units, one coordinated water molecule and a nine-coordinate metal center, forming a monoanionic complex. To neutralize the charge system, there is an *n*-tetrabutylammonium cation, in addition to four crystallization water molecules (**1**) or one water molecule and three crystallization DMSO molecules (**3**) (Figure 1). Selected bond lengths and angles for **1–4** are listed in Tables S1–S4.

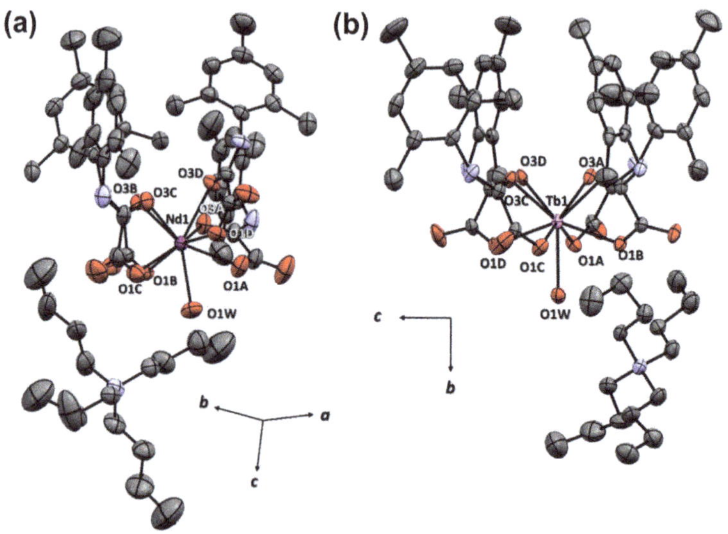


**Figure 1.** ORTEP view of the crystal structures of (**a**) **1** and (**b**) **3**. Crystallization water molecules and hydrogen atoms have been omitted for clarity, and ellipsoids represent 50% probability levels.

The coordination sphere of the nine-coordinate metal center in **1** and **3** consists of eight carbonyl oxygen atoms from the carboxylate amide groups of the four chelating oxamate ligands, as well as the O1W atom from the coordinated water molecule (Figure 2). According to the SHAPE software (version 2.0) [30], the coordination geometry of **2–4** corresponds to the CSAPR-9 (spherical capped square antiprism, $C_{4v}$), while **1** corresponds to the MFF-9 (Muffin, $C_s$) (Table S9). The values of the SHAPE parameters [S(MFF) = 1.640 (**1**), and S(CSAPR) = 0.379(**2**)/0.322(**3**)/0.383(**4**)] indicate that distortion concerning the ideal CSAPR-9 symmetry increases in the following order: **3** < **2** < **4**.

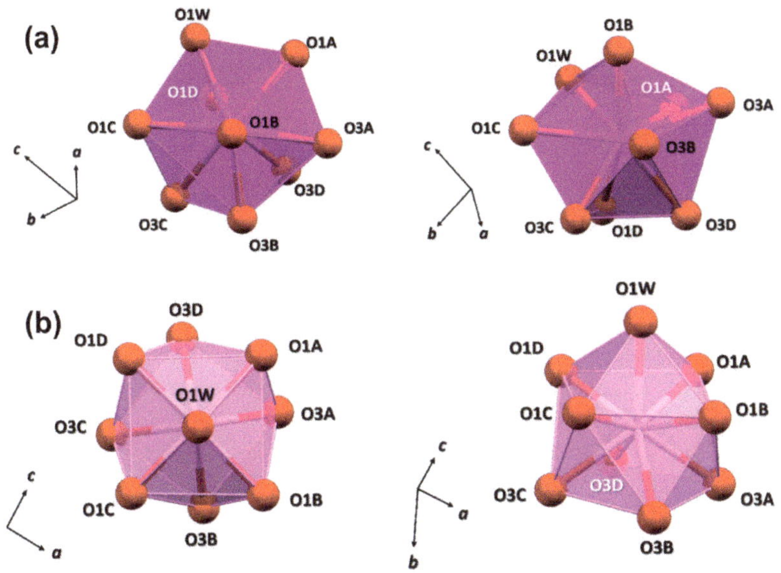


**Figure 2.** (**a**) Top and side views of the coordination polyhedron around the Nd1 metal center of **1** with the atom numbering scheme for the donor atoms, showing the distorted muffin geometry. The O1D/O3C/O3D atoms compose the triangular base, and the O1A/O1C/O3A/O3B/O1w atoms compose the pentagonal plane with the O1B atom in the capped position. (**b**) Top and side views of the coordination polyhedron around the Tb1 metal center of **3** with the atom numbering scheme for the donor atoms, showing the distorted capped square antiprism geometry. The O3A/O3B/O3C/O3D atoms compose the square base of the prism, and the O1A/O1B/O1C/O1D atoms compose the quadratic plane of the capped face with the O1W atom in the capped position.

The four units are pointed in the same direction in both **1** and **3**, forming a calixarene-like species around the metal center (Figure 1). In the crystal lattice of **3**, the adjacent mononuclear units interact with each other through self-complementary double hydrogen bonds between the N–H and C=O groups from each of the four Htmpa ligands leading to supramolecular layers growing within the *ac* (**3**) plane (Figure 3b). These layers are assembled glide-plane symmetry-related molecules, which are placed alternately pointing up and down into the hydrogen-bonded chains grown along the [101] and [−101] directions for the bi-dimensional layer. The shortest intermetallic distances between neighboring hydrogen-bonded units along the [101] and [−101] are 11.158 and 11.206 Å (**2**), 11.128 and 11.177 Å (**3**) and 11.179 and 11.211 Å (**4**). In **1**, the two-dimensional layer is also assembled, but it is moved onto the *ab* plane. The self-complementary double hydrogen bonds between the N–H and C=O groups also occur in this layer, but only along the [100] direction wherein inversion symmetry molecules are alternately placed as in **3**. The shortest two distinct intrachain intermetallic distances along [100] are 11.466 and 11.443 Å in **1**. However, different from **3**, there are no self-complementary double hydrogen bonds between the N–H and C=O groups in the other chain growing [010] direction responsible for mounting the dimensional layer. In this pattern, there are only N–H$\cdots$O=C hydrogen bonds between neighboring translation symmetry-related molecules, which, therefore, are not alternately placed but always pointing towards the same side. The shortest two distinct intrachain intermetallic distances along [010] are 12.665 and 11.976 Å in **1**.

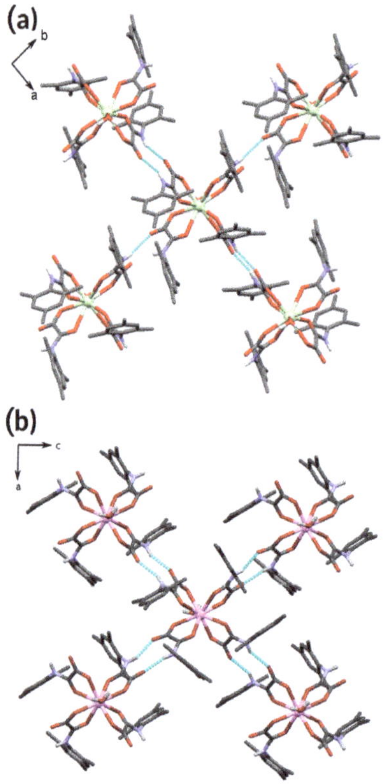


**Figure 3.** (**a**) View of the supramolecular layers of hydrogen-bonded mononuclear units of **1** along the crystallographic *ab* plane. (**b**) View of the supramolecular layers of hydrogen-bonded mononuclear units of **3** along the crystallographic *ac* plane, showing the presence of additional hydrogen-bonded DMSO molecules. Cyan dashed lines represent hydrogen bonds. CH hydrogen atoms, remaining solvent molecules and tetrabutylammonium cations were omitted for the sake of clarity.

These intralayer hydrogen-bonded supramolecular synthons are classified by Etter as $R_2^2$ (10) (Figure 3), except for that responsible for assembling the chain along the [010] in **1**, which is as defined as $C(8)$ [31]. The layers of **1** and **3** are packed differently, including the distinct disposition of tetrabutylammonium cations among them (Figure 4), which are involved in weaker non-classical hydrogen bonds and vdW interactions with the molecules of the complexes. In addition, there is a complex pattern of interlayer hydrogen bonding interactions between coordinated and crystallization DMSO and water molecules governing the 3D self-assembly of the system. Selected intra- and interlayer hydrogen bond lengths and angles for **1–4** are listed in Tables S5–S8.

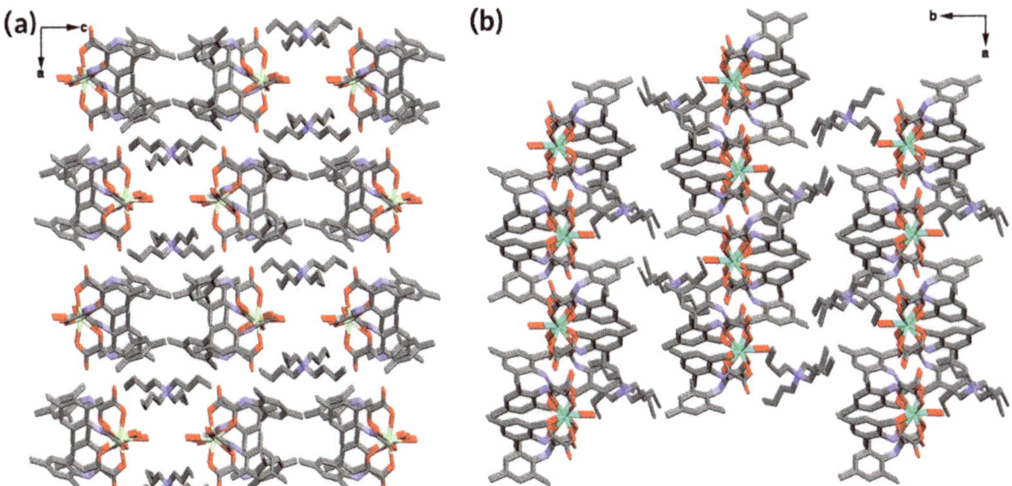


**Figure 4.** Views perpendicular to the layers growing onto the (**a**) *ab* plane in **1** and (**b**) *ac* plane in **3**. Three layers are depicted in each panel, while hydrogen atoms and solvent molecules were omitted for the sake of clarity. Color codes: carbon, gray; nitrogen, light blue; oxygen, red; Nd, light green; and terbium, green.

### 3.2. Static Magnetic Properties

Static direct current (dc) magnetic properties for **1–4** are presented in Figure 5 as the $\chi_M T$ vs. $T$ and $M$ vs. $H$ curves ($\chi_M$ and $M$ being the molar dc magnetic susceptibility and magnetization per mononuclear unit). The $M$ vs. $H/T$ curves are shown in Supplementary Materials (Figure S10). Table 2 summarizes all magnetic data for **1–4**.

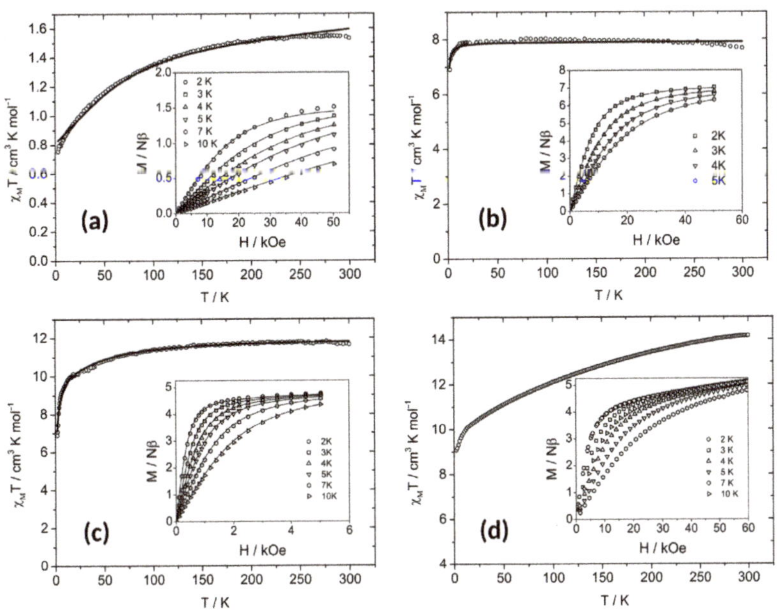


**Figure 5.** $\chi_M T$ vs. $T$ curves for **1–4** (**a–d**). Inset: $M$ vs. $H$ curves. Solid lines represent the best-fit curves according to the text.

**Table 2.** Experimental and calculated magnetic dc data for compounds **1–4**.

| Complex | Theory | | | | | $\chi_M T$ (cm$^3$ K mol$^{-1}$) Calculated | $\chi_M T$ (cm$^3$ K mol$^{-1}$) Experimental | Fitting Parameters | | | | $\Delta E$ [1] (cm$^{-1}$) |
|---|---|---|---|---|---|---|---|---|---|---|---|---|
| | $g$ | $S$ | $L$ | $J$ | Free Ion Term | | | $\lambda$ (cm$^{-1}$) | $\Delta$ (cm$^{-1}$) | $g$ | $zJ$ (cm$^{-1}$) | |
| **1** | 8/11 | 3/2 | 6 | 9/2 | $^4I_{9/2}$ | 1.64 | 1.53 | +258.94 | +55.45 | | | 1639 |
| **2** | 2 | 7/2 | 0 | 7/2 | $^8S_{7/2}$ | 7.88 | 7.20 | | | 2.0 | −0.01 | |
| **3** | 3/2 | 3 | 3 | 6 | $^7F_6$ | 11.82 | 11.65 | −387.29 | −48.64 | | −0.02 | 1936 |
| **4** | 4/3 | 5/2 | 5 | 15/2 | $^6H_{15/2}$ | 14.17 | 14.16 | | | | | |

[1] $\Delta E = -J|\lambda|$.

For the neodymium(III) complex **1**, a $\chi_M T$ value of 1.53 cm$^3$ K mol$^{-1}$ (Figure 5a) is observed at room temperature, which is slightly below the expected value for a $Nd^{3+}$ ion with a $^4I_{9/2}$ ground state, well-separated from the first $^4I_{11/2}$ excited state, in a free ion approximation (1.64 cm$^3$ K mol$^{-1}$, with $J = 9/2$ and $g_J = 8/11$). Cooling the sample, the $\chi_M T$ value decreases sharply, reaching 0.75 cm$^3$ K mol$^{-1}$ at 2.90 K. Under the action of LF, the $^4I_9/_2$ state splits into 5 Kramer doublets, and depending on the symmetry of this field, the $m_J = \pm 9/2$ or $\pm 1/2$ states could be the ground state. Hence, the maximum magnetization value of 1.51 Nβ at 2 K is smaller than expected for a decet ground state ($M_{sat} = g_J J N\beta = 3.27$ Nβ with $J = 9/2$ and $g_J = 8/11$) (inset of Figure 5a), being however close to that observed experimentally for other magnetically isolated $Nd^{3+}$ ions in Nd-oxamate compounds [17,32]. The reduced magnetization curves show an almost complete overlap of the curves (Figure S10a). Due to the symmetry of compound **1**, it was possible to use the Hamiltonian below (Equation (1)) to simulate the magnetic properties of this compound [33]:

$$H = -\lambda LS + \Delta\left[L_z^2 - L^2\right] + (-\kappa L + 2S)\beta H \tag{1}$$

where $\lambda$ is the spin-orbit coupling parameter; $\Delta$, energy gap between the $M_L$ values; $\kappa$ = orbital reduction parameter ($\kappa = 1$ was fixed); $\beta$ is Bohr magneton; $L$, $S$ and $H$ represent the angular operator, the spin operator and the magnetic field vector, respectively. Simultaneous fitting of $\chi_M T$ vs. $T$ and $M$ vs. $H$ curves using PHI software (version 3.1.6) [34] yielded the following values: $\lambda = 258.94$ cm$^{-1}$ and $\Delta = +55.45$ cm$^{-1}$ (Table 2). The calculated curves match the experimental data over the temperature range (solid lines in Figure 5a). The $\lambda$, $\Delta$ and $\Delta E$ (the separation between the ground and the first excited states) values are within the range observed for similar $Nd^{3+}$ compounds (see Table 2) [33,35]. Simulations with negative axial distortion values were performed, showing a very slight decrease in the $\chi_M T$ product with cooling, which is inconsistent with the fact that compound **1** shows an $\chi_M T$ product at low temperatures close to 0.8 cm$^3$ K mol$^{-1}$, a value expected for a ground state with $m_J = \pm 1/2$ [35].

For isotropic gadolinium(III) complex **2**, the $\chi_M T$ vs. $T$ curve follows a Curie law over an extensive temperature range, with a $\chi_M T$ value at room temperature of 7.67 cm$^3$ K mol$^{-1}$, slightly lower than the expected value for a $Gd^{3+}$ ion with a $^8S_{7/2}$ ground state (7.88 cm$^3$ K mol$^{-1}$ with $S = 7/2$ and $g = 2$) (Figure 5b). Upon cooling, the $\chi_M T$ product remains constant until approximately 13 K, after which a decrease is observed, reaching a value of 6.32 cm$^3$ K mol$^{-1}$ at 2.0 K. The slight decrease in the $\chi_M T$ at low temperatures may occur due to weak intermolecular antiferromagnetic interactions. The reduced magnetization curves show an almost complete overlap (inset of Figure S10b). A saturation effect is observed with a magnetization value of 7.05 Nβ at 50 kOe, close to the expected value for isotropic $Gd^{3+}$ ions ($M_{sat} = 7$ Nβ with $S = 7/2$ and $g = 2$). Using the PHI software, simultaneous fitting of the $\chi_M T$ vs. $T$ and $M$ vs. $H$ was performed through a mean-field model considering an effective intermolecular interaction ($zJ$) between the $Gd^{3+}$ ions. The calculated curves match the experimental data over the temperature range and, particularly, the decrease in $\chi_M T$ at low temperatures (solid lines in Figure 5a). The calculated value of $g = 2.0$ agrees with the expected value for this isotropic ion ($g = 2.0$). The calculated $zJ$ value of −0.01 cm$^{-1}$ is consistent with a small but non-negligible interaction, likely a dipolar interaction.

For terbium(III) complex **3**, $\chi_M T$ is 11.65 $cm^3$ K $mol^{-1}$ at room temperature and gradually decreases until around 50 K, when the decrease becomes steeper, reaching 6.89 $cm^3$ K $mol^{-1}$ at 2.0 K (Figure 5c). The expected $\chi_M T$ value for $Tb^{3+}$ ions with a $^7F_6$ ground state is well-separated from the first $^7F_5$ excited state in a free-ion approximation is 11.88 $cm^3$ K $mol^{-1}$ ($S = 3$, $J = 6$ and $g_J = 3/2$). The decrease in $\chi_M T$ upon cooling from room temperature is due to the depopulation of higher energy $m_J$ states, resulting from the LF splitting. Hence, the reduced magnetization curves do not overlap (inset of Figure S10) but show apparent saturation with a maximum value of 4.76 Nβ at 2.0 K and 5.0 T. This value is smaller than expected for a tredecet ground state ($M_{sat}$ = 9 Nβ with $J = 6$ and $g_J = 3/2$) (inset of Figure 5c). The experimental magnetic data were fitted using the previously presented Hamiltonian (Equation (1)). This fit showed good agreement between the experimental and theoretical data. The calculated $\lambda$ value of $-387.29$ $cm^{-1}$ is consistent with related terbium(III) compounds [33]. The calculated $\Delta$ parameter of $-48.64$ $cm^{-1}$ is within the range observed in similar terbium(III) compounds and has a negative sign, confirming an $m_J = \pm 6$ (ground state). To improve the fit at low temperatures, a $zJ$ component was added, yielding a value of $-0.023$ $cm^{-1}$, like that observed for **2**.

For dysprosium(III) complex **4**, a decrease in the $\chi_M T$ is 14.16 $cm^3$ K $mol^{-1}$ at room temperature. According to the free ion approximation model, this value is very close to the expected one for a $Dy^{3+}$ ion with a $^6H_{15/2}$ ground state, well-separated from the first $^6H_{13/2}$ excited state (14.17 $cm^3$ K $mol^{-1}$ with $J = 15/2$ and $g_J = 4/3$). As the sample is cooled, a gradual decrease in the $\chi_M T$ is observed (Figure 5d), reaching 10.00 $cm^3$ K $mol^{-1}$ at approximately 13 K. This smooth decrease in the $\chi_M T$ product is characteristic of the depopulation of higher energy $m_J$ states resulting from the LF splitting. Below 10 K, a more pronounced decrease is observed; however, in this case, this effect may be related to weak intermolecular antiferromagnetic interactions, already observed and quantified in compounds **2** and **3**. The reduced magnetization curves do not overlap (Figure S10), and no saturation effects are observed, even at the lowest temperature and highest field. The maximum $M$ value of 5.3 Nβ at 2.0 K and 7.0 T is smaller than expected for a hexadecet ground state ($M_{sat}$ = 10 Nβ with $J = 15/2$ and $g_J = 4/3$).

Based on the crystal structure and the formal charge of the atoms in the structure, it is possible to determine the easy magnetization axis in compounds containing dysprosium(III) ions and its energy barrier through the electrostatic model (Table S10) [36]. In the case of **4**, this axis lies in the *xz* plane (crystallographic *ac* plane), intersecting these axes at an angle close to 45 degrees (Figure 6a). By observing the crystal packing of **4**, it can be noted that there is an alternation in the directions of the anisotropy axis, which justifies the presence of antiferromagnetic interactions in this compound and possibly in the others presented in this contribution (Figure 6b). However, despite the simplicity of providing the easy axis of magnetization using the Magellan software [36], high-level quantum chemical calculations such as Complete Active Space Self-Consistent Field (CASSCF calculations) could give further insights into the effects of crystal field, the nature of the Dy–O bond, and the covalent factors influencing the magnetic properties of these compounds.

### 3.3. Dynamic Magnetic Properties

Dynamic alternating current (ac) magnetic properties for **1–4** are presented in Figures 7–10, respectively, in the form of the $\chi_M'$ and $\chi_M''$ vs. $\nu$ curves ($\chi_M'$ and $\chi_M''$ being the in-phase and out-of-phase ac magnetic susceptibilities per mononuclear unit, and $\nu$ is the frequency). Table 3 summarizes all magnetic ac data for **1–4** obtained using CC-FIT2 software v5.8.0 [37,38].

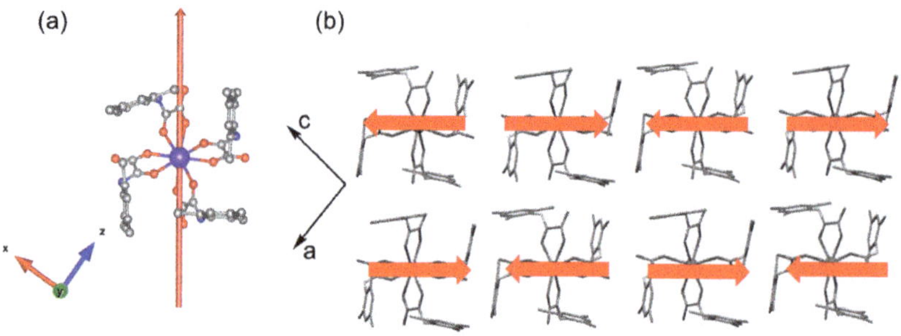


**Figure 6.** (**a**) Representation of the easy magnetization axis of dysprosium(III) complex **4** as a red arrow in each single-ion magnet (color code: Dy in violet, N in blue, oxygen in red, carbon in grey). (**b**) The easy magnetization axis is represented by red arrows in the crystal packing of **4** along the crystallographic *ac* direction. For clarity, the hydrogen atoms, solvent molecules and counterions were omitted.

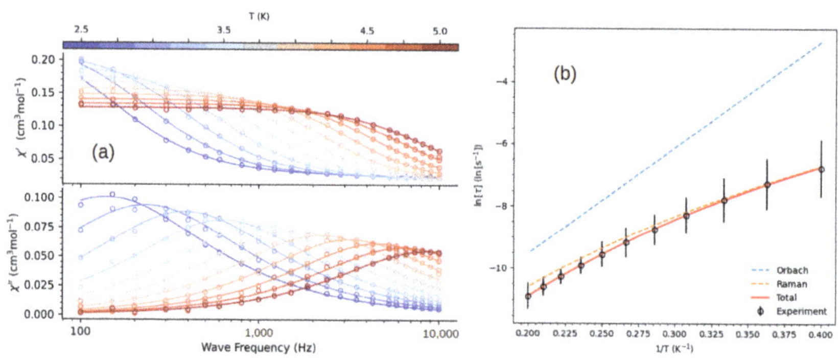


**Figure 7.** (**a**) $\chi_M{}'$ and $\chi_M{}''$ vs. $\nu$ curves of **1** under 1.0 kOe dc magnetic field. (**b**) Arrhenius plot of **1** under 1.0 kOe dc magnetic field. Solid lines represent fits to the data (see text).

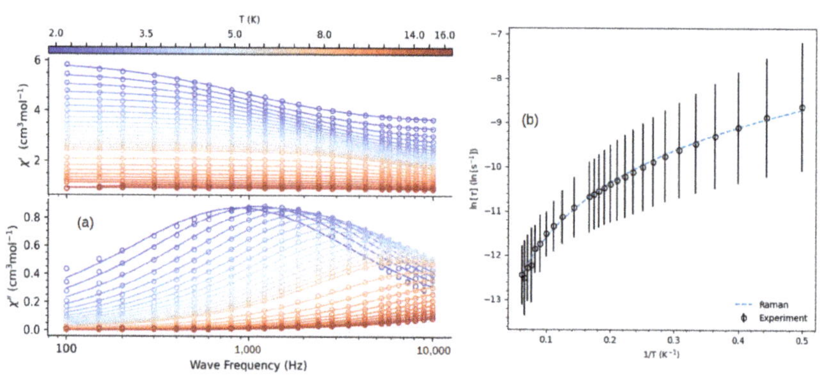


**Figure 8.** (**a**) $\chi_M{}'$ and $\chi_M{}''$ vs. $\nu$ curves of **2** under 1.0 kOe dc magnetic field. (**b**) Arrhenius plot of **2** under a 1.0 kOe dc magnetic field. Solid lines represent fits to the data (see text).

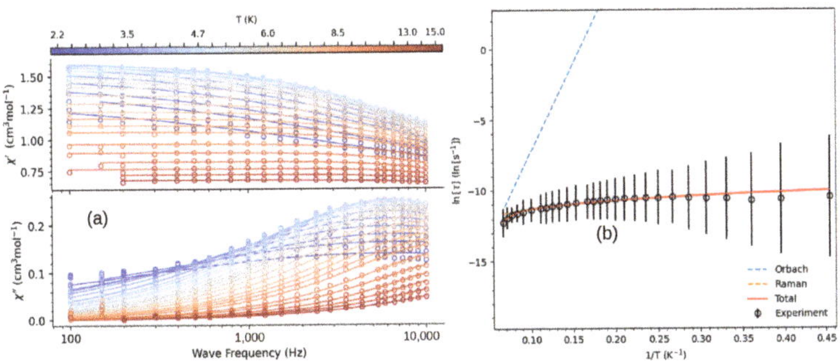


**Figure 9.** (**a**) $\chi_M{}'$ and $\chi_M{}''$ vs. $\nu$ curves of **3** under 5.0 kOe dc magnetic field. (**b**) Arrhenius plot of **3** under 5.0 kOe dc magnetic field. Solid lines represent fits to the data (see text).

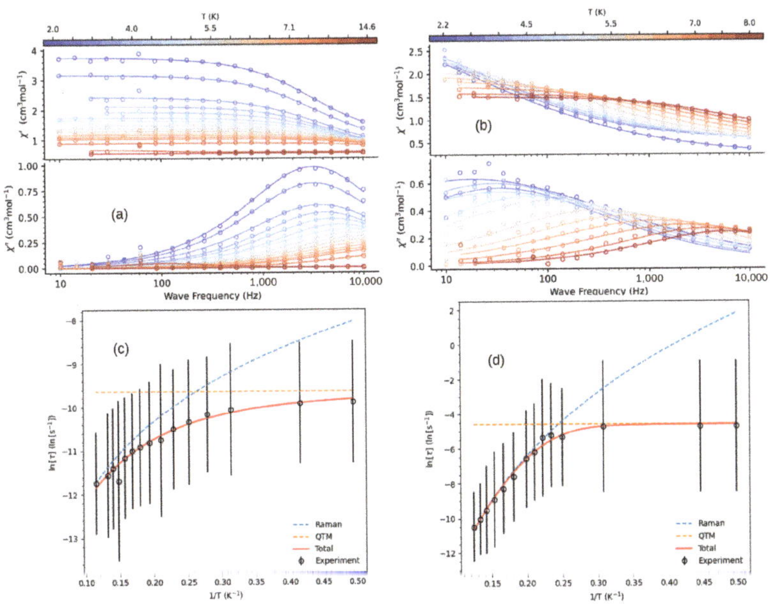


**Figure 10.** $\chi_M{}'$ and $\chi_M{}''$ vs. $\nu$ curves of compound **4** under (**a**) zero dc magnetic field and (**b**) 1.0 kOe dc magnetic field. Arrhenius plots of **4** (**c**) under zero dc field and (**d**) 1.0 kOe dc field. Solid lines represent fits to the data (see text).

Under zero dc field, frequency-dependent $\chi_M{}'$ and $\chi_M{}''$ signals were observed in **4** with $\chi_M{}''$ maxima at temperatures ranging from 2 to 6 K, indicating the occurrence of slow magnetic relaxation, as shown in Figure 10a. Upon applying a static dc field of 1.0 kOe, the temperature at which $\chi_M{}''$ appears shifts to 8 K, as illustrated in Figure 10b. The $\chi_M{}'$ and $\chi_M{}''$ vs. $\nu$ curves were fitted to the generalized Debye model [39] to extract temperature-dependent relaxation times ($\tau$). The resulting $\tau$ versus $T^{-1}$ (so-called Arrhenius plots) are shown in Figure 10c for a zero-dc field and Figure 10d for a dc field of 1.0 kOe. All parameters determined from these fitting processes, such as the isothermal and thermal susceptibility, $\chi_S$ and $\chi_T$, relaxation time and its distribution, $\tau$ and $\alpha$, are summarized in Tables S11 and S12. The distribution of relaxation times is relatively narrow, with $\alpha$ close to 0.2 at zero dc field (Table S11). For a dc field of 1.0 kOe, $\alpha$ approaches the value of 0.5 at a low temperature (Table S12).

**Table 3.** Selected ac magneto-structural data for some mononuclear oxamate-containing Ln-SMMs and Ln-SIMs.

| Compound | Ln···Ln [a] (Å) | Ln | CN [b] | Sym. [c] | $H_{DC}$ (kOe) | $U_{eff}/k_B$ (K) | $\tau_0$ (s) | $\tau_{QTM}$ (s) | $C$ ($s^{-1}$ $K^{-1}$) | $n$ | $A$ ($s^{-1}$ $K^{-n}$) | Process | Ref. |
|---|---|---|---|---|---|---|---|---|---|---|---|---|---|
| **1** | 11.466 | Nd | 9 | $C_s$ | 1.0 | 34(2) | $8.0(9) \times 10^{-8}$ | | 4.87(3) | 5.6(2) | | Orbach + Raman | This work |
| **2** | 11.158 | Gd | 9 | $C_{4v}$ | 1.0 | | | | 1746(12) | 1.80(2) | | Bottleneck | This work |
| **3** | 11.128 | Tb | 9 | $C_{4v}$ | 5.0 | 21(5) × 10 | $1.2(5) \times 10^{-11}$ | | | | 889(3) × 10 | Orbach + Raman + Direct | This work |
| **4** | 11.179 | Dy | 9 | $C_{4v}$ | 0 | | | $6.6(2) \times 10^{-5}$ | 520(32) | 2.5(3) | | Raman + QTM | This work |
| | | | | | 1.0 | | | $9.06(7) \times 10^{-3}$ | $3.2(3) \times 10^{-4}$ | 9.0(3) | | Raman + QTM | This work |
| $n$-$Bu_4N[Nd(L1)_4(DMSO)]_2 \cdot 3H_2O$ | 11.169 | Nd | 9 | $D_{3h}$ | 1.0 | 16.7(5) [d] | $1(3) \times 10^{-4}$ | | 0.005(4) | 4.7(2) | | Orbach + Raman | [19] |
| $n$-$Bu_4N[Nd(L2)_4(DMSO)]_2 \cdot 2H_2O$ | 11.228 | Nd | 9 | $D_{3h}$ | 1.0 | 4.9(1) [d] | $9.5(7) \times 10^{-4}$ | | 0.03(4) | 11(1) | | Orbach + Raman | [19] |
| $[Nd(L3)_3(DMSO)_2]_n$ | 8.8709(3) | Nd | 9 | $C_{4v}$ | 1.0 | 42.7(2) | $2.3(1) \times 10^{-8}$ | | 26.7(1) | 3.9(1) | | Orbach + Raman | [34] |
| $Na[Gd(L4)_4(H_2O)]_2 \cdot 2H_2O$ | 7.3014(7) | Gd | 9 | $C_{4v}$ | 1.0 | 13(1) | $6.1 \times 10^{-6}$ | | $2(1) \times 10^{-5}$ | 3 | 2.9(1) | Direct + Raman | [14] |
| $n$-$Bu_4N[Gd(L1)_4(DMSO)]_2 \cdot 2H_2O$ | 11.0272(6) | Gd | 9 | $D_{3h}$ | 1.0 | | | | $4.5(5) \times 10^{3}$ | 1.6(1) | | Bottleneck | [18] |
| $n$-$Bu_4N[Gd(L2)_4(DMSO)]_2 \cdot 3H_2O$ | 10.6005(7) | Gd | 9 | $D_{3h}$ | 1.0 | | | | $1.1(1) \times 10^{3}$ | 1.8(1) | | Bottleneck | [18] |
| $\{Gd_2(L5)_3(H_2O)_5 \cdot H_2O\}_n$ | 10.163(3) | Gd | 8 | $T_d$ | 1.0 | 3.27 [d] | $5.4 \times 10^{-6}$ | | | | | Orbach | [35] |
| $[Gd(L6)_3(DMSO)_2]_n \cdot nH_2O$ | 6.702 | Gd | 9 | $C_{4v}$ | 1.0 | | | | | 1.9(2) | 125.89 | Bottleneck | [17] |
| | 6.702 | Gd | 9 | $C_{4v}$ | 2.5 | | | | | 2.1(1) | 39.81 | Bottleneck | [17] |
| $Me_4N[Dy(L7)_4]_2CH_3CN$ | 9.032(1) | Dy | 8 | $D_{2d}$ | 1.0 | 93.66 [d] | $2.88(8) \times 10^{-10}$ | | | | | Orbach | [13] |
| $Na[Dy(L4)_4(H_2O)]_2 \cdot 2H_2O$ | 7.2881(4) | Dy | 9 | $C_{4v}$ | 1.0 | 16(1) | $1.4(9) \times 10^{-8}$ | | 2(5) | 9 | | Orbach + Raman | [14] |
| $n$-$Bu_4N[Tb(L1)_4(DMSO)]_2 \cdot 2H_2O$ | 10.9998(5) | Tb | 9 | $D_{3h}$ | 2.0 | | | $2.04 \times 10^{-4}$ | 20(6) | 3.5(2) | | Raman + QTM | [18] |
| $n$-$Bu_4N[Tb(L2)_4(DMSO)]_2 \cdot 3H_2O$ | 10.6121(2) | Tb | 9 | $D_{3h}$ | 1.0 | | | | 6.6(9) | 3.8(2) | $13.7(3) \times 10^{1}$ | Raman + direct | [18] |

Ligand abbreviations: L1 = *N*-4-(chorophenyl)oxamate; L2 = *N*-4-(fluorophenyl)oxamate; L3 = *N*-(4-methylphenyl)oxamate; L4 = *N*-4-(hydroxyphenyl)oxamate; L5 = *N*-(4-carboxyphenyl)oxamic acid; L6 = *N*-(phenyl)oxamate; L7 = *N*-(2,6-dimethylphenyl)oxamic acid. [a] Shortest intermolecular Ln···Ln distance. [b] Coordination number of the Ln ion. [c] Symmetry around the Ln ion according to the continuous shape measurements [30]. [d] Original values were converted from $cm^{-1}$ to K for comparative purposes.

Relaxation processes of Dy-based SIMs have been analyzed, considering a combination of Orbach relaxation process at higher temperatures, Raman at intermediate temperatures and quantum tunneling of magnetization (QTM) at low temperatures [5]. Consequently, the data obtained for **4** were analyzed using Equation (2) below.

$$\tau^{-1} = CT^{n} + \tau_0^{-1}\exp(-U_{eff}/k_BT) + \tau_{QTM}^{-1} \quad (2)$$

The first term of the right side of Equation (2) represents the Raman process, the second term represents the Orbach process and the third term represents the QTM relaxation processes. A combination of Raman and QTM processes better described the data. For the zero dc field, resulting in the best-fit parameters as C = 520(32) $s^{-1}$ $K^{-n}$, $n$ = 2.5(3) and $\tau_{QTM}$ = 6.6(2) × $10^{-5}$ s. Under a 1.0 kOe dc applied field, the best-fit parameters were $C$ = 3.2(3) × $10^{-4}$ $s^{-1}$ $K^{-n}$, $n$ = 9.0(3) and $\tau_{QTM}$ = 9.8(1) × $10^{-3}$ s. These parameter values are consistent with the energy barrier observed for $Dy^{3+}$ ions with the same coordination number [40].

Unlike **4**, compounds **1**–**3** show no $\chi_M''$ signal under zero dc magnetic field above 2.0 K, indicating fast relaxation processes through QTM. Applying a static dc field, however, induces slow magnetic relaxation in **1**–**3**, as shown in Figures 7a, 8a and 9a. Relaxation times for **1**–**3** were determined by fitting the $\chi_M'$ and $\chi_M''$ vs. $\nu$ curves to the generalized Debye model [39]. The resulting Arrhenius plots are shown in Figures 7b, 8b and 9b, respectively. The determined values of $\chi_S$ and $\chi_T$, $\tau$ and $\alpha$ for **1**–**3** are summarized in Tables S13–S16.

For **1**, the relaxation time was accurately determined within the 2–5 K temperature range. The distribution of relaxation times was narrow, with $\alpha$ close to 0.1 (Table S12). The Arrhenius plot was fitted considering Raman and Orbach processes, as reported previously [41,42]. The best fit parameters are $U_{eff}$ = 34(2) K, $\tau_0$ = 8.0(9) × $10^{-8}$ s, C = 4.87(3) $s^{-1}$ $K^{-n}$, $n$ = 5.6(2). These results are consistent with values reported by other authors for Nd-based molecular material [41,42]. The direct process ($\tau^{-1} = A\,H^{n1}T$) was ruled out for this complex as the field-dependent results deviated significantly from the typically exponential values ($n_1$ = 2–4) described by other researchers, as shown in Figure S11 [43].

For **2**, under an applied field of 1.0 kOe, the SMR was observed within the temperature range of 2–16 K, as shown in Figure 8a. The distribution of relaxation times is narrow, reaching a maximum value at 2 K with $\alpha$ close to 0.2 (see Table S14). Therefore, the Arrhenius plot of **2** (Figure 8b) was fitted by considering the Raman process, resulting in the following values: $C$ = 1746(12) $s^{-1}$ $K^{-n}$ and $n$ = 1.80(2). These values suggest that the magnetic relaxation mechanism in the $Gd^{3+}$ compound is associated with the bottleneck effect, as expected for $n$ values near 2 [12–16]. This is consistent with findings reported by other authors for field-induced Gd-based SIMs [12,13,15,16,44].

Although compound **3** showed frequency dependence in out-of-phase magnetic susceptibility under a 1.0 kOe magnetic field (Figure S12, Table S15), peaks in the $\chi_M''$ vs. $\nu$ curves were only observed under a 5.0 kOe dc magnetic field, as shown in Figure 9a. The distribution of relaxation times is broad at low temperatures ($\alpha$ value close to 0.6 at 2 K), suggesting multiple relaxation processes occurring within this temperature range. Above 5.0 K, the distribution is relatively narrow, where $\alpha$ begins to decrease from a value close to 0.2, as shown in Table S16. The Arrhenius plot of **3** (Figure 9b) was fitted considering the Orbach, Raman, Direct and QTM processes [12,39]. The best fit was obtained considering the Orbach and Direct processes, resulting in $U_{eff}$ = 21(5) × 10 K, $\tau_0$ = 1.2(5) × $10^{-11}$ s, $A$ = 889(3) × 10 $s^{-1}$ $K^{-1}$. These values are consistent with findings reported by other authors for field-induced slow magnetic relaxation for Tb-based SIMs [45].

Table 3 compares the results of mononuclear complexes and coordination polymers based on oxamate containing Kramer $Ln^{III}$ ions (Ln = Nd, Dy, and Gd) and non-Kramer $Ln^{III}$ ions (Tb) in low coordinate that exhibited SMR phenomena. Additionally, it is worth emphasizing that an ideal point group symmetry such as $C_{\infty v}$, $D_{4d}$, $S_8$, $D_{5h}$, $D_{6d}$ and $D_{\infty h}$ could suppress the unwanted QTM process [5]. Once these precise geometries are unlikely to be realized in molecular complexes in general, taking into account the oxamate derivatives (Table 3). The suppression of QTM phenomena can be achieved by diluting the paramagnetic complexes within an isostructural diamagnetic matrix, which can reduce the dipolar interactions and/or by applying an external static dc magnetic field. Only **4** exhibits the SMR at zero dc field, while all other oxamate compounds synthesized up to date (Table 3) behave as field-induced SMMs. In the case of **4**, the Raman and QTM mechanisms are dominant. Dysprosium(III) ions are anisotropic; thus, the observation of the SRM phenomena is expected for Dy-complexes. The fitting results show that the QTM occurs for $Dy^{3+}$ complexes in some extensions, even with an application of an $H_{dc}$ of 1.0 kOe, which is not strong enough to suppress the QTM. Additionally, the extension of the QTM in this complex does not preclude the observation of SMR phenomena in **4**. The energy of the barrier for spin reversal was obtained in the Magellan software (Table S10). The minimum energy of the crystalline structure was calculated as $-187$ cm$^{-1}$, while the minimum energy barrier for the spin reversal was determined as +22.9 cm$^{-1}$. This value is very close to the experimental one determined through the fits of magnetic data (+23.63 cm$^{-1}$ or 34 K, see Table 3), indicating that the model employed describes the energy values with reasonable precision [36].

Neodymium(III)–oxamate compounds exhibited the Orbach and Raman mechanism that accounts for the SMR phenomena, as observed for all Nd-complexes of Table 3, even with different local symmetries around the metal ion and ligands containing donor [*N*-4-(methyl)oxamate] or electron-withdrawing groups [*N*-4-(chlorophenyl)oxamate and (*N*-4-(fluorophenyl)oxamate)]. Comparing the value of the energy barrier of **1** and the other compounds of Table 3, the energy barrier values are higher for complexes containing only oxamate-donor ligands. Comparing the energy barrier of **1** and **4** with $H_{dc} \neq 0$, due to the higher anisotropy and $m_J$ states of **4**, a higher energy barrier for **4** is achieved.

In contrast to $Nd^{3+}$, $Tb^{3+}$ and $Dy^{3+}$ ions, $Gd^{3+}$ ions are isotropic, and the SMR is not expected for such complexes. However, a very low magnetic anisotropy with a low value of zero-field splitting ($D$~0.1 cm$^{-1}$) can be found experimentally for gadolinium(III) compounds. With the application of a magnetic field, the asymmetric surrounding of the gadolinium(III) ions can increase the $D$ value parameter, which can cause a split in the ground state doublet ($m_s = \pm 7/2$) and, consequently, the admixture with the ground and some excited states. Consequently, the SMR cannot be explained in many cases by considering the conventional over-barrier mechanisms or suppression of tunneling, which occurs with anisotropic systems. When the SMR phenomena are observed in $Gd^{3+}$ complexes, the interaction between the magnetic ion and its environment should be considered. Focusing on the case of nine-coordinate gadolinium(III) ions, the phonon bottleneck mechanism is predominant, observed for **2** and the other three compounds from Table 3, with $D_{3h}$ and $C_4$ symmetries. The phonon bottleneck effect occurs when the dc field is applied, suppressing QTM relaxation. It arises when the energy from lattice modes generated by relaxing spins, transitioning from excited to ground states, cannot be dissipated into the thermal reservoir at a sufficiently fast rate. However, for $\{Gd_2(L5)_3(H_2O)_5 \cdot H_2O\}_n$ (L5 = *N*-(4-carboxyphenyl)oxamic acid), the mechanism involved is Orbach, while for $Na[Gd(L4)_4(H_2O)]_2 \cdot 2H_2O$ (L4 = *N*-4-(hydroxyphenyl)oxamate) it is attributable to direct and Raman processes.

In the case of terbium(III) complex **3**, a non-Kramer ion, the SMR is not expected without an external magnetic field [5]. Moreover, the application of a small external magnetic field can induce SMR; therefore, most terbium(III) complexes exhibit behavior characteristic of field-induced single-molecule magnets (SMMs) or field-induced single-ion magnets (SIMs). Besides the external field, an ideal molecular symmetry causing appropriate local symmetry around the metal centre can also cause SMR in mononuclear Tb compounds. In the case of **4**, the fit of the Arrhenius plot reveals that the magnetic dynamics follow the power law corresponding to Orbach and direct processes. Comparing this compound with other Tb–oxamate complexes, it is noticed that **3** has the highest energy barrier. Probably, this result is associated with the higher value of the magnetic field to promote the SMR.

## 4. Conclusions

In summary, we synthesized four lanthanide(III) complexes with the ligand *N*-(2,4,6-trimethylphenyl)oxamate featuring a calixarene-like structure. The gadolinium(III), terbium(III) and dysprosium(III) complexes are nine-coordinated and isostructural, with CSAPR-9 symmetry, while the neodymium(III) complex, with a nine-coordinate (MFF-9) geometry, is not. The dysprosium(III) complex **4** is the first example of an oxamate-based Ln-SIM without applying a dc field. The neodymium(III), gadolinium(III) and terbium(III) complexes behave instead as field-induced Ln-SIMs. These results suggest that by improving the geometry around the metal center and using oblate ions, it would be possible to improve the performance of oxamate-based Ln SIMs, reaching higher temperatures and higher energy barriers for spin reversal.

**Supplementary Materials:** The following supporting information can be downloaded at: https://www.mdpi.com/article/10.3390/magnetochemistry10120103/s1, Figure S1: FTIR spectra of **1–4**; Figure S2: TG (solid line) and DTA (dashed line) curves for **1** in synthetic air atmosphere; Figure S3: TG (solid line) and DTA (dashed line) curves for **2** in synthetic air atmosphere; Figure S4: TG (solid line) and DTA (dashed line) curves for **3** in synthetic air atmosphere; Figure S5: TG (solid line) and DTA (dashed line) curves for **3** in synthetic air atmosphere; Figure S6: Experimental and simulated PXRD patterns of **1**; Figure S7: Experimental and simulated PXRD patterns of **2**; Figure S8: Experimental and simulated PXRD patterns of **3**; Figure S9: Experimental and simulated PXRD patterns of **4**; Figure S10: Reduced magnetization curves of **1–4** ($H_{dc}$ = 0.2 kOe). Figure S11: Field dependence of the direct process coefficient for **1**. (a) Fitted considering Orbach + Direct processes, and (b) Raman + Direct processes. Figure S12: In-phase and out-of-phase magnetic susceptibilities of **3** were obtained for a 1.0 kOe dc magnetic field. Table S1: Selected bond lengths (Å) and angles (°) for complex **1**. Table S2: Selected bond lengths (Å) and angles (°) for complex **2**. Table S3: Selected bond lengths (Å) and angles (°) for complex **3**. Table S4: Selected bond lengths (Å) and angles (°) for complex **4**. Table S5: Hydrogen bonds for complex **1** [Å and °]. Table S6: Hydrogen bonds for complex **2** [Å and °]. Table S7: Hydrogen bonds for complex **3** [Å and °]. Table S8: Hydrogen bonds for complex **4** [Å and °]. Table S9: CShM calculations of the coordination environment at lanthanide(III) ion in all complexes of $LnL_9$ coordination sphere of **1–4** [30]. Table S10: Energy values were obtained through the Magellan software [36]. Table S11: Debye parameters of **4** from the AC susceptibility data measured with zero dc magnetic field. Table S12: Debye parameters of **4** from the ac susceptibility data were measured with a 1.0 kOe dc magnetic field. Table S13: Debye parameters of **1** from the ac susceptibility data were measured with a 1.0 kOe dc magnetic field. Table S14: Debye parameters of **2** from the ac susceptibility data were measured with a 1.0 kOe dc magnetic field. Table S15: Debye parameters of **3** from the ac susceptibility data were measured with a 1.0 kOe dc magnetic field. Table S16: Debye parameters of **3** from the ac susceptibility data were measured with a 5.0 kOe dc magnetic field.

**Author Contributions:** Investigation and methodology, writing original draft, T.T.d.C.; investigation, J.H.d.A.-N. and M.E.A.; validation, writing review and editing, F.T.M.; investigation, writing original draft, E.F.P., D.L.M., W.C.N. and N.M.; resources, project administration, writing—review and editing, F.L., M.J.; conceptualization, funding acquisition, supervising, writing review and editing, C.L.M.P. All authors have read and agreed to the published version of the manuscript.

**Funding:** This research was funded by Conselho Nacional de Desenvolvimento Científico e Tecnológico (CNPq), the Coordenação de Aperfeiçoamento de Pessoal de Nível Superior (CAPES), the Bioanalytical Facility NEPS-DQ/UFMG, the Fundação de Amparo à Pesquisa do Estado de Minas Gerais, the Fundo de Amparo à Pesquisa do Estado do Rio de Janeiro, the Spanish MINECO (Project PID201935-GB-I00) and Unidad de Excelencia Maria de Maeztu CEX2019-000919 for financial support.

**Data Availability Statement:** All data of this manuscript are available in the Supplementary Materials. Crystal structures were deposited with the Cambridge Crystallographic Data Center with numbers 2380352 (**1**), 2380353 (**2**), 2380351 (**3**) and 2380354 (**4**) and are available under request.

**Conflicts of Interest:** The authors declare no conflicts of interest to disclose.

## References


1. Bartolomé, E.; Arauzo, A.; Luzón, J.; Bartolomé, J.; Bartolomé, F. Magnetic Relaxation of Lanthanide-Based Molecular Magnets. In *Handbook of Magnetic Materials*; North-Holland: Amsterdam, The Netherlands, 2017; Volume 26, pp. 1–289. [CrossRef]
2. Ferrando-Soria, J.; Vallejo, J.; Castellano, M.; Martínez-Lillo, J.; Pardo, E.; Cano, J.; Castro, I.; Lloret, F.; Ruiz-García, R.; Julve, M. Molecular Magnetism, Quo Vadis? A Historical Perspective from a Coordination Chemist Viewpoint. *Coord. Chem. Rev.* **2017**, *339*, 17–103. [CrossRef]
3. Rabelo, R.; Stiriba, S.E.; Cangussu, D.; Pereira, C.L.M.; Moliner, N.; Ruiz-García, R.; Cano, J.; Faus, J.; Journaux, Y.; Julve, M. When Molecular Magnetism Meets Supramolecular Chemistry: Multifunctional and Multiresponsive Dicopper(Ii) Metallacyclophanes as Proof-of-Concept for Single-Molecule Spintronics and Quantum Computing Technologies? *Magnetochemistry* **2020**, *6*, 69. [CrossRef]
4. Ishikawa, N. Simultaneous Determination of Ligand-Field Parameters of Isostructural Lanthanide Complexes by Multidimensional Optimization. *J. Phys. Chem. A* **2003**, *107*, 5831–5835. [CrossRef]
5. Parmar, V.S.; Mills, D.P.; Winpenny, R.E.P. Mononuclear Dysprosium Alkoxide and Aryloxide Single-Molecule Magnets. *Chem. Eur. J.* **2021**, *27*, 7625–7645. [CrossRef]
6. Blagg, R.J.; Ungur, L.; Tuna, F.; Speak, J.; Comar, P.; Collison, D.; Wernsdorfer, W.; McInnes, E.J.L.; Chibotaru, L.F.; Winpenny, R.E.P. Magnetic Relaxation Pathways in Lanthanide Single-Molecule Magnets. *Nat. Chem.* **2013**, *5*, 673–678. [CrossRef]
7. Rinehart, J.D.; Long, J.R. Exploiting Single-Ion Anisotropy in the Design of f-Element Single-Molecule Magnets. *Chem. Sci.* **2011**, *2*, 2078. [CrossRef]
8. Zheng, Y.Z.; Zhou, G.J.; Zheng, Z.; Winpenny, R.E.P. Molecule-Based Magnetic Coolers. *Chem. Soc. Rev.* **2014**, *43*, 1462–1475. [CrossRef]
9. Bünzli, J.-C.G.; Chauvin, A.-S. Lanthanides in Solar Energy Conversion. In *Handbook on the Physics and Chemistry of Rare Earths*; North-Holland: Amsterdam, The Netherlands, 2014; Volume 44, pp. 169–281. [CrossRef]
10. Bünzli, J.C.G. On the Design of Highly Luminescent Lanthanide Complexes. *Coord. Chem. Rev.* **2015**, *293–294*, 19–47. [CrossRef]
11. Fortea-Pérez, F.R.; Vallejo, J.; Julve, M.; Lloret, F.; De Munno, G.; Armentano, D.; Pardo, E. Slow Magnetic Relaxation in a Hydrogen-Bonded 2D Array of Mononuclear Dysprosium(III) Oxamates. *Inorg. Chem.* **2013**, *52*, 4777–4779. [CrossRef]
12. da Cunha, T.T.; Barbosa, V.M.M.; Oliveira, W.X.C.; Pinheiro, C.B.; Pedroso, E.F.; Nunes, W.C.; Pereira, C.L.M. Slow Magnetic Relaxation in Mononuclear Gadolinium(III) and Dysprosium(III) Oxamato Complexes. *Polyhedron* **2019**, *169*, 102–113. [CrossRef]
13. Mayans, J.; Escuer, A. Correlating the Axial Zero Field Splitting with the Slow Magnetic Relaxation in $Gd^{III}$ SIMs. *Chem. Commun.* **2021**, *57*, 721–724. [CrossRef] [PubMed]
14. da Cunha, T.T.; Barbosa, V.M.M.; Oliveira, W.X.C.; Pedroso, E.F.; García, D.M.A.; Nunes, W.C.; Pereira, C.L.M. Field-Induced Slow Magnetic Relaxation of a Six-Coordinate Mononuclear Manganese(II) and Cobalt(II) Oxamate Complexes. *Inorg. Chem.* **2020**, *59*, 12983–12987. [CrossRef] [PubMed]
15. da Silveira, C.O.C.; Oliveira, W.X.C.; da Silva Júnior, E.N.; Alvarenga, M.E.; Martins, F.T.; Gatto, C.C.; Pinheiro, C.B.; Pedroso, E.F.; Silva, J.P.O.; Marques, L.F.; et al. Photoluminescence and Magnetic Properties of Isostructural Europium(III), Gadolinium(III) and Terbium(III) Oxamate-Based Coordination Polymers. *Dalton Trans.* **2024**, *53*, 14995–15009. [CrossRef] [PubMed]
16. Vaz, R.C.A.; Esteves, I.O.; Oliveira, W.X.C.; Honorato, J.; Martins, F.T.; Marques, L.F.; dos Santos, G.L.; Freire, R.O.; Jesus, L.T.; Pedroso, E.F.; et al. Mononuclear Lanthanide(III)-Oxamate Complexes as New Photoluminescent Field-Induced Single-Molecule Magnets: Solid-State Photophysical and Magnetic Properties. *Dalton Trans.* **2020**, *49*, 16106–16124. [CrossRef]
17. Vaz, R.C.A.; Esteves, I.O.; Oliveira, W.X.C.; Honorato, J.; Martins, F.T.; da Silva Júnior, E.N.; de C. A. Valente, D.; Cardozo, T.M.; Horta, B.A.C.; Mariano, D.L.; et al. Lanthanide(III)-Oxamato Complexes Containing $Nd^{3+}$ and $Ho^{3+}$: Crystal Structures, Magnetic Properties, and Ab Initio Calculations. *CrystEngComm* **2022**, *24*, 6628–6641. [CrossRef]
18. Wang, J.; Jing, Y.; Cui, M.; Lu, Y.; Ouyang, Z.; Shao, C.; Wang, Z.; Song, Y. Spin Qubit in a 2D $Gd^{III}$ $Na^{I}$ -Based Oxamato Supramolecular Coordination Framework. *Chem. Eur. J.* **2023**, *29*, e202301771. [CrossRef]
19. Pardo, E.; Ruiz-García, R.; Lloret, F.; Faus, J.; Julve, M.; Journaux, Y.; Delgado, F.; Ruiz-Pérez, C. Cobalt(II)-Copper(II) Bimetallic Chains as a New Class of Single-Chain Magnets. *Adv. Mater.* **2004**, *16*, 1597–1600. [CrossRef]
20. Araujo Junior, C.R.; Oliveira, W.X.C.; Pinheiro, C.B.; Pedroso, E.F.; Nunes, W.C.; de Almeida, A.A.; Knobel, M.; Julve, M.; Pereira, C.L.M. Crystal Structure and Cryomagnetic Study of a Mononuclear Erbium(III) Oxamate Inclusion Complex. *Acta Crystallogr. C Struct. Chem.* **2024**, *C80*, 349–356. [CrossRef]
21. Macrae, C.F.; Sovago, I.; Cottrell, S.J.; Galek, P.T.A.; McCabe, P.; Pidcock, E.; Platings, M.; Shields, G.P.; Stevens, J.S.; Towler, M.; et al. Mercury 4.0: From Visualization to Analysis, Design and Prediction. *J. Appl. Crystallogr.* **2020**, *53*, 226–235. [CrossRef]
22. Sheldrick, G.M. A Short History of *SHELX*. *Acta. Crystallogr. A* **2008**, *64*, 112–122. [CrossRef]
23. Dolomanov, O.V.; Bourhis, L.J.; Gildea, R.J.; Howard, J.A.K.; Puschmann, H. *OLEX2*: A Complete Structure Solution, Refinement and Analysis Program. *J. Appl. Crystallogr.* **2009**, *42*, 339–341. [CrossRef]
24. Rigaku, O.D. *CrysAlis PRO*; Rigaku Corporation: Oxford, UK, 2018.
25. Palatinus, L.; Chapuis, G. SUPERFLIP—A Computer Program for the Solution of Crystal Structures by Charge Flipping in Arbitrary Dimensions. *J. Appl. Crystallogr.* **2007**, *40*, 786–790. [CrossRef]
26. Sheldrick, G.M. Crystal Structure Refinement with SHELXL. *Acta Crystallogr. C Struct. Chem.* **2015**, *71*, 3–8. [CrossRef]
27. Johnson, C.K. *Crystallographic Computing Techniques*; Ahmed, F., Hall, S.R., Huber, C.P., Eds.; Munksgaard: Copenhagen, Denmark, 1970; pp. 207–220.

28. Pereira, C.L.M.; Doriguetto, A.C.; Konzen, C.; Meira-Belo, L.C.; Leitão, U.A.; Fernandes, N.G.; Mascarenhas, Y.P.; Ellena, J.; Brandl, A.L.; Knobel, M.; et al. A Crystalline Phase Transition and Optical Properties in a $Co^{II}Cu^{II}$ Oxamato-Bridged Ferrimagnetic Chain. *Eur. J. Inorg. Chem.* **2005**, *2005*, 5018–5025. [CrossRef]
29. da Cunha, T.T.; da Silveira, C.O.C.; Barbosa, V.M.M.; Oliveira, W.X.C.; da Silva Júnior, E.N.; Ferreira, F.F.; Pedroso, E.F.; Pereira, C.L.M. Ferromagnetic Coupling in a Dicopper(II) Oxamate Complex Bridged by Carboxylate Groups. *CrystEngComm* **2021**, *23*, 1885–1897. [CrossRef]
30. Lunell, M.; Casanova, D.; Cirera, J.; Alemany, P.; Alvarez, S. *SHAPE*, version 2.0. 2010.
31. Etter, M.C. Encoding and Decoding Hydrogen-Bond Patterns of Organic Compounds. *Acc. Chem. Res.* **1990**, *23*, 120–126. [CrossRef]
32. Diogo, E.B.T.; da Silva Júnior, E.N.; Oliveira, W.X.C.; Stumpf, H.O.; Fabris, F.; de Almeida, A.A.; Knobel, M.; Ferreira, F.F.; Nunes, W.C.; Pedroso, E.F.; et al. Isostructural Oxamate Complexes with Visible Luminescence ($Eu^{3+}$) and Field-Induced Single-Molecule Magnet ($Nd^{3+}$). *Chem. Asian J.* **2024**, *24*, e202400887. [CrossRef]
33. de Oliveira Maciel, J.W.; Lemes, M.A.; Valdo, A.K.; Rabelo, R.; Martins, F.T.; Queiroz Maia, L.J.; de Santana, R.C.; Lloret, F.; Julve, M.; Cangussu, D. Europium(III), Terbium(III), and Gadolinium(III) Oxamato-Based Coordination Polymers: Visible Luminescence and Slow Magnetic Relaxation. *Inorg. Chem.* **2021**, *60*, 6176–6190. [CrossRef]
34. Chilton, N.F.; Anderson, R.P.; Turner, L.D.; Soncini, A.; Murray, K.S. PHI: A Powerful New Program for the Analysis of Anisotropic Monomeric and Exchange-Coupled Polynuclear d- and f-Block Complexes. *J. Comput. Chem.* **2013**, *34*, 1164–1175. [CrossRef] [PubMed]
35. Ungur, L.; Chibotaru, L.F. Computational Modelling of the Magnetic Properties of Lanthanide Compounds. In *Lanthanides and Actinides in Molecular Magnetism*; Wiley: Weinheim, Germany, 2015; pp. 153–184. [CrossRef]
36. Chilton, N.F.; Collison, D.; McInnes, E.J.L.; Winpenny, R.E.P.; Soncini, A. An Electrostatic Model for the Determination of Magnetic Anisotropy in Dysprosium Complexes. *Nat. Commun.* **2013**, *4*, 2551. [CrossRef] [PubMed]
37. Reta, D.; Chilton, N.F. Uncertainty Estimates for Magnetic Relaxation Times and Magnetic Relaxation Parameters. *Phys. Chem. Chem. Phys.* **2019**, *21*, 23567–23575. [CrossRef]
38. Blackmore, W.J.A.; Gransbury, G.K.; Evans, P.; Kragskow, J.G.C.; Mills, D.P.; Chilton, N.F. Characterisation of Magnetic Relaxation on Extremely Long Timescales. *Phys. Chem. Chem. Phys.* **2023**, *25*, 16735–16744. [CrossRef]
39. Cole, K.S.; Cole, R.H. Dispersion and Absorption in Dielectrics, I. Alternating Current Characteristics. *J. Chem. Phys.* **1941**, *9*, 341–351. [CrossRef]
40. Hu, B.; Xi, J.; Cen, P.; Guo, Y.; Ding, Y.; Qin, Y.; Zhang, Y.-Q.; Liu, X. A Mononuclear Nine-Coordinated Dy(III) Complex Exhibiting Field-Induced Single-Ion Magnetism Behaviour. *RSC Adv.* **2022**, *12*, 13992–13998. [CrossRef]
41. Wada, H.; Ooka, S.; Iwasawa, D.; Hasegawa, M.; Kajiwara, T. Slow Magnetic Relaxation of Lanthanide(III) Complexes with a Helical Ligand. *Magnetochemistry* **2016**, *2*, 43. [CrossRef]
42. Wada, H.; Ooka, S.; Yamamura, T.; Kajiwara, T. Light Lanthanide Complexes with Crown Ether and Its Aza Derivative Which Show Slow Magnetic Relaxation Behaviors. *Inorg. Chem.* **2017**, *56*, 147–155. [CrossRef]
43. Liddle, S.T.; Van Slageren, J. Improving F-Element Single Molecule Magnets. *Chem. Soc. Rev.* **2015**, *44*, 6655–6669. [CrossRef]
44. Holmberg, R.J.; Ho, L.T.A.; Ungur, L.; Korobkov, I.; Chibotaru, L.F.; Murugesu, M. Observation of Unusual Slow-Relaxation of the Magnetisation in a Gd-EDTA Chelate. *Dalton Trans.* **2015**, *44*, 20321–20325. [CrossRef] [PubMed]
45. Katoh, K.; Yasuda, N.; Damjanović, M.; Wernsdorfer, W.; Breedlove, B.K.; Yamashita, M. Manipulation of the Coordination Geometry along the $C_4$ Rotation Axis in a Dinuclear $Tb^{3+}$ Triple-Decker Complex via a Supramolecular Approach. *Chem. Eur. J.* **2020**, *26*, 4805–4815. [CrossRef] [PubMed]

magnetochemistry

MDPI

*Article*

# Engineering Mononuclear Ln(III) Complexes with a Pseudo-Macrocyclic Hexadentate $N_4O_2$ Schiff Base Ligand Exhibiting Slow Magnetic Relaxation †

**Ismael Francisco Diaz-Ortega [1,2], Yating Ye [1], Jesus Jover [3], Eliseo Ruiz [3,*], Enrique Colacio [1,*] and Juan Manuel Herrera [1,*]**

1 Departamento de Química Inorgánica, Facultad de Ciencias, Universidad de Granada and Unidad de Excelencia en Química (UEQ), Avenida Fuentenueva s/n, 18071 Granada, Spain; ifdior@ual.es (I.F.D.-O.); yatingwsy@ugr.es (Y.Y.)

2 Departamento de Química y Física-CIESOL, Universidad de Almería, Carretera Sacramento s/n, 04120 Almería, Spain

3 Departament de Química Inorgánica and Institut de Recerca de Química Teórica i Computacional, Universitat de Barcelona, Diagonal 645, 08028 Barcelona, Spain; jesus.jover@qi.ub.edu

* Correspondence: eliseo.ruiz@qi.ub.edu (E.R.); ecolacio@ugr.es (E.C.); jmherrera@ugr.es (J.M.H.)

† This article is a tribute to Professors Miguel Julve and Francesc Lloret on the occasion of their retirement (Miguel sadly passed away very recently). They can be considered two of the most active, outstanding, and brilliant researchers in the field of Molecular Magnetism based on Coordination Complexes have achieved key milestones in this field over the last forty-five years and, consequently, contributed decisively to its development. Both of them and their collaborators at the University of Valencia and beyond, by subtly playing with the tools offered by Coordination Chemistry, have deliberately designed and prepared new and interesting molecular materials with aesthetic structures and outstanding magnetic properties. In addition, Miguel and Paco are very nice and friendly people who always helped many of us to walk along the exciting and attractive road of Molecular Magnetism. Thank you very much for your friendship, support, and teachings.

**Abstract:** We report here the synthesis of a series of nine coordinated mononuclear $Ln^{III}$ complexes $[LnL_1Cl_2(DMF)]Cl{\cdot}2.5DMF$ and $[LnL^1(L_2)_2]Cl{\cdot}4CH_3OH$ ($Ln^{III}$ = $Gd^{III}$, $Dy^{III}$, $Er^{III}$ and $Yb^{III}$, $HL_2$ = 9-anthracenecarboxylic acid), where L1 is a hexadentate $N_4O_2$ Schiff base ligand prepared from the condensation of 1,10-phenanthroline-2,9-dicarbaldehyde and semicarbazone. The X-ray crystal structures of these complexes show the $Ln^{III}$ ions to possess $LnN_4O_2Cl_2$ and $LnN_4O_4$ coordination spheres, which can be considered to be derived from a hexagonal bipyramidal geometry, with the ligand in the equatorial plane and the anions (chloride or 9-antracenecarboxylate) in axial positions, which undergo distortion after coordination of either a molecule of DMF or a bidentate coordination of the 9-anthracenecarboxxylate ligand. All these compounds exhibit field-induced slow magnetization relaxation (SMR). The absence of SMR at zero field due to QTM, as well as the processes involved in the magnetic relaxation under a field of 0.1 T, have been justified on the basis of theoretical calculations and the distortion of the respective coordination spheres. The severe discrepancy between the calculated and experimental thermal energy barriers for the $Dy^{III}$ complexes seems to indicate that the relaxation occurs with the contribution of spin–vibrational coupling, which is favored by the flexibility of the ligand.

**Keywords:** single-molecule magnets; lanthanides; molecular magnetism

## 1. Introduction

The field of molecular magnetism experienced a major boost with the discovery of single-molecule magnets (SMMs) [1–10]. These nanoscale molecule-based magnets are paramagnetic metal complexes exhibiting both quantum (quantum tunneling of the magnetization (QTM), quantum coherence, and quantum oscillations) and classical (slow relaxation of the magnetization, leading to magnetic hysteresis and remnant magnetization at zero field below the commonly named blocking temperature, $T_B$) properties. Owing to this intriguing association of physical properties, SMMs possess great potential for creating a new generation of ultrahigh-density data storage, spin-qubits for quantum computing, and spintronic devices [1–11]. The SMM behavior is tied to the existence of a large magnetic moment, a bistable ground state (a classical binary memory requires two degenerate states between which magnetization can be switched), and magnetic anisotropy. The latter issue is of paramount importance for observing SMM behavior, because the blocking of the magnetization below $T_B$ is due to an energy barrier that mainly depends on the magnetic anisotropy, which in turn arises from the combined action of spin–orbit coupling and crystal field effects [1–10]. In view of this, it is not surprising that the use of $Ln^{III}$ ions to construct SMMs allowed this area of research to enter into a fascinating and fruitful new stage of development.

Numerous examples of SMMs containing $Dy^{III}$, $Tb^{III}$, $Er^{III}$, $Ho^{III}$ and even $Nd^{III}$, $Tm^{III}$ and $Yb^{III}$ lanthanide ions have been reported so far [1–18]. Among these systems, those of mononuclear nature, also named Mononuclear SMMs (MSMMs) or Single-Ion Magnets (SIMs), are the best choice for achieving high-performance SMMs. This is because in these mononuclear complexes, the magnetic anisotropy, which depends on the crystal field (CF), can be deliberately modulated by playing with the coordination environment around the lanthanide metal ion, which ultimately depends on the electronic and steric effects of the ligands. In connection with this, experimental and theoretical studies carried out on $Ln^{III}$-based SIMs have shown that for oblate ions ($Dy^{III}$, $Tb^{III}$, $Ho^{III}$ and $Nd^{III}$), their axial anisotropy is maximized when the crystal field is axial (repulsive contacts are minimized), which can be reached with weak equatorial and strong axial CF components. For prolate lanthanide metal ions such as $Er^{III}$, $Nd^{III}$ and $Yb^{III}$, weak axial and strong equatorial CF components are required to generate the anisotropic energy barrier [19–23]. In spite of that, it is of paramount importance to engineer molecular vibrations to control the spin lifetime of SMM complexes [24]. The best SMM properties have been observed for $Dy^{III}$-based SIMs because of the large magnetic anisotropy and Kramers nature of the $Dy^{III}$ ion. In this case, the strong axial anisotropy stabilizes the largest $M_J = \pm 15/2$ ground state and leads to a larger separation from the excited states [1–23]. Thus, dysprosium sandwich metallocenes with large axial (easy-axis) anisotropy and negligible rhombicity exhibit $T_B$ and $U_{eff}$ values as high as 1540 $cm^{-1}$ and 80 K, respectively [25–27]. Nevertheless, the record is held by the mixed-valence $Ln^{III}$-$Ln^{II}$ dinuclear metallocene complex $[Dy_2I_3(CpiPr_5)_2]$ with $U_{eff}$ = 2347 K and $T_B$ = 80 K. In this compound, the collinearity of the anisotropy axes and the strong 4f-radical Ising exchange interaction result in a large separation of the first and excited states and the suppression of QTM [28].

Taking into consideration that the ultimate goal in this field of research is the incorporation of high-efficiency SMMs in devices for technological applications, these systems must be stable at ambient conditions, and this is not the case for $Ln^{III}$-based metallocenes. In view of this, thermal-, air- and humidity-stable monometallic complexes with axial symmetry, such as pentagonal and hexagonal bipyramids, have been designed with oblate $Ln^{III}$ ions, some of which present insignificant transverse anisotropy, suppressed QTM and high-performance SMM behavior [29–34]. A good strategy for achieving this kind of system is using macrocyclic ligands containing weak donor atoms coordinated in the equatorial positions, whereas donor atoms with high electronic density are strongly bonded in axial positions. The magneto-structural properties of axial mononuclear SMMs containing macrocyclic ligands at the waist have recently been reported, demonstrating the efficacy of the proposed approach [35,36]. Moreover, complexes containing macrocyclic ligands generally show great stability at ambient conditions. Pseudo-macrocyclic polydentate ligands would also be a good choice for constructing axial SIMs stable at ambient conditions. In fact, some examples of $Dy^{III}$-based PBPY-7 SMMs have been reported so far [37–41]. However, to the best of our knowledge, there are no reported examples of HBPY-8 SMMs containing hexacoordinate pseudomacrocycles as ligands.

In this paper, we have focused on the challenge of synthesizing a new series of $Ln^{III}$-based SMMs showing pseudo-HBPY-8 geometry, using the hexadentate pseudo-macrocyclic ligand 2,2′-((1,10-phenanthroline-2,9-diyl)bis(methanylylidene)]-bis(hydrazin-1-carboxamide) (**$L_1$** in Scheme 1), which presents a high tendency to bond the $Ln^{III}$ centers around their equatorial coordination positions. This ligand has been previously used to prepare $Eu^{III}$, $Tb^{III}$, $Lu^{III}$ and $Y^{III}$ derivatives [42], some of which show interesting properties as anion sensors.

The crystal structures showed that ligand $L_1$ encapsulated the $Ln^{III}$ ions in their cleft with additional solvent molecules and/or counterions occupying the axial coordination positions. Firstly, we prepared a series of complexes of the formula $[LnL_1Cl_2(DMF)]Cl{\cdot}2.5DMF$, [Ln = Yb(**1a**), Er(**2a**), Dy(**3a**), Gd(**4a**)], which are isostructural to the complex $[Tb(L_1)Cl_2(DMF)]Cl$ previously reported by Albrecht et al. [11]. Additionally, we prepared a second series of complexes of the formula $[LnL_1(L_2)_2]Cl{\cdot}4CH_3OH$, [Ln = Yb(**1b**), Er(**2b**), Dy(**3b**), Gd(**4b**)], where the axially coordinated chloride anions and DMF molecules were replaced by the strongly coordinating ligand 9-anthracene carboxylate in order to generate an axially stressed ligand field around the $Ln^{III}$ centers (Scheme 1). Here, we report the X-ray structures, dc and ac magnetic properties, ab initio calculations and luminescence properties of these two series of compounds. The aim of this work is not only to prepare a new series of potential $Ln^{III}$-SMMs but also to study how this ligand replacement affects the magnetic properties.

## 2. Experimental

### 2.1. Materials and General Methods

All chemical reactants and solvents were obtained from commercial sources (Merck, Burlington, MA, USA) and used as received. Ligand $L_1$ was prepared as described previously by Albrecht et al. [42].

#### Synthesis of the Complexes

**$[LnL_1Cl_2(DMF)]Cl{\cdot}2.5DMF$, [Ln = Yb(1a), Er(2a), Dy(3a), Gd(4a)]:** To a suspension of $L_1$ (0.2 mmol) in 10 mL of methanol, one molar equivalent of the corresponding Ln(III) chloride salt dissolved in a minimal amount of the same solvent was added. The resulting solution was gently heated for 10 min and then dried under a vacuum, resulting in the precipitation of an orange solid. The slow diffusion of diethyl ether into a DMF solution of these precipitates gave a good yield of single crystals of complexes **1a–4a**.

**Yb(1a):** Yield: 64%; elemental anal. for $C_{26.5}H_{38.5}Cl_3N_{11.5}YbO_{5.5}$: C, 35.94; N, 18.19; H, 4.38; found: C, 36.51; N, 18.00; H, 4.14.

**Er(2a):** Yield: 67%; elemental anal. for $C_{26.5}H_{38.5}Cl_3N_{11.5}ErO_{5.5}$: C, 36.18; N, 18.31; H, 4.41; found: C, 36.40; N, 18.74; H, 4.51.

**Dy(3a):** Yield: 65%; elemental anal. for $C_{26.5}H_{38.5}Cl_3N_{11.5}DyO_{5.5}$: C, 36.37; N, 18.41; H, 4.43; found: C, 36.79; N, 18.27; H, 4.72.

**Gd(4a):** Yield: 69%; elemental anal. for $C_{26.5}H_{38.5}Cl_3N_{11.5}GdO_{5.5}$: C, 36.59; N, 18.52; H, 4.46; found: C, 37.29; N, 18.34; H, 4.56.

**$[LnL_1(L_2)_2]Cl{\cdot}4CH_3OH$, [Ln = Yb(1b), Er(2b), Dy(3b), Gd(4b)]:** To a suspension of $L_1$ (0.2 mmol) in 10 mL of methanol, one molar equivalent of the corresponding Ln(III) chloride salt dissolved in a minimal amount of the same solvent was added. To the resulting solution, two equivalents of 9-anthracenecarboxylic acid ($L_2$, 0.4 mmol) and triethylamine (0.4 mmol) dissolved in 5 mL of methanol were added dropwise with stirring. The resulting orange solution was reduced in volume to about 5 mL and then diethyl ether was allowed to diffuse. Well-formed single crystals of complexes 1b–4b appeared after 48–72 h.

**Yb(1b)**: Yield: 80%; elemental anal. for $C_{50}H_{47}ClN_8YbO_{10}$: C, 53.21; N, 9.93; H, 4.20; found: C, 53.36; N, 10.30; H, 4.00.

**Er(2b)**: Yield: 81%; elemental anal. for $C_{50}H_{47}ClN_8ErO_{10}$: C, 53.49; N, 9.98; H, 4.22; found: C, 53.68; N, 10.35; H, 3.98.

**Dy(3b)**: Yield: 82%; elemental anal. for $C_{50}H_{47}ClN_8DyO_{10}$: C, 53.72; N, 10.02; H, 4.24; found: C, 53.30; N, 10.38; H, 4.63.

**Gd(4b)**: Yield: 81%; elemental anal. for $C_{50}H_{47}ClN_8GdO_{10}$: C, 53.94; N, 10.07; H, 4.26; found: C, 54.26; N, 10.47; H, 3.94.

### 2.2. Physical Measurements

Elemental analyses were carried out at the Centro de Instrumentación Científica of the University of Granada with a Fisons-Carlo Erba EA 1108 Elemental Analyser. Direct (dc) and alternating (ac) current susceptibility measurements in different applied statics fields were performed with an oscillating ac field of 3.5 Oe and ac frequencies ranging from 50 to 1400 Hz with a Quantum Design SQUID MPMS XL-5 device. Ac magnetic susceptibility measurements in the range of 1000–10,000 Hz were carried out with a Quantum Design Physical Property Measurement System in an oscillating ac field of 5 Oe.

### 2.3. Computational Methodology

Low-energy spectra and g factors of the eight lowest Kramer's doublets were obtained by means of Complete Active Space Self-Consistent Field and Restricted Active Space State Interaction (CASSCF + RASSI) calculations, as implemented in the open MOLCAS 18.09 software package [43]. The method is divided into two steps: (1) CASSCF calculations for three different multiplicities (sextet, quartet and doublet); (2) the effect of spin–orbit coupling on the basis of the converged wavefunctions obtained in the previous step is included by the RASSI method. Spin Hamiltonian parameters (such as g factors) can be calculated from the wavefunctions resulting from the state interaction step employing the SINGLE ANISO program [44] implemented in the open MOLCAS 18.09 package. We included 21, 128 and 98 roots for the sextet, quartet and doublet $Dy^{III}$ CASSCF calculations, 35 quartets and 112 doublet states were included for the $Er^{III}$ systems, and 7 doublet states were used for the $Yb^{III}$ calculations. The employed basis set has the following contractions: Dy[9s8p6d4f3g2h], Er[9s8p6d4f3g2h], Yb[9s8p6d4f3g2h], O[4s3p2d1f], N[4s3p2d1f], Cl[5s4p2d1f], C[3s2p], H[2s]. The structure of the model was extracted from the corresponding X-ray structure without any ligand truncation.

### 2.4. Single-Crystal Structure Determination

Suitable crystals of complexes **1a–4a** and **1b–4b** were mounted on a Bruker D8 Venture (Mo K$\alpha$ radiation, $\lambda$ = 0.71073 Å, Photon 100 CMOS detector). Details of the crystals, data collection and refinement parameters are shown in Tables S1 and S2. Once the data were processed (raw integration, merging of equivalent reflections and empirical correction of the absorption), the structures were solved by either Patterson or direct methods and refined by full-matrix least-squares on weighted $F^2$ values using the SHELX–2014 suite of programs [45] integrated in Olex2 [46]. Selected bond angles and distances can be found in Tables S3–S5. For structures **1b–3b**, several solvent molecules in the asymmetric unit were found to be disordered over multiple positions and could therefore not be modeled satisfactorily. They were removed from the electron density map using the OLEX solvent MASK command. These structures crystallize in the non-centrosymmetric *Cc* space group. The calculated Flack parameters reflect the presence of racemic twinning. CCDC numbers 2392736–2392743 contain the supplementary crystallographic data for this article. These data are provided free of charge by the Cambridge Crystallographic Data Centre.

## 3. Results and Discussion

### 3.1. Syntheses and Crystal Structures

The synthetic procedures for the preparation of complexes **1a–4a** and **1b–4b** are shown in Scheme 1. For complexes **1a–4a**, the reaction of L1 with the corresponding LnCl3 salt in methanol afforded in all cases microcrystalline orange solids soluble in dimethylformamide (DMF). When diethyl ether was slowly diffused into concentrated DMF solutions of such solids, the complexes appeared as single crystals in good yields. With respect to complexes 1b–4b, single crystals of these complexes were obtained by slow diffusion of diethyl ether into a methanolic solution containing L1, 9-anthracenecarboxylic acid (L2), triethylamine and the corresponding LnIII chloride salt in a 1:2:2:1 molar ratio.

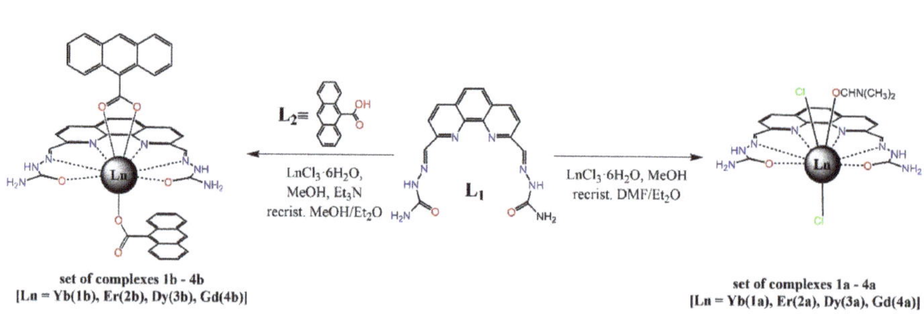


**Scheme 1.** Synthesis of complexes **1a–4a** and **1b–4b**.

X-ray studies revealed that complexes **1a–4a** are isostructural to the $Tb^{III}$ analog complex described previously by Albrecht et al. [42]. The complexes crystallize in the triclinic crystallographic system, the *P-1* space group. The lanthanide ions exhibit an $n = 9$ coordination number, with the hexadentate $L_1$ ligand enveloping the $Ln^{III}$ centers, creating an equatorial $Ln(N_4O_2)$ fragment (Figure 1a). Ln-$N_{L1}$ bond distances are slightly longer than those of Ln–$O_{L1}$ (see Table S5). Within the Ln–$O_{L1}$ bond distances, Ln–O1 (ranging between 2.527 Å for **1a** and 2.551 Å for **4a**) is significantly longer than Ln–O2 [2.342 Å (**1a**)–2.391 Å (**4a**)]. Bond distances decrease as the ionic radii of the $Ln^{III}$ centers decrease due to lanthanide contraction. $L_1$ is slightly twisted with a deviation from planarity that decreases as the $Ln^{III}$ ionic radius raises. To illustrate this writhing, we have considered a plane containing atoms Ln1, N4 and N5 (Figure S1a). O1 (above the plane) and O2 (below) are separated from this plane by a distance ranging between 1.194 Å (**1a**) and 1.185 Å (**1b**) and between 0.573 Å (**1a**) and 0.488 Å (**4a**), respectively. The coordination around the $Ln^{III}$ centers is completed by two chloride counterions (Cl1 and Cl2) axially coordinated at opposite sides of $L_1$ and one oxygen atom (O3) that belongs to a molecule of DMF. According to the continuous-shape-measures method (CShMs) [47,48], the coordination sphere geometry around the $Yb^{III}$ center in **1a** is close to ideal $C_s$ (muffin) and $C_{2v}$ (hula-hoop) geometries (CShMs values of 3.773 and 3.787, respectively). As the ionic radius of the $Ln^{III}$ center increases in the rest of the series (**2a–4a**), the coordination geometries around the $Ln^{III}$ centers progressively deviate from a muffin-like ($C_s$) polyhedron and approach a hula-hoop coordination environment (Table S6). In all the cases, several DMF molecules of crystallization and a third chloride counterion, which ensures electroneutrality, are present in the structures.

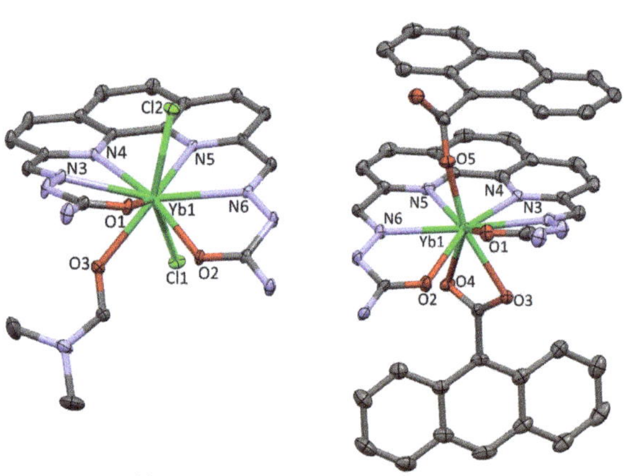


(a) (b)

**Figure 1.** X-ray structures of complexes **1a** (**a**) and **1b** (**b**). Hydrogen atoms, crystallization solvent molecules and non-coordinated counterions have been omitted for the sake of clarity. Ellipsoids are drawn at 50% probability.

Complexes **1b–4b** are isostructural and crystallize in the monoclinic *Cc* space group. The main structural motif in this set of complexes is not very different from that observed for **1a–4a**. The ligand $L_1$ also embraces the $Ln^{III}$ centers describing a wristed equatorial plane (see Figure 1b). In these cases, however, the axial coordination positions around the Ln centers are occupied by two $L_2$ ligands via their carboxylate groups. On one side of the $LnL_1$ platform, the carboxylate group acts as a bidentate ligand (O2 and O3 atoms), while that on the opposite side is monodentate (O5). The $Ln^{III}$ centers exhibit a $LnO_5N_4$ coordination sphere with a geometry intermediate between ideal hula-hoop ($C_{2v}$) and spherically relaxed capped cube ($C_{4v}$) polyhedra (see Table S6). As the ionic radius of the $Ln^{III}$ ion decreases, the $C_{2v}$-CShMs parameter increases, and the geometry becomes closer to $C_{4v}$. For the $LnL_1$ unit, Ln–$N_{L1}$ bond distances are, as expected, longer than those of Ln-$O_{L1}$ (see Table S5). As in the previous series, the twist of the ligand $L_1$ can be illustrated by considering the distances between atoms O1 and O2 and the mean plane defined by atoms N4–Ln1–N5 (Figure S1b). These distances range between 1.041 Å (**1b**) and 1.003 Å (**3b**) and between 0.787 Å (**1b**) and 0.745 Å (**4b**) for O1 and O2, respectively. Compared to Ln–Cl bond distances in **1a–4a**, axial Ln–$O_{L2}$ bond distances are significantly shorter (see Table S5). In particular, the Ln–O5 bond distances (corresponding to the monodentate carboxylate group) are significantly shorter than those of Ln–O3 and Ln–O4 (which correspond to the bidentate carboxylate group). In general, all bond distances increase slightly as the size of the $Ln^{III}$ center increases due to lanthanide contraction. Several methanol and water crystallization molecules exist in the structure connected between them and with the $[LnL_1(L_2)_2]$ entity through hydrogen bonds.

### 3.2. Magnetic Properties

The thermal variations of the $\chi_M T$ product ($\chi_M$ is the magnetic susceptibility measured under an applied *dc* magnetic field of 0.1 T) in the temperature range 2 K–300 K for complexes **1a–4a** and **1b–4b** are presented in Figure S2 and the numerical data are listed in Table 1. At room temperature, the $\chi_M T$ values are close to those expected for magnetically isolated $Ln^{III}$ ions under the free-ion approximation. For $Gd^{III}$ derivatives **4a** and **4b**, the $\chi_M T$ products remain almost constant throughout the range of temperatures studied, as expected for such an isotropic ion with negligible intermolecular magnetic exchange interactions and zero-field splitting (ZFS). Conversely, for complexes **1a–3a** and **1b–3b**, the $\chi_M T$ product gradually decreases as the temperature falls to 2 K, which is due to the thermal depopulation of the excited $m_J$ sublevels arising from the ligand field splitting of the LS ground terms $^2F_{7/2}$, $^4I_{15/2}$ and $^6H_{15/2}$ of the $Yb^{III}$ (**1a** and **1b**), $Er^{III}$ (**2a** and **2b**) and $Dy^{III}$ (**3a** and **3b**) ions, respectively.

**Table 1.** *DC* magnetic data for the complexes studied in this work.

| Complex | $Ln^{III}$ Ground State | Theoretical $\chi_M T$ [$cm^3$ K $mol^{-1}$] [a] | Experimental $\chi_M T_{300\,K}/\chi_M T_{2K}$ [$cm^3$ K $mol^{-1}$] | Theoretical $M_{sat}$ [$N\mu_B$] [b] | Experimental $M_{sat}$ [$N\mu_B$] (T = 2K, H = 5 T) | CASSCF $\chi_M T$ 300 K/2 K [$cm^3$ K $mol^{-1}$] | CASSCF $M_{sat}$ [$N\mu_B$] (T = 2 K, H = 5 T) |
|---|---|---|---|---|---|---|---|
| **1a**<br>**1b** | $Yb^{III}$, $^2F_{7/2}$, $g_J = 8/7$ | 2.57 | 2.39/1.11<br>2.44/1.38 | 4 | 1.65<br>1.94 | 2.49/1.85<br>2.43/1.65 | 1.92<br>1.86 |
| **2a**<br>**2b** | $Er^{III}$, $^4I_{15/2}$, $g_J = 6/5$ | 11.48 | 11.45/5.44<br>11.8/6.98 | 9 | 4.02<br>5.4 | 11.26/7.95<br>11.18/6.73 | 4.43<br>4.22 |
| **3a**<br>**3b** | $Dy^{III}$, $^6H_{15/2}$, $g_J = 4/3$ | 14.17 | 14.35/10.92<br>14.42/10.60 | 10 | 5.56<br>5.73 | 13.93/11.86<br>13.83/12.27 | 5.06<br>5.02 |
| **4a**<br>**4b** | $Gd^{III}$, $^8S_{7/2}$, $g_J = 2$ | 7 | 7.55/7.36<br>7.65/7.39 | 7 | 7.23<br>7.41 | | |

[a] $\chi_M T = \frac{N\beta^2}{3K}\left\{g_J^2\, J(J+1)\right\}$; [b] $M = NgJ\mu_B$; $J = L + S$; $g_J = \frac{3}{2} + \frac{S_T(S_T+1)-L(L+1)}{2J(J+1)}$.

The $M$ vs. H plots at $T = 2$ K (Figure S2b,d) for all the complexes studied in this work show a relatively sharp increase in the magnetization at fields below 1.5 T. At higher fields, the increase in the magnetization is smoother, almost reaching saturation at H = 5 T. For the $Gd^{III}$ derivatives (**4a** and **4b**), the saturation value is very close to that expected for an isotropic S = 7/2 system. For the rest of the complexes, the $M_{sat}$ values are lower than those calculated (Table 1), which can be mainly attributed to crystal field effects giving rise to significant magnetic anisotropy [49,50].

The $M_{sat}$ values for the $Dy^{III}$ complexes (**3a** and **3b**), as well as the value of $\chi_M'T$ product where all the temperature- and frequency-dependent curves converge, which is close to that expected for an isolated $M_J = \pm 15/2$ ground state (12.5 $cm^3$ $mol^{-1}$ K), indicate the axial nature of the ground state (Figure S3). In the case of compound **3a**, the $\chi_M'T$ value (13.9 $cm^3$ $mol^{-1}$ K) is higher than expected, which can be due to a slight orientation of the microcrystals in the magnetic field.

### 3.3. Dynamic Magnetic Properties

In order to determine if the $Yb^{III}$, $Er^{III}$ and $Dy^{III}$ complexes presented in this work exhibit slow magnetic relaxation properties, *ac* dynamic susceptibility measurements were taken as a function of temperature and frequency. Initially, the measurements were carried out in the absence of an external dc magnetic field. Under these conditions, none of the complexes exhibited SMM behavior, probably due to the fast relaxation of the magnetization through quantum tunneling (QTM). When a *dc* external field of 0.1 T was applied (this was the field leading in all cases to the larger relaxation time at 4 K; see Figures S5–S8), frequency- and temperature-dependent out-of-phase signals appeared, thus indicating the existence of slow magnetic relaxation in these compounds. For the $Yb^{III}$ derivatives **1a** and **1b**, the $\chi_M''$ vs. $T$ plots (Figure 2, right) show the frequency dependence of $\chi_M''$ with maxima between 4.0 K (1000 Hz) and 5.7 K (10,000 Hz) and between 2.4 K (1000 Hz) and 3.9 K (10,000 Hz), respectively. At low temperatures, the $\chi_M''$ signals do not clearly tend to zero, which could be indicative that the QTM has not been fully canceled. The Cole–Cole plots (Figure 2, right) in the range 2.0 K–5.0 K displayed semicircular shapes with $\alpha$ values between 0.27 (2.0 K) and 0.038 (5.0 K) and between 0.106 (2.0 K) and 0.201 (5.0 K) for **1a** and **1b**, respectively. Considering that $\alpha$ values close to zero indicate a single relaxation process, whereas $\alpha = 1$ corresponds to an infinitely wide distribution of relaxation times, the $\alpha$ values observed for both complexes at low temperatures suggested the existence of several competitive relaxation processes. The temperature dependence of the relaxation times was extracted from the fitting of the frequency dependence of $\chi_M''$ to the Debye model (Figure S8). The $\tau^{-1}$ vs. $T$ is generally described by the following equation:

$$\tau^{-1} = AT + \tau_{QTM}^{-1} + BT^n + \tau_0 \exp^{\left(-\frac{U_{eff}}{k_B T}\right)} \quad (1)$$

where the two first terms represent the direct and QTM relaxation processes, respectively, whereas the third and fourth terms describe Raman and Orbach relaxation processes. Taking into consideration that the magnetic relaxation in $Yb^{III}$ complexes generally takes place through a Raman process [17,51–53], the extracted $\tau$ vs. $T$ data were fitted to a combination of Raman and QTM (second and third terms in Equation (1)). An excellent fit was obtained with magnetic parameters B, n and $\tau_{QTM}^{-1}$ given in Table 2. Although the magnetic relaxation does not take place through a thermally activated Orbach process, for comparative purposes, we have phenomenologically extracted the energy barriers ($U_{eff}$) for the reversal of the magnetization and the relaxation times ($\tau_0$) from the fit of the high-temperature $\tau$ vs. $T$ data to an Arrhenius plot (Table 2).

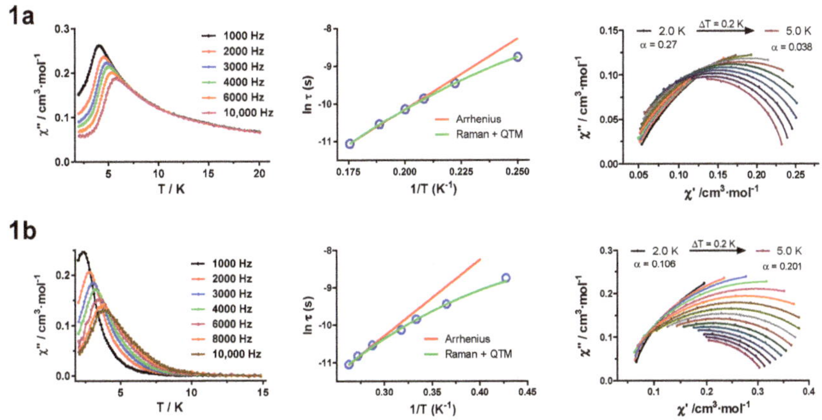


**Figure 2.** $Yb^{III}$ complexes **1a** (**top**) and **1b** (**bottom**): thermal dependence of the out-of-phase $\chi_M''$ signal under a 1000 Oe dc field (**left**). Arrhenius plots of the relaxation time under a 1000 Oe dc field (**middle**; red and green lines represent the best fit of the experimental data to the Arrhenius equation and to the Raman + QTM relaxation processes, respectively). Cole–Cole plots under $H_{dc}$ = 1000 Oe (**right**).

**Table 2.** $U_{eff}$ and $\tau_0$ values for the reported complexes studied in this work.

| Compound | Orbach ($H_{dc}$ = 0.1 T) [a] | | Orbach/Raman/QTM | | | | |
|---|---|---|---|---|---|---|---|
| | $U_{eff}$ [K] | $\tau_0$ [s] | B ($s^{-1}\,K^{-n}$) | $n$ | $\tau_{QTM}$ [s] | $U_{eff}$ [K] | $\tau_0$ [s] |
| **1a** | 37.3 | $2.3 \times 10^{-8}$ | 0.18 [d] | 7.3 [d] | $5.9 \times 10^{-4}$ [d] | | |
| **1b** | 20.3 | $7.9 \times 10^{-8}$ | 41.6 [d] | 5.4 [d] | $3.8 \times 10^{-4}$ [d] | | |
| **2a** | 39.4 | $2.8 \times 10^{-9}$ | 0.02 [e] | 9 | $2.3 \times 10^{-4}$ [e] | 43.3 [e] | $2.2 \times 10^{-9}$ [e] |
| **2b** | 21.1 | $1.3 \times 10^{-7}$ | 0.03 [e] | 9 | $3.2 \times 10^{-3}$ [e] | 16.8 [e] | $7.9 \times 10^{-7}$ [e] |
| **3a** | ≈8.9 [b] | $\approx 3.3 \times 10^{-6}$ [b] | ≈0.002 [c] | ≈2.5 [c] | | | |
| **3b** | ≈5.3 [b] | $\approx 8.5 \times 10^{-6}$ [b] | ≈0.001 [c] | ≈1.7 [c] | | | |

[a] Calculated from the high-temperature data of the Arrhenius plot. [b] Using Equation (3). [c] Using Equation (4). [d] Using a combination of Raman and QTM in Equation (1). [e] Using a combination of Orbach, Raman and QTM in Equation (1).

The magnetic parameters extracted from this fit indicate that the replacement of the axial chloride and DMF ligands by the carboxylate groups of the anthracene carboxylate ligand gives rise to a moderate acceleration of the magnetization relaxation.

The $Er^{III}$ complexes **2a** and **2b** also exhibited field-induced SMM behavior with neat frequency-dependent $\chi_M''$ maxima that appeared between 3.1 K (1000 Hz) and 4.6 K (10,000 Hz) and between 2.5 K (1,000 Hz) and 4.4 K (10, Hz) for complexes **2a** and **2b,** respectively (Figure 3, left). $U_{eff}$ and $\tau_0$ were extracted for both complexes by fitting the calculated relaxation times $\tau$ at different temperatures to an Arrhenius law and are listed in Table 2. The deviation of the data from linearity observed at low temperatures (Figure 3, middle) suggests the existence of several competitive relaxation processes. In agreement with this, the Cole–Cole plots in the range 3.0 K–5.0 K display semicircular shapes with $\alpha$ values between 0.14 (3.0 K) and 0.04 (5.0 K) and between 0.104 (2.0 K) and 0.04 (5.0 K) for **2a** and **2b**, respectively (Figure 3, right). As in the case of the $Yb^{III}$ complexes, the $\chi_M''$ signals do not clearly tend to zero below the maxima, which suggests that the QTM has not been fully suppressed by the field. The best fit was obtained by considering that only Orbach, Raman and QTM processes (fourth, second and third terms in Equation (1), respectively) contribute simultaneously to magnetic relaxation. In order to avoid overparametrization, the n parameter for the Raman process was fixed to the expected value for a Kramers ion ($n$ = 9) [54]. A very good fit was obtained with the parameters given in Table 2. Compared to the simple Arrhenius law, as usual, higher $U_{eff}$ and lower $\tau_0$ values were obtained from the fit of the experimental data to this equation. Here, again, the magnetization relaxation is slightly accelerated when the DMF molecule and the chloride counterions axially coordinated to the $Er^{III}$ centers are replaced by anthracene carboxylate groups (Table 2). It is worth noting that a good-quality fit can also be obtained for **2a** and **2b** when Equation (1) is used (see Table 2). However, bad-quality fits were obtained in both cases by using a combination of Raman and Orbach processes.

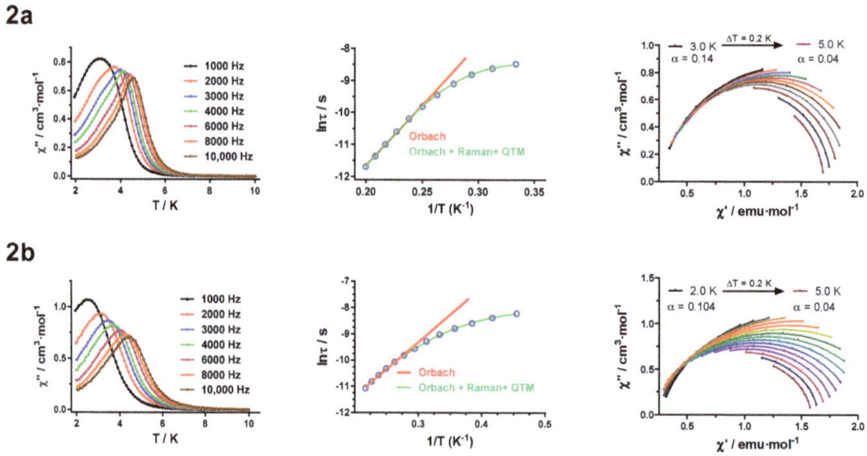


**Figure 3.** $Er^{III}$ complexes **2a** and **2b**: thermal dependence of the out-of-phase $\chi_M''$ signal under a 1000 Oe dc field (**Left**). Arrhenius plots of the relaxation time under a 1000 Oe dc field (**middle**; red and green lines represent the best fit of the experimental data to the Arrhenius equation and to the Orbach + QTM relaxation processes, respectively). Cole–Cole plots under $H_{dc}$ = 1000 Oe (**right**).

Regarding the $Dy^{III}$ derivatives, complex **3a** also exhibits field-induced slow relaxation of the magnetization, but without maxima at temperatures above 2 K. Complex **3b** presents maxima in the $\chi_M''$ signals only at frequencies higher than 6000 Hz (Figure 4). In view of this, we can affirm that either the energy barrier for the reversal of the magnetization is too small to be blocked at temperatures above 2 K or that fast QTM relaxation exists.

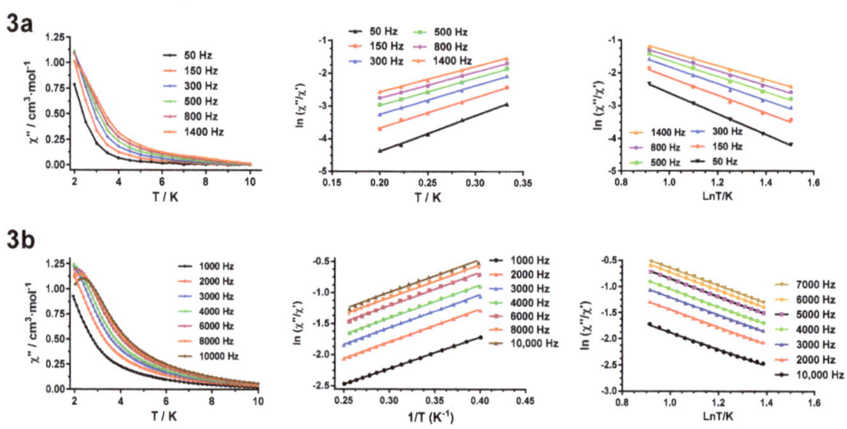


**Figure 4.** $Dy^{III}$ complexes **3a** and **3b**: thermal dependence of the out-of-phase $\chi_M''$ signal under a 1000 Oe dc field (**Left**). Plots of $\ln(\chi_M''/\chi_M')$ vs. $T^{-1}$ (**middle**; solid lines represent the best fit to Equation (3)). Plots of $\ln(\chi_M''/\chi_M')$ vs. $\ln T$ (**right**; solid lines represent the best fit to Equation (4)).

The ac data for **3a** and **3b** could not be fitted satisfactorily, neither by a Debye model nor by a generalized Debye model including a distribution of relaxation times, because the relaxation rates were too close to or beyond the experimental limit of our equipment of 10 kHz. To overcome this problem, and to make a quantitative comparison with the $Yb^{III}$ and $Er^{III}$ derivatives, we have used an alternative approach to obtain the relaxation parameters from the ac data, which is based on the fact that the ratio between the out-of-phase and in-phase ac susceptibility can be expressed in an approximate manner as (Equation (2)) [55],

$$\chi''_M/\chi''_M = 2\pi f\tau \quad (2)$$

where $f$ is the frequency of the ac magnetic field.

The substitution in this equation of the relaxation time ($\tau$) by its expression for each relaxation mechanism (Orbach or Raman) allows us to determine the respective relaxation parameters. If we assume that, hypothetically, the relaxation occurs entirely through an Orbach relaxation mechanism, for which $\tau = \tau_0 \cdot \exp(U_{eff}/k_B T)$, after applying natural logarithms, the equation becomes as follows:

$$\ln\left(\frac{\chi''_M}{\chi'_M}\right) = \ln(2\pi f\tau_0) - \frac{U_{eff}}{k_B T} \quad (3)$$

The effective energy barrier $U_{eff}$ and $\tau_0$ could be approximately estimated by fitting the experimental $\chi''/\chi'$ data in the high-frequency region to this equation (Figure 4). Indeed, the extracted $U_{eff}$ and $\tau_0$ values for these complexes (Table 2) are significantly smaller than those calculated for the $Yb^{III}$ and $Er^{III}$ derivatives. Nevertheless, the $U_{eff}$ values should be taken as phenomenological and may not represent the actual relaxation processes. In the presence of a magnetic field, direct/Raman processes should dominate in the studied low-temperature range. Bearing this in mind, we have also fitted the $\chi''/\chi'$ data to the following equation:

$$\ln\left(\frac{\chi''_M}{\chi'_M}\right) = \ln(2\pi f C) - n(\ln T) \quad (4)$$

which was obtained by replacing $\tau$ in Equation (2) with its expression for Raman/direct processes ($\tau = CT^{-n}$). The data in the 3 K–5 K range for **3a** and those in the range 2.5–4 K for **3b** were fitted to Equation (4) using frequencies in the ranges 50–1,400 Hz and 1000–10,000 Hz, respectively, leading to the parameters $C \approx 1.6 \times 10^{-3}$ s and $n \approx 2.5$ for **3a** and $C \approx 0.0007$ and $n \approx 1.70$ for **3b** (Figure 4).

In both cases, the inverse of the relaxation times follows a temperature dependence close to $T^2$, which could be ascribed to a resonance phonon trapping (RPT) mechanism. This process occurs when there are phonon-bottleneck effects, so that the energy of the lattice modes generated by the relaxing spins cannot be liberated fast enough into a thermal reservoir. These phonons can be reabsorbed by other spins, and the effective relaxation time becomes longer.

### 3.4. Ab Initio Calculations

In order to rationalize the experimental magnetization relaxation and to estimate the values of the thermal energy barrier for complexes **1a–3a** and **1b–3b**, we performed ab initio CASSCF (Complete-Active Space Self-Consistent Field)/RASSI + SO/SINGLE-ANISO calculations on their X-ray structures using open MOLCAS 18.09 [43]. The energy for the first four KDs and their g values for all the compounds are given in Table S7, whereas the g tensors for the ground KD and the energy of the first excited KD for these complexes ($\Delta$) are given in Table 3. The calculated orientation of the main anisotropy axis and the ab initio computed blocking barrier for the magnetization reversal are given in Figures 5 and 6, respectively. The computed crystal field parameters are gathered in Table S8, whereas the calculated susceptibility and magnetization data derived from the ab initio calculations are given in Tables S9 and S10.

**Table 3.** Computed first excitation energies after including the spin–orbit effects (Δ) and $g$ values for the ground-state Kramers' doublet of complexes **1a**/**1b**–**3a**/**3b**.

| Complex | Δ ($cm^{-1}$) | $g_{xx}$, $g_{yy}$, $g_{zz}$ |
|---|---|---|
| **1a** | 216.8 | 0.07, 0.12, 7.68 |
| **1b** | 257.9 | 0.52, 1.10, 7.16 |
| **2a** | 54.6 | 0.43, 0.71, 15.81 |
| **2b** | 64.1 | 0.74, 1.34, 14.52 |
| **3a** | 118.3 | 0.17, 0.36, 19.42 |
| **3b** | 241.7 | 0.03, 0.04, 19.81 |

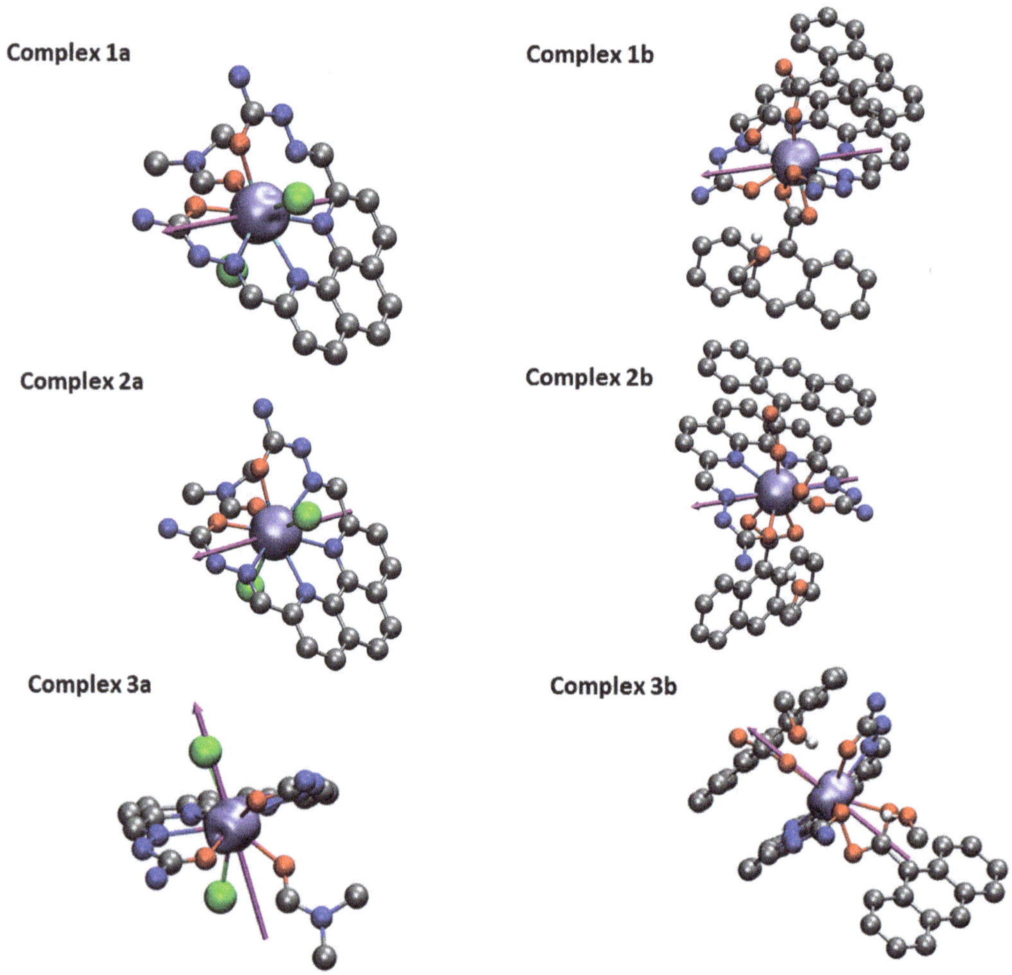


**Figure 5.** CASSCF computed orientation of the main magnetization axis ($g_{zz}$) along the beta-spin density for the $Yb^{III}$ complexes (**top**), $Er^{III}$ complexes (**middle**) and $Dy^{III}$ complexes (**bottom**).

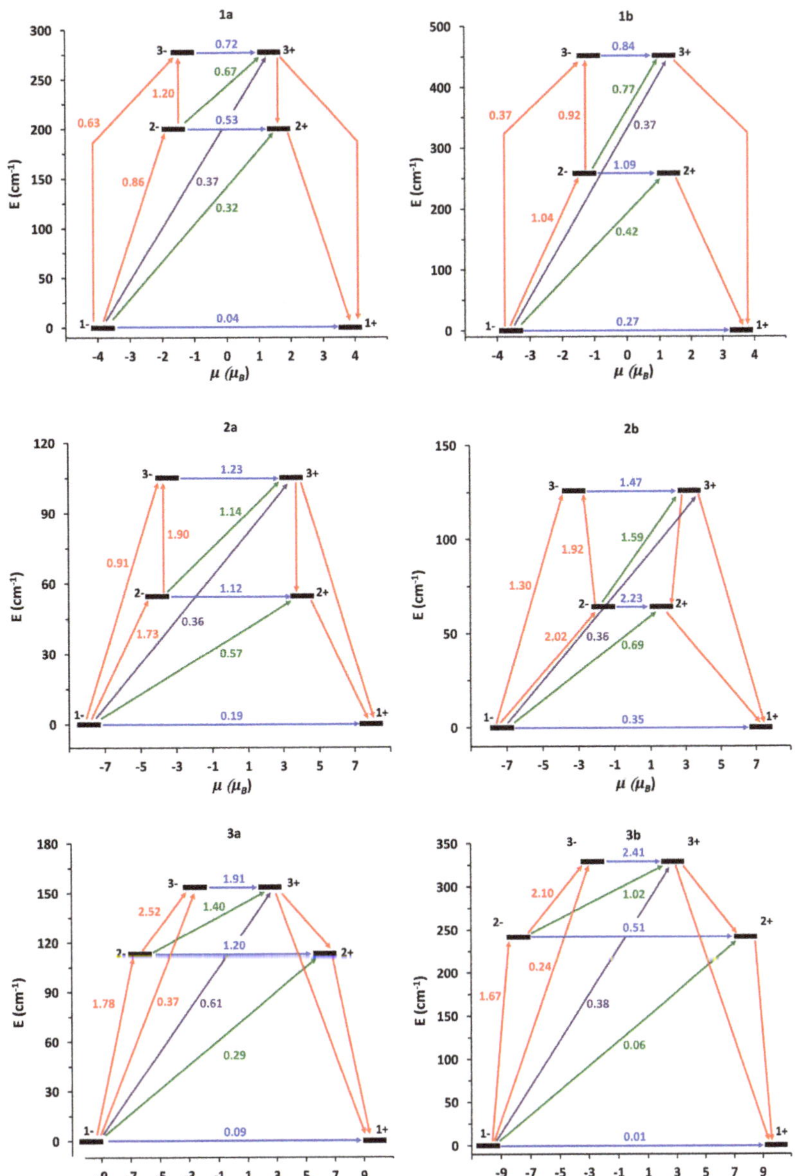


**Figure 6.** Ab initio blocking barrier for the $Yb^{III}$ complexes (**top**), $Er^{III}$ complexes (**middle**) and $Dy^{III}$ complexes (**bottom**). The blue, red and green arrows represent spin-phonon relaxation mechanisms and their transition magnetic moment matrix elements.

With regard to the $Yb^{III}$ complexes, the main anisotropic axis for **1a** is almost included to the $DyN_4O_2$ mean plane and practically perpendicular with the pseudo-$C_6$ axis, whereas for **1b**, the main anisotropy axis is located within the plane (Figure 5, top). In **1a**, the shortest distances Yb–O2 and Yb–O3 are close to the mean plane of the pentadentate ligand, and therefore, the prolate electronic density is directed toward the axial positions perpendicular to the $DyN_4O_2$ mean plane (close to the Yb–Cl bonds), thus reducing electronic repulsions. However, in the case of **1b**, the shortest Yb–O distances are by far those in axial or close to axial positions, and therefore, the electronic density of the $Yb^{III}$ ion spreads out on the ligand plane, placing the main anisotropic axis in the plane and oriented along the bisectrix of the N6–Yb–O2 angle. The CASSCF-computed beta-spin density presents the prolate-like shape expected for a pure $M_J = \pm 7/2$ ground state with an almost spherical density with some holes in the region of the Yb–Cl bonds (for **1a**) due to the unique 4f empty orbital. The $U_{eff}$ values calculated through the first excited state of 216.8 and 257.9 $cm^{-1}$ are significantly larger than the experimental ones (extracted from the ac susceptibility measurements in the presence of a small magnetic field of 0.1 T, see Table 2). The calculated transition magnetic moments between the ground-state KD are close to zero with values of 0.04 and 0.27, respectively (Figure 6, top), indicating a weak QTM relaxation mechanism. Therefore, an efficient spin-relaxation mechanism without QTM has been predicted in $Yb^{III}$ complexes, in agreement with our experimental observation [17,51–53]. These results suggest that spin relaxation occurs through phonon-activate processes, with the Raman, direct mechanisms or local-mode spin-relaxation being predominant, as generally observed for reported $Yb^{III}$ field-induced mononuclear SMMs [53,56].

The ab initio calculations carried out on the $Er^{III}$ complexes **2a** and **2b** show that the main anisotropic axis (Figure 5, middle) is close to be included in the mean plane of the hexadentate $N_4O_2$ ligand similar to those of the Yb(III) systems. This orientation is adopted to reduce the crystal field repulsion on the prolate electron density distribution of the $Er^{III}$ ion with the short distances with the axial ligands. In this case, the mismatching between the orientation of the main magnetic axis and the axial direction of the coordination geometry can lead to a strong transverse magnetic anisotropy and fast magnetic relaxation at zero field [57]. In accordance with this hypothesis, the computed blocking barrier through the first excited state for **2a** and **2b**, respectively, is smaller than that for the $Dy^{III}$ counterparts. Moreover, the computed transition magnetic moments for these compounds between the ground KD of 0.19 and 0.35, respectively (Figure 6, middle), point out the existence of QTM and fast magnetic relaxation. This behavior is in agreement with the absence of slow relaxation in **2a** and **2b**. After quenching the QTM by applying a small *dc* static magnetic field, the experimental $U_{eff}$ values extracted for **2a** and **2b** are only slightly lower than those calculated, which may be due to limitations inherent to the method, spin–vibronic coupling and fast direct processes.

These complexes **3a** and **3b** show a highly anisotropic ground state ($M_J = \pm 15/2$) with strong axiality ($g_{zz}$~20 and $g_{xx}$ and $g_{yy}$~0). This is in agreement with the fact that for lower-symmetry complexes, like **3a** and **3b**, the ground KD state is mostly axial with $M_J = \pm 15/2$, whereas the orientation of the anisotropy axis depends on the symmetry and CF strength of the $Dy^{III}$ coordination sphere. Nevertheless, this ground state is not of pure Ising type as the $g_{xx}$ and $g_{yy}$ tensor components are small but significant. The CASSCF-computed beta-spin density does not present the characteristic disk-like shape expected for a pure $M_J = \pm 15/2$ ground state, which points out that the ground state is instead mixed with other excited states. The spin density for **3a** and **3b** spreads out along the $DyN_4O_2$ coordination plane, and thus, the anisotropy axis is oriented perpendicular to this plane (Figure 5, bottom). It is worth noting that this orientation of the anisotropy axis in **3a** and **3b** is perpendicular to that observed for complexes with analogous hydrazone pentadentate ligands and PBPY-7 geometry [37–41]. This difference could be due to the fact that in the latter, the pentadentate ligand is deprotonated, leading to Dy–O distances in the equatorial plane that are significantly shorter than those found for **3a** and **3b**, where the ligand is found in a non-deprotonated form. In order to reduce the repulsion between the large CF in the equatorial plane of these PBPY-7 complexes with deprotonated hydrazone pentadentate ligands and the oblate electron density distribution of the $Dy^{III}$ ground state, the anisotropic axis is located near the $DyN_3O_2$ plane. Additionally, the distortion of the $Dy^{III}$ coordination sphere in **3a** and **3b**, where oxygen donor atoms with short Dy–O distances belonging to the hexadentate ligand and to the DMF molecule are located above and below the $DyN_4O_2$ plane, could increase the contribution to the axial crystal field with respect to the situation where these donor atoms are much closer to this plane, as observed in hydrazone-based PBPY-7 complexes. For **3a** and **3b**, the orientation of the main magnetic axis matches well with the Dy–Cl direction in **3a** and with the Dy–O5 direction in **3b**, which are close to the axial direction of the coordination geometry, thus favoring the uniaxial anisotropy. The ab initio computed blocking barriers for the magnetization reversal for **3a** and **3b** are given in Figure 6 (bottom). The computed transverse magnetic moments for these compounds between the ground KD of 0.09 and 0.01 are near the threshold value of 0.1 generally assumed for efficient magnetic relaxation. Nevertheless, the accepted value for complete suppression of the QTM is $<10^{-6}$ [39]. Therefore, a non-negligible contribution of QTM is expected for these compounds, which would explain why **3a** and **3b** do not show slow magnetic relaxation at zero field. The computed transverse magnetic moments for the thermally assisted quantum tunneling of magnetization (TA-QTM) through the first excited state of 1.20 and 0.51 for complexes **3a** and **3b**, respectively, are larger than those connecting the ground KD, which suggest that, in the presence of a magnetic field, magnetic relaxation should take place predominantly through the first excited KD doublet with $U_{calc}$ values of 118.3 and 241.7 $cm^{-1}$ for **3a** and **3b**, respectively. The larger $U_{calc}$ value for **3b** could be due to the fact that the Dy-O5 bond distance, involving the oxygen donor atom of the monodentate carboxylate ligand in the axial position, is comparatively short (2.224 Å), thus favoring a stronger thermal energy barrier for **3b**. The large difference between the calculated and experimental thermal energy barrier could be due to dipolar intermolecular interactions that induce QTM, hyperfine interactions and spin–vibrational coupling [1–10,24], which were not considered in the ab initio calculations. Even in the presence of a small dc field, the slow relaxation is quite weak (the out-of-phase signal steadily increases when the temperature decreases), thus indicating that the QTM is not fully suppressed and/or the spin lifetime is severely restricted due to the coupling between the low-energy vibrational modes (the most populated at the low experimental temperatures) and the spin, thus inducing spin-relaxation at low temperatures through anharmonic Orbach and/or Raman mechanisms [29]. The flexible first coordination sphere could be responsible for low-energy vibration that provokes magnetization relaxation in the reported field-induced mononuclear SMMs [58,59].

Finally, we have also computed crystal field parameters (Table S8) to analyze the QTM between the ground-state KDs. The corresponding crystal field Hamiltonian is given as $H_{CF} = B_k^q O_k^q$, where $B_k^q$ is the crystal field parameter while $O_k^q$ is the Steven's operator. Larger non-axial $B_k^q$ (where $q \neq 0$, and $k$ = 2, 4, 6) terms than the axial $B_k^q$ (where $q$ = 0, and $k$ = 2, 4, 6) terms are found to favor the QTM process. For all the reported complexes, both the axial and non-axial terms are found to be equal in strength, suggesting non-negligible relaxation via ground KDs, which agrees with the experimental results.

## 4. Conclusions

The above results allow us to extract the following conclusions: (i) The $N_4O_2$ hexadentate pseudo-macrocyclic Schiff base ligand (L1), prepared from phenanthroline dialdehyde and semicarbazide, does not form the expected mononuclear $Ln^{III}$ complexes ($Ln^{III}$ = $Gd^{III}$, $Dy^{III}$, $Er^{III}$ and $Yb^{III}$) with hexagonal bipyramidal geometry (HBPY-8), where the ligand L1 would occupy the equatorial positions and two monodentate anions (either chloride or 9-anthracenecarboxylato) the axial positions. Instead, nine coordinated complexes were formed by the additional coordination of either a DMF molecule or the oxygen atom of a bidentate 9-anthracenecarboxylato anion, thus distorting the coordination sphere from the HBPY-8 geometry. (ii) None of the nine coordinated $Ln^{III}$ complexes exhibit slow relaxations at zero field due to QTM, which has been supported by ab initio calculations. The QTM results from the transverse anisotropy were generated by the distortion of the coordination sphere and, in the case of the $Er^{III}$ complexes, also by the mismatching of the main magnetic axis, which is located close to the mean plane of the coordinated pseudo-macrocyclic ligand, and the pseudoaxial direction of the coordination geometry. (iii) After the partial suppression of the QTM, by application of a small optimal dc field, all the complexes exhibit slow relaxation of the magnetization, which takes place through a combination of Raman and QTM for the $Yb^{III}$ complexes, through a combination of Orbach, Raman and QTM for the $Er^{III}$ complexes, and through a resonance-trapping mechanism for the $Dy^{III}$ counterparts. (iv) The fact that the thermal energy barriers ($U_{eff}$) theoretically calculated for the $Dy^{III}$ complexes are moderately high (118.3 and 241.7 $cm^{-1}$), whereas those extracted from experimental data are much smaller (a few $cm^{-1}$), seems to indicate that other relaxation processes, such the coupling between spin and vibrational molecular modes arising from the flexibility of the first coordination sphere, could govern the magnetization relaxation in these field-induced mononuclear SMMs.

In view of the above considerations and with the aim of disposing of very efficient $Dy^{III}$-based HBPY-8 SMMs with the ligand L1, we have planned to use bulky neutral (tri alkyl/aryl phosphine oxides) or anionic monodentate ligands (aromatic alkoxides, trialkyl or triaryl alkoxides, or silanoxides), with high electronic density on the oxygen donor atom, which after coordination in axial positions prevent steric hindrance in the coordination of additional ligands to the HBPY-8 coordination sphere.

**Supplementary Materials:** The following supporting information can be downloaded at: https://www.mdpi.com/article/10.3390/magnetochemistry10120104/s1, Figure S1: Molecular Representation, Figures S2–S8 Magnetic Measurements; Tables S1–S2: Crystallographic Data, Tables S3–S5: Bond Distances, Table S6: Continuous Shape Measures, Tables S7–S10 CASSCF results.

**Author Contributions:** Conceptualization, E.C. and J.M.H.; methodology, I.F.D.-O., Y.Y. and J.M.H.; crystallography, Y.Y. and J.M.H.; magnetic study and analysis, I.F.D.-O., E.R. and E.C.; Theory, J.J. and E.R.; validation, J.M.H., E.R. and E.C.; formal analysis, I.F.D.-O., Y.Y. and J.J.; resources, J.M.H., E.R. and E.C.; data curation, J.M.H., E.C. and E.R.; writing—original draft preparation, J.M.H., E.C. and E.R.; writing—review and editing, J.M.H., E.R. and E.C.; visualization, J.M.H. and E.C.; supervision, J.M.H., E.C. and E.R.; funding acquisition, J.M.H., E.C. and E.R. All authors have read and agreed to the published version of the manuscript.

**Funding:** Financial support from the Ministerio de Ciencia e Innovación, Grant PID2022-138090NB-C21 funded by MICIU/AEI/10.13039/501100011033; ERDF/EU, Junta de Andalucía (FQM-195 and project I + D + i P20_00692) and the University of Granada is gratefully acknowledged. ER thanks Ministerio de Ciencia e Innovación (project PID2022-138090NB-C21, PID2021-122464NB-I00, TED2021-129593B-I00, Maria de Maeztu CEX2021-001202-M), and Generalitat Catalunya for ICREA academia and the 2021-SGR-00286 grant for financial support, and CSUC for computational resources.

**Institutional Review Board Statement:** Not applicable.

**Informed Consent Statement:** Not applicable.

**Data Availability Statement:** The original contributions presented in this study are included in the article and Supplementary Materials. Further inquiries can be directed to the corresponding authors.

**Conflicts of Interest:** The authors declare no conflicts of interest.

## References

1. Gatteschi, D.; Sessoli, R. Quantum Tunneling of Magnetization and Related Phenomena in Molecular Materials. *Angew. Chem. Int. Ed.* **2003**, *42*, 268–297. [CrossRef]
2. Gatteschi, D.; Sessoli, R.; Villain, J. *Molecular Nanomagnets*; Oxford University Press: Oxford, UK, 2006.
3. Bartolomé, J.; Luis, F.; Fernández, J.F. *Molecular Magnets: Physics and Applications*; Springer: Berlin/Heidelberg, Germany, 2014.
4. Craig, G.A.; Murrie, M. 3d single-ion magnets. *Chem. Soc. Rev.* **2015**, *44*, 2135–2147. [CrossRef]
5. Ming, Y.-S.; Jiang, S.D.; Wang, B.-W.; Gao, S. Understanding the Magnetic Anisotropy toward Single-Ion Magnets. *Acc. Chem. Res.* **2016**, *49*, 2381–2389. [CrossRef]
6. Day, A.; Kalita, P.; Chandrasekhar, V.V. Lanthanide(III)-Based Single-Ion Magnets. *ACS Omega* **2018**, *3*, 9462–9475. [CrossRef]
7. Fen, M.; Tong, M.-L. Single Ion Magnets from 3d to 5f: Developments and Strategies. *Chem. Eur. J.* **2018**, *24*, 7574–7594.
8. Zabala-Lekuona, A.; Seco, J.M.; Colacio, E. Single-Molecule Magnets: From Mn12-ac to dysprosium metallocenes, a travel in time. *Coord. Chem. Rev.* **2021**, *441*, 213984. [CrossRef]
9. Chilton, N.F. Molecular Magnetism. *Annu. Rev. Mat. Res.* **2022**, *52*, 79–101. [CrossRef]

10. Yin, X.; Deng, L.; Ruan, L.; Wu, Y.; Luo, F.; Qin, G.; Han, X.; Zhang, X. Recent Progress for Single-Molecule Magnets Based on Rare Earth Elements. *Materials* **2023**, *16*, 3568. [CrossRef]
11. Moreno-Pineda, E.; Wernsdorfer, W. Measuring molecular magnets for quantum technologies. *Nat. Rev. Phys.* **2021**, *3*, 645–659. [CrossRef]
12. Chen, Y.-C.; Liu, J.-L.; Wernsdorfer, W.; Liu, D.; Chibotaru, L.F.; Chen, X.-M.; Tong, M.-L. Hyperfine-Interaction-Driven Suppression of Quantum Tunneling at Zero Field in a Holmium(III) Single-Ion Magnet. *Angew. Chem. Int. Ed.* **2017**, *56*, 4996–5000. [CrossRef]
13. Chen, Y.-C.; Huang, X.-S.; Liu, J.-L.; Tong, M.L. Magnetic Dynamics of a Neodymium(III) Single-Ion Magnet. *Inorg. Chem.* **2018**, *57*, 11782–11784. [CrossRef] [PubMed]
14. Regincós Martí, E.; Canaj, A.; Sharma, T.; Celmina, A.; Wilson, C.; Rajaraman, G.; Murrie, M. Importance of an Axial LnIII-F Bond across the Lanthanide Series and Single-Molecule Magnet Behavior in the Ce and Nd Analogues. *Inorg. Chem.* **2022**, *61*, 9906–9917. [CrossRef] [PubMed]
15. Huang, G.; Calvez, G.; Suffren, Y.; Daiguebonne, C.; Freslon, S.; Guillou, O.; Bernot, K. Closing the Circle of the Lanthanide-Murexide Series: Single-Molecule Magnet Behavior and Near-Infrared Emission of the NdIII Derivative. *Magnetochemistry* **2018**, *4*, 44. [CrossRef]
16. Meng, Y.-S.; Qiao, Y.-S.; Zhang, Y.-Q.; Jiang, S.-D.; Meng, Z.-S.; Wang, B.-W.; Wang, Z.-M.; Gao, S. Can Non-Kramers TmIII Mononuclear Molecules be Single-Molecule Magnets (SMMs)? *Chem. Eur. J.* **2016**, *22*, 4704–4708. [CrossRef] [PubMed]
17. Fondo, M.; Corredoira-Vázquez, J.; García-Deibe, A.M.; Sanmartín-Matalobos, J.; Amoza, M.; Botas, A.M.P.; Ferreira, R.A.S.; Carlos, L.D.; Colacio, E. Field-induced slow magnetic relaxation and luminescence thermometry in a mononuclear ytterbium complex. *Inorg. Chem. Front.* **2020**, *7*, 3019–3029. [CrossRef]
18. Borah, A.; Murugavel, R. Magnetic relaxation in single-ion magnets formed by less-studied lanthanide ions Ce(III), Nd(III), Gd(III), Ho(III), Tm(II/III) and Yb(III). *Coord. Chem. Rev.* **2022**, *453*, 214288. [CrossRef]
19. Liddle, S.T.; van Slageren, J. Improving f-element single molecule magnets. *Chem. Soc. Rev.* **2015**, *44*, 6655–6669. [CrossRef] [PubMed]
20. Rinehart, J.D.; Long, J.R. Exploiting single-ion anisotropy in the design of f-element single-molecule magnets. *Chem. Sci.* **2011**, *2*, 2078–2085. [CrossRef]
21. Chilton, N.F. Design criteria for high-temperature single-molecule magnets. *Inorg. Chem.* **2015**, *54*, 2097–2099. [CrossRef] [PubMed]
22. Chilton, N.F.; Goodwin, C.A.P.; Mills, D.P.; Winpenny, R.E.P. The first near-linear bis(amide) f-block complex: A blueprint for a high temperature single molecule magnet. *Chem. Commun.* **2015**, *51*, 101–103. [CrossRef] [PubMed]
23. Ungur, L.; Chibotaru, L.F. Strategies toward High-Temperature Lanthanide-Based Single-Molecule Magnets. *Inorg. Chem.* **2016**, *55*, 10043–10056. [CrossRef] [PubMed]
24. Briganti, M.; Santanni, F.; Tesi, L.; Totti, F.; Sessoli, R.; Lunghi, A. A Complete Ab Initio View of Orbach and Raman Spin-Lattice Relaxation in a Dysprosium Coordination Compound. *J. Am. Chem. Soc.* **2021**, *143*, 13633–13645. [CrossRef]
25. Goodwin, C.A.P.; Ortu, F.; Reta, D.; Chilton, N.F.; Mills, D.P. Molecular magnetic hysteresis at 60 kelvin in dysprosocenium. *Nature* **2017**, *548*, 439–442. [CrossRef] [PubMed]
26. Guo, F.S.; Day, B.M.; Chen, Y.C.; Tong, M.L.; Mansikkamäki, A.; Layfield, R.A. A Dysprosium Metallocene Single-Molecule Magnet Functioning at the Axial Limit. *Angew. Chem. Int. Ed.* **2017**, *56*, 11445–11449. [CrossRef] [PubMed]
27. Guo, F.S.; Day, B.M.; Chen, Y.C.; Tong, M.L.; Mansikkamäki, A.; Layfield, R.A. Magnetic hysteresis up to 80 kelvin in a dysprosium metallocene single-molecule magnet. *Science* **2018**, *362*, 1400–1403. [CrossRef]
28. Gould, C.A.; McClain, K.R.; Reta, D.; Kragskow, J.G.C.; Marchiori, D.A.; Lachman, E.; Choi, E.-S.; Analytis, J.G.; Britt, R.D.; Chilton, N.F.; et al. Ultrahard magnetism from mixed-valence dilanthanide complexes with metal-metal bonding. *Science* **2022**, *375*, 198–202. [CrossRef]
29. Dey, S.; Sharma, T.; Rajaraman, G. Unravelling the role of spin–vibrational coupling in designing high-performance pentagonal bipyramidal Dy(III) single ion magnets. *Chem. Sci.* **2024**, *15*, 6465–6477. [CrossRef] [PubMed]
30. Zhang, B.; Cheng, Z.; Wu, Y.; Chen, L.; Jing, R.; Cai, X.; Jiang, C.; Zhang, Y.-Q.; Yuan, A.; Cui, H.-H.; et al. Pseudo-mono-axial ligand fields that support high energy barriers in triangular dodecahedral Dy(III) single-ion magnets. *Chem. Sci.* **2022**, *13*, 13231–13240. [CrossRef]
31. Jia, S.; Zhu, X.; Yin, B.; Dong, Y.; Sun, A.; Li, D.-f. Macrocyclic Hexagonal Bipyramidal Dy(III)-Based Single-Molecule Magnets with a D6h Symmetry. *Cryst. Growth Des.* **2023**, *23*, 6967–6973. [CrossRef]
32. Zhu, Z.; Zhao, C.; Feng, T.; Liu, X.; Ying, X.; Li, X.-L.; Zhang, Y.-Q.; Tang, J. Air-Stable Chiral Single-Molecule Magnets with Record Anisotropy Barrier Exceeding 1800 K. *J. Am. Chem. Soc.* **2021**, *143*, 10077–10082. [CrossRef]
33. Acharya, J.; Kalita, P.; Chandrasekhar, V. High-Coordinate Mononuclear Ln(III) Complexes: Synthetic Strategies and Magnetic Properties. *Magnetochemistry* **2021**, *7*, 1. [CrossRef]
34. Kalita, P.; Acharya, J.; Chandrasekhar, V. Mononuclear pentagonal bipyramidal Ln(III) complexes: Syntheses and magnetic properties. *J. Magn. Magn. Mater.* **2020**, *498*, 166098. [CrossRef]
35. Gil, Y.; Castro-Alvarez, A.; Fuentealba, P.; Spodine, E.; Aravena, D. Lanthanide SMMs Based on Belt Macrocycles: Recent Advances and General Trends. *Chem. Eur. J.* **2022**, *28*, e202200336. [CrossRef]

36. Corredoira-Vázquez, J.; González-Barreira, C.; Oreiro-Martínez, P.; García-Deibe, A.M.; Sanmartín-Matalobos, J.; Fondo, M. Synthesis and applications of lanthanoid complexes of pentadentate and hexadentate $N_5$ and $N_6$ macrocycles: A review. *J. Rare Earths* **2024**, *42*, 1–15. [CrossRef]
37. Bar, A.K.; Kalita, P.; Sutter, J.-P.; Chandrasekhar, V. Pentagonal-Bipyramid Ln(III) Complexes Exhibiting Single-Ion-Magnet Behavior: A Rational Synthetic Approach for a Rigid Equatorial Plane. *Inorg. Chem.* **2018**, *57*, 2398–2401. [CrossRef]
38. Kalita, P.; Ahmed, N.; Bar, A.K.; Dey, S.; Jana, A.; Rajaraman, G.; Sutter, J.-P.; Chandrasekhar, V. Pentagonal Bipyramidal Ln(III) Complexes Containing an Axial Phosphine Oxide Ligand: Field-induced Single-ion Magnetism Behavior of the Dy(III) Analogues. *Inorg. Chem.* **2020**, *59*, 6603–6612. [CrossRef]
39. Kalita, P.; Ahmed, N.; Moorthy, S.; Béreau, V.; Bar, A.K.; Kumar, P.; Nayak, P.; Sutter, J.P.; Singh, S.K.; Chandrasekhar, V. Slow magnetic relaxation in a homoaxially phosphine oxide coordinated pentagonal bipyramidal Dy(III) complex. *Dalton Trans.* **2023**, *52*, 2804–2815. [CrossRef] [PubMed]
40. Singh, V.; Das, D.; Anga, S.; Sutter, J.P.; Chandrasekhar, V.; Kumar, A.K. Rigid $N_3O_2$-Pentadentate Ligand-Assisted Octacoordinate Mononuclear Ln(III) Complexes: Syntheses, Characterization, and Slow Magnetization Relaxation. *ACS Omega* **2022**, *7*, 25881–25890. [CrossRef] [PubMed]
41. Kalita, P.; Kumari, K.; Umar, P.; Kumar, V.; Singh, S.K.; Rogez, G.; Chandrasekhar, V. Eight-coordinate mono- and dinuclear Dy(III) complexes containing a rigid equatorial plane and an anisobidentate carboxylate ligand in the axial position: Synthesis, structure and magnetism. *Dalton Trans.* **2024**, *53*, 10521–10535. [CrossRef]
42. Nadella, S.; Selvakumar, P.M.; Suresh, E.; Subramanian, P.S.; Albrecht, M.; Giese, M.; Fröhlich, R. Lanthanide(III) Complexes of Bis-semicarbazone and Bis-imine-Substituted Phenanthroline Ligands: Solid-State Structures, Photophysical Properties, and Anion Sensing. *Chem. Eur. J.* **2012**, *18*, 16784–16792. [CrossRef]
43. Fdez Galván, I.; Vacher, M.; Alavi, A.; Angeli, C.; Aquilante, F.; Autschbach, J.; Bao, J.J.; Bokarev, S.I.; Bogdanov, N.A.; Carlson, R.K.; et al. OpenMolcas: From Source Code to Insight. *J. Chem. Theor. Comput.* **2019**, *15*, 5925–5964. [CrossRef]
44. Chibotaru, L.F.; Ungur, L.J. Ab initio calculation of anisotropic magnetic properties of complexes. I. Unique definition of pseudospin Hamiltonians and their derivation. *Chem. Phys.* **2012**, *137*, 064112–064122. [CrossRef]
45. Sheldrick, G.M.A. A short history of SHELX. *Acta Crystallogr.* **2008**, *64*, 112–122. [CrossRef]
46. Dolomanov, O.V.; Bourhis, L.J.; Gildea, R.J.; Howard, J.A.K.; Puschmann, H. OLEX2: A complete structure solution, refinement and analysis program. *J. Appl. Crystallogr.* **2009**, *42*, 339–341. [CrossRef]
47. Alvarez, S.; Alemany, P.; Casanova, D.; Cirera, J.; Llunell, M.; Avnir, D. Shape maps and polyhedral interconversion paths in transition metal chemistry. *Coord. Chem. Rev.* **2005**, *249*, 1693–1708. [CrossRef]
48. Cirera, J.; Ruiz, E.; Alvarez, S. Continuous Shape Measures as a Stereochemical Tool in Organometallic Chemistry. *Organometallics* **2005**, *24*, 1556–1562. [CrossRef]
49. Oyarzabal, I.; Ruiz, J.; Ruiz, E.; Aravena, D.; Seco, J.M.; Colacio, E. Increasing the effective energy barrier promoted by the change of a counteranion in a Zn-Dy-Zn SMM: Slow relaxation via the second excited state. *Chem. Commun.* **2015**, *51*, 12353–12356. [CrossRef] [PubMed]
50. Costes, J.P.; Titos-Padilla, S.; Oyarzabal, I.; Gupta, T.; Duhayon, C.; Rajaraman, G.; Colacio, E. Analysis of the Role of Peripheral Ligands Coordinated to $Zn^{II}$ in Enhancing the Energy Barrier in Luminescent Linear Trinuclear Zn-Dy-Zn Single-Molecule Magnets. *Chem. Eur. J.* **2015**, *21*, 15785–15796. [CrossRef]
51. Ruiz, J.; Lorusso, G.; Evangelisti, M.; Brechin, E.K.; Pope, S.J.A.; Colacio, E. Closely-Related $Zn^{II}{}_2Ln^{III}{}_2$ Complexes ($Ln^{III}$ = Gd, Yb) with Either Magnetic Refrigerant or Luminescent Single-Molecule Magnet Properties. *Inorg. Chem.* **2014**, *53*, 3586–3594. [CrossRef]
52. Liu, J.-L.; Yuan, K.; Leng, J.-D.; Ungur, L.; Wernsdorfer, W.; Guo, F.-S.; Chibotaru, L.F.; Tong, M.-L. A Six-Coordinate Ytterbium Complex Exhibiting Easy-Plane Anisotropy and Field-Induced Single-Ion Magnet Behavior. *Inorg. Chem.* **2012**, *51*, 8538–8544. [CrossRef] [PubMed]
53. Guégan, F.; Jung, J.; Le Guennic, B.; Riobé, F.; Maury, O.; Gillon, B.; Jacquot, J.-F.; Guyot, Y.; Morell, C.; Luneau, D. Evidencing under-barrier phenomena in a Yb(III) SMM: A joint luminescence/neutron diffraction/SQUID study. *Inorg. Chem. Front.* **2019**, *6*, 3152–3157. [CrossRef]
54. Abragam, A.; Bleaney, B. *Electron Paramagnetic Resonance of Transition Ions*; Clarendon Press: Oxford, UK, 1970.
55. Bartolomé, J.; Filoti, G.; Kuncser, V.; Schinteie, G.; Mereacre, V.; Anson, C.E.; Powell, A.K.; Prodius, D.; Turta, C. Magnetostructural correlations in the tetranuclear series of {$Fe_3LnO_2$} butterfly core clusters: Magnetic and Mössbauer spectroscopic study. *Phys. Rev. B* **2009**, *80*, 014430. [CrossRef]
56. Jain, N.; Houard, F.; Marchal, R.; Cordier, M.; Le Guennic, B.; Suffren, Y.; Sarazin, Y.; Bernot, K. Luminescence and Single-Molecule Magnet Properties in Ideal Symmetry Compounds: Example of a Near-Planar Tricoordinate Ytterbium(III) Amide. *ChemistryEurope* **2024**, *2*, e202400062. [CrossRef]
57. Wang, F.; Gong, H.-W.; Zhang, Y.; Xue, A.-Q.; Zhu, W.-H.; Zhang, Y.-Q.; Huang, Z.-N.; Sun, H.-L.; Liu, B.; Fang, Y.-Y.; et al. The comparative studies on the magnetic relaxation behaviour of the axially-elongated pentagonal-bipyramidal dysprosium and erbium ions in similar one-dimensional chain structures. *Dalton Trans.* **2021**, *50*, 8736–8745. [CrossRef]

58. Lunghi, A.; Totti, F.; Sanvito, S.; Sessoli, R. Intramolecular origin of spin-phonon coupling in slow-relaxing molecular magnets. *Chem. Sci.* **2017**, *8*, 6051–6059. [CrossRef]
59. Gu, L.; Wu, R. Origins of slow relaxation in Single-molecule Magnets. *Phys. Rev. Lett.* **2020**, *125*, 117203. [CrossRef] [PubMed]

*magnetochemistry*

MDPI

*Review*

# Exploring Spin-Crossover Cobalt(II) Single-Ion Magnets as Multifunctional and Multiresponsive Magnetic Devices: Advancements and Prospects in Molecular Spintronics and Quantum Computing Technologies †

**Renato Rabelo [1,2], Luminita M. Toma [1], Abdeslem Bentama [3], Salah-Eddine Stiriba [1], Rafael Ruiz-García [1] and Joan Cano [1,*]**

1 Instituto de Ciencia Molecular (ICMol), Universitat de València, 46980 Paterna, València, Spain; renato_rabelo@ufg.br (R.R.); lumimtoma@gmail.com (L.M.T.); salah.stiriba@uv.es (S.-E.S.); rafael.ruiz@uv.es (R.R.-G.)
2 Instituto de Química, Universidade Federal de Goiás, Goiânia 74690-900, Brazil
3 Laboratoire de Chimie Organique Appliquée, Faculté des Sciences Techniques de Fès, Université Sidi Mohammed Ben Abdellah, Fès 30000, Morocco; abdeslem.bentama@usmba.ac.ma
* Correspondence: joan.cano@uv.es

† Dedicated to Professors Miguel Julve and Francesc Lloret, two outstanding chemists, excellent teachers, best friends, and colleagues on the occasion of their retirement and, particularly, to the memory of Miguel Julve, who passed away in July 2024.

**Abstract:** Spin-crossover (SCO) and single-ion magnets (SIMs), or their mixed SCO-SIM derivatives, are a convenient solution in the evolution from molecular magnetism toward molecular spintronics and quantum computing. Herein, we report on the current trends and future directions on the use of mononuclear six-coordinate $Co^{II}$ SCO-SIM complexes with potential opto-, electro-, or chemo-active 2,6-pyridinediimine (PDI)- and 2,2′:6′,2′-terpyridine (TERPY)-type ligands as archetypical examples of multifunctional and multiresponsive magnetic devices for applications in molecular spintronics and quantum computing technologies. This unique class of spin-crossover cobalt(II) molecular nanomagnets is particularly well suited for addressing and scaling on different supports, like metal molecular junctions or carbon nanomaterials (CNMs) and metal–organic frameworks (MOFs) or metal-covalent organic frameworks (MCOFs), in order to measure the single-molecule electron transport and quantum coherence properties, which are two major challenges in single-molecule spintronics (SMS) and quantum information processing (QIP).

**Keywords:** cobalt complexes; coordination chemistry; dynamic molecular systems; ligand design; non-innocent ligands; single-ion magnets; spin-crossover; molecular magnetism; molecular spintronics; quantum computing

## 1. Introduction and Background

Faster, smaller, cheaper—this has been the trend in computing for the last decades. Since 1947, electronics and computing have experienced a booming revolution with the invention of the primary examples of working transistors based on germanium and silicon materials and their subsequent developments as integrated systems. These technologies guided us to what is now known as the digital revolution. Unfortunately, one of the established bases of the digital revolution—Intel co-founder Gordon Moore's prediction that the density of transistors in integrated systems would be doubled every two years, known as the Moore's Law [1]—is nearing its end. In 1959, the physicist and Nobel prize winner Richard Feynman and his iconic phrase, "There's plenty of room at the bottom", granted thoughtful insights into and warmed up the nanotechnology and nanoscience fields [2]. Feynman's forward-thinking ideas suggested that nanotechnology could enable a proper toolkit to take the miniaturization to atom scale, offering a pathway to enhance system functionality and overcome current limitations.

Spintronics is a promising way to miniaturize electronic circuits, but also a more efficient technology that might replace the current device, the complementary metal-oxide-semiconductor (CMOS), by leading to highly integrated systems. This novel approach relies on the use of spin (and, thus, its associated magnetic moment) instead of only the electrical charge for device functionality. The electron spin is already being applied in computing to read data in hard disk drives based on the giant magnetoresistance (GMR) effect, discovered by Nobel Prize winners Albert Fert and Peter Grunberg [3,4]. Nevertheless, beyond that, once the electron spin is a two-level system, it is an inherent candidate for realizing a quantum bit (qubit) in quantum information processing (QIP) [5].

A variety of qubit-based algorithms have been developed since the 1990s. A priori, these algorithms can handle very complex calculations that are not available in current classical computers. In 2019, IBM unveiled the first quantum computer designed for commercial use, the IBM Q System One, which operates with twenty qubits. Recently, Google claimed the achievement of quantum supremacy with its 54-qubit Sycamore processor, which could perform, in two-hundred seconds, a calculation that would have taken ten thousand years to the most powerful non-quantum computer [6]. Unlike the IBM Q System One and Google's Sycamore systems that use superconducting qubits, other proposals for a quantum computer emerged using photons, named Jiuzhang, which generates up to 76 output photons and allows for the quantum cloning of two-phonon entangled states for the first time [7,8].

As the race toward quantum supremacy intensifies and electronics nears the limits of silicon, the development and implementation of spintronic components are becoming increasingly crucial. These technologies promise enhanced information processing performance and higher-density storage at lower costs, following the familiar trend of faster, smaller, and cheaper devices [9,10]. To date, spintronic devices have relied on conventional materials like inorganic metals and semiconductors. However, the prospect of a spin-based quantum computing machine is no longer far-fetched [11–13].

A new field, known as single-molecule spintronics (SMS), has emerged, combining spintronic concepts with the unique quantum properties of molecular systems at the nanoscale [14–29]. SMS offers several key advantages:

(i) It provides an ideal platform for studying and understanding the fundamental principles of spintronics;
(ii) It enables the creation of devices governed by quantum mechanics;
(iii) It achieves the ultimate level of miniaturization for spintronic devices;
(iv) It leverages a broad range of chemical tools to optimize quantum properties for various device architectures and applications, and, most importantly;
(v) It paves the way for retrieving information from dynamic devices integrated into computers or sensors, facilitating a clear and efficient information read-out process.

Mononuclear spin-crossover (SCO) [30–47] and single-molecule magnet (SMM) compounds, also known as single-ion magnets (SIMs) [48–60], represent the smallest magnetic units in molecular magnetism (MM). Due to their dynamic molecular and quantum behavior, these compounds have promising applications in quantum computing and high-density information storage. Their multifunctional, multiresponsive, and bistable properties, along with their nanoscale size, ease of manipulation, and versatile addressing options, make SCO and SIM compounds ideal synthetic and theoretical models for fundamental studies on the phenomena that rule SMS and QIP [61–73]. Indeed, both classes of molecules are strong candidates for active components in molecular spintronic circuits and quantum computers.

### *1.1. SCO Complexes as Spin Quantum Nanodevices in SMS*

Using a single molecule as a spin carrier represents the ultimate level of miniaturization for an SMS device. SCO molecules have garnered attention due to their ability to respond to multiple stimuli, making them well suited as conductive elements within nanoscale structures, known as molecular junctions. The most basic device configuration consists of a source and drain electrode to establish electron flux. A more complex setup, the transistor, incorporates an additional gate electrode to regulate electron flow through the molecule by applying a potential. Notably, integrating SCO molecules into these devices has yielded promising results.

In an experiment conducted by Park et al., thiol-functionalized terpyridine-cobalt(II) complexes with the formula $[Co(TERPY\{CH_2\}_nSH)_2]^{2+}$, where $n = 0$ or 5, molecules that are part of the well-known family of octahedral spin-crossover (SCO) cobalt(II) imine complexes [35] connect two gold electrodes [61]. The study demonstrated that altering the number of methylene groups in the ligand could control electron transport properties. When the low-spin (LS) cobalt(II) complex ($S = 1/2$) with the short linker ($n = 0$) was anchored to the gold electrodes, a Kondo peak appeared, evincing a strong coupling regime. In contrast, using a longer carbon chain ($n = 5$) as the linker resulted in weak coupling, where the Coulomb blockade dominated the behavior. Subsequently, the parent $[Co(TERPY\text{-}SH)_2]^{2+}$ complex was connected to two mobile gold electrodes [62]. The mechanical manipulation of the junction enabled spin control by altering molecular symmetry, leading to the splitting of the Kondo peak as the molecule was stretched.

The authors concluded that the observed Kondo resonance arises from a triplet spin ground state originating from the reduction in the metal to a high-spin (HS) cobalt(I) complex ($S = 1$) as the current flows through the molecule and not the molecular changes linked to the HS/LS transition on the original cobalt(II) complex, as previously seen in related iron(II) SCO complexes with redox-active ligands such as 2,6-bis(pyrazol-1-yl)pyridine (DPZPY) and 3,5-bis(pyrazol-1-yl)-4,4′-bipyridine (DPZBPY) [63]. In the last case, the authors suggested that the observed Kondo peak splitting, as a function of gate voltage, was due to changes in the charge of the two ancillary ligands, modifying the spin state but not the metal oxidation state, as supported by theoretical calculations on the $[Fe(DPZPY)_2]^{2+}$ model complex [63]. This charge-induced spin-crossover was associated with the transformation of the LS iron(II) ($S = 0$) complex into the antiferromagnetically coupled, doubly reduced π-radical anion HS iron(II) ($S = 2$) species, $[Fe^{II}(DPZBPY^{\bullet -})_2]$, with a resulting $S = 1$ state.

Harzman et al. reported another innovative method for controlling electron transport in single-molecule spintronic devices, as illustrated in Figure 1 [64]. The novelty lies in the sensitivity of the spin state to the spatial arrangement of the ligands and the overall distortion of the asymmetric $[Fe(TERPY\text{-}PhSAc)(TERPY\text{-}PhXY)]^{2+}$ complex (Figure 1a), where two sulfur-containing groups modify one TERPY-PhSAc ligand to allow the complex to be anchoring to the surface of gold electrodes, while the other TERPY-PhXY ligand containing two polar groups remains free to rotate. The strong ligand field pairs the electrons, resulting in a diamagnetic LS ground state for the iron(II) complex in its almost ideal geometry (Figure 1b, top). However, mechanical distortion of the coordination environment reduces symmetry and weakens the ligand field, making $e_g$ orbitals accessible, which results in a paramagnetic HS ground state (Figure 1b, bottom). The application of an electric field polarizes the free TERPY ligand, causing a reduction on the symmetry of the octahedral coordination sphere, which alters the conductance across the junction as a bias-dependent spin-switching mechanism of control. In contrast, when the SCO iron(II) $[Fe(S\text{-}BPP)_2](ClO_4)_2$ complex [S-BPP = (S)-(4-{[2,6-(dipyrazol-1-yl)pyrid-4-yl]ethynyl}phenyl)ethanethioate] was integrated into gold nanoparticle arrays, the resistance exhibited a minimum around 200 K, according to a higher molecular resistance of the HS state [65].

Remarkable results were also achieved by addressing the individual *trans*-bis(thiocyanate) iron(II) SCO molecules $[FeL_2(NCS)_2]$ {L = 3-(2-pyridyl)[1-3]triazolo [1,5-a]pyridine (TZPY) and *N,N′*-bis(1-pyridin-2-ylethylidene)-2,2-dimethylpropane-1,3-diamine (BEPYDMP)} between a gold surface and the ferromagnetic nickel tip of the scanning tunneling microscope (STM), as depicted in Figure 2 [67,69]. In their seminal work, Aragonès et al. found out that, when subjected to a spin-polarized current, the device exhibited spin state-dependent electron transport, in which the TZPY ligand acts like a spin filter, and the HS complex increases the single-molecule conductance as the polarization of the nickel electrode shifted from spin-up (α) to spin-down (β) [69]. In contrast, the electrical conductance remained unchanged when the diamagnetic LS iron(II) complex with the BEPYDMP ligand was concerned [67]. These results agreed with earlier theoretical predictions by Aravena and Ruiz, whose anticipated higher conductance for the aforementioned paramagnetic HS iron(II) complex with the TZPY ligand [66].

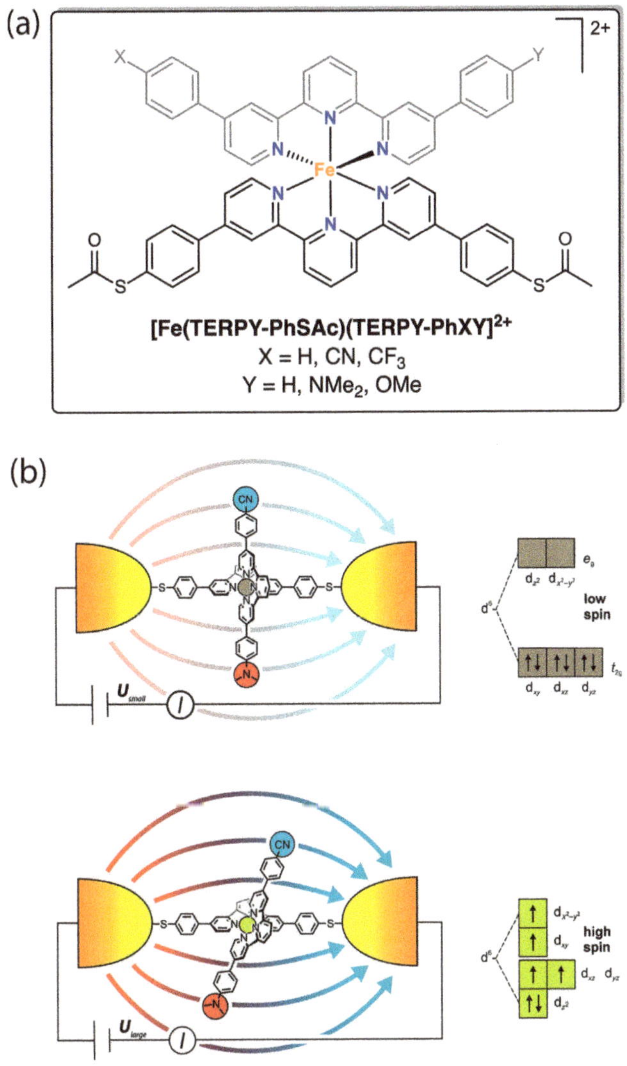


**Figure 1.** (**a**) The general chemical structure of the series of polar heteroleptic iron(II) complexes of formula [Fe(TERPY-PhSAc)(TERPY-PhXY)]$^{2+}$ as a prototype of the electric field-effect spin quantum transistor. (**b**) An illustration of the single-molecule junction constructed from the iron(II) complex connecting two gold electrodes at a small, applied voltage (top) and the distorted HS complex due to the alignment of the ligand TERPY-PhCN(NMe$_2$) to the applied electric field. Adapted with permission from Ref. [64].

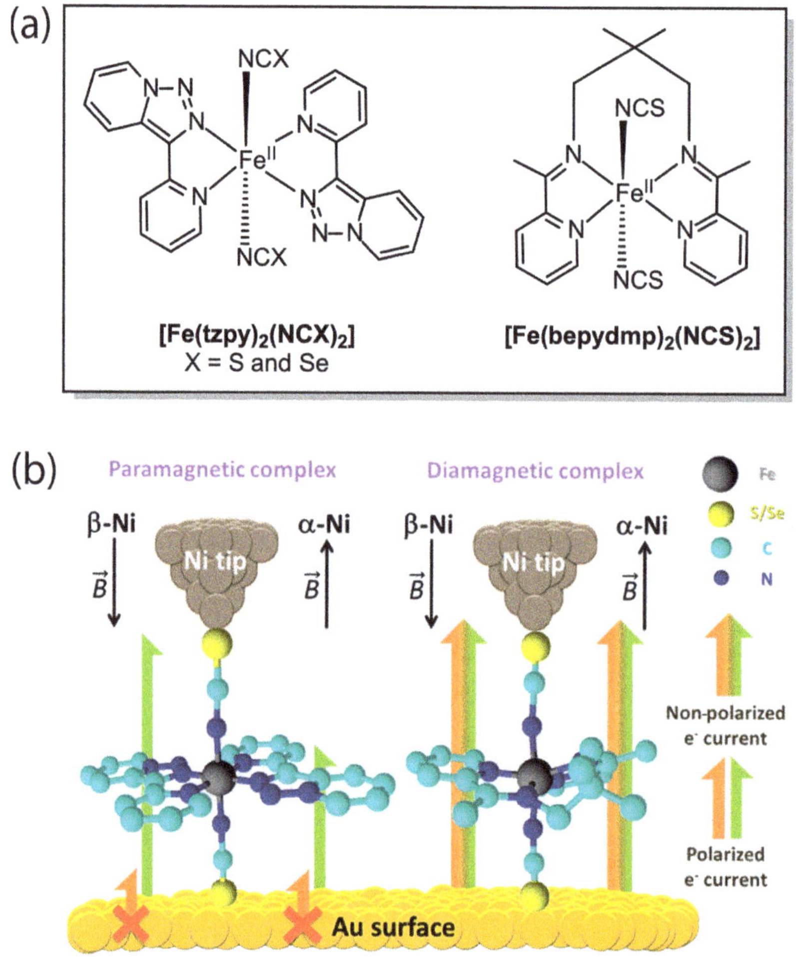


**Figure 2.** (**a**) General chemical structures of the mononuclear iron(II) complexes $[Fe(TZPY)_2(NCX)_2]$ (X = S and Se) and $[Fe(BEPYDMP)_2(NCS)_2]$ as a prototype of the spin quantum filter. (**b**) An illustration of the conductance switching of the spin-dependent electron transport in a real single-molecule junction constructed from $[Fe(TZPY)_2(NCX)_2]$ (X = S and Se) and $[Fe(BEPYDMP)_2(NCS)_2]$ connected between the gold surface and the ferromagnetic nickel tip of the STM with opposite spin polarizations. Adapted with permission from Ref. [67].

Graphene is an attractive component for hybrid device materials due to its exceptional surface sensitivity to environmental dielectric properties [68]. In this context, the electrical read-out of light-induced spin transition was investigated by building a thin film of the SCO iron(II) complex $[Fe(HBDMPZ)_2]$ [HBDMPZ = hydrotris(3-5-dimethylpyrazole)borate] on a graphene sensing layer. The switch from diamagnetic LS to paramagnetic HS increased the conductance of the graphene channel under both light irradiation and temperature stimulus [69]. These findings highlight the critical role of single molecules in addressing the search for new functionalities that would give rise to future nanoscale spintronic devices.

### 1.2. SIMs as Spin Quantum Nanodevices in SMS

Organic molecules have been extensively used in the design of promising electronic devices, leading to important advancements in applications such as semiconductors, light-emitting diodes (LEDs), photovoltaic cells, spin valves, and field-effect transistors (FETs) [70]. Since the discovery of GMR in purely inorganic materials, the study of electron transport, combined with magnetic properties associated with the electron spin, has pushed molecular electronics into a newly emerging area with vast possibilities: molecular spintronics. Some examples of spintronic uses of SIMs as quantum spin transistors and spin valves are detailed hereunder.

An outstanding example of a spintronic transistor is the oxidized double-decker $[Tb(PC)_2]$ SIM molecule (PC = phthalocyanine) absorbed on a gold surface, as shown in Figure 3 [71]. This device exhibits a Kondo resonance peak in electrical conductance at zero-bias voltage ($V = 0$), which indicates transistor-like behavior. This Kondo effect arises from the unpaired spin on a $\pi$-orbital of the oxidized PC ligand since the large magnetic anisotropy of the $Tb^{III}$ ion prevents the degeneration of states necessary to produce Kondo signals. Moreover, controlled current pulses can significantly alter the conformation of the SIM, rotating the upper PC ligand of $[Tb(PC)_2]$ from the ideal azimuthal angle ($\theta$) of 45 to 30° (Figure 3a). This conformational change reversibly commutes the spin, switching the Kondo state ON and OFF (Figure 3b).

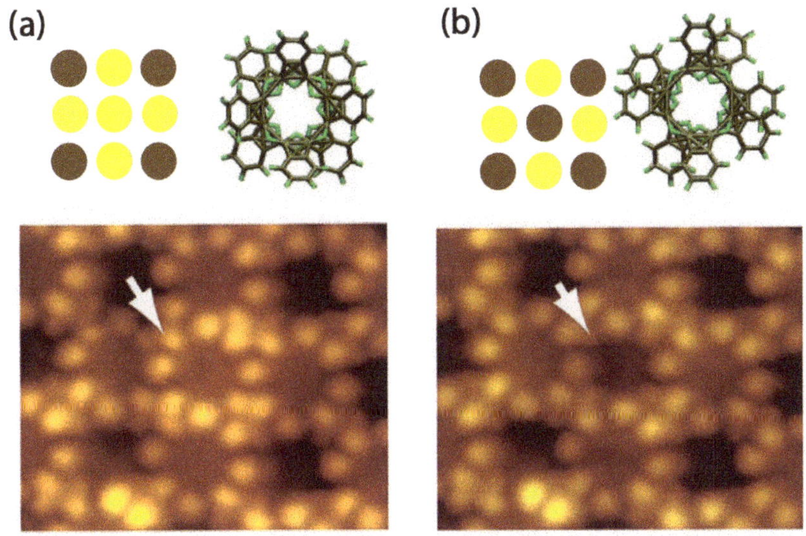


**Figure 3.** STM images of the conversion of the mononuclear terbium(III) complex $[Tb(PC)_2]$ from $\theta = 45°$ (**a**) to $\theta = 30°$ (**b**) by applying a current pulse as a prototype of the spin quantum transistor. An arrow marks the target molecule, and magnified images are shown in the bottom panel. Changes in the contrast and the top view of the center molecule are schematically illustrated. Adapted with permission from Ref. [71].

Taking advantage of the GMR, spin valves, a key component of modern hard drives, offer another possible solution for memory device implementation. Traditionally, these devices utilize metal layers. However, recent research has explored a novel design incorporating SIMs or their supramolecular arrays coupled to a single-walled carbon nanotube (SWCNT), as illustrated in Figure 4 [72,73]. The chosen SIM was the mononuclear terbium(III) derivative $[Tb(PC)(Hx_6BuOpyrPC)]$ accounting for one modified pyrene and six hexyl groups into one of the two PC rings, serving as anchoring points (Figure 4a). Notably, both the energy barrier and relaxation time increased upon anchoring, likely due to suppressing the intermolecular interactions between the SIM units [72]. The device consists of a source and drain connected by an SWCNT, to which the SIM randomly binds through supramolecular interactions (Figure 4b) [73]. The hypothesis was that the giant magnetic anisotropy of the SIM could affect electron transport through the SWCNT. Experimental results confirmed that, when a magnetic field is applied in the plane of the device and then reversed, magnetoresistance ratios rise to 300% at temperatures below 1 K. A simplified model suggests that two SIM molecules (A and B) in the channel generate an effective energy barrier when their spin moments are antiparallel, an effective energy barrier forms, hindering charge flow (Figure 4b, top); however, if the spin moment of B switches under the magnetic field, high conductance is restored across the SWCNT (Figure 4b, bottom).

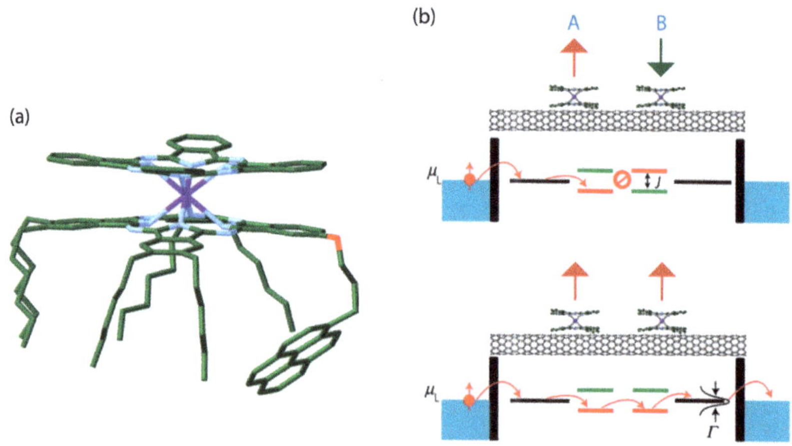


**Figure 4.** (**a**) The chemical structure of the mononuclear terbium(III) complex [Tb(PC)(Hx$_6$BuOpyr PC)] as a prototype of the spin quantum valve. (**b**) An illustration of the magnetoresistance switching mechanism for an SWCNT with two anchored 4f SIMs. The thick red and green arrows are the localized spins of the 4f SIMs, while the thin red arrows represent the spin-polarized conduction electrons across the SWCNT. Adapted with permission from Ref. [73].

## 2. Current Trends and Future Directions

A wide array of SCO systems based on mononuclear iron(II) and cobalt(II) complexes is well known and has served as an essential platform for understanding the SCO phenomenon [34,35,74–87]. Meanwhile, cobalt(II) SIMs have garnered significant attention due to the inherent high magnetic anisotropy of the HS $Co^{II}$ ion and its extensive coordination chemistry [81,88–96]. However, the coexistence of both bistabilities—SCO and SIMs—in a single system remains rare. In first-row transition metal complexes, SIM behavior typically manifests only at very low temperatures, while thermally induced HS/LS transitions occur at higher temperatures, resulting in a diamagnetic or paramagnetic LS state ($S$ = 0 or 1/2 for $Fe^{II}$ and $Co^{II}$ ions) at low temperatures. To date, only a few studies report the coexistence of SCO and SIM behavior in cobalt(II)-based systems [97–108]. This coexistence is not possible for the $Fe^{II}$ ion with a diamagnetic LS state, but it could happen in the HS state. Thus, only three known SCO-SIM systems have been observed in metastable, magnetically anisotropic HS excited states ($S^*$ = 2), achieved through light irradiation via a light-induced excited spin-state trapping (LIESST) process [109–111].

Multifunctional and multiresponsive cobalt(II)-based SCO-SIMs have gained renewed interest in the ongoing transition from MM to SMS and QIP [112]. The cornerstone of this advancement lies in designing and synthesizing novel classes of potentially opto-, electro-, and chemo-active ligands and, therefore, capable of allowing precise control and switching of the spin transition and magnetic relaxation dynamics on metal centers.

### 2.1. $Co^{II}$ SCO-SIMs as Spin Quantum Nanodevices in SMS and QIP

As we continue going further on the road from MM to SMS, our journey has led us to some new insightful outcomes that we will share and discuss below since they will undoubtedly help to reinforce the consolidated MM field while also offering alternative pathways in the emerging SMS domain. Hence, because of their unusual combination of chemical (redox and Brønsted or Lewis acidity) and physical (magnetic and optical or luminescent) properties arising from metal and ligand components, $Co^{II}$-PDI and $Co^{II}$-TERPY complexes are well suited for the design of multifunctional, multiresponsive advanced magnetic materials for SMS nanotechnology. In particular, they emerge as promising candidates for spintronic nanodevices, like spin quantum capacitors and switches.

#### 2.1.1. Molecular Spin Quantum Capacitors and Supercapacitors

Transition metal complexes containing non-innocent electroactive ligands have been extensively studied as single-molecule spintronic devices for charge storage applications [113–115]. The ability to access multiple ligand-based electron reductions (up to four) and one metal-based oxidation makes $Co^{II}$-PDI complexes, featuring electron-donating (methyl, methoxy, dimethylamine, thiomethyl) or electron-withdrawing groups (iodo, bromo, chloro) at the *para*, *meta*, or *ortho* positions of the terminal phenyl rings, promising candidates for use as molecular spin quantum capacitors storing and delivering charge in spintronic circuits, as depicted in Figure 5 [107]. Additionally, their high reversibility allows for repeated charging and discharging cycles without compromising performance.

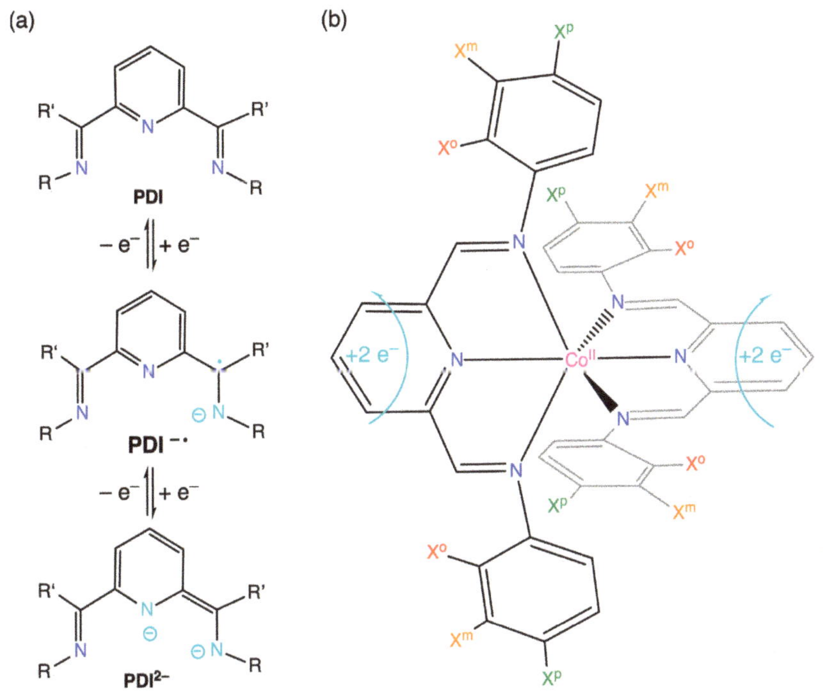


**Figure 5.** (**a**) Redox equilibria of PDI derivatives. (**b**) A schematic illustration of the potentiality of charge storage/delivery of the redox-active phenyl-substituted $Co^{II}$-PDI complexes proposed as a new class of molecular spin quantum capacitors.

Building on this concept, we will explore the idea of supercapacitors, which are relevant to molecules capable of storing energy and facilitating rapid, stable faradaic processes, such as charge and discharge cycles, to deliver substantial amounts of charge during circuit operations [116,117]. We propose the anthraquinone-substituted $Co^{II}$-PDI complex as a promising candidate, as illustrated in Figure 6. Anthraquinone (AQ) derivatives are well known for their $\pi$-acceptor character and redox chemistry, comprising quinone, semiquinone radical, and quinolate forms (Figure 6a) [118,119]. These anthraquinone moieties can serve as efficient multielectron reservoirs to develop a new class of charge storage spintronic devices: molecular spin quantum supercapacitors. Coupled with the PDI ligand, they could enhance the reduction capacity by two electrons per anthraquinone group, potentially allowing up to twelve ligand-based electron reductions within the $Co^{II}$-PDI complex (Figure 6b).

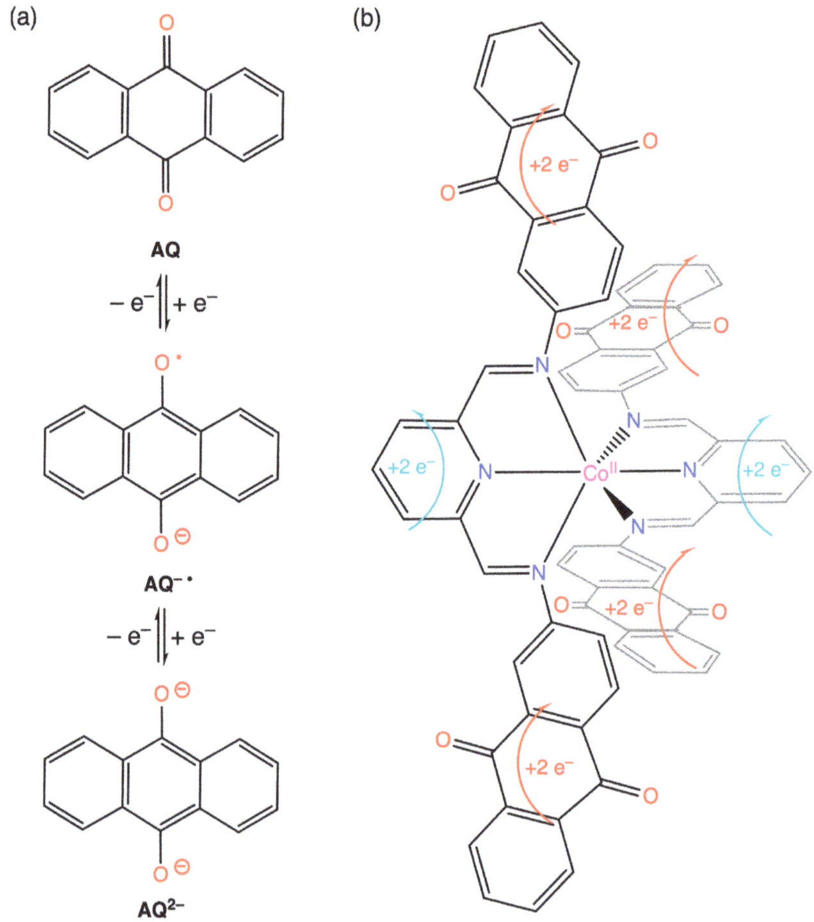


**Figure 6.** (**a**) Redox equilibria of anthraquinone in aprotic media. (**b**) A schematic illustration of the potentiality of charge storage/delivery of the redox-active anthraquinone-substituted $Co^{II}$-PDI complex proposed as a new class of molecular spin quantum supercapacitors.

2.1.2. Molecular Spin Quantum Chemo-, Electro-, or Photo-Switches

Molecular switches are essential components in any highly integrated molecular spintronic circuit, responsible for allowing and interrupting electron transport through the molecule by altering the molecular electronic state. Aware of this need, we propose the development of molecular switches whose spin state (LS or HS) and spin dynamics [fast (FR) or slow relaxation (SR)] can be triggered by means of chemical, electrochemical, and photonic stimuli, referred to as molecular spin quantum chemo- or electro- and photo-switches.

The ON–OFF electro-switching behavior of the *para*-methoxyphenyl-substituted $Co^{II}$-PDI complex, detailed in Figure 7, demonstrates how this $Co^{II}$ SCO-SIM can work as a room-temperature molecular spin quantum electro-switch [104]. This complex can reversibly switch between the slow-relaxing paramagnetic LS $Co^{II}$ ion (ON) and the diamagnetic LS $Co^{III}$ ion (OFF) through chemical or electrochemical redox processes.

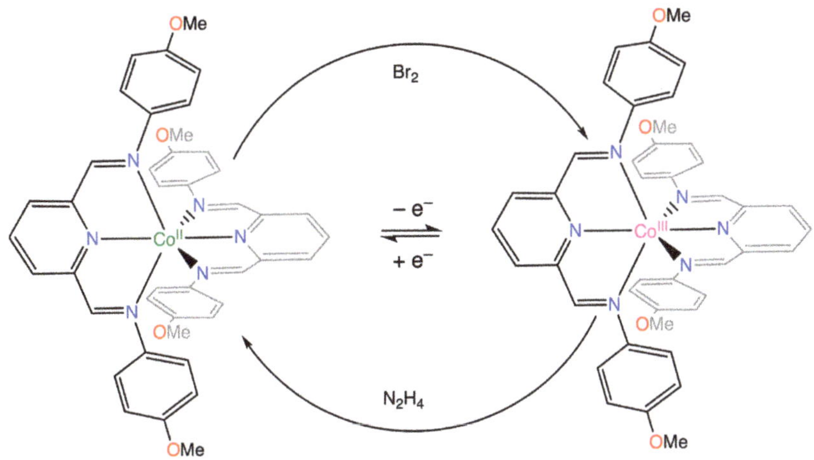


**Figure 7.** Reversible redox switching of *p*-methoxyphenyl-substituted $Co^{II/III}$-PDI complexes under chemical or electrochemical oxidation and reduction, as a new class of molecular spin quantum electro-switch.

The chemo-switching observed in the benzoic/benzoate-substituted $Co^{II}$-TERPY complexes inspired the second molecular spin quantum switch design, as illustrated in Figure 8 [105]. On them, we demonstrated how a single input, specifically a proton, can switch the redox, luminescent, and spin dynamic properties of a single molecule, while minimally affecting the SCO properties. A progressive enhancement of the luminescence in the ultraviolet (UV) region occurs upon deprotonation, concomitantly with a color change from orange to red, and is accompanied by a partial reversibility lost for the one-electron oxidation of the paramagnetic LS $Co^{II}$ ion to the diamagnetic LS $Co^{III}$ ion upon protonation. The dynamic equilibrium between the FS-protonated and SR-deprotonated forms of the LS $Co^{II}$ ion illustrates well the whole potential of this advanced new class of pH-responsive electric, optical, and magnetic switches in the pursuit of multifunctional molecules that can serve as multiresponsive switches in highly integrated spintronic circuits.

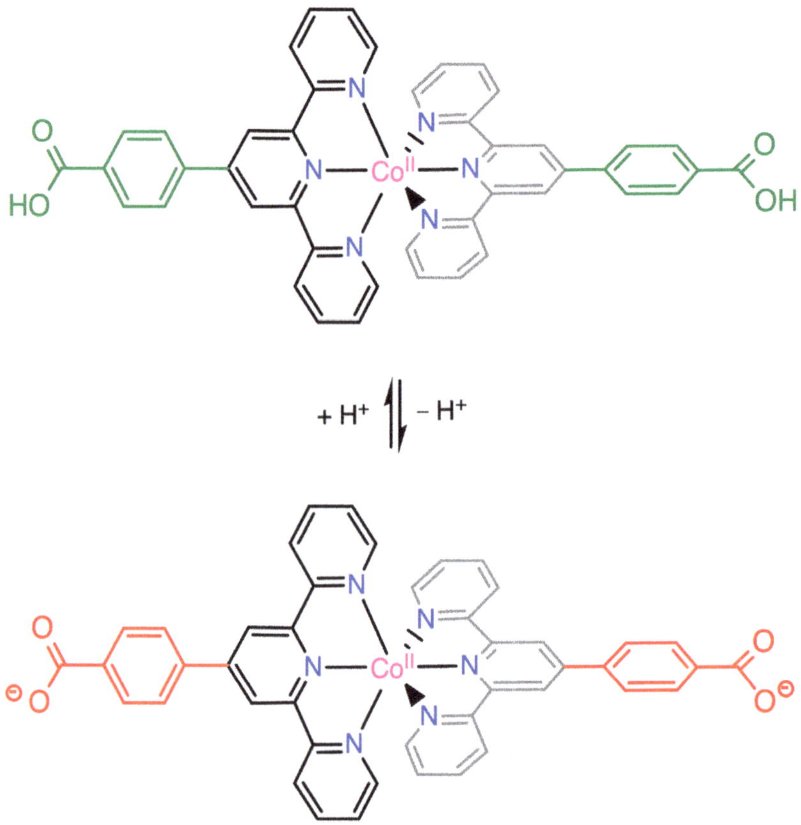


**Figure 8.** Reversible chemical switching of *p*-benzoic/benzoate-substituted $Co^{II}$-TERPY complexes under deprotonation and protonation, as a new class of molecular spin quantum chemo-switch.

The photo-control of magnetic properties along this family of cobalt(II)-TERPY complexes constitutes a major challenge for SMS, where SCO and photochemical properties converge, as well as SIM ones. Incorporating photoactive styryl or azophenyl substituents in the $Co^{II}$-TERPY complex, we can explore an innovative design of molecular switches focused on the well-known *cis-trans* (*Z-E*) photoisomerization reaction of styrene and azobenzene derivatives, as well as the duality of the magnetic relaxation dynamics of the $Co^{II}$ ion depending on its electronic configuration, as illustrated in Figure 9. The LS OFF state ($S = 1/2$) and the HS ON state ($S = 3/2$) are a consequence of the changes in the ligand field strength caused by the variation in the aromaticity between planar *E* and non-planar *Z* configurations [120]. Our current efforts focus on further exploring this alternative ligand-driven, light-induced spin change (LD-LISC) route on these related pairs of photoactive stilbene- or azobenzene-substituted $Co^{II}$-TERPY complexes, provided that a significant variation between the LS and HS population can be observed upon UV and visible light irradiation.

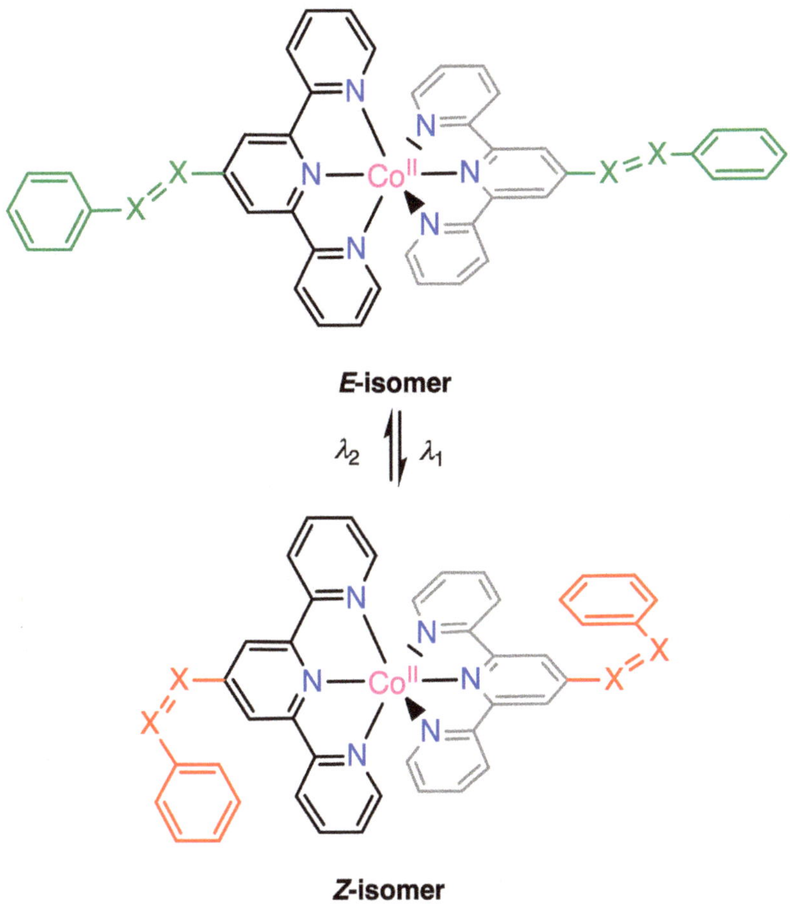


**Figure 9.** Photoisomerisation *cis-trans* (*Z-E*) equilibria of *p*-styryl- (X = CH) and *p*-azobenzene-substituted (X = N) $Co^{II}$-PDI complexes under UV and visible light irradiation, as a new class of molecular spin quantum photo-switches.

2.1.3. Chemo- and Electro-Switchable Molecular Spin Quantum Machines

We observed notable similarities between our $Co^{II}$-PDI complexes and the precursors of classical [2]catenane molecular machines, as illustrated in Figure 10. Both structures feature a mutually perpendicular arrangement of the coordinating-group PDI fragments, including reactive sites that enable the double-cyclization reaction.

The least risky strategy for synthesizing $Co^{II}$-PDI [2]catenanes begins with a pre-synthesized PDI macrocycle prior to the template reaction (Figure 10a). Then, by mixing the PDI macrocycle, $Co^{II}$ ion, and open-chain PDI ligand in a 1:1:1 ratio, the only complex that forms is the desired intermediate (Figure 10b). In the final step, the open-chain fragment converts into another PDI macrocycle, resulting in two interlocked PDI rings [121,122]. At this stage, different approaches can be pursued depending on the functionality chosen for the potentially switchable molecular spin quantum machine. The versatility of these new systems is vast; for instance, the synthesis of a metal-based electro-switch using mixed TERPY/PDI macrocycles would integrate both platforms of cobalt(II) SCO-SIM complexes explored above (Figure 10c). Hopefully, the interplay could be triggered by either a chemical or electrochemical oxidation on the metal ion, causing a sliding motion that enhances the stabilization of the $Co^{III}$ ion in the TERPY coordination environment.

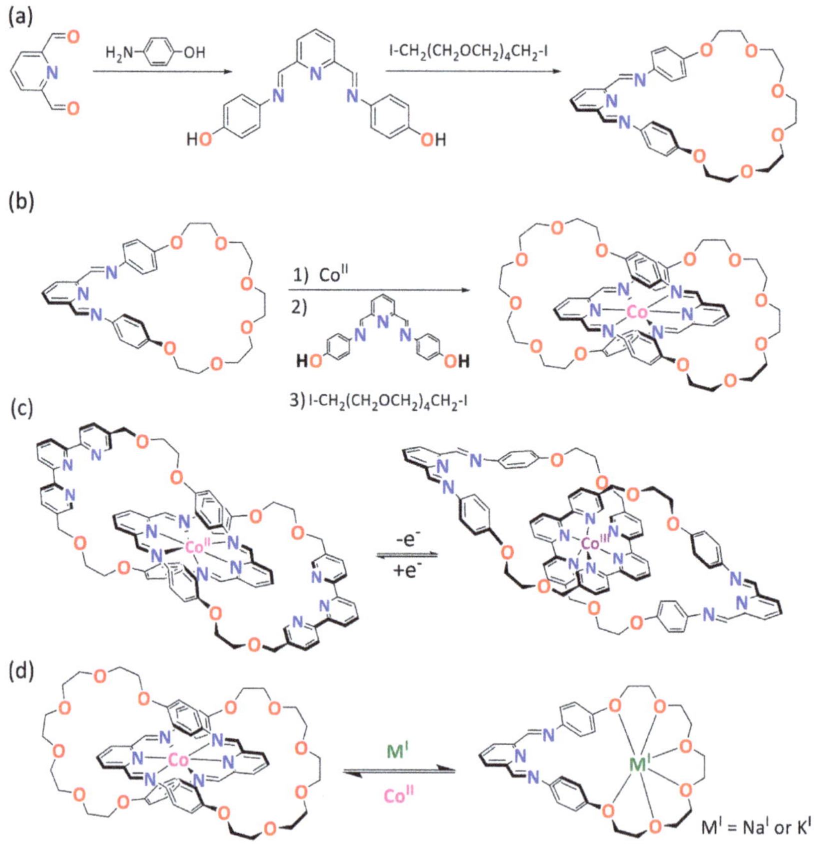

**Figure 10.** Proposed synthetic routes for (**a**) functionalized PDI macrocyclic and (**b**) catenane-type $Co^{II}$-PDI SCO-SIM complexes through a metal-directed template and double-cyclization reactions, leading to chemo- and photo-switchable molecular spin quantum machines. (**c**) The electrochemically or (**d**) chemically triggered sliding motion on mixed TERPY/PDI or pure PDI, catenane-type $Co^{II}$ SCO-SIMs.

Ligand-based chemo-switches using these polyether-functionalized PDI macrocycles could interact with $Na^+$ and $K^+$ alkali cations, causing a sliding motion of the catenane and the subsequent release of the coordinated $Co^{II}$ ion (Figure 10d). This feature evokes catenanes to work as molecular playgrounds for $Na^I/K^I$ and $Co^{II}$ ions. When the $Co^{II}$ ion is present, both SCO and SIM phenomena would be in the ON state, while high concentrations of alkali metal ions lead to the OFF state.

This innovative design may open up the path to a new class of mechanically bonded, catenane-type $Co^{II}$ SCO-SIMs as unique examples of dynamic molecular systems with multiple switchable properties triggered by internal molecular motion.

### 2.2. Addressing of $Co^{II}$ SCO-SIMs as Spin Quantum Nanodevices in SMS

The high expectations for this novel class of $Co^{II}$ SCO-SIMs in data storage and processing stem from the potential to create spintronic devices in which these molecules are uniformly organized across different inorganic and organic supports (e.g., silver, silicon, or graphene) while preserving their singular magnetic behavior [123–128].

The current challenge in the SMS area is addressing individual magnetic molecules into functional devices to study their spin-dependent electron transport properties. Basically, there are three approaches to address molecules for electron transport:

(i) Electrode-based junctions: Placing molecules between two electrodes and measuring the conductance across the device.
(ii) Scanning tunneling microscopy (STM): Depositing molecules on a surface and using an STM tip as a lead to measure the electron transport properties across the individual molecule.
(iii) Nanostructure integration: Anchoring molecules to carbon-based nanostructures through supramolecular interactions, using molecules for their spin, while electron transport occurs through the quantum dot or the carbon nanostructure.

Recent research has yielded promising prototypes of spintronic devices based on either SCO or SIM molecules from these three strategies (see the Introduction and Background Section).

Due to their nanometer size and ease of handling, the $Co^{II}$ SCO-SIM compounds that constitute the leitmotif of this perspective review appear as suitable candidates for the study of spin dependence of the electron transport through single molecules, envisaging the magnetic control of conduction properties or vice versa, the electric control of magnetic properties. Let us briefly anticipate some of the expected results as spin quantum transistors, filters, or valves for single-molecule electron transport, which can be achieved with this variety of SCO molecular nanomagnets when they are connected to two gold electrodes, attached to a gold surface, integrated in a SWCNT, or anchored on a nanographene (NG).

2.2.1. Molecular Spin Quantum Transistors in Gold Molecular Junctions

The observed capacitor-like redox behavior in the class of cobalt(II)-PDI SCO-SIMs allows us to propose them as prototypes of electric field-effect spin quantum transistors (EFE-SQTs) in SMS [64]. So, for instance, the 4-thiomethylphenyl-substituted $Co^{II}$-4-MeSPhPDI complex is a very appealing candidate due to its strong affinity for gold electrodes via the sulfur atoms of the thiomethyl substituents, as illustrated in Figure 11 [107]. Two opposite thiomethyl substituents from one ligand would connect to the source and drain electrodes, establishing the electron flux (Figure 11a). Depending on the applied electric voltage, dramatic changes in the electrical conductances can be then presumed when passing from the HS ($S$ = 3/2) or LS ($S$ = 1/2) cobalt(II) complex, $[Co^{II}(4\text{-MeSPhPDI})_2]^{2+}$ to the antiferro- ($S$ = 1/2) or ferromagnetically coupled ($S$ = 3/2) metal-diradical reduced species, $[Co^{II}(4\text{-MeSPhPDI}^{\bullet-})_2]$, as previously reported for the related $Fe^{II}$ DPZBPY SCO complex [63].

As with most transition metal-based SIMs, the presence of a fast quantum tunneling of magnetization (QTM) in our cobalt(II) SCO-SIMs under a zero applied *dc* magnetic field offers an alternative and highly efficient relaxation mechanism that severely limits their use in high-density information storage devices. Otherwise, the considerable decrease in the magnetic relaxation rate with an increasing external magnetic field observed in most LS cobalt(II) SIMs could be exploited to develop a new class of magnetic field-effect spin quantum transistors (MFE-SQTs) for applications in SMS [106]. In contrast to conventional electronic devices controlled by an electric field, the electrical current flow, in such cases, would be governed by a magnetic field. Indeed, different electrical conductances can be expected for FR and SR LS $Co^{II}$ ions, occurring in the absence or presence of an external magnetic field. Although this approach to operating on a device involves a greater degree of difficulty than using an electric current due to the challenge of confining a magnetic field to a nanometric or smaller region, Bogani and Ruben demonstrated that this was possible by creating supramolecular SVQs [73].

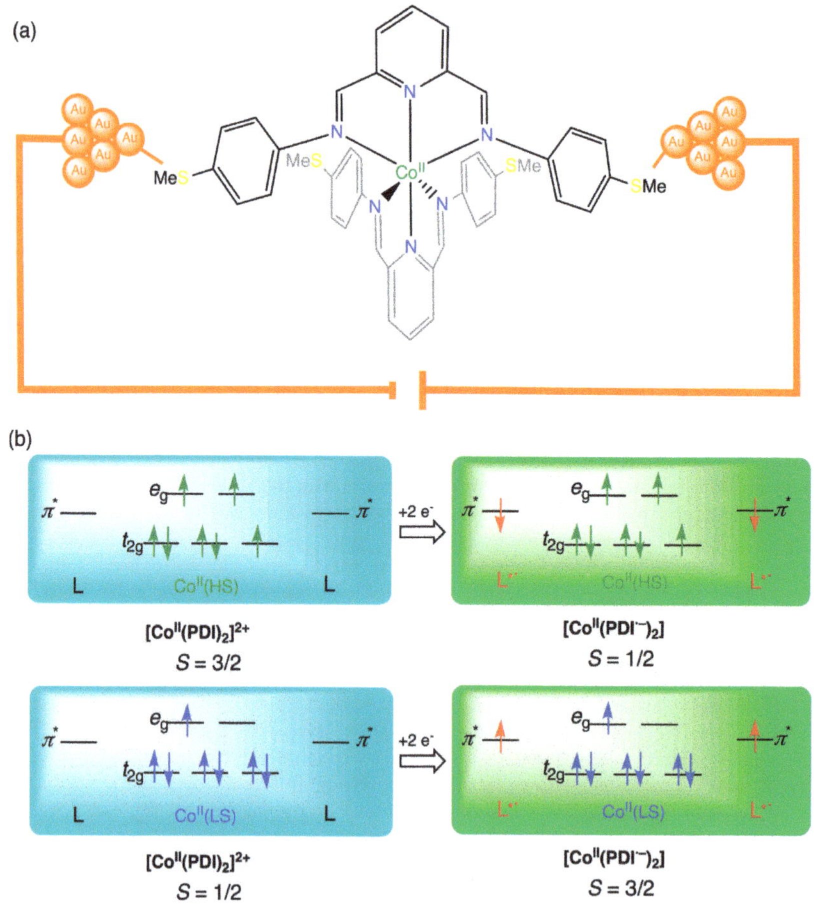


**Figure 11.** (**a**) An illustration of the proposed molecular junction device constructed by the 4-thiomethylphenyl-substituted $Co^{II}$-PDI complex connecting two gold electrodes as a prototype of the electric field-effect spin quantum transistor. (**b**) A simplified energy level diagram of the interaction between the d-type metal orbitals and the $\pi$*-type ligand orbitals for the HS (**top**) and LS (**bottom**) $Co^{II}$ forms of the cationic $[Co^{II}(PDI)_2]^{2+}$ complex and the corresponding doubly reduced neutral $[Co^{II}(PDI^{\bullet-})_2]$ species, illustrating the charge-induced electro-switching magnetic behavior.

### 2.2.2. Molecular Spin Quantum Filters in Gold Surfaces

Within this context, the electroactive $Co^{II}$-PDI complexes with electron-donor thiomethyl substituents on the terminal phenyl rings offer a promising platform for developing a new class of spin quantum filters (SQFs), which is due to several factors:

(i) The sulfur atoms have a large affinity for various metallic substrates.
(ii) The spin dynamic properties can be tuned when an external magnetic field is applied, similar to the behavior of bulk materials.
(iii) The electrochemical activity associated with metal oxidation can work as an ON–OFF switch for the spin filter device.

Consequently, while the paramagnetic HS $Co^{II}$ ion in the ON state would filter spin-up ($\alpha$) electrons, the diamagnetic LS $Co^{III}$ ion in the OFF state would lack any selectivity in distinguishing spin-polarized electrons, as previously reported for the isoelectronic diamagnetic LS $Fe^{II}$-BEPYDMP complex (see Figure 2) [67].

The chiral-induced spin-selectivity (CISS) effect links chiral symmetry directly to the electron spin within molecules, evident through phenomena such as electron transmission and displacement currents driven by charge polarization that subsequently induce spin polarization [129]. Consequently, the traditional definition of the OFF state needs to change when considering chiral molecules; what was previously deemed OFF now forms the foundation for chiral spin filters, thereby expanding the scope of spintronic applications. However, the previous OFF state is expected to show an attenuated spin polarization concerning the ON state [130], which can be used to design memristors that are resistance-switching memory devices capable of mimicking biological synaptic functions in artificial neural networks [131].

### 2.2.3. Molecular Spin Quantum Valves in Carbon-Based Nanostructures

Hybrid carbon-based materials play a crucial role in enhancing and thoroughly exploring the spin dynamic properties of SIMs. Recognizing this potential, Bogani and Ruben proposed anchoring a lanthanide-based SIM onto an SWCNT [70,71]. Achieving this required a rational-designed SIM incorporating anchoring groups capable of engaging in supramolecular $\pi$–$\pi$ interactions with the SWCNT (see Figure 4). In their pioneering work, Bogani and Ruben successfully self-assembled a heteroleptic derivative of the archetypical double-decker [Tb(PC)$_2$], bearing one modified pyrene and six hexyl groups, onto the SWCNT (see Figure 4) [72]. From these results, a new molecular spintronic device emerged: the supramolecular spin quantum valve (SQV) [73].

Mononuclear Co$^{II}$-PDI complexes with oligoacene (naphthalene and anthracene) substituents would be suitable for preparing these hybrid metallocarbon nanotube and nanographene architectures. However, a more advantageous option might be Co$^{II}$-PDI derivatives incorporating extended aromatic anchoring groups, like pyrene or its derivatives, along with flexible aliphatic linkers, as depicted in Figure 12.

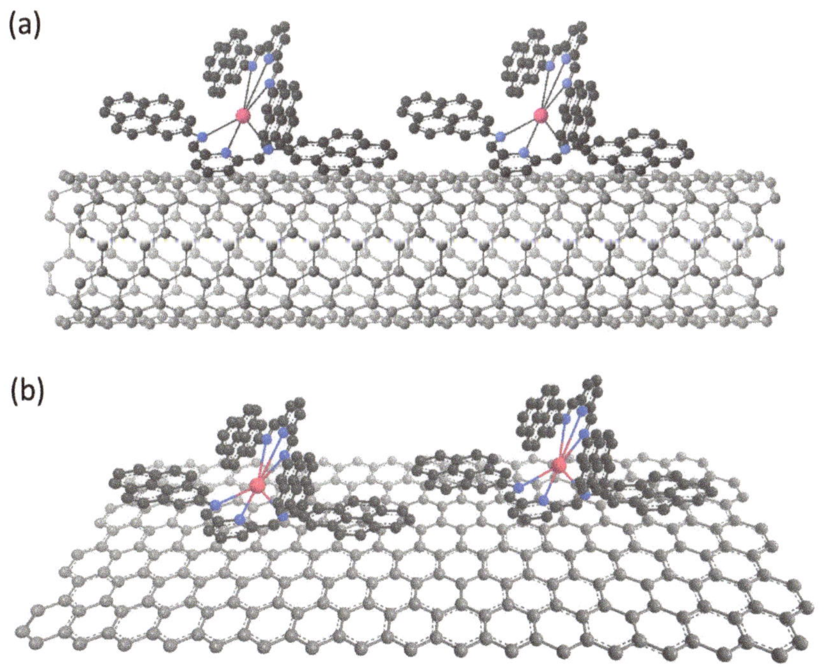


**Figure 12.** Illustration of the supramolecular carbon-based nanotubes (**a**) and nanographenes (**b**) with pyrene-substituted Co$^{II}$-PDI complexes as prototypes of molecular spin quantum valves.

### 2.3. Scaling of $Co^{II}$ SCO-SIMs as Spin Quantum Nanodevices in QIP

A molecular approach to quantum computing nanotechnologies can be likened to a three-tiered process, analogous to the bottom-up strategy employed in supramolecular chemistry. Each step progressively increases complexity, culminating in constructing the desired material. These stages can be summarized as follows:

(i) Designing qubits with exceptional quantum coherence;
(ii) Scaling up to implement double qubit-based quantum gates (QGs);
(iii) Conceiving a quantum computing device as a multiqubit 2D array capable of storing and processing quantum information.

While significant strides have been made in designing and controlling qubits, both individually and within QGs, the final stage, representing the ultimate objective, remains largely unexplored.

In the quest to minimize quantum decoherence, several promising candidates for molecular-based qubits have emerged, incorporating many first-row transition metal ions from vanadium to copper [132–139]. These include the octahedral vanadium(IV), iron(III), and chromium(III) complexes of formula $[V(C_8S_8)_3]^{2-}$, $[Fe(C_5O_5)_3]^{3-}$, and $[Cr(C_3S_5)_2]^{3-}$, in which only nuclear spin-free elements (C, O, and S) compose the ligands [132–134]. This feature, combined with the low abundance of nuclear spin-active isotopes in these elements, grants long coherence times ($T_2$ or $T_m$) once nuclear spins are the main source of magnetic noise and thus decoherence. Changing the chemical matrix to a spin-free solvent like $CS_2$ or $SO_2$ made it possible to increase $T_2$ up to 675 μs at 10 K for $[V(C_8S_8)_3]^{2-}$ [132]. In contrast, the related square-pyramidal oxovanadium(IV) complex, $[VO(C_8S_8)_2]^{2-}$, exhibited a slight decrease in $T_2$ to 104 μs at 10 K in $SO_2$ [135]. Other remarkable candidates for the qubit include the square-pyramidal oxovanadium(IV)-phthalocyanine [VO(PC)] and square-planar copper(II)-maleonitriledithiolate $[Cu(MNT)_2]^{2-}$ complexes, owing $T_2$ values of 1 μs at room temperature, despite placing donor or nuclear spin-active isotope (N) atoms close to the metal center [136,137].

Scalability is essential in the development of spin-based quantum computing. This term, closely related to entanglement, refers to the ability to increase the number of qubits in arrays while preserving the capacity to manipulate and read out information without compromising quantum coherence. This topic is a challenging step in the development of a spin-based quantum computer. Each array has its own advantages and disadvantages, making it essential to have a versatile and easy-to-handle qubit platform that may be processed in various architectures, such as our $Co^{II}$-TERPY and $Co^{II}$-PDI complexes, illustrated in Figure 13. To follow, we provide some attempts to exemplify the potential and versatility in designing, scaling, and addressing cobalt(II) SCO-SIMs discussed in this review as building components for quantum computing technologies. In this context, new constructions could be achieved through diverse supramolecular cobalt(II) SCO-SIM architectures obtained through ligand design.

Our results may open a pathway toward the design of operative molecular spin qubits or qubit arrays projecting the physical implementation of a new class of multifunctional electrical, optical, chemical, and magnetic materials for QIP. After identifying the key design criteria to improve the SCO and SIM performance of our mononuclear cobalt(II) complexes, two critical challenges should be addressed to achieve this goal:

(i) Pinpointing the primary sources of quantum decoherence in cobalt(II) SCO-SIMs;
(ii) Building supramolecular arrays of cobalt(II) SCO-SIMs with increasing complexity, featuring non-negligible feeble and potentially switchable intermolecular interactions;
(iii) Effectively addressing suitable cobalt(II) SCO-SIMs and their supramolecular arrays on surfaces and other solid supports.

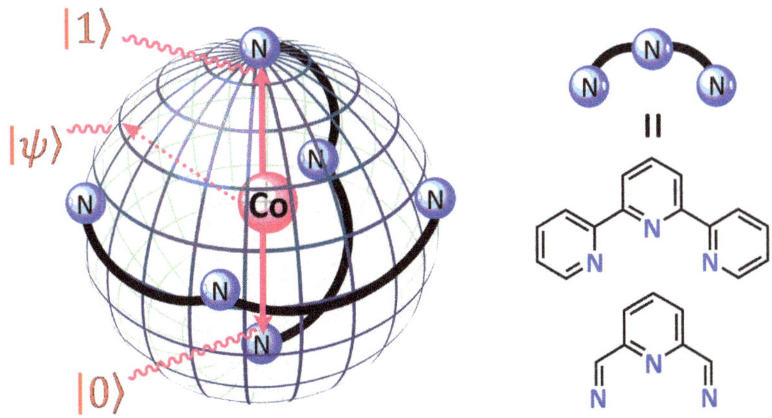


**Figure 13.** Selected TERPY and PDI platforms for the creation of a new chemical library of $Co^{II}$ SCO-SIM molecules as candidates to molecular spin quantum bits.

2.3.1. Molecular Spin Quantum Bits

In cobalt(II) SCO-SIMs, the LS $Co^{II}$ ion is a genuine two-level magnetic quantum system represented by the $m_S = +1/2$ and $-1/2$ states, which can play the role of the superposed 0 and 1 states in a qubit for quantum computing applications. In contrast, due to significant first-order spin–orbit coupling (SOC), the HS state possesses an effective doublet ground spin state ($S_{eff} = 1/2$) coming from the ground $m_S = \pm1/2$ or $m_S = \pm3/2$ Kramers doublet for an easy-plane ($D > 0$) or easy-axis ($D < 0$) magnetic anisotropy, respectively. As a matter of fact, the slow magnetic relaxation behavior, along with this family of SCO cobalt(II)-TERPY and PDI complexes, is dominated by single or double Raman mechanisms between the $m_S = +1/2$ and $-1/2$ states [LS or HS ($D > 0$)] or $m_S = +3/2$ and $-3/2$ states [HS ($D < 0$)] through a virtual excited state, with no participation of Orbach one through the excited $m_S = +1/2$ and $-1/2$ states in the latter case, and occasionally with additional temperature-independent Intra-Kramer [LS or HS ($D > 0$)] or QTM mechanisms [HS ($D < 0$)] [105–107].

Typically, the magnetic relaxation time for a doublet spin ground state is equivalent to the spin-lattice time ($T_1 = \tau$), which acts as a limiting factor for the phase memory time ($T_m \leq T_1$). Our octahedral LS $Co^{II}$-TERPY and $Co^{II}$-PDI complexes exhibit promising $T_1$ values at 2.0 K, ranging from 100 μs to 1 ms [105–107]. However, these values are slightly lower than those reported for some square-planar LS cobalt(II) complexes (1.3 and 11.1 ms at 2.0 K), specifically $[Co(TMC)(CH_3CN)]^{2+}$ and [Co(PC)], where TMC = 1,4,7,10-tetramethyl-1,4,7,10-tetraazacyclododecane [101,138]. However, complete pulsed EPR studies should be performed on both LS and HS complexes to characterize and assess their suitability as new candidates for spin qubits.

Recently, our research group has focused on pulsed EPR studies of octahedral HS cobalt(II) complex $[Co(DMPHEN)_2(SAL)]^+$ (DMPHEN = 2,9-dimethyl-1-10-phenantroline; SAL = salicylaldehyde) with a positive $D$ value ($S = 3/2$) [139]. In this case, due to the strong SOC contribution, the $T_1$ rapidly decreases as the temperature increases. Yet, the $T_m$ value for this complex is somehow comparable to other qubit candidates (around 1 μs at 2.0 K) [132–137], which is a good relaxation time for this low temperature. Notably, the dc and ac magnetic or cw-EPR characteristics of this complex are similar to those of our HS cobalt(II)-PDI complexes with *p*-clorophenyl and *p*-thiomethylphenyl substituents (see Figure 5), positioning them as potential candidates for molecular spin quantum bits.

2.3.2. Optically, Chemically or Electrochemically, and Electrically or Magnetically Addressable Molecular Spin Quantum Bits

These distinctive multifunctional and multiresponsive $Co^{II}$ SCO-SIMs appear as up-and-coming candidates for the multiple (optical, chemical or electrochemical, and electrical or magnetic) qubits addressing in future QIP applications. The results herein presented constitute an important step in the pursuit of advanced addressable qubits and the controlled resetting of the quantum gate.

All the compounds discussed here are attractive for developing advanced qubits with electrical readout capabilities. The electro-switching from the paramagnetic LS/HS $Co^{II}$ state to the diamagnetic $Co^{III}$ state of the active metal center offers an excellent opportunity to erase and reinitiate the qubit at the conclusion and beginning of information processing. Furthermore, when exposed to a suitable combination of two or more stimuli—such as an external magnetic field, electrical voltage at the gate, light irradiation, and pH—$Co^{II}$ SCO-SIM compounds can act as quantum logic gates processing all the inputs into a single output.

In the search for ideal qubits, there is a consensus that the all optically addressable qubits are particularly advantageous for several reasons:

(i) Light is an environmentally green stimulus;
(ii) They can be remotely controlled with precision timing and positioning;
(iii) They enable straightforward single-spin readout and initialization;
(iv) They can be orthogonally coupled with other switchable entities in a single-spin configuration to build complex logic multitaskers.

This appealing need within the field of quantum computing field moved us to invest efforts in creating and studying a family of optically active $Co^{II}$ SCO-SIMs by simply incorporating oligoacene luminophore groups, such as naphthalene or anthracene, into our $Co^{II}$-PDI platform, as illustrated in Figure 14. A preliminary study of their photophysical activity in solution reveals significant metal-promoted quenching of the ligand luminescence upon UV light irradiation at room temperature, likely due to a ligand-to-metal charge transfer (LMCT) process that leads to a short-lived excited state of the $Co^{II}$ ion, analogous to the LIESST excited state, which is not readily accessible in our $Co^{II}$ SCO-SIM [140].

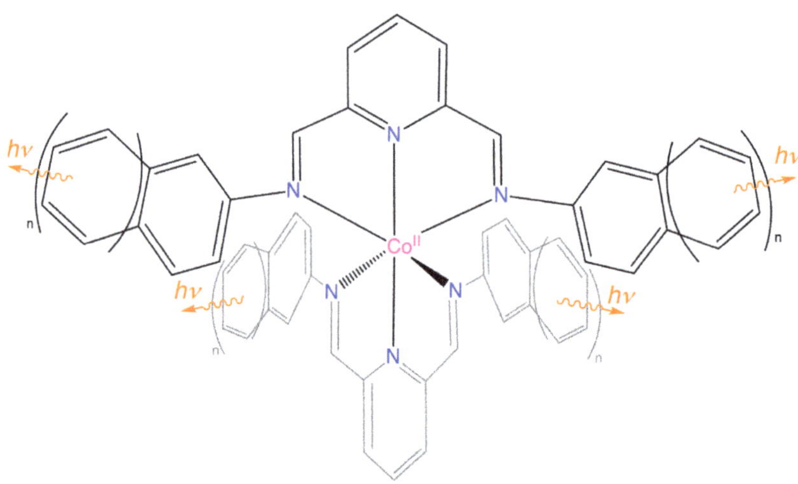


**Figure 14.** A schematic illustration of the potentiality of luminescent 2-naphthalene- (n = 2) and 2-anthracene-substituted (n = 3) $Co^{II}$-PDI complexes proposed as a new class of optically addressable molecular spin quantum bits.

In this regard, the parent $Co^{II}$-TERPY complex exhibits ultrafast Jahn-Teller (JT) photoswitching behavior in aqueous solution, transitioning between axially elongated LS and axially compressed HS forms on the femtosecond timescale upon optical metal-to-ligand charge transfer (MLCT) excitation at room temperature under deep violet-blue light [108]. The photo-control of magnetic properties along this family of cobalt(II)-TERPY complexes constitutes a major challenge for SMS, where SCO and photophysical properties converge, as do SIM ones. Moreover, this finding opens new exciting perspectives for implementing logic schemes in QIP through the optical control of quantum coherence (QC) properties of multiresponsive $Co^{II}$ SCO-SIMs as prototypes of optically addressable qubits, provided that the lifetimes of the photogenerated metastable states are sufficiently long relative to the typical manipulation time of about 10 ns for a qubit.

#### 2.3.3. Exotic Arrays of Molecular Spin Quantum Bits in Soft Matter

Once most sources of quantum decoherence in solution or a solid state (diluted solid solutions) are identified, further investigations into $Co^{II}$ SCO-SIMs in soft media will benefit their potential applications in QIP.

Although some questions about the SCO and SIM behaviors in crystalline solids remain unclear, these topics are still largely studied. This is different when it comes to soft matter. Bulk assemblies such as colloids, liquid crystals, polymers, vesicles, and Langmuir–Blodgett (LB) films incorporating SCO and SIM behaviors remain largely uncharted territory. However, the few works reported in this area provided promising results regarding improving SCO behavior. Examples of this feature include the LB films based on an amphiphilic iron(II) complex, $[Fe([CH_3(CH_2)_{16}]_2BBY)_2(NCS)_2]$ ($[CH_3(CH_2)_{16}]_2BBY$ = 4,4′-diheptadecyl-2,2′-bipyridine), which exhibits thermally assisted cycle-dependent SCO [141,142]. This atypical behavior is attributed to the melting of the alkyl chain groups, aiding in reorganizing the structure to favor the SCO process. Additionally, Sun et al. prepared SCO vesicular nanospheres based on the series of neutral iron(II) complexes with 2,2′-bipyridine-4,4′-dicarboxamide ligands bearing different alkyl group substituents, ranging from propyl to cetyl ($R_2BPYCA$ with R = $CH_3(CH_2)_n$ and n = 2–15), represented by the general formula $[Fe(BPZ)_2(R_2BPYCA)]$ [BPZ = dihydrobis(1-pyrazolyl)borate] [143]. The spin transition temperature of the vesicular nanospheres was significantly shifted from 160 K for the bulk material to room temperature [143].

Another intriguing example is the cobalt(II) complex $[Co(C_{16}\text{-TERPY})_2]^{2+}$, which evinces an unexpected reverse spin transition triggered by a structural vitreous phase transition [144]; however, the possible SIM behavior based on the reverse HS state was unfortunately not studied. In this context, mononuclear $Co^{II}$-PDI and TERPY complexes with long polyalkyl substituents offer a suitable $Co^{II}$ SCO-SIM platform for studying both phenomena and, ultimately, for testing their qubit performance in soft matter for the first time, as depicted in Figure 15.

#### 2.3.4. Molecular Spin Quantum Bits in Porous MOF-Based Materials

An original approach to address individual spin-crossover cobalt(II) molecular nanomagnets within matrixes would be to take advantage of the combined host–guest chemistry and geometrical features of metal–organic frameworks (MOFs) to produce a new class of hybrid materials. MOFs have been extensively studied for their applications in gas storage and separation, drug delivery, molecular recognition, transport and catalysis, as well as for their use as a template for the growth of nanoparticles and the encapsulation of a wide variety of functional moieties [145–149].

Owing to their porosity and rich host–guest chemistry, MOFs are excellent playgrounds for designing multifunctional and multiresponsive hybrid materials by hosting $Co^{II}$ SCO-SIMs on pre-existent MOF matrices. This setup allows for the investigation of quantum decoherence in a confined environment of the adsorbed $Co^{II}$ SCO-SIM, the results of which will be compared with those obtained in solution, a solid state, or soft matter to optimize their performance as qubits for future applications in QIP.

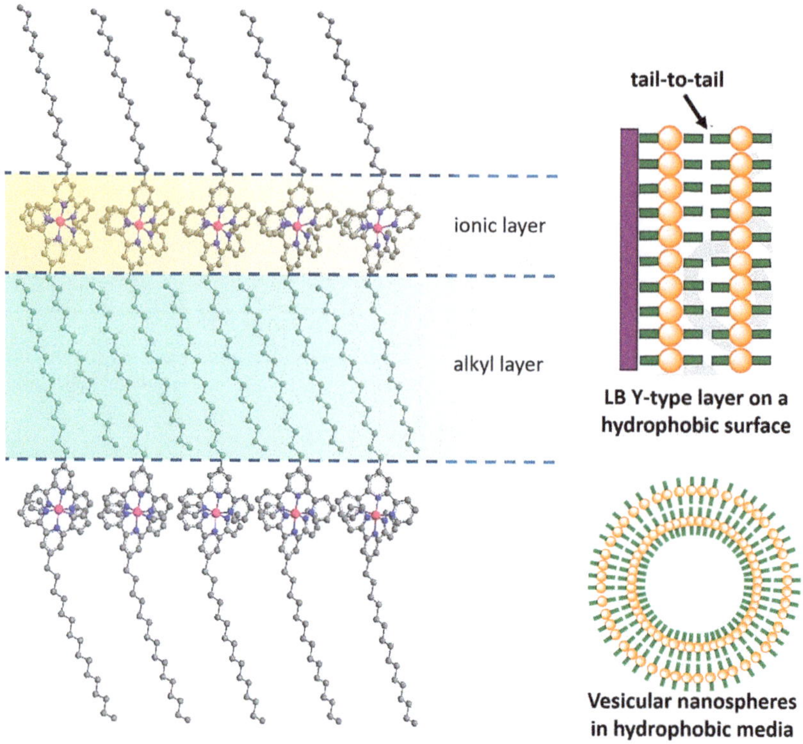


**Figure 15.** Examples of LB films and vesicular nanospheres from $Co^{II}$ SCO-SIMs with long polyalkyl-substituted TERPY ligands as building blocks for exotic arrays of entangled and potentially switchable molecular spin quantum qubits.

Additionally, the long-range magnetic ordering from the magnetic MOF host can influence the slow magnetic relaxation of the incorporated $Co^{II}$ SCO-SIM guests, resulting in a synergistic interplay of both magnetic phenomena. Previously, our research group reported the partial incorporation of cationic iron(III)-Schiff's base SCO and manganese(III)-porphyrin SIM complexes into the large channels of an anionic oxamato-based copper(II)-manganese(II) 3D MOF of formula $Na_4\{Mn_4[Cu_2(TMMPBA)_2]_3\}\cdot 60H_2O$ [TMMPBA = *N*,*N*′-2,4,6-trimethyl-1,3-phenylenebis(oxamate)] via cation-exchange single-crystal-to-single-crystal (SC-SC) processes [150–152]. The resulting host–guest hybrid materials exhibited hysteretic SCO behavior or internal field-induced partial quenching of the QTM of the iron(III) and manganese(III) guests, respectively [151,152]. These features constitute unique proofs of the synergistic effects between the spin dynamics of the guests and the long-range magnetic ordering of the host, demonstrating that this class of porous magnets is a well-suited platform for the controlled addressing of SIMs.

There again, our preliminary studies revealed that the parent cationic $Co^{II}$ SCO-SIM complex $[Co^{II}(PhPDI)_2]^{2+}$ could be hosted within the large pores of the aforementioned anionic ferrimagnetic copper(II)-manganese(II) MOF, as illustrated in Figure 16 [140]. Energy-dispersive X-ray (EDX) analysis indicates that all exchangeable $Na^{I}$ ions were successfully replaced by the new $[Co^{II}(PhPDI)_2]^{2+}$ guest over the course of one week. This promising result offers a new perspective on how SCO and SIM behaviors could be modulated by addressing the spin carriers as hybrid materials. Moving forward, we plan to investigate quantum decoherence through pulsed EPR experiments on partially exchanged diluted samples of the $[Co^{II}(PhPDI)_2]^{2+}$ complex adsorbed into a diamagnetic nickel(II)-magnesium(II) MOF derivative.

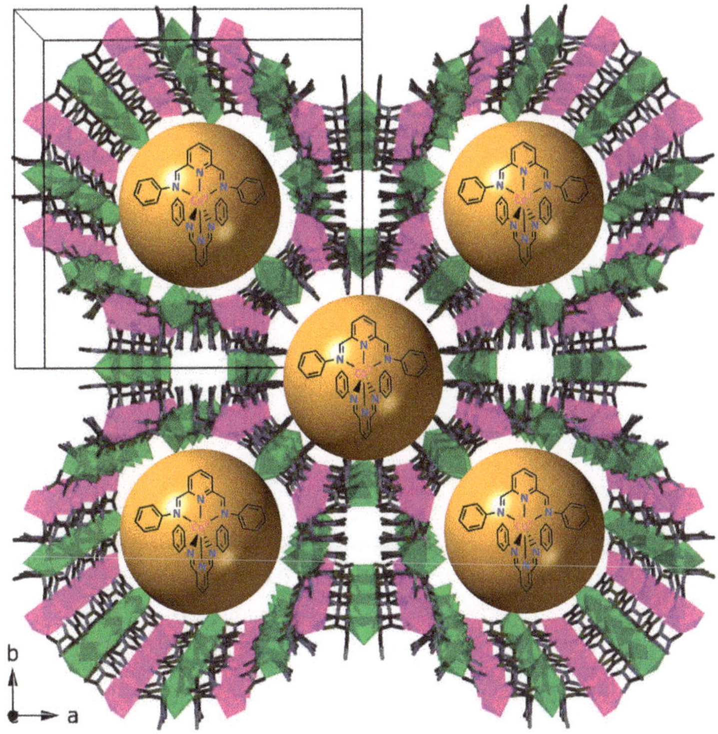

**Figure 16.** Illustrative representation of the chemisorption of the $Co^{II}$-PDI complexes into the giant pores of the anionic MOF of formula $\{Mn_4[Cu_2(TMMPBA)_2]_3\}^{4-}$ as building blocks for arrays of molecular spin quantum qubits.

#### 2.3.5. MOFs and MCOFs as Entangled and Potentially Switchable Ordered Arrays of Molecular Spin Quantum Bits

Designing and synthesizing weakly interacting (entangled) and potentially switchable ordered arrays of qubits is mandatory for constructing a quantum computer, as depicted in Figure 17. Hence, the Sycamore processor, a commercial quantum computer realized under Feynman's vision, is based on microscopic superconducting circuits arranged in a square grid array of 54 qubits (53 operatives), each connected to its four nearest neighbors with appropriate couplers [6]. Inspired by this achievement, we believe that the challenges of spin-based quantum computing machines could be addressed shortly by applying the principles of reticular chemistry (Figure 17) [153].

MOFs made of metal ions or small metal clusters serving as secondary building units (SBUs) linked by a diverse range of organic spacers are particularly well suited for this purpose [154,155]. We have demonstrated that the bottom-up "complex as ligand" approach is an effective method for creating new mixed-metal MOFs and, in some instances, the resulting thin films of these MOFs [156,157]. In fact, metal complexes with additional coordinating groups function as ligands (metalloligands) toward other metal ions, forming high-dimensional 2D and 3D MOF structures.

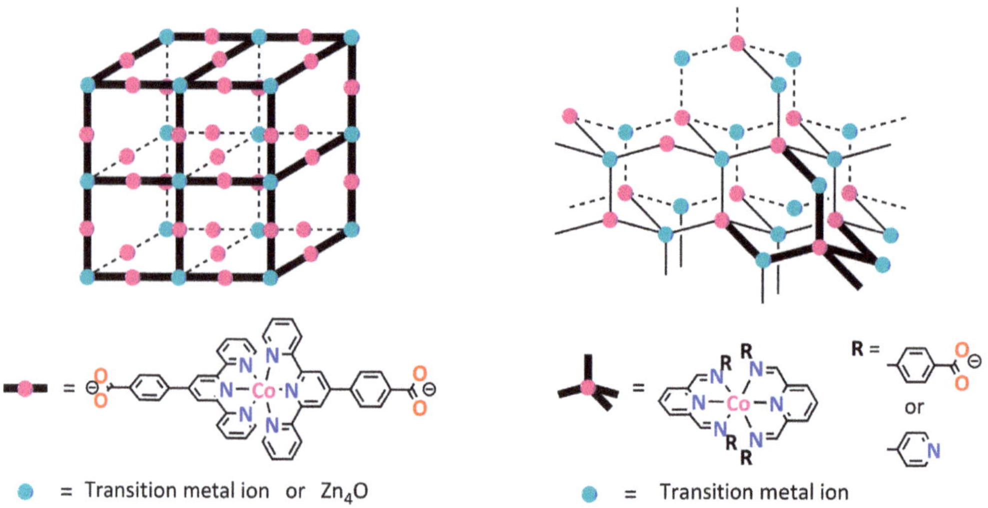


**Figure 17.** Illustration of the proposed 3D MOFs of cubic or diamond topologies built from mononuclear cobalt(II) SCO-SIM complexes with benzoate- or pyridine-substituted TERPY and PDI ligands as linkers towards transition metal ions or clusters as SBUs for arrays of entangled and potentially switchable molecular spin quantum qubits.

In this regard, the $Co^{II}$-TERPY and $Co^{II}$-PDI SCO-SIM complexes discussed in this review are promising metalloligands for constructing mixed-metal MOFs since, by modifying peripheral substituents, these complexes can be tailored to match the Lewis acidity of different metal ions, leading to nD (where n = 2 or 3) SCO-SIM networks. Oxygen-donor benzoate groups coordinate to hard ions like lanthanides, while nitrogen-donor groups from pyridines are appropriate for interacting with many first-row transition metals of interest. In addition, the different directional nature of these metalloligands favors the formation of different 3D MOF topologies, such as cubic (for $Co^{II}$-TERPY) or diamond-type (for $Co^{II}$-PDI), where the cobalt(II) complexes occupy the edges, and the second metal ion (or SBU) occupies the nodes (see Figure 17). It is worth noting that our benzoate-substituted TERPY cobalt(II) complex (see Figure 8), significantly longer than the conventional aromatic dicarboxylates used in Yaghi's MOFs [154,155], would enhance the porosity of the new material.

Covalent organic frameworks (COFs) resulting from the linking of organic building units through well-established organic reactions (e.g., C=N imine formation), and, eventually, their metal-containing derivatives, so-called metal-covalent organic frameworks (MCOFs), are now emerging as propitious alternatives to MOFs, representing unique examples of porous crystalline hybrid materials [158,159]. Among COFs, those with interlaced structures—created through the self-assembly of molecular organic building blocks into linear organic strands woven in two or three dimensions—mean a major challenge for organic chemists. A clever approach to achieving interlaced structures involves using metal ions with specific stereochemical preferences to template the polymerization reaction, resulting in an ordered woven pattern in the MCOF, provided that the organic building blocks include the required coordination donor groups [160,161].

Recently, Yaghi and co-workers have synthesized a unique example of an imine-based 3D MCOF built from mononuclear cobalt(II) complexes with perpendicularly oriented aminophenyl-substituted PDI ligands as nodes of the woven framework, as depicted in Figure 18 [161]. The imine reaction synthesis not only facilitates the control of nucleation and crystal growth of the cobalt(II)-based woven MCOF, leading to uniform nanocrystals through microwave-assisted reactions (Figure 18a), but also allows for the straightforward preparation of oriented thin films of 190 nm thickness on silicon substrates (Figure 18b). This unique class of woven MCOFs, comprising arrays of weakly interacting ("entangled") and potentially switchable $Co^{II}$ SCO-SIMs exhibiting sufficiently long quantum decoherence relaxation times, would offer an excellent opportunity to test them as prototypes of multiple qubit arrays for the construction of molecular logic systems, like spin quantum processors (SQPs).

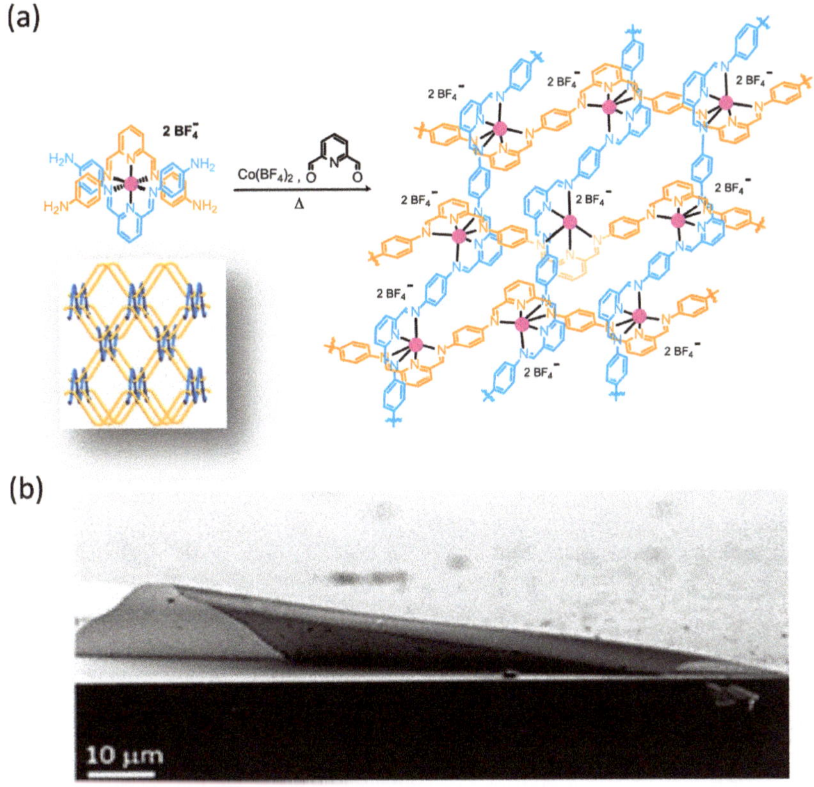


**Figure 18.** (**a**) An illustration of the 3D MCOF of diamond topology built from mononuclear cobalt(II) SCO-SIM complexes with phenylamine-substituted PDI ligands as nodes of the woven-like framework (see boxed structure) as a prototype of the molecular spin quantum processor. (**b**) A microphotograph of the oriented MCOF thin film showing high uniformity and small thickness. Adapted with permission from Ref. [161].

## 3. Conclusions and Outlook

"Every adventure requires a first step." While this may sound cliché, it holds, particularly in science. The results gathered in this perspective review mark the initial steps of our research group in exploring multifunctional and multiresponsive magnetic devices based on spin-crossover molecular nanomagnets within the emerging fields of molecular spintronics and quantum computing. Although the search for some relationship (synergy) between the SCO and SIM properties is a challenging task, given the different temperature regions for each phenomenon, other alternatives exist that we explored along this family of cobalt(II)-TERPY and PDI SCO-SIM complexes. Hence, both SCO and SIM behaviors are sensible to other stimuli besides temperature, whether chemical (pH or analytes) or physical (light or electron flow), which allow the control of the spin state and spin relaxation dynamics simultaneously so that a spin modulation of the spin relaxation dynamics is made possible. These results are also a key motivation for us to continue pursuing this challenging multidisciplinary research project where SCO and SIMs converge with redox and photophysical or photochemical properties. Indeed, most proposals concerning the scaling and addressing of spin-crossover cobalt(II) molecular nanomagnets as prototypes of single-molecule quantum spintronics and quantum computing devices feel more like daydreams or aspirations than tangible achievements. In the coming years, we aim to turn some of these dream worlds into reality, particularly in the physical realization of quantum information processing. In this respect, this perspective review serves as a message in a bottle for materials scientists with large expertise in manipulating and measuring single-molecule electron transport properties and those skilled in processing and addressing molecules on different supports, as well as for theoretical physicists who may find our work relevant for conceptualizing and scaling molecules as spin quantum bits in future quantum computers.

## 4. Epilog

*"There is plenty of room at the bottom"*. With all due respect, we would like to offer our own interpretation of Richard Feynman's famous phrase, reflecting our perspective on the opportunities presented in this review. Our intention is not to undermine the original meaning of his invaluable insights, which have inspired numerous fields within Molecular Sciences, but rather to share the additional reflections that this statement evokes for us.

Molecular Sciences have indeed come a long way, and the remarkable advancements in this area have led us to a heightened awareness of its potential. It is widely acknowledged that nanoscience and nanotechnology have only scratched the surface of what is possible. These fields are highly competitive and evolve rapidly, often leading to a focus on the most significant or recent findings—the "top of the tip". In this pursuit, many valuable results are overlooked simply because they were reported "too early". Consequently, a lack of knowledge and appropriate techniques at the time has resulted in a wealth of findings being lost to scientific obscurity, which we refer to as "the bottom". These results are still waiting for new, enthusiastic scientists to recognize their worth and recontextualize them with the current understanding and advanced technologies. This indicates that there is indeed plenty of room at the bottom.

We have thus realized that a new multifunctional system does not always need to arise from developing novel properties. Instead, by looking into "the bottom", new materials may be created from creative designs that combine several well-known properties, establishing synergies between them. In this perspective review, we embraced the simple (but not effortless) idea of using two $Co^{II}N_6$ platforms with TERPY- and related PDI-type ligands, known for decades, to successfully obtain a new chemical library of spin-crossover molecular nanomagnets. Their chemical modulation and switchable behavior concerning both spin-crossover and the slow relaxation of magnetization phenomena illustrate the whole potential of this new class of multifunctional and multiresponsive magnetic molecules in molecular spintronics and quantum computing.

**Author Contributions:** This manuscript was written with the contributions of all authors. All authors have read and agreed to the published version of the manuscript.

**Funding:** This work was supported by the Spanish MINECO (Projects PID2019-109735GB-I00 and Unidad de Excelencia María de Maeztu CEX2019-000919-M) and the Generalitat Valenciana (AICO/2020/183 and AICO/2021/295). R. R. thanks the Generalitat Valenciana and CAPES/BRASIL for the doctoral (GRISOLIAP/2017/057) and postdoctoral (88887.798611/2022-00) grants.

**Conflicts of Interest:** The authors declare no conflicts of interest.

**List of Abbreviations:** AQ, anthraquinone; BEPYDMP, *N,N′*-bis(1-pyridin-2-ylethylidene)-2,2-dimethylpropane-1,3-diamine; BPYCA, 2,2′-bipyridine-4,4′-dicarboxamide; BPZ, dihydrobis(1-pyrazolyl)borate; CISS, chiral-induced spin-selectivity; CMOS, complementary metal-oxide-semiconductor; CNM, carbon nanomaterial; COF, covalent organic framework; DMPHEN, 2,9-dimethyl-1-10-phenantroline; DFT, density functional theory; DPZBPY, 3,5-bis(pyrazol-1-yl)-4,4′-

bipyridine; DPZPY, 2,6-bis(pyrazol-1-yl)pyridine; EDX, energy-dispersive X-ray; EFE-SQT, electric field-effect spin quantum transistor; FET, field-effect transistor; FR, fast relaxation; GMR, giant magnetoresistance; HBDMPZ, hydrotris(3-5-dimethylpyrazole)borate; HS, high-spin; JT, Jahn-Teller; LB, Langmuir-Blodgett; LD-LISC, ligand-driven light-induced spin change; LED, light-emitting diode; LIESST, light-induced excited spin-state trapping; LMCT, ligand-to-metal charge transfer; LS, low-spin; MCOF, metal-covalent organic framework; MFE-SQT, magnetic field-effect spin quantum transistor; MLCT, metal-to-ligand charge transfer; MM, molecular magnetism; MNT, maleonitriledithiolate; MOF, metal–organic framework; PC, phthalocyanine; PDI, 2,6-pyridinediimine; QC, quantum coherence; QG, quantum gate; QIP, quantum information processing; QTM, quantum tunneling of magnetization; qubit, quantum bit; SAL, salicylaldehyde; *S*-BPP, (*S*)-(4-{[2,6-(dipyrazol-1-yl)pyrid-4-yl]ethynyl}phenyl)ethanethioate; SBU, secondary building unit; SCO, spin-crossover; SC-SC, single-crystal-to-single-crystal; SIM, single-ion magnet; SMM, single-molecule magnet; SMS, single-molecule spintronics; SOC, spin–orbit coupling; SQP, spin quantum processor; STM, scanning tunneling microscope; SQF, spin quantum filter; SQV, spin quantum valve; SR, slow relaxation; SWCNT, single-walled carbon nanotube; TERPY, 2,2′:6′,2′-terpyridine; TMC, 1,4,7,10-tetramethyl-1,4,7,10-tetraazacyclododecane; TMMPBA, *N,N*′-2,4,6-trimethyl-1,3-phenylenebis(oxamate); TZPY, 3-(2-pyridyl)[1-3]triazolo [1,5-a]pyridine; and UV, ultraviolet.

## References


1. Moore, G.E. Cramming more components onto integrated circuits. *Electron. Mag.* **1965**, *38*, 114–117. [CrossRef]
2. Feynman, R.P. There's plenty of room at the bottom: An invitation to enter a new field of physics. *Eng. Sci.* **1960**, *23*, 22–36.
3. Baibich, M.N.; Broto, J.M.; Fert, A.; Nguyen Van Dau, F.; Petroff, F.; Etienne, P.; Creuzet, G.; Friederich, A.; Chazelas, J. Giant magnetoresistance of (001)Fe/(001)Cr magnetic superlattices. *Phys. Rev. Lett.* **1988**, *61*, 2472–2475. [CrossRef] [PubMed]
4. Binash, G.; Grünberg, P.; Saurenbach, F.; Zinn, W. Enhanced magnetoresistance in layered magnetic structures with antiferromagnetic interlayer exchange. *Phys. Rev. B* **1989**, *39*, 4828–4830. [CrossRef]
5. Awschalom, D.D.; Bassett, L.C.; Dzurak, A.S.; Hu, E.L.; Petta, J.R. Quantum spintronics: Engineering and manipulating atom-like spins in semiconductors. *Science* **2013**, *339*, 1174–1179. [CrossRef]
6. Arute, F.; Arya, K.; Babbush, R.; Bacon, D.; Bardin, J.C.; Barends, R.; Biswas, R.; Boixo, S.; Brandao, F.G.S.L.; Buell, D.A.; et al. Quantum supremacy using a programmable superconducting processor. *Nature* **2019**, *574*, 505–510. [CrossRef] [PubMed]
7. Zhong, H.-S.; Wang, H.; Deng, Y.-H.; Chen, M.-C.; Peng, L.-C.; Luo, Y.-H.; Qin, J.; Wu, D.; Ding, X.; Hu, P.; et al. Quantum computational advantage using photons. *Science* **2020**, *370*, 1460–1463. [CrossRef] [PubMed]
8. Peng, L.-C.; Wu, D.; Zhong, H.-S.; Luo, Y.-H.; Li, Y.; Hu, Y.; Jiang, X.; Chen, M.-C.; Li, L.; Liu, N.-L.; et al. Cloning of quantum entanglement. *Phys. Rev. Lett.* **2020**, *125*, 210502. [CrossRef]
9. Coronado, E. Molecular magnetism: From chemical design to spin control in molecules, materials and devices. *Nat. Rev. Mater.* **2020**, *5*, 87–104. [CrossRef]
10. Fert, A. Nobel lecture: Origin, development, and future of spintronics. *Rev. Mod. Phys.* **2008**, *8*, 1517–1530. [CrossRef]
11. Shor, P.W. Algorithms for quantum computation: Discrete logarithms and factoring. In Proceedings of the 35nd Annual Symposium on Foundations of Computer Science, Santa Fe, NM, USA, 20–22 November 1994; Goldwasser, S., Ed.; IEEE Computer Society Press: Los Alamitos, CA, USA, 1994; pp. 124–134.
12. Grover, L.K. Quantum computers can search arbitrarily large databases by a single query. *Phys. Rev. Lett.* **1997**, *79*, 4709–4712. [CrossRef]
13. Kloeffel, C.; Loss, D. Prospects for spin-based quantum computing in quantum dots. *Annu. Rev. Condens. Matter Phys.* **2013**, *4*, 51–81. [CrossRef]
14. Sanvito, S. Molecular spintronics. *Chem. Soc. Rev.* **2011**, *40*, 3336–3355. [CrossRef]
15. Sanvito, S. Injecting and controlling spins in organic materials. *J. Mater. Chem.* **2007**, *17*, 4455–4459. [CrossRef]
16. Affronte, M.; Troiani, F.; Ghirri, A.; Candini, A.; Evangelisti, M.; Corradini, V.; Carretta, S.; Santini, P.; Amoretti, G.; Tuna, F.; et al. Single molecule magnets for quantum computation. *J. Phys. D Appl. Phys.* **2007**, *40*, 2999–3004. [CrossRef]
17. Bogani, L.; Wernsdorfer, W. Molecular spintronics using single-molecule magnets. *Nat. Mater.* **2008**, *7*, 179–186. [CrossRef] [PubMed]
18. Winpenny, R.E.P. Quantum information processing using molecular nanomagnets as qubits. *Angew. Chem. Int. Ed.* **2008**, *47*, 7992–7994. [CrossRef]
19. Lehmann, J.; Gaita-Ariño, A.; Coronado, E.; Loss, D. Quantum computing with molecular spin systems. *J. Mater. Chem.* **2009**, *19*, 1672–1677. [CrossRef]
20. Stamp, P.C.E.; Gaita-Ariño, A. Spin-based quantum computers made by chemistry: Hows and whys. *J. Mater. Chem.* **2009**, *19*, 1718–1730. [CrossRef]
21. Affronte, M. Molecular nanomagnets for information technologies. *J. Mater Chem.* **2009**, *19*, 1731–1737. [CrossRef]
22. Ardavan, A.; Blundell, S.J. Storing quantum information in chemically engineered nanoscale magnets. *J. Mater. Chem.* **2009**, *19*, 1754–1760. [CrossRef]

23. Dei, A.; Gatteschi, D. Molecular (nano)magnets as test grounds of quantum mechanics. *Angew. Chem. Int. Ed.* **2011**, *50*, 11852–11858. [CrossRef]
24. Timco, G.A.; Faust, T.B.; Tuna, F.; Winpenny, R.E.P. Linking heterometallic rings for quantum information processing and amusement. *Chem. Soc. Rev.* **2011**, *40*, 3067–3075. [CrossRef]
25. Troiani, F.; Affronte, M. Molecular spins for quantum information technologies. *Chem. Soc. Rev.* **2011**, *40*, 3119–3129. [CrossRef]
26. Aromí, G.; Aguilà, D.; Gamez, P.; Luis, F.; Roubeau, O. Design of magnetic coordination complexes for quantum computing. *Chem. Soc. Rev.* **2012**, *41*, 537–546. [CrossRef] [PubMed]
27. Ghirri, A.; Troiani, F.; Affronte, M. Quantum computation with molecular nanomagnets: Achievements, challenges, and new trends. *Struct. Bond.* **2015**, *164*, 383–430.
28. Sessoli, R. Toward the quantum computer: Magnetic molecules back in the race. *ACS Cent. Sci.* **2015**, *1*, 473–474. [CrossRef]
29. Gaita-Ariño, A.; Luis, F.; Hill, S.; Coronado, E. Molecular spins for quantum computation. *Nat. Chem.* **2019**, *11*, 301–309. [CrossRef]
30. Figgins, P.E.; Busch, D.H. Complexes of iron(II), cobalt(II) and nickel(II) with biacetyl-bis-methylimine, 2-pyridinal-methylimine and 2,6-pyridindial-bis-methylimine. *J. Am. Chem. Soc.* **1960**, *82*, 820–824. [CrossRef]
31. Goodwin, H.A. Spin transitions in six-coordinate iron(II) complexes. *Coord. Chem. Rev.* **1976**, *18*, 293–325. [CrossRef]
32. Gütlich, P. Spin crossover in iron(II) complexes. *Struct. Bond.* **1981**, *44*, 83–195.
33. König, E.; Ritter, G.; Kulshreshtha, S.K. The nature of spin-state transitions in solid complexes of iron(II) and the interpretation of some associated phenomena. *Chem. Rev.* **1985**, *85*, 219–234. [CrossRef]
34. Boillot, M.L.; Sour, A.; Delhaes, P.; Mingotaud, C.; Soyer, H. A photomagnetic effect for controlling spin states of iron(II) complexes in molecular materials. *Coord. Chem. Rev.* **1999**, *190–192*, 47–59. [CrossRef]
35. Krivokapic, I.; Zerara, M.; Daku, M.L.; Vargas, A.; Enachescu, C.; Ambrus, C.; Tregenna-Piggott, P.; Amstutz, N.; Krausz, E.; Hauser, A. Spin-crossover in cobalt(II) imine complexes. *Coord. Chem. Rev.* **2007**, *251*, 364–378. [CrossRef]
36. Real, J.A.; Gaspar, A.B.; Muñoz, M.C. Thermal, pressure and light switchable spin-crossover materials. *Dalton Trans.* **2005**, 2062–2079. [CrossRef] [PubMed]
37. Bousseksou, A.; Molnár, G.; Salmon, L.; Nicolazzi, W. Molecular spin crossover phenomenon: Recent achievements and prospects. *Chem. Soc. Rev.* **2011**, *40*, 3313–3335. [CrossRef] [PubMed]
38. Ruiz, E. Charge transport properties of spin crossover systems. *Phys. Chem. Chem. Phys.* **2014**, *16*, 14–22. [CrossRef] [PubMed]
39. Bertoni, R.; Cammarata, M.; Lorenc, M.; Matar, S.F.; Matar, J.-F.; Létard, J.F.; Lemke, E.; Collet, H.T. Ultrafast light-induced spin-state trapping photophysics investigated in $Fe(phen)_2(NCS)_2$ spin-crossover crystal. *Acc. Chem. Res.* **2015**, *48*, 774–781. [CrossRef]
40. Harding, D.J.; Harding, P.; Phonsri, W. Spin crossover in iron(II) complexes. *Coord. Chem. Rev.* **2016**, *313*, 38–61. [CrossRef]
41. Khusniyarov, M.M. How to switch spin-crossover metal complexes at constant room temperature. *Chem. Eur. J.* **2016**, *22*, 15178–15191. [CrossRef]
42. Halcrow, M.A. Structure: Function relationships in molecular spin-crossover complexes. *Chem. Soc. Rev.* **2011**, *40*, 4119–4142. [CrossRef] [PubMed]
43. Halcrow, M.A. Spin-crossover compounds with wide thermal hysteresis. *Chem. Lett.* **2014**, *43*, 1178–1188. [CrossRef]
44. Halcrow, M.A. The effect of ligand design on metal ion spin state-Lessons from spin crossover complexes. *Crystals* **2016**, *6*, 58. [CrossRef]
45. Mallah, T.; Cavallini, M. Surfaces, thin films and patterning of spin crossover compounds. *C. R. Chimie* **2018**, *21*, 1270–1286. [CrossRef]
46. Molnár, G.; Rat, S.; Salmon, L.; Nicolazzi, W.; Bousseksou, A. Spin crossover nanomaterials: From fundamental concepts to devices. *Adv. Mater.* **2018**, *30*, 17003862. [CrossRef]
47. Molnár, G.; Mikolasek, M.; Ridier, K.; Fahs, A.; Nicolazzi, W.; Bousseksou, A. Molecular spin crossover materials: Review of the lattice dynamical properties. *Ann. Phys.* **2019**, *531*, 1900076. [CrossRef]
48. Cirera, J.; Ruiz, E.; Alvarez, S.; Neese, F.; Kortus, J. How to build molecules with large magnetic anisotropy. *Chem. Eur. J.* **2009**, *15*, 4078–4087.
49. Neese, F.; Pantazis, D.A. What is not required to make a single molecule magnet. *Faraday Discuss.* **2011**, *148*, 229–238. [PubMed]
50. Atanasov, M.; Aravena, D.; Suturina, E.; Bill, E.; Maganas, D.; Neese, F. First principles approach to the electronic structure, magnetic anisotropy and spin relaxation in mononuclear 3d-transition metal single molecule magnets. *Coord. Chem. Rev.* **2015**, *289–290*, 177–214.
51. Gomez-Coca, S.; Aravena, D.; Morales, R.; Ruiz, E. Large magnetic anisotropy in mononuclear metal complexes. *Coord. Chem. Rev.* **2015**, *289–290*, 379–392. [CrossRef]
52. Chibotaru, L.F. Theoretical understanding of anisotropy in molecular nanomagnets. *Struct. Bond.* **2015**, *164*, 185–229.
53. Craig, G.A.; Murrie, M. 3d Single-ion magnets. *Chem. Soc. Rev.* **2015**, *44*, 2135–2147. [CrossRef] [PubMed]
54. Frost, J.M.; Harriman, K.L.M.; Murugesu, M. The rise of 3-d single-ion magnets in molecular magnetism: Towards materials from molecules? *Chem. Sci.* **2016**, *7*, 2470–2491. [CrossRef]
55. Bar, A.K.; Pichon, C.; Sutter, J.-P. Magnetic anisotropy in two- to eight-coordinated transition-metal complexes: Recent developments in molecular magnetism. *Coord. Chem. Rev.* **2016**, *308*, 346–380. [CrossRef]
56. Meng, Y.-S.; Jiang, S.-D.; Wang, B.-W.; Gao, S. Understanding the magnetic anisotropy toward single-ion magnets. *Acc. Chem. Res.* **2016**, *49*, 2381–2389. [CrossRef]

57. Feng, M.; Tong, M.-L. Single-ion magnets from 3d to 5f: Developments and strategies. *Chem. Eur. J.* **2018**, *24*, 7574–7594. [CrossRef] [PubMed]
58. Moreno-Pineda, E.; Wernsdorfer, W. Measuring molecular magnets for quantum technologies. *Nat. Rev. Phys.* **2021**, *3*, 645–659. [CrossRef]
59. Zabala-Lekuona, A.; Seco, J.M.; Colacio, E. Single-molecule magnets: From Mn12-ac to dysprosium metallocenes. A travel in time. *Coord. Chem. Rev.* **2021**, *441*, 213984. [CrossRef]
60. Chilton, N.F. Molecular Magnetism. *Annu. Rev. Matter. Res.* **2022**, *52*, 79–101. [CrossRef]
61. Park, J.; Pasupathy, A.N.; Goldsmith, J.I.; Chang, C.; Yaish, Y.; Petta, J.R.; Rinkoski, M.; Sethna, J.P.; Abruña, H.D.; McEuen, P.L.; et al. Coulomb blockade and the kondo effect in single-atom transistors. *Nat. Mater.* **2002**, *417*, 722–725. [CrossRef] [PubMed]
62. Parks, J.J.; Champagne, A.R.; Costi, T.A.; Shum, W.W.; Pasupathy, A.N.; Neuscamman, E.; Flores-Torres, S.; Cornaglia, P.S.; Aligia, A.A.; Balseiro, C.A.; et al. Mechanical control of spin states in spin-1 molecules and the underscreened kondo effect. *Science* **2010**, *328*, 1370–1373. [CrossRef]
63. Meded, V.; Bagrets, A.; Fink, K.; Chandrasekar, R.; Ruben, M.; Evers, F.; Bernand-Mantel, A.; Seldenthuis, J.S.; Beukman, A.; van der Zant, H.S.J. Electrical control over the Fe(II) spin crossover in a single molecule: Theory and experiment. *Phys. Rev. B* **2011**, *83*, 245415. [CrossRef]
64. Harzmann, G.D.; Frisenda, R.; van der Zant, H.S.J.; Mayor, M. Single-molecules spin switch based on voltage-triggered distortion of the coordination sphere. *Angew. Chem. Int. Ed.* **2015**, *54*, 13425–13430. [CrossRef]
65. Devid, E.J.; Martinho, P.N.; Kamalakar, M.V.; Salitros, I.; Prendergast, Ú.; Dayen, J.-F.; Meded, V.; Lemma, T.; González-Pietro, R.; Evers, F.; et al. Spin transition in arrays of gold nanoparticles and spin crossover molecules. *ACS Nano* **2015**, *9*, 4486–4507. [CrossRef]
66. Aravena, D.; Ruiz, E. Coherent transport through spin-crossover single molecules. *J. Am. Chem. Soc.* **2012**, *134*, 777–779. [CrossRef]
67. Aragonès, A.C.; Aravena, D.; Cerdá, J.I.; Acís-Castillo, Z.; Li, H.; Real, J.A.; Sanz, F.; Hihath, J.; Ruiz, E.; Díez-Pérez, I. Large conductance switching in a single-molecule device through room temperature spin-dependent transport. *Nano Lett.* **2016**, *16*, 218–226. [CrossRef] [PubMed]
68. Katsnelson, M.I. Graphene: Carbon in two dimensions. *Mater. Today* **2007**, *10*, 20–27. [CrossRef]
69. Konstantinov, N.; Tauzin, A.; Noumbé, U.N.; Dragoe, D.; Kundys, B.; Majjad, H.; Brosseau, A.; Lenertz, M.; Singh, A.; Berciaud, S.; et al. Electrical read-out of light-induced spin transition in thin film spin crossover/graphene heterostructures. *J. Mater. Chem. C* **2021**, *9*, 2712–2720. [CrossRef]
70. Lee, C.W.; Kim, O.Y.; Lee, J.Y. Organic materials for organic electronic devices. *J. Ind. Eng. Chem.* **2014**, *20*, 1198–1208. [CrossRef]
71. Komeda, T.; Isshiki, H.; Liu, J.; Zhang, Y.-F.; Lorente, N.; Katoh, K.; Breedlove, B.K.; Yamashita, M. Observation and electric current control of a local spin in a single-molecule magnet. *Nat. Commun.* **2011**, *2*, 217–223. [CrossRef] [PubMed]
72. Kyatskaya, S.; Galán-Mascarós, J.R.; Bogani, L.; Hennrich, F.; Kappes, M.; Wernsdorfer, W.; Ruben, M. Anchoring of rare-earth-based single-molecule magnets on single-walled carbon nanotubes. *J. Am. Chem. Soc.* **2009**, *131*, 15143–15151. [CrossRef] [PubMed]
73. Urdampilleta, M.; Klyatskaya, S.; Cleuziou, J.-P.; Ruben, M.; Wernsdorfer, W. Supramolecular spin valves. *Nat. Mater.* **2011**, *10*, 502–506. [CrossRef] [PubMed]
74. Muñoz, M.C.; Real, J.A. Thermo-, piezo-, photo- and chemo-switchable spin crossover iron(II)-metallocyanate based coordination polymers. *Coord. Chem. Rev.* **2011**, *255*, 2068–2093. [CrossRef]
75. Goodwin, H.A. Spin crossover in cobalt(II) systems. *Top. Curr. Chem.* **2004**, *234*, 23–47.
76. Cowan, M.G.; Olguin, J.; Narayanaswamy, S.; Tallon, J.L.; Brooker, S. Reversible switching of a cobalt complex by thermal, pressure, and electrochemical stimuli: Abrupt, complete, hysteretic spin crossover. *J. Am. Chem. Soc.* **2012**, *134*, 2892–2894. [CrossRef] [PubMed]
77. Kahn, O. *Molecular Magnetism*; VCH Publishers: New York, NY, USA, 1993.
78. Ishikawa, N.; Sugita, M.; Ishikawa, T.; Koshihara, S.Y.; Kaizu, Y. Lanthanide double-decker complexes functioning as magnets at the single-molecular level. *J. Am. Chem. Soc.* **2003**, *125*, 8694–8695. [CrossRef] [PubMed]
79. Ishikawa, N.; Sugita, M.; Okubo, T.; Tanaka, N.; Lino, T.; Kaizu, Y. Determination of ligand-field parameters and f-electronic structures of double-decker bis(phthalocyaninato)lanthanide complexes. *Inorg. Chem.* **2003**, *42*, 2440–2446. [CrossRef] [PubMed]
80. Ishikawa, N.; Sugita, M.; Ishikawa, T.; Koshihara, S.-Y.; Kaizu, Y. Mononuclear lanthanide complexes with a long magnetisation relaxation time at high temperatures: A new category of magnets at the single-molecular level. *J. Phys. Chem. B* **2004**, *108*, 11265–11271. [CrossRef]
81. Vallejo, J.; Castro, I.; Ruiz-García, R.; Cano, J.; Julve, M.; Lloret, F.; De Munno, G.; Wernsdorfer, W.; Pardo, E. Field-induced slow magnetic relaxation in a six-coordinate mononuclear cobalt(II) complex with a positive anisotropy. *J. Am. Chem. Soc.* **2012**, *134*, 15704–15707. [CrossRef] [PubMed]
82. Gaspar, A.B.; Ksenofontov, V.; Seredyuk, M.; Gutlich, P. Multifunctionality in spin crossover materials. *Coord. Chem Rev.* **2005**, *249*, 2661–2676. [CrossRef]
83. Bousseksou, A.; Molnár, G.; Demont, P.; Menegotto, J. Observation of a thermal hysteresis loop in the dielectric constant of spin crossover complexes: Towards molecular memory devices. *J. Mater. Chem.* **2003**, *13*, 2069–2071. [CrossRef]
84. Kumar, K.S.; Ruben, M. Emerging trends in spin crossover (SCO) based functional materials and devices. *Coord. Chem. Rev.* **2017**, *346*, 176–205. [CrossRef]

85. Gutlich, P.; Hauser, A. Thermal and light-induced spin crossover in iron(II) complexes. *Coord. Chem. Rev.* **1990**, *97*, 1–22. [CrossRef]
86. Hayami, S.; Komatsu, Y.; Shimizu, T.; Kamihata, H.; Lee, Y.H. Spin-crossover in cobalt(II) compounds containing terpyridine and its derivatives. *Coord. Chem. Rev.* **2011**, *255*, 1981–1990. [CrossRef]
87. Galet, A.; Gaspar, A.B.; Muñoz, M.C.; Real, J.A. Influence of the counterion and the solvent molecules in the spin crossover system $[Co(4\text{-terpyridone})_2]X_p{\cdot}nH_2O$. *Inorg. Chem.* **2006**, *45*, 4413–4422. [CrossRef]
88. Murrie, M. Cobalt(II) single-molecule magnets. *Chem. Soc. Rev.* **2010**, *39*, 1986–1995. [CrossRef]
89. Ding, Y.-S.; Deng, Y.-F.; Zheng, Y.-Z. The rise of single-ion magnets as spin qubits. *Magnetochemistry* **2016**, *2*, 40. [CrossRef]
90. Zadrozny, J.M.; Long, J.R. Slow magnetic relaxation at zero field in the tetrahedral complex $[Co(SPh)_4]^{2-}$. *J. Am. Chem. Soc.* **2011**, *133*, 20732–20734. [CrossRef] [PubMed]
91. Lloret, F.; Julve, M.; Cano, J.; Ruiz-García, R.; Pardo, E. Magnetic properties of six-coordinated high-spin cobalt(II) complexes: Theoretical background and its application. *Inorg. Chim. Acta* **2008**, *361*, 3432–3445. [CrossRef]
92. Herchel, R.; Váhovská, L.; Potocnák, I.; Trávnícek, Z. Slow magnetic relaxation in octahedral cobalt(II) field-induced single-ion magnet with positive axial and large rhombic anisotropy. *Inorg. Chem.* **2014**, *53*, 5896–5898. [CrossRef] [PubMed]
93. Roy, S.; Oyarzabal, I.; Vallejo, J.; Cano, J.; Colacio, E.; Bauza, A.; Frontera, A.; Kirillov, A.M.; Drew, M.G.B.; Das, S. Two polymorphic forms of a six-coordinate mononuclear cobalt(II) complex with easy-plane anisotropy: Structural features, theoretical calculations, and field-induced slow relaxation of the magnetisation. *Inorg. Chem.* **2016**, *55*, 8502–8513. [CrossRef] [PubMed]
94. Ding, Z.-Y.; Meng, Y.-S.; Xiao, Y.; Zhang, Y.-Q.; Zhu, Y.-Y.; Gao, S. Probing the influence of molecular symmetry on the magnetic anisotropy of octahedral cobalt(II) complexes. *Inorg. Chem. Front.* **2017**, *4*, 1909–1916. [CrossRef]
95. Wu, Y.; Tian, D.; Ferrando-Soria, J.; Cano, J.; Yin, L.; Ouyang, Z.; Wang, Z.; Luo, S.; Liu, X.; Pardo, E. Modulation of the magnetic anisotropy of octahedral cobalt(II) single-ion magnets by fine-tuning the axial coordination microenvironment. *Inorg. Chem. Front.* **2019**, *6*, 848–856. [CrossRef]
96. Cen, P.; Yuan, W.; Tan, M.; Chen, B.; Song, W.; Wang, Z.; Liu, X. Field-induced slow magnetic relaxation in an octahedral high-spin Co(II) complex. *Inorg. Chem. Commun.* **2019**, *99*, 195–198. [CrossRef]
97. Palion-Gazda, J.; Machura, B.; Kruszynski, R.; Grancha, T.; Moliner, N.; Lloret, F.; Julve, M. Spin crossover in double salts containing six- and four-coordinate cobalt(II) ions. *Inorg. Chem.* **2017**, *56*, 6281–6296. [CrossRef] [PubMed]
98. Shao, D.; Deng, L.-D.; Shi, L.; Wu, D.-Q.; Wei, X.-Q.; Yang, S.-R.; Wang, X.-Y. Slow magnetic relaxation and spin crossover behaviour in a bicomponent ion-pair cobalt(II) complex. *Eur. J. Inorg. Chem.* **2017**, *2017*, 3862–3867. [CrossRef]
99. Cui, H.-H.; Wang, J.; Chen, X.-T.; Xue, Z.-L. Slow magnetic relaxation in five-coordinate spin-crossover cobalt(II) complexes. *Chem. Commun.* **2017**, *53*, 9304–9307. [CrossRef] [PubMed]
100. Chen, L.; Song, J.; Zhao, W.; Yi, G.; Zhou, Z.; Yuan, A.; Song, Y.; Wang, Z.; Ouyang, Z.-W. A mononuclear five-coordinate Co(II) single molecule magnet with a spin crossover between the $S = 1/2$ and $3/2$ states. *Dalton Trans.* **2018**, *47*, 16596–16602. [CrossRef]
101. Xu, M.-X.; Liu, Z.; Dong, B.-W.; Cui, H.-H.; Wang, Y.-X.; Su, J.; Wang, Z.; Song, Y.; Chen, X.-T.; Jiang, S.-D.; et al. Single-crystal study of a low spin Co(II) molecular qubit: Observation of anisotropic rabi cycles. *Inorg. Chem.* **2019**, *58*, 2330–2335. [CrossRef] [PubMed]
102. Shao, D.; Shi, L.; Yin, J.; Wang, B.-L.; Wang, Z.-X.; Zhang, Y.-Q.; Wang, X.-Y. Reversible on-off switching of both spin crossover and single-molecule magnet behaviours via a crystal-to-crystal transformation. *Chem. Sci.* **2018**, *9*, 7986–7991. [CrossRef] [PubMed]
103. Kobayashi, F.; Komatsumaru, Y.; Akiyoshi, R.; Nakamura, M.; Zhang, Y.; Lindoy, L.F.; Hayami, S. Water molecule-induced reversible magnetic switching in a bis-terpyridine cobalt(II) complex exhibiting coexistence of spin crossover and orbital transition behaviors. *Inorg. Chem.* **2020**, *59*, 16843–16852. [CrossRef] [PubMed]
104. Rabelo, R.; Toma, L.; Moliner, N.; Julve, M.; Lloret, F.; Pasán, J.; Ruiz-Pérez, C.; Ruiz-García, R.; Cano, J. Electro-switching of the single-molecule magnet behaviour in an octahedral spin crossover cobalt(II) complex with a redox-active pyridinediimine ligand. *Chem. Commun.* **2020**, *56*, 12242–12245. [CrossRef]
105. Rabelo, R.; Toma, L.; Moliner, N.; Julve, M.; Lloret, F.; Inclán, M.; García-España, E.; Pasán, J.; Ruiz-García, R.; Cano, J. pH-Switching of the luminescent, redox, and magnetic properties in a spin crossover cobalt(II) molecular nanomagnet. *Chem. Sci.* **2023**, *14*, 8850. [CrossRef] [PubMed]
106. Rabelo, R.; Toma, L.; Julve, M.; Lloret, F.; Pasán, J.; Cangussu, D.; Ruiz-García, R.; Cano, J. How the spin state tunes the slow magnetic relaxation field dependence in spin crossover cobalt(II) complexes. *Dalton Trans.* **2024**, *53*, 5507–5520. [CrossRef] [PubMed]
107. Rabelo, R.; Toma, L.; Julve, M.; Lloret, F.; Pasán, J.; Cangussu, D.; Ruiz-García, R.; Cano, J. Multi-electron transfer and field-induced slow magnetic relaxation in spin crossover cobalt(II) complexes: Structure-function correlations. *Inorg. Chem. Front.* **2024**, *11*, 6028–6043. [CrossRef]
108. Canton, S.E.; Biednov, M.; Pápai, M.; Lima, F.A.; Choi, T.-K.; Otte, F.; Jiang, Y.; Frankenberger, P.; Knoll, M.; Zalden, P.; et al. Ultrafast Jahn-Teller photoswitching in cobalt single-ion magnets. *Adv. Sci.* **2023**, *10*, 2206880. [CrossRef]
109. Urtizberea, A.; Roubeau, O. Switchable slow relaxation of magnetization in the native low temperature phase of a cooperative spin crossover compound. *Chem. Sci.* **2017**, *8*, 2290–2295. [CrossRef] [PubMed]
110. Feng, X.; Mathonière, C.; Jeon, I.-R.; Rouzières, M.; Ozarowski, A.; Aubrey, M.L.; Gonzalez, M.I.; Clérac, R.; Long, J.R. Tristability in a light-actuated single-molecule magnet. *J. Am. Chem. Soc.* **2013**, *135*, 15880–15884. [CrossRef]
111. Mathonière, C.; Lin, H.-J.; Siretanu, D.; Clérac, R.; Smith, J.M. Photoinduced single-molecule magnet properties in a four-coordinate iron(II) spin crossover complex. *J. Am. Chem. Soc.* **2013**, *135*, 19083–19086. [CrossRef] [PubMed]

112. Ferrando-Soria, J.; Vallejo, J.; Castellano, M.; Martínez-Lillo, J.; Pardo, E.; Cano, J.; Castro, I.; Lloret, F.; Ruiz-García, R.; Julve, M. Molecular magnetism, quo vadis? A historical perspective from a coordination chemist viewpoint. *Coord. Chem. Rev.* **2017**, *339*, 17–103. [CrossRef]
113. Ruben, M.; Rojo, J.; Romero-Salguero, F.J.; Uppadine, L.H.; Lehn, J.-M. Grid-type metal ion architectures: Functional metallosupramolecular arrays. *Angew. Chem. Int. Ed.* **2004**, *43*, 3644–3662. [CrossRef] [PubMed]
114. Moriuchi, T.; Hirao, T. Design and redox function of conjugated complexes with polyanilines or quinonediimines. *Acc. Chem. Res.* **2012**, *45*, 347–360. [CrossRef]
115. Mansoor, I.F.; Wozniak, D.I.; Wu, Y.; Lipke, M.C. A delocalized cobaltoviologen with seven reversibly accessible redox states and highly tunable electrochromic behaviour. *Chem. Commun.* **2020**, *56*, 13864–13867. [CrossRef] [PubMed]
116. Liu, Q.; Liu, X.; Shi, C.; Zhang, Y.; Feng, X.; Cheng, M.-L.; Su, S.; Gu, J. A copper-based layered coordination polymer: Synthesis, magnetic properties and electrochemical performance in supercapacitors. *Dalton Trans.* **2015**, *44*, 19175–19184. [CrossRef]
117. Zhang, Q.; Uchaker, E.; Candelaria, S.L.; Cao, G. Nanomaterials for energy conversion and storage. *Chem. Soc. Rev.* **2013**, *42*, 3127–3171. [CrossRef] [PubMed]
118. Durfee, W.S.; Pierpont, C.G. Complexes of quinone-functionalised chelating ligands for multiple electron/proton transfer reduction reactions. *Inorg. Chem.* **1993**, *32*, 493–494. [CrossRef]
119. Malwitz, J.; Alkire, N.; Durfee, W. Quinone-functionalized transition metal complexes for multi-electron transfer. the electrochemistry of a bis-naphthoquinone-functionalized schiff base Ni(II) complex. *J. Coord. Chem.* **2010**, *55*, 641–650. [CrossRef]
120. Roux, C.; Zarembowitch, J.; Gallois, B.; Granier, T.; Claude, R. Toward ligand-driven light-induced spin changing. influence of the configuration of 4-styrylpyridine (stpy) on the magnetic properties of $Fe^{II}(stpy)_4(NCS)_2$ complexes. Crystal Structures of the spin-crossover species Fe(*trans*-stpy)$_4$(NCS)$_2$ and of the high-spin species Fe(*cis*-stpy)$_4$(NCS)$_2$. *Inorg. Chem.* **1994**, *33*, 2273–2279.
121. Dietrich-Buchecker, C.O.; Sauvage, J.-P.; Kintzinger, J.-P. Une nouvelle famille de molécules: Les métallo-caténanes. *Tetrahedron Lett.* **1983**, *24*, 5095–5098. [CrossRef]
122. Mohr, B.; Sauvage, J.-P.; Grubbs, R.H.; Weck, M. High-yield synthesis of [2]catenanes by intramolecular ring-closing metathesis. *Angew. Chem. Int. Ed.* **1997**, *36*, 1308–1310. [CrossRef]
123. Hruby, J.; Vavrecková, S.; Masaryk, L.; Sojka, A.; Navarro-Giraldo, J.; Bartos, M.; Herchel, R.; Moncol, J.; Nemec, I.; Neugebauer, P. Deposition of tetracoordinate Co(II) complex with chalcone ligands on graphene. *Molecules* **2020**, *25*, 5021. [CrossRef] [PubMed]
124. Hruby, J.; Dvorák, D.; Squillantini, L.; Mannini, M.; van Slageren, J.; Herchel, R.; Nemec, I.; Neugebauer, P. Co(II)-based single-ion magnets with 1,1′-ferrocenediyl-bis(diphenylphosphine) metalloligands. *Dalton Trans.* **2020**, *49*, 11697–11707. [CrossRef] [PubMed]
125. Juráková, J.; Midlíková, J.D.; Hruby, J.; Kliuikov, A.; Santana, V.T.; Pavlik, J.; Moncol, J.; Cizmár, E.; Orlita, M.; Mohelsky, I.; et al. Pentaccordinate cobalt(II) single ion magnets with pendant alkyl chains: Shall we go for chloride or bromide? *Inorg. Chem. Front.* **2022**, *9*, 1179–1194. [CrossRef]
126. Juráková, J.; Fellner, O.F.; Schlittenhardt, S.; Vavrecková, S.; Nemec, I.; Herchel, R.; Cizmár, E.; Santana, V.T.; Orlita, M.; Gentili, D.; et al. Neutral cobalt(II)-bis(benzimidazole)pyridine field-induced single-ion magnets for surface deposition. *Inorg. Chem. Front.* **2023**, *10*, 5406–5419. [CrossRef]
127. Navarro-Giraldo, J.; Hruby, J.; Vavrecková, S.; Fellner, O.F.; Havlicek, L.; Henry, D.; de Silva, S.; Herchel, R.; Bartos, M.; Salitros, I.; et al. Tetracoordinate Co(II) complexes with semi-coordination as stable single-ion magnets for deposition on graphene. *Phys. Chem. Chem. Phys.* **2023**, *25*, 29516–29530. [CrossRef]
128. García-López, V.; Giaconi, N.; Poggini, L.; Calbo, J.; Juhin, A.; Cortigiani, B.; Herrero-Martín, J.; Ortí, E.; Mannini, M.; Clemente-León, M.; et al. Spin-crossover grafted monolayer of a Co(II) terpyridine derivative functionalized with carboxylic acid groups. *Adv. Func. Mater.* **2023**, *33*, 2300351. [CrossRef]
129. Bloom, B.P.; Paltiel, Y.; Naaman, R.; Waldeck, D.H. Chiral Induced Spin Selectivity. *Chem. Rev.* **2024**, *124*, 1950–1991. [CrossRef] [PubMed]
130. Torres-Cavanillas, R.; Escorzia-Ariza, G.; Brotons-Alcázar, I.; Sanchis-Gual, R.; Mondal, P.C.; Rosaleny, L.E.; Giménez-Santamarina, S.; Sessolo, M.; Galbiati, M.; Tatay, S.; et al. Reinforced Room-Temperature Spin Filtering in Chiral Paramagnetic Metallopeptides. *J. Am. Chem. Soc.* **2020**, *142*, 17572–17580. [CrossRef] [PubMed]
131. Cardona-Serra, S.; Rosaleny, L.E.; Giménez-Santamarina, S.; Martínez-Gil, L.; Gaita-Ariño, A. Towards peptide-based tunable multistate memristive materials. *Phys. Chem. Chem. Phys.* **2021**, *23*, 1802–1810. [CrossRef] [PubMed]
132. Zadrozny, J.M.; Niklas, J.; Poluektov, O.G.; Freedman, D.E. Multiple quantum coherences form hyperfine transitions in a vanadium(IV) complex. *J. Am. Chem. Soc.* **2014**, *136*, 15841–15844. [CrossRef]
133. Zadrozny, J.M.; Freedman, D.E. Qubit control limited by spin-lattice relaxation in a nuclear spin-free iron(III) complex. *Inorg. Chem.* **2015**, *54*, 12027–12031. [CrossRef] [PubMed]
134. Fataftah, M.S.; Zadrozny, J.M.; Coste, S.C.; Graham, M.J.; Rogers, D.M.; Freedman, D.E. Employing forbidden transitions as qubits in a nuclear spin-free chromium complex. *J. Am. Chem. Soc.* **2016**, *138*, 1344–1348. [CrossRef]
135. Yu, C.J.; Graham, M.J.; Zadrozny, J.M.; Niklas, J.; Krzyaniak, M.; Wasielewski, M.R.; Poluektov, O.G.; Freedman, D.E. Long coherence times in nuclear spin-free vanadyl qubits. *J. Am. Chem. Soc.* **2016**, *138*, 14678–14685. [CrossRef]
136. Bader, K.; Dengler, D.; Lenz, S.; Endeward, B.; Jiang, S.-D.; Neugebauer, P.; van Slageren, J. Room temperature quantum coherence in a potential molecular qubit. *Nat. Commun.* **2014**, *5*, 5304–5308. [CrossRef]

137. Atzori, M.; Tesi, L.; Morra, E.; Chiesa, M.; Sorace, L.; Sessoli, R. Room-temperature quantum coherence and rabi oscillations in vanadyl phthalocyanine: Toward multifunctional molecular spin qubits. *J. Am. Chem. Soc.* **2016**, *138*, 2154–2157. [CrossRef]
138. Bader, K.; Winkler, M.; van Slageren, J. Tuning of molecular qubits: Very long coherence and spin–lattice relaxation times. *Chem. Commun.* **2016**, *52*, 3623–3626. [CrossRef]
139. Carbonell-Vilar, J.M. Estudio de Propiedades Magnéticas en Compuestos de Coordinación Multifuncionales. Ph.D. Thesis, Universitat de València, Valencia, Spain, 2021.
140. Rabelo, R. Spin Crossover Cobalt(II) Molecular Nanomagnets as Dynamic Molecular Systems. Ph.D. Thesis, Universitat de València, Valencia, Spain, 2022.
141. Soyer, H.; Mingotaud, C.; Boillot, M.L.; Delhaes, P. Spin crossover of a langmuir−blodgett film based on an amphiphilic iron(II) complex. *Langmuir* **1998**, *14*, 5890–5895. [CrossRef]
142. Soyer, H.; Mingotaud, C.; Boillot, M.L.; Delhaes, P. Spin-crossover complex stabilized on a formamide/water subphase. *Thin Solid Films* **1998**, *329*, 435–438. [CrossRef]
143. Luo, Y.-H.; Liu, Q.-L.; Yang, L.-J.; Sun, Y.; Wang, J.-W.; You, C.-Q.; Sun, B.-W. Magnetic observation of above room-temperature spin transition in vesicular nano-spheres. *J. Mater. Chem. C* **2016**, *4*, 8061–8069. [CrossRef]
144. Hayami, S.; Shigeyoshi, Y.; Akita, M.; Inoue, K.; Kato, K.; Osaka, K.; Takata, M.; Kawajiri, R.; Mitani, T.; Maeda, Y. Reverse spin transition triggered by a structural phase transition. *Angew. Chem. Int. Ed.* **2005**, *44*, 4899–4903. [CrossRef] [PubMed]
145. Kitagawa, S.; Kitaura, R.; Noro, S.I. Functional porous coordination polymers. *Angew. Chem. Int. Ed.* **2004**, *43*, 2334–2375. [CrossRef] [PubMed]
146. Kitagawa, S.; Matsuda, R. Chemistry of coordination space of porous coordination polymers. *Coord. Chem. Rev.* **2007**, *251*, 2490–2509. [CrossRef]
147. Férey, G.; Serre, C. Large breathing effects in three-dimensional porous hybrid matter: Facts, analyses, rules and consequences. *Chem. Soc. Rev.* **2009**, *38*, 1380–1399. [CrossRef] [PubMed]
148. Mon, M.; Bruno, R.; Ferrando-Soria, J.; Armentano, D.; Pardo, E. Metal–organic framework technologies for water remediation: Towards a sustainable ecosystem. *J. Mat. Chem. A* **2018**, *6*, 4912–4947. [CrossRef]
149. Viciano-Chumillas, M.; Mon, M.; Ferrando-Soria, J.; Corma, A.; Leyva-Pérez, A.; Armentano, D.; Pardo, E. Metal–organic frameworks as chemical nanoreactors: Synthesis and stabilisation of catalytically active metal species in confined spaces. *Acc. Chem. Res.* **2020**, *53*, 520–531. [CrossRef]
150. Ferrando-Soria, J.; Ruiz-García, R.; Cano, J.; Stiriba, S.-E.; Vallejo, J.; Castro, I.; Julve, M.; Lloret, F.; Amorós, P.; Pasán, J.; et al. Reversible solvatomagnetic switching in a spongelike manganese(II)–copper(II) 3D open framework with a pillared square/octagonal layer architecture. *Chem. Eur. J.* **2012**, *18*, 1608–1617. [CrossRef] [PubMed]
151. Abhervé, A.; Grancha, T.; Ferrando-Soria, J.; Clemente-León, M.; Coronado, E.; Waerenborgh, J.C.; Lloret, F.; Pardo, E. Spin-crossover complex encapsulation within a magnetic metal–organic framework. *Chem. Commun.* **2016**, *52*, 7360–7363. [CrossRef] [PubMed]
152. Mon, M.; Pascual-Álvarez, A.; Grancha, T.; Cano, J.; Ferrando-Soria, J.; Lloret, F.; Gascon, J.; Pasán, J.; Armentano, D.; Pardo, E. Solid-state molecular nanomagnet inclusion into a magnetic metal–organic framework: Interplay of the magnetic properties. *Chem. Eur. J.* **2016**, *22*, 539–545. [CrossRef]
153. Graham, M.J.; Zadrozny, J.M.; Fataftah, M.S.; Freedman, D.E. Forging solid-state qubit design principles in a molecular furnace. *Chem. Mater.* **2017**, *29*, 1885–1897. [CrossRef]
154. Yaghi, O.M.; O'Keeffe, M.; Eddaoudi, M.; Chae, K.; Kim, J.; Ockwing, N.W. Reticular synthesis and the design of new materials. *Nature* **2003**, *423*, 705–714. [CrossRef]
155. Long, J.R.; Yaghi, O.M. The pervasive chemistry of metal–organic frameworks. *Chem. Soc. Rev.* **2009**, *38*, 1213–1214. [CrossRef]
156. Castellano, M.; Ferrando-Soria, J.; Moliner, N.; Cano, J.; Julve, M.; Lloret, F. Growth of thin films of single-chain magnets on functionalized silicon surfaces. *J. Coord. Chem.* **2018**, *71*, 725–736. [CrossRef]
157. Castellano, M.; Moliner, N.; Cano, J.; Lloret, F. Magnetic phase thransition and magneic bistability in oxamato-based $Co^{II}Cu^{II}$ bimetallic MOF thin films. *Polyhedron* **2019**, *170*, 7–11. [CrossRef]
158. Geng, K.; He, T.; Liu, R.; Dalapati, S.; Tan, K.T.; Li, Z.; Tao, S.; Gong, Y.; Jiang, Q.; Jiang, D. Covalent organic frameworks: Design, synthesis, and functions. *Chem. Rev.* **2020**, *120*, 8814–8933. [CrossRef] [PubMed]
159. Dong, J.; Han, X.; Liu, Y.; Li, H.; Cui, Y. Metal–Covalent Organic Frameworks (MCOFs): A bridge between metal–organic frameworks and covalent organic frameworks. *Angew. Chem. Int. Ed.* **2020**, *59*, 13722–13733. [CrossRef] [PubMed]
160. August, D.P.; Dryfe, R.A.W.; Haigh, S.J.; Kent, P.R.C.; Leigh, D.A.; Lemonnier, J.-F.; Li, Z.; Muryn, C.A.; Leoni, I.; Song, Y.; et al. Self-assembly of a layered two-dimensional molecularly woven fabric. *Nature* **2020**, *588*, 429–435. [CrossRef]
161. Zao, Y.; Guo, L.; Gándara, F.; Ma, Y.; Liu, Z.; Zhu, C.; Lyu, H.; Trickett, C.A.; Kasputin, E.A.; Terasaki, O.; et al. A synthetic route for crystals of woven structures, uniform nanocrystals, and thin films of imine covalent organic frameworks. *J. Am. Chem. Soc.* **2017**, *139*, 13166–13172. [CrossRef] [PubMed]

*magnetochemistry*

MDPI

*Article*

# Uncovering the Mechanisms of Long-Range Magnetic Order in $[Mn(mal)(H_2O)]_n$: Insights from Microscopic and Macroscopic Magnetic Analysis †

**Fernando S. Delgado [1], Laura Cañadillas-Delgado [2,*], Juan Rodríguez-Carvajal [2], Óscar Fabelo [2,*] and Jorge Pasán [3,*]**

[1] Laboratorio de Rayos X y Materiales Moleculares, Departamento de Física, Universidad de La Laguna, 38204 La Laguna, Tenerife, Spain
[2] Diffraction Group, Institut Laue Langevin, 38000 Grenoble, France; jrc@ill.fr
[3] Laboratorio de Materiales para Análisis Químico (MAT4LL), Departamento de Química, Universidad de La Laguna, 38204 La Laguna, Tenerife, Spain
* Correspondence: lcd@ill.fr (L.C.-D.); fabelo@ill.fr (Ó.F.); jpasang@ull.edu.es (J.P.)
† With this paper, we wish to pay a heartfelt tribute to Paco and Miguel on the occasion of their retirement. Fernando, Laura, Oscar, and Jorge are deeply thankful for the invaluable lessons we learned from Paco and Miguel, not only in the fields of Molecular Magnetism and Coordination Chemistry, but also in mentoring and their example of humanity. Tragically, during the preparation of this manuscript, we received the devastating news of the passing of Miguel. In his loss, we mourn not only an exceptional scientist, but also a cherished friend and a truly remarkable person.

**Abstract:** In this study, we investigate the magnetic properties of the molecular compound $[Mn(mal)(H_2O)]_n$ (mal = dianion of malonic acid) by integrating microscopic and macroscopic characterization, combining unpolarized neutron diffraction and magnetometry measurements. Neutron diffraction, though non-commonly applied to molecular compounds, proved essential for fully resolving the magnetic structure, as well as overcoming challenges such as hydrogen-related incoherent scattering and difficulties in accurately locating light atoms. Our neutron data provided critical structural details, including the precise location of hydrogen atoms, especially those associated with crystallization water molecules. By conducting low-temperature measurements below the magnetic ordering temperature, we identified the correct Shubnikov space group ($Pc'a2_1'$) and established a magnetic model consistent with the observed weak ferromagnetism. Our findings reveal that the compound presents a spin-canted structure with a weak ferromagnetic signal along the *b*-axis. This signal originates primarily from antisymmetric exchange interactions rather than single-ion anisotropy, consistent with the isotropic nature of the Mn(II) ($^6A_{1g}$) ground state. The combined neutron diffraction and magnetometry results provide a comprehensive understanding of how structural and symmetry factors influence the magnetic properties of malonate-based manganese compounds.

**Keywords:** molecular materials (malonates); X-ray diffraction; neutron diffraction; magnetometry measurements

## 1. Introduction

The field of molecular magnetism has been extensively studied, particularly by renowned coordination chemists such as Prof. Miguel Julve and Prof. Francesc Lloret. Through strategic modifications of the coordination environment of metal ions, they have successfully transformed and tailored the magnetic properties of materials [1]. These materials differ from traditional solid-state magnets due to their unique structural and bonding characteristics, which strongly influence their magnetic behavior. Molecular magnets are categorized by their dimensionality and magnetic properties, spanning a diverse range of systems: from 0D single-molecule magnets (SMMs) that show slow magnetic relaxation and bistability at low temperatures to low-dimensional systems like 1D single-chain magnets and 2D molecular layers. Additionally, molecular materials can be constructed as 3D coordination polymers, enabling the design of tailored magnetic functionalities [2,3].

The magnetic properties of molecular materials are strongly influenced by their electronic structure and the interactions between magnetic centers. These interactions are governed by several mechanisms, including double exchange, orthogonal magnetic orbitals, spin-polarization effects, and others [4]. Malonate coordination complexes are viewed as an extension of evergreen oxalate chemistry [5–12]; however, the three-carbon-atom skeleton of malonic acid introduces more flexibility than expected. Therefore, the magnetic properties of malonate compounds, particularly those incorporating transition metals, exhibit a notable versatility driven mainly by their structural configurations. Of particular interest is that the carboxylate bridge tends to appear in an out-of-plane configuration, leading to dihedral angles between the equatorial planes of bridging metal ions close to orthogonality, and thus yields weak ferro- (usually for Cu(II)-containing compounds) or antiferromagnetic interactions (for Co(II), Ni(II), or Mn(II)). The inherent flexibility of the malonate ligand has been extensively leveraged to tune magnetic behavior. Variations in the conformation of the malonate ligand and substituents can significantly impact crystal packing, thereby modulating the magnetic properties of these complexes [13]. For instance, different alkyl or aryl substitutions can tweak the crystal structure, affecting magnetic interactions [14].

The two-dimensional coordination networks adopted for metal transition malonates can facilitate magnetic interactions over longer distances, leading in some cases to long-range antiferromagnetic order, as observed in $Na_2M(H_2C_3O_4)_2{\cdot}2H_2O$ with M = Mn(II), Fe(II), Co(II), and Ni(II), and in the malonate compound described in this communication [15]. Focusing on Mn(II) compounds, whose magnetic behavior can be described by a magnetic chain model, the primary exchange coupling between Mn(II) atoms exhibits a weak antiferromagnetic character. The isotropic nature of the Mn(II) ($^6A_1g$) ground state allows the bridging modes of the malonate ligand to play a significant role in influencing magnetic coupling [14,16–20]. Additionally, other factors can affect magnetic interactions, such as spin-polarization effects on peripheral ligands, particularly those with $sp^2$-hybridized carbon sites, which can enhance the transmission of magnetic interactions over considerable distances [21–26].

The analysis of the magnetic properties of these compounds is usually based on macroscopic magnetometer measurements (isothermal magnetization curves, thermal dependence of magnetization, *ac* susceptibility ($\chi'$), out-of-phase susceptibility ($\chi''$), *dc* susceptibility, or relaxation measurements, among others), and generally, the actual magnetic structure with the correct alignment of the magnetic moments in each atom remains unknown. Neutron diffraction experiments are currently the most reliable approach for unambiguously resolving magnetic structures, including those of coordination compounds. However, certain prerequisites are essential to achieving a reliable model. For neutron powder diffraction, a sample of at least 1 g is typically required due to the weak interaction of neutrons with matter. Furthermore, low hydrogen content (or the use of deuterated samples) is critical, as hydrogen contributes to incoherent scattering, potentially obscuring the typically low-intensity magnetic reflections. In contrast, single-crystal studies demand less stringent sample requirements. Despite the relatively low number of magnetic structures reported for coordination polymers, previous studies demonstrate that neutron diffraction techniques are a valuable tool for advancing our understanding of magnetic molecular materials. These findings show that the challenges associated with such systems can be effectively addressed [27–46]. The data obtained from the magnetic structure not only support macroscopic analyses, but also facilitate a comprehensive understanding of the magnetic behavior in complex systems.

In this article, we present the magnetic structure of the malonate-based Mn(II) compound $[Mn(mal)(H_2O)]_n$, which exhibits a weak ferromagnetic signal at low temperatures after the onset of antiferromagnetic order, as observed in the macroscopic measurements. To determine the accurate position of light atoms and the magnetic structure below the Néel temperature, we perform low-temperature neutron diffraction experiments. Additionally, we conducted magnetic symmetry analysis to identify all possible Shubnikov magnetic space groups and refined the resulting models against the experimental data. By combining macroscopic measurements with neutron diffraction data, we successfully determined the magnetic structure of this compound.

## 2. Materials and Methods

Manganese(II) acetate tetrahydrate (from Merck, Burlington, MA, USA) and malonic acid (from Sigma Aldrich, Burlington, MA, USA) were used as received. Elemental analyses (C, H, N) were performed on an EA 1108 CHNS-O microanalytical analyzer.

### 2.1. Synthesis of $[Mn(mal)(H_2O)]_n$

The compound was synthesized by reacting an aqueous solution (15 mL) of manganese(II) acetate (8 mmol, 1.96 g) with a separate solution (10 mL) of malonic acid (8 mmol, 0.823 g) at 80 °C. This process resulted in the formation of pinkish crystallites, which were collected by filtration, washed with water and ethanol, and air-dried, with a yield of approximately 95% (1.46 g). Light-pink single crystals suitable for X-ray diffraction and magnetic measurements were obtained. The elemental analysis revealed the following composition: C, 18.76; H, 3.05. Calc. for $C_3H_6O_6Mn$: C, 18.66; H, 3.11%. Infrared spectroscopy (KBr, $cm^{-1}$) $\nu_{as}$(COO) 1669 s, 1570vs; $\nu_s$(COO) 1449 m, 1380 s, 1284 m; $\delta$(OCO) 811 m and 722 m.

### 2.2. X-Ray and Neutron Diffraction

The X-ray powder diffraction (XRD) pattern of the $[Mn(mal)(H_2O)]_n$ compound was recorded using a Panalytical X'Pert PRO diffractometer (Worcestershire, UK) operated with Cu-K$\alpha$ radiation $\lambda_1$ = 1.54056 Å, $\lambda_2$ = 1.54439 Å, and a PiXCel 1D detector operating with a Bragg–Brentano geometry. The data collection was carried out in the 5–80° 2-theta range in 0.02° steps, with a total acquisition time of 15 min.

A clear, light-pink, almost colorless prism-like crystal of 0.1 × 0.15 × 0.15 $mm^3$ was used for single-crystal diffraction at 100 K using a Ag microsource ($\lambda$ = 0.56086 Å) on a dual-source Bruker D8 VENTURE (Bruker Coorporation, Karlsruhe, Germany) equipped with a Photon III detector. For the purpose of collecting a full data set of reflections, the crystal was put on loops and covered with paratone oil. The crystallographic data and experimental details of the structural analysis are summarized in Table 1. A total of 578 frames were collected. The frames were integrated with the Bruker SAINT software package (version V8.40B) using a narrow-frame algorithm. Data were corrected for absorption effects using the Multi-Scan method (SADABS). The structure was solved to locate all the non-H atoms, which were refined anisotropically using the Bruker SHELXTL software package (version 2018/2) and the non-centrosymmetric space group $Pca2_1$, with Z = 4 for the formula unit $C_3H_6MnO_6$. The structure was refined using SHELXL-2019/1, using full-matrix least squares on the $F^2$ procedure [47,48]. H atoms from the malonate ligand were placed in the calculated positions and constrained to ride on their parent atoms. H atoms from the coordinated water molecules were placed in the maximum density compatible with the probable H-bond contacts and fixed with bond distance restraints. The final anisotropic full-matrix least squares refinement on $F^2$ with 107 variables converged at $R_1$ = 2.27% for the observed data and $wR_2$ = 5.42% for all data.

Neutron diffraction experiments were conducted at the Orphée reactor facilities, Laboratoire Léon Brillouin, Saclay, France. High-resolution neutron powder data collected in the paramagnetic phase were obtained using the G4.2 diffractometer (high resolution (HR), featuring 70 detectors with Soller collimators) with a wavelength of 2.34246 Å. The step increment in $2\theta$ was set to 0.1°. The *HR* G4.2 data, which extended from 3° to 160°, were utilized for structural and magnetic refinements. Additionally, neutron diffraction patterns as a function of temperature from 1.5 to 5.0 K were recorded using the G4.1 high-flux powder diffractometer with a 2.428 Å wavelength obtained from a graphite monochromator. The powdered samples were contained in a vanadium cylindrical container (8 mm of diameter). All powder patterns were refined using the Rietveld method, as implemented in the FullProf program (October 2024) [49].

**Table 1.** Crystal data and structure refinement for $[Mn(mal)(H_2O)]_n$.

| | | |
|---|---|---|
| **Chemical formula** | $C_3H_6MnO_6$ | |
| **Formula weight** | 193.02 g/mol | |
| **Temperature** | 100(2) K | |
| **Wavelength** | 0.56086 Å | |
| **Crystal size** | $0.100 \times 0.150 \times 0.150$ mm$^3$ | |
| **Crystal system** | orthorhombic | |
| **Space group** | $Pca2_1$ | |
| **Unit cell dimensions** | $a = 9.6321(6)$ Å<br>$b = 7.3554(5)$ Å<br>$c = 8.2307(5)$ Å | $\alpha = 90°$<br>$\beta = 90°$<br>$\gamma = 90°$ |
| **Volume** | 583.13(6) Å$^3$ | |
| **Z** | 4 | |
| **Theta range for data collection** | 2.19 to 28.14° | |
| **Index ranges** | $-15 \leq h \leq 16, -9 \leq k \leq 12, -12 \leq l \leq 13$ | |
| **Reflections collected** | 11,840 | |
| **Independent reflections** | 2745 [$R$(int) = 0.0373] | |
| **Data/restraints/parameters** | 2745/7/107 | |
| **Goodness-of-fit on $F^2$** | 1.052 | |
| **$\Delta/\sigma_{max}$** | 0.001 | |
| **Final $R$ indices** | 2683 data; $I > 2\sigma(I)$<br>$R_1 = 0.0227, wR_2 = 0.0537$<br>all data<br>$R_1 = 0.0233, wR_2 = 0.0542$ | |

## 3. Results

### 3.1. Crystal Structure of $[Mn(mal)(H_2O)]_n$

The crystal structure of this compound has previously been reported in several publications [50–52]. In this manuscript, we present a summary of its key structural details to assist the reader in following the subsequent discussions.

The structure of $[Mn(mal)(H_2O)]_n$ consists of layers of trans-diaquamanganese(II) units bridged by carboxylate–malonate groups in both *anti–anti* and *anti–syn* coordination modes (see Figure 1). These layers are stacked along the *c*-direction following an *ABAB* sequence (see Figure 2). The manganese (Mn) atoms are octahedrally coordinated by four carboxylate oxygen atoms from three different malonate ions and two *trans*-water oxygen atoms. Each malonate ligand coordinates to three different Mn(II) ions, forming a 4,4-square two-dimensional network parallel to the (001) plane. The four carboxylate oxygens—O(1), O(3), O(2a), and O(4b) ($a = x - 1/2, -y + 1, z$ and $b = x - 1/2, -y, z$)—fill the equatorial positions around the manganese centers. This coordination results in significant planarity within the layers, characterized by a very short interlayer Mn···Mn distance of 5.3502(4) Å. The Mn···Mn separations through the *anti–anti* and *anti–syn* carboxylate bridges measure 5.7997(4) and 6.3403(4) Å, respectively. An extensive network of hydrogen bonds holds the polymeric layers together. All four hydrogen atoms from the two water molecules are involved in hydrogen bonding, enhancing the stability of the structure (see Figure 1b).

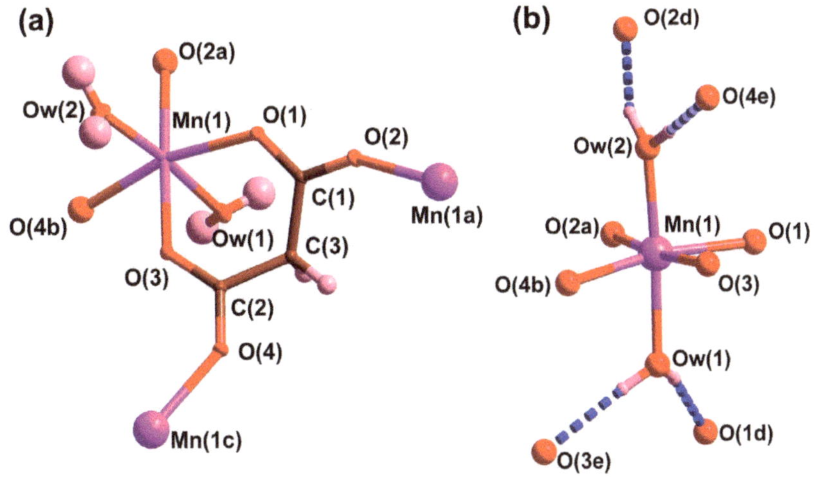


**Figure 1.** (**a**) Detailed view of the coordination mode of the malonate ligand in $[Mn(mal)(H_2O)]_n$. The coordination environment of the manganese atom in the asymmetric unit has been completed to enhance understanding of the structure. Atoms in the asymmetric unit are represented as ellipsoids with a 50% probability. (**b**) Coordination environment around Mn with blue-dashed lines indicates hydrogen bonds according to the neutron-refined model at 5 K. Symmetry operators: (a) = $-1/2 + x$, $1 - y$, $z$; (b) = $-1/2 + x$, $-y$, $z$ (c) = $2 - x$, $1 - y$, $-1/2 + z$; (d) = $2 - x$, $-y$, $-1/2 + z$; (e) = $5/2 - x$, $y$, $1/2 + z$.

The polymeric network of $[Mn(mal)(H_2O)]_n$ is quite distinct from those of other manganese carboxylates, such as formate, acetate, propionate, citrate, ethylenediaminetetraacetate, malate, and gluconate, among others. Although most of the carboxylates of Mn(II) have the tendency to crystallize as polymeric structures, the layered structure of $[Mn(mal)(H_2O)]_n$ is not common. However, it remains as an intrinsic feature of malonate-based compounds. Although with slight modifications, the (4,4)-planes can also be observed in other malonates such as $Na_2M(mal)_2{\cdot}2H_2O$, with M = Mn, Co, and Ni [15], as well as in compounds where a rod-like ligand has been introduced to modulate the interplanar distance (e.g., 1,2-bis(4-pyridyl)ethylene, 1,2-bis(4-pyridyl)ethane, 4,4′-bipyridine...) [53,54]. The bond lengths and angles in the malonate groups are similar to those previously reported, with C–O lengths being nearly equal, with values ranging from 1.251(2) to 1.263(2) Å. The malonate ligands exhibit an envelope conformation, common in such compounds. The C(1)–C(3)–C(2) angle is measured at 115.5(8)°, which is significantly distorted from the ideal tetrahedral geometry (109.5°). The chelating character of the malonate ligand gives rise to a six-membered chelate ring that twists the planes of the two carboxylate groups closer to the plane of the carbon atom chain, resulting in an angle of 36.2(7)° and 45.5(9)° for each carboxylate group with respect to the planes containing the C-atom chains.

To summarize, the crystal structure of $[Mn(mal)(H_2O)]_n$ exhibits a well-defined layered architecture where Mn atoms are octahedrally coordinated, with two water molecules occupying the apical positions and the remaining four positions filled by oxygen atoms from the carboxylate groups. The metal atoms are linked via *anti–anti* and *anti–syn* carboxylate bridges, forming distinct 4,4 layers. The magnetic interactions across these bridges play a pivotal role in defining the magnetic properties of the compound, while weaker interlayer interactions, due to the short Mn···Mn distance between the layers and well-oriented H-bonds, support long-range magnetic order at low temperatures.

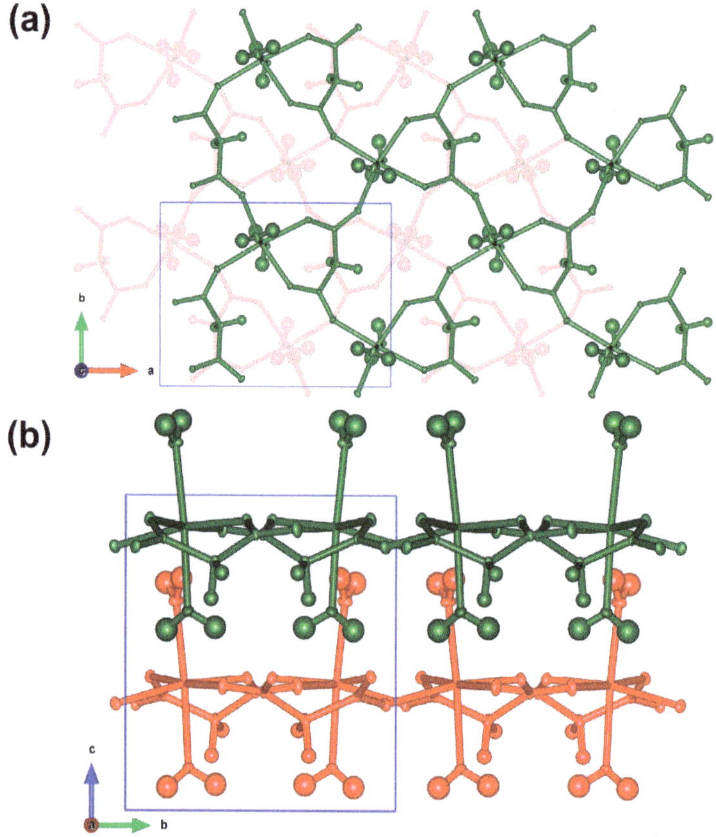


**Figure 2.** (**a**) View along the *c*-axis and (**b**) view along the *a*-axis of the refined structural model for $[Mn(mal)(H_2O)]_n$ at 5 K, obtained using neutron powder diffraction data from the G4.2 diffractometer. For clarity, alternating layers stacked along the *c*-axis are shown in red and green to illustrate the *ABAB* stacking sequence.

### 3.2. Magnetic Measurements of $[Mn(mal)(H_2O)]_n$

As with the crystal structure section, the magnetic properties of this compound have been described in detail in reference [52]; here, we will only highlight the key features to improve the clarity of the manuscript.

Magnetic measurements in the form of $\chi_M T$ vs. $T$, in the temperature range from RT to 3 K, show the typical behavior of an antiferromagnetic system, with a smooth decrease in $\chi_M T$ from RT to 100 K. A more rapid decrease below this temperature shows that the exchange coupling interactions are more important than the thermal effects. This antiferromagnetic behavior is also confirmed in the temperature dependence of the susceptibility curve, where a well-defined peak is observed. However, an abrupt increase in susceptibility is observed below ca. 3 K (see Figure 3).

As discussed in the previous section, the crystal structure of $[Mn(mal)(H_2O)]_n$ consists of planar layers where manganese(II) ions are connected by carboxylate groups in alternating *anti–anti* and *anti–syn* coordination modes. Both configurations can mediate significant magnetic coupling between paramagnetic centers, though the *anti–syn* configuration produces considerably weaker interactions than the *anti–anti* configuration [55]. Considering this, the temperature dependence of the magnetic susceptibility was fitted using Fisher's expression for a chain of interacting spins (S = 5/2) [56], with an additional mean-field correction to account for interchain interactions. This fitting yielded magnetic exchange couplings across the *anti–anti* and *anti–syn* carboxylate bridges, both of which were antiferromagnetic in nature, with $J = -0.64(1)$ and $j = -0.0075$ cm$^{-1}$.

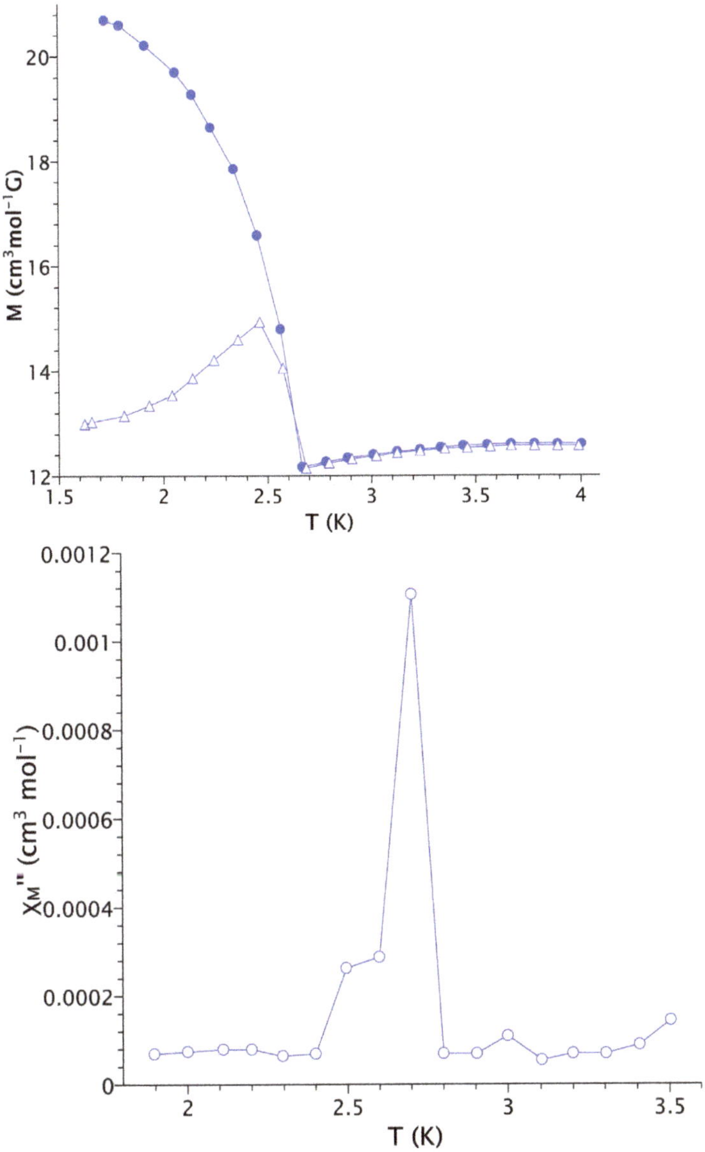


**Figure 3.** (**Top**) Temperature dependence of the magnetization for $[Mn(mal)(H_2O)_2]_n$, showing zero-field-cooled (open triangles) and field-cooled (solid circles) measurements. Solid lines are included as a visual guide to aid in identifying trends. (**Bottom**) Temperature dependence of out-of-phase *ac* susceptibility, $\chi_M$, measured under an oscillating field of 1 G at a frequency of 333 Hz.

This simple model provides a good fit to the experimental data from 3 K up to room temperature. However, the abrupt increase in susceptibility observed below 3 K cannot be adequately captured. Furthermore, the relatively small values obtained for interchain interactions ($j$) may be attributed to competing antiferromagnetic and ferromagnetic interactions, with the latter driving the susceptibility increase below 3 K.

To rule out any artifacts or the presence of a paramagnetic impurity that could account for the increase in susceptibility below 3 K, field-cooled (FC) and zero-field-cooled (ZFC) magnetization curves were collected (see Figure 3). The FC magnetization curve exhibits a sharp increase at temperatures below 2.8 K, approaching saturation at lower temperatures. In contrast, the ZFC magnetization curve rises with increasing temperature, reaching a maximum at $T$ = 2.6 K, and at $T$ = 2.75 K, its value aligns with that of the FC curve. From these experiments, the ordering temperature can be established more precisely as 2.7(1) K. The three-dimensional nature of this magnetic signal was confirmed by *ac* magnetic susceptibility measurements. The out-of-phase magnetic susceptibility, $\chi''$, exhibits a distinct peak at 2.7 K, with no frequency dependence observed in the range of 100 to 1500 Hz. This observation rules out any dynamic effects as the origin of this signal.

### 3.3. Neutron Diffraction

The initial structural model was derived from single-crystal measurements conducted at 100 K. The purity of the sample used for the neutron diffraction experiments was assessed using X-ray powder diffraction. In the X-ray powder measurement, the intensity of the observed peaks aligned well with the calculated pattern derived from the single-crystal diffraction model. However, the positions of the hydrogen atoms in the two water molecules occupying the apical positions of the Mn(II) atom were determined using difference Fourier maps derived from X-ray data. Due to the weak diffraction power of the hydrogen atoms in the X-ray analysis, the obtained positions may contain significant inaccuracies. To further evaluate the crystal structure of the sample, a diffraction pattern was collected using the G4.2 instrument in the paramagnetic phase (5 K). A Rietveld refinement was performed based on our initial model to refine the cell parameters and atomic positions, including those of the lighter atoms (see Figure 4). The final fit is depicted in Figure 5. We confirmed that the crystal structure corresponds to the expected orthorhombic structure (space group $Pca2_1$), with unit cell parameters ($a$ = 9.6323(2) Å, $b$ = 7.3452 (2) Å, $c$ = 8.2109(2) Å, $V$ = 580.93(2) Å$^3$) in agreement with those values reported in the literature [50–52] and in our X-ray measurement at 100 K.

The temperature dependence of the neutron diffraction patterns was collected by the G4.1 instrument and is presented in Figure 6. Below the Néel temperature ($T_N$), new reflections appear, which were indexed using the K-search program within the FullProf suite [49]. The most probable propagation vector, $\mathbf{k}$ = (0, 0, 0), accurately accounts for all observed magnetic reflections. To explore possible magnetic structures compatible with the symmetry of the parent crystal structure of $[Mn(mal)(H_2O)]_n$, which is described in the paramagnetic space group $Pca2_1.1'$ and the $\mathbf{k}$ = 0 propagation vector, we utilized the MAXMAGN tool on the Bilbao Crystallographic Server [57].

There are four maximal Shubnikov space groups: $Pca2_1.1$ (#29.99), $Pc'a'2_1$ (#29.103), $Pca'2_1'$ (#29.102), and $Pc'a2_1'$ (#29.101). Each of these models corresponds to an irreducible representation, from $\Gamma_1$ to $\Gamma_4$, describing four different Shubnikov magnetic space groups. In all four magnetic structures, the Mn(II) atoms occupy the 4*a* Wyckoff position, meaning that no orbital splitting occurs within these orthorhombic space groups. In other words, the four magnetic moments in the unit cell are generated from a single site by applying symmetry operators. Additionally, the orientation of the magnetic moments in each model is not symmetry-restricted, allowing for the refinement of all three components of $\mathbf{m} = (m_x, m_y, m_z)$.

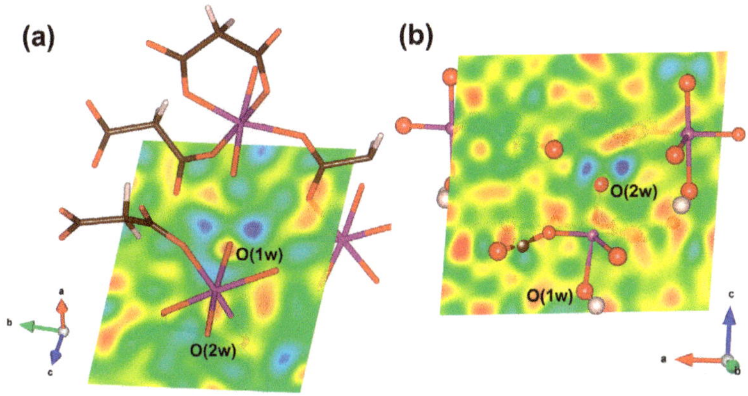


**Figure 4.** (**a**) View of the Fourier map around O(1w), derived from low-temperature (5 K) neutron diffraction data, using the X-ray model (without hydrogen atoms of the crystallization water) as the starting point. (**b**) Fourier map around O(2w), following the placement of hydrogen atoms on O(1w).

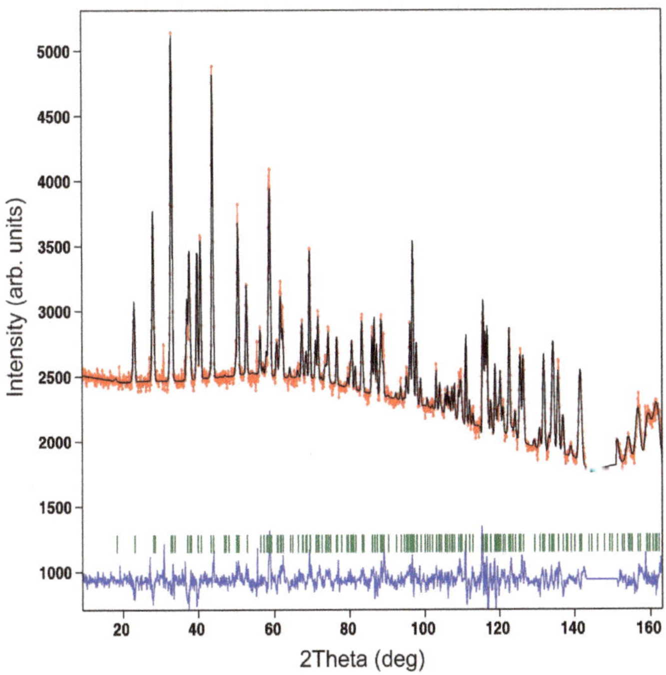


**Figure 5.** Rietveld refinement of $[Mn(mal)(H_2O)]_n$ collected at 5 K, measured on the G4.2 diffractometer. Experimental data and fitted profiles are represented by red and black lines, respectively, while the blue line indicates the residuals of the refinement. Vertical green ticks mark the positions of the Bragg reflections. The gap between 143° and 151° corresponds to an excluded region where contributions from the sample environment become visible. The final Bragg R-factor obtained for the paramagnetic phase in the last refinement cycle is 9.2%.

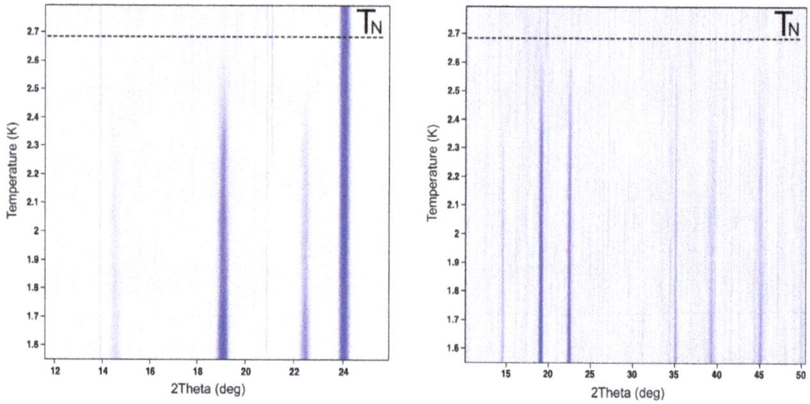


(**a**) (**b**)

**Figure 6.** (**a**) Thermodiffractogram of $[Mn(mal)(H_2O)]_n$ collected on the G4.1 instrument. (**b**) Thermodiffractogram of the magnetic signal, calculated as the difference between the pattern measured at different temperatures and the one measured at 5 K in the paramagnetic phase (difference pattern). The Néel temperature is highlighted with a black-dashed line in both figures.

Hereafter, we will describe the four maximal magnetic structures. To facilitate understanding, we will describe the orientation of the magnetic moments using the two carboxylate bridges (*anti–anti* and *anti–syn*) as a reference, considering that the *anti–anti* bridge represents the main magnetic interaction.

The magnetic structure described by the $Pca2_1.1$ Shubnikov space group can be visualized as a layered arrangement extended within the *ab*-plane, with layers stacked along the *c*-axis in an *ABAB* sequence. This stacking sequence arises from the antiferromagnetic coupling between adjacent layers. Within the layers, both carboxylate bridges mediate antiferromagnetic interactions between the Mn(II) ions in the *a*- and *c*-directions, while allowing a ferromagnetic component along the *b*-axis. However, this ferromagnetic component is counterbalanced by the adjacent layer, resulting in an overall antiferromagnetic structure.

This model adequately fits the magnetic reflections; however, it results in a purely antiferromagnetic structure. The fit of the experimental data indicates that the $m_y$ component along the *b*-axis is significantly smaller than those along the *a*- and *c*-axes, being nearly equal to its error bar. The fit reveals that the magnetic component along the *a*-axis is the dominant one. Fixing $m_y$ to zero yields a comparable fit, improving the statistics slightly. Nevertheless, the model described by the Shubnikov group $Pca2_1.1$ is strictly AF and, therefore, incompatible with the weak ferromagnetic signal observed in the macroscopic measurements. Consequently, this model is ruled out.

The magnetic structures described in the $Pc'a'2_1$ and $Pca'2_1'$ Shubnikov space groups can be discarded after refinement against the experimental model. Furthermore, only $Pc'a'2_1$ allows for a global ferromagnetic signal, being compatible with the magnetometer measurements. Within the layers, the carboxylate bridges facilitate antiferromagnetic coupling in the *a*- and *b*-directions while permitting a ferromagnetic component in the *c*-direction. This component is not counterbalanced due to the ferromagnetic coupling between the layers; therefore, this model allows a global ferromagnetic signal along the *c*-axis.

The last magnetic structure is described by the $Pc'a2_1'$ Shubnikov space group (see Figure 7). As in the previous case, the magnetic interactions between adjacent layers are ferromagnetic. Within the layers, the magnetic moments are antiferromagnetically coupled in the *a*- and *c*-directions while allowing a ferromagnetic signal along the *b*-axis. This ferromagnetic component is globally conserved due to the ferromagnetic interlayer interactions. The refinement against the experimental data gives a good fit, with the main component of the magnetic moment along the *c*-axis. Although there are not any restraints in terms of symmetry in the magnetic moment components, the $m_x$ component is zero within the experimental error; therefore, it will be fixed to zero in subsequent refinements. The $m_x$ component of the magnetic moment has the strongest effect on the (011) reflection, which shows no evidence of magnetic contribution between the pattern collected at 1.5 and 5 K, corresponding to the ordered and paramagnetic state, respectively. Thus, fixing $m_x$ to zero is justified.

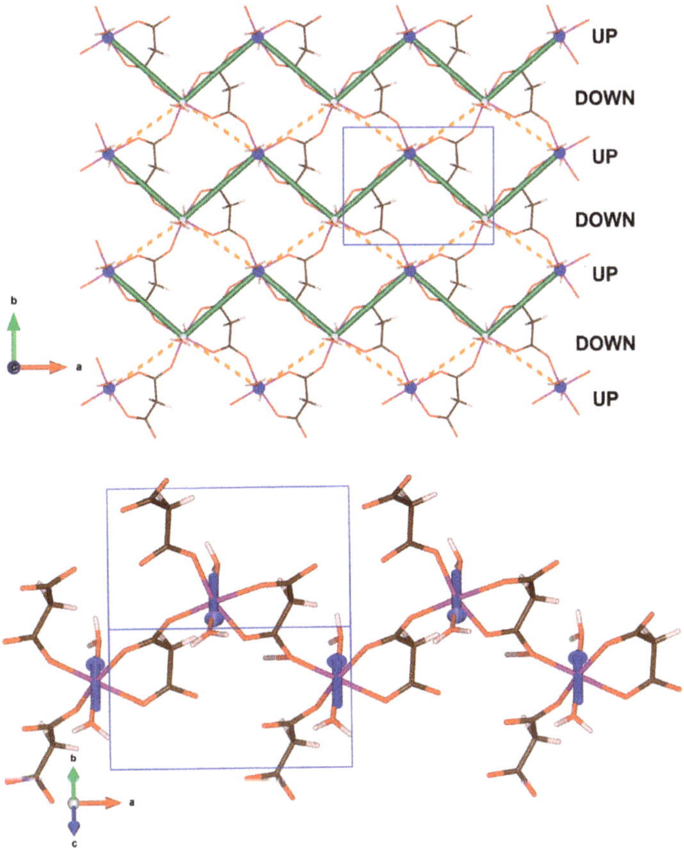


**Figure 7.** Magnetic structure of $[Mn(mal)(H_2O)]_n$ compound in the $Pc'a2_1'$ magnetic space group. (**Top**) View along the *c*-axis of one of the layers of $[Mn(mal)(H_2O)]_n$ extended in the *ab*-plane. The unit cell is indicated in blue for reference. The main magnetic interactions between Mn ions via *anti–anti* exchange (*J*) are shown with a green bond, while the secondary interaction (*j*) is depicted with an orange-dashed line. The magnetic moments are represented in blue. (**Bottom**) Detailed view in the [011] direction of the magnetic chain formed by Mn atoms linked through carboxylate bridges in an *anti–anti* configuration.

The $m_y$ component along the *b*-axis is therefore responsible for the weak ferromagnetism observed in the macroscopic measurements. The refined value for the magnetic moment is 4.30(3) $\mu_B$, which is slightly lower than what would be expected from a spin-only contribution for a high-spin Mn(II) ion. However, this reduction may arise from the fact that the magnetic moments are not fully saturated, as the ordering temperature (2.7 K) is close to the lowest temperature achieved (1.5 K). From the evolution of the magnetic moments, there is no clear evidence of saturation in the magnetization derived from the neutron fit at different temperatures.

The refined value of the $m_y$ component is notably weaker than the component along the *c*-axis and presents a large error bar (0.73(3) $\mu_B$). The reflections most influenced by the $m_y$ component are (110), (200), and (111), which exhibit both nuclear and magnetic intensity contributions. Therefore, to accurately determine $m_y$, we opt to work with the difference pattern, calculated as the difference between the measurements taken at 1.5 K, below $T_N$, and those taken at 5 K in the paramagnetic phase. This approach enables us to isolate the magnetic contribution (see Figure 8).

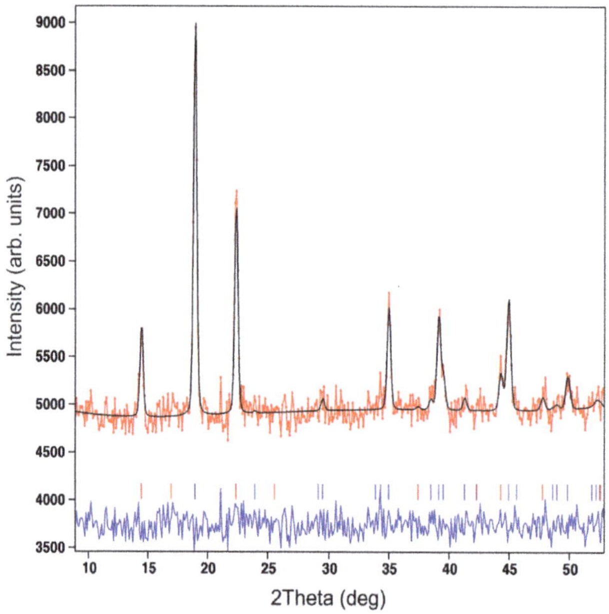


**Figure 8.** Rietveld refinement of the difference pattern (1.5–5 K) for [Mn(mal)($H_2O$)]n, measured on the G4.1 diffractometer, was performed using the *Pc′a*$2_1$*′* Shubnikov space group (Γ4), as described in the main text. The refinement corresponds to a model where $m_x$ is fixed to zero and $m_y$ is fixed to 0.05 $\mu_B$ (for further details, see the main text). The resulting Bragg R-factor for the magnetic reflections is 10.6%. Experimental data and the fitted profiles are shown by the red and black lines, respectively, with the blue line representing the residuals of the refinement. Red vertical ticks mark pure magnetic reflections, while blue ticks denote both magnetic and nuclear Bragg peaks (note that the nuclear contribution is zero due to the difference pattern).

Among the three reflections that are most sensitive to the $m_y$ component, the (200) reflection, which is perpendicular to the magnetic moment component, is likely the most significant reference reflection, as even minor changes in the $m_y$ component significantly impact its intensity. After removing the nuclear contribution, this reflection becomes solely sensitive to the $m_y$ component of the magnetic moment. The (200) reflection is located very close to the (111) reflection; thus, the magnetic contribution to the (200) reflection should appear as a shoulder to the left of the (111) peak. Unfortunately, the measurements are quite noisy, and the profiles of these two reflections are not well defined, preventing us from accurately determining the $m_y$ component. However, the magnetometry measurements indicate that the magnetic structure has a small ferromagnetic contribution, requiring that $m_y \neq 0$. Thus, while fixing the $m_y$ component at zero also yields a good fit, we conclude that it must be different from zero, although our current measurements do not have the precision to determine its exact value. Based on our fit, we determine a canting angle of 9(3)°. However, magnetometry measurements suggest that the actual canting angle is less than 1° (approximately 0.6°). Using this angle as a reference, we estimate the magnetic component along the $b$-axis to be about 0.05 $\mu_B$, which is significantly below the precision of our measurements. However, for consistency, this value is used as a fixed parameter for the $m_y$ component in the last refinement cycle. It is worth noting that the refinement statistics for the magnetic model described in the $Pc'a2_1'$ Shubnikov space group are the lowest among the four possible maximal space groups. The refined magnetic structure can be seen in Figure 9.

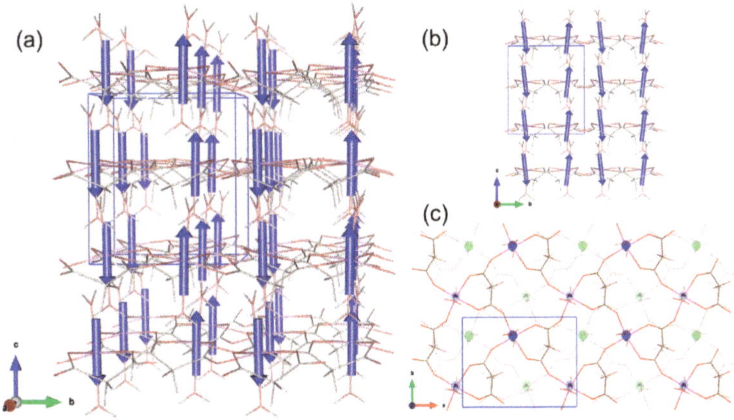


**Figure 9.** (**a**) Refined magnetic structure of $[Mn(mal)(H_2O)]_n$, with magnetic moments depicted as blue arrows. The nuclear structure is shown in transparent mode for clarity. (**b**) View along the $a$-axis, where magnetic moment components along the $b$-axis (representing the weak ferromagnetic component—see main text) are magnified by a factor of 10 for visibility. (**c**) View along the $c$-axis, with the $m_y$ component similarly enhanced as in (**b**). Different stacking layers along the $c$-axis are shown with varying transparencies, and magnetic moments in each layer are represented in alternating blue and green hues to improve visual distinction.

The refined magnetic moment was fitted in the critical region ($|T - T_N|/T_N < 0.1$) using a power law of $M \propto |T - T_N|^{\beta}$(red solid line in Figure 10), giving rise to a critical exponent $\beta$ of 0.34(2). This number should be interpreted with caution, given the limited number of experimental points available for fitting in the critical region. Nevertheless, the obtained value aligns well with the expected value for a 3D Heisenberg model ($\beta$ = 0.365).

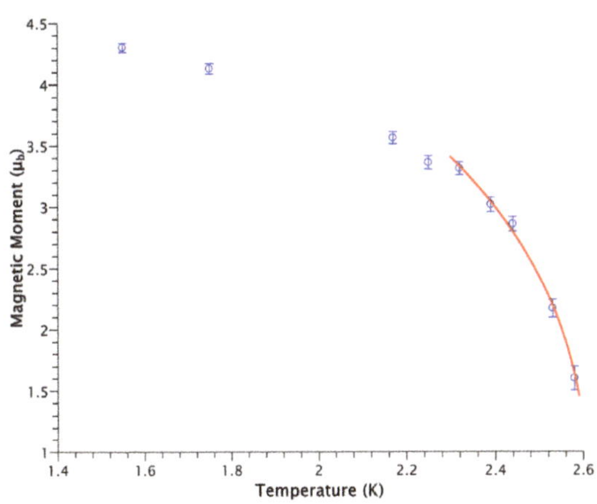


**Figure 10.** Evolution of the magnetic moment as a function of temperature. The solid line represents the power law fit of the data in the critical region (see main text).

## 4. Discussion

The ability of carboxylate bridges to mediate magnetically between paramagnetic centers is well known. For the case of malonates and R-malonates, the main exchange coupling between paramagnetic centers is mainly antiferromagnetic for Mn-, Co-, and Ni-based compounds, whereas it becomes ferromagnetic in the case of Cu [14,58–61].

The crystal structures of the substituted malonate manganese(II) compounds exhibit big differences to the layered $[Mn(mal)(H_2O)]_n$ compound. However, when a rod-like bidentate co-ligand is introduced in the framework, the system commonly adopts an *anti–syn* carboxylate bridged (4,4)-layer, further connected through the co-ligand to give three-dimensional structures. This situation occur in the following compounds: $[Mn_2(Etmal)(H_2O)_2(L)]_n$ with Etmal = ethylmalonate and L = 1,2-bis(4-pyridyl)ethane, 4,4′-bipyridyl, and 1,2-bis(4-pyridyl)ethylene; $[Mn_2(Rmal)_2(H_2O)_2(4,4'\text{-azobispyridine})]_n$ with Rmal = methylmalonate, dimethylmalonate, and butylmalonate; and $[Mn(\text{phenylmalonate})(H_2O)(4,4'\text{-azobispyridine})]_n$ [53,54]. Interestingly, the layers in these latter cases are corrugated, and the carboxylate bridge exhibits only the *anti–syn* conformation. The dihedral angles among the equatorial planes of adjacent Mn(II) ions are in the range of 83° to 89°. The magnetic behavior in all these examples is weakly antiferromagnetic, as is expected for poorly overlapping molecular orbitals.

Among the malonate-based manganese(II) compounds that exhibit weak ferromagnetism, $[Mn(Et_2mal)(H_2O)]_n$ ($Et_2mal$ = diethylmalonate) has magnetic properties similar to those of $[Mn(mal)(H_2O)]_n$ [54]. However, $[Mn(Et_2mal)(H_2O)]_n$ does not exhibit the layered structure of the $[Mn(mal)(H_2O)]_n$ compound. The main difference arises from the steric effects of the ethyl groups that preclude the formation of (4,4)-layers. The two-dimensional structure of $[Mn(Et_2mal)(H_2O)]_n$ consists of chains of $\mu$-aqua and double *syn–syn* carboxylate-bridged manganese(II) ions running along the crystallographic *b*-axis, which are connected through $Et_2mal$ ligands to form neutral layers in the *bc* plane. The magnetic behavior above the long-range magnetic order can be described using a model of uniform Mn(II) chains bridged by a single aqua and two *syn–syn* carboxylate ligands. Consequently, as occurs with the $[Mn(mal)(H_2O)]_n$ compound, the susceptibility data can be fitted using the Fisher model [56]. The least squares best-fit parameters yield a $J$ value of $-3.69(4)$ cm$^{-1}$, which is higher than the obtained value for the $[Mn(mal)(H_2O)]_n$ compound ($-0.64(1)$ cm$^{-1}$). This increase in $J$ most likely results from the presence of two distinct antiferromagnetic exchange pathways: $\mu$-$\kappa$O:$\kappa$O and the *syn–syn* carboxylate bridge. The strength of this exchange coupling also explains the significant rise in the ordering temperature, which is almost 30 K above that of the $[Mn(mal)(H_2O)]_n$ compound. Although the magnetic structure of this compound remains unreported, only two of the possible maximal space groups (*Cm′ca′* and *Cm′c′a*) yield net ferromagnetic components under the condition of a $\mathbf{k} = 0$ propagation vector, which is needed for the emergence of net ferromagnetism. Moreover, in both Shubnikov space groups, the ferromagnetic signal arises from weak spin canting, with the main difference being the orientation of this ferromagnetic signal. Thus, the canting angle, estimated from the reported saturation magnetization curve, is 0.95°, a value comparable to that derived from the magnetometry curves in the $[Mn(mal)(H_2O)]_n$ compound.

Among transition metal malonate compounds, the only family studied microscopically using neutron diffraction are $Na_2M(mal)_2{\cdot}2H_2O$ compounds with M = Mn, Fe, Co, and Ni [15]. Their crystal structure and magnetic behavior share similarities with those of the $[Mn(mal)(H_2O)]_n$ compound. Below, we provide a brief comparison of these two compounds.

The crystal structures of the $Na_2M(mal)_2{\cdot}2H_2O$ compounds [15,27,28], which include transition metals such as Mn, Fe, Co, Ni, and Cu, are isostructural and can be described as sheets composed of metal–oxygen octahedra linked by tridentate malonate ligands, alternating with sodium cations and crystallization water molecules. The transition metal environment creates a slightly distorted octahedron composed of six oxygen atoms from four malonate ligands, with two acting as bidentate and two as monodentate. Of the two carboxylate groups in the malonate ligand, only one acts as a bridge between adjacent metal atoms in an *anti–anti* configuration, which is relatively uncommon in transition metal malonate compounds. The space between the layers is occupied by sodium cations and water molecules, with the sodium ions situated in an octahedral coordination, surrounded by four oxygen atoms from carboxylate groups and two from water molecules.

While the macroscopic magnetic properties of transition metal molecular-based compounds are routinely studied, magnetic structure determination using neutron diffraction is rare, with only one study on malonate compounds reported [15,27,28]. The large spacing between transition metal ions often restricts long-range magnetic ordering, and hydrogen-rich ligands introduce substantial diffuse scattering, complicating both neutron diffraction experiments and their analysis. Although the deuteration of ligands can mitigate this issue, this is often complex and expensive, making such studies even less common. Consequently, there are very few reports in the literature on coordination polymers or metal–organic frameworks [20–45,51].

The $[Na_2Mn(mal)_2]{\cdot}2H_2O$ compound exhibits antiferromagnetic ordering around 7 K. The magnetic structure, determined by neutron diffraction, reveals antiferromagnetic layers with moments aligned along the *c*-axis. At 1.5 K, the magnetic moment of Mn(II) is 4.42 $\mu_B$, slightly below the expected spin-only value of 5 $\mu_B$, possibly due to electron delocalization toward the surrounding oxygen atoms [46] or zero-point quantum fluctuations in the ground state. In the context of an ideal isotropic Heisenberg exchange for a model characterized by a uniform ground state, the influence of fluctuations is typically negligible. Additionally, the magnetic moment value obtained from neutron diffraction refinement closely aligns with that measured at the same temperature for the compound $[Mn(mal)(H_2O)]_n$, which is approximately 4.30(3) $\mu_B$. The magnetic interactions within layers are primarily antiferromagnetic, mediated by carboxylate ligands in an *anti–anti* configuration. The interlayer interactions, though notably weaker, are ferromagnetic in this compound and contribute to the establishment of long-range magnetic order. Therefore, both manganese compounds exhibit layered antiferromagnetic structures, but the key differences lay in the orientation and coupling of magnetic moments within and between layers. In $[Na_2Mn(mal)_2]{\cdot}2H_2O$, the moments are aligned strictly along the *c*-axis and show no ferromagnetic component, leading to a purely antiferromagnetic structure. In contrast, $[Mn(mal)(H_2O)]_n$ displays a more complex coupling, where moments along the *b*-axis are not fully compensated, resulting in a net ferromagnetic component. Furthermore, while both compounds exhibit similar reductions in their magnetic moments compared to the theoretical spin-only value, the $[Mn(mal)(H_2O)]_n$ structure allows for a ferromagnetic signal due to weak spin canting within the *ab*-layers. This difference in behavior is attributed to the competition between *anti–anti* and *anti–syn* interactions in $[Mn(mal)(H_2O)]_n$.

## 5. Conclusions

This study presents a microscopic magnetic investigation of the $[Mn(mal)(H_2O)]_n$ compound, underscoring the valuable role of neutron diffraction—a method rarely employed for molecular compounds—in revealing its unique magnetic properties. Challenges in this approach included inaccuracies in locating light atoms via X-ray diffraction and substantial incoherent scattering from hydrogen atoms in the neutron powder diffraction due to the significant hydrogen content. Despite these limitations, the neutron diffraction data allowed for the completion of the structural model, including the precise positions of hydrogen atoms, in particular, via crystallization water molecules.

Low-temperature measurements conducted below the ordering temperature facilitated the elucidation of the magnetic structure. By combining magnetometry with neutron diffraction data, we determined the correct Shubnikov space group and constructed a magnetic model aligned with the observed spin-canting behavior. The resulting zero-field magnetic structure is consistent with both the *ac* and *dc* magnetic measurements and aligns well with the neutron diffraction data.

The observed spin canting in $[Mn(mal)(H_2O)]_n$ appears to be primarily driven by antisymmetric exchange interactions rather than single-ion anisotropy, in agreement with the isotropic character of the six-coordinated manganese(II) ($^6A_1g$) ground state. The acentric

$Pc'a2_{1'}$ Shubnikov space group further supports the spin-canted structure, with antisymmetric exchange as the primary mechanism behind the weak ferromagnetic signal. This robust experimental approach, despite the limited number of magnetic centers per unit cell and the resulting weak magnetic reflections, facilitates a deeper understanding of how structural and symmetry-related factors in malonate-based manganese compounds affect their magnetic properties, particularly in terms of the emergence of weak ferromagnetism through spin canting.

**Supplementary Materials:** The following supporting information can be downloaded at: https://www.mdpi.com/article/10.3390/magnetochemistry10120109/s1, Supplementary 1: X-ray diffraction data at 100 K in the format of a CIF file (also available in the CCDC database under identification number 2403191); Supplementary 2: Magnetic model obtained from the 1.5 K refinement in the format of MCIF (also available in the MAGNDATA database on the Bilbao Crystallographic Server.)

**Author Contributions:** Conceptualization, F.S.D.; methodology, F.S.D., L.C.-D., J.R.-C., Ó.F. and J.P.; investigation, L.C.-D., Ó.F. and J.P.; data curation, J.R.-C., L.C.-D. and Ó.F.; writing—original draft preparation, F.S.D.; writing—review and editing, L.C.-D., Ó.F. and J.P.; funding acquisition, L.C.-D., Ó.F. and J.P. All authors have read and agreed to the published version of the manuscript.

**Funding:** This research was funded by Ministerio de Ciencia, Innovación y Universidades, grant number: PID2023-147246-OB-I00.

**Institutional Review Board Statement:** Not applicable.

**Informed Consent Statement:** Not applicable.

**Data Availability Statement:** The original contributions presented in this study are included in the article/Supplementary Materials. Further inquiries can be directed to the corresponding authors.

**Acknowledgments:** The authors would like to thank Gilles André and the G4.1 and G4.2 instrument technicians for their valuable support during the measurements. The authors also gratefully acknowledge the Laboratoire Léon Brillouin for the beam time allocated.

**Conflicts of Interest:** The authors declare no conflicts of interest.

## References

1. Ferrando-Soria, J.; Vallejo, J.; Castellano, M.; Martínez-Lillo, J.; Pardo, E.; Cano, J.; Castro, I.; Lloret, F.; Ruiz-García, R.; Julve, M. Molecular magnetism, quo vadis? A historical perspective from a coordination chemist viewpoint. *Coord. Chem. Rev.* **2017**, *339*, 17–103. [CrossRef]
2. Bałanda, M.; Fitta, M. Molecular Magnets. *Crystals* **2019**, *9*, 132. [CrossRef]
3. Coronado, E.; Gatteschi, D. Trends and challenges in molecule-based magnetic materials. *J. Mater. Chem.* **2006**, *16*, 2513–2515. [CrossRef]
4. Glaser, T. Rational design of single-molecule magnets: A supramolecular approach. *Chem. Commun.* **2011**, *47*, 116–130. [CrossRef] [PubMed]
5. Marinescu, G.; Andruh, M.; Lloret, F.; Julve, M. Bis(oxalato)chromium(III) complexes: Versatile tectons in designing heterometallic coordination compounds. *Coord. Chem. Rev.* **2011**, *255*, 161–185. [CrossRef]
6. Armentano, D.; Mastropietro, T.F.; Munno, G.; Rossi, P.; Lloret, F.; Julve, M. New extended magnetic systems based on oxalate and iron(III) ions. *Inorg. Chem.* **2008**, *47*, 3772–3786. [CrossRef]
7. De Munno, G.; Ruiz, R.; Lloret, F.; Faus, J.; Sessoli, R.; Julve, M. Oxalate and 2,2′-Bipyrimidine as Useful Tools in Designing Layered Compounds. *Inorg. Chem.* **1995**, *34*, 408–411. [CrossRef]
8. Lloret, F.; Julve, M.; Faus, J.; Solans, X.; Journaux, Y.; Morgenstern-Badarau, I. Synthesis and magnetic properties of binuclear iron(III) complexes with oxalate, 2,5-dihydroxy-1,4-benzoquinone dianion, and squarate as bridging ligands. Crystal structure of (μ-1,3-squarato)bis[(*N*,*N*′-ethylenebis(salicylideneaminato))(methanol)iron(III)]. *Inorg. Chem.* **1990**, *29*, 2232–2237. [CrossRef]
9. Cangussu, D.; Stumpf, H.O.; Adams, H.; Thomas, J.A.; Lloret, F.; Julve, M. Oxalate, squarate and croconate complexes with bis(2-pyrimidylcarbonyl)amidatecopper(II): Synthesis, crystal structures and magnetic properties. *Inorg. Chim. Acta* **2005**, *358*, 2292–2302. [CrossRef]
10. Román, P.; Guzmán-Miralles, C.; Luque, A.; Beitia, J.I.; Cano, J.; Lloret, F.; Julve, M.; Alvarez, S. Influence of the Peripheral Ligand Atoms on the Exchange Interaction in Oxalato-Bridged Nickel(II) Complexes: An Orbital Model. Crystal Structures and Magnetic Properties of $(H_3dien)_2[Ni_2(ox)_5]\cdot 12H_2O$ and $[Ni_2(dien)_2(H_2O)_2(ox)]Cl_2$. *Inorg. Chem.* **1996**, *35*, 3741–3751. [CrossRef] [PubMed]

11. Castillo, O.; Luque, A.; Román, P.; Lloret, F.; Julve, M. Syntheses, Crystal Structures, and Magnetic Properties of One-Dimensional Oxalato-Bridged Co(II), Ni(II), and Cu(II) Complexes with n-Aminopyridine (n = 2−4) as Terminal Ligand. *Inorg. Chem.* **2001**, *40*, 5526–5535. [CrossRef]
12. Dinca, A.S.; Shova, S.; Ion, A.E.; Maxim, C.; Lloret, F.; Julve, M.; Andruh, M. Ascorbic acid decomposition into oxalate ions: A simple synthetic route towards oxalato-bridged heterometallic 3d–4f clusters. *Dalton Trans.* **2015**, *44*, 7148–7151. [CrossRef]
13. Rodríguez-Martín, Y.; Hernández-Molina, M.; Delgado, F.S.; Pasán, J.; Ruiz-Pérez, C.; Sanchiz, J.; Lloret, F.; Julve, M. Structural versatility of the malonate ligand as a tool for crystal engineering in the design of molecular magnets. *CrytsEngComm* **2002**, *4*, 522–535. [CrossRef]
14. Déniz, M.; Hernández-Rodríguez, I.; Pasán, J.; Fabelo, O.; Cañadillas-Delgado, L.; Vallejo, J.; Julve, M.; Lloret, F.; Ruiz-Pérez, C. Syntheses, crystal structures and magnetic properties of five new manganese(II) complexes: Influence of the conformation of different alkyl/aryl substituted malonate ligands on the crystal packing. *CrystEngComm* **2014**, *16*, 2766–2778. [CrossRef]
15. Rousse, G.; Radtke, G.; Klein, Y.; Ahouari, H. Long-range antiferromagnetic order in malonate-based compounds $Na_2M(H_2C_3O_4)_2{\cdot}2H_2O$ (M = Mn, Fe, Co, Ni). *Dalton Trans.* **2016**, *45*, 2536–2548. [CrossRef]
16. Tan, Y.-X.; He, Y.-P.; Zhang, Y.; Zheng, Y.-J.; Zhang, J. Solvent controlled assembly of four Mn(ii)-2,5-thiophenedicarboxylate frameworks with rod-packing architectures and magnetic properties. *CrystEngComm* **2013**, *15*, 6009–6014. [CrossRef]
17. Tian, C.-B.; Chen, R.-P.; He, C.; Li, W.-J.; Wei, Q.; Zhang, X.-D.; Du, S.-W. Reversible crystal-to-amorphous-to-crystal phase transition and a large magnetocaloric effect in a spongelike metal organic framework material. *Chem. Commun.* **2014**, *50*, 1915–1917. [CrossRef] [PubMed]
18. Tian, C.-B.; He, C.; Han, Y.-H.; Wei, Q.; Li, Q.-P.; Lin, P.; Du, S.-W. Four New MnII Inorganic–Organic Hybrid Frameworks with Diverse Inorganic Magnetic Chain's Sequences: Syntheses, Structures, Magnetic, NLO, and Dielectric Properties. *Inorg. Chem.* **2015**, *54*, 2560–2571. [CrossRef] [PubMed]
19. Zhou, X.; Liu, P.; Huang, W.-H.; Kang, M.; Wang, Y.-Y.; Shi, Q.-Z. Solvents influence on sizes of channels in three fry topological Mn(II)-MOFs based on metal–carboxylate chains: Syntheses, structures and magnetic properties. *CrystEngComm* **2013**, *15*, 8125–8132. [CrossRef]
20. Su, F.; Zhou, C.-Y.; Han, C.; Wu, L.-T.; Wu, X.; Sun, L.; Su, J.; Feng, S.-S.; Lu, L.-P.; Zhu, M.-L. Binuclear $Mn^{2+}$ complexes of a biphenyltetracarboxylic acid with variable N-donor ligands: Syntheses, structures, and magnetic properties. *CrystEngComm* **2018**, *20*, 1818–1831. [CrossRef]
21. Lloret, F.; De Munno, G.; Julve, M.; Cano, J.; Ruiz, R.; Caneschi, A. Spin Polarization and Ferromagnetism in Two-Dimensional Sheetlike Cobalt(II) Polymers: $[Co(L)_2(NCS)_2]$ (L = Pyrimidine or Pyrazine). *Angew. Chem. Int. Ed. Engl.* **1998**, *37*, 135–138. [CrossRef]
22. Dougan, J.; Xue, Z.L. Polarized neutron diffraction and its application to spin density studies. *Sci. China Ser. B Chem.* **2009**, *12*, 2083–2095. [CrossRef]
23. Nakayama, K.; Ishida, T.; Takayama, R.; Hashizume, D.; Yasui, M.; Iwasaki, F.; Nogami, T. Pyrimidine-Bridged Cobalt(II) Complexes with a Chiral 3-D Network Showing Weak Ferromagnetism. *Chem. Lett.* **1998**, *27*, 497–498. [CrossRef]
24. Ishida, T.; Kawakami, T.; Mitsubori, S.; Nogami, T.; Yamaguchi, K.; Iwamura, H. Antiferromagnetic coupling of transition metal spins across pyrimidine and pyrazine bridges in dinuclear manganese(II), cobalt(II), nickel(II) and copper(II) 1,1,1,5,5,5-hexafluoropentane-2,4-dionate complexes. *J. Chem. Soc. Dalton Trans.* **2002**, 3177–3186. [CrossRef]
25. Doi, Y.; Ishida, T.; Nogami, T. Metal-Azide-Pyrimidine Complexes $M(N_3)_2$(pm) with a Three-Dimensional Network Showing Weak Ferromagnetism for M = Mn and Fe and Antiferromagnetism for M = Co and Ni. *Bull. Chem. Soc. Jpn.* **2002**, *75*, 2455–2461. [CrossRef]
26. Kusaka, T.; Ishida, T.; Hashizune, D.; Iwasaki, F.; Nogami, T. Low-temperature magnets $M[N(CN)_2]_2$ (pyrimidine) (M = Fe and Co) with a 3-D network. *Chem. Lett.* **2000**, *29*, 1146–1147. [CrossRef]
27. Ghoshal, D.; Kumar Maji, T.; Mallah, T.; Lu, T.-H.; Mostafa, G.; Chaudhuri, N.R. Synthesis of a metal–dicarboxylate hybrid with three-dimensional Na–O–Cu connectivity: Structure, magnetic property and controlled solid state thermolysis leading to CuO nanorod. *Inorg. Chim. Acta* **2005**, *358*, 1027–1033. [CrossRef]
28. Ahouari, H.; Rousse, G.; Klein, Y.; Chotard, J.-N.; Sougrati, M.T.; Recham, N.; Tarascon, J.-M. Synthesis, structure and electrochemical properties of metal malonate $Na_2M(H_2C_3O_4)_2{\cdot}nH_2O$ (n = 0, 2) compounds and comparison with oxalate $Na_2M_2(C_2O_4)_3{\cdot}2H_2O$ compounds. *Solid State Sci.* **2015**, *42*, 6–13. [CrossRef]
29. Cañadillas-Delgado, L.; Fabelo, O.; Rodriguez-Velamazan, J.A.; Lemee-Cailleau, M.-H.; Mason, S.A.; Pardo, E.; Lloret, F.; Zhao, J.-P.; Bu, X.-H.; Simonet, V.; et al. The Role of Order–Disorder Transitions in the Quest for Molecular Multiferroics: Structural and Magnetic Neutron Studies of a Mixed Valence Iron(II)–Iron(III) Formate Framework. *J. Am. Chem. Soc.* **2012**, *134*, 19772–19781. [CrossRef]
30. Sibille, R.; Lhotel, E.; Mazet, T.; Malaman, B.; Ritter, C.; Ban, V.; François, M. Magnetic structure and dynamics of a strongly one-dimensional cobalt(II) metal-organic framework. *Phys. Rev. B* **2014**, *89*, 104413. [CrossRef]
31. Lappas, A.; Wills, A.S.; Green, M.A.; Prassides, K.; Kurmoo, M. Magnetic ordering in the rutile molecular magnets $M^{II}[N(CN)_2]_2$ ($M$ = Ni, Co, Fe, Mn, $Ni_{0.5}Co_{0.5}$, and $Ni_{0.5}Fe_{0.5}$). *Phys. Rev. B* **2003**, *67*, 144406. [CrossRef]
32. Mazzuca, L.; Cañadillas-Delgado, L.; Rodríguez-Velamazán, J.A.; Fabelo, O.; Scarrozza, M.; Stroppa, A.; Picozzi, S.; Zhao, J.-P.; Bu, X.-H.; Rodríguez-Carvajal, J. Magnetic Structures of Heterometallic M(II)-M(III) Formate Compounds. *Inorg. Chem.* **2016**, *56*, 197–207. [CrossRef] [PubMed]

33. Viswanathan, M.; Bhat, S.G.; Bera, A.K.; Rodríguez-Carvajal, J. Neutron Diffraction Study on the Magnetic Structure of the Promised Multiferroic Hybrid Perovskite [$C(ND_2)_3$]$Cu(DCOO)_3$ and Its Centrosymmetric Analogues. *J. Phys. Chem. C* **2019**, *123*, 18551–18559. [CrossRef]
34. Mazzuca, L.; Cañadillas-Delgado, L.; Fabelo, O.; Rodríguez-Velamazán, J.A.; Luzón, J.; Vallcorba, O.; Simonet, V.; Colin, C.V.; Rodríguez-Carvajal, J. Microscopic Insights on the Multiferroic Perovskite-Like [$CH_3NH_3$][$Co(COOH)_3$] Compound. *Chem. Eur. J.* **2018**, *24*, 388–399. [CrossRef] [PubMed]
35. Fabelo, O.; Cañadillas-Delgado, L.; Pasán, J.; Díaz-Gallifa, P.; Ruiz-Pérez, C.; Lloret, F.; Julve, M.; Puente Orench, I.; Campo, J.; Rodríguez-Carvajal, J. Neutron Diffraction Studies of the Molecular Compound [$Co_2$(bta)]$_n$ ($H_4$bta = 1,2,4,5-Benzenetetracarboxylic Acid): In the Quest of Canted Ferromagnetism. *Inorg. Chem.* **2013**, *52*, 12818–12827. [CrossRef]
36. Calder, S.; Baral, R.; Narayanan, N.; Sanjeewa, L.D. Magnetic order in the two-dimensional metal-organic framework manganese pyrazinecarboxylate with Mn-Mn dimers. *Phys. Rev. Mater.* **2023**, *7*, 124408. [CrossRef]
37. Geers, M.; Lee, J.Y.; Ling, S.; Fabelo, O.; Cañadillas-Delgado, L.; Cliffe, M.J. Non-collinear magnetism in the post-perovskite thiocyanate frameworks $CsM(NCS)_3$. *Chem. Sci.* **2023**, *14*, 3531–3540. [CrossRef]
38. Diaz-Gallifa, P.; Fabelo, O.; Pasán, J.; Cañadillas-Delgado, L.; Rodríguez-Carvajal, J.; Lloret, F.; Julve, M.; Ruiz-Pérez, C. Synthesis, Crystal Structure, and Magnetic Characterization of the Three-Dimensional Compound [$Co_2$(cbut)($H_2O$)$_3$]$_n$ ($H_4$cbut = 1,2,3,4-Cyclobutanetetracarboxylic Acid). *Inorg. Chem.* **2014**, *53*, 5674–5683. [CrossRef] [PubMed]
39. Greenfield, J.T.; Garlea, V.O.; Kamali, S.; Chen, M.; Kovnir, K. Synthesis, crystal growth, structural and magnetic characterization of $NH_4MCl_2$(HCOO), M = (Fe, Co, Ni). *J. Solid State Chem.* **2016**, *236*, 222–229. [CrossRef]
40. Cliffe, M.J.; Lee, J.; Paddison, J.A.M.; Schott, S.; Mukherjee, P.; Gaultois, M.W.; Manuel, P.; Sirringhaus, H.; Dutton, S.E.; Grey, C.P. Low-dimensional quantum magnetism in $Cu(NCS)_2$: A molecular framework material. *Phys. Rev. B* **2018**, *97*, 144421. [CrossRef]
41. Rousse, G.; Rodríguez-Carvajal, J. Oxalate-mediated long-range antiferromagnetism order in $Fe_2(C_2O_4)3{\cdot}4H_2O$. *Dalton Trans.* **2016**, *45*, 14311–14319. [CrossRef]
42. Kubus, M.; Lanza, A.; Scatena, R.; Dos Santos, L.H.R.; Wehinger, B.; Casati, N.; Fiolka, C.; Keller, L.; Macchi, P.; Ruegg, C.; et al. Quasi-2D Heisenberg Antiferromagnets [$CuX(pyz)_2$]($BF_4$) with X = Cl and Br. *Inorg. Chem.* **2018**, *57*, 4934–4943. [CrossRef] [PubMed]
43. Manson, J.L.; Huang, Q.-Z.; Brown, C.M.; Lynn, J.W.; Stone, M.B.; Singleton, J.; Xiao, F. Magnetic Structure and Exchange Interactions in Quasi-One-Dimensional $MnCl_2(urea)_2$. *Inorg. Chem.* **2015**, *54*, 11897–11905. [CrossRef]
44. Cañadillas-Delgado, L.; Mazzuca, L.; Fabelo, O.; Rodríguez-Carvajal, J.; Petricek, V. Experimental Evidence of the Coexistence of Proper Magnetic and Structural Incommensurability on the [$CH_3NH_3$][$Ni(COOH)_3$] Compound. *Inorg. Chem.* **2020**, *59*, 17896–17905. [CrossRef] [PubMed]
45. Geers, M.; Fabelo, O.; Cliffe, M.; Cañadillas-Delgado, L. Tuning structural modulation and magnetic properties in metal–organic coordination polymers [$CH_3NH_3$]$Co_xNi_{1-x}(HCOO)_3$. *IUCrJ* **2024**, *11*, 910–920. [CrossRef]
46. Cañadillas-Delgado, L.; Fabelo, O.; Rodríguez-Velamazán, J.A.; Stunault, A.; Zhao, J.-P.; Bu, X.-H.; Rodríguez-Carvajal, J. Spin-density studies of the multiferroic metal-organic compound [$NH_2(CH_3)_2$][$Fe^{III}Fe^{II}(HCOO)_6$]. *IUCrJ* **2020**, *7*, 803–813. [CrossRef]
47. Burla, M.C.; Caliandro, R.; Camalli, M.; Carrozzini, B.; Cascarano, G.L.; De Caro, L.; Giacovazzo, C.; Polidori, G.; Spagna, R. SIR2004: An improved tool for crystal structure determination and refinement. *J. Appl. Crystallogr.* **2005**, *38*, 381–388. [CrossRef]
48. Sheldrick, G.M. A short history of SHELX. *Acta Crystallogr. Sect. A* **2008**, *64*, 112–122. [CrossRef]
49. Rodríguez-Carvajal, J. Recent advances in magnetic structure determination by neutron powder diffraction. *Phys. B* **1993**, *192*, 55–69. [CrossRef]
50. Gupta, M.P.; Chand, P. The crystal structure of manganese malonate dehydrate $C_3H_2O_4Mn{\cdot}2H_2O$. *Curr. Sci.* **1977**, *46*, 557.
51. Lis, T.; Matuszewski, J. Manganese(II) malonate dihydrate: A reinvestigation. *Acta Crystallogr. Sect. B Struct. Crystallogr. Cryst. Chem.* **1979**, *35*, 2212–2214. [CrossRef]
52. Rodriguez-Martín, Y.; Hernández-Molina, M.; Sanchiz, J.; Ruiz-Pérez, C.; Lloret, F.; Julve, M. Crystal structures and magnetic properties of two- and three-dimensional malonato-bridged manganese(II) complexes. *Dalton Trans.* **2003**, 2359–2365. [CrossRef]
53. Deniz, M.; Hernandez-Rodriguez, I.; Pasan, J.; Fabelo, O.; Canadillas-Delgado, L.; Yuste, C.; Julve, M.; Lloret, F.; Ruiz-Perez, C. Pillaring Role of 4,4′-Azobis(pyridine) in Substituted Malonate-Containing Manganese(II) Complexes: Syntheses, Crystal Structures, and Magnetic Properties. *Cryst. Growth Des.* **2012**, *12*, 4505–4518. [CrossRef]
54. Deniz, M.; Pasan, J.; Ferrando-Soria, J.; Fabelo, O.; Canadillas-Delgado, L.; Yuste, C.; Julve, M.; Cano, J.; Ruiz-Perez, C. Synthesis, Structural Analysis, and Magnetic Properties of Ethylmalonate-Manganese(II) Complexes. *Inorg. Chem.* **2011**, *50*, 10765–10776. [CrossRef] [PubMed]
55. Colacio, E.; Domínguez-Vera, J.M.; Ghazi, M.; Kivekkäs, R.; Moreno, J.M.; Romerosa, A.; Ruiz, J. Structure and magnetic properties of a syn-anti carboxylate bridged linear trinuclear copper(II) complex with ferromagnetic exchange interaction. *Inorg. Chem.* **1993**, *212*, 115–121. [CrossRef]
56. Fisher, M.E. Magnetism in One-Dimensional Systems-The Heisenberg Model for Infinite Spin. *Am. J. Phys.* **1964**, *32*, 343–346. [CrossRef]
57. Perez-Mato, J.M.; Gallego, S.V.; Tasci, E.S.; Elcoro, L.; de la Flor, G.; Aroyo, M.I. Symmetry-based computational tools for magnetic crystallography. *Annu. Rev. Mater. Res.* **2015**, *45*, 217–248. [CrossRef]

58. Pasán, J.; Delgado, F.S.; Rodríguez-Martín, Y.; Hernández-Molina, M.; Ruiz-Pérez, C.; Sanchiz, J.; Lloret, F.; Julve, M. Malonate-based copper(II) coordination compounds: Ferromagnetic coupling controlled by dicarboxylates. *Polyhedron* **2003**, *22*, 2143–2153. [CrossRef]
59. Su, F.; Lu, L.; Feng, S.; Zhu, M.; Gao, Z.; Dong, Y. Synthesis, structures and magnetic properties in 3d-electron-rich isostructural complexes based on chains with sole syn–anti carboxylate bridges. *Dalton Trans.* **2015**, *44*, 7213–7222. [CrossRef] [PubMed]
60. Colacio, E.; Dominguez-Vera, J.M.; Ghazi, M.; Kivekäs, R.; Klinga, M.; Moreno, J.M. Singly anti-anti Carboxylate-Bridged Zig-Zag Chain Complexes from a Carboxylate-Containing Tridentate Schiff Base Ligand and $M(hfac)_2$ [M = $Mn^{II}$, $Ni^{II}$, and $Cu^{II}$]: Synthesis, Crystal Structure, and Magnetic Properties. *Eur. J. Inorg. Chem.* **1999**, *1999*, 441–446. [CrossRef]
61. Delgado, F.S.; Sanchiz, J.; Ruiz-Pérez, C.; Lloret, F.; Julve, M. Design of High-Dimensional Copper(II) Malonate Complexes with Exo-Polydentate N-Donor Ligands. *Inorg. Chem.* **2003**, *42*, 5938–5948. [CrossRef] [PubMed]

*magnetochemistry*

MDPI

*Article*

# Slow Relaxation of the Magnetisation in a Two-Dimensional Metal–Organic Framework with a Layered Square Lattice †

**Samia Benmansour *, Christian Cerezo-Navarrete and Carlos J. Gómez-García ***

Departamento de Química Inorgánica, Universidad de Valencia, C/Dr. Moliner 50, 46100 Burjasot, Valencia, Spain; chcena@alumni.uv.es

* Correspondence: sam.ben@uv.es (S.B.); carlos.gomez@uv.es (C.J.G.-G.)

† This article is dedicated to the long and outstanding scientific careers of Miguel Julve and Paco Lloret, two pioneers in the fields of magnetochemistry and molecular magnetism. Sadly, Miguel left us on 9 July 2024. We have not only lost a brilliant chemist and colleague, whose desk door was also open to anyone needing help, an outstanding teacher, highly esteemed by his students, but, even more, a close friend with a fine sense of humour, a fast mind and a generosity that we will never forget.

**Abstract:** Herein, we present the synthesis and structural characterisation of two layered MOFs with the asymmetric ligand 3-chloro,6-cyano-2,5-dihydroxy-1,4-benzoquinone dianion ($C_6O_4(CN)Cl^{2-}$ = chlorocyananilato). These compounds, formulated as $(H_3O)[Eu(C_6O_4(CN)Cl)_2(H_2O)]\cdot 34H_2O$ (**1**) and $(H_3O)[Dy(C_6O_4(CN)Cl)_2(H_2O)]\cdot 44H_2O$ (**2**), are isostructural and show a (4,4)-layered square structure with the crystallisation water molecules located between the layers. The lanthanoid ions are surrounded by four bis-bidentate chlorocyananilato ligands that connect each $Ln^{III}$ centre with other four, giving rise to square cavities formed by $Ln^{III}$ centres in the vertices and chlorocyananilato ligands as the sides. There is an additional coordinated water molecule that occupies the caped position of the capped square antiprismatic coordination geometry around the $Ln^{III}$ centres. The magnetic properties show the presence of a field-induced slow relaxation of the magnetisation in the $Dy^{III}$ derivative at low temperatures that follows Direct and Orbach relaxation mechanisms with an energy barrier of 36(3) K.

**Keywords:** slow magnetisation relaxation; metal–organic frameworks; europium coordination polymer; dysprosium coordination polymer; layered square lattice

## 1. Introduction

The last decade has witnessed a growing interest in the field of lanthanoid-containing metal–organic frameworks (MOFs) and coordination polymers (CPs). This interest has been fuelled by the numerous potential applications due to their interesting properties for gas and solvent adsorption/absorption and exchange. Additionally, they may also present interesting magnetic and luminescent properties, making them perfect candidates for the preparation of magnetic and optical sensors of gases, contaminants and different chemical species [1–3].

Although the number of ligands used to prepare these MOFs and CPs is very large, most of them are O-donors (given the high oxophilicity of $Ln^{III}$ ions) [4]. Among the many O-donor ligands used to prepare these MOFs and CPs, the derivatives of the 3,6-disubstituted-2,5-dihydroxy-1,4-benzoquinone dianion $(C_6O_4X_2)^{2-}$ (anilato-type ligands, Scheme 1a) are becoming very common due to properties as: (i) The different coordination modes from simple mono- and bidentate (*1kO* and $1k^2O,O'$), to monodentate–bidentate

(*1kO;2k²O′,O″*), bis-bidentate (*1k²O,O′;2k²O″,O‴*, Scheme 1c) and other more complex coordination modes, such as (*1k²O,O′;2k²O″,O‴;3kO″*) [5,6]. (ii) The possibility of connecting two metal ions when acting as bis-bidentate linear bridges (Scheme 1c), giving rise to several dozens of MOFs and CPs [6,7]. (iii) The possibility of modifying the magnetic coupling between transition metal centres connected by these ligands by simply changing their X groups [8]. (iv) The magnetic isolation provided by these ligands when connecting lanthanoid metal ions (due to the extremely weak overlap with the 4f orbitals). This magnetic isolation has resulted in several examples of lanthanoid-containing MOFs and CPs with slow relaxation of the magnetisation [9–18]. (v) The electrochemical activity, as these ligands can be reduced by one or two electrons to their semiquinone and catecholate forms, giving rise to stronger magnetic couplings and ordering temperatures [19]. (vi) A final interesting property is their topological equivalence with the oxalato ligand ($C_2O_4^{2-}$). This fact allows the prediction of the coordination geometries, structural types and dimensionalities to be expected when combining anilato ligands with transition metals and lanthanoids. This equivalence has been confirmed in many monomeric tris-anilato complexes [20,21] and also in extended 1D, 2D and 3D lattices [5,6,22]. The only difference is the larger size of the cavities and channels, given the larger size of anilato compared with oxalate.

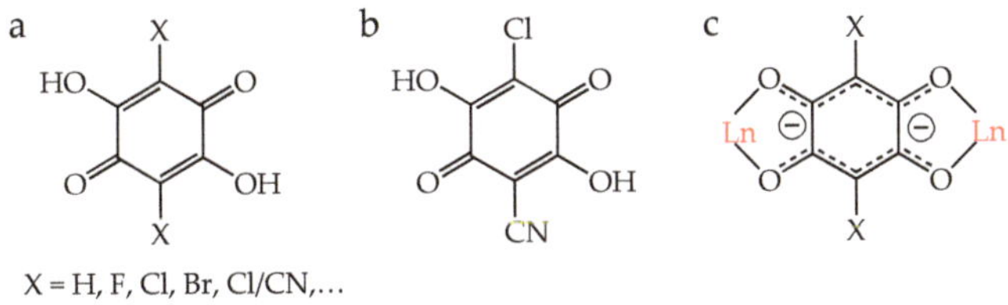


**Scheme 1.** (**a**) The 3,6-disubstituted anilato derivatives ($H_2C_6O_4X_2$). (**b**) The ligand chlorocyananilato used in this work (**c**) The bis-bidentate coordination modes (*1k²O,O′;2k²O″,O‴*) of the anilato ligands present in compounds **1** and **2**.

Although there are many anilato-type ligands, most of the reported Ln-containing anilato lattices have been prepared with symmetrical derivatives with X = H, Cl and Br (Scheme 1a) [9]. Only a few examples have been reported with other derivatives, such as nitranilato (X = $NO_2$) [23], tert-butylanilato (X = *t*-Bu) [24], or even triazolylanilato [17]. There are also a few reported lattices with the asymmetric chlorocyananilato (C = Cl/CN). Most of them are neutral layers formulated as $[Ln_2(C_6O_4(CN)Cl)_3(G)_n]\cdot mG$ with G = dimethylformamide (dmf) [25–28] or dimethylsulfoxide (dmso) [25,26,28–31], although there is also a couple of discrete dimers with the same formula, but G = water [29]. The layered compounds show distorted (6,3)-gon hexagonal honeycomb structures [25–28,30]. In around half of them, the hexagons are so distorted that they resemble rectangles showing either a brick wall [25,26,28,29] or a herringbone disposition [25,27]. In addition to these (6,3)-gon layered structures, there are a total of three examples with the unusual (4,4)-square lattice [25,28,32]. In this case, the Ln:anilato stoichiometry is 1:2 and, therefore, the lattices are anionic. The cations are $H_3O^+$ [25], $NEt_2H_2^+$ [28], or $PPh_4^+$ [32]. Although very unusual, this (4,4)-square lattice had already been observed by Robson et al. [22] in the heterometallic lattice with s- and p-block metals: $(Et_4N)_2[Sn^{IV}Ca^{II}(C_6O_4Cl_2)_4]$.

Here, we report two rare examples of (4,4) square lattices obtained with $Eu^{III}$ and $Dy^{III}$, the asymmetric chlorocyananilato ligand (Scheme 1b) and water as solvent. These compounds, formulated as $(H_3O)[Eu(C_6O_4(CN)Cl)_2(H_2O)]\cdot 34H_2O$ (**1**) and $(H_3O)[Dy(C_6O_4(CN)Cl)_2(H_2O)]\cdot 44H_2O$ (**2**), contain nona-coordinated $Ln^{III}$ ions with a regular caped square antiprism geometry and show a 2D square lattice. Compound **2** is a solvate of compound DIFMAJ [25], previously reported by us with only four water molecules. In compound **2,** we obtain a total of 44 delocalised crystallisation water molecules based on the electron density found in the voids with the mask procedure implanted in Olex2 (see below). The magnetic properties of the $Dy^{III}$ derivative have not been reported; therefore, here we perform a complete study of its magnetic properties, including AC measurements that show the presence of a field-induced slow relaxation of the magnetisation (SRM). Compound **2** is, therefore, the first example of SRM in any anilato-based square 2D lattice. We also report the magnetic properties of the $Eu^{III}$ derivative.

## 2. Results and Discussion

### 2.1. Synthesis of $(H_3O)[Eu(C_6O_4(CN)Cl)_2(H_2O)]\cdot 34H_2O$ (1) and $(H_3O)[Dy(C_6O_4(CN)Cl)_2(H_2O)]\cdot 44H_2O$ (2)

The synthesis of both compounds was performed by a slow diffusion method that allows for the obtaining of good-quality single crystals for **2** and of poorer quality for **1**. The difference in the size of the lanthanoid ions and the hydration degree may be the origin of the difference in the quality of the single crystals. X-ray powder diffraction indicates that the sample of compound **2** used for the magnetic characterization corresponds with the simulated from the single crystal X-ray structure determination (Figure S1, Supporting Information).

### 2.2. Structure of $(H_3O)[Eu(C_6O_4(CN)Cl)_2(H_2O)]\cdot 34H_2O$ (1) and $(H_3O)[Dy(C_6O_4(CN)Cl)_2(H_2O)]\cdot 44H_2O$ (2)

Both compounds are isomorphous and, since the structure of a solvate of compound **2** was already reported by us, we will concentrate on the description of the structure of compound **1**. Both compounds crystallise in the monoclinic *P2/n* space group (#13). The asymmetric unit contains a $Ln^{III}$ ion (Eu in **1** and Dy in **2**) located on the $C_2$ axis, two halves of chlorocyananilato ligands on general positions and a coordinated water molecule, also located on the $C_2$ axis (Figure 1a). The formula unit is, therefore, $[Ln(C_6O_4(CN)Cl)_2(H_2O)]^-$. The solvent mask routine shows the presence of a total of 347 and 453 electrons, corresponding to 35 and 45 water molecules, delocalised on voids of 1329.9 Å$^3$ and 1326.0 Å$^3$ per formula unit in **1** and **2**, respectively. As the formula unit is monoanionic and there is no cation present in the interlayer space, we have to assume that one of the 35 (in **1**) or 45 (in **2**) crystallisation water molecules must be protonated; therefore, we can formulate compounds **1** and **2** as: $(H_3O)[Eu(C_6O_4(CN)Cl)_2(H_2O)]\cdot 34H_2O$ (**1**) and $(H_3O)[Dy(C_6O_4(CN)Cl)_2(H_2O)]\cdot 44H_2O$ (**2**). As is customary in this asymmetric chlorocyananilato ligand, the Cl and CN groups are delocalised over the two possible positions with occupancy of 0.5 due to the presence of the two possible orientations of the ligand with equal probability (Figure 1a).

Each $Eu^{III}$ ion is connected to four other $Eu^{III}$ ions through four bis-bidentate chlorocyananilate ligands (Figure 1b). This unusual connectivity of the $Eu^{III}$ centres results in a 2D-(4,4) square lattice, where the $Eu^{III}$ ions are located in the vertices of a square cavity and the chlorocyananilato ligands constitute the sides of the squares (Figure 2a). The slightly corrugated anionic $[Eu(C_6O_4(CN)Cl)_2(H_2O)]^-$ layers are parallel to the *ac* plane and are stacked in an eclipsed conformation along the *b* direction (Figure 2b). The $H_3O^+$ cation and the 34 (in **1**) and 44 (in **2**) crystallisation water molecules are located in the interlayer space. The interlayer distance, corresponding to the *b* parameter, is 13.885(6) Å in **1** and 14.0496(8) Å in **2**. The larger interlayer space in **2** agrees with the larger number of crystallisation water molecules found in **2**. This result suggests the possibility of increasing/decreasing the number of crystallisation water molecules by changing the synthetic conditions. Attempts are in progress to modify the amount of water molecules in the inter-layer space and even to exchange them with other solvents.

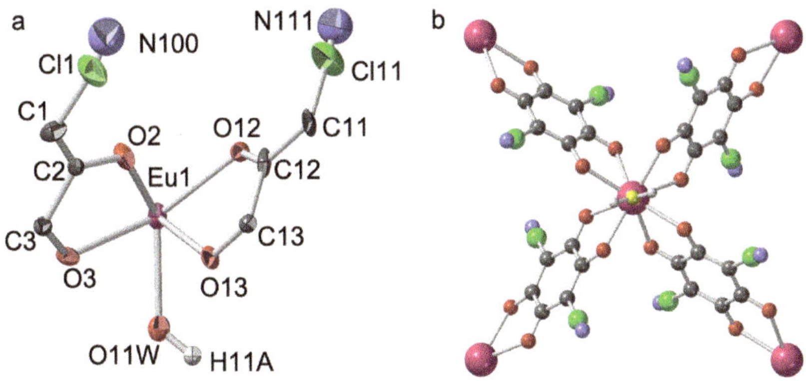


**Figure 1.** (**a**) Ortep view of the asymmetric unit of compound **1** with the labelling scheme (ellipsoids drawn at 50% probability). (**b**) Connectivity of the $Eu^{III}$ centres in compound **1**. Colour code in (**b**): Eu = pink, O = red, Cl = green, N = blue, C = grey, H = white and $O_{water}$ = yellow.

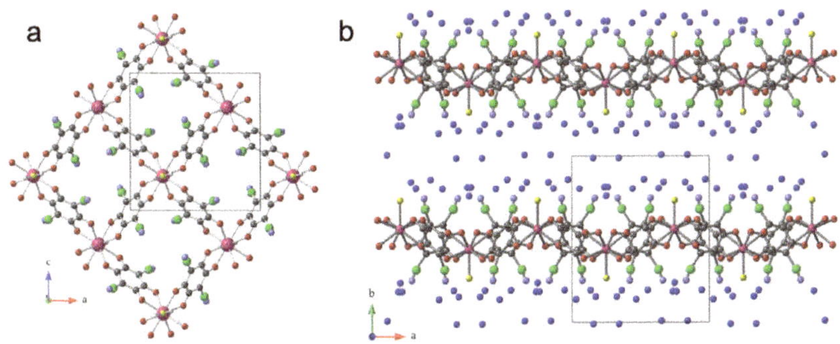


**Figure 2.** (**a**) Top view, along the *b* axis, of a layer of compound **1**. (**b**) Side view of the slightly corrugated anionic layers $[Dy(C_6O_4(CN)Cl)_2(H_2O)]^-$ in **2** alternating with layers containing the water molecules (dark blue, drawn with the DIFMAJ cif file). Colour code: Eu/Dy = pink, O = red, Cl = green, N = blue, C = grey, H = white, $O_{cryst\ water}$ = dark blue and $O_{coordt\ water}$ = yellow.

As mentioned above, the $Ln^{III}$ ions are surrounded by four bidentate chlorocyananilato ligands and complete their nona-coordination with a water molecule. The coordination environment around the $Ln^{III}$ ions can be described as a quite regular caped square antiprism (Figure 3a), as confirmed by the low shape parameters (Table 1) [33]. Each of the four bidentate chlorocyananilato ligands occupies one position of the base and one of the caped square face (Figure 3a) pointing out radially from the $Ln^{III}$ centre (Figure 1b), whereas the coordinated water molecule occupies the caped position (Figure 3a). The square cavities are quite regular (Figure 2a), as shown by the two very close Eu–Eu distances along the sides of the square (8.736 Å and 8.766 Å) and the Eu–Eu–Eu angles (89.16° and 86.33°). The $Dy^{III}$ derivative shows a similar regularity (Dy–Dy distances of 8.686 Å and 8.723 Å and Dy–Dy–Dy angles of 89.23° and 86.46°). The lack of planarity is confirmed by the sum of the four Ln–Ln–Ln angles (350.98° in **1** and 351.38° in **2**), both below 360°.

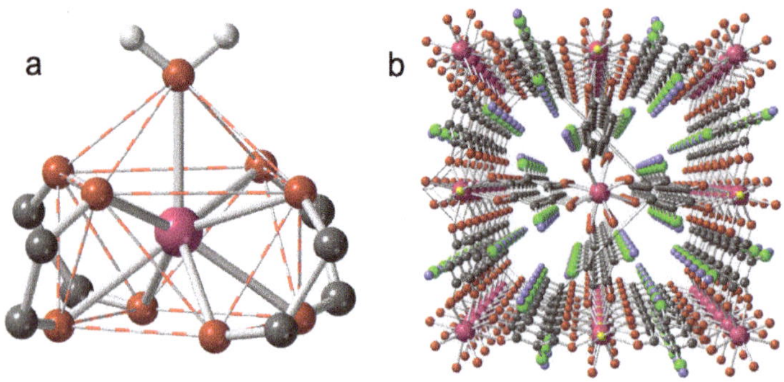


**Figure 3.** (**a**) View of the coordination environment around the $Eu^{III}$ ion in Compound **1**. (**b**) Perspective view of four square channels in Compound **1**. Colour code: Eu = pink, O = red, Cl = green, N = blue, C = grey, H = white.

**Table 1.** Continuous SHAPE measurement (CShM) values of the 13 possible coordination geometries for the $Ln^{III}$ ions with coordination number nine in compounds $(H_3O)[Eu(C_6O_4(CN)Cl)_2(H_2O)]\cdot34H_2O$ (**1**) and $(H_3O)[Dy(C_6O_4(CN)Cl)_2(H_2O)]\cdot44H_2O$ (**2**) [34].

| Geometry | Symmetry | 1 | 2 |
|---|---|---|---|
| EP-9 | D9h | 37.102 | 37.988 |
| OPY-9 | C8v | 18.863 | 19.698 |
| HBPY-9 | D7h | 19.891 | 19.821 |
| JTC-9 | C3v | 17.633 | 17.417 |
| JCCU-9 | C4v | 9.755 | 9.324 |
| CCU-9 | C4v | 8.022 | 7.786 |
| JCSAPR-9 | C4v | 1.704 | 1.464 |
| **CSAPR-9** | **C4v** | **0.393** | **0.306** |
| JTCTPR-9 | D3h | 3.465 | 3.238 |
| TCTPR-9 | D3h | 1.278 | 1.259 |
| JTDIC-9 | C3v | 12.487 | 12.546 |
| HH-9 | C2v | 12.576 | 12.399 |
| MFF-9 | Cs | 1.304 | 1.182 |

EP-9 = Enneagon; OPY-9 = Octagonal pyramid; HBPY-9 = Heptagonal bipyramid; JTC-9 = Triangular cupola (J3) = Trivacant cuboctahedron; JCCU-9 = Capped cube (Elongated square pyramid, J8); CCU-9 = Capped cube; JCSAPR-9 = Capped square antiprism (Gyroelongated square pyramid J10); **CSAPR-9 = Capped square antiprism**; JTCTPR-9 = Tricapped trigonal prism (J51); TCTPR-9 = Tricapped trigonal prism; JTDIC-9 = Tridiminished icosahedron (J63); HH-9 = Hula-hoop; MFF-9 = Muffin. The lowest value is highlighted in bold.

Since the $Eu^{III}$ ion is located on a $C_2$ axis, there are only four different Eu–$O_{anilato}$ bond distances, all very similar, in the range 2.423(4)–2.441(5) Å (Table S1, Supporting Information). The Eu–$O_{water}$ bond distance is slightly shorter (2.402(7) Å). The longer Eu–$O_{anilato}$ distances may be attributed to the tension of the five-membered chelate ring of the anilato ligand. Given the smaller size of $Dy^{III}$ compared to $Eu^{III}$, the Dy–$O_{anilato}$ bond distances are slightly shorter in compound **2** (in the range 2.392(4)–2.411(4) Å, Table S3, Supporting Information). However, the Dy–$O_{water}$ bond distance is slightly longer (2.433(6) Å) than the Dy–$O_{anilato}$ ones. This elongation may be due to the formation of stronger H-bonds between the H atoms of the coordinated water molecule and the crystallisation ones present in the interlayer space. Since the Dy derivative contains a higher density of water molecules, they must be closer to the coordinated water molecule (which points to the interlayer space, Figure 2b), favouring the H-bonding with it.

The anionic layers are packed in an eclipsed way, giving rise to square channels running along the *b* direction (Figure 3b).

### 2.3. *Magnetic Properties of* $(H_3O)[Eu(C_6O_4(CN)Cl)_2(H_2O)]\cdot34H_2O$ (**1**)

The thermal variation of the product of the molar magnetic susceptibility per $Eu^{III}$ ion times the temperature ($\chi_m T$) for compound **1** shows a value of around 1.3 $cm^3$ K $mol^{-1}$ at 300 K (Figure S2). When compound **1** is cooled, $\chi_m T$ decreases and reaches a value close to zero at low temperatures (Figure 4).

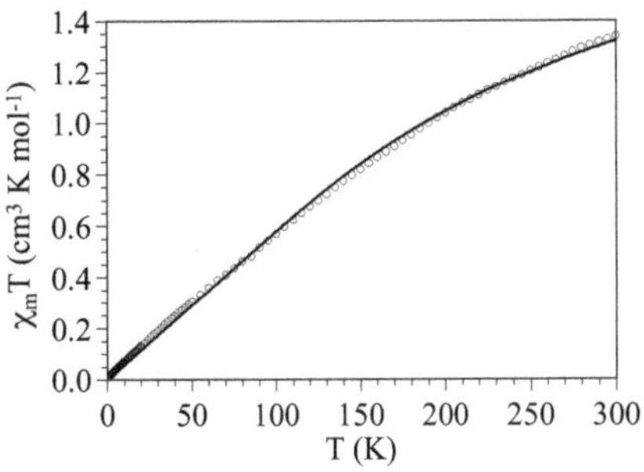


**Figure 4.** Thermal variation of $\chi_m T$ of compound **1**. Solid line is the best fit to Equation (1).

This behaviour is the expected one for an isolated $Eu^{III}$ ion, despite having a non-magnetic $^7F_0$ ground state. This magnetic moment observed in $Eu^{III}$ compounds is due to the thermal population of the excited $^7F_J$ states (J = 1–6) that results from the splitting of the $^7F$ ground term due to the first-order spin-orbit coupling (λ). Usually, the two first excited states ($^7F_1$ and $^7F_2$) are close enough to the ground state and they can be populated at room temperature [35]. Accordingly, we have fitted the magnetic susceptibility data of Compound **1** to the expression [36]:

$$\chi_m T = (N\beta^2)/(3kx)\cdot[24 + (27x/2 - 3/2)e^{-x} + (135x/2 - 5/2)e^{-3x} + (189x - 7/2)e^{-6x} + (405x - 9/2)e^{-10x} + (1485x/2 - 11/2)e^{-15x} + (2457x/2 - 13/2)e^{-21x}]/[1 + 3e^{-x} + 5e^{-3x} + 7e^{-6x} + 9e^{-10x} + 11e^{-15x} + 13e^{-21x}] \quad (1)$$

with $x = \lambda/kT$.

As can be seen in Figure 4, this equation reproduces satisfactorily the magnetic data of compound **1** with λ = 359(1) $cm^{-1}$ (solid line in Figure 4). This value is very similar to those reported for other $Eu^{III}$ compounds [36–47].

### 2.4. *Magnetic Properties of* $(H_3O)[Dy(C_6O_4(CN)Cl)_2(H_2O)]\cdot44H_2O$ (**2**)

The thermal variation of the $\chi_m T$ product per $Dy^{III}$ ion for compound **2** shows a value of around 14.4 $cm^3$ K $mol^{-1}$ at 300 K (Figure S2). This value is close to the expected one for one magnetically isolated $Dy^{III}$ ion with a ground multiplet $^6H_{15/2}$ and g = 4/3 [35]. The $\chi_m T$ product smoothly decreases when compound **2** is cooled and reaches a value close to 10.5 $cm^3$ K $mol^{-1}$ at 2 K. This behaviour is not due to the presence of an antiferromagnetic coupling through the anilato bridge but rather to the depopulation of the sub-levels produced by the ligand field, as has been observed in $Dy^{III}$ latices connected by similar anilato bridges [9,12–14].

The lack of any noticeable Dy···Dy interaction and the previous results in other $Dy^{III}$-anilato lattices, prompted us to measure the AC susceptibility at low temperatures to look for a possible slow relaxation of the magnetisation. These AC measurements show the presence of a frequency-dependent out-of-phase signal ($\chi''_m$) at low temperatures only when a DC field is applied (Figure 5a), indicating the presence of a field-induced SRM. The lack of SMR when no DC field is applied suggests the presence of a very efficient relaxation of the magnetisation through a quantum tunnelling mechanism, which is suppressed when a DC field is applied. The intensity of the $\chi''_m$ signal increases and the position of the maximum shifts to lower frequencies as the DC field increases from 0 to ca. 80 mT. For values above ca. 80 mT, the $\chi''_m$ value decreases and the maximum shifts to higher frequencies. This initial shift to lower frequencies is confirmed by the increase in the relaxation times obtained by fitting the frequency dependence to the Debye model at each DC magnetic field [48]. For fields above ca. 80 mT, the shift to higher frequencies corresponds to the decrease observed in the relaxation times (Figure 5b). The fit of the relaxation time as a function of the magnetic field with Equation (2) shows that the maximum relaxation time is obtained for a DC field of around 80 mT. Accordingly, we have performed a detailed study at different temperatures of the frequency dependence of the out-of-phase signal with an applied DC field of 80 mT (Figure 6a) [49].

$$\tau^{-1} = AH^nT + \frac{B_1}{1 + B_2H^2} + D \quad (2)$$

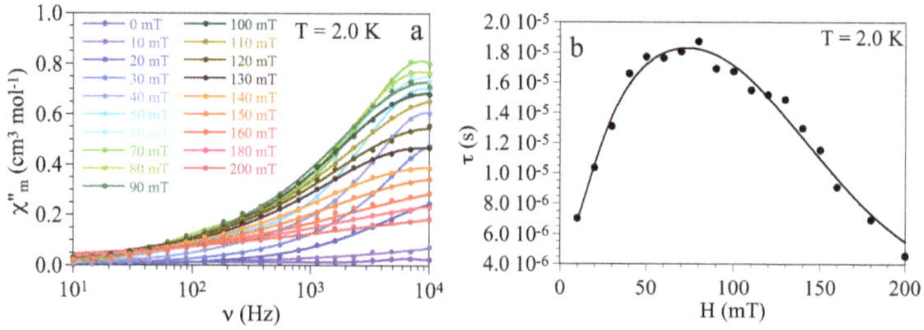

**Figure 5.** (**a**) Frequency-dependence of $\chi''_m$ at 2.0 K with different DC fields for **2**. Solid lines are the best fit to the Debye model. (**b**) Dependence of the relaxation time ($\tau$) with the DC magnetic field for compound **2** at 2.0 K. Solid line is the best fit to Equation (2).

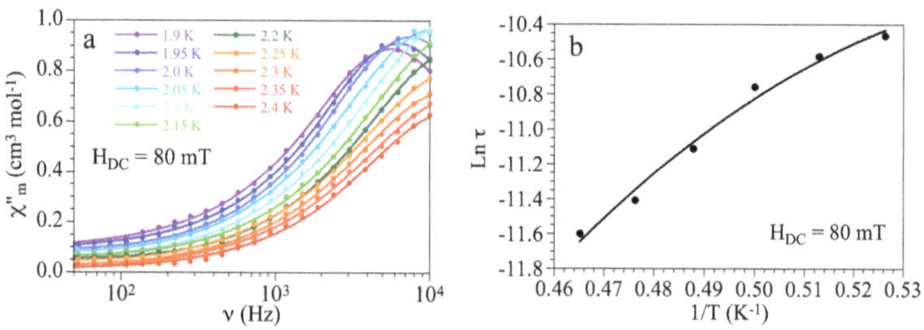

**Figure 6.** (**a**) Frequency-dependence of $\chi''_m$ at different temperatures with an applied DC field of 80 mT for compound **2**. Solid lines are the best fit to the Debye model. (**b**) Arrhenius plot of the relaxation times with a DC field of 80 mT. The solid line is the best fit to a model with Orbach and Direct relaxation mechanisms (Equation (3)).

In this equation, the parameters A, $B_1$, $B_2$ and D, are constants corresponding to the four different relaxation mechanisms: direct (A), quantum tunnelling ($B_1$ and $B_2$) and the field-independent Raman and Orbach mechanisms (D). H is the applied DC magnetic field and n is equal to 4 for Kramers ions (as $Dy^{III}$).

The frequency-dependence of $\chi''_m$ with an applied DC field of 80 mT shows, at 1.9 K, a maximum at around 8 kHz that shifts to higher frequencies as the temperature increases (Figure 6a). The in-phase signal, $\chi'_m$, shows a sigmoidal shape with an inflexion point at around 8 kHz that shifts to higher frequencies as the temperature increases (Figure S3). The fit of these data to the Debye model [48] allows for the determination of the relaxation time at each temperature. The Arrhenius plot (ln$\tau$ vs. $1/T$, Figure 6b) for compound

**2** at 80 mT shows an almost linear dependence with a curvature at low temperatures, suggesting the presence of a thermally activated relaxation mechanism (Orbach type, first term in Equation (3)) plus a Raman and/or direct mechanisms (second and third terms, respectively in Equation (3)) [50]. Attempts to fit the data of Figure 6b to all the possible mechanisms of Equation (3) show that the best fit is obtained including only the Orbach and Direct mechanisms (first and third terms in Equation (3)) with $\tau_0$ = 5.4(2) $\times$ $10^{-13}$ s, $U_{eff}$ = 36(3) K and $AH^2$ = 1.3(4) $\times$ $10^4$ $K^{-1}$ (solid line in Figure 6b). If we used the data obtained by fitting $\chi'_m$ only below 2.15 K, we obtain similar values: $\tau_0$ = 6.7(2) $\times$ $10^{-13}$ s, $U_{eff}$ = 40(8) K and $AH^2$ = 1.3(3) $\times$ $10^4$ $K^{-1}$ (solid line in Figure S4).

Note that, given the high frequency of the maxima, the temperature range studied is small and the data of the relaxation times are not very reliable above 2.4 K, as well as the corresponding fit of the $\ln\tau$ vs. $1/T$ plot.

$$\tau^{-1} = \tau_0^{-1}\exp\left(-\frac{U_{eff}}{kT}\right) + CT^n + AH^2T \qquad (3)$$

Although compound **2** is the only one to show SMR with a (4,4)-square lattice where the $Dy^{III}$ ion is surrounded by four anilato ligands, we have attempted to establish a magneto-structural correlation between the energy barrier ($U_{eff}$) and different structural parameters as: (i) the coordination number, (ii) the coordination geometry, (iii) the distortions of the coordination geometry and (iv) the electronegativity of the X group of the anilato ligand.

Table 2 summarises the twenty $Dy^{III}$-anilato lattices with SMR reported to date, including compound **2**. Among them, sixteen are 2D lattices and only four are 3D diamond lattices, although interestingly, in all these 3D lattices, the $Dy^{III}$ is coordinated to only three anilato bis-bidentate ligands and the fourth bridge corresponds to an ethylene glycol bridge [12], to a double-oxido bridge [18], or to a triazole of the anilato ligand. Among the sixteen 2D lattices, fifteen show a (6,3)-hexagonal honeycomb structure and only one (compound **2**) shows a (4,4)-square lattice. In seven of the fifteen hexagonal lattices, the hexagons are so distorted that they appear as rectangles, with a brick-wall disposition in four cases and with a herringbone disposition in the remaining three.

**Table 2.** Structural and magnetic parameters of all the structurally and magnetically characterised Dy-anilato coordination polymers, including compound **2**.

| CCDC | X [a] | Solvent [b] | Structure [c] | Geom-c.n. [d] | CSh [e] | $H_{dc}$ (mT) | $U_{eff}$ (K) | Ref. |
|---|---|---|---|---|---|---|---|---|
| GEXGID | H | eg | 3D-dia | TCTPR-9 | 0.562 | 60<br>100 | 49(1)/25(3) [g]<br>36(1)/16(3) [g] | [12] |
| GEXGOJ | Cl | eg | 2D-brick | TCTPR-9 | 0.508 | 50<br>100 | 27(2)/19(4) [g]<br>46(2)/32(1) [g] | [12] |
| GEXGUP | Cl | eg/$H_2O$ | 3D-dia | TCTPR-9 | 0.695 | 60 | 37(1) | [12] |
| GEXHAW | Br | eg/MeOH | 2D-hb | TCTPR-9 | 0.860 | 70 | 29.4(2) | [12] |
| GEXHEA | Br | Eg | 2D-brick | CSAPR-9 | 0.513 | 0<br>50 | 40.2(8)<br>39(1)/48(3) [g] | [12] |
| GEXHIE | Br | eg | 2D-brick | CSAPR-9 | 0.617 [h]<br>1.102 [h] | 60 | 20(2)<br>47(3) [g] | [12] |
| JUSBOR | Cl | MeOH | 3D-dia | TDD-8 | 0.617 | 200 | 175(9) | [18] |
| JUSBIL | Cl | dmf | 2D-hex | SAPR-8 | 0.406 | 200 | 145(7) | [18] |
| SOKZAX | Br | $H_2O$ | 2D-hex | TCTPR-9 | 0.595 | 100 | 9.6(1) | [15] |
| LUTROK | Br | dmf | 2D-hb | CSAPR-9 | 0.450 | 100 | 11.4(6)/36(1) [g] | [15] |
| NIDFUE | Br | dmso | 2D-hex | TDD-8 | 1.079 | 100 | 22.8(4) | [15] |

**Table 2.** *Cont.*

| CCDC | X [a] | Solvent [b] | Structure [c] | Geom-c.n. [d] | CSh [e] | $H_{dc}$ (mT) | $U_{eff}$ (K) | Ref. |
|---|---|---|---|---|---|---|---|---|
| NOQBON [f] | Br | dmso | 2D-hb | CSAPR-9 | 0.453 | 0 | 40.9 | [29] |
| UNOCUY | H | dmso/$H_2O$ | 2D-hex | TDD-8 | 0.902 | 60 | 17.1(9) | [14] |
| UNODAF | Cl | dmso | 2D-hex | TDD-8 | 1.208 | 100 | 31.5(7) | [14] |
| XOYTUD | ClCN | dmso | 2D-brick | CSAPR-9 | 0.365 | 100 | 21(3) | [14,28] |
| NULPIX | Cl | fma | 2D-hex | CSAPR-9 | 0.649 | 120 | 30(2) | [11] |
| NULPOD | Br | fma/$H_2O$ | 2D-hex | CSAPR-9 | 0.388 | 120 | 41(1) | [11] |
| NULPUJ | Cl | fma/$H_2O$ | 2D-hex | CSAPR-9 | 0.378 | 70 | 42(7) | [11] |
| VOTTAD | trz | $H_2O$ | 3D | TCTPR-9 | 0.520 | 100 | 25.7 | [17] |
| **2** | Cl/CN | $H_2O$ | 2D-4,4 | CSAPR-9 | 0.306 | 80 | 36(3) | This work |

[a] trz = triazolyl; [b] eg = ethylene glycol, dmf = dimethylformamide, dmso = dimethylsulfoxide; fma = formamide; [c] dia = diamantine, brick = brick-wall (6,3), hb = herringbone (6,3), hex = hexagonal (6,3); [d] TCTPR = tri-capped trigonal prism, CSAPR = capped square antiprism, TDD = triangular dodecahedron, SAPR = square antiprism; c.n. = coordination number; [e] Continuous shape parameter [33]; [f] 2% Dy-doped Eu compound; [g] there are two relaxation processes; [h] there are two independent $Dy^{III}$ centres.

If we focus on the coordination number and geometry, we can see that there are fifteen lattices where the $Dy^{III}$ ions show a coordination number of nine (ten with a tri-capped trigonal prism coordination geometry, TCTPR-9 and five with a capped square antiprism geometry, CSAPR-9). There are also five lattices with a coordination number of eight (four with a triangular dodecahedron geometry, TDD-8, and one with a square antiprism geometry, SAPR-8). As can be seen in Table 2, there is no correlation between the $U_{eff}$ values and any of the above-mentioned structural features (dimensionality, shape and distortion of the cavities, coordination number or coordination geometry).

It is well known that the energy barrier is related to the coordination geometry and the anisotropy of the lanthanoid ions and that certain geometries may suppress the quantum tunnelling and facilitate the single-ion magnet behaviour [51,52]. Additionally, the modulation of the ligand field and the distortions of the coordination geometry by playing with the solvents and co-ligands may also affect the slow magnetic relaxation behaviour in lanthanoid lattices and compounds [12,15,53–56]. Accordingly, we have tried to establish a correlation between $U_{eff}$ and the distortion of the coordination geometry, measured by the continuous shape parameter [33]. Figure 7a shows the dependence of the $U_{eff}$ value for all the reported lattices on the shape parameter for all compounds, grouped by coordination geometry. As can be seen, there is no clear correlation between the distortion of the geometry and the value of $U_{eff}$ for any coordination geometry, indicating that the effective energy barrier could not be correlated with only this parameter. Note that we removed compounds JUSBOR and JUSBIL because, in these compounds, the $U_{eff}$ value was determined using only the data at high temperatures [18].

Finally, in order to check if the electronegativity of the X group of the anilato ligand influences the energy barrier, we also plotted the value of $U_{eff}$ as a function of the electronegativity of the X group (Figure 7b). Again, there is no correlation between the effective energy barrier and the electronegativity of the X groups in the anilato ligand, suggesting, as expected, that the interaction between the anilato O-donor atoms and the $Dy^{III}$ ions is negligible in these series.

In conclusion, although compound **2** is unique in this series of Dy-anilato lattices with SMR, its energy barrier ($U_{eff}$ = 36(3) K) is within the normal range (10–49 K, Table 2) found in all these anilato-Dy lattices (except for two examples whose values were obtained from the fit of the high-temperature data only) [18]. This fact supports the conclusion that the effective energy barrier in these Dy-anilato lattices is not very sensitive to the exact environment surrounding the $Dy^{III}$ ion (at least for coordination numbers of eight and nine and quite regular coordination geometries).

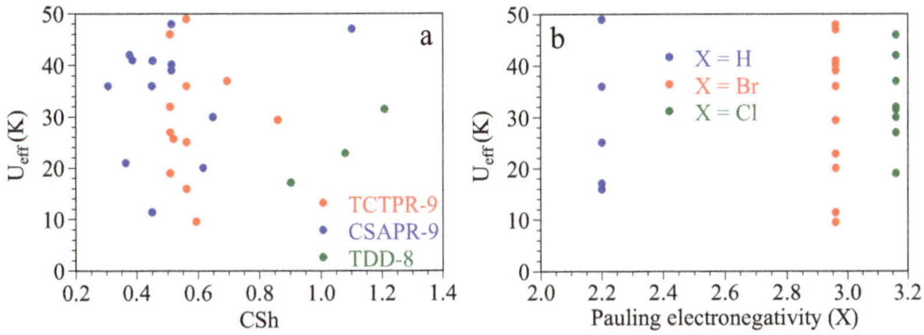


**Figure 7.** (**a**) Effective energy barrier ($U_{eff}$) as a function of the distortion from the ideal geometry (CSh) for the three main geometries: TCTPR-9 = tri-capped trigonal prism, CSAPR-9 = capped-square antiprism and TDD-8 = triangular dodecahedron in all the reported Dy-anilato lattices with SMR. (**b**) Effective energy barrier ($U_{eff}$) as a function of the electronegativity of the X group of the anilato ligand.

## 3. Materials and Methods

**General Remarks.** $(Ph_4P)_2[C_6O_4(CN)Cl]\cdot 2H_2O$ was synthesised as previously reported [57]. Reagents and solvents were purchased from Thermofisher Scientific Alfa-Aesar (Ward Hill, MA, USA) and used as received.

**Synthesis of $(H_3O)[Eu(C_6O_4(CN)Cl)_2(H_2O)]\cdot 34H_2O$ (1).** Single crystals of compound **1** were obtained by carefully layering in thin tubes, at room temperature, a solution of $(Ph_4P)_2[C_6O_4(CN)Cl]\cdot 2H_2O$ (27.4 mg, 0.03 mmol) in 7.5 mL of methanol on top of a solution of $Eu(NO_3)_3\cdot 6H_2O$ (13.4 mg, 0.03 mmol) in 7.5 mL of $H_2O$. The solutions were separated with a layer of water. The tubes were allowed to stand for about 7 days to obtain red single crystals suitable for X-ray diffraction that were filtered and air-dried.

**Synthesis of $(H_3O)[Dy(C_6O_4(CN)Cl)_2(H_2O)]\cdot 44H_2O$ (2).** Compound **2** was prepared as **1**, but using $Dy(NO_3)_3\cdot 5H_2O$ (10.5 mg, 0.03 mmol) instead of $Eu(NO_3)_3\cdot 6H_2O$. The tubes were allowed to stand for about 10 days to obtain pink single crystals suitable for X-ray diffraction that were filtered and air-dried.

### *3.1. Magnetic Susceptibility Measurements*

Variable temperature susceptibility measurements were carried out for compounds **1** and **2** in the temperature range 2–300 K with an applied magnetic field of 100 mT on ground polycrystalline samples using a Quantum Design (San Diego, CA, USA) MPMS XL-5 SQUID magnetometer. The susceptibility data were corrected for the sample holder and the diamagnetic contributions of the sample, using Pascal's constants [58]. AC susceptibility measurements were performed on the same sample of compound **2** in the temperature range 1.9–2.4 K with different applied DC fields in the frequency range 10 Hz–10 kHz using a Quantum Design (San Diego, CA, USA) PPMS-9 equipment.

### *3.2. X-Ray Diffraction*

Suitable single crystals of compounds **1** and **2** with dimensions $0.09 \times 0.05 \times 0.03$ mm$^3$ and $0.20 \times 0.10 \times 0.06$ mm$^3$, respectively, were selected and mounted on a loop using a viscous paratone oil and transferred directly to the cold nitrogen stream for data collection on a

Supernova Agilent Technologies diffractometer equipped with a graphite-monochromated Enhanced (Mo) X-ray Source (λ = 0.71073 Å).

X-ray data were collected at 120 K using $\omega$ scans with Mo $K_\alpha$ radiation. The program CrysAlis$^{Pro}$ was used for unit cell determinations and data reduction [59]. The structure was solved with ShelXT using Intrinsic Phasing and by using Olex2 1.5. as the graphical interface [60]. The model was refined with ShelXL 2018/3 [61] using full-matrix least-squares minimisation on $F^2$.

Empirical absorption correction was performed using spherical harmonics, implemented in the SCALE3 ABSPACK scaling algorithm. All non-hydrogen atoms were refined anisotropically. Hydrogen atom positions were calculated geometrically and refined using the riding model. Data collection and refinement parameters are provided in Table S5.

A solvent mask was calculated for compounds **1** and **2**. This analysis shows 346.8 electrons in a volume of 1329.9 Å$^3$ in one void for **1** and 453 electrons in a volume of 1326.0 Å$^3$ in one void for **2** per unit cell This is consistent with the presence of 35 and 45 $H_2O$ molecules per formula unit in **1** and **2**, respectively.

CCDC-2401611 and 2402436 contain the supplementary crystallographic data for **1** and **2**, respectively. These data can be obtained free of charge from The Cambridge Crystallographic Data Centre via www.ccdc.cam.ac.uk/data_request/cif.

## 4. Conclusions

We have prepared and characterised two isostructural MOFs with a very unusual square (4,4) layered structure containing lanthanoid ions and the asymmetric chlorocyananilato ligand. These compounds, formulated as $(H_3O)[Eu(C_6O_4(CN)Cl)_2(H_2O)]\cdot 34H_2O$ (**1**) and $(H_3O)[Dy(C_6O_4(CN)Cl)_2(H_2O)]\cdot 44H_2O$ (**2**), contain lanthanoid ions surrounded by four bis-bidentate chlorocyananilato ligands, in contrast to most reported structures, where the lanthanoids are surrounded by only three bidentate anilato ligands [9]. The $Ln^{III}$ ions complete their caped square antiprismatic coordination geometry with a water molecule located in the caped position, pointing to the interlayer space. The unusual coordination around the $Ln^{III}$ ions results in a slightly corrugated 4,4 lattice formed by square cavities, where the $Ln^{III}$ ions occupy the vertices of the squares and are connected to other four through the bis-bidentate chlorocyananilato ligands, which form the sides of the squares. The anionic layers are packed in an eclipsed way and are separated by layers containing the crystallisation water molecules and one oxonium cation.

The magnetic properties of the $Dy^{III}$ derivative show the presence of a field-induced slow relaxation of the magnetisation at low temperatures. The magnetisation relaxes following Direct and Orbach mechanisms with an energy barrier of 36(3) K, which is within the normal range observed in all the reported $Dy^{III}$ 2D and 3D lattices with anilato-type ligands.

**Supplementary Materials:** The following supporting information can be downloaded at: https://www.mdpi.com/article/10.3390/magnetochemistry11010001/s1, Figure S1. Simulated and experimental X-ray powder diffractograms for compound $(H_3O)[Dy(C_6O_4(CN)Cl)_2(H_2O)]\cdot 44H_2O$ (**2**). Figure S2. Thermal variation of the $\chi_m T$ product for **2**. Figure S3. Frequency dependence of $\chi'_m$ at different temperatures with a DC field of 80 mT for **2**. Figure S4. Arrhenius plot of the relaxation times (τ) obtained from the fit of $\chi'_m$, with a DC field of 80 mT for 2. Table S1. Main bond lengths in $(H_3O)[Eu(C_6O_4(CN)Cl)_2(H_2O)]\cdot 34H_2O$ (**1**). Table S2. Main bond angles (°) in $(H_3O)[Eu(C_6O_4(CN)Cl)_2(H_2O)]\cdot 34H_2O$ (**1**). Table S3. Main bond lengths in $(H_3O)[Dy(C_6O_4(CN)Cl)_2(H_2O)]\cdot 44H_2O$ (**2**). Table S4. Main bond angles (°) in $(H_3O)[Dy(C_6O_4(CN)Cl)_2(H_2O)]\cdot 44H_2O$ (**2**). Table S5. Crystal data for compounds $(H_3O)[Eu(C_6O_4(CN)Cl)_2(H_2O)]\cdot 34H_2O$ (**1**) and $(H_3O)[Dy(C_6O_4(CN)Cl)_2(H_2O)]\cdot 44H_2O$ (**2**).

**Author Contributions:** Conceptualisation, S.B. and C.J.G.-G.; formal analysis, S.B., C.C.-N. and C.J.G.-G.; investigation, S.B., C.C.-N. and C.J.G.-G.; writing—original draft preparation, S.B. and C.J.G.-G.; writing—review and editing, S.B., C.C.-N. and C.J.G.-G. All authors have read and agreed to publish this version of the manuscript.

**Funding:** This study forms part of the Advanced Materials program and was supported by the Spanish MCIN with funding from European Union NextGeneration EU (PRTR-C17.I1) and the Generalitat Valenciana (project MFA-2022-057). We also thank the Grant PID2021-125907NB-I00 funded by MCIN/AEI/10.13039/501100011033 and by "ERDF A way of making Europe" and project CIPROM-2022-060 from the Generalitat Valenciana, for financial support.

**Institutional Review Board Statement:** Not applicable.

**Informed Consent Statement:** Not applicable.

**Data Availability Statement:** The original contributions presented in this study are included in the article/Supplementary Materials. Further inquiries can be directed to the corresponding authors.

**Conflicts of Interest:** The authors declare no conflicts of interest.

## References

1. Fordham, S.; Wang, X.; Bosch, M.; Zhou, H. Lanthanide metal-organic frameworks: Syntheses, properties, and potential applications. In *Lanthanide Metal-Organic Frameworks;* Springer: Berlin/Heidelberg, Germany, 2015; pp. 1–27.
2. Zhang, S.; Wang, J.; Zhang, Y.; Ma, J.; Huang, L.; Yu, S.; Chen, L.; Song, G.; Qiu, M.; Wang, X. Applications of water-stable metal-organic frameworks in the removal of water pollutants: A review. *Environ. Pollut.* **2021**, *291*, 118076. [CrossRef]
3. Liu, X.; Fu, W.; Bouwman, E. One-step growth of lanthanoid metal–organic framework (MOF) films under solvothermal conditions for temperature sensing. *Chem. Commun.* **2016**, *52*, 6926–6929. [CrossRef] [PubMed]
4. Kepp, K.P. A quantitative scale of oxophilicity and thiophilicity. *Inorg. Chem.* **2016**, *55*, 9461–9470. [CrossRef] [PubMed]
5. Benmansour, S.; Valles-Garcia, C.; Gomez-Claramunt, P.; Minguez Espallargas, G.; Gomez-Garcia, C.J. 2d and 3d anilato-based heterometallic M (I) M (III) lattices: The missing link. *Inorg. Chem.* **2015**, *54*, 5410–5418. [CrossRef]
6. Kitagawa, S.; Kawata, S. Coordination compounds of 1,4-dihydroxybenzoquinone and its homologues. Structures and properties. *Coord. Chem. Rev.* **2002**, *224*, 11–34. [CrossRef]
7. Mercuri, M.L.; Congiu, F.; Concas, G.; Sahadevan, S.A. Recent Advances on Anilato-Based Molecular Materials with Magnetic and/or Conducting Properties. *Magnetochemistry* **2017**, *3*, 17. [CrossRef]
8. Atzori, M.; Benmansour, S.; Mínguez Espallargas, G.; Clemente-León, M.; Abhervé, A.; Gómez-Claramunt, P.; Coronado, E.; Artizzu, F.; Sessini, E.; Deplano, P.; et al. A Family of Layered Chiral Porous Magnets Exhibiting Tunable Ordering Temperatures. *Inorg. Chem.* **2013**, *52*, 10031–10040. [CrossRef] [PubMed]
9. Benmansour, S.; Gómez-García, C.J. Lanthanoid-anilato complexes and lattices. *Magnetochemistry* **2020**, *6*, 71. [CrossRef]
10. Benmansour, S.; Segura-Andreu, M.; Gómez-García, C.J. Tb and Er dimers based on an asymmetric anilato ligand with slow magnetic relaxation. *Inorg. Chim. Acta* **2024**, *571*, 122189. [CrossRef]
11. Benmansour, S.; Pintado-Zaldo, C.; Casal-García, S.H.; Martínez-Ponce, J.; Gómez-García, C.J. Anilato-based Coordination Polymers with Slow Relaxation of the Magnetization: Role of the Synthetic Method and Anilato ligand. *Chem. Eur. J.* **2024**, *30*, e202400410. [CrossRef]
12. Benmansour, S.; Pintado-Zaldo, C.; Martínez-Ponce, J.; Hernández-Paredes, A.; Valero-Martínez, A.; Gómez-Benmansour, M.; Gómez-García, C.J. The Versatility of Ethylene Glycol to Tune the Dimensionality and Magnetic Properties in $Dy^{III}$-Anilato-Based Single-Ion Magnets. *Cryst. Growth Des.* **2023**, *23*, 1269–1280. [CrossRef] [PubMed]
13. Benmansour, S.; Gómez-García, C.J.; Hernandez-Paredes, A. The Complete Series of Lanthanoid-Chloranilato Lattices with Dimethylsulfoxide: Role of the Lanthanoid Size on the Coordination Number and Crystal Structure. *Crystals* **2022**, *12*, 261. [CrossRef]
14. Benmansour, S.; Hernandez-Paredes, A.; Bayona-Andres, M.; Gómez-García, C.J. Slow Relaxation of the Magnetization in Anilato-Based Dy(III) 2D Lattices. *Molecules* **2021**, *26*, 1190. [CrossRef] [PubMed]
15. Benmansour, S.; Hernandez-Paredes, A.; Mondal, A.; Lopez Martinez, G.; Canet-Ferrer, J.; Konar, S.; Gómez-García, C.J. Slow relaxation of the magnetization, reversible solvent exchange and luminescence in 2D anilato-based frameworks. *Chem. Commun.* **2020**, *56*, 9862–9865. [CrossRef] [PubMed]

16. Monni, N.; Baldoví, J.J.; García-López, V.; Oggianu, M.; Cadoni, E.; Quochi, F.; Clemente-León, M.; Mercuri, M.L.; Coronado, E. Reversible tuning of luminescence and magnetism in a structurally flexible erbium–anilato MOF. *Chem. Sci.* **2022**, *13*, 7419–7428. [CrossRef] [PubMed]
17. Monni, N.; Dey, S.; Garcia-Lopez, V.; Oggianu, M.; Baldovi, J.J.; Mercuri, M.L.; Clemente-Leon, M.; Coronado, E. Tunable SIM properties in a family of 3D anilato-based lanthanide-MOFs. *Inorg. Chem. Front.* **2024**, *11*, 5913–5923. [CrossRef]
18. Mondal, A.; Roy, S.; Konar, S. Remarkable Energy Barrier for Magnetization Reversal in 3D and 2D Dysprosium-Chloranilate-Based Coordination Polymers. *Chem. Eur. J.* **2020**, *26*, 8774–8783. [CrossRef]
19. Jeon, I.; Negru, B.; Duyne, R.P.V.; Harris, T.D. A 2D Semiquinone Radical-Containing Microporous Magnet with Solvent-Induced Switching from $T_c$ = 26 to 80 K. *J. Am. Chem. Soc.* **2015**, *137*, 15699–15702. [CrossRef] [PubMed]
20. Atzori, M.; Artizzu, F.; Sessini, E.; Marchio, L.; Loche, D.; Serpe, A.; Deplano, P.; Concas, G.; Pop, F.; Avarvari, N.; et al. Halogen-bonding in a new family of tris(haloanilato)metallate(iii) magnetic molecular building blocks. *Dalton Trans.* **2014**, *43*, 7006–7019. [CrossRef] [PubMed]
21. Benmansour, S.; Gómez-Claramunt, P.; Vallés-García, C.; Mínguez Espallargas, G.; Gómez García, C.J. Key Role of the Cation in the Crystallization of Chiral Tris(Anilato)Metalate Magnetic Anions. *Cryst. Growth Des.* **2016**, *16*, 518–526. [CrossRef]
22. Abrahams, B.F.; Grannas, M.J.; Hudson, T.A.; Hughes, S.A.; Pranoto, N.H.; Robson, R. Synthesis, structure and host-guest properties of $(Et_4N)_2[Sn^{IV}Ca^{II}$(chloranilate)$_4]$, a new type of robust microporous coordination polymer with a 2D square grid structure. *Dalton Trans.* **2011**, *40*, 12242–12247. [CrossRef] [PubMed]
23. Benmansour, S.; López-Martínez, G.; Canet-Ferrer, J.; Gómez-García, C.J. A Family of Lanthanoid Dimers with Nitroanilato Bridges. *Magnetochemistry* **2016**, *2*, 32. [CrossRef]
24. Kharitonov, A.D.; Trofimova, O.Y.; Meshcheryakova, I.N.; Fukin, G.K.; Khrizanforov, M.N.; Budnikova, Y.H.; Bogomyakov, A.S.; Aysin, R.R.; Kovalenko, K.A.; Piskunov, A.V. 2D-metal–organic coordination polymers of lanthanides (La (III), Pr (III) and Nd (III)) with redox-active dioxolene bridging ligands. *CrystEngComm* **2020**, *22*, 4675–4679. [CrossRef]
25. Gomez-Claramunt, P.; Benmansour, S.; Hernandez-Paredes, A.; Cerezo-Navarrete, C.; Rodriguez-Fernandez, C.; Canet-Ferrer, J.; Cantarero, A.; Gómez-García, C.J. Tuning the Structure and Properties of Lanthanoid Coordination Polymers with an Asymmetric Anilato Ligand. *Magnetochemistry* **2018**, *4*, 6. [CrossRef]
26. Benmansour, S.; Hernandez-Paredes, A.; Gómez-García, C.J. Two Dimensional Magnetic Coordination Polymers Formed by Lanthanoids and Chlorocyananilato. *Magnetochemistry* **2018**, *4*, 58. [CrossRef]
27. Sahadevan, S.A.; Monni, N.; Abherve, A.; Marongiu, D.; Sarritzu, V.; Sestu, N.; Saba, M.; Mura, A.; Bongiovanni, G.; Cannas, C.; et al. Nanosheets of Two-Dimensional Neutral Coordination Polymers Based on Near-Infrared-Emitting Lanthanides and a Chlorocyananilate Ligand. *Chem. Mater.* **2018**, *30*, 6575–6586. [CrossRef]
28. Sahadevan, S.A.; Monni, N.; Abherve, A.; Cosquer, G.; Oggianu, M.; Ennas, G.; Yamashita, M.; Avarvari, N.; Mercuri, M.L. Dysprosium Chlorocyanoanilate-Based 2D-Layered Coordination Polymers. *Inorg. Chem.* **2019**, *58*, 13988–13998. [CrossRef] [PubMed]
29. Hernandez-Paredes, A.; Cerezo-Navarrete, C.; Gómez-García, C.J.; Benmansour, S. Slow relaxation in doped coordination polymers and dimers based on lanthanoids and anilato ligands. *Polyhedron* **2019**, *170*, 476–485. [CrossRef]
30. Artizzu, F.; Atzori, M.; Liu, J.; Mara, D.; Van Hecke, K.; Van Deun, R. Solution-processable Yb/Er 2D-layered metallorganic frameworks with high NIR-emission quantum yields. *J. Mater. Chem. C* **2019**, *7*, 11207–11214. [CrossRef]
31. Oggianu, M.; Abherve, A.; Marongiu, D.; Quochi, F.; Galan-Mascaros, J.R.; Bertolotti, F.; Masciocchi, N.; Avarvari, N.; Mercuri, M.L. Terbium and Europium Chlorocyananilate-Based 2D Coordination Polymers. *Molecules* **2023**, *28*, 6453. [CrossRef]
32. Hua, C. Structure directing effects of the tetraphenylphosphonium cation upon the formation of lanthanoid anilate coordination polymers. *Eur. J. Inorg. Chem.* **2024**, *27*, e202400026. [CrossRef]
33. Alvarez, S. Distortion Pathways of Transition Metal Coordination Polyhedra Induced by Chelating Topology. *Chem. Rev.* **2015**, *115*, 13447–13483. [CrossRef]
34. Ruiz-Martínez, A.; Casanova, D.; Alvarez, S. Polyhedral structures with an odd number of vertices: Nine-coordinate metal compounds. *Chem. Eur. J.* **2008**, *14*, 1291–1303. [CrossRef]
35. Sorace, L.; Gatteschi, D. Electronic Structure and Magnetic Properties of Lanthanide Molecular Complexes. In *Lanthanides and Actinides in Molecular Magnetism*; Layfield, R.A., Murugesu, M., Eds.; Wiley-VCH: Weinheim, Germany, 2015; Volume 1, pp. 1–25.
36. Andruh, M.; Bakalbassis, E.; Kahn, O.; Trombe, J.C.; Porcher, P. Structure, spectroscopic and magnetic properties of rare earth metal (III) derivatives with the 2-formyl-4-methyl-6-(N-(2-pyridylethyl) formimidoyl) phenol ligand. *Inorg. Chem.* **1993**, *32*, 1616–1622. [CrossRef]
37. Thompson, L.; Legendziewicz, J.; Cybinska, J.; Pan, L.; Brennessel, W. Structure, photophysics and magnetism of a europium mixed complex, Eu(HFAA)$_3$bipy.$H_2O$, in the solid state and solution. *J. Alloys Comp.* **2002**, *341*, 312–322. [CrossRef]

38. de Oliveira Maciel, J.W.; Lemes, M.A.; Valdo, A.K.; Rabelo, R.; Martins, F.T.; Queiroz Maia, L.J.; de Santana, R.C.; Lloret, F.; Julve, M.; Cangussu, D. Europium(III), Terbium(III), and Gadolinium(III) Oxamato-Based Coordination Polymers: Visible Luminescence and Slow Magnetic Relaxation. *Inorg. Chem.* **2021**, *60*, 6176–6190. [CrossRef] [PubMed]
39. Mičová, R.; Bielková, Z.; Rajnák, C.; Titiš, J.; Moncol', J.; Bieńko, A.; Boča, R. Magnetic properties of a europium(III) complex–possible multiplet crossover. *Dalton Trans.* **2024**, *53*, 1492–1496. [CrossRef] [PubMed]
40. Vostrikova, K.E.; Sukhikh, T.S.; Lavrov, A.N. The Synthesis, Crystal Structure, and Magnetic Properties of Mono-Scorpionate Eu (III) Complexes. *Inorganics* **2023**, *11*, 418. [CrossRef]
41. Arauzo, A.; Lazarescu, A.; Shova, S.; Bartolomé, E.; Cases, R.; Luzón, J.; Bartolomé, J.; Turta, C. Structural and magnetic properties of some lanthanide (Ln= Eu (III), Gd (III) and Nd (III)) cyanoacetate polymers: Field-induced slow magnetic relaxation in the Gd and Nd substitutions. *Dalton Trans.* **2014**, *43*, 12342–12356. [CrossRef]
42. Wu, M.; Wang, M.; Guo, S.; Zheng, F.; Chen, H.; Jiang, X.; Liu, G.; Guo, G.; Huang, J. Photoluminescent and magnetic properties of a series of lanthanide coordination polymers with 1 H-tetrazolate-5-formic acid. *Cryst. Growth Des.* **2011**, *11*, 372–381. [CrossRef]
43. Li, Y.; Zheng, F.; Liu, X.; Zou, W.; Guo; Lu, C.; Huang, J. Crystal structures and magnetic and luminescent properties of a series of homodinuclear lanthanide complexes with 4-cyanobenzoic ligand. *Inorg. Chem.* **2006**, *45*, 6308–6316. [CrossRef]
44. Xu, N.; Wang, C.; Shi, W.; Yan, S.; Cheng, P.; Liao, D. Magnetic and Luminescent Properties of Sm, Eu, Tb, and Dy Coordination Polymers with 2-Hydroxynicotinic Acid. *Eur. J. Inorg. Chem.* **2011**, *2011*, 2387–2393. [CrossRef]
45. Vaz, R.C.; Esteves, I.O.; Oliveira, W.X.; Honorato, J.; Martins, F.T.; Marques, L.F.; Dos Santos, G.L.; Freire, R.O.; Jesus, L.T.; Pedroso, E.F. Mononuclear lanthanide(III)-oxamate complexes as new photoluminescent field-induced single-molecule magnets: Solid-state photophysical and magnetic properties. *Dalton Trans.* **2020**, *49*, 16106–16124. [CrossRef]
46. Chen, J.; Carpenter, S.H.; Fetrow, T.V.; Mengell, J.; Kirk, M.L.; Tondreau, A.M. Magnetism Studies of Bis(acyl)phosphide-Supported $Eu_3$ and $Eu_2$ Complexes. *Inorg. Chem.* **2022**, *61*, 18466–18475. [CrossRef] [PubMed]
47. Shintoyo, S.; Fujinami, T.; Matsumoto, N.; Tsuchimoto, M.; Weselski, M.; Bieńko, A.; Mrozinski, J. Synthesis, crystal structure, luminescent and magnetic properties of europium (III) and terbium (III) complexes with a bidentate benzoate and a tripod N7 ligand containing three imidazole, $[Ln^{III}(H_3L)\ benzoate](ClO_4)_2{\cdot}H_2O{\cdot}2MeOH$ ($Ln^{III} = Eu^{III}$ and $Tb^{III}$). *Polyhedron* **2015**, *91*, 28–34. [CrossRef]
48. Aubin, S.M.; Sun, Z.; Pardi, L.; Krzystek, J.; Folting, K.; Brunel, L.; Rheingold, A.L.; Christou, G.; Hendrickson, D.N. Reduced anionic $Mn_{12}$ molecules with half-integer ground states as single-molecule magnets. *Inorg. Chem.* **1999**, *38*, 5329–5340. [CrossRef]
49. Dey, A.; Kalita, P.; Chandrasekhar, V. Lanthanide (III)-based single-ion magnets. *ACS Omega* **2018**, *3*, 9462–9475. [CrossRef]
50. Demir, S.; Zadrozny, J.M.; Long, J.R. Large Spin-Relaxation Barriers for the Low-Symmetry Organolanthanide Complexes $[Cp^*{}_2Ln(BPh_4)]$($Cp^*$ = pentamethylcyclopentadienyl; Ln = Tb, Dy). *Chem. Eur. J.* **2014**, *20*, 9524–9529. [CrossRef] [PubMed]
51. Lin, P.-H.; Burchell, T.J.; Clérac, R.; Murugesu, M. Dinuclear dysprosium(III) single-molecule magnets with a large anisotropic barrier. *Angew. Chem. Int. Ed.* **2008**, *47*, 8848–8851. [CrossRef] [PubMed]
52. Wu, H.; Li, M.; Xia, Z.; Montigaud, V.; Cador, O.; Le Guennic, B.; Ke, H.; Wang, W.; Xie, G.; Chen, S. High temperature quantum tunnelling of magnetization and thousand kelvin anisotropy barrier in a $Dy_2$ single-molecule magnet. *Chem. Commun.* **2021**, *57*, 371–374. [CrossRef]
53. Liu, K.; Li, H.; Zhang, X.; Shi, W.; Cheng, P. Constraining and tuning the coordination geometry of a lanthanide ion in metal–organic frameworks: Approach toward a single-molecule magnet. *Inorg. Chem.* **2015**, *54*, 10224–10231. [CrossRef] [PubMed]
54. Xin, Y.; Wang, J.; Zychowicz, M.; Zakrzewski, J.J.; Nakabayashi, K.; Sieklucka, B.; Chorazy, S.; Ohkoshi, S. Dehydration–Hydration Switching of Single-Molecule Magnet Behavior and Visible Photoluminescence in a Cyanido-Bridged $Dy^{III}Co^{III}$ Framework. *J. Am. Chem. Soc.* **2019**, *141*, 18211–18220. [CrossRef]
55. Huizi-Rayo, U.; Zabala-Lekuona, A.; Terenzi, A.; Cruz, C.M.; Cuerva, J.M.; Rodríguez-Diéguez, A.; García, J.A.; Seco, J.M.; San Sebastian, E.; Cepeda, J. Influence of thermally induced structural transformations on the magnetic and luminescence properties of tartrate-based chiral lanthanide organic-frameworks. *J. Mater. Chem. C* **2020**, *8*, 8243–8256. [CrossRef]
56. Manna, F.; Oggianu, M.; Avarvari, N.; Mercuri, M.L. Lanthanide-Based Metal-Organic Frameworks with Single-Molecule Magnet Properties. *Magnetochemistry* **2023**, *9*, 190. [CrossRef]
57. Atzori, M.; Artizzu, F.; Marchiò, L.; Loche, D.; Caneschi, A.; Serpe, A.; Deplano, P.; Avarvari, N.; Mercuri, M.L. Switching-on luminescence in anilate-based molecular materials. *Dalton Trans.* **2015**, *44*, 15786–15802. [CrossRef] [PubMed]
58. Bain, G.A.; Berry, J.F. Diamagnetic corrections and Pascal's constants. *J. Chem. Educ.* **2008**, *85*, 532–536. [CrossRef]
59. *Oxford Diffraction CRYSALISPRO*, version 171.33.55; Agilent: Oxfordshire, UK, 2010.

60. Dolomanov, O.V.; Bourhis, L.J.; Gildea, R.J.; Howard, J.A.K.; Puschmann, H. OLEX2: A complete structure solution, refinement and analysis program. *J. Appl. Cryst.* **2009**, *42*, 339–341. [CrossRef]
61. Sheldrick, G.M. SHELXT–Integrated space-group and crystal-structure determination. *Acta Crystallogr.* **2015**, *71*, 3–8. [CrossRef]

magnetochemistry

MDPI

*Article*

# Chemical and Structural Versatility in the Copper/2,2′-Bipyrimidine/Iodide System: A Regular Alternating Mixed-Valent Cu(II)-Cu(I) Chain Showing Unusually Similar Metal Coordination Environments †

**Nadia Marino [1,*], Francesc Lloret [2], Miguel Julve [2] and Giovanni De Munno [1]**

1 Dipartimento di Chimica e Tecnologie Chimiche, Ponte P. Bucci, 14/c, Università della Calabria, 87036 Arcavacata di Rende, Cosenza, Italy
2 Instituto de Ciencia Molecular (ICMol), Universitat de València, c/José Beltrán 2, E-46980 Paterna, València, Spain
* Correspondence: nadia.marino@unical.it

† Dedicated to two one-of-a-kind friends, Francesc Lloret, and Miguel Julve, who sadly passed away prematurely, in July 2024.

**Abstract:** The reaction in reagent grade acetone of copper(II) nitrate hexahydrate, 2,2′-bipyrimidine (bpm) and potassium iodide in a 1:2:2 molar ratio afforded three different products: an unreduced Cu(II) species, a fully reduced Cu(I) species and a mixed-valent Cu(II)/Cu(I) species. Of these, only the unreduced Cu(II) complex of formula $[Cu^{II}(bpm)_3](I_3)(I)$ (**1**) could be structurally characterized, the other two products being initially only isolated as amorphous powders. X-ray quality, beautifully shaped, *quasi*-black prismatic crystals of compound **2**, namely $\{[Cu^{I}(I_3)Cu^{II}(I)(bpm)_2](I_3)\}_n$, and brick-reddish parallelepipeds of compound **3**, namely $\{[Cu^{I}_2(\mu\text{-}I)_2(bpm)]\}_n$, were successively obtained through the slow diffusion in H-shaped tubes of aqueous solutions of the three reagents, after extensive optimization of the crystallization conditions. Compound **1** consists of a rare tris(2,2′-bipyrimidine)copper(II) monomeric dication, charge balanced by both iodide and triiodide anions. Compound **3**, whose structure as well as optical and photocatalytic properties were recently disclosed, consists of a regular alternating $\mu$-bpm/di-$\mu$-iodide copper(I) chain. Finally, compound **2** consists of a rare, regular alternating mixed-valent Cu(II)-Cu(I) $\mu$-bpm copper chain, showing unusual similarities in the metal coordination environment. The magnetic properties of compound **2** remarkably reveal a very weak antiferromagnetic coupling between the paramagnetic Cu(II) ions which are well separated both intra- and inter-chain.

**Keywords:** copper(II) partial reduction; iodide; triiodide; mixed-valent species; crystal structure

## 1. Introduction

The interest in inorganic exchange–coupled systems has led to many synthetic endeavors and magnetostructural studies in the past decades [1]. In this field of research, the scientific contribution of Miguel Julve and Francesc Lloret is no less than outstanding [2–5].

The work presented herein aimed to give a modest tribute to these two brilliant researchers on the occasion of their retirement. Now that Miguel is dead, we would also like to remember him by briefly sharing some aspects of our personal journey together. There has been a very close scientific collaboration between the research group at the University of Calabria (De Munno) and that at the University of Valencia (Julve, Lloret), which lasted for almost forty years, also owing to the similar character of the protagonists and their true and profound friendship. Thanks to the pooling of different but complementary skills and backgrounds, such as structural chemistry and molecular magnetism, this collaboration has led to about 140 joint articles in peer-reviewed international scientific journals. It is hard not to remember with pride and nostalgia the early days, especially. In the period 1993–1998 alone, for example, six articles were published in Angewandte Chemie [6–11]. At the beginning, the scientific partnership was based on the magnetic and structural study of 2,2′-bipyrimidine (bpm) compounds. For this reason, the present commemorative contribution also focuses on bpm-containing species. As a bis-chelating ligand, its ability to mediate magnetic interactions between transition metal ions separated by distances ca. 5.5 A is now well known but, at the time the collaboration was born, there was only one structurally characterized bpm bridged compound, obtained in Prof. De Munno's laboratory [12]. Over the years, more than 80 compounds containing bpm have been published by the two research groups. These comprise mononuclear, dinuclear and polynuclear species, as well as mono-, two- and three-dimensional polymers [6–10,12–28]. In multiple cases, they have shown fascinating architecture and interesting magnetic features. Several reports are also known concerning the biological [29,30], photophysical, electrochemical and catalytical properties of bpm-containing compounds [23,31–38], proving the interest toward this widely used ligand is not limited to molecular magnetism.

At a certain point in time, we become involved in the study of halides and pseudohalides of paramagnetic metal ions containing bpm, whose synthesis and characterization appear very attractive from a structural and magnetic point of view. According to data reported in the literature, while the magnetic interaction through bpm is mostly antiferromagnetic, that mediated by bridging halogen/pseudoalogen can vary from antiferro- to weak or strong ferromagnetic [9,13,17,25,39–44]. Restricting ourselves to bridging bpm and halides Cu(II) systems, and excluding mixed-metal [45] or further mixed-ligand species [39], only three complexes result characterized to date, all displaying large antiferromagnetic behavior: two bidimensional polymers of general formula $[Cu_2(bpm)X_4]_n$ (X = Cl, Br) [13] and one chain compound of formula $[\{Cu_2(\mu\text{-bpm})(H_2O)_2(\mu\text{-F})_2F_2\}_n$ [25]. With the iodide ion, to the best of our knowledge and according to a recent search within the Cambridge structural Database (CCDC 2024; ConQuest version 2024.3.0), no bpm complexes of Cu(II) have ever been reported, which prompted us to undertake the study presented herein. In this work, we focus on the versatility of the copper/2,2′-bipyrimidine/iodide system, which has led to the isolation and characterization of an unreduced Cu(II) species, a mixed-valent Cu(II)/Cu(I) species and a fully reduced Cu(I) species under the same, mild experimental conditions, starting from a Cu(II) salt, potassium iodide and bpm. The three complexes have formulae, respectively, $[Cu^{II}(bpm)_3](I_3)(I)$ (**1**), $\{[Cu^{I}I_3\ Cu^{II}I(bpm)_2](I_3)\}_n$ (**2**) and $\{[Cu^{I}_2(\mu\text{-I})_2(bpm)]\}_n$ (**3**). A rationalized synthetic approach is presented, which proved pivotal for the obtaining of complexes **2** and **3**, each as clean and almost quantitative product of two slightly distinct crystallization strategies. The crystal structure of complex **3**, of broad interest for its optical and photocatalytic properties, has been recently evaluated by Hou et al. [38], CCDC refcode ZIJWEX. For this reason, only the new synthetic route to **3** is discussed herein. Although the leitmotiv of this work is both synthetic and structural, of course, we would not have achieved justice to honoring our friends Paco and Miguel without some magnetic measurements. They analyzed and conducted the magneto–structural correlation study themselves on compound **2**, validating its nature as that of a quite unique, mixed-valent, perfectly alternating Cu(I)-Cu(II) chain based on bpm bridges.

## 2. Materials and Methods

All chemicals were purchased from commercial sources and used as received without further purification. Infrared spectra were recorded on a PerkinElmer 1750 FT-IR spectrometer (PerkinElmer, Waltham, MA, USA) as KBr pellets in the 4000–400 $cm^{-1}$ region. The relative intensity of reported signals is defined as s = strong, br = broad, sh = sharp, m = medium, and w = weak. Elemental analysis (C, H, N) was performed by the Microanalytical Service of the Università della Calabria. Variable-temperature (2.0–300 K) magnetic susceptibility and variable-field (0–5 T) magnetization measurements were carried out with a SQUID magnetometer under an applied field of 0.1 T ($T \geq 25$ K). A small magnetic field of 0.025 T was used in the low-temperature region ($T < 25$ K) to prevent saturation effects. The experimental magnetic susceptibility data were corrected for the diamagnetic contributions of the constituent atoms and the sample holder, as well as for the temperature-independent paramagnetism (tip) of the $Cu^{II}$ ion ($60 \times 10^{-6}$ $cm^3$ $mol^{-1}$).

### 2.1. Preparation of the Complexes

$[Cu^{II}(bpm)_3](I_3)(I)$ (**1**), $\{[Cu^{I}(I_3)Cu^{II}(I)(bpm)_2](I_3)\}_n$ (**2**) and $\{[Cu^{I}_2(\mu\text{-}I)_2(bpm)]\}_n$ (**3**).

$Cu(NO_3)_2{\cdot}6H_2O$ (0.2 mmol) and bpm (0.4 mmol) were dissolved in reagent grade acetone (50 mL), giving a dark green solution. To this solution, KI (0.4 mmol) was added as solid, slowly and under continuous stirring. Upon dissolution of KI, the solution color turned brown-yellowish, and a brick-red precipitate appeared simultaneously. This was filtered and the solution was left to evaporate slowly at room temperature. After a few hours, a new precipitate separated from the solution, exhibiting a darker color this time. The observation of this second precipitate fraction under the microscope revealed the presence of both brick-red and very dark, *quasi* black, grains in it. After filtering out this second solid fraction, amber-yellow parallelepipeds of compound **1** appeared in ca. 2 days on standing of the reaction solution at room temperature. Yield: ca. 20% based on Cu(II). Anal. Calcd. for $C_{24}H_{18}CuI_4N_{12}$ (**1**): C, 27.57; H, 1.74; N, 16.07. Found: C, 27.48; H, 1.65; N, 16.15. IR (KBr/$cm^{-1}$): 3550 (br), 3410 (br), 3090 (w), 1568 (sh), 1552 (sh,s), 1449 (m), 1403 (sh,s), 1235 (m), 1145 (sh, m), 1098 (m), 1004 (sh, m), 802 (sh), 800 (sh), 750 (sh,s), 687 (sh), 653 (sh).

Various attempts to dissolve and recrystallize the two initial amorphous precipitate fractions failed, but their analysis through comparative infrared spectroscopy later revealed the first fraction being compound 3, while the second fraction being a mixture of compounds **2** and **3**. In fact, single crystals of compounds **2** and **3** were successively obtained through slow diffusion in H-shaped tubes. Firstly, an H-shaped tube was prepared as follows. One arm was filled with a dark green aqueous solution (2 mL) of $Cu(NO_3)_2{\cdot}6H_2O$ (0.2 mmol) and bpm (0.1 mmol) and the other with a colorless aqueous solution of KI (0.4 mmol, so that, overall, the synthetic Cu(II):bpm:KI molar ratio was 2:1:4), layering both arms slowly with water. This method afforded, after a few weeks, very dark crystals of compound **2**, their surface being extensively covered by much smaller, reddish crystals of compound **3**, however. Clean, homogeneous batches of beautifully shaped, X-ray quality *quasi*-black prisms and reddish parallelepipeds of compound **2** and **3**, respectively, could be finally grown in H-shaped tubes, following the same procedure indicated above, by adjusting the synthetic Cu(II):bpm:KI molar ratio to either 2:1:2 (for **2**) or 2:1:1 (for **3**), respectively. Yield: ca. 60% (**2**) and 75% (**3**) based on Cu(II). Anal. Calcd. for $C_{16}H_{12}Cu_2I_7N_8$ (**2**): C, 14.43; H, 0.91; N, 8.41. Found: C, 14.20; H, 0.89; N, 8.50. Anal. Calcd. for $C_8H_6Cu_2I_2N_4$ (**3**): C, 17.82; H, 1.12; N, 10.39. Found: C, 17.74; H, 1.05; N, 10.42. IR (KBr/$cm^{-1}$) for **2**: 3048 (m), 1595 (w), 1557 (sh, s), 1471 (w), 1403 (sh, s), 1329 (w), 1298 (w), 1215 (w), 1012 (m), 831 (sh), 818 (sh), 740 (sh,s), 668 (sh). IR (KBr/$cm^{-1}$) for **3**: 3050 (sh, m), 1558 (sh, s), 1407 (sh, s), 1140 (w), 1011 (sh, w), 832 (sh, m), 751 (sh, s), 681(sh, m), 667 (sh, m). Notes:

sample purity of crystalline batches of **1–3** obtained through the optimized synthetic procedures was assessed first by eye and then confirmed by elemental analyses; spectroscopic analysis (FT-IR) was conducted over the entire course of the synthetic exploration briefly summarized above. The most characteristic band in the IR spectra of complex **1–3** is the strong and sharp band around 1580 $cm^{-1}$, diagnostic of a coordinated bpm ligand. For **1**, this band appears as a quasi-symmetric doublet, which is typically associated with the presence of terminal chelating bpm, while the same band appears as a sharp and strongly asymmetric doublet in the IR spectra of complexes **2** and **3**, as expected [25,41]. Indeed, the IR spectra of **2** and **3** are very similar, although each band is sharper in the monovalent (**3**) vs. mixed-valent (**2**) species.

### 2.2. X-Ray Data Collection and Structure Refinement

X-ray diffraction data on single crystals of **1–3** were collected at room temperature with a Bruker-Nonius X8-APEXII CCD area detector system by using graphite-monochromated Mo-K$\alpha$ radiation (Bruker, Billerica, MA, USA) ($\lambda$ = 0.71073 Å). The data were processed through the *SAINT* [46] reduction and *SADABS* [47] absorption software. The structures were solved by standard direct methods and subsequently completed by Fourier recycling by using the SHELXTL software packages [48]. The obtained models were refined with the 2019/3 version of SHELXL against $F^2$ on all data by full-matrix least squares.

All non-hydrogen atoms of the bpm ligands were set in calculated positions and refined isotropically using the riding model. The final geometrical calculations and graphical manipulations were performed using the XP utility of SHELX 2019-3 [48] and the software Diamond (version 4.6.8) [49]. A summary of the crystallographic data and structure refinement for **1–3** is given in Table 1. Main bond lengths and angles for **1** and **2** are shown in Tables 2 and 3, respectively (for a detailed structural description of compound **3** see ref. [38]). Crystallographic data have been deposited at the Cambridge Crystallographic Data Centre and have been assigned CCDC reference numbers 2419535 (**1**), 2419536 (**2**) and 2425651 (**3**). These data can be obtained free of charge from the Cambridge Crystallographic Data Center via https://www.ccdc.cam.ac.uk/structures/Home.

**Table 1.** Summary of crystallographic data for $[Cu^{II}(bpm)_3](I_3)(I)$ (**1**), $\{[Cu^{I}I_3\,Cu^{II}I(bpm)_2](I_3)\}_n$ (**2**) and $\{[Cu^{I}_2(\mu\text{-}I)_2(bpm)]\}_n$ (**3**).

| | 1 | 2 | 3 |
|---|---|---|---|
| Formula | $C_{24}H_{18}CuI_4N_{12}$ | $C_{16}H_{12}Cu_2I_7N_8$ | $C_8H_6Cu_2I_2N_4$ |
| $F_w$/g mol$^{-1}$ | 1045.64 | 1331.72 | 539.05 |
| Crystal system | triclinic | orthorhombic | monoclinic |
| Space group | *P*-1 | *Pna*2$_1$ | *C*2/*m* |
| *a*/Å | 10.7606(3) | 22.9658(6) | 12.7554(9) |
| *b*/Å | 11.3856(3) | 10.8435(3) | 8.1499(9) |
| *c*/Å | 13.6376(3) | 11.7187(3) | 6.2362(6) |
| $\alpha$/° | 88.454(1) | 90 | 90 |
| $\beta$/° | 87.043(1) | 90 | 102.737(4) |
| $\gamma$/° | 72.620(1) | 90 | 90 |
| *V*/Å$^3$ | 1592.29(7) | 2918.3(1) | 632.33(10) |
| *Z* | 2 | 4 | 2 |
| $D_c$/g cm$^{-3}$ | 2.181 | 3.031 | 2.831 |
| *F*(000) | 974 | 2372 | 492 |
| $\mu$(Mo-K$\alpha$)mm$^{-1}$ | 4.600 | 8.884 | 8.222 |
| Refl. Collected | 22,922 | 60,622 | 5095 |
| Refl. Indep. [Rint] | 7847 [0.0248] | 9912 [0.0258] | 839 [0.0715] |
| Reflec. Obs. [I > 2$\sigma$(I)] | 5551 | 8616 | 586 |

**Table 1.** *Cont.*

| | 1 | 2 | 3 |
|---|---|---|---|
| Data/restraints/parameters | 7847/0/370 | 9912/1/299 | 839/0/42 |
| Completeness to $\theta$ = 25.242° | 99.2% | 100% | 100% |
| *GOOF* [c] | 1.038 | 1.055 | 1.115 |
| $R_1$ [a] [I > 2$\sigma$(I)](all) | 0.0350(0.0559) | 0.0332(0.0411) | 0.0427(0.0507) |
| $wR_2$ [b] [I > 2$\sigma$(I)](all) | 0.0803(0.0901) | 0.0780(0.0819) | 0.1016(0.1083) |
| Largest diff. peak/hole/eÅ$^{-3}$ | 1.202/−1.224 | 1.523/−1.431 | 2.384/−1.217 |

[a] $R_1 = \sum(|F_o| - |F_c|)/\sum|F_o|$. [b] $wR_2 = [\sum w(F_o^2 - F_c^2)^2/\sum w(F_o^2)^2]^{1/2}$. [c] $S = [\sum w(|F_o| - |F_c|)^2/(N_o - N_p)]^{1/2}$.

## 3. Results and Discussion

### *3.1. Description of the Structures*

#### 3.1.1. $[Cu^{II}(bpm)_3](I_3)(I)$ (**1**)

Compound **1** crystallizes in the triclinic space group *P*-1 and consists of a racemic mixture of Δ- and Λ-isomers of a monomeric $[Cu^{II}(bpm)_3]^{2+}$ cation (see Figure 1), charge-balanced by both iodide and triiodide anions. The metal ion in **1** is six-coordinate by the N atoms of three crystallographically independent bpm molecules (noted as A, B and C), adopting an axially elongated octahedral surrounding, as confirmed by continuous shape measures (CShM) [50] (see Table 2 for selected bond distances and angles).

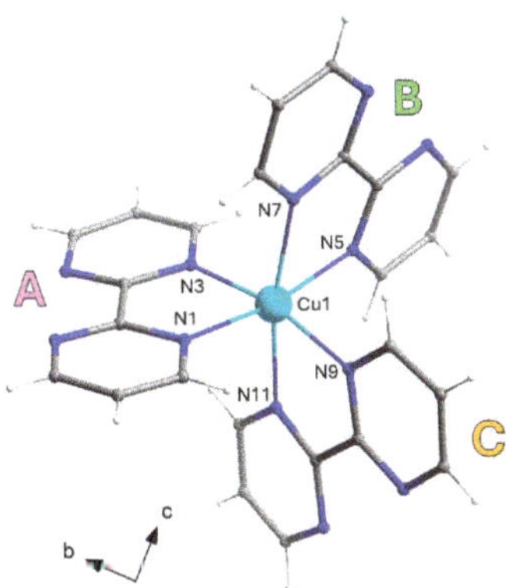


**Figure 1.** Molecular structure of the monomeric $[Cu^{II}(bpm)_3]^{2+}$ cationic unit in **1**, with selected labels. The Δ-isomer has been arbitrarily chosen as representative of the asymmetric unit; the Λ-isomer is generated by the inversion center. The three crystallographically independent bpm molecules are noted as A, B, and C. [Color codes: Cu(II), cyan; C, gray; N, blue].

**Table 2.** Selected bond distances [Å] and angles [°] for **1**.

| | | | |
|---|---|---|---|
| Cu1-N1 | 2.096(4) | Cu1-N7 | 2.108(4) |
| Cu1-N5 | 2.236(4) | Cu1-N9 | 2.060(3) |
| Cu1-N3 | 2.044(3) | Cu1-N11 | 2.244(3) |
| I1-I2 | 2.9835(5) | I2-I3 | 2.8679(5) |
| N3-Cu1-N9 | 166.29(13) | N1-Cu1-N5 | 96.31(14) |
| N3-Cu1-N1 | 78.84(13) | N7-Cu1-N5 | 76.12(14) |
| N9-Cu1-N1 | 95.67(12) | N3-Cu1-N11 | 91.34(12) |
| N3-Cu1-N7 | 95.05(13) | N9-Cu1-N11 | 76.22(12) |
| N9-Cu1-N7 | 92.02(12) | N1-Cu1-N11 | 92.20(12) |
| N1-Cu1-N7 | 169.86(12) | N7-Cu1-N11 | 96.06(13) |
| N3-Cu1-N5 | 95.93(12) | N5-Cu1-N11 | 169.73(13) |
| N9-Cu1-N5 | 97.16(12) | I(1)-I(2)-I(3) | 177.896(16) |

The three chelating bpm molecules are rather planar, with mean deviations from planarity laying in the range 0.020(4)–0.049(4) Å, and they form dihedral angles of 79.2(1), 86.6(1) and 85.9(1)° between each other. Each bpm self-interacts with a symmetry-equivalent molecule, following A···Aa, B···Bb and C···Cc *off-set* π-π interactions paths [(a) $-x$, $-y + 2$, $-z$; (b) $-x$, $-y + 1$, $-z + 1$; (c) $-x$, $-y + 1$, -$z$, see Figure 2a], and establishing interplanar distances of ca. 3.35, 3.30 and 3.15 Å, respectively. In this way, each Δ-isomer is surrounded by three Λ- isomers and vice versa, with metal–metal distances in the range 8.3–9.4 Å [Cu1···Cu1a, 9.380(1); Cu1···Cu1b, 8.327(1); Cu1···Cu1c, 8.307(1) Å]. The resulting motif is overall half-way between the brick-wall and the honeycomb (6,3) net, as shown in Figure 2a, with the supramolecular layers growing parallel to the crystallographic *bc* plane. Interestingly, the iodide anions form an ideally identical 2D anionic net, with consistent I···I separations [8.105(1)–9.520(1) Å]. The cationic and anionic hexameric motifs are eclipsed and they alternate regularly along the crystallographic *a* axis, forming *pseudo*-hexagonal cavities (Figure 2b).

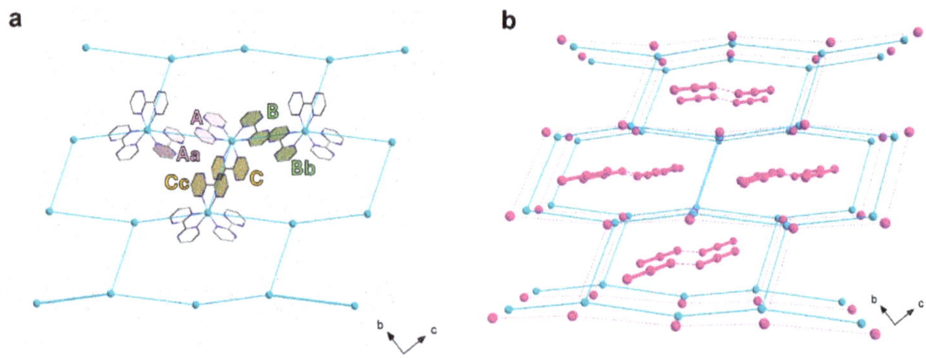


**Figure 2.** (**a**) View along the crystallographic *a* axis of one cationic layer in **1**, highlighting the bpm interaction paths and the topological 2D network of Cu(II) ions. (**b**) Perspective view of the alternating cationic and anionic topological 2D nets in **1**, highlighting the content of the supramolecular *pseudo*-hexagonal cavities. [Color codes: Cu(II), cyan; C, gray; N, blue; I, pink. Solid and dotted lines are only a guide for the eyes].

A close look into the alternating layers reveals they are not well separated. Rather, numerous CH···I contacts between the bpm ligand and the iodide anions (with H···I distances in the range 3.0–3.5 Å) are observed, which likely contribute to the formation of the ordered anionic hexagonal net (Figure 3a). The triiodide anions, which are fairly symmetrical but deviate somewhat from linearity [I1-I2, 2.984(1) Å; I2-I3, 2.868(1) Å; I1-I2-I3, 162.1(1)°], form dimers reminiscent of a $I_6^{2-}$ anion [51–53], by means of I···I contacts that fall in the typical range of the so-called secondary halogen interactions [I3···I3d, 3.698(7) Å; I2-I3···I3d, 161.4(2)°; (d) = $-x + 2$, $-y$, $-z + 1$]. These anionic triiodide pairs are segregated within the supramolecular *pseudo*-hexagonal cavities by means of several, weak CH···I (in the range ~3.1–3.5 Å) and N···I (in the range ~3.8–4.0 Å) contacts (Figure 3b,c). No iodide-triiodide contacts are noted in the structure, instead.

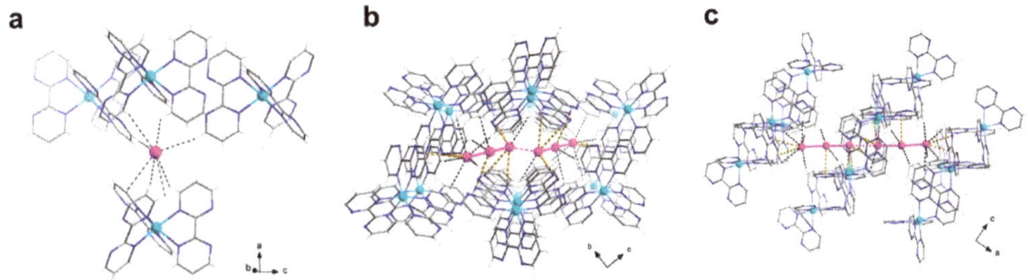


**Figure 3.** (**a**) A view of the iodide anions close surrounding in **1**, highlighting the several CH···I contacts (dashed black lines) established with the bpm molecules of vicinal complex cations. (**b**,**c**) Perspective views, along either the crystallographic *a* (**b**) or *b* (**c**) axes, of a portion of the 3D packing in **1**, showing the triiodide pairs close surrounding within the *pseudo*-hexagonal cavities in **1**. The weak CH···I and N···I contacts contributing to the anion segregation within the channels are drawn as black and yellow dashed lines, respectively. [Color codes: Cu(II), cyan; C, gray; N, blue; H, white; I, pink].

### 3.1.2. $\{[Cu^I(I_3)Cu^{II}(I)(bpm)_2](I_3)\}_n$ (**2**)

Compound **2** crystallizes in the orthorhombic space group $Pna2_1$ with two crystallographically independent copper atoms in the asymmetric unit, Cu1 and Cu2. It consists of chains of alternating Cu1 and Cu2 ions connected by two distinct bpm bridges (denoted A and B), with triiodide or iodide anions completing the coordination sphere of the two metal ions (Figure 4).

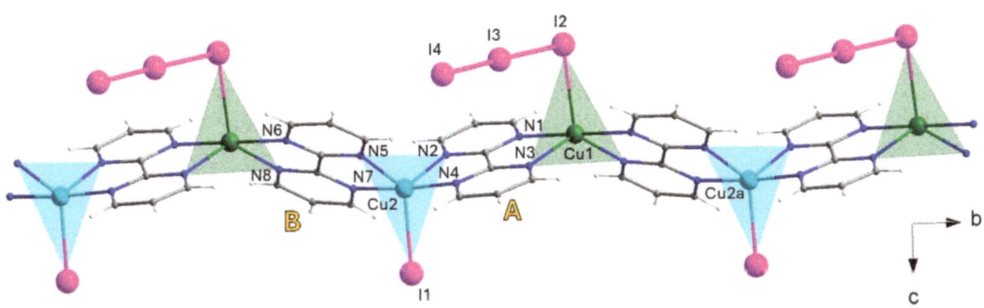


**Figure 4.** A fragment of the cationic alternating copper(I)–copper(II) chain growing along the crystallographic *b* axis in **2**, with selected labels, also showing the equatorial planes at both metal centers along the chain [Symmetry code: (a) *x*, 1 + *y*, *z*. Color codes: Cu(II), cyan; Cu(I), green; C, gray; N, blue; H, white; I, pink.].

**Table 3.** Selected bond distances [Å] and angles [°] for **2**[a].

| | | | |
|---|---|---|---|
| Cu1-N1 | 2.240(6) | Cu2-N2 | 2.140(5) |
| Cu1-N3 | 2.106(6) | Cu2-N4 | 1.993(5) |
| Cu1-N6a | 2.260(6) | Cu2-N5 | 2.121(5) |
| Cu1-N8a | 2.097(5) | Cu2-N7 | 1.985(5) |
| Cu1-I2 | 2.7907(16) | Cu2-I1 | 2.6179(10) |
| I2-I3 [b] | 3.0321(11) | I5-I6 [c] | 2.9291(10) |
| I3-I4 [b] | 2.8704(11) | I6-I7 [c] | 2.9448(10) |

**Table 3.** *Cont.*

| | | | |
|---|---|---|---|
| N8a-Cu1-N3 | 128.8(2) | N5-Cu2-N2 | 122.7(2) |
| N8a-Cu1-N1 | 103.2(2) | N4-Cu2-N5 | 98.1(2) |
| N3-Cu1-N1 | 77.7(2) | N4-Cu2-N2 | 80.0(2) |
| N8a-Cu1-N6a | 77.2(2) | N7-Cu2-N5 | 80.0(2) |
| N3-Cu1-N6a | 101.6(2) | N7-Cu2-N2 | 97.49(19) |
| N1-Cu1-N6a | 179.2(2) | N7-Cu2-N4 | 175.5(2) |
| N8a-Cu1-I2 | 123.84(17) | N5-Cu2-I1 | 122.10(16) |
| N3-Cu1-I2 | 107.29(16) | N2-Cu2-I1 | 115.23(15) |
| N1-Cu1-I2 | 89.59(17) | N4-Cu2-I1 | 90.99(17) |
| N6a-Cu1-I2 | 90.72(17) | N7-Cu2-I1 | 93.52(16) |
| I2-I3-I4 [b] | 176.71(3) | I5-I6-I7 [c] | 176.45(2) |

[a] Symmetry code: (a) $x$, $1 + y$, $z$. [b] Coordinated triiodide anion. [c] Uncoordinated triiodide anion.

The presence of an uncoordinated extra triiodide anion in the asymmetric unit indicates that the chain is cationic, meaning **2** must be a mixed valence copper(I/II) compound. Both metal ions in **2** are five-coordinate, with four N atoms of two bridging bpm molecules and a $I_3^-$ or $I^-$ anion building, in either case, a distorted {CuN4I} trigonal bipyramidal coordination arrangement (as confirmed by CShM [50]), where the triiodide (at Cu1) or iodide (at Cu2) anions occupies one of the three equatorial positions. Despite the striking likeliness of the two metal surroundings, the average Cu-N and Cu-I bond lengths are somewhat longer for Cu1 [2.176(6) and 2.791(2) Å, respectively] than for Cu2 [2.060(5) and 2.618(1) Å, respectively]. A closer look at the two coordination polyhedral shows that the two axial Cu-N bond lengths are longer than the two equatorial Cu-N bond lengths for Cu1, while it is the opposite for Cu2 (see Table 3). Both these findings suggest that Cu1 and Cu2 could be considered Cu(I) and Cu(II) centers [54,55], respectively, with crystallographic full occupancy.

The alternating Cu(I)/Cu(II) chains develop in the direction of the crystallographic $b$ axis. The equatorial planes of adjacent Cu(I) and Cu(II) polyhedra are essentially coplanar, but rotated by about 174° (see Figure 4). The two crystallographically independent bis-chelating bpm molecules are reasonably planar and they form a dihedral angle of 47.9(1)° between each other. The metal–metal distance through either the A bpm or the B bpm corresponds to about 5.64 Å, which results in Cu1···Cu1a and Cu2···Cu2a intra-chain separations of ca. 10.844(1) Å [(a) $x$, $1 + y$, $z$].

In the $ac$ plane, the chains are arranged according to a *quasi*-ideal square net (Figure 5a), whose edges correspond to the smallest inter-chain Cu1-Cu1 and Cu2-Cu2 distances [Cu1···Cu1b, 8.127(2) Å; Cu1···Cu1c, 9.898(1) Å; Cu2···Cu2c, 9.810(1) Å; Cu2···Cu2d, 8.232(1) Å; (b) $1 - x$, $2 - y$, $-0.5 + z$; (c) $0.5 - x$, $0.5 + y$, $0.5 + z$; (d) $1 - x$, $1 - y$, $0.5 + z$]. Among those, the shorter separations [Cu1···Cu1b and Cu2···Cu2d, ca. 8 Å] are associated with weak *off-set* $\pi$-$\pi$ stacking interactions established between an A bpm on one chain and a B bpm on the other (dihedral angle of 4.7°, interplanar distance of 3.4 Å); for the longer distances [Cu1···Cu1c and Cu2···Cu2c, ca. 10 Å], the aromatic interaction is even less relevant as the dihedral angle between the ligands would be 22° (Figure 5b).

Concerning the anions, the two crystallographically independent triiodide groups in **2** are structurally not equivalent. The uncoordinated anion is fairly linear (with a I5-I6-I7 angle of 176.45(2)°) and symmetrical (with I5-I6 and I6-I7 bond distances of 2.929(1) and 2.945(1) Å), while the Cu(I)-coordinated anion is linear (with a I2-I3-I4 angle of 176.71(3)°) but much less symmetrical (with I2-I3 and I3-I4 bond distances of 3.032(1) and 2.870(1) Å), as a consequence of coordination. They both participate in an extended, herringbone-like polyiodide 2D supramolecular network, with triiodide-triiodide contacts slightly below 4 Å, and weaker iodide-triiodide contacts in the range 4.2–4.3 Å. There is an apparent alternation of cationic (copper-bpm backbones) and anionic (coordinated iodide/coordinated and uncoordinated triiodide) regions in the crystal packing of **2**, along the crystallographic *c* axis (Figure 5a). Truly, the copper chains are well separated along the *a* and *c* axes, but not halfway between them, where they interpenetrate the iodide-triiodide net through the mentioned moderate to weak *off-set* $\pi$-$\pi$ contacts between bpm molecules placed above and beyond the anionic layer (Figure 5a,b).

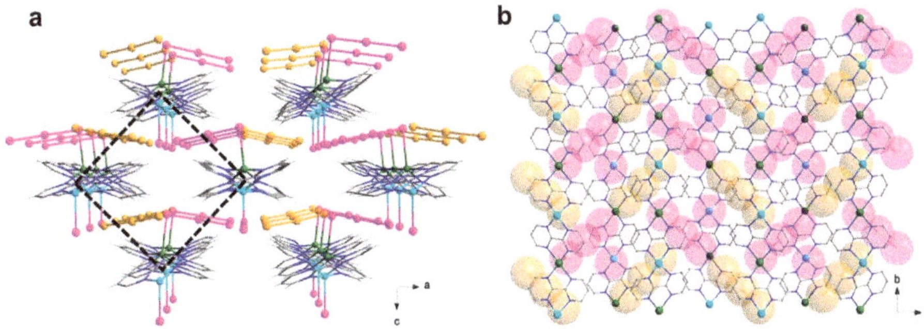


**Figure 5.** (**a**) Perspective view, along the crystallographic *b* axis, of a fragment of the crystal packing of **2**, highlighting the square arrangement of the copper(I/II) chains in the *ac* plane, and the apparent alternation between anionic and cationic regions along the *c* axis. (**b**) View of a fragment of the crystal packing of **2** along the *c* axis, showing the bpm-bpm contacts taking place through the 'holes' of the anionic layer (I atoms are depicted as van der Waals spheres). [Color codes: Cu(II), cyan; Cu(I), green; C, gray; N, blue; I(coordinated), pink; I(uncoordinated), gold].

### 3.2. Magnetic Properties

The magnetic properties of **1** and **2** in the form of the $\chi_M T$ vs. $T$ plot ($\chi_M$ being the magnetic susceptibility per formula unit) are shown in Figure 6. At room temperature, the $\chi_M T$ value of 0.41 cm$^3$ mol$^{-1}$ K is that expected for a magnetically isolated spin doublet from the 3d$^9$ Cu(II) ion ($S$ = 1/2). This value remains constant upon cooling until 5 (**1**) and 50 K (**2**); then, it decreases slightly to reach 40 (**1**) and 0.32 cm$^3$ mol$^{-1}$ K (**2**) at 2.0 K. No maximum in the susceptibility curve is observed in any case. The magnetic behavior of **1** is typical of a mononuclear copper(II) complex with almost negligible intermolecular antiferromagnetic interactions, while that of **2** is indicative of the occurrence of a very weak, but non-negligible, antiferromagnetic interaction between the paramagnetic Cu(II) ions, which can be due to both inter- or intrachain interactions. In both cases, however, the closest Cu(II)-Cu(II) distances are very large (10.84 and 8.23 Å for the shortest intra- and inter-chain distances, respectively).

Hence, the fit of the magnetic susceptibility data for **1** and **2** through the Curie–Weiss law [Equation (1) with $S$ = 1/2] gives $\theta$ = –0.08 (**1**) and –0.61 K (**2**) with $g$ = 2.11 (**1**) and 2.12 (**2**), where $\theta$ is the Weiss temperature and $g$ is the average Landé factor of the Cu$^{II}$ ion. The absolute $\theta$ value for **1** is certainly negligible ($|\theta| < 0.1$ K), while that of **2** is very small but non-negligible. This is in agreement with the $M$ vs. $H$ plots for **1** and **2** ($M$ being the molar magnetization per formula unit and $H$ the applied magnetic field) at 2.0 K (insets of Figure 6a,b). In fact, the isothermal magnetization curve exactly superposes for **1**, but slightly deviates for **2**, from the Brillouin function for one isolated doublet ($S$ = 1/2) state with $g$ = 2.11 (**1**) and 2.12 (**2**) obtained from the fit of the experimental data (solid line in the insets of Figure 6a,b).

$$\chi_M = [N\beta^2 g^2/3k(T-\theta)]S(S+1) \quad (1)$$

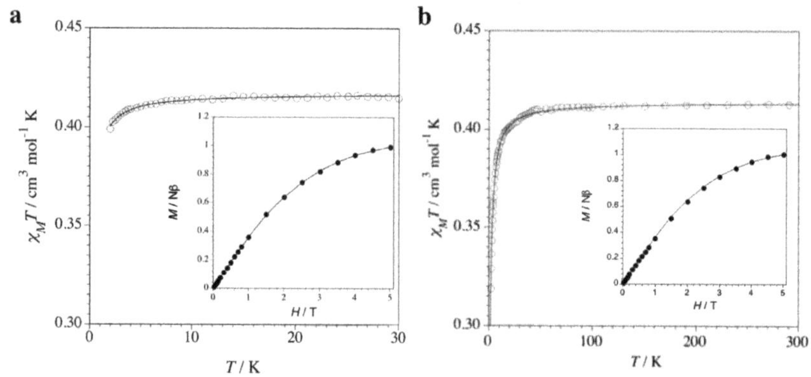


**Figure 6.** Temperature dependence of $\chi_M T$ for **1** (**a**) and **2** (**b**). The solid lines are the best-fit curves through Equations (1) and (2), respectively (see text). The insets show the field dependence of $M$ at 2.0 K. The solid lines are the simulated curves from the Brillouin function for an $S = 1/2$ with $g = 2.16$ (**1**) and 2.12 (**2**) (see text).

Because of the one-dimensional character of **2**, and given the diamagnetic nature of the $3d^{10}$ Cu(I) ions, from a magnetic point of view, we can describe compound **2** as an antiferromagnetic coupled uniform chain of spin doublets through the spin Hamiltonian $H = -J \sum_{i = 1-n} S_i \cdot S_{i+1}$. So, we can alternatively analyze the magnetic susceptibility data through the numerical expression proposed by Bonner and Fisher [Equation (2) with $x = |J|/kT$] [56].

$$\chi_M T = (N\beta^2 g^2/k)(0.25 + 0.074975x + 0.075235x^2)/(1.0 + 0.9931x + 0.172135x^2 + 0.757825x^3) \quad (2)$$

In such expression, $|J|$ is the absolute value of the exchange coupling parameter between adjacent Cu(II) ions into the chain, $g$ is the average Landé factor and $N$, $\beta$ and $k$ have their usual meaning. In the case of **2**, neglecting the inter-chain interactions, least-squares fit of the magnetic susceptibility to Equation (2) leads to $J = -0.650(3)$ cm$^{-1}$ and $g = 2.120(5)$. The calculated plot matches well the experimental data in the whole temperature range explored. The absolute $J$ value is very small, as expected for such a long bridge: N-C-N(bpm)-Cu(I)-N-C-N(bpm). In fact, the antiferromagnetic exchange interactions between the unpaired electrons occupying the 3d($z^2$)-type magnetic orbitals of the paramagnetic trigonal bipyramidal Cu(II) ions, and partially delocalized on the N atoms of one of the two pyrimidine rings of the bpm bridges, are expected to be very weak, although not negligible, because of their small orbital overlap through the long orbital pathway involving the fully occupied 3d($z^2$)-type orbital of appropriate symmetry of the diamagnetic trigonal bipyramidal Cu(I) ions (Scheme 1a). Interestingly, a different magnetic behavior would result from a square pyramidal coordination environment at the metal atoms along the bipym-bridged mixed-valent copper(I)–copper(II) chain (Scheme 1b). In this hypothetical case, a stronger orbital overlap would occur between the unpaired electrons occupying the 3d($x^2$–$y^2$)-type magnetic orbitals of the paramagnetic square pyramidal Cu(II) ions, and largely delocalized on the four N atoms of the two pyrimidine rings of the bpm bridges, across the orbital pathway involving the fully occupied 3d($x^2$–$y^2$)-type orbital of appropriate symmetry of the diamagnetic square bipyramidal Cu(I) ions. This so-called "orbital reversal" phenomenon would instead lead to a moderate antiferromagnetic intrachain coupling.

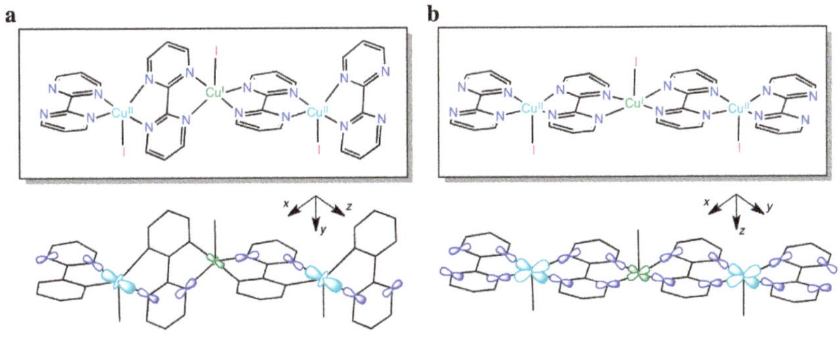

**Scheme 1.** Illustration of the intra-chain orbital pathway for the electron exchange magnetic interactions operating in **2** for the actual trigonal bibyramidal (**a**) or hypothetical square pyramidal (**b**) coordination environment at the metal atoms along the bipym-bridged mixed-valent copper(I)–copper(II) chain, showing the orientation of the molecular axes centered on each metal atom.

To our knowledge, there are a few examples of polymetallic complexes with a non-negligible, weak to moderate magnetic coupling, either antiferro- or ferromagnetic, between paramagnetic metal centers (e.g., $Cu^{II}$ or $Cr^{III}$) mediated by a diamagnetic metal ion (e.g., $Zn^{II}$, $Pd^{II}$, or $Nb^{V}$) [57–62].

## 4. Discussion

Cu(I) and Cu(I)-Cu(II) mixed-valent complexes have been generally obtained by using Cu(I) as precursor, while few examples are reported for the in situ total or partial reduction of Cu(II) to Cu(I) [63,64]. For instance, it was shown that pyridine derivatives under hydrothermal conditions can reduce the starting Cu(II) material to afford mixed-valent Cu(I)-Cu(II) species [65,66]. In general, the in situ generation of reagents is an attractive topic of coordination chemistry [63,64], especially if mild synthetic conditions can be used and, above all, when compounds whose preparation would be hampered otherwise become easily accessible instead.

In this work, we have explored the copper/2,2′-bipyrimidine/iodide system, in aqueous media and at room temperature, with the intent of combining the well-established reducing properties of the iodide ion towards copper(II) in water with the protective effect of ligand coordination, to make the metal oxidation state tunable at our wish. Our approach afforded systematically, with 100% reproducibility, compounds **2** and **3**, at first together, and later as the only products, thanks to an extended optimization of the crystallization procedure. Our findings can be at least partially rationalized as follows.

When the reduction of Cu(II) to Cu(I) by iodide ions is conducted in dilute aqueous solution conditions and in defect of iodide, the formation of the triiodide anion (from $2Cu^{2+} + 4I^{-} \rightarrow 2\ CuI_{(s)} + I_2$ first, and then $I_2 + I^{-} \rightarrow I_3{}^{-}$) is not competitive and, in the presence of the bpm ligand, the fully reduced compound **3** can be easily accessed, not just as microcrystalline precipitate, but rather as X-ray quality crystals, and in optimum yield (cfr. Ref. [38] for a previously described synthesis of **2** directly from CuI and bpm in acetonitrile or mixed water/acetonitrile/acetone media). When the formation of triiodide anions is not sufficiently hampered synthetically, they do form, consistently, and, in the presence of the bpm ligand, the reduction of the Cu(II) ions does not proceed to completeness. In this case, the mixed-valent species **2**, whose crystal structure comprises two triiodide anions every Cu(I/II) metal pair, can be afforded instead. The peculiarities of compound **2** are numerous. First, the five-coordination geometry is not very common for the copper(I) ion; at present, it occurs in about the1.3% of all copper(I) compounds deposited in the Cambridge Structural Database (CCDC 2024; ConQuest version 2024.3.0), and the percentage drastically lowers if we exclude those comprising tetra- or penta-dentate ligands. More importantly, to the best of our knowledge, **2** represents the first example of structurally characterized alternating Cu(I)/Cu(II) chain, where the two metal ions exhibit the same coordination environment and a strikingly similar coordination chromophore [$N_4I$]. Let us now focus on complex **1**. Although the metal ion in **1** is in its +2 oxidation state, the presence of triiodide as co-counteranion requires that a partial Cu(II) $\rightarrow$ Cu(I) reduction process must happen in the reaction vessel. Thus, while **2** could be considered, in some way, a solid-state reaction intermediate, **1** cannot be considered a 'precursor' of compounds **2** and **3**. We were never able to form **1** in pure water, even after extensive optimization of the synthetic parameters. Rather, the compound was obtained from acetone, only after the formation of **2** and **3** as insoluble powders. This suggests that, when the reaction media is not ideal for the slow self-assembly of the mixed valent, Cu(I)/Cu(II) chain of **2**, complex **1** becomes accessible, thanks to the simultaneous presence in solution of an excess of the bpm ligand as well as of residual iodide and triiodide anions. Interestingly, several attempts to intentionally form a tris(bpm)copper(II) dication with all sort of different counteranions, made over years also by us, were never successful. Thus, while the [bis(bpm)copper(II)$H_2O]^{2+}$ species is somewhat known [14,15], and the analog tris(2,2′-bipyridine)copper(II) species are well documented [67–69], to the best of our knowledge the complex cation in **1** represents the first tris(bpm)copper(II) species reported to date.

## 5. Conclusions

In this work, we have demonstrated the possibility to obtain three different products, i.e., the complexes of formulae $[Cu^{II}(bpm)_3](I_3)(I)$ (**1**), $\{[Cu^{I}(I_3)Cu^{II}(I)(bpm)_2](I_3)\}_n$ (**2**) and $\{[Cu^{I}_2(\mu\text{-}I)_2(bpm)]\}_n$ (**3**), by using copper(II)nitrate hexahydrate, 2,2′-bipyrimidine (bpm) and potassium iodide as starting reagents, in mild conditions, in reagent grade acetone or aqueous media. To our knowledge, Cu(I)- or mixed Cu(I)/Cu(II)-halides frameworks, prepared by the in situ reduction of Cu(II) precursors with the bpm ligand and subsequent self-assembly, have not been reported to date. Compound **2** was of the most interesting nature. X-ray single crystal analysis revealed for it the occurrence of two crystallographically distinct copper centers, regularly alternating within a $\mu$-bpm bridged chain structure, charge balanced by three monovalent anions (two of which coordinated). Despite both metal centers having *quasi*-identical pentacoordination surroundings, it was possible to assign undoubtedly formal +1 oxidation state to one copper ion and +2 to the other, based on the Cu-N and Cu-I bond distances. The magnetic analysis of **2** revealed a very weak antiferromagnetic coupling between the Cu(II) ions, which can be due to both inter- or intrachain interactions [the intrachain Cu(II)-Cu(II) separation, through the diamagnetic Cu(I) center, is 10.8435(3) Å; the shortest interchain Cu(II)-Cu(II) distance is 8.232(1) Å, and it is associated with moderate *off-set* $\pi$-$\pi$ stacking interactions between bpm ligands]. In either case, the magnetic measurements were in agreement with the hypothesis, based on the structural observations, that in **2** the Cu(I) and Cu(II) metal sites alternate regularly; i.e., the Cu(II) ions are well isolated from each other's. Compound **2** can be regarded as a solid-state reaction intermediate. The numerous I···I contacts established between the coordinated and uncoordinated iodide and triiodide anions reasonably account for its great stability and striking synthetic reproducibility.

**Author Contributions:** N.M. and G.D.M. conceived the project and designed the experiments; N.M. synthesized and characterized the complexes, and carried out the crystallographic study; F.L. and M.J. performed the magnetic study; N.M., G.D.M., F.L. and M.J. discussed the results and agreed on their interpretation, exactly as discussed herein; N.M., G.D.M., F.L. and M.J. wrote the manuscript (original draft); N.M. and G.D.M. revised and edited the manuscript. All authors have read and agreed to the published version of the manuscript, except for M.J., who sadly is no longer with us.

**Funding:** This work was supported by the Italian MUR through the ElectroLight4Value project (PRIN 2020, 2020927WY3).

**Data Availability Statement:** The data used in this study are included in the article. Further inquiries can be directed to the corresponding author.

**Acknowledgments:** Special thanks are due to Rafael Ruiz-García (ICMol, Valencia) for his unselfish support with the magnetic measurements on compound **1** and with the revision of Section 3.2.

**Conflicts of Interest:** The authors declare no conflicts of interest. The funders had no role in the design of the study; in the collection, analyses, or interpretation of data; in the writing of the manuscript; or in the decision to publish the results.

## References

1. Ferrando-Soria, J.; Vallejo, J.; Castellano, M.; Martínez-Lillo, J.; Pardo, E.; Cano, J.; Castro, I.; Lloret, F.; Ruiz-García, R.; Julve, M. Molecular magnetism, quo vadis? A historical perspective from a coordination chemist viewpoint. *Coord. Chem. Rev.* **2017**, *339*, 17–103. [CrossRef]
2. Marinescu, G.; Andruh, M.; Lloret, F.; Julve, M. Bis(oxalato)chromium(III) complexes: Versatile tectons in designing heterometallic coordination compounds. *Coord. Chem. Rev.* **2011**, *255*, 161–185. [CrossRef]
3. Martínez-Lillo, J.; Faus, J.; Lloret, F.; Julve, M. Towards multifunctional magnetic systems through molecular-programmed self-assembly of Re(IV) metalloligands. *Coord. Chem. Rev.* **2015**, *289–290*, 215–237. [CrossRef]
4. Castro, I.; Barros, W.P.; Calatayud, M.L.; Lloret, F.; Marino, N.; De Munno, G.; Stumpf, H.O.; Ruiz-García, R.; Julve, M. Dicopper(II) pyrazolenophanes: Ligand effects on their structures and magnetic properties. *Coord. Chem. Rev.* **2016**, *315*, 135–152. [CrossRef]
5. Castro, I.; Calatayud, M.L.; Orts-Arroyo, M.; Marino, N.; De Munno, G.; Lloret, F.; Ruiz-García, R.; Julve, M. Oxalato as polyatomic coordination center and magnetic coupler in copper(II)-polypyrazole inverse polynuclear complexes and coordination polymers. *Coord. Chem. Rev.* **2022**, *471*, 214730–214737. [CrossRef]
6. De Munno, G.; Julve, M.; Nicolò, F.; Lloret, F.; Faus, J.; Ruiz, R.; Sinn, E. 2,2′-Bipyrimidineoxalatocopper(II) Complexes: From the Mononuclear Complex to the 2D Sheetlike Polymer. *Angew. Chem. Int. Ed. Engl.* **1993**, *32*, 613–615. [CrossRef]
7. Julve, M.; Lloret, F.; Faus, J.; De Munno, G.; Verdaguer, M.; Caneschi, A. Alternating Ferro- and Antiferromagnetic Interactions in a Chainlike $Cu^{II}$ Coordination Polymer. *Angew. Chem. Int. Ed. Engl.* **1993**, *32*, 1046–1048. [CrossRef]
8. Real, J.; De Munno, G.; Chiappetta, R.; Julve, M.; Lloret, F.; Journaux, Y.; Colin, J.-C.; Blondin, G. Structural and Magnetic Characterization of a Novel Heptanuclear Hydroxo-Bridged Copper(II) Cluster of the Dicubane Type. *Angew. Chem. Int. Ed. Engl.* **1994**, *33*, 1184–1186. [CrossRef]
9. De Munno, G.; Julve, M.; Viau, G.; Lloret, F.; Faus, J.; Viterbo, D. Azido and 2,2′-Bipirimidine Ligands as Useful Tools in Designing Two- and Three-Dimensional Manganese(II) Networks. *Angew. Chem. Int. Ed. Engl.* **1996**, *35*, 1807–1810. [CrossRef]
10. De Munno, G.; Poerio, T.; Viau, G.; Julve, M.; Lloret, F. Ferromagnetic Coupling in the Bis($\mu$-end-on-azido)iron(III) Dinuclear Complex Anion of $[Fe^{II}(bpym)_3]_2[Fe^{III}_2(N_3)_{10}]\cdot 2H_2O$. *Angew. Chem. Int. Ed. Engl.* **1997**, *36*, 1459–1461. [CrossRef]
11. Lloret, F.; De Munno, G.; Julve, M.; Cano, J.; Ruiz, R.; Caneschi, A. Spin Polarization and Ferromagnetism in 2D Sheetlike Cobalt(II) Polymers: $[Co(L)_2(SCN)_2]$ (L = Pyrimidine or Pyrazine). *Angew. Chem. Int. Ed. Engl.* **1998**, *37*, 135–138. [CrossRef]
12. De Munno, G.; Bruno, G. Catena-poly{$\mu$-(2,2′-bypirimidine-*N,N′,N″,N‴*)-[(nitrato-O,O′)copper(II)]-di-($\mu$-nitrato-$\mu$-O)-[(nitrato-*O,O′*)copper(II)]}, $[Cu_2(NO_3)_4(C_8H_6N_4)]$. *Acta Crystallogr.* **1984**, *40*, 2030–2032. [CrossRef]
13. Julve, M.; De Munno, G.; Bruno, G.; Verdaguer, M. Synthesis, Structure and Magnetic Properties of $\mu$-(2,2′-Bipyrimidine-*N,N′,N″,N‴*) Copper(II) Complexes. *Inorg. Chem.* **1988**, *27*, 3160–3165. [CrossRef]
14. De Munno, G.; Bruno, G.; Julve, M.; Romeo, M. Synthesis and Structure of Aquabis(2,2′-bipyrimidine-*N,N′*)copper(II) hexafluorophosphate Dihydrate. *Acta Crystallogr.* **1990**, *46*, 1828–1830. [CrossRef]
15. De Munno, G.; Julve, M.; Verdaguer, M.; Bruno, G. Bonding Flexibility of 2,2′-Bipyrimidine (bpm): Symmetry and Magnetism of Three Copper(II) Complexes with Different Cu:bpm Ratios. *Inorg. Chem.* **1993**, *32*, 2215–2220. [CrossRef]
16. De Munno, G.; Julve, M.; Lloret, F.; Cano, J.; Caneschi, A. Magneto-Structural Effects of the Jahn-Teller Distortions on 2,2′-Bipyrimidine (bpm)-Bridged Dinuclear Copper(II) Complexes: Crystal Structures and Magnetic Properties of $[Cu_2(bpm)(H_2O)_4(SO_4)_2]\cdot 3H_2O$ and $[Cu_2(bpm)(H_2O)_8](SO_4)_2\cdot 2H_2O$. *Inorg. Chem.* **1995**, *34*, 2048–2053. [CrossRef]
17. De Munno, G.; Poerio, T.; Viaux, G.; Julve, M.; Lloret, F.; Journaux, J.; Rivière, E. New magnetic behaviour of honeycomb layered compounds. Crystal structure of [M2(bipym)(N3)] (M = CoII, FeII; bipym = 2,2′-bipyrimidine). *Chem. Commun.* **1996**, *22*, 2587–2588. [CrossRef]
18. De Munno, G.; Julve, M.; Real, J.A. Synthesis and crystal structure of the low-spin iron(II) complex $[Fe(bpym)_3](ClO_4)_2\cdot 1/4H_2O$. *Inorganica Chim. Acta* **1997**, *255*, 185–188. [CrossRef]

19. De Munno, G.; Poerio, T.; Julve, M.; Lloret, F.; Faus, J.; Caneschi, A. Syntheses, crystal structures and magnetic properties of one-, two- and three-dimensional 2,2′-bipyrimidine-containig copper(II) complexes. *J. Chem. Soc. Dalton Trans.* **1998**, *10*, 1679–1685. [CrossRef]
20. De Munno, G.; Ventura, W.; Viau, G.; Lloret, F.; Faus, J.; Julve, M. Structural Characterization and Magnetic Properties of the First 2,2′-Bipyrimidine (bpm)-containing Iron(III) Complexes. *Inorg. Chem.* **1998**, *37*, 1458–1464. [CrossRef]
21. De Munno, G.; Armentano, D.; Julve, M.; Lloret, F.; Lescouëzec, R.; Faus, J. Two-dimensional Assembling of (2,2′-Bipirimidine)bis(oxalato)chromate(III) Units through Alkaline Cations. *Inorg. Chem.* **1999**, *38*, 2234–2237. [CrossRef] [PubMed]
22. Armentano, D.; De Munno, G.; Lloret, F.; Julve, M. *Inorg. Chem.* Novel Three-Dimensional Cage Assembly of a $\mu_4$-Carbonato Bridged Cobalt(II) Compound $[Co_2(bpm)(H_2O)_2(CO_3)(OH)]NO_3{\cdot}4H_2O$. *Inorg. Chem.* **1999**, *38*, 3744–3747. [CrossRef] [PubMed]
23. Chiozzone, R.; Gonzáles, R.; Kremer, C.; Cerdá, M.F.; Armentano, D.; De Munno, G.; Martínez-Lillo, J.; Faus, J. A novel series of rhenium-bipyrimidine complexes: Synthesis, crystal structure and electrochemical properties. *Dalton Trans.* **2007**, *6*, 653–660. [CrossRef] [PubMed]
24. Martínez-Lillo, J.; Armentano, D.; De Munno, G.; Cano, J.; Lloret, F.; Julve, M.; Faus, J. First Magnetostructural Study on a Heterodinuclear 2,2′-Bipyrimidine-Bridged Complex. *Inorg. Chem.* **2011**, *50*, 12405–12407. [CrossRef]
25. Marino, N.; Armentano, D.; De Munno, G.; Cano, J.; Lloret, F.; Julve, M. Synthesis, Structure, and Magnetic Properties of Regular Alternating μ-bpm/di-μ-X Copper (II) Chains (bpm= 2, 2′-bipyrimidine; X= OH, F). *Inorg. Chem.* **2012**, *51*, 4323–4334. [CrossRef]
26. Marino, N.; Armentano, D.; De Munno, G.; Lloret, F.; Cano, J.; Julve, M. Toward a better understanding of honeycomb alternating magnetic networks. *Dalton Trans.* **2015**, *44*, 11040–11051. [CrossRef]
27. Martínez-Lillo, J.; Armentano, D.; Fortea-Pérez, F.R.; Stiriba, S.-E.; De Munno, G.; Lloret, F.; Julve, M.; Faus, J. A Chiral, Photoluminescent and Spin-Canted $\{Cu^{I}Re^{IV}{}_2\}_n$ Branched Chain. *Inorg. Chem.* **2015**, *54*, 4594–4596. [CrossRef]
28. Marino, N.; Mastropietro, T.F.; Armentano, D.; De Munno, G.; Doyle, R.P.; Lloret, F.; Julve, M. Spin canting in an unprecedented three-dimensional pyrophosphate- and 2,2′-bipyrimidine-bridged cobalt(II) framework. *Dalton Trans.* **2008**, *54*, 5152–5154. [CrossRef]
29. Martínez-Lillo, J.; Mastropietro, T.F.; Lappano, R.; Madeo, A.; Alberto, M.E.; Russo, N.; Maggiolini, M.; De Munno, G. Rhenium(IV) compounds inducing apoptosis in cancer cells. *Chem. Commun.* **2011**, *47*, 5283–5285. [CrossRef]
30. Betanzos-Lara, S.; Novakova, O.; Deeth, R.J.; Pizarro, A.M.; Clarkson, G.J.; Liskova, B.; Bravec, V.; PJSadler, P.J.; Habtemariam, A. Bipyrimidine ruthenium(II) arene complexes: Structure, reactivity and cytotoxicity. *J. Biol. Inorg. Chem.* **2012**, *17*, 1033–1051. [CrossRef]
31. Kojima, T.; Morimoto, T.; Miyazaki, S.; Fukuzumi, S. Ruthenium(II) Pyridylamine Complexes with Diimine Ligands Showing Reversible Photochemical and Thermal Structural Change. *Chem Eur. J.* **2008**, *14*, 8904–8915. [CrossRef] [PubMed]
32. Ishizuka, T.; Itogawa, M.; Shimomura, H.; Shiota, Y.; Kotani, H.; Yoshizawa, K.; Kojima, T. Redox properties of a bipyrimidine-bridged dinuclear ruthenium(II) complex. *Inorg. Chem. Commun.* **2020**, *120*, 108150. [CrossRef]
33. Henke, W.C.; Kerr, T.A.; Sheridan, T.R.; Henling, L.M.; Takase, M.K.; Day, V.W.; Gray, H.B.; Blakemore, J.D. Synthesis, structural studies, and redox chemistry of bimetallic $[Mn(CO)_3]$ and $[Re(CO)_3]$ complexes. *Dalton Trans.* **2021**, *50*, 2746–2756. [CrossRef] [PubMed]
34. Kaim, W.; Dogan, A.; Wanner, M.; Klein, A.; Tiritiris, I.; Schleid, T.; Stufkens, D.J.; Snoeck, T.L.; McInnes, E.J.L.; Fiedler, J.; et al. Reduced and Excited States of (bpym)$[PtCl_2]_n$ (bpym = 2,2′-Bipyrimidine; $n$ = 1, 2): Experiments and DFT Calculations. *Inorg. Chem.* **2002**, *41*, 4139–4148. [CrossRef]
35. Real, J.A.; Gaspar, A.B.; Muñoz, M.C.; Gütlich, P.; Ksenofontov, V.; Spiering, H. Bipyrimidine-bridged dinuclear iron (II) spin crossover compounds. *Top. Curr. Chem.* **2004**, *233*, 167–193. [CrossRef]
36. Vogler, C.; Hausen, H.-D.; Kaim, W.; Kohlmann, S.; Kramer, H.E.A.; Rieker, J. Copper(I)-Assisted Formation of an "Organic" Sandwich Structure: Structural Prerequisites for Luminescence of the Dinuclear Complexes $[(\mu\text{-Bipyrimidine})\{Cu(PR_3)_2\}_2]X_2$. *Angew. Chem. Int. Ed.* **1989**, *28*, 1659–1660. [CrossRef]
37. Sieger, M.; Vogler, C.; Klein, A.; Knödler, A.; Wanner, M.; Fiedler, J.; Záliš, S.; Snoeck, T.I.; Kaim, W. Geometrical and Electronic Structures of Dinuclear Complex Ions $\{(\mu\text{-bpym})[Cu(EAr_3)_2]_2\}^{2+}$ with Intramolecular "Organic Sandwich" Formation (E = P or As; Ar = Aryl; bpym = 2,2′-Bipyrimidine). *Inorg. Chem.* **2005**, *44*, 4637–4643. [CrossRef]
38. Hou, F.; Powell, M.; Dougherty, D.B.; Sommer, R.D.; Maggard, P.A. Tunable Optical and Photocatalytic Properties of Low-Dimensional Copper(I)-Iodide Hybrids Using Coordinating Organic Ligands. *Cryst. Growth Des.* **2018**, *18*, 5406–5416. [CrossRef]
39. Decurtins, S.; Schmalle, H.W.; Schneuwly, P.; Zheng, L.-M.; Ensling, J.; Hauser, A. Synthesis, Crystal Structure, Optical and Magnetic Properties of a Novel Two-Dimensional Copper(II) Network formed conjointly with $\mu$-Bipyrimidine, $\mu$-Oxalato, and $\mu$-Chloro Ligands. *Inorg. Chem.* **1995**, *34*, 5501–5506. [CrossRef]
40. Ikotun, O.F.; Ouellette, W.; Lloret, F.; Julve, M.; Doyle, R.P. Synthesis, X-ray Structure, Thermal and Magnetic Behavior of $[(bipy)_2Ni_2(\mu\text{-}Cl)_2Cl_2(H_2O)_2]$: The First Neutral Ferromagnetically Coupled Six-Coordinate Dichlorido-Bridged Nickel(II) Dimer. *Eur. J. Inorg. Chem.* **2007**, *2007*, 2083–2088. [CrossRef]

41. Julve, M.; Verdaguer, M.; De Munno, G.; Real, J.A.; Bruno, G. Synthesis, crystal structure, and magnetic properties of ($\mu$-bipyrimidine)(cyanato)copper(II) and -(thiocyanato)copper(II) complexes. *Inorg. Chem.* **1993**, *32*, 795–802. [CrossRef]
42. Cortés, R.; Lezama, L.; Rojo, T.; Arriortua, M.I.; Pizarro, J.L.; Solans, X. Alternating Ferro- and Antiferromagnetic Interactions in a $Mn^{II}$ Chain with Alternating End-On and End-to-End Bridging Azido Ligands. *Angew. Chem. Int. Ed. Engl.* **1994**, *33*, 2488–2489. [CrossRef]
43. Cortés, R.; Drillon, M.; Solans, X.; Lezama, L.; Rojo, T. Alternating Ferromagnetic−Antiferromagnetic Interactions in a Manganese(II)−Azido One-Dimensional Compound: $[Mn(bipy)(N_3)_2]$. *Inorg. Chem.* **1997**, *36*, 677–683. [CrossRef]
44. Viau, G.; Lombardi, M.G.; De Munno, D.; Julve, M.; Lloret, F.; Faus, J.; Caneschi, A.; Clemente-Juan, J.M. The azido ligand: A useful tool in designing chain compounds exhibiting alternating ferro- and antiferro-magnetic interactions. *Chem. Commun.* **1997**, *13*, 1195–1196. [CrossRef]
45. Armentano, D.; Sanchis-Perucho, A.; Rojas-Dottib, C.; Martínez-Lillo, J. Halogen halogen interactions in the self-assembly of one-dimensional 2,2′-bipyrimidine-based $Cu^{II}Re^{IV}$ systems. *CrystEngComm* **2018**, *20*, 4575–4581. [CrossRef]
46. *SAINT*, version 6.45; Bruker Analytical X-Ray Systems Inc.: Madison, WI, USA, 2003.
47. *SADABS*, version 2.03; Bruker AXS Inc.: Madison, WI, USA, 2000.
48. Sheldrick, G.M. Crystal structure refinement with SHELXL. *Acta Cryst.* **2015**, *71*, 3–8. [CrossRef]
49. Diamond (version 4.6.8)—Crystal and Molecular Structure Visualization, Crystal Impact—Dr. H. Putz & Dr. K. Brandenburg GbR, Kreuzherrenstr. 102, 53227 Bonn, Germany. Available online: https://www.crystalimpact.de/diamond (accessed on 30 January 2025).
50. *SHAPE*, version 2.1; Electronic Structure Group, Universitat de Barcelona: Barcelona, Spain, 2013.
51. Blake, A.J.; Li, W.-S.; Lippolis, V.; Schröder, M.; Devillanova, F.A.; Gould, R.O.; Parsons, S.; Radek, C. Template self-assembly of polyiodide networks. *Chem. Soc. Rev.* **1998**, *27*, 195–206. [CrossRef]
52. Svensson, P.H.; Kloo, L. Synthesis, Structure, and Bonding in Polyiodide and Metal Iodide-Iodine Systems. *Chem. Rev.* **2003**, *103*, 1649–1684. [CrossRef]
53. van Megen, M.; Reiss, G.J. $I_6^{2-}$ Anion Composed of Two Asymmetric Triiodide Moieties: A Competition between Halogen and Hydrogen Bond. *Inorganics* **2013**, *1*, 3–13. [CrossRef]
54. Kulkarni, P.; Padhye, S.; Sinn, E.; Anson, C.E.; Powell, A.K. Comparative studies on copper(I) complexes: Synthesis, X-ray crystallography and electrochemical properties of [CuI (dafone)nX] complexes (dafone=/4,5-diaza-fluoren-9-one, X=/Br, I, SCN). *Inorg. Chim. Acta* **2002**, *332*, 167–175. [CrossRef]
55. Horn, C.; Ali, B.; Dance, I.; Scudder, M.; Craig, D. Crystal supramolecularity: Extended aryl embraces in dimorphs of [Cu(1,10-phen)2I]I3. *CrystEngComm* **2000**, *2*, 6–15. [CrossRef]
56. Kahn, O. *Molecular Magnetism*; VCH Publishers: New York, NY, USA, 1993.
57. Ruiz, R.; Julve, M.; Faus, J.; Lloret, F.; Muñoz, M.C.; Journaux, Y.; Bois, C. Ferromagnetic Coupling between Copper(II) Centers through the Diamagnetic Zinc(II) Ion: Crystal Structure and Magnetic Properties of $[Cu_2Zn(Hdmg)_2(dmg)_2(H_2O)]\cdot 0.5H_2dmg\cdot H_2O$ ($H_2dmg$ = Dimethylglyoxime). *Inorg. Chem.* **1997**, *36*, 3434–3439. [CrossRef] [PubMed]
58. Buvaylo, E.A.; Kokozay, V.N.; Vassilyeva, O.Y.; Skelton, B.W., Jezierska, J.; Brunel, L.C.; Ozarowski, A. Ozarowski. A Cu–Zn–Cu–Zn heterometallomacrocycle shows significant antiferromagnetic coupling between paramagnetic centres mediated by diamagnetic metal. *Chem. Commun.* **2005**, *39*, 4976–4978. [CrossRef] [PubMed]
59. Semenaka, V.V.; Nesterova, O.V.; Kokozay, V.N.; Dyakonenko, V.V.; Zubatyuk, R.I.; Shishkin, O.V.; Boča, R.; Jezierska, J.; Ozarowski, A. Ozarowski. $Cr^{III}-Cr^{III}$ Interactions in Two Alkoxo-Bridged Heterometallic $Zn_2Cr_2$ Complexes Self-Assembled from Zinc Oxide, Reinecke's Salt, and Diethanolamine. *Inorg. Chem.* **2010**, *49*, 5460–5471. [CrossRef]
60. Oliveira, W.X.; Ribeiro, M.A.; Pinheiro, C.B.; da Costa, M.M.; Fontes, A.P.S.; Nunes, W.C.; Cangussu, D.; Julve, M.; Stumpf, H.O.; Pereira, C.L. Palladium(II)–Copper(II) Assembling with Bis(2-pyridylcarbonyl)amidate and Bis(oxamate) Type Ligands. *Cryst. Growth Des.* **2015**, *15*, 1325–1335. [CrossRef]
61. Oliveira, W.X.C.; Ribeiro, M.A.; Pinheiro, C.B.; da Costa, M.M.; Fontes, A.P.S.; Nunes, W.C.; Cangussu, D.; Julve, M.; Stumpf, H.O.; Pereira, C.L.M. Experimental and Theoretical Investigation of the Anti-Ferromagnetic Coupling of $Cr^{III}$ Ions through Diamagnetic $-O-Nb^{V}-O-$ Bridges. *Inorg. Chem.* **2017**, *56*, 6879–6889. [CrossRef]
62. Jurić, M.; Dubraja, L.A.; Popović, J.; Molčanov, K.; Torić, F.; Pajić, D.; Lončarić, I. From a square core to square opening: Structural diversity and magnetic properties of the oxo-bridged $[Cr^{III}Nb^{V}]$ complexes. *Dalton Trans.* **2018**, *47*, 4183–4190. [CrossRef]
63. Lu, J.Y. Crystal engineering of Cu-containing metal–organic coordination polymers under hydrothermal conditions. *Coord. Chem. Rev.* **2003**, *246*, 327–347. [CrossRef]
64. Zhang, J.-P.; Zheng, S.-L.; Huang, X.-C.; Chen, X.-M. Two Unprecedented 3-Connected Three-Dimensional Networks of Copper(I) Triazolates: In Situ Formation of Ligands by Cycloaddition of Nitriles and Ammonia. *Angew. Chem. Int. Ed.* **2004**, *43*, 206–209. [CrossRef]

65. Tao, J.; Zhang, Y.; Tong, M.-L.; Chen, X.-M.; Yuen, T.; Lin, C.L.; Huang, X.; Li, J. A mixed-valence copper coordination polymer generated by hydrothermal metal/ligand redox reactions. *Chem. Commun.* **2002**, *13*, 1342. [CrossRef]
66. Zhang, X.-M.; Tong, M.-L.; Chen, X.-M. Hydroxylation of N-Heterocycle Ligands Observed in Two Unusual Mixed-Valence $Cu^{I}/Cu^{II}$ Complexes. *Angew. Chem. Int. Ed.* **2002**, *41*, 1029–1031. [CrossRef]
67. Chamayou, J.A.A.-C.; Biswas, C.; Janiak, C.; Ghosh, A. Tris(2,2′-bipyridine-κ$^2$N,N′)copper(II) bis(tetrafluoridoborate). *Acta Cryst.* **2007**, *63*, m1936–m1937. [CrossRef]
68. Xu, F.; You, W.; Huang, W. Tris(2,20-bipyridyl-j2 N,N0)copper(II) sulfate 7.5-hydrate. *Acta Cryst.* **2009**, *5*, m129–m130. [CrossRef]
69. Murphy, B.; Aljabri, M.; Ahmed, A.M.; Murphy, G.; Hathaway, B.J.; Light, M.E.; Geilbrich, T.; Hursthouse, M.B. Structural systematics of the $[Cu(chelate)_3][Y]_2$ series. An interesting crystallographic structural insight involving vibronic coupling and the Jahn–Teller effect (JTE). The syntheses and low temperature crystal structures of tris(2,2′bipyridyl)copper(ii) tetraphenylborate and tris(2,2′bipyridyl)zinc(ii) tetraphenylborate. *Dalton Trans.* **2006**, *2*, 357–367. [CrossRef]

magnetochemistry

MDPI

*Article*

# Slow Relaxation of Magnetization and Magnetocaloric Effects in One-Dimensional Oxamato-Based Lanthanide(III) Coordination Polymers †

**Jhonny W. Maciel [1], Lucas H. G. Kalinke [2], Renato Rabelo [1], Meiry E. Alvarenga [1], Felipe Terra Martins [1], Nicolás Moliner [3] and Danielle Cangussu [1,*]**

1 Instituto de Química, Universidade Federal de Goiás, 74690-900 Goiânia, GO, Brazil; jhonny_willians@ufg.br (J.W.M.); renato_rabelo@ufg.br (R.R.); meiryalvarenga@yahoo.com.br (M.E.A.); felipe@ufg.br (F.T.M.)

2 Instituto Federal de Goiás—IFG, Campus Anápolis, 75131-457 Anápolis, GO, Brazil; lucas.kalinke@ifg.edu.br

3 Instituto de Ciencia Molecular (ICMol), Universitat de València, 46980 Paterna, Spain; fernando.moliner@uv.es

* Correspondence: danielle_cangussu@ufg.br

† We dedicate this article to Miguel Julve and Francisco Lloret, two exceptional professors and researchers whose brilliant careers were devoted to exploring coordination chemistry and studying magnetic properties. Their passion for guiding students and nurturing colleagues around the world stands as the biggest testament to their generosity and dedication. We are deeply grateful for the priceless knowledge and wisdom they shared over more than 20 years of collaboration. In loving memory of Miguel, our dear friend and colleague, whose legacy will continue to inspire us all.

**Abstract:** Herein, we present the synthesis and characterization of a series of isostructural lanthanide(III) compounds with the *N*-(4-carboxyphenyl)oxamic acid ($H_3$pcpa) ligand of the general formula as $\{[Ln_2(Hpcpa)_3(H_2O)_5]\}_n$ [Ln = Dy(III) **1**, Ho(III) **2**, Er(III) **3**]. The structure of **3** consists of neutral zig–zag chains of Er(III) ions, with $Hpcpa^{2-}$ ligands acting as bridges in a bidentate/monodentate coordination mode with five water molecules achieving the eight-coordination around the two Er(III) ions within the repeating bis(carboxylate)-bridged dinuclear units along the chain. The magnetic and magnetocaloric properties were studied for **1–3**. Compound **1** presents a field-induced slow relaxation of the magnetization with a "reciprocating thermal behavior" below 5 K for $H = 0.25$ T, while **2** shows maxima of the magnetic entropy from 3 up to 6 K for $\Delta H > 2$ T.

**Keywords:** coordination chemistry; oxamato ligand; lanthanide; magnetic properties; magnetocaloric properties

## 1. Introduction

The development of new multifunctional molecular materials (MMMs) is an important field of research due to the ability of these materials to combine different structures and chemical or physical properties for technological innovation. Coordination chemistry offers suitable alternatives for building MMMs for this purpose, such as coordination polymers (CPs). There are already many studies of porosity, good absorption capacity [1,2], and optical [3] and magnetic or magnetocaloric properties [4–6] where the CPs are used. The studies of CPs with 4*f* metal ions have been widely explored due to the enormous potential of working with MMMs that combine, for example, structural diversity with optical and magnetic/magnetocaloric properties [3,7]. The lanthanide(III) ions can show high coordination numbers, which allows different structures topologies for coordination polymers resulting from interaction with different organic ligands. The lanthanide(III) CPs can show good optical and magnetic/magnetocaloric properties depending on the choice of metal ion and organic ligand. These materials are interesting for applications like optoelectronics, spintronics, and cryogenic magnetic refrigeration [8].

Aromatic polyoxamato ligands have been amply used to obtain CPs with 3*d* metals. The great advantage of this type of ligand is the ability to build CPs with predictable structures and tunable magnetic properties; the ligand can efficiently transmit the magnetic coupling between neighboring metal atoms [9–12]. Many compounds with 4*f* metal ions and aromatic-substituted monooxamato ligands have been reported in the literature, most of which use ligands with monoprotonated oxamato groups. A mononuclear lanthanide(III) oxamato compound exhibiting a single-ion magnet (SIM) behavior is $Me_4N[Dy^{III}(H\text{-}2,6\text{-}dmpa)_4]\cdot 2CH_3CN$ [2,6-dmpa = *N*-2,6-dimethylphenyloxamate]; the hydrogen-bond-mediated self-assembly of this compound in the solid state provides a first example of 2D hydrogen-bonded polymer with a herringbone net topology [9]. Other examples employing the tetrabutylammonium salt of the *N*-(4-chlorophenyl)oxamate ligand (4-Clpa) yielded a series of mononuclear lanthanide(III) complexes of general formula $NBu_4[Ln^{III}(H\text{-}4Clpa)_4(dmso)]$, where Ln = Y, La, Nd, Eu, Gd, Tb, Dy, and Ho. Among these, the Nd, Gd, Tb, and Dy compounds exhibited a field-induced slow relaxation of magnetization typical of SIM behavior [13–15]. Using the sodium salt of the *N*-4-hydroxyphenyloxamate ligand (4-OHpa) towards the Ln(III) ions resulted in a series of a 2D $Na^{I}Ln^{III}$ supramolecular coordination networks, where Ln = Eu, Gd, Dy, and Ho, and only the Gd(III) compound presented a rare slow relaxation of the magnetization [16]. The simpler *N*-phenyloxamate ligand (pa) recently afforded 1D $Ln^{III}$ CPs, where Ln = Eu, Gd, and Tb. These are examples of CPs featuring an aryl-monoxamato ligand without coordinating groups attached to the aromatic ring [17].

Our group's extensive work with the coordination chemistry of the proligand *N*-(4-carboxyphenyl)oxamic acid ($H_3pcpa$) towards *p*-, *d*-, and *f*-block metal ions has led to the synthesis of a discrete $Fe^{III}$ complex in addition to $Pb^{II}$, $Mn^{II}$, $Co^{II}$, $Zn^{II}$, $Cd^{II}$, $Eu^{III}$, $Tb^{III}$, and $Gd^{III}$ CPs [18–20]. Their crystal structures revealed diverse coordination and bridging modes of the anionic $Hpcpa^{2-}$ ligand (Scheme 1), highlighting its remarkable versatility in self-assembling CPs with various metal ions.

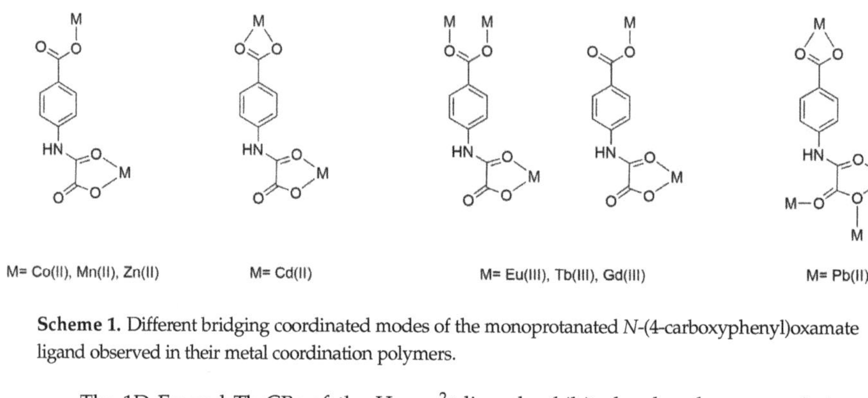

**Scheme 1.** Different bridging coordinated modes of the monoprotanated *N*-(4-carboxyphenyl)oxamate ligand observed in their metal coordination polymers.

The 1D Eu and Tb CPs of the $Hpcpa^{2-}$ ligand exhibited red and green emission, respectively, with the Tb compound showing a higher quantum yield due to the oxamate ligand's superior sensitization of $Tb^{III}$ ions. In contrast, the Gd analog displays slow magnetic relaxation under an applied field, classifying it as a field-induced single-molecule magnet (SMM) [19]. Focusing on the magnetic and magnetocaloric properties, we present a related series of Ln(III) CPs (Ln = Dy, Ho, and Er) with the $Hpcpa^{2-}$ ligand. Their synthesis, chemical and structural characterization, and the analysis of their magnetic and magnetocaloric properties are reported herein. In this regard, highly anisotropic holmium(III), dysprosium(III), and erbium(III) ions with $^5I_8$, $^6H_{15/2}$, and $^4I_{15/2}$ ground states, respectively, are particularly well suited to test their magnetocaloric efficiency, when compared with the isotropic gadolinium(III) ion with an orbital-free $^8S_{7/2}$ ground state ($J = 7/2$ with $S = 7/2$ and $L = 0$). They possess the highest value of total angular momentum along the lanthanide(III) series because of the first-order spin-orbit coupling (SOC) effects of the $4f^9$ $Dy^{III}$ ($J = 15/2$ with $S = 5/2$ and $L = 5$), $4f^{10}$ $Ho^{III}$ ($J = 8$ with $S = 2$ and $L = 6$), and $4f^{11}$ $Er^{III}$ ions ($J = 15/2$ with $S = 3/2$ and $L = 6$).

## 2. Experimental Section

### 2.1. General Information

All manipulations were carried out under aerobic conditions. Chemicals were purchased from Sigma-Aldrich and used without further purification. The synthesis of the *N*-(4-carboxyphenyl)oxamic acid ethyl ester ($EtH_2pcpa$) and $Na_2Hpcpa$ were performed following the previous report [18–21]. Elemental analysis (C, H, and N) was performed by Fisons EA1108. Solid-state infrared spectra were recorded on a Perkin Elmer Spectrum 400 FT-IR/FT-FIR spectrometer using KBr in the 400–4000 $cm^{-1}$ range, and 1H NMR were conducted by the Microanalytical Service of the Central de Análises Multiusuária of Federal University of Goiás. Thermogravimetric (TG) analysis was performed using samples obtained from the polycrystalline synthesis. The analyses were carried out at the Regional Center for Technological Development and Innovation (CRTI) on a Netzsch STA 449 F3 Nevio instrument over a temperature range of 30 to 650 °C, with a heating rate of 10 °C per minute under a synthetic air atmosphere. Powder X-ray diffraction (PXRD) data were collected for polycrystalline powders at CRTI using a Bruker D8 Discover powder diffractometer, using Cu-Kα radiation at a voltage of 40 kV and a current of 40 mA in the 2θ range 3.00–50.00° with a step-size of 0.0151°. The samples were kept at 15 rpm rotations during a measurement.

### 2.2. Synthesis

$\{[Ln_2(Hpcpa)_3(H_2O)_5]\}_n$ [Ln = $Dy^{III}$ (**1**), $Ho^{III}$ (**2**) and $Er^{III}$ (**3**)] were prepared by two different methods. The first method involved slow diffusion techniques in H-shaped tubes as follows: aqueous solutions of $Dy(NO_3)_3$ (0.1 mmol, 0.0348 g), $HoCl_3{\cdot}6H_2O$ (0.1 mmol, 0.0379 g), and $Er(NO_3)_3{\cdot}5H_2O$ (0.1 mmol, 0.0443 g) were placed in one arm of the corresponding H-tube, whereas an aqueous solution of $Na_2Hpcpa$ (0.1 mmol, 0.037 g) was placed in the other arm. Water was added dropwise to fill the H-tube, which was covered with parafilm and allowed to stand at room temperature. All tubes yielded crystals; however, not all were suitable for single-crystal X-ray diffraction analysis. The second method involved mixing a solution of the rare-earth metal salt with a solution of the ligand salt in a 2:3 metal-to-ligand molar ratio. This process resulted in a polycrystalline solid, which was filtered and dried under vacuum. The yields were 78%, 81%, and 83% for **1–3**, respectively. Elemental analysis (C,H,N) Calcd for $C_{27}H_{25}Dy_2N_3O_{20}$ (**1**): C 31.29; H, 2.43; N, 4.05. Found: C 30.74; H, 2.72; N, 4.04%. Anal. Calcd for $C_{27}H_{25}Ho_2N_3O_{20}$ (**2**): C, 31.14; H, 2.42; N, 4.04. Found: C 30.62; H, 2.61; N, 3.94%. Anal. Calcd for $C_{27}H_{25}Er_2N_3O_{20}$ (**3**): C 31.00; H, 2.41; N, 4.02. Found: C 30.47; H, 2.59; N, 4.05%. Selected IR data (KBr disk, $cm^{-1}$): 3425(br), 1651(s), 1604(s), 1537(s), 1412(s), 1311(w), 1182(w), 862(w), and 786(m) for **1**; 3419(br), 1651(s), 1604(s), 1538(s), 1415(s), 1311(w), 1182(w), 862(w), and 786(m) for **2**. 3434(br), 1650(s), 1604(s), 1539(s), 1415(s), 1311(w), 1182(w), 862(w), and 786(m) for **3**.

### 2.3. Crystallographic Data Collection and Refinement

All reflections for single crystals of **3** were collected on a Bruker-AXS Kappa Duo diffractometer (Billerica, MA, USA) equipped with an APEX II CCD detector with either Mo Kα (λ = 0.71073 Å). The structure was solved by direct methods through SHELXS-2014 [22] of phase retrieval and then refined by full-matrix least-squares on $F^2$ with anisotropic thermal parameters for all non-hydrogen atoms, while the hydrogen atoms were stereochemically positioned and their displacement parameters fixed and set to isotropic [$U_{iso}$ (H) = 1.2Ueq (CH or NH) or 1.5Ueq (water)]. In addition, some scattered peaks with residual electron density were still observed in the difference Fourier map located in a void for **3**. Attempts were made to assign these peaks as nonstoichiometric water molecules; however, checkcif alerts arise, such as short distances with other atoms, like coordination water molecules. Therefore, PLATON/SQUEEZE [23] was used to remove their contribution to the diffraction data. The Mercury [24] and ORTEP-3 [25] programs were used within the WinGX [26] software version 4.0 package to prepare artwork representations. Crystallographic data and refinement parameters of **3** are summarized in Table 1, whereas selected bond distances and angles are listed in Table S2 and Table S3, respectively. All crystallographic data are available from the CCDC. The CCDC number is 2417034 (**3**).

**Table 1.** X-ray diffraction data collection and refinement parameters for **3**.

| | 3 |
|---|---|
| Empirical formula | $C_{27}H_{25}N_3O_{20}Er_2$ |
| Formula weight | 1046.02 |
| Crystal system | Monoclinic |
| Space group | Cc |
| $a$/Å | 24.290(4) |
| $b$/Å | 19.873(4) |
| $c$/Å | 7.4327(13) |
| β/° | 104.998(5) |
| V/Å$^3$ | 3465.7(11) |
| Z | 4 |
| T/K | 296 |
| Radiation | Mo Kα |
| Wavelength/Å | 0.71073 |
| $\rho$/g cm$^{-3}$ | 2.005 |
| μ//mm$^{-1}$ | 4.897 |
| Symmetry factor (Rint) | 0.1171 |
| Completeness to θmax/% | 98.70 |
| $F$(000) | 2016.0 |
| Refined parameters | 272 |
| Reflections collected | 6351 |
| R [a], wR [b] [I > 2s(I)] | 0.1087, 0.3155 |
| R [a], wR [b] (tall data) | 0.1502 |
| S [c] | 1.127 |
| $\rho_{max}$ and $\rho_{min}/e$Å$^{-1}$ | 2.3400/–4.2300 |
| CCDC | 2417034 |

[a] R = S | |F$_o$| − |F$_c$| | /S |F$_o$|. [b] wR = $[S\,w(|F_o|^2 - |F_c|^2)^2/S\,w|F_o|^2]^{1/2}$. [c] S = $[S\,w(|F_o|^2 - |F_c|^2)^2/(n_o - np)]^{1/2}$, where $w \propto 1/s$, $n_o$ = observed, and $n_p$ = fitted parameters.

### 2.4. Magnetic Measurements

Variable-temperature (2–300 K) direct-current (dc) magnetic susceptibility measurements under an applied magnetic field of 0.5 T and variable-temperature (2–10 K) and variable-field (0–5 T) magnetization measurements were carried out using a Quantum Design Superconducting Quantum Interference Device (SQUID) magnetometer.

Variable-temperature (2–16 K) and variable-field (0–0.25 T) alternating-current (ac) magnetic susceptibility measurements under a ±5.0 Oe oscillating field at frequencies in the range 0.1–10 kHz were performed with a Quantum Design Physical Property Measurement System (PPMS). The magnetic measurements, performed on powdered samples embedded in eicosane to prevent any crystal reorientation, were corrected for the diamagnetism of the constituent atoms and the sample holder.

## 3. Results

### *3.1. Syntheses and General Physicochemical Characterization of* **1–3**

The preparation of **1–3** was performed by slow diffusion in the H tube that yielded X-ray quality single crystals for **3** and polycrystalline samples for **1** and **2**. All of them were also prepared using the precipitation method with good yields. The infrared spectra of the complexes can be seen in Figure S1. All compounds present characteristic bands of the ligand such as stretching of the C=O amide bond and C–H and C=C phenyl bonds. They also show strong absorptions in the high-frequency region above 3400 $cm^{-1}$ due to stretching vibrations of the O–H bond of crystallization and coordination water molecules involved in hydrogen bonding. At the same region, superimposed on the broad O–H stretching band, it is possible to observe a narrow band of N–H stretching, thus being an indication that the nitrogen of the oxamate group is not involved in metal coordination. The PXRD patterns for the polycrystalline samples of **1–3** (Figure 1) indicated that all compounds were isostructural, possessing the same structure of **3** described below, and confirmed the phase purity of the compounds.

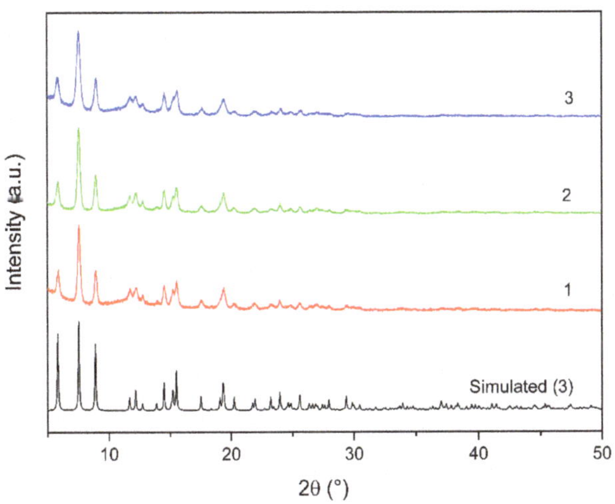


**Figure 1.** Experimental PXRD patterns for **1–3** and simulated one for **3**.

The TG curves (Figure S2) of **1–3** are similar. They lose about 10% of their weight up to 170 °C, which can be attributed to the loss of six water molecules [calculated/found = 10.2/10.9 (**1**), 10.2/10.5, (**2**) and 10.1/9.7 (**3**)]. This first mass loss is followed by an almost plateau up to 350 °C. Two major mass losses occur between 350 and 450 °C due to the decomposition of the anhydrous phases. The final waste from thermal decomposition can be attributed to the respective lanthanide oxides ($Ln_2O_3$).

The electronic spectra in the UV-vis-NIR region of **1–3** were collected at room temperature (Figure S3). The diffuse reflectance spectra display characteristic absorption lines of 4f transitions [27–29] of the $Dy^{III}$, $Ho^{III}$ and $Er^{III}$ ions [**1**: $^6H_{15/6} \rightarrow {}^6H_{7/6} + {}^6F_{9/2}$ (1106 nm), $^6H_{15/6} \rightarrow {}^6F_{7/2}$ (910 nm), $^6H_{15/6} \rightarrow {}^6F_{5/2}$ (807 nm), $^6H_{15/6} \rightarrow {}^6F_{3/2}$ (756nm), $^6H_{15/6} \rightarrow {}^4F_{9/2}$ (474 nm), $^6H_{15/6} \rightarrow {}^4I_{15/2}$ (463 nm), and $^6H_{15/6} \rightarrow {}^4G_{11/2}$ (427 nm); **2**: $^5I_8 \rightarrow {}^5I_6$ (1159 nm), $^5I_8 \rightarrow {}^5I_5$ (897 nm), $^5I_8 \rightarrow {}^5I_4$ (753 nm), $^5I_8 \rightarrow {}^5F_5$ (643 nm), $^5I_8 \rightarrow {}^5S_2$ (538 nm), $^5I_8 \rightarrow {}^5F_4$ (486 nm), $^5I_8 \rightarrow {}^5F_3$ (474 nm), $^5I_8 \rightarrow {}^5F_2$ (468 nm), $^5I_8 \rightarrow {}^5G_6$ (450 nm), $^5I_8 \rightarrow {}^5G_5$ (418 nm), $^5I_8 \rightarrow {}^5G_4$ (386 nm), and $^5I_8 \rightarrow {}^3K_7$ (382 nm); **3**: $^4I_{15/2} \rightarrow {}^4I_{11/2}$ (980 nm), $^4I_{15/2} \rightarrow {}^4I_{9/2}$ (802 nm), $^4I_{15/2} \rightarrow {}^4F_{9/2}$ (652 nm), $^4I_{15/2} \rightarrow {}^4S_{3/2}$ (544 nm), $^4I_{15/2} \rightarrow {}^4H_{11/2}$ (520 nm), $^4I_{15/2} \rightarrow {}^4F_{7/2}$ (488 nm), $^4I_{15/2} \rightarrow {}^4F_{5/2}$ (451 nm), $^4I_{15/2} \rightarrow {}^4F_{3/2}$ (444 nm), $^4I_{15/2} \rightarrow {}^4H_{9/2}$ (407 nm), and $^4I_{15/2} \rightarrow {}^4G_{11/2}$ (378 nm)].

### 3.2. *Description of the Structure of* 3

The crystallographic results reveal that **3** crystallizes in the monoclinic *Cc* space group and forms zig–zag chains parallel to the crystallographic b axis [101] direction, as shown in Figure 2. The asymmetric dinuclear unit, shown in Figure 2a, consists of two crystallographically independent $Er^{III}$ ions (Er1 and Er2), three anionic $Hpcpa^{2-}$ ligands, and five coordinated water molecules. The Er1 and Er2 ions exhibit eight-coordination, with three (Er1) or two (Er2) water molecules, four (Er1) or three (Er2) carboxylate oxygen atoms, and one (Er1) or two (Er2) amide oxygen atoms, defining a geometry intermediate between triangular dodecahedron ($D_{2d}$), square antiprism ($D_{4d}$), and biaugmented trigonal prism ($C_{2v}$), as determined by the SHAPE program (see Table S1). The Er1−O bond lengths range from 2.24(3) to 2.63(3) Å, while the Er2−O bond lengths range from 2.27(3) to 2.58(3) Å (see Table S2). Additionally, the $Er^{III}$−O (water) and $Er^{III}$−O (oxamate) distances are consistent with previously reported structures. The O−$Er^{III}$−O bond angles range from 60.4(8)° to 147.9(10)° (see Table S2). These angles were comparable with related compounds reported in the literature.

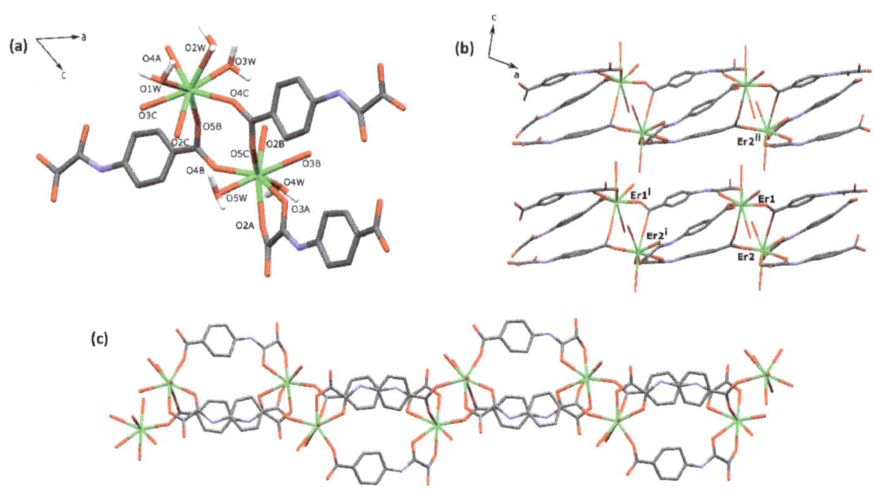


**Figure 2.** (**a**) A view of the asymmetric unit of **3** with the atom numbering of the metal environment. (**b**) A view of the arrangement of two neighboring chains parallel to the *ac* plane. (**c**) A perspective view of a fragment of the zigzag chain of **3** running along the crystallographic *a*-axis. Color code: C (gray), N (blue), O (red), Er (green), and H (white). Symmetry code: (i) = $\frac{1}{2} + x$, $\frac{1}{2}$ $y$, $\frac{1}{2} + z$; (ii) = $x$, $-y$, $\frac{1}{2} + z$.

Double *syn–syn* carboxylate(oxamate) groups connect Er1 and Er2 into pairs within each chain, the erbium–erbium separation being 4.961(4) Å for Er1···Er2. These dinuclear units are, in turn, interlinked by three $Hpcpa^{2-}$ ligands, one of them through the μ-κ²O,O′:κO″ coordination mode and the other two across the $\mu_3$-κ²O,O′:κO″:κO‴ one, resulting in neutral 1D CPs (Figure 2b). Two root mean square deviation (rmsd) planes were fitted through the atoms that connect the erbium cations across the oxamate bridge in the double syn–syn conformation cited above. The planes were defined as A and B and fitted through the C6B,C9B,O4B,O5B and C6C,C9C,O4C,O5C sets of atoms, respectively.

Within the resulting triple-stranded moieties, the values of the metal–metal distances through these extended μ-κ²O,O′:κO″- and $\mu_3$-κ²O,O′:κO″:κO‴-$Hpcpa^{2-}$ bridges are 10.152(2), 12.227(5), and 12.220(6) Å for Er1···Er2$^{i}$, Er1···Er1$^{i}$ and Er2···Er2$^{i}$ [symmetry operations ($^{i}$): = $\frac{1}{2} + x$, $\frac{1}{2} - y$, $\frac{1}{2} + z$], respectively (Figure 2c). Additionally, the interaction between two parallel chains through hydrogen bonds and π–π interactions results in a supramolecular 2D network (Figure 3) in which the shortest interlayer contact between the erbium atoms is 6.075(3) Å [Er1 and Er2$^{ii}$, symmetry operations ($^{ii}$): = $x$, $-y$, $\frac{1}{2} + z$]. These π–π interactions occur along the crystallographic *c* axis with values of the distance between centroids of 3.717 (2) Å (see Figure 3a).

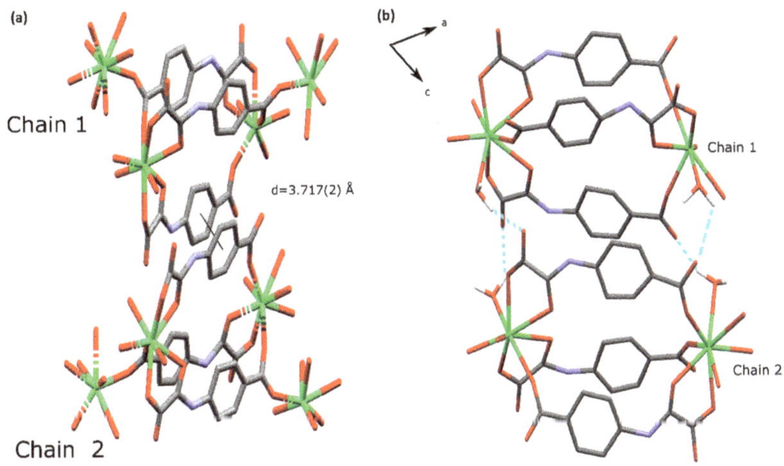


**Figure 3.** Intermolecular interactions between two parallel chains of **3** through (**a**) π-π stacking showing the distance between the centroid of the aromatic rings along the crystallographic *c*-axis and (**b**) hydrogen bonds between coordinated water molecules and oxygen atoms from the two $Hpcpa^{2-}$ bridges along *b*-axis. Some hydrogen atoms were omitted for clarity. Dashed cyan and black lines represent hydrogen bonds and π-π interactions, respectively.

The interaction between two chains is established by hydrogen bonds interactions, which occur along the *b* and *c* axes. The connection between two chains through O–H···O hydrogen bonds along the *b* axis can be observed in Figure 3. These intermolecular interactions involve oxygen atoms belonging to coordinated water molecules (O1W, O2W, O3W, O4W, and O5W) and oxamate oxygen atoms (O1B and O2B). The connection between the chains is also observed along the *c* axis (Figure 3b). These interactions involve coordinated water molecules (O1W, O2W, O4W, and O5W) and oxamate oxygen atoms (O1A, O2A, O5A, and O2C). The molecular contacts include O–H···O interactions that are mainly established by water molecules and oxamate oxygen atoms. The geometric parameters of the hydrogen bond interactions for **3** are summarized in Table S3.

### 3.3. *Magnetic Properties of* **1–3**

The dc magnetic properties of **1–3** in the form of the $\chi_M T$ vs. $T$ plots [where $\chi_M$ represents the dc magnetic susceptibility per two lanthanide(III) ions] are presented in Figure S4. At 300 K, the $\chi_M T$ values are 27.0 (**1**), 27.3 (**2**), and 21.5 cm$^3$ mol$^{-1}$ K (**3**). Such values are close to the calculated value for two magnetically isolated Ln$^{III}$ ions with $^5I_8$ (Ln = Ho), $^6H_{15/2}$ (Ln = Dy), and $^4I_{15/2}$ (Ln = Er) ground states, well separated from the low-lying excited states. Then, $\chi_M T$ decreases with decreasing temperature due to the thermal depopulation of the $M_J$ states resulting from the ligand field (LF) splitting, reaching values of 15.1 (**1**), 10.0 (**2**) and 9.3 cm$^3$ mol$^{-1}$ K (**3**) at 1.9 K. However, the occurrence of antiferromagnetic interactions between the Ln$^{III}$ ions through the double *syn–syn* carboxylate bridge cannot be discarded [16]. The normalized isothermal magnetization (*M* vs. *H/T*) curves do not superimpose below 10 K in agreement with the strong magnetic anisotropy of these ions, as shown in the inset of Figure S4.

The ac magnetic properties of **1–3** were investigated under different applied dc magnetic fields (*H* = 0, 0.1 and 0.25 T). Their behaviors are shown in Figures 4, S5 and S6 in the form of $\chi_M'$ and $\chi_M''$ vs. $T$ or $\nu$ plots [$\chi_M'$ and $\chi_M''$ being the in-phase and out-of-phase ac molar magnetic susceptibilities per two lanthanide(III) ions, $T$ the temperature, and $\nu$ the frequency of the oscillating ac field]. Even though the static dc magnetic properties of **1–3** are similar, that is not the case concerning their dynamic ac magnetic properties. Neither the frequency dependence of $\chi_M'$ nor a $\chi_M''$ signal was observed for **2**. However, slow magnetic relaxation phenomena occur in **1** and **3**. While in **3**, only field-induced incipient signals of $\chi_M'$ and $\chi_M''$ are observed by applying a *H*, **1** exhibits frequency- and temperature dependence of the $\chi_M'$ and $\chi_M''$ even in the absence of *H*, which varies significantly by changing the *T*, $\nu$, or *H*. Only incipient $\chi_M''$ signals are observed for **1** under *H* = 0 or 0.1 T [Figure 4a,b(right)], while $\chi_M''$ maxima arise under *H* of 0.25 T [Figure 4c(right)]. For instance, these maxima slightly shift toward higher temperatures with increasing $\nu$ from 0.1 to 1 kHz but revert with the further increase in $\nu$ up to 10 kHz. This uncommon behavior noted for the $\chi_M''$ vs. $T$ plots was described by Boča and collaborators as a phenomenological "reciprocating thermal behavior" due to a "strange" relaxation mechanism [30].

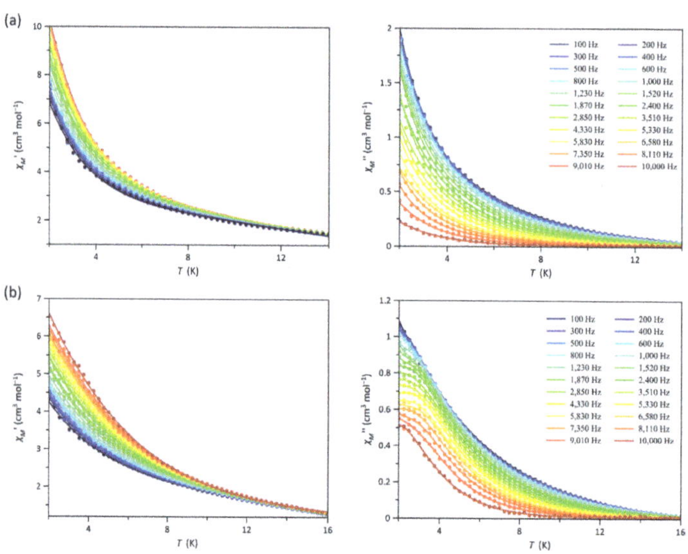


**Figure 4.** *Cont.*

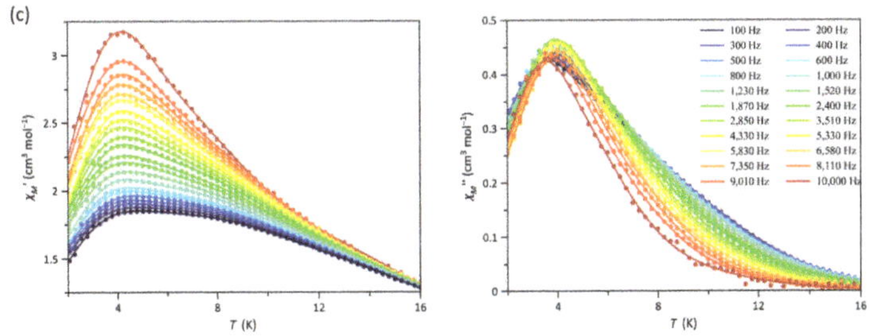


**Figure 4.** Temperature dependence of $\chi_M{}'$ (**left**) and $\chi_M{}''$ (**right**) for **1** at a 5.0 G oscillating field in the frequency range of 0.1–10 kHz (red to blue) in the absence of applied static magnetic fields (**a**) and under 0.1 (**b**) and 0.25 T (**c**). The solid lines are only eye guides.

Unfortunately, the incipient nature of the data prevented accurate analysis of most of the ac magnetic measurements. Thus, only the magnetic relaxation time ($\tau$) values for **1** at $H = 0.25$ T were extracted by fitting the $\chi_M{}'$ and $\chi_M{}''$ vs. $\nu$ plots through the generalized Debye model [31]. They are shown as an Arrhenius plot in Figure 5. At higher temperatures (above 5 K), the experimental data follow an Arrhenius law of a thermally activated magnetic relaxation process referring to the climbing of the energy barrier to spin reversal. However, below 5 K, the relaxation process speeds up again upon cooling, and the "strange" mechanism becomes a more efficient path for the spin reversal. So, the experimental relaxation times can be satisfactorily simulated through a model that considers a two-phonon Orbach mechanism [$\tau^{-1} = \tau_0{}^{-1} \exp(-U_{eff}/k_BT)$] plus a "strange" one [$\tau^{-1} = FT^{-k}$, with $k > 0$]. The best-fit values of $\tau_0$ and $U_{eff}$ [$1.50(3) \times 10^{-6}$ s and 14.81(13) $cm^{-1}$] are within the range found in other Dy(III) compounds [32]. Meanwhile, the found values of $F$ and $k$ are $1.96(4) \times 10^6$ $K^k$ $s^{-1}$ and 4.2(2), respectively. Only two examples were reported in the literature: a dinuclear $\{Ni^{II}Gd^{III}\}$ complex with a Schiff's Base ligand and a $Dy^{III}$ chain represented by the formulas $[Ni(\mathit{vanen})Gd(H_2O)Cl_3]$ and $[Dy(DPOA)(NO_3)(H_2O)_2]_n \cdot nH_2O$. For the Dy compound, the reported parameters related to the "strange" mechanism [$\tau^{-1} = FT^{-k}$, with $k > 0$] are $F = 1.0 \times 10^4$ $K^k$ $s^{-1}$ and $k = 2.8$, values comparable to those found for compound **1** in our work [30]. Still, it is hard to argue on those "strange" parameters since, to date, this type of relaxation mechanism has only a phenomenological meaning.

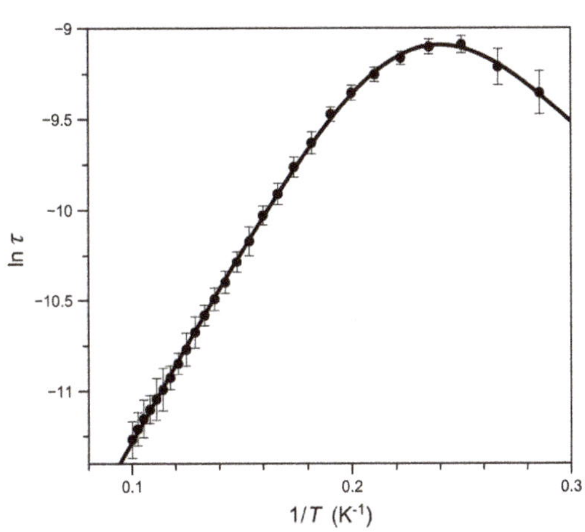


**Figure 5.** Arrhenius plots for **1** under applied dc magnetic field of 0.25 T. The solid line is the best-fit curve (see text).

### 3.4. *Magnetocaloric Properties of* **1–3**

The magnetocaloric effect (MCE) of **1–3** has been investigated by variable-temperature ($T$ = 2–10 K) and variable-field ($H$ = 0–5 T) magnetization measurements (see Section 2). The field and temperature dependence of the magnetic entropy change ($\Delta S_m$) upon connecting the magnetic field ($\Delta H = H - H_0$ with $H_0$ = 0) has been estimated from the $M$ vs. $H$ and $T$ plots ($M$ being the gravimetric magnetization) according to the Maxwell equation (Figure 6), as reported elsewhere [33,34].

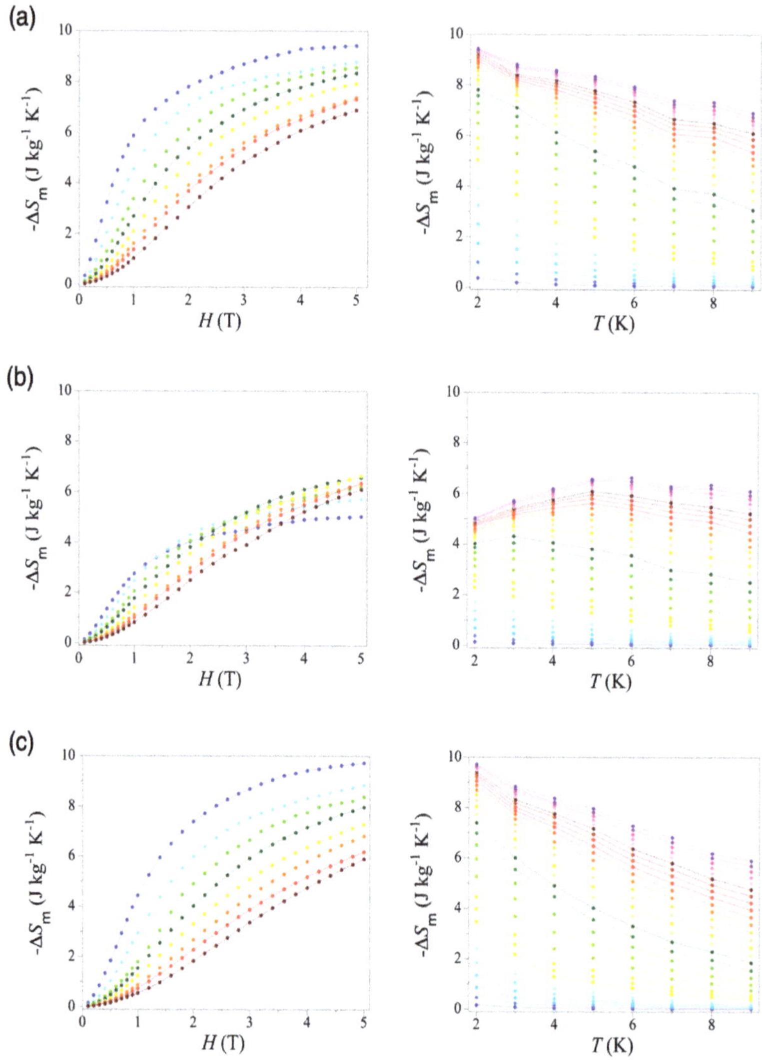


**Figure 6.** Field (**left**) and temperature (**right**) dependence of $-\Delta S_m$ for **1** (**a**), **2** (**b**), and **3** (**c**) in the temperature and field range of 2.0–10 K (blue to red) and 0–5 T (blue to purple), respectively. The solid lines are only eye guides.

The $-\Delta S_m$ vs. $H$ and $T$ plots show a monotonic increase with decreasing the temperature or increasing the field for **1** and **3** (Figure 6a and c, respectively), whereas **2** evidences a decrease upon cooling further for $T < 5$ K with the progressive development of a maximum from 3 up to 6 K for $\Delta H > 2$ T (Figure 6b). Interestingly, the $-\Delta S_m$ maxima for **2** occur for $\Delta H > 2$ T, whereas it does not show slow magnetic relaxation effects above 2 K. In contrast, no $-\Delta S_m$ maxima are observed for **1** and **3** for $\Delta H < 5$ T, evidencing thus that the blockage of the magnetization that occurs at low temperatures under the application of a weak *dc* magnetic field [$T_B < 4.5$ (**1**) and 3.5 K (**3**) for $H = 0.025$ T] does not preclude the observation of a significant MCE along this series.

The maximum $-\Delta S_m$ value in gravimetric units for $\Delta H = 5$ T increases from **2** to **1** and **3** [9.4 (**1**)/6.6 (**2**)/9.7 J kg$^{-1}$ K$^{-1}$ (**3**) at $T = 2$ (**1**)/6 (**2**)/2 K (**3**)], being comparable to those of the gadolinium(III)–gallium(III) garnet used as commercial cryogenic material [$-\Delta S_m = 6.5$ and 28.0 J kg$^{-1}$ K$^{-1}$ at $T = 2$ and 10 K, respectively, for $\Delta H = 8$ T] [35]. Moreover, the corresponding $-\Delta S_m$ value for $\Delta H = 1$ T [5.8 (**1**)/1.4 (**2**)/4.4 J kg$^{-1}$ K$^{-1}$ (**3**) at $T = 2$ (**1**)/6 (**2**)/2 K (**3**)] reaches up to 62 (**1**), 21 (**2**), and 45% (**3**) of the maximum $-\Delta S_m$ value. This relatively weak magnetic field is close enough to the stronger field to be achieved with the standard permanent magnets used commercially (ranging from 0.5 to 1 T for ceramic magnets and up to 1.4 T for neodymium ones). These two features (high magnetic entropy changes and large slope of the corresponding isothermal magnetic entropy variation curves) are mandatory for potential applications in CMR.

The overall enhancement of the MCE in the order **2** < **1** ≈ **3** is likely explained by the larger magnetic anisotropy of the $4f^{10}$ $Ho^{III}$ ion ($^5I_8$) compared to the $4f^9$ $Dy^{III}$ ($^6H_{15/2}$) and $4f^{11}$ $Er^{III}$ ($^4I_{15/2}$) ions, despite the larger $J$ value in the former case. Indeed, the maximum $-\Delta S_m$ values in molar units [9.7 (**1**)/6.9 (**2**)/10.1 J mol$^{-1}$ K$^{-1}$ (**3**) at $T = 2$ (**1**)/6 (**2**)/2 K (**3**) for $\Delta H = 5$ T] are somewhat lower than the limiting values for two magnetically isolated $Ln^{III}$ ions with no zero-field splitting (*zfs*) [$-\Delta S_m = 2R\ln(2J + 1) = 46.1$ (Dy)/47.1 (Ho)/46.1 J mol$^{-1}$ K$^{-1}$ (Er) with $J = 15/2$ (Dy), 8 (Ho), and 15/2 (Er)]. By comparison, the $-\Delta S_m$ values in molar units (per mol of lanthanide atom) [3.0 (**1**)/0.7 (**2**)/2.3 J mol$^{-1}$ K$^{-1}$ (**3**) at $T = 2$ (**1**)/6 (**2**)/2 K (**3**) for $\Delta H = 1$ T] are similar to those reported for the gadolinium(III)–gallium(III) garnet and its dysprosium(III)-substituted derivatives of general formula $(Dy_xGd_{1-x})_3Ga_5O_{12}$ [$x = 0$ (GGG), 0.5 (DGGG), and 1 (DGG)] [$-\Delta S_m = 3.1$–3.3 J mol$^{-1}$ K$^{-1}$ at $T = 5$ K for $\Delta H = 2$ T] [36].

In this respect, the MCE efficiencies for **1**–**3** are comparable to those of related magnetically anisotropic lanthanide(III) metal–organic frameworks (MOFs) and its mixed gadolinium(III)-substituted heterometallic derivatives of general formula $[(Ln_xGd_{1-x})(HCO_2)_3]_n$, $[(Ln_xGd_{1-x})(HCO_4)]_n$, $[(Ln_xGd_{1-x})_3(ad)_{4.5}(DMF)_2]_n$, and $[(Ln_xGd_{1-x})_3(cb)_4(NO_3)(DMF)_y]_n$ [ad = adipate and cb = 1,7-di(4-carboxyphenyl)-1,7-dicarba-*closo*-dodecaborane with Ln = Eu, Tb, Dy, Ho, Er ($x = 0$–1)], which have been recently proposed as efficient molecular cryomagnetic refrigerants [37–41] in place of traditional purely inorganic intermetallic alloys and oxides, such as mixed lanthanide-transition metal antimonides and tellurides, $Ln_6MSb_2$ and $Ln_6MTe_2$ (Ln = Gd, Tb, and Dy; M = Mn, Fe, Co, and Ni), or lanthanide–gallium garnets, $Ln_3Ga_5O_{12}$ (Ln = Gd, Tb, Dy, and Nd) [42,43].

## 4. Conclusions

The magnetic properties of **1**–**3** are very variable. Although the dc magnetic susceptibility measurements reveal similar trends for all compounds, the ac magnetic properties diverge, with only compounds **1** and **3** displaying slow magnetic relaxation phenomena. The temperature- and frequency-dependent ac susceptibility behavior in **1** is especially remarkable due to a "strange" relaxation mechanism, which has been phenomenologically described but not fully understood to date. This isostructural series of oxamato-based $Ln^{III}$

CPs also exhibits weak (Ln = Ho) to moderate (Ln = Dy and Er) MCE with magnetic entropy change maxima at relatively high temperatures (Ln = Ho), far above He liquefaction.

These findings reinforce the potential of lanthanide-based compounds as efficient molecular cryomagnetic refrigerants, comparable to traditional materials such as gadolinium–gallium garnets, as new optional materials for magnetic refrigeration technology. Further exploration of the "strange" relaxation mechanism and its impact could lead to new insights in understanding some enigmatic features of the slow magnetic relaxation phenomena.

**Supplementary Materials:** The following supporting information can be downloaded at: https://www.mdpi.com/article/10.3390/magnetochemistry11040023/s1. Table S1. Analyses of coordination geometry for each LnIII ion in 3. Table S2. Selected bond lengths (Å) and angles (°) for 3. Table S3. Hydrogen bonds in 3. Figure S1. Infrared spectra of 1–3 (top to bottom). Figure S2. Termogravimetric curves for (red line) 1, (green line) 2 and (blue line) 3. Figure S3. Powder diffuse reflectance spectra of 1–3. Figure S4. Temperature dependence of $\chi_M T$ for 1–3 (a-c). The inset shows the M vs H/T plots in the temperature range of 2.0–10 K. Figure S5. Temperature dependence of $\chi_M'$ (a) and $\chi_M''$ (b) for 3 at a 5.0 G oscillating field in the frequency range of 0.5–10.0 kHz (yellow to blue) under an applied static magnetic field of 0.25 T. Figure S6. Frequency dependence of $\chi_M'$ (a) and $\chi_M''$ (b) and Argand plots (c) for 1 at a 5.0 G oscillating field in the temperature range of 3.5–10 K under an applied static magnetic field of 0.25 T and frequencies of 0.1–10 kHz (gray to brown).

**Author Contributions:** Conceptualization, D.C.; methodology, J.W.M. and L.H.G.K.; formal analysis, J.W.M., F.T.M., M.E.A., N.M. and R.R.; investigation, J.W.M., L.H.G.K. and R.R.; data curation, J.W.M., M.E.A. and R.R.; writing—original draft preparation, D.C., J.W.M., L.H.G.K. and R.R.; writing—review and editing, J.W.M. and R.R. All authors have read and agreed to the published version of the manuscript.

**Funding:** This research was funded by CAPES n° 88887798611/2022-00, CNPq n° 305691/2022-8 and FAPEG project n° 202410267000958.

**Institutional Review Board Statement:** Not applicable.

**Informed Consent Statement:** Not applicable.

**Data Availability Statement:** The original contributions presented in this work are included in the article and Supplementary Materials. Further inquiries can be directed to the corresponding authors.

**Acknowledgments:** J.W.M. thanks CAPES for the doctoral grant.

**Conflicts of Interest:** The authors declare no conflicts of interest.

## References

1. Wang, T.; Lin, E.; Peng, Y.L.; Chen, Y.; Cheng, P.; Zhang, Z. Rational Design and Synthesis of Ultramicroporous Metal-Organic Frameworks for Gas Separation. *Coord. Chem. Rev.* **2020**, *423*, 213485. [CrossRef]
2. Wang, C.; Liu, X.; Keser Demir, N.; Chen, J.P.; Li, K. Applications of Water Stable Metal–Organic Frameworks. *Chem. Soc. Rev.* **2016**, *45*, 5107–5134. [CrossRef]
3. Liu, J.Q.; Luo, Z.D.; Pan, Y.; Kumar Singh, A.; Trivedi, M.; Kumar, A. Recent Developments in Luminescent Coordination Polymers: Designing Strategies, Sensing Application and Theoretical Evidences. *Coord. Chem. Rev.* **2020**, *406*, 213145. [CrossRef]
4. Zheng, Y.-Z.; Zheng, Z.; Chen, X.-M. A Symbol Approach for Classification of Molecule-Based Magnetic Materials Exemplified by Coordination Polymers of Metal Carboxylates. *Coord. Chem. Rev.* **2014**, *258–259*, 1–15. [CrossRef]
5. Journaux, Y.; Ferrando-Soria, J.; Pardo, E.; Ruiz-Garcia, R.; Julve, M.; Lloret, F.; Cano, J.; Li, Y.; Lisnard, L.; Yu, P.; et al. Design of Magnetic Coordination Polymers Built from Polyoxalamide Ligands: A Thirty Year Story. *Eur. J. Inorg. Chem.* **2018**, *2018*, 228–247. [CrossRef]
6. Leong, W.L.; Vittal, J.J. One-Dimensional Coordination Polymers: Complexity and Diversity in Structures, Properties, and Applications. *Chem. Rev.* **2011**, *111*, 688–764. [CrossRef]

7. Marin, R.; Brunet, G.; Murugesu, M. Shining New Light on Multifunctional Lanthanide Single-Molecule Magnets. *Angew. Chem. Int. Ed.* **2021**, *60*, 1728–1746. [CrossRef]
8. Edelmann, F.T. Lanthanide Amidinates and Guanidinates in Catalysis and Materials Science: A Continuing Success Story. *Chem. Soc. Rev.* **2012**, *41*, 7657. [CrossRef]
9. Ferrando-Soria, J.; Pasán, J.; Ruiz-Pérez, C.; Journaux, Y.; Julve, M.; Lloret, F.; Cano, J.; Pardo, E. Spin Control in Oxamato-Based Manganese(II)-Copper(II) Coordination Polymers with Brick-Wall Layer Architectures. *Inorg. Chem.* **2011**, *50*, 8694–8696. [CrossRef]
10. Ferrando-Soria, J.; Grancha, T.; Julve, M.; Cano, J.; Lloret, F.; Journaux, Y.; Pasán, J.; Ruiz-Pérez, C.; Pardo, E. Ligand Effects on the Dimensionality of Oxamato-Bridged Mixed-Metal Open-Framework Magnets. *Chem. Commun.* **2012**, *48*, 3539. [CrossRef]
11. Grancha, T.; Acosta, A.; Cano, J.; Ferrando-Soria, J.; Seoane, B.; Gascon, J.; Pasán, J.; Armentano, D.; Pardo, E. Cation Exchange in Dynamic 3D Porous Magnets: Improvement of the Physical Properties. *Inorg. Chem.* **2015**, *54*, 10834–10840. [CrossRef]
12. Abhervé, A.; Grancha, T.; Ferrando-Soria, J.; Clemente-León, M.; Coronado, E.; Waerenborgh, J.C.; Lloret, F.; Pardo, E. Spin-Crossover Complex Encapsulation within a Magnetic Metal-Organic Framework. *Chem. Commun.* **2016**, *52*, 7360–7363. [CrossRef]
13. Fortea-Pérez, F.R.; Vallejo, J.; Julve, M.; Lloret, F.; De Munno, G.; Armentano, D.; Pardo, E. Slow Magnetic Relaxation in a Hydrogen-Bonded 2D Array of Mononuclear Dysprosium(III) Oxamates. *Inorg. Chem.* **2013**, *52*, 4777–4779. [CrossRef]
14. Vaz, R.C.A.; Esteves, I.O.; Oliveira, W.X.C.; Honorato, J.; Martins, F.T.; da Silva Júnior, E.N.; Valente, D.d.C.A.; Cardozo, T.M.; Horta, B.A.C.; Mariano, D.L.; et al. Lanthanide(III)-Oxamato Complexes Containing $Nd^{3+}$ and $Ho^{3+}$: Crystal Structures, Magnetic Properties, and Ab Initio Calculations. *CrystEngComm* **2022**, *24*, 6628–6641. [CrossRef]
15. Vaz, R.C.A.; Esteves, I.O.; Oliveira, W.X.C.; Honorato, J.; Martins, F.T.; Marques, L.F.; Dos Santos, G.L.; Freire, R.O.; Jesus, L.T.; Pedroso, E.F.; et al. Mononuclear Lanthanide(III)-Oxamate Complexes as New Photoluminescent Field-Induced Single-Molecule Magnets: Solid-State Photophysical and Magnetic Properties. *Dalton Trans.* **2020**, *49*, 16106–16124. [CrossRef]
16. da Cunha, T.T.; Barbosa, V.M.M.; Oliveira, W.X.C.; Pinheiro, C.B.; Pedroso, E.F.; Nunes, W.C.; Pereira, C.L.M. Slow Magnetic Relaxation in Mononuclear Gadolinium(III) and Dysprosium(III) Oxamato Complexes. *Polyhedron* **2019**, *169*, 102–113. [CrossRef]
17. da Silveira, C.O.C.; Oliveira, W.X.C.; da Silva Júnior, E.N.; Alvarenga, M.E.; Martins, F.T.; Gatto, C.C.; Pinheiro, C.B.; Pedroso, E.F.; Silva, J.P.O.; Marques, L.F.; et al. Photoluminescence and Magnetic Properties of Isostructural Europium(III), Gadolinium(III) and Terbium(III) Oxamate-Based Coordination Polymers. *Dalton Trans.* **2024**, *53*, 14995–15009. [CrossRef]
18. Maciel, J.W.; de Souza, J.J.S.; Valdo, A.K.; Martins, F.T.; Guimarães, F.F.; de Santana, R.C.; Maia, L.J.Q.; Cangussu, D. Investigating Solid-State Luminescent Properties of New Cadmium(II) and Lead(II) Oxamato-Based Coordination Polymers. *J. Braz. Chem. Soc.* **2024**, *35*, e-20240072. [CrossRef]
19. De Oliveira Maciel, J.W.; Lemes, M.A.; Valdo, A.K.; Rabelo, R.; Martins, F.T.; Queiroz Maia, L.J.; De Santana, R.C.; Lloret, F.; Julve, M.; Cangussu, D. Europium(III), Terbium(III), and Gadolinium(III) Oxamato-Based Coordination Polymers: Visible Luminescence and Slow Magnetic Relaxation. *Inorg. Chem.* **2021**, *60*, 6176–6190. [CrossRef]
20. Maciel, J.W.; Kalinke, L.H.G.; Valdo, A.K.; Martins, F.T.; Rabelo, R.; Moliner, N.; Cano, J.; Julve, M.; Lloret, F.; Cangussu, D. New Metal-Organic Systems with a Functionalized Oxamate-Type Ligand and $Mn^{II}$, $Fe^{II}$, $Cu^{II}$ and $Zn^{II}$. *J. Braz. Chem. Soc.* **2019**, *30*, 2413–2429. [CrossRef]
21. Oliveira, T.L.; Kalinke, L.H.G.; Mascarenhas, E.J.; Castro, R.; Martins, F.T.; Sabino, J.R.; Stumpf, H.O.; Ferrando, J.; Julve, M.; Lloret, F.; et al. Cobalt(II) and Copper(II) Assembling through a Functionalized Oxamate-Type Ligand. *Polyhedron* **2014**, *81*, 105–114. [CrossRef]
22. Sheldrick, G.M. Crystal Structure Refinement with SHELXL. *Acta Crystallogr. C Struct. Chem.* **2015**, *71*, 3–8. [CrossRef]
23. Spek, A.L. Single-Crystal Structure Validation with the Program PLATON. *J. Appl. Crystallogr.* **2003**, *36*, 7–13. [CrossRef]
24. Macrae, C.F.; Sovago, I.; Cottrell, S.J.; Galek, P.T.A.; McCabe, P.; Pidcock, E.; Platings, M.; Shields, G.P.; Stevens, J.S.; Towler, M.; et al. *Mercury 4.0*: From Visualization to Analysis, Design and Prediction. *J. Appl. Crystallogr.* **2020**, *53*, 226–235. [CrossRef]
25. Farrugia, L.J. *ORTEP-3;* School of Chemistry, University of Glasgow: Glasgow, Scotland, 2012.
26. Farrugia, L.J. WinGX and ORTEP for Windows: An Update. *J. Appl. Crystallogr.* **2012**, *45*, 849–854. [CrossRef]
27. Gruber, J.B.; Zandi, B.; Valiev, U.V.; Rakhimov, S.A. Energy Levels of $Dy^{3+}(^4f_9)$ in Orthoaluminate Crystals. *J. Appl. Phys.* **2003**, *94*, 1030–1034. [CrossRef]
28. Caspers, H.H.; Rast, H.E.; Fry, J.L. Absorption, Fluorescence, and Energy Levels of $Ho^{3+}$ in $LaF_3$. *J. Chem. Phys.* **1970**, *53*, 3208–3216. [CrossRef]
29. Krupke, W.F.; Grdbee, J.B. Absorption and Fluorescence Spectra of $Er^{3+}(^4f_{11})$ in $LaF_3$. *J. Chem. Phys.* **1963**, *39*, 1024–1030. [CrossRef]
30. Rajnák, C.; Boča, R. Reciprocating Thermal Behavior in the Family of Single Ion Magnets. *Coord. Chem. Rev.* **2021**, *436*, 213808. [CrossRef]

31. Cole, K.S.; Cole, R.H. Dispersion and Absorption in Dielectrics I. Alternating Current Characteristics. *J. Chem. Phys.* **1941**, *9*, 341–351. [CrossRef]
32. Sahu, P.K.; Konar, S. Dysprosium (III) Based One-Dimensional Zigzag Chain Exhibiting Slow Magnetic Relaxation. *Inorganica Chim. Acta* **2024**, *572*, 122255. [CrossRef]
33. Orts-Arroyo, M.; Rabelo, R.; Carrasco-Berlanga, A.; Moliner, N.; Cano, J.; Julve, M.; Lloret, F.; De Munno, G.; Ruiz-García, R.; Mayans, J.; et al. Field-Induced Slow Magnetic Relaxation and Magnetocaloric Effects in an Oxalato-Bridged Gadolinium(III)-Based 2D MOF. *Dalton Trans.* **2021**, *50*, 3801–3805. [CrossRef] [PubMed]
34. El Alouani Dahmouni, N.; Orts-Arroyo, M.; Sanchis-Perucho, A.; Moliner, N.; Mayans, J.; Pacheco, M.; Castro, I.; De Munno, G.; Marino, N.; Ruiz-García, R.; et al. Magnetocaloric Efficiency Tuning through Solvent-Triggered 3D to 2D Interconversion in Holmium(III)-Based Dynamic MOFs. *Chem. Commun.* **2024**, *60*, 7451–7454. [CrossRef] [PubMed]
35. Daudin, B.; Lagnier, R.; Salce, B. Thermodynamic Properties of the Gadolinium Gallium Garnet, $Gd_3Ga_5O_{12}$, between 0.05 and 25 K. *J. Magn. Magn. Mater.* **1982**, *27*, 315–322. [CrossRef]
36. Numazawa, T.; Kamiya, K.; Okano, T.; Matsumoto, K. Magneto Caloric Effect in $(Dy_xGd+)_3Ga_5O_{12}$ for Adiabatic Demagnetization Refrigeration. *Physica B Condens. Matter* **2003**, *329–333*, 1656–1657. [CrossRef]
37. Saines, P.J.; Paddison, J.A.M.; Thygesen, P.M.M.; Tucker, M.G. Searching beyond Gd for Magnetocaloric Frameworks: Magnetic Properties and Interactions of the $Ln(HCO_2)_3$ Series. *Mater. Horiz.* **2015**, *2*, 528–535. [CrossRef]
38. Dixey, R.J.C.; Saines, P.J. Optimization of the Magnetocaloric Effect in Low Applied Magnetic Fields in $LnOHCO_3$ Frameworks. *Inorg. Chem.* **2018**, *57*, 12543–12551. [CrossRef]
39. Doheny, P.W.; Cassidy, S.J.; Saines, P.J. Investigations of the Magnetocaloric and Thermal Expansion Properties of the $Ln_3(Adipate)_{4.5}(DMF)_2$(Ln = Gd-Er) Framework Series. *Inorg. Chem.* **2022**, *61*, 4957–4964. [CrossRef] [PubMed]
40. Li, Z.; Arauzo, A.; Giner Planas, J.; Bartolomé, E. Magnetic Properties and Magnetocaloric Effect of Ln = Dy, Tb Carborane-Based Metal–Organic Frameworks. *Dalton Trans.* **2024**, *53*, 8969–8979. [CrossRef]
41. Li, Z.; Arauzo, A.; Roscini, C.; Planas, J.G.; Bartolomé, E. Multifunctional Self-Refrigerated Multivariate {GdLn} (Ln = Dy, Tb, Tb/Eu) Metal–Organic Frameworks. *J. Mater. Chem. A Mater.* **2024**, *12*, 21971–21986. [CrossRef]
42. Herrero, A.; Aseguinolaza, I.R.; Oleaga, A.; Garcia-Adeva, A.J.; Apiñaniz, E.; Garshev, A.V.; Yapaskurt, V.O.; Morozkin, A.V. Selecting Optimal $R_6TX_2$ Intermetallics (R = Gd, Tb, Dy; T = Mn, Fe, Co, Ni; X = Sb, Te) for Magnetic Refrigeration. *Dalton Trans.* **2023**, *52*, 5780–5797. [CrossRef]
43. Kleinhans, M.; Eibensteiner, K.; Leiner, J.C.; Resch, C.; Worch, L.; Wilde, M.A.; Spallek, J.; Regnat, A.; Pfleiderer, C. Magnetocaloric Properties of $R_3Ga_5O_{12}$ (R=Tb, Gd, Nd, Dy). *Phys. Rev. Appl.* **2023**, *19*, 014038. [CrossRef]

magnetochemistry

Article

# Lanthanoid Coordination Polymers Based on Homoditopic Picolinate Ligands: Synthesis, Structure and Magnetic Properties †

Verónica Jornet-Mollá, Carlos J. Gómez-García *, Miquel J. Dolz-Lozano and Francisco M. Romero *

Departament de Química Inorgànica, Universitat de València, C/Dr. Moliner, 50, 46100 Burjassot, Spain; miquel.dolz@uv.es (M.J.D.-L.)

* Correspondence: carlos.gomez@uv.es (C.J.G.-G.); fmrm@uv.es (F.M.R.)

† This work is dedicated to Professors Miguel Julve and Francisco Lloret, two outstanding teachers and brilliant scientists in the field of coordination chemistry. With sincere gratitude for their friendship and support. Sadly, Professor Julve passed away on July 2024. Miguel was gifted with a quick mind, a generous heart and uncommon energy. His wisdom and constant inspiration will never be forgotten.

**Abstract:** A ditopic ligand ($H_2\mathbf{L_1}$), containing picolinate subunits segmented by ethynylene bridges, has been used in the synthesis of a series of isostructural coordination polymers, formulated as $[(CH_3)_2NH_2][Ln(\mathbf{L_1})_2]\cdot H_2O\cdot CH_3COOH$, where Ln = Eu (**1**), Gd (**2**), Tb (**3**), Dy (**4**) and Ho (**5**). The single-crystal structures show that these compounds crystallise in the orthorhombic $Pna2_1$ space group and form 3D anionic lattices with triangular cavities. AC magnetic susceptibility measurements show that the Gd, Tb and Dy derivatives (**2–4**) present a slow relaxation in their magnetisation under an applied DC magnetic field. The detailed study of the AC susceptibility in compounds **2** and **4** shows that they relax following direct and Orbach mechanisms under these conditions. The Dy derivative (**4**) retains this behaviour in the absence of an external field, relaxing via quantum tunnelling and Orbach mechanisms. Compound **2** is one of the very few reported Gd(III) compounds showing slow relaxation in its magnetisation.

**Keywords:** picolinate; lanthanoid; coordination polymers; magnetic properties; slow relaxation

## 1. Introduction

There is, nowadays, a strong interest in the synthesis and physical properties of lanthanoid coordination polymers (Ln-CPs) [1–3]. The reasons for this are manifold: (i) their structural versatility in comparison with transition metals, with higher coordination numbers leading easily to extended frameworks of different dimensionalities, including 3D porous structures [4–8]. (ii) Their fascinating photophysical and magnetic properties, with potential applications in different fields, such as luminescent (bio)sensors and magnetic materials for molecular spintronics and quantum computing [9,10].

Due to the internal character of the 4*f* electron shell, the magnetism of lanthanoid coordination complexes has traditionally drawn less attention with respect to 3*d* metal ions. This situation was reversed after the report of single-molecule magnetism (SMM) in the lanthanoid bis-phthalocyaninato anions, $[LnPc_2]^-$ (Ln = Dy, Tb; Pc = phthalocyanine), reported by Ishikawa and coworkers [11,12]. In these compounds, the key to SMM behaviour is the large local magnetic anisotropy that arises from the splitting of the $m_J$ sublevels by the crystal field, giving rise to a high energy barrier for magnetisation and slow spin dynamics [13–15]. Most studies concern mononuclear Dy(III) compounds [16,17], but examples of other nuclearities and cations also exist [18–23], including Ln-CPs of varying dimensionality [24,25]. In this latter case, the exchange interactions between neighbouring lanthanoid cations can lead to SCM (single-chain magnet) behaviour, especially when dealing with heterometallic 3d–4f systems [26–30]. In most other systems, however, the exchange interactions between lanthanoid ions are rather weak and the slow magnetic relaxation observed is mainly due to single-ion anisotropy effects [31].

Lanthanoid single-ion magnetic anisotropy arises from the interaction of the ground spin-orbit coupled $J$ level with the electrostatic crystal field. This results in a splitting of the $m_J$ states in such a way that the energy gaps between the different $\pm m_J$ doublets (resulting in different orientations of the spins) are very sensitive to the coordination environment. Large energy gaps can be obtained by optimizing the strength and symmetry of the ligand field with respect to the shape of the electronic density of the free metal ions. This results in high energy barriers for magnetisation reversal, giving rise to slow relaxation effects and magnetic hysteresis. As described by Long and coworkers, when the shape of the free-ion electron density is oblate, as for $Tb^{3+}$, $Dy^{3+}$ and $Ho^{3+}$ cations, an axial crystal field is needed to maximise the magnetic anisotropy [32]. The family of dysprosium metallocenes, with its ligand electron density concentrated above and below the $xy$ plane, are excellent examples of SMM behaviour [33,34]. Instead, ions such as $Er^{3+}$, having a prolate electron density, would perform better in an equatorial crystal field. It has been suggested that SMMs based on lanthanoid coordination polymers may be of interest since the lanthanoid coordination sphere can be constrained by the bridging linkers and the crystal symmetry, yielding new materials with improved magnetic properties [24,35]. Further, by using different bridging ligands, different compounds can be obtained that are essentially identical in their first coordination spheres but differ markedly in the way the lanthanoid ions assemble in the crystal lattice, giving some hints about the importance of dipolar interactions in fine-tuning magnetic behaviour [36].

In the quest for interesting bridging ligands, our attention has been focused on chelating and anionic units, which provide thermodynamical stability and charge compensation. Previous examples of coordination polymers containing units of this type involve tetraanionic bis-bidentate dioxidobenzenedicarboxylate ligands [37–39] and catecholates [40]. Surprisingly, despite the fact that the picolinate ligand is a universal chelating anionic entity, only a few polytopic picolinate ligands have been reported [41–44], and their use in metal–organic frameworks are underexplored. We have reported on a family of polytopic picolinate ligands bridged by ethynylene links, which provide rigidity, linear connectivity and electron delocalisation [45,46]. Coordination polymers containing these ligands, in combination with transition metal ions, have also been described. More recently, knowing the ability of picolinate ligands to act as photosensitisers for lanthanoid luminescence [47,48], we reported on Eu(III) CPs based on the ethynylene-bridged bis-picolinate ligand 5,5′-(ethyne-1,2-diyl)dipicolinate ($\mathbf{L_1}^{2-}$, Figure 1), a strongly luminescent material that can be useful in the detection of several species [49]. We herein describe the whole family of isostructural compounds of formulae $[(CH_3)_2NH_2][Ln(\mathbf{L_1})_2]\cdot H_2O\cdot CH_3COOH$ (Ln = Eu (**1**), Gd (**2**), Tb (**3**), Dy (**4**), Ho (**5**)), including their synthesis, structural and magnetic characterisation.

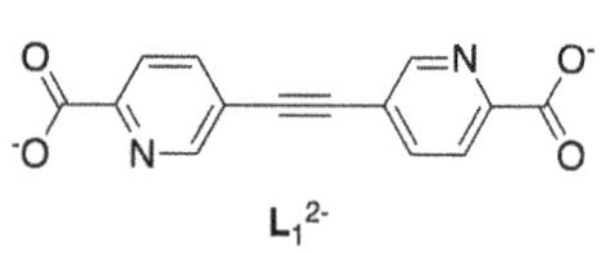


**Figure 1.** The segmented picolinate-based ligand $\mathbf{L_1}^{2-}$.

## 2. Materials and Methods

All chemicals and solvents were used as received. Ligand $\mathbf{L_1}^{2-}$ was synthesised as previously described [45].

*Synthesis of [$(CH_3)_2NH_2$][Ln($\mathbf{L_1}$)$_2$]·$H_2O$·$CH_3COOH$ (**1–5**).* A solution of 5,5′-(ethyne-1,2-diyl)dipicolinic acid (15 mg, 0.056 mmol) and $Ln(NO_3)_3{\cdot}6H_2O$ (0.056 mol) in 4.5 mL of DMF/$H_2O$ (8:1 ratio) was placed in a glass vial and stirred for 30 min until a solution was formed. Then, acetic acid was added, and the mixture was placed in an oven at 130 °C for 3 days. After cooling to room temperature at a cooling rate of 0.2 K·min$^{-1}$, the liquid phase was decanted, and colourless aggregated crystals from compounds **1–5** were obtained. Details concerning the yields, elemental analyses and IR spectra are given in the Supplementary Materials.

*Single crystal X-ray diffraction.* The suitable crystals of **1–5** were coated with paratone N oil, fixed on a small fibre loop and mounted on an Oxford Diffraction Supernova diffractometer equipped with a graphite-monochromated Enhance Mo X-Ray Source (λ = 0.71073 Å) at 120 K. The data collection routines, unit cell refinements and data processing were carried out using the CrysAlisPro v38.46 software package [50]. The structures were solved using SHELXT 2018/2 via the WinGX v2018.3 graphical interface [51] and refined using SHELXL-2018/3 [52]. Although X-ray experiments (powder and single-crystal) unambiguously support the isostructural character of the whole lanthanoid series, satisfactory data for complete structural characterization could only be obtained for the Eu (**1**) (reported previously) and Tb (**3**) derivatives. All non-hydrogen atoms were refined anisotropically for **1** (DELU and SIMU restraints were applied to C and O atoms of the acetic acid molecules to allow their anisotropic refinement). For **3**, all non-hydrogen atoms were refined anisotropically, except those concerning the acetic acid molecule and O1W. H atoms on carbon atoms were included at calculated positions and refined with a riding model with relative isotropic displacement parameters. Instead, H atoms on solvent molecules and amine H atoms on dimethylammonium cations were found using Fourier difference maps and refined with constrained isotropic thermal parameters, except for the H atoms of O1W. CCDC 2068875 and CCDC 2420909 contain the supplementary crystallographic data for **1** and **3**, respectively. These data were provided free of charge by the Cambridge Crystallographic Data Centre.

*Powder X-ray Diffraction (PXRD).* PXRD measurements for compounds **1–5** were collected using Cu Kα radiation (λ = 1.54056 Å) at room temperature and in a $2\theta$ range from 2 to 40°. Polycrystalline samples were lightly ground in an agate mortar and filled into a 0.5 mm borosilicate capillary prior to being mounted and aligned on an Empyrean PANalytical powder diffractometer. Simulated diffractograms were obtained from single-crystal X-ray data using the CrystalDiffract software.

*Magnetic susceptibility measurements.* Variable temperature susceptibility measurements were carried out for compounds **1–5** in the temperature range of 2–300 K with an applied magnetic field of 100 mT on ground polycrystalline samples, using a Quantum Design (San Diego, CA, USA) MPMS XL-5 SQUID magnetometer. The susceptibility data were corrected for the sample holder and the diamagnetic contributions of the sample using Pascal's constants [53]. AC susceptibility measurements were performed on the same samples for compounds **2–5** at low temperatures with different applied DC fields in a frequency range of 10 Hz–10 kHz, using Quantum Design (San Diego, CA, USA) PPMS-9 equipment.

*Other characterization techniques.* Thermogravimetric analyses were performed on a Mettler-Toledo TGA/SDTA/851e apparatus under an $N_2$ atmosphere at a scan rate of 10 K·min$^{-1}$. IR transmission measurements were performed from the powdered samples at room temperature in an FT-IR spectrometer (Bruker, alpha II) equipped with an attenuated total reflection (ATR) accessory in the range of 400–4000 $cm^{-1}$. C, H and N elemental analyses were measured in a CE INSTRUMENTS 1110 EA elemental analyser (SCSIE, Universitat de València).

## 3. Results and Discussion

### 3.1. Synthesis and Characterization

Coordination polymers **1–5** were prepared by direct reaction between the corresponding lanthanoid nitrate and ligand $H_2\mathbf{L_1}$ under solvothermal conditions. Although important efforts have been devoted to obtaining single crystals of the erbium derivative, its isolation was not possible. The addition of acetic acid as a modulator was necessary in order to obtain single crystals for X-ray structural analysis. These could also be obtained from the corresponding methyl diesters, $Me_2\mathbf{L_1}$. In this case, the modulator of choice was formic acid.

In order to confirm the phase purity of the bulk materials, powder X-ray diffraction measurements were performed (Figure S1). The experimental patterns recorded at room temperature for compounds **1–5** were compared with the simulated diffractogram for the Tb derivative **3** obtained from single-crystal data at 120 K. There is a perfect agreement between both subsets of patterns, discarding the presence of crystalline impurities in the materials.

The thermogravimetric analysis (Figure S2) of the different compounds is very similar, exhibiting a first weight loss between room temperature and 390 K that is associated with the release of water. A second step ascribed to the loss of acetic acid is detected between 425 and 590 K. Above this temperature, the samples start to decompose.

### 3.2. Structural Properties

Compounds **1–5** are isostructural and crystallise in the orthorhombic $Pna2_1$ space group. Their asymmetric unit contains an $Ln^{3+}$ cation, two crystallographically independent $\mathbf{L_1}^{2-}$ dianions, one dimethylammonium cation and crystallization solvent molecules (one water molecule and one acetic acid). The Flack parameter is close to zero (Table 1), implying that the model has the correct absolute structure [54].

The presence of dimethylammonium cations is required in order to balance the negative charges of the host framework. They arise from the hydrolysis of dimethylformamide, used as a solvent during the synthesis. In the following, the structure of the Tb compound (**3**) is described. A comparison with the structure of the Eu derivative (**1**) is given in Tables S1 and S2.

The terbium cation is located in a general position and exhibits a non-coordinated tricapped trigonal prismatic (TTP) geometry (Figure 2). It is coordinated by four picolinate subunits in a chelating mode. One of these picolinate moieties acts also as a $\mu$-carboxylato bridging ligand and occupies the ninth position in the lanthanoid coordination sphere. The Tb–O distances lie in the 2.33–2.39 Å range, much shorter than the Tb–N1 distances, which comprise between 2.58 and 2.68 Å (Table S1). This is in agreement with the reported values for Ln(III) complexes based on picolinate ligands and is due to the strong oxophilicity of the $Ln^{3+}$ cations [55]. The triangular faces of the tricapped trigonal prism are defined by O3O6N2 and O1O5O7 atoms, whereas the capping positions over each rectangular face are occupied by N1, N3 and N4 nitrogen atoms. The angle between the trigonal faces is 175.3°, indicating a small deviation from the ideal parallel arrangement.

**Table 1.** X-ray crystal data for **1** and **3**.

| | **1** | **3** |
|---|---|---|
| chemical formula | $C_{32}H_{26}EuN_5O_{11}$ | $C_{32}H_{26}N_5O_{11}Tb$ |
| $a$ (Å) | 19.6440(2) | 19.70340(10) |
| $b$ (Å) | 12.35500(10) | 12.32080(10) |
| $c$ (Å) | 13.21490(10) | 13.14510(10) |
| $\alpha$ (deg) | 90.00 | 90.00 |
| $\beta$ (deg) | 90.00 | 90.00 |
| $\gamma$ (deg) | 90.00 | 90.00 |
| $V$ (Å$^3$) | 3207.28(5) | 3191.13(4) |
| $T$ (K) | 119.7(8) | 120.00(10) |
| $Z$ | 4 | 4 |
| $M_r$ (g/mol) | 808.54 | 815.50 |
| crystal system | orthorhombic | orthorhombic |
| space group | $Pna2_1$ (No. 33) | $Pna2_1$ (No. 33) |
| crystal dimensions (mm) | 0.124 0.105 0.073 | 0.180 0.136 0.122 |
| $\mu$ (Mo $K\alpha$) (mm$^{-1}$) | 2.025 | 2.286 |
| $\lambda$ (Å) | 0.71073 | 0.71073 |
| density (Mg/m$^3$) | 1.674 | 1.697 |
| index ranges for $h, k, l$ | −25/24, −16/16, −16/17 | −25/25, −16/15, −17/17 |
| $\theta$ range (deg) | 3.298 to 28.107 | 3.307 to 27.710 |
| goodness-of-fit on $F^2$ | 1.082 | 1.115 |
| reflns collected | 118251 | 116487 |
| independent reflns ($R_{int}$) | 7383 (0.0975) | 7172 (0.0435) |
| data/restraints/parameters | 7383/33/453 | 7172/6/422 |
| $R1$, $wR2$ [$I > 2\sigma(I)$] [1] | 0.0389, 0.0765 | 0.0308, 0.0714 |
| $R1$, $wR2$ (all data) [1] | 0.0627, 0.0883 | 0.0356, 0.0747 |
| absolute structure parameter | −0.039(6) | −0.026(4) |

[1] $R1 = \sum(|F_o| - |F_c|)/\sum|F_o|$; $wR2 = [\sum[w(F_o^2 - F_c^2)^2]/\sum[wF_o^4]]^{1/2}$.

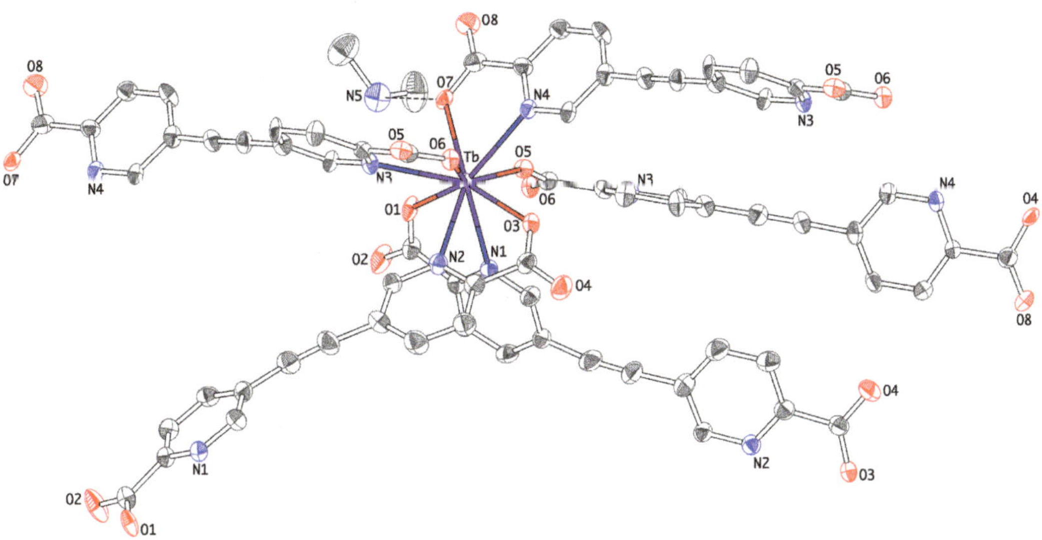


**Figure 2.** Thermal ellipsoid plot of the $Tb^{3+}$ coordination environment in **3**.

Continuous shape measure (CShM) calculations using SHAPE 2.1 confirm that the coordination geometries of **1** and **3** are slightly distorted tricapped trigonal prisms (with minimum CShM calculated values corresponding to $D_{3h}$ symmetry, Table S2) [56–58]. A

CShM calculation of a coordination sphere indicates its deviation from an ideal structure, regardless of its size and orientation, and is given by the following expression:

$$S(G) = min\frac{\sum_{k=1}^{N}|Q_k - P_k|}{\sum_{k=1}^{N}|Q_k - Q_0|} \tag{1}$$

The vectors $Q_k$ ($k$ = 1, 2. . .$N$) indicate the positions of the $N$ donor atoms defining the coordination sphere, and $G$ is a specific and perfect symmetry group with coordinates $P_k$ ($k$ = 1, 2. . .$N$). The distances between the vertices of the two objects are calculated until a set of $P_k$ coordinates is found that minimises $S(G)$ (the denominator is a size normalisation factor, $Q_0$ being the coordinate vector of the centre of mass in the investigated structure). The lower the CShM parameter, the better the agreement between the investigated and reference coordination geometries [59]. According to Equation (1), CShM measures must lie within the range of $100 \geq S \geq 0$. In compounds **1** and **3**, values of 3.882 and 3.891, respectively, indicate a close agreement with the ideal TTP geometry ($D_{3h}$ symmetry).

The two independent $\mathbf{L_1}^{2-}$ bis-picolinate ligands adopt different coordination modes. Ligand A acts as a terminal (N1,O1);(N2,O3) bis-chelating ligand, adopting a *transoid* arrangement of the two picolinate moieties and connecting two $Tb^{3+}$ complexes separated by the ethynylene spacer. Ligand B behaves also as a ditopic ligand, coordinating to two distant $Tb^{3+}$ cations related by a translation along the $y$ direction, but one of the picolinate subunits (N3O6) acts as a tridentate ligand in an (N,O);O mode, connecting two lanthanoid metal ions by *anti, anti*-carboxylate (O5O6) bridges that organise zigzag chains propagating along the $c$ axis. The intrachain terbium–terbium distance is 6.6564(7) Å (Figure 3a). This connectivity results in layers of Tb(III) complexes that run parallel to the bc plane.

The layers are further connected by ligand A, binding two $Tb^{3+}$ cations separated by a distance of 12.8146(6) Å. This yields a 3D anionic lattice showing triangular cavities (Figure 3b). Dimethylammonium countercations sit within these cavities and are hydrogen-bonded to a coordinating picolinate oxygen atom (H-bond distance N5···O7: 2.741(9) Å) and a water molecule (H-bond distance N5···O1W: 2.704(14) Å). On the other hand, the acetic acid molecule forms a short–strong H-bond with a non-coordinated oxygen atom (H-bond distance O9···O4: 2.486(19) Å).

Both anions deviate considerably from a planar structure. Ligand A adopts a *transoid* conformation, with a dihedral angle between the two pyridine rings close to 27°. In ligand B, however, the two pyridine rings are almost perpendicular to each other, with a dihedral angle of about 80°. This may be attributed to the fact that this ligand binds to a third $Ln^{3+}$ cation via the *anti, anti*-carboxylate (O5O6) bridge. For the same reason, the torsion angle involving the O5 carboxylate oxygen atom (12.59°) is higher than those obtained for the other carboxylate moieties (0.82–8.82°). With respect to the bending of the triple bond, it is similar in both ligands, with a maximum deviation of 8.23° to the ideal –C–C≡C– and –C≡C–C– angles of 180°. Table S3 lists the relevant angles for comparing the crystal structures of **1** and **3**.

Although compounds **1**–**5** are isostructural, it can be observed that, as the atomic number increases, cell volume decreases. This is a clear signature of the lanthanoid contraction effect (Figure S3).

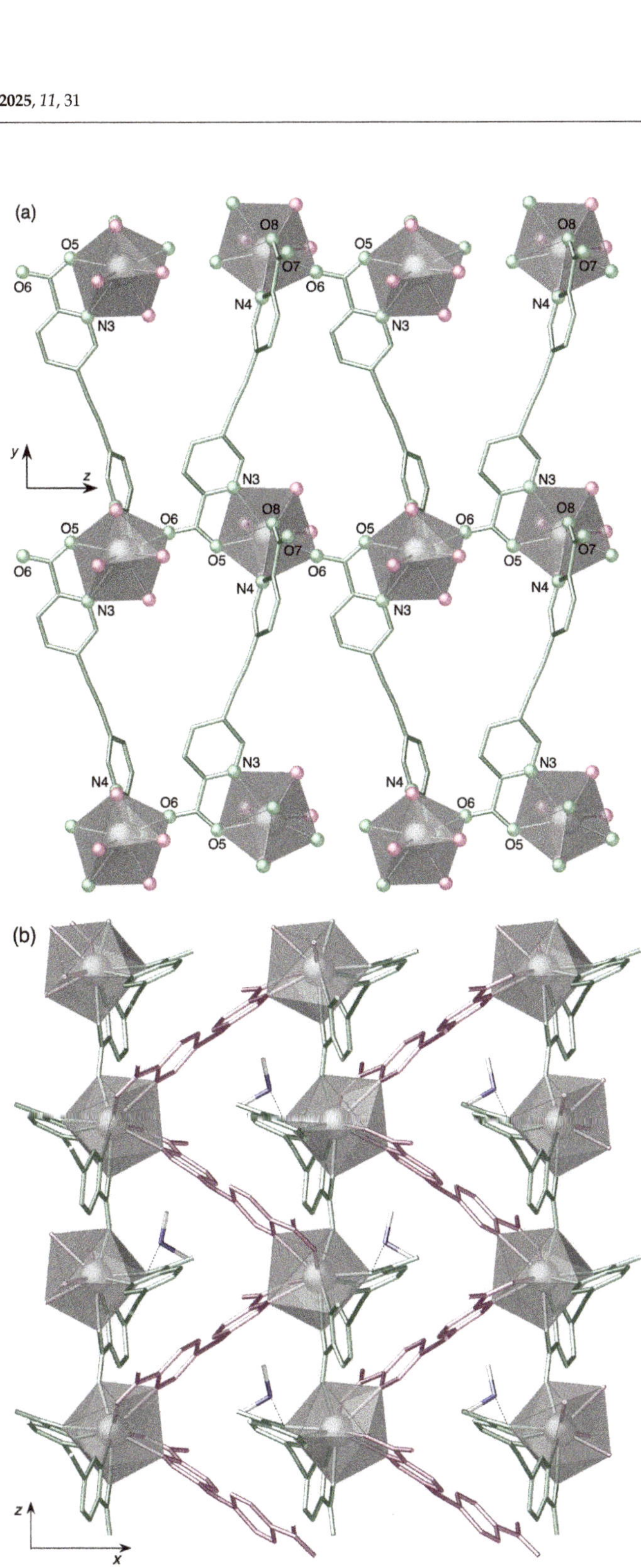


**Figure 3.** (**a**) View of the crystal structure of **3** along the $x$ direction, showing the layers defined by ligand B (green). (**b**) Projection of the crystal structure of **3** onto the $xz$ plane, showing the connections between layers mediated by ligand A (pink) and the presence of triangular voids containing dimethylammonium cations (blue). H atoms have been omitted for clarity.

### 3.3. Magnetic Properties

Magnetic susceptibility measurements were performed on the bulk samples **1–5**, in a static dc magnetic field of 100 mT. The thermal variation in the $\chi_m T$ product ($\chi_m$ = molar magnetic susceptibility) in the 2–300 K temperature range (Figure 4) shows constant values at high temperatures, in agreement with those expected for the $Ln^{3+}$ ions in their ground states (Table 2) [60]. The exception of this is the europium derivative (**1**), with a value of $\chi_m T$ at room temperature equal to 1.55 emu·K·mol$^{-1}$, showing a steady decrease in the $\chi_m T$ signal upon cooling until the signal practically vanishes at 2 K, as expected for the diamagnetic $^7F_0$ ground state of $Eu^{3+}$ ions. The non-zero value observed at room temperature indicates the thermal population of excited $^7F_J$ ($J$ = 1–6) multiplet levels resulting from the splitting of the $^7F$ ground term by first-order spin-orbit coupling ($\lambda$). In fact, the two first excited states ($^7F_1$ and $^7F_2$) are quite close to the ground state, and they can be populated at room temperature. Accordingly, we can fit the magnetic susceptibility data for compound **1** to a simple expression that is only dependent on the first-order spin-orbit coupling parameter ($\lambda$), as follows:

$$\chi_m T = \frac{N\beta^2}{3kTx}\,\frac{24 + \left(\frac{27x}{2} - \frac{3}{2}\right)e^{-x} + \left(\frac{135x}{2} - \frac{5}{2}\right)e^{-3x} + \left(189x - \frac{7}{2}\right)e^{-6x} + \left(405x - \frac{9}{2}\right)e^{-10x} + \left(\frac{1485x}{2} - \frac{11}{2}\right)e^{-15x} + \left(\frac{2457x}{2} - \frac{13}{2}\right)e^{-21x}}{1 + 3e^{-x} + 5e^{-3x} + 7e^{-6x} + 9e^{-10x} + 11e^{-15x} + 13e^{-21x}} \quad (2)$$

with $x = \lambda/kT$. Equation (2) very satisfactorily reproduces the magnetic data for compound **1** with $\lambda$ = 389(2) cm$^{-1}$ (solid line in Figure 4). This value is similar to those reported in other $Eu^{III}$ compounds [61–63].

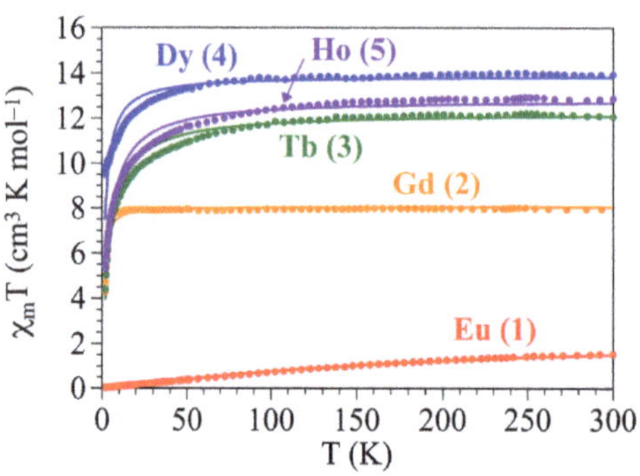


**Figure 4.** Thermal variation in the $\chi_m T$ product for compounds **1–5**. The solid lines are the fits to a simple model with weak intermolecular interactions (see text).

**Table 2.** Experimental and expected $\chi_m T$ values (emu·K·mol$^{-1}$) at 300 K and magnetic parameters extracted from the fit of the DC magnetic data for **1–5**.

| | **1** | **2** | **3** | **4** | **5** |
|---|---|---|---|---|---|
| Experimental $\chi_m T$ | 1.55 | 7.86 | 12.05 | 13.65 | 14.04 |
| Expected $\chi_m T$ | 0 (1.50) | 7.87 | 11.82 | 14.05 | 14.04 |
| $\lambda$ (cm$^{-1}$) | 389(2) | | | | |
| Experimental $g$ | | 2.023(4) | 1.528(2) | 1.301(2) | 1.195(2) |
| Expected $g$ | | 2.0 | 1.5 | 1.33 | 1.20 |
| $\lvert D \rvert$ (cm$^{-1}$) | | 0.27(3) | 0.15(1) | 1.35(6) | 1.14(2) |
| $zj$ (cm$^{-1}$) | | −0.018(2) | −0.091(2) | −0.032(1) | −0.079(1) |

The $\chi_m T$ values of the isotropic $Gd^{3+}$ derivative (**2**) remain essentially constant in the whole temperature range and decrease abruptly only at temperatures below 10 K. This behaviour indicates that, despite the presence of carboxylate bridges between neighbouring $Ln^{3+}$ ions, the exchange interactions are very weak [64,65]. In the other compounds (**3–5**), the decrease is more gradual and may be ascribed to the progressively thermal depopulation in the highest ground-state $Ln^{3+}$ sublevels that arise from the splitting of the ground term by the ligand field.

Given the lack of significant interactions, we have fitted the $\chi_m T$ product for compounds **2–5** to a simple model, including a ZFS contribution (*D*) and a very weak antiferromagnetic coupling (*zj*), modelled using the mean-field approximation with the program PHI (solid lines in Figure 4) [66]. The parameters found in the fit are displayed in Table 2.

The good magnetic isolation provided by the picolinate ligands prompted us to measure the AC magnetic susceptibility at low temperatures for **2–5** in different applied DC fields to look for slow relaxation in the magnetisation (SRM). With the exception of the Ho derivative (**5**), the compounds show the presence of a frequency-dependent out-of-phase signal ($\chi''_m$) at low temperatures when a DC field is applied (Figures S4–S7). Interestingly, for the Dy derivative (**4**), we could observe an out-of-phase signal even in the absence of a DC field, although with a lower intensity (Figure S6). The lack of SRM in **2–3** when no DC field is applied is due to the presence of an efficient relaxation in the magnetization through a quantum tunnelling mechanism. This fast relaxation is suppressed in a DC field. As it can be seen in Figures S4–S6, the intensity of the $\chi''_m$ signal initially increases with an increase in the magnetic field and shows two different maxima, one at low frequencies (slow relaxation process with a relaxation time $\tau_1$), and a second one at high frequencies (fast relaxation process, $\tau_2$). In the Gd (**2**) and Dy (**4**) derivatives, we can clearly observe how the maximum of the fast relaxation process ($\tau_2$) shifts to lower frequencies and increases in intensity, reaching a minimum frequency and a maximum intensity for DC fields of around 120 mT in **2** and 80 mT in **4** (Figures S4 and S6, respectively). In the Tb derivative (**3**), the two maxima appear outside the frequency range of our equipment, although we could determine that the maximum intensity is reached for DC values around 140 mT (Figure S5).

The fitting of the frequency dependence of $\chi''_m$ to a Debye model for the two relaxation processes (Equation (3), solid lines in Figures S4–S6) gives us the relaxation times ($\tau$) for each applied DC field [67]. Note that, since the maxima of the first relaxation process lie outside the measured frequency range, $\tau_1$ values are unreliable in all cases. This applies also to the $\tau_2$ value for compound **3**. The field dependence of $\tau_2$ for compounds **2** and **4** show, as expected, maxima at values around 120 and 80 mT, respectively (Figures S8 and S9). Despite that there are seven adjustable parameters in Equation (3), the isothermal ($\chi_{T1}$ and $\chi_{T2}$) and adiabatic ($\chi_S$) susceptibilities, as well as the relaxation times ($\tau_1$ and $\tau_2$), can be easily estimated from the in-phase signal ($\chi'_m$) values at different frequencies and from the position of the low- and high-frequency maxima, respectively. Note also that, although there is only one crystallographically independent $Ln^{3+}$ ion in compounds **2–4**, the presence of two different relaxation processes has already been observed in other lanthanoid-based lattices and has been attributed to the presence of relaxation pathways via excited states [68].

$$\chi''(\omega) = (\chi_{T1} - \chi_S)\frac{(\omega\tau_1)^{1-\alpha_1}\cos\left(\frac{\pi\alpha_1}{2}\right)}{1+2(\omega\tau_1)^{1-\alpha_1}\sin\left(\frac{\pi\alpha_1}{2}\right)+(\omega\tau_1)^{2-2\alpha_1}} + (\chi_{T2} - \chi_{T1})\frac{(\omega\tau_2)^{1-\alpha_2}\cos\left(\frac{\pi\alpha_2}{2}\right)}{1+2(\omega\tau_2)^{1-\alpha_2}\sin\left(\frac{\pi\alpha_2}{2}\right)+(\omega\tau_2)^{2-2\alpha_2}} \quad (3)$$

The relaxation times obtained from the fit to the Debye model increase as the DC field increases, reaching maxima at around 120 mT for the Gd derivative (Figure S8) and 80 mT for the Dy compound (Figure S9). This behaviour can be well reproduced with the general model (Equation (4)) [69], using the parameters displayed in Table S4 (solid lines in Figures S8 and S9).

$$\tau^{-1} = AH^{n}T + \frac{B_1}{1 + B_2H^2} + D \tag{4}$$

In this equation, A, $B_1$, $B_2$ and D are parameters corresponding to the different relaxation mechanisms: direct (A), quantum tunnelling ($B_1$ and $B_2$), and Raman and Orbach (D, since they are field-independent). $H$ is the applied DC magnetic field and n is equal to 4 or 2 for Kramers (Gd and Dy) and non-Kramers (Tb) ions, respectively.

A detailed study of the frequency dependence of $\chi''_m$ at different temperatures was performed under a constant DC field set at the values where relaxation time $\tau_2$ and signal intensity are the highest. The frequency dependence of $\chi''_m$ in an applied DC field of 120 mT for the Gd derivative (**2**) shows at 1.9 K a maximum of around 740 Hz that shifts to higher frequencies as the temperature increases (Figure 5a). The fit of this data to the Debye model for the two relaxation processes (Equation (3)), gives us the relaxation times at each temperature. These data, displayed in an Arrhenius plot (ln$\tau$ vs. $1/T$, Figure 5b), show linear dependence at high temperatures for **2**, with a clear curvature at lower temperatures. This result suggests the presence of an Orbach thermally activated relaxation mechanism (second term in Equation (5)), with additional Raman and/or direct mechanisms (third and fourth terms, respectively, in Equation (5)) [70]. Attempts to fit the Arrhenius plot for compound **2** to all the possible mechanisms in Equation (5) show that the best fit is obtained including only the Orbach and direct mechanisms (second and fourth terms in Equation (5)), using the parameters displayed in Table S5 (solid line in Figure 5b).

$$\tau^{-1} = \tau_{QT}^{-1} + \tau_0^{-1}\exp\left(-\frac{U_{eff}}{kT}\right) + CT^{n} + AH^{4}T \tag{5}$$

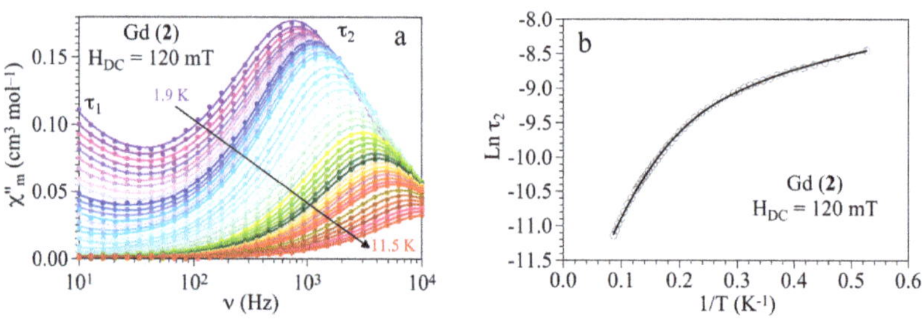


**Figure 5.** (**a**) Frequency dependence of $\chi''_m$ for the Gd derivative (**2**) in a DC field of 120 mT at different temperatures. The solid lines are the best fit for the Debye model for the two processes (Equation (3)). (**b**) Arrhenius plot of the relaxation times for **2** in a DC field of 120 mT. The solid line is the best fit for a model with direct and Orbach relaxation mechanisms (Equation (5)).

The Tb derivative (**3**) also shows two different relaxation processes but none of them exhibit a corresponding maximum in the frequency range measured (Figure S10). Therefore, the relaxation times obtained with the Debye model, including the two relaxation processes (Equation (3)), are unreliable and preclude the extraction of relevant parameters from the Arrhenius plot.

Finally, for the Dy derivative (**4**), we studied the thermal dependence of the out-of-phase signal in both zero and 80 mT DC fields. In both cases, at 1.9 K, we observe maxima in the $\chi''_m$ signal at 3200 Hz, with the signal in $H_{DC}$ = 80 mT being stronger (Figure 6). As the temperature is increased under the zero DC field, the maximum remains almost unchanged, in the range of 1.9–5.2 K (Figure 6a), and shifts to higher frequencies above around 5.4 K

(Figure S11). For $H_{DC}$ = 80 mT, when the temperature is increased, the maximum shifts very fast to higher frequencies (Figure 6b).

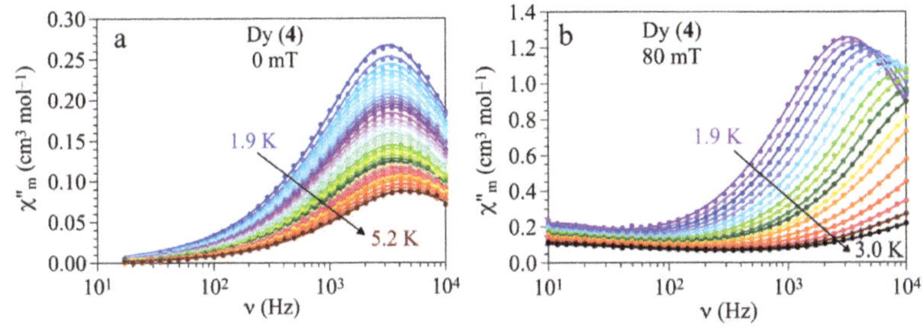


**Figure 6.** (**a**) Frequency dependence of $\chi''_m$ for the Dy derivative (**4**) at different temperatures in zero DC field. (**b**) Same plot in a DC field of 80 mT. The solid lines are the best fit for the Debye model for the two processes (Equation (3)).

This behaviour indicates that in a zero static field, the predominant mechanism for the relaxation of the magnetisation at low temperatures is quantum tunnelling, which is temperature independent, whereas, at higher temperatures, there is a thermally activated mechanism (Orbach type, see below). As expected, when $H_{DC}$ = 80 mT, the quantum tunnelling is suppressed and the magnetisation relaxes through temperature-dependent mechanisms (direct and Orbach, see below).

As in compound **2**, we fitted the data to a Debye model for the two relaxation processes (Equation (3)) to obtain the relaxation times of the fast relaxation process ($\tau_2$) at each temperature. The Arrhenius plot (ln$\tau$ versus $1/T$, Figure 7), in a zero static field, shows an almost temperature-independent behaviour at low temperatures and a curvature as the temperature increases, reaching linear behaviour at higher temperatures. This behaviour suggests the presence of quantum tunnelling and Orbach relaxation mechanisms (first and second terms in Equation (5)). Accordingly, we have fit the data with these two terms to obtain the parameters gathered in Table S5 (solid line in Figure 7a). When $H_{DC}$ = 80 mT, the Arrhenius plot shows linear behaviour with a slight curvature at low temperatures, indicative of the presence of Orbach, direct, and/or Raman mechanisms. Attempts to fit the data with the three possible mechanisms (quantum tunnelling was discarded under the presence of an applied DC field) show that the data can be fitted to Orbach and direct mechanisms, using the parameters shown in Table S5 (solid line in Figure 7b).

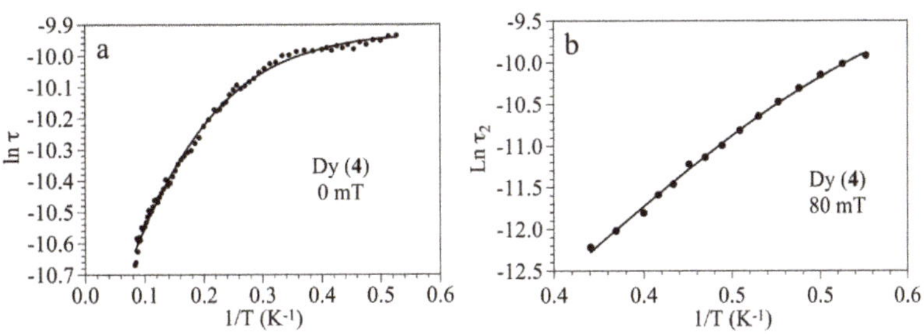


**Figure 7.** (**a**) Arrhenius plot of the relaxation times for the Dy derivative (**4**) in a zero static field. (**b**) The same plot in a DC field of 80 mT. The solid lines are the best fits to the model (see text).

The frequency dependence of the in-phase signals ($\chi'_m$) for compounds **2–4** show an inflexion point corresponding to the maxima in the out-of-phase signals (Figures S12–S16).

The Cole–Cole plots ($\chi''_m$ vs. $\chi'_m$, Figures S17–S21) show the typical semi-circular shapes in all cases, with one or two semicircles depending on the number of relaxation processes observed in the out-of-phase plots.

It is interesting to note that **2** is one of the few reported Gd(III) lattices showing slow relaxation in its magnetisation (SRM). Thus, although there are hundreds of reports on Dy(III) compounds showing SRM [71–74], the number of compounds with Gd(III) showing SRM is, by far, much lower. This is due to the fact that the $Gd^{3+}$ ion has a half-filled $4f^7$ configuration with a ground multiplet $^8S_{7/2}$ with no orbital contribution. In the first stage, slow relaxation in Gd(III) complexes has been ascribed to a phonon-bottleneck effect [75]. This picture assumes that the wavelength of the resonant phonon is larger than the separation between $Gd^{3+}$ ions, a situation that is possible in our case. Indeed, the first reports on SRM for Gd(III) compounds invariably are related to polymeric structures showing relatively short Gd–Gd distances [31,76–78], although, recently, well-isolated mononuclear Gd(III) compounds exhibiting SRM have been reported [79] and it has been argued that zero-field splitting in strongly distorted environments could account for the efficiency of the resonant phonon trapping [80]. Thus, the observation of SRM for Gd(III) compounds might be more common than initially expected, as the recent literature provides more examples of this behaviour. An extensive list of materials is given in Table S6 [81–95].

## 4. Conclusions

A segmented ligand $\mathbf{L_1}^{2-}$, containing two picolinate anions connected by ethynylene bridges, has been used in the synthesis of five isostructural 3D anionic lattices with triangular cavities, formulated as $[(CH_3)_2NH_2][Ln(\mathbf{L_1})_2]\cdot H_2O\cdot CH_3COOH$, where Ln = Eu (**1**), Gd (**2**), Tb (**3**), Dy (**4**) and Ho (**5**). AC magnetic measurements show that **2–4** present slow relaxation when a DC magnetic field is applied, whereas the Dy derivative (**4**) exhibits slow relaxation in its magnetisation, even in the absence of a static field. The AC susceptibility in compounds **2** and **4** shows that, in both compounds, the dynamic properties follow direct and Orbach mechanisms in a DC field. When this field is switched off, magnetic relaxation in **4** follows quantum tunnelling and Orbach mechanisms. Compound **2** constitutes one of the very rare examples of slow relaxation in a Gd(III) compound.

**Supplementary Materials:** The following supporting information can be downloaded at: https://www.mdpi.com/article/10.3390/magnetochemistry11040031/s1. Details concerning the synthesis and characterisation of **1–5**; Table S1: relevant bond distances (Å) in the crystal structures of **1** and **3**; Table S2: continuous shape measures (CShM) for **1** and **3** using SHAPE 2.1; Table S3: relevant angles (deg) in the crystal structures of **1** and **3**; Table S4: magnetic parameters obtained from the fit of the field dependence of the relaxation times at 1.9 K (Equation (4)); Table S5: magnetic parameters obtained from the fit of the Arrhenius plot to the general model (Equation (5)); Figure S1: powder X-ray diffractograms for compounds **1–5**. The simulation of the diffractogram for **3** from single-crystal data is offered for comparison; Figure S2: thermogravimetric analysis for compounds **1–5**; Figure S3: variation in the unit cell volume with increasing atomic numbers (Z) for **1–5**; Figure S4: frequency dependence of the out-of-phase susceptibility ($\chi''_m$) for the Gd derivative (**2**) at 1.9 K in different applied DC fields; Figure S5: frequency dependence of the out-of-phase susceptibility ($\chi''_m$) for the Tb derivative (**3**) at 1.9 K in different applied DC fields; Figure S6: frequency dependence of the out-of-phase susceptibility ($\chi''_m$) for the Dy derivative (**4**) at 1.9 K in different applied DC fields; Figure S7: frequency dependence of the out-of-phase susceptibility ($\chi''_m$) for the Ho derivative (**5**) at 1.9 K in different applied DC fields; Figure S8: field dependence of the relaxation time for the fast relaxation process ($\tau_2$) in the Gd derivative (**2**) at 1.9 K; Figure S9: field dependence of the relaxation time for the fast relaxation process ($\tau_2$) in the Dy derivative (**4**) at 1.9 K; Figure S10: frequency dependence of $\chi''_m$ for the Tb derivative (**3**) in a DC field of 140 mT at different temperatures;

Figure S11: frequency dependence of $\chi''_m$ for the Dy derivative (**4**) in a zero DC field at different temperatures above 5.4 K; Figure S12: frequency dependence of $\chi'_m$ for the Gd derivative (**2**) with $H_{DC}$ = 120 mT, in the temperature range of 1.9–11.5 K; Figure S13: frequency dependence of $\chi'_m$ for the Tb derivative (**3**) with $H_{DC}$ = 140 mT, in the temperature range of 1.9–4.8 K; Figure S14: frequency dependence of $\chi'_m$ for the Dy derivative (**4**) with $H_{DC}$ = 0 mT, in the temperature range of 1.9–5.2 K; Figure S15: frequency dependence of $\chi'_m$ for the Dy derivative (**4**) with $H_{DC}$ = 0 mT, in the temperature range of 5.4–12.0 K; Figure S16: frequency dependence of $\chi'_m$ for the Dy derivative (**4**) with $H_{DC}$ = 80 mT, in the temperature range of 1.9–3.0 K; Figure S17: Cole-Cole plot for the Gd derivative (**2**) with $H_{DC}$ = 120 mT, in the temperature range of 1.9–11.5 K; Figure S18: Cole-Cole plot for the Tb derivative (**3**) with $H_{DC}$ = 140 mT, in the temperature range of 1.9–4.8 K; Figure S19: Cole-Cole plot for the Dy derivative (**4**) with $H_{DC}$ = 0 mT, in the temperature range of 1.9–5.2 K; Figure S20: Cole-Cole plot for the Dy derivative (**4**) with $H_{DC}$ = 0 mT, in the temperature range of 5.4–12.0 K; Figure S21: Cole-Cole plot for the Dy derivative (**4**) with $H_{DC}$ = 80 mT, in the temperature range of 1.9–3.0 K.

**Author Contributions:** Conceptualization, F.M.R.; formal analysis, V.J.-M., C.J.G.-G., M.J.D.-L. and F.M.R.; investigation, V.J.-M., C.J.G.-G., M.J.D.-L. and F.M.R.; writing—original draft preparation, C.J.G.-G. and F.M.R.; writing—review and editing, V.J.-M., C.J.G.-G., M.J.D.-L. and F.M.R. All authors have read and agreed to the published version of the manuscript.

**Funding:** This research forms part of the Advanced Materials program and was funded by the Spanish MCIN, with funding from the European Union NextGeneration EU (PRTR-C17.I1) and the Generalitat Valenciana (project MFA-2022-057). We also express thanks for Grant PID2021-125907NB-I00, funded by the MCIN/AEI/10.13039/501100011033, as well as "ERDF A way of making Europe" and project CIPROM-2022-060 from the Generalitat Valenciana, for financial support.

**Institutional Review Board Statement:** Not applicable.

**Informed Consent Statement:** Not applicable.

**Data Availability Statement:** The data in this contribution will be provided upon request to the corresponding author.

**Conflicts of Interest:** The authors declare no conflicts of interest.

## References

1. Yan, B. Lanthanide-functionalized metal–organic framework hybrid systems to create multiple luminescent centers for chemical sensing. *Acc. Chem. Res.* **2017**, *50*, 2789–2798. [CrossRef] [PubMed]
2. Shi, Y.; Song, M.-M.; Tao, D.-L.; Bo, Q.-B. Diverse lanthanide coordination polymers with 3,3′-dimethylcyclopropane-1,2-dicarboxylate ligand: Synthesis, crystal structure, and properties. *ACS Omega* **2018**, *3*, 12122–12131. [CrossRef]
3. Feng, X.; Shang, Y.; Zhang, H.; Li, R.; Wang, W.; Zhang, D.; Wang, L.; Li, Z. Enhanced luminescence and tunable magnetic properties of lanthanide coordination polymers based on fluorine substitution and phenanthroline ligand. *RSC Adv.* **2019**, *9*, 16328–16338. [CrossRef]
4. Ruiz-Muelle, A.B.; García- García, A.; García-Valdivia, A.A.; Oyarzabal, I.; Cepeda, J.; Seco, J.M.; Colacio, E.; Rodríguez-Diéguez, A.; Fernández, I. Design and synthesis of a family of 1D-lanthanide-coordination polymers showing luminescence and slow relaxation of the magnetization. *Dalton Trans.* **2018**, *47*, 12783–12794. [CrossRef] [PubMed]
5. Liu, J.-H.; Zhang, R.-T.; Zhang, J.; Zhao, D.; Li, X.-X.; Sun, Y.-Q.; Zheng, S.-T. A series of 3D porous lanthanide-substituted polyoxometalate frameworks based on rare hexadecahedral {$Ln_6W_8O_{28}$} heterometallic cage-shaped clusters. *Inorg. Chem.* **2019**, *58*, 14734–14740. [CrossRef]
6. Zhang, S.; Shi, W.; Li, L.; Duan, E.; Cheng, P. Lanthanide coordination polymers with "fsy-type" topology based on 4, 4′-azobenzoic acid: Syntheses, crystal structures, and magnetic properties. *Inorg. Chem.* **2014**, *53*, 10340–10346. [CrossRef]
7. Ma, F.; Sun, R.; Sun, A.-H.; Xiong, J.; Sun, H.-L.; Gao, S. Regulating the structural dimensionality and dynamic properties of a porous dysprosium coordination polymer through solvent molecules. *Inorg. Chem. Front.* **2020**, *7*, 930–938. [CrossRef]
8. Blais, C.; Calvez, G.; Suffren, Y.; Daiguebonne, C.; Paranthoen, C.; Bazin, E.; Freslon, S.; Bernot, K.; Guillou, O. Luminance and brightness: Application to lanthanide-based coordination polymers. *Inorg. Chem.* **2022**, *48*, 19588–19596. [CrossRef]
9. Bogani, L.; Wernsdorfer, W. Molecular spintronics using single-molecule magnets. *Nat. Mater.* **2008**, *7*, 179–186. [CrossRef]

10. Urdampilleta, M.; Nguyen, N.V.; Cleuziou, J.P.; Klyatskaya, S.; Ruben, M.; Wernsdorfer, W. Molecular quantum spintronics: Supramolecular spin valves based on single-molecule magnets and carbon nanotubes. *Int. J. Mol. Sci.* **2011**, *12*, 6656–6667. [CrossRef]
11. Ishikawa, N.; Sugita, M.; Ishikawa, T.; Koshihara, S.; Kaizu, Y. Lanthanide double-decker complexes functioning as magnets at the single-molecular level. *J. Am. Chem. Soc.* **2003**, *125*, 8694–8695. [CrossRef] [PubMed]
12. Ishikawa, N.; Sugita, M.; Ishikawa, T.; Koshihara, S.; Kaizu, Y. Mononuclear lanthanide complexes with a long magnetization relaxation time at high temperatures: A new category of magnets at the single-molecular level. *J. Phys. Chem. B* **2004**, *108*, 11265–11271. [CrossRef]
13. Woodruff, D.N.; Winpenny, R.E.P.; Layfield, R.A. Lanthanide single-molecule magnets. *Chem. Rev.* **2013**, *113*, 5110–5148. [CrossRef]
14. Ungur, L.; Chibotaru, L.F. Strategies toward high-temperature lanthanide-based single-molecule magnets. *Inorg. Chem.* **2016**, *55*, 10043–10056. [CrossRef] [PubMed]
15. Bernot, K. Get under the umbrella: A comprehensive gateway for researchers on lanthanide-based single-molecule magnets. *Eur. J. Inorg. Chem.* **2023**, *26*, e202300336. [CrossRef]
16. Zhang, P.; Guo, Y.-N.; Tang, J. Recent advances in dysprosium-based single molecule magnets: Structural overview and synthetic strategies. *Coord. Chem. Rev.* **2013**, *257*, 1728–1763. [CrossRef]
17. AlDamen, M.A.; Clemente-Juan, J.M.; Coronado, E.; Martí-Gastaldo, C.; Gaita-Ariño, A. Mononuclear lanthanide single-molecule magnets based on polyoxometalates. *J. Am. Chem. Soc.* **2008**, *130*, 8874–8875. [CrossRef] [PubMed]
18. Meihaus, K.R.; Long, J.R. Magnetic blocking at 10 K and a dipolar-mediated avalanche in salts of the bis($\eta^8$-cyclooctatetraenide) complex $[Er(COT)_2]^-$. *J. Am. Chem. Soc.* **2013**, *135*, 17952–17957. [CrossRef]
19. Le Roy, J.J.; Jeletic, M.; Gorelsky, S.I.; Korobkov, I.; Ungur, L.; Chibotaru, L.F.; Murugesu, M. An organometallic building block approach to produce a multidecker 4*f* single-molecule magnet. *J. Am. Chem. Soc.* **2013**, *135*, 3502–3510. [CrossRef]
20. Tang, J.; Hewitt, I.; Madhu, N.T.; Chastanet, G.; Wernsdorfer, W.; Anson, C.E.; Benelli, C.; Sessoli, R.; Powell, A.K. Dysprosium triangles showing single-molecule magnet behavior of thermally excited spin states. *Angew. Chem. Int. Ed.* **2006**, *45*, 1729–1733. [CrossRef]
21. Wang, J.; Sun, C.-Y.; Zheng, Q.; Wang, D.-Q.; Chen, Y.-T.; Ju, J.-F.; Sun, T.-M.; Cui, Y.; Ding, Y.; Tang, Y.-F. Lanthanide single-molecule magnets: Synthetic strategy, structures, properties and recent advances. *Chem. Asian J.* **2023**, *18*, e202201297. [PubMed]
22. Dolinar, B.S.; Alexandropoulos, D.I.; Vignesh, K.R.; James, T.; Dunbar, K.R. Lanthanide triangles supported by radical bridging ligands. *J. Am. Chem. Soc.* **2018**, *140*, 908–911. [PubMed]
23. Boulon, M.-E.; Cucinotta, G.; Luzon, J.; Degl'Innocenti, C.; Perfetti, M.; Bernot, K.; Calvez, G.; Caneschi, A.; Sessoli, R. Magnetic anisotropy and spin-parity effect along the series of lanthanide complexes with DOTA. *Angew. Chem. Int. Ed.* **2013**, *52*, 350–354. [CrossRef]
24. Wan, Q.; Wakizaka, M.; Yamashita, M. Single-ion magnetism behaviors in lanthanide (III) based coordination frameworks. *Inorg. Chem. Front.* **2023**, *10*, 5212–5224.
25. Huang, G.; Fernandez-Garcia, G.; Badiane, I.; Camarra, M.; Freslon, S.; Guillou, O.; Daiguebonne, C.; Totti, F.; Cador, O.; Guizouarn, T.; et al. Magnetic slow relaxation in a metal–organic framework made of chains of ferromagnetically coupled single-molecule magnets. *Chem. Eur. J.* **2018**, *24*, 6983–6991. [CrossRef] [PubMed]
26. Bernot, K.; Bogani, L.; Caneschi, A.; Gatteschi, D.; Sessoli, R. A family of rare-earth-based single chain magnets: Playing with anisotropy. *J. Am. Chem. Soc.* **2006**, *128*, 7947–7956.
27. Gheorghe, R.; Madalan, A.M.; Costes, J.-P.; Wernsdorfer, W.; Andruh, M. A heterotrimetallic 3*d*–3*d*′–4*f* single chain magnet constructed from anisotropic high-spin 3*d*–4*f* nodes and paramagnetic spacers. *Dalton Trans.* **2010**, *39*, 4734–4736.
28. Paulovič, J.; Cimpoesu, F.; Ferbinteanu, M.; Hirao, K. Mechanism of ferromagnetic coupling in copper (II)-gadolinium (III) complexes. *J. Am. Chem. Soc.* **2004**, *126*, 3321–3331.
29. Singh, S.K.; Beg, M.F.; Rajaraman, G. Role of magnetic exchange interactions in the magnetization relaxation of {3*d*–4*f*} single-molecule magnets: A theoretical perspective. *Chem. Eur. J.* **2016**, *22*, 672–680.
30. Dey, A.; Acharya, J.; Chandrasekhar, V. Heterometallic 3*d*–4*f* complexes as single-molecule magnets. *Chem. Asian J.* **2019**, *14*, 4433–4453.
31. Girginova, P.I.; Pereira, L.C.J.; Coutinho, J.T.; Santos, I.C.; Almeida, M. Slow magnetic relaxation in lanthanide ladder type coordination polymers. *Dalton Trans.* **2014**, *43*, 1897–1905. [CrossRef] [PubMed]
32. Rinehart, J.D.; Long, J.R. Exploiting single-ion anisotropy in the design of f-element single-molecule magnets. *Chem. Sci.* **2011**, *2*, 2078–2085. [CrossRef]
33. Guo, F.-S.; Day, B.M.; Chen, Y.-C.; Tong, M.-L.; Mansikkamäki, A.; Layfield, R.A. Magnetic hysteresis up to 80 kelvin in a dysprosium metallocene single-molecule magnet. *Science* **2018**, *362*, 1400–1403. [CrossRef]

34. Tarannum, I.; Moorthy, S.; Singh, S.K. Understanding electrostatics and covalency effects in highly anisotropic organometallic sandwich dysprosium complexes [$Dy(C_mR_m)_2$] (where R = H, $SiH_3$, $CH_3$ and m = 4 to 9): A computational perspective. *Dalton Trans.* **2023**, *52*, 15576–15589. [CrossRef] [PubMed]
35. Liu, K.; Zhang, X.; Meng, X.; Shi, W.; Cheng, P.; Powell, A.K. Constraining the coordination geometries of lanthanide centers and magnetic building blocks in frameworks: A new strategy for molecular nanomagnets. *Chem. Soc. Rev.* **2016**, *45*, 2423–2439. [CrossRef]
36. Zhang, X.; Vieru, V.; Feng, X.; Liu, J.-L.; Zhang, Z.; Na, B.; Shi, W.; Wang, B.-W.; Powell, A.K.; Chibotaru, L.F.; et al. Influence of guest exchange on the magnetization dynamics of dilanthanide single-molecule-magnet nodes within a metal–organic framework. *Angew. Chem. Int. Ed.* **2015**, *54*, 9861–9865. [CrossRef]
37. Caskey, S.R.; Wong-Foy, A.G.; Matzger, A.J. Dramatic tuning of carbon dioxide uptake via metal substitution in a coordination polymer with cylindrical pores. *J. Am. Chem. Soc.* **2008**, *130*, 10870–10871. [CrossRef] [PubMed]
38. Bloch, E.D.; Murray, L.J.; Queen, W.L.; Chavan, S.; Maximoff, S.N.; Bigi, J.P.; Krishna, R.; Peterson, V.K.; Grandjean, F.; Long, G.J.; et al. Selective binding of $O_2$ over $N_2$ in a redox–active metal–organic framework with open iron(II) coordination sites. *J. Am. Chem. Soc.* **2011**, *133*, 14814–14822. [CrossRef]
39. Kapelewski, M.T.; Geier, S.J.; Hudson, M.R.; Stück, D.; Mason, J.A.; Nelson, J.N.; Xiao, D.J.; Hulvey, Z.; Gilmour, E.; FitzGerald, S.A.; et al. $M_2$(*m*-dobdc) (M = Mg, Mn, Fe, Co, Ni) metal–organic frameworks exhibiting increased charge density and enhanced $H_2$ binding at the open metal sites. *J. Am. Chem. Soc.* **2014**, *136*, 12119–12129. [CrossRef]
40. Hmadeh, M.; Lu, Z.; Liu, Z.; Gándara, F.; Furukawa, H.; Wan, S.; Augustyn, V.; Chang, R.; Liao, L.; Zhou, F.; et al. New porous crystals of extended metal-catecholates. *Chem. Mater.* **2012**, *24*, 3511–3513. [CrossRef]
41. Park, B.G.; Pink, M.; Lee, D. Fluorogenic N,O-chelates built on a $C_2$-symmetric aryleneethynylene platform: Spectroscopic and structural consequences of conformational preorganization and ligand denticity. *J. Organomet. Chem.* **2011**, *696*, 4039–4045. [CrossRef]
42. Kodanko, J.J.; Lippard, S.J. Synthesis and characterization of a ditriflate-bridged, diiron(II) complex with *syn*-N-donor ligands: [$Fe_2(\mu\text{-OTf})_2(PIC_2DET)_2$]$(BARF)_2$. *Inorg. Chim. Acta* **2008**, *361*, 894–900.
43. Kodanko, J.J.; Xu, D.; Song, D.; Lippard, S.J. Iron substitution for sodium in a carboxylate-bridged, heterodinuclear sodium–iron Complex. *J. Am. Chem. Soc.* **2005**, *127*, 16004–16005. [CrossRef]
44. Severance, R.C.; Rountree, E.S.; Smith, M.D.; zur Loye, H.-C. Ligand-based luminescence in lead-containing complexes: The effect of conjugated organic ligands on fluorescence. *Solid State Sci.* **2012**, *14*, 1512–1519.
45. Jornet-Mollá, V.; Romero, F.M. Synthesis of rigid ethynyl-bridged polytopic picolinate ligands for MOF applications. *Tetrahedron Lett.* **2015**, *56*, 6120–6122. [CrossRef]
46. Jornet-Mollá, V.; Martín-Mezquita, C.; Giménez-Saiz, C.; Romero, F.M. Zinc (II) picolinate-based coordination polymers as luminescent sensors of $Fe^{3+}$ ions and nitroaromatic compounds. *Inorg. Chim. Acta* **2022**, *538*, 120993. [CrossRef]
47. Bünzli, J.-C.G.; Piguet, C. Taking advantage of luminescent lanthanide ions. *Chem. Soc. Rev.* **2005**, *34*, 1048–1077.
48. Lamture, J.B.; Zhou, Z.H.; Kumar, A.S.; Wensel, T.G. Luminescence properties of terbium (III) complexes with 4-substituted dipicolinic acid analogs. *Inorg. Chem.* **1995**, *34*, 864–869. [CrossRef]
49. Jornet-Mollá, V.; Dreessen, C.; Romero, F.M. Robust lanthanoid picolinate-based coordination polymers for luminescence and sensing applications. *Inorg. Chem.* **2021**, *60*, 10572–10584. [CrossRef]
50. *CrysAlisPro v38.46*; Oxford Diffraction Ltd.: Abingdon, UK, 2017.
51. Farrugia, L.J. WinGX suite for small-molecule single-crystal crystallography. *J. Appl. Crystallogr.* **1999**, *32*, 837–838. [CrossRef]
52. Sheldrick, G.M. Crystal structure refinement with SHELXL. *Acta Cryst.* **2015**, *C71*, 3–8.
53. Bain, G.A.; Berry, J.F. Diamagnetic corrections and Pascal's constants. *J. Chem. Educ.* **2008**, *85*, 532–536. [CrossRef]
54. Flack, H.D. On enantiomorph-polarity estimation. *Acta Cryst.* **1983**, *A39*, 876–881. [CrossRef]
55. Wu, Y.-P.; Xu, G.-W.; Dong, W.-W.; Zhao, J.; Li, D.-S.; Zhang, J.; Bu, X. Anionic lanthanide MOFs as a platform for iron-selective sensing, systematic color tuning, and efficient nanoparticle catalysis. *Inorg. Chem.* **2017**, *56*, 1402–1411. [CrossRef]
56. Casanova, D.; Alemany, P.; Bofill, J.M.; Alvarez, S. Shape and symmetry of heptacoordinate transition-metal complexes: Structural trends. *Chem. Eur. J.* **2003**, *9*, 1281–1295. [CrossRef]
57. Alvarez, S.; Alemany, P.; Casanova, D.; Cirera, J.; Llunell, M.; Avnir, D. Shape maps and polyhedral interconversion paths in transition metal chemistry. *Coord. Chem. Rev.* **2005**, *249*, 1693–1708. [CrossRef]
58. Llunell, M.; Casanova, D.; Cirera, J.; Bofill, J.M.; Alemany, P.; Alvarez, S. *SHAPE*, version 2.1; Universitat de Barcelona: Barcelona, Spain, 2013.
59. Alvarez, S.; Avnir, D.; Llunell, M.; Pinsky, M. Continuous symmetry maps and shape classification. The case of six-coordinated metal compounds. *New J. Chem.* **2002**, *26*, 996–1009. [CrossRef]

60. Sorace, L.; Gatteschi, D. Electronic Structure and Magnetic Properties of Lanthanide Molecular Complexes. In *Lanthanides and Actinides in Molecular Magnetism*; Wiley-VCH: Hoboken, NJ, USA, 2015; pp. 1–26.
61. de Oliveira Maciel, J.W.; Lemes, M.A.; Valdo, A.K.; Rabelo, R.; Martins, F.T.; Queiroz Maia, L.J.; de Santana, R.C.; Lloret, F.; Julve, M.; Cangussu, D. Europium(III), terbium(III), and gadolinium(III) oxamato-based coordination polymers: Visible luminescence and slow magnetic relaxation. *Inorg. Chem.* **2021**, *60*, 6176–6190. [CrossRef]
62. Vostrikova, K.E.; Sukhikh, T.S.; Lavrov, A.N. The synthesis, crystal structure, and magnetic properties of mono-scorpionate Eu(III) complexes. *Inorganics* **2023**, *11*, 418. [CrossRef]
63. Benmansour, S.; Cerezo-Navarrete, C.; Gómez-García, C.J. Slow relaxation of the magnetisation in a two-dimensional metal–organic framework with a layered square lattice. *Magnetochemistry* **2025**, *11*, 1. [CrossRef]
64. AlDamen, M.A.; Cardona-Serra, S.; Clemente-Juan, J.M.; Coronado, E.; Gaita-Ariño, A.; Martí-Gastaldo, C.; Luis, F.; Montero, O. Mononuclear lanthanide single molecule magnets based on the polyoxometalates $[Ln(W_5O_{18})_2]^{9-}$ and $[Ln(\beta_2\text{-}SiW_{11}O_{39})_2]^{13-}$ ($Ln^{III}$ = Tb, Dy, Ho, Er, Tm, and Yb). *Inorg. Chem.* **2009**, *48*, 3467–3479. [CrossRef]
65. Kalita, P.; Malakar, A.; Goura, J.; Nayak, S.; Herrera, J.M.; Colacio, E.; Chandrasekhar, V. Mononuclear lanthanide complexes assembled from a tridentate NNO donor ligand: Design of a $Dy^{III}$ single-ion magnet. *Dalton Trans.* **2019**, *48*, 4857–4866.
66. Chilton, N.F.; Anderson, R.P.; Turner, L.D.; Soncini, A.; Murray, K.S. PHI: A powerful new program for the analysis of anisotropic monomeric and exchange-coupled polynuclear *d*- and *f*-block complexes. *J. Comput. Chem.* **2013**, *34*, 1164–1175.
67. Dolai, M.; Ali, M.; Titiš, J.; Boča, R. Cu(II)–Dy(III) and Co(III)–Dy(III) based single molecule magnets with multiple slow magnetic relaxation processes in the Cu(II)–Dy(III) complex. *Dalton Trans.* **2015**, *44*, 13242–13249. [PubMed]
68. Blagg, R.J.; Ungur, L.; Tuna, F.; Speak, J.; Comar, P.; Collison, D.; Wernsdorfer, W.; McInnes, E.J.L.; Chibotaru, L.; Winpenny, R.E.P. Magnetic relaxation pathways in lanthanide single-molecule magnets. *Nature Chem.* **2013**, *5*, 673–678.
69. Wang, M.-H.; Tsai, M.-Y.; Su, Y.-C.; Chiu, S.-T.; Lin, P.-H.; Long, J. Zero-field single-molecule magnet behavior in a series of dinuclear dysprosium(III) complexes based on benzothiazolyl-based ligands and $\beta$-diketonates. *Cryst. Growth Des.* **2024**, *24*, 422–431.
70. Demir, S.; Zadrozny, J.M.; Long, J.R. Large spin-relaxation barriers for the low-symmetry organolanthanide complexes $[Cp^*_2Ln(BPh_4)]$ (Cp*=pentamethylcyclopentadienyl; Ln=Tb, Dy). *Chem. Eur. J.* **2014**, *20*, 9524–9529.
71. Zhu, Z.; Guo, M.; Li, X.; Tang, J. Molecular magnetism of lanthanide: Advances and perspectives. *Coord. Chem. Rev.* **2019**, *378*, 350–364.
72. Parmar, V.S.; Mills, D.P.; Winpenny, R.E.P. Mononuclear dysprosium alkoxide and aryloxide single-molecule magnets. *Chem. Eur. J.* **2021**, *27*, 7625–7645. [CrossRef]
73. Zabala-Lekuona, A.; Manuel Seco, J.; Colacio, E. Single-molecule magnets: From $Mn_{12}$-ac to dysprosium metallocenes, a travel in time. *Coord. Chem. Rev.* **2021**, *441*, 213984.
74. Ashebr, T.G.; Li, H.; Ying, X.; Li, X.-L.; Zhao, C.; Liu, S.; Tang, J. Emerging trends on designing high-performance dysprosium(III) single-molecule magnets. *ACS Mater. Lett.* **2022**, *4*, 307–319. [CrossRef]
75. Orendáč, M.; Sedláková, L.; Čizmár, E.; Orendáčova, A.; Feher, A.; Zvyagin, S.A.; Wosnitza, J.; Zhu, W.H.; Wang, Z.M.; Gao, S. Spin relaxation and resonant phonon trapping in $[Gd_2(fum)_3(H_2O)_4]\cdot 3H_2O$. *Phys. Rev. B* **2010**, *81*, 214410. [CrossRef]
76. Rossin, A.; Giambastiani, G.; Peruzzini, M.; Sessoli, R. Amine-templated polymeric lanthanide formates: Synthesis, characterization, and applications in luminescence and magnetism. *Inorg. Chem.* **2012**, *51*, 6962–6968. [CrossRef] [PubMed]
77. Orts-Arroyo, M.; Rabelo, R.; Carrasco-Berlanga, A.; Moliner, N.; Cano, J.; Julve, M.; Lloret, F.; De Munno, G.; Ruiz-García, R.; Mayans, J.; et al. Field-induced slow magnetic relaxation and magnetocaloric effects in an oxalato-bridged gadolinium(III)-based 2D MOF. *Dalton Trans.* **2021**, *50*, 3801–3805. [CrossRef] [PubMed]
78. Orts-Arroyo, M.; Sanchis-Perucho, A.; Moliner, N.; Castro, I.; Lloret, F.; Martínez-Lillo, J. One-dimensional gadolinium(III) complexes based on alpha-and beta-amino acids exhibiting field-induced slow relaxation of magnetization. *Inorganics* **2022**, *10*, 32. [CrossRef]
79. Chen, Y.C.; Peng, Y.-Y.; Liu, J.-L.; Tong, M.-L. Field-induced slow magnetic relaxation in a mononuclear Gd (III) complex. *Inorg. Chem. Commun.* **2019**, *107*, 107449. [CrossRef]
80. Mayans, J.; Escuer, A. Correlating the axial zero field splitting with the slow magnetic relaxation in $Gd^{III}$ SIMs. *Chem. Commun.* **2021**, *57*, 721–724. [CrossRef]
81. Liu, B.; Wang, B.; Wang, Z. Static field induced magnetic relaxations in dinuclear lanthanide compounds of $[phen_2Ln_2(HCOO)_4(HCOO)_{2-2x}(NO_3)_{2x}]$ (1, Ln = Gd and $x$ = 0.52; 2, Ln = Er and $x$ = 0.90; phen = 1,10-phenanthroline). *Sci. China Chem.* **2012**, *55*, 926–933. [CrossRef]
82. Martínez-Pérez, M.J.; Cardona-Serra, S.; Schlegel, C.; Moro, F.; Alonso, P.J.; Prima-García, H.; Clemente-Juan, J.M.; Evangelisti, M.; Gaita-Ariño, A.; Sesé, J.; et al. Gd-based single-ion magnets with tunable magnetic anisotropy: Molecular design of spin qubits. *Phys. Rev. Lett.* **2012**, *108*, 247213. [CrossRef]

83. Arauzo, A.; Lazarescu, A.; Shova, S.; Bartolomé, E.; Cases, R.; Luzón, J.; Bartolomé, J.; Turta, C. Structural and magnetic properties of some lanthanide (Ln = Eu(III), Gd(III) and Nd(III)) cyanoacetate polymers: Field-induced slow magnetic relaxation in the Gd and Nd substitutions. *Dalton Trans.* **2014**, *43*, 12342–12356. [CrossRef]
84. Holmberg, R.J.; Anh Ho, L.T.; Ungur, L.; Korobkov, I.; Chibotaru, L.F.; Murugesu, M. Observation of unusual slow-relaxation of the magnetisation in a Gd-EDTA chelate. *Dalton Trans.* **2015**, *44*, 20321–20325. [CrossRef] [PubMed]
85. Yoshida, T.; Cosquer, G.; Izuogu, D.C.; Ohtsu, H.; Kawano, M.; Lan, Y.; Wernsdorfer, W.; Nojiri, H.; Breedlove, B.K.; Yamashita, M. Field-induced slow magnetic relaxation of $Gd^{III}$ complex with a Pt–Gd heterometallic bond. *Chem. Eur. J.* **2017**, *23*, 4551–4556. [CrossRef] [PubMed]
86. Izuogu, D.C.; Yoshida, T.; Zhang, H.; Cosquer, G.; Katoh, K.; Ogata, S.; Hasegawa, M.; Nojiri, H.; Damjanović, M.; Wernsdorfer, W.; et al. Slow magnetic relaxation in a palladium–gadolinium complex induced by electron density donation from the palladium ion. *Chem. Eur. J.* **2018**, *24*, 9285–9294. [CrossRef] [PubMed]
87. da Cunha, T.T.; Barbosa, V.M.M.; Oliveira, W.X.C.; Pinheiro, C.B.; Pedroso, E.F.; Nunes, W.C.; Pereira, C.L.M. Slow magnetic relaxation in mononuclear gadolinium(III) and dysprosium(III) oxamato complexes. *Polyhedron* **2019**, *169*, 102–113. [CrossRef]
88. Vaz, R.C.A.; Esteves, I.O.; Oliveira, W.X.C.; Honorato, J.; Martins, F.T.; Marques, L.F.; dos Santos, G.L.; Freire, R.O.; Jesus, L.T.; Pedroso, E.F.; et al. Mononuclear lanthanide(III)-oxamate complexes as new photoluminescent field-induced single-molecule magnets: Solid-state photophysical and magnetic properties. *Dalton Trans.* **2020**, *49*, 16106–16124. [CrossRef]
89. Handzlik, G.; Magott, M.; Arczyński, M.; Sheveleva, A.M.; Tuna, F.; Sarewicz, M.; Osyczka, A.; Rams, M.; Vieru, V.; Chibotaru, L.F.; et al. Magnetization dynamics and coherent spin manipulation of a propeller Gd(III) complex with the smallest helicene ligand. *J. Phys. Chem. Lett.* **2020**, *11*, 1508–1515. [CrossRef]
90. Brzozowska, M.; Handzlik, G.; Zychowicz, M.; Pinkowicz, D. Low-coordinate dinuclear dysprosium(III) single molecule magnets utilizing LiCl as bridging moieties and tris(amido)amine as blocking ligands. *Magnetochemistry* **2021**, *7*, 125. [CrossRef]
91. Tkáč, V.; Tarasenko, R.; Doroshenko, A.; Kavečansky, V.; Čizmár, E.; Orendáčová, A.; Smolko, R.; Černák, J.; Orendáč, M. Reciprocating thermal behavior in a magnetic field induced slow spin relaxation of $[Gd_2(H_2O)_6(C_2O_4)_3]\cdot 2.5H_2O$. *Solid State Sci.* **2023**, *136*, 107105. [CrossRef]
92. Masuda, Y.; Sakata, S.; Kayahara, S.; Irie, N.; Kofu, M.; Kono, Y.; Sakakibara, T.; Horii, Y.; Kajiwara, T. Slow magnetic relaxation of linear trinuclear M(II)–Gd(III)–M(II) complexes with $D_3$ point group symmetry (M(II) = Zn(II) and Mg(II)). *J. Phys. Chem. C* **2023**, *127*, 3295–3306. [CrossRef]
93. Kumar, S.; Gabarró Riera, G.; Arauzo, A.; Hruby, J.; Hill, S.; Bogani, L.; Rubio-Zuazo, J.; Jover, J.; Bartolomé, E.; Sañudo, E.C. On-surface magnetocaloric effect for a van der Waals Gd(III) 2D MOF grown on Si. *J. Mater. Chem. A* **2024**, *12*, 6269–6279. [CrossRef]
94. da Silveira, C.O.C.; Oliveira, W.X.C.; da Silva Júnior, E.N.; Alvarenga, M.E.; Martins, F.T.; Gatto, C.C.; Pinheiro, C.B.; Pedroso, E.F.; Silva, J.P.O.; Marques, L.F.; et al. Photoluminescence and magnetic properties of isostructural europium(III), gadolinium(III) and terbium(III) oxamate-based coordination polymers. *Dalton Trans.* **2024**, *53*, 14995–15009. [CrossRef] [PubMed]
95. Orts-Arroyo, M.; Ten-Esteve, A.; Ginés-Cárdenas, S.; Cerdá-Alberich, L.; Martí-Bonmatí, L.; Martínez-Lillo, J. A gadolinium(III) complex based on pyridoxine molecule with single-ion magnet and magnetic resonance imaging properties. *Int. J. Mol. Sci.* **2024**, *25*, 2112. [CrossRef] [PubMed]

magnetochemistry

MDPI

*Article*

# Cobalt(II) and Nickel(II) Cubane $\{M_4O_4\}$ Complexes Derived from Di-2-pyridyl Ketone and Benzoate: Syntheses, Structure and Magnetic Properties †

**Carolina Pejo [1], Santiago Valiero [1], Carlos Rojas-Dotti [1], Guilherme P. Guedes [2], Joan Cano [3], Miguel A. Novak [4], Raúl Chiozzone [1], Maria G. F. Vaz [2,*] and Ricardo González [1,*]**

1 Facultad de Química, Universidad de la República, General Flores 2124, Montevideo UY-11800, Uruguay; cpejo@fq.edu.uy (C.P.); svaliero@fq.edu.uy (S.V.); carlosrojas@fq.edu.uy (C.R.-D.); rchiozzo@fq.edu.uy (R.C.)

2 Instituto de Química, Universidade Federal Fluminense, Niterói 24020-150, RJ, Brazil; guilherme_guedes@id.uff.br

3 Instituto de Ciencia Molecular (ICMol), Universitat de València, 46980 Paterna, València, Spain; joan.cano@uv.es

4 Instituto de Física, Universidade Federal Fluminense, Niterói 24210-240, RJ, Brazil; miguelnovak@id.uff.br

* Correspondence: mariavaz@id.uff.br (M.G.F.V.); rgonzale@fq.edu.uy (R.G.)

† With this article, we would like to contribute to the well-deserved tribute to the extensive and outstanding academic careers of Paco and Miguel (sadly, Miguel passed away in July 2024), brilliant researchers who made fundamental contributions to coordination chemistry and molecular magnetism. Beyond this we also celebrate them as the extraordinary individuals—extremely generous and always willing to give a helping hand. We are deeply grateful for their friendship, unwavering support, collaboration and the enriching and joyful moments we shared. Thank you Paco and Miguel.

**Abstract:** Two tetranuclear complexes were obtained by a self-assembly process employing di-2-pyridyl ketone ($(py)_2CO$), benzoate and $M(NO_3)_2$ (M = Co and Ni). The compounds $[M_4\{(py)_2C(OH)O\}_4(O_2CPh)_4]$, where $\{(py)_2C(OH)O\}^-$ is the monoanion of the *gem*-diol form of $(py)_2CO$, were characterized through single-crystal X-ray diffraction and magnetic measurements. Structural analysis revealed that both complexes possess a $[M_4O_4]$ cubane-like core. A two-*J* model and magnetic anisotropy were employed to analyze the magnetic properties of both compounds. These studies indicate the presence of dominant ferromagnetic interactions within both tetranuclear cores. DFT and CASSCF/NEVPT2 calculations were also performed to support the fitting of experimental magnetic data.

**Keywords:** cobalt; nickel; crystal structures; cubane complexes; magnetic properties; DFT and CASSCF/NEVPT2 calculations

## 1. Introduction

Polynuclear metal complexes have attracted interest in the context of molecular magnetism for more than three decades. Early research primarily focused on the study of magnetic interactions in dinuclear complexes; however, the discovery of slow relaxation of magnetization in $Mn_{12}$ [1] triggered the search for novel coordination clusters with diverse topologies and single-molecule magnet (SMM) behavior [2–4]. Serendipity has played a key role as a synthetic strategy that leads to remarkable structural diversity in magnetic systems. This synthetic approach typically involves the use of flexible polydentate organic ligands capable of coordinating two or more metal ions. These ligands often contain hydroxyl groups that can adopt different coordination modes. Since the primary ligand usually does not saturate the coordination sphere of the metal, the solvent molecules or auxiliary organic or inorganic ligands incorporated during synthesis may also coordinate to the metal. Carboxylates are frequently employed as ancillary ligands owing to their ability to act as both terminal and bridging, with a variety of ligation modes. This versatility adds additional flexibility to the mixed-ligand systems. Although this flexibility complicates predictions regarding the possible outcomes of synthesis, serendipitous self-assembly has allowed researchers to obtain a vast number of coordination clusters with a large diversity of spin topologies and nuclearities. These clusters range from trinuclear structures to systems comprising over tens of metal ions, involving first-row transition metals and lanthanide elements [5–16]. Among this diverse group of compounds, tetranuclear complexes featuring the $M_4O_4$ cubane core are one of the most reported structures. All first-row transition metals form this type of isolated cubane complexes, nickel being the most prominent. The CCSD database lists over 270 entries for tetranuclear compounds featuring the $Ni_4O_4$ core. Cobalt has also enriched the library of oxo cubane complexes, for which around 130 $Co_4O_4$ compounds can be found in the CCSD database. The observation of SMM behavior in some of these Ni and Co complexes has stimulated the extensive study of magnetic properties in these $M_4O_4$ families of compounds [17–30]. The accidental orthogonality of magnetic orbitals in $M_4O_4$ cubes may lead to high-spin ground states, which contribute to a significant barrier to magnetization reversal ($U$), since the upper limit of $U$ is given by $S^2|D|$ and $(S^2 - 1/4)|D|$ for integer and half-integer spin, respectively. The molecular spin ground state and the anisotropy of the system are influenced by the nature of the ligand system. Zero-field splitting (zfs) anisotropy associated with the individual $M^{II}$ sites and the exchange coupling between them are affected by changes in the coordination sphere of the $M^{II}$ ions and the $M-O-M$ bond angles. For $Ni_4O_4$ compounds, a clear correlation between the sign and magnitude of the magnetic exchange constant ($J$) and the M–O–M bond angle has been established [19,31,32]. On the other hand, although several magnetic studies have been conducted on $Co_4O_4$ compounds, to the best of our knowledge, magnetostructural correlations have not been reported.

Slight variations in the Ni–O–Ni bond angles can induce significant changes in the magnitude and sign of magnetic interactions, potentially altering the spin ground state. Such changes can occur, for example, through the exchange of coordinated solvent molecules [32] or even by more subtle modifications, such as variations in the peripheral halide atoms of the ligands [33]. Also, changes in coordinated solvent molecules or the substitution of a peripheral H atom with a $CH_3$ group may switch the SMM behavior in cubane complexes [28,32].

Di-2-pyridyl ketone, $(py)_2CO$, has been proven highly effective in the synthesis of coordination clusters. In the presence of a metal ion, the carbonyl group can undergo addition reactions with different nucleophilic agents such as water or alcohols (ROH), and as a result, the species $(py)_2C(OR)(OH)$ (R = H, alkyl, aryl) are obtained (Scheme 1). Upon the deprotonation of the hydroxyl groups, the resulting dianionic $\{(py)_2C(O)_2\}^{2-}$ or the monoanionic $\{(py)_2C(OR)(O)\}^{-}$ species can function as flexible N, N′, O chelating ligands. This flexibility arises from the ability of the negatively charged oxygen atom to bridge two ($\mu_2$) or three ($\mu_3$) metal ions.

Several polynuclear complexes with nuclearities as high as twenty-six metal centers [34] and diverse topologies have been obtained starting with $(py)_2CO$ [35]. For nickel and cobalt, di-2-pyridyl ketone-based complexes with nuclearities of up to eleven and nine, respectively, have been observed, both with acetate as a coligand [36,37]. Particularly, nine different tetranuclear $Ni_4O_4$ cubane compounds based on $(py)_2CO$—both neutral and cationic—can be found in the CCSD, with different auxiliary ligands such as hydroxylamine, dicyanamide, nitrate, sulfate, acetate, water and methanol. Magnetic studies have been conducted on five of these complexes, revealing competing ferromagnetic and antiferromagnetic interactions, resulting in $S = 0$ or $S = 2$ spin ground states [38–42]. Regarding di-2-pyridyl ketone-based $Co_4O_4$ cubane compounds, the CCSD shows seven complexes—both neutral and cationic—in which acetate/water, chloride, acetate and nitrate act as auxiliary ligands. Magnetic properties have been measured for only four of them. Although no data fitting was performed, competing ferro- and antiferromagnetic interactions can be inferred for the compounds $[Co_4\{(py)_2C(OH)O\}_4(NO_3)_3(H_2O)]NO_3$, $[Co_4\{(py)_2C(OH)O\}_4(O_2CMe)_3(H_2O)](ClO_4)\cdot 1.8H_2O$ and $[Co_4\{(py)_2C(OH)O\}_4(O_2CMe)_4]\cdot H_2O$ [37,39,43]. An intermediate $S = 3$ spin ground state was suggested for the latter, while a second hydrated form, $[Co_4\{(py)_2C(OH)O\}_4(O_2CMe)_4]\cdot 7H_2O$, shows antiferromagnetic interactions leading to a non-magnetic ground state [44].

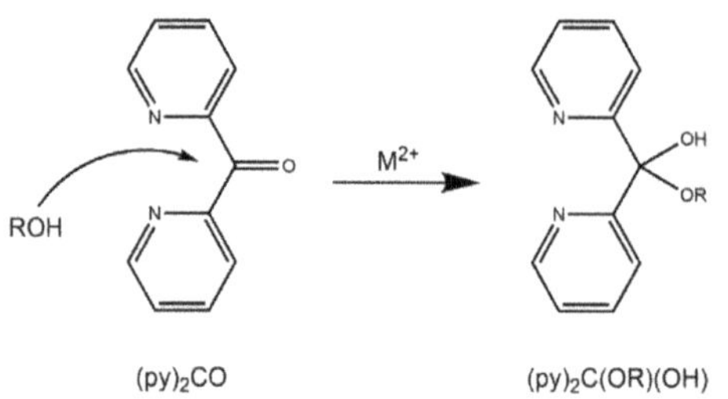


**Scheme 1.** Di-2-pyridyl ketone $(py)_2CO$ and its neutral *gem*-diol, $(py)_2C(OH)_2$, and hemiketal, $(py)_2C(OR)(OH)$, forms.

Based on the previous discussion about the potential structural diversity and magnetic properties of these systems and taking into account the limited number of studies in which benzoate ($PhCO_2^-$) serves as an auxiliary ligand in $(py)_2CO$-based coordination clusters, we explored the chemistry of $M^{II}/(py)_2CO/PhCO_2^-$ systems, where M = Ni and Co. In this paper, we report the synthesis, crystal structures and both experimental and theoretical studies of magnetic properties of compounds $[M_4\{(py)_2C(OH)O\}_4(O_2CPh)_4]$, where M = Co (**1**) and Ni (**2**). Both compounds present a $M_4O_4$ cubane core, which exhibits predominant ferromagnetic interactions, leading to high-spin ground states. *Broken-symmetry* DFT and CASSCF/NEVPT2 calculations support these results.

## 2. Materials and Methods

### 2.1. Materials and Characterization

All reagents and solvents were purchased from commercial sources and used without further purification. Infrared spectra were recorded with a Shimadzu IR Prestige-21 FTIR spectrometer using KBr pellets from 4000 to 400 $cm^{-1}$. Elemental analyses of carbon, hydrogen and nitrogen were performed on a Thermo Flash 2000 analyzer.

Magnetic measurements were carried out on polycrystalline samples of **1** and **2** using a Cryogenics SX600 SQUID magnetometer from 2 to 280 K under an applied magnetic field of 1 and 0.1 kOe for **1** and **2**, respectively. The device was calibrated with YFe garnet NIST reference samples. Diamagnetic corrections were estimated from Pascal's constants, and experimental susceptibilities were also corrected for the magnetization of the sample holder.

### 2.2. Synthesis of $[Co_4\{(py)_2C(O)(OH)\}_4(O_2CPh)_4]$ (1) and $[Ni_4\{(py)_2C(O)(OH)\}_4(O_2CPh)_4]$ (2)

Compound **1** was obtained by adding 73 mg of $Co(NO_3)_2\cdot 6H_2O$ (0.25 mmol) to 10 mL of methanol solution containing $NaO_2CPh$ (0.5 mmol) and 46 mg of $(py)_2CO$ (0.25 mmol). The resulting solution was filtered off and layered with toluene (10 mL). Well-shaped rhombic blue–purple crystals were obtained after 30 days and were filtered off, washed with ethanol and left to dry in air. Yield: 41 mg (34%). *Anal. Calc.* for $[Co_4\{(py)_2C(O)(OH)\}_4(O_2CPh)_4]$ (**2**): C: 56.71; H: 3.69; N: 7.35(%). Found: C: 56.65; H:3.59; N: 7.36 (%). IR ($\nu/cm^{-1}$, KBr): 3494(m), 3416(m), 2800 (m), 1601 (s), 1560 (s), 1387 (s), 1225 (m), 1121 (m), 1071 (m), 1044 (m), 953 (w), 803 (w), 771 (w), 723 (m), 683 (m), 595 (w), 533 (w), 515 (w), 429 (w).

Single crystals of compound **2** were obtained by the same procedure followed for **1** but replacing $Co(NO_3)_2{\cdot}6H_2O$ with $Ni(NO_3)_2{\cdot}6H_2O$. Yield: 50 mg (40%). *Anal. Calc.* for $[Ni_4\{(py)_2C(O)(OH)\}_4(O_2CPh)_4]$ (**1**): C: 56.74; H: 3.69; N: 7.35(%). Found: C: 56.50; H:3.61; N: 7.32 (%). IR ($\nu/cm^{-1}$, KBr): 3500 (m), 3419 (m), 2800 (m), 1601 (s), 1560 (s), 1384 (s), 1223 (m), 1125 (m), 1080 (s), 1045 (m), 1022 (m), 951 (m), 803 (m), 769 (m), 728 (s), 684 (m), 589 (w), 533 (w), 515 (w), 429 (w), 417 (w), 400 (w).

### 2.3. X-Ray Data Collection and Structure Refinement

Single-crystal X-ray diffraction data were collected on an Xcalibur Atlas Gemini Ultra (**1**) or Bruker KAPPA CCD (**2**) diffractometer using graphite-monochromated Mo K$\alpha$ radiation ($\lambda$ = 0.71073 Å) at room temperature. For **1**, CRYSALISPRO software [45] was used to determine the unit cell parameters, the integration of the collected reflections and the absorption correction. The structure solutions and full-matrix least-squares refinements based on $F^2$ were performed with the SHELXS and SHELXL programs [46]. For **2**, the final unit cell parameters were determined from all reflections obtained with the DIRAX program [47,48], whereas the integration of the collected reflections was performed using the EVALCCD program [49]. Absorption correction using equivalent reflections was performed with the SADABS program [47]. All non-hydrogen atoms were refined anisotropically, while hydrogen atoms were set in calculated positions and refined as riding atoms. Details of the crystal structure, collection and refinement are gathered in Table 1. CCDC numbers 2417037-2417038 contain the supplementary crystallographic data for this article. These data are provided free of charge by the Cambridge Crystallographic Data Centre.

### 2.4. Computational Methodology

All calculations were performed in the ORCA software suite version 5.0.3 [50] using the experimental crystal structure as input. The Def2-tzvp basis set was used for all atoms [51].

Magnetic coupling constants ($J$) were evaluated by DFT calculations using the Gaussian version of the B3LYP hybrid functional [52,53]. The evaluation of coulomb and exchange two-electron integrals were accelerated using the resolution of identity (RI) and chains of spheres (COSX) [54] approximations with the def2/J auxiliary basis set [55]. A polarizable continuum model (PCM) was introduced in the calculations with parameters corresponding to acetonitrile to reduce the delocalization error typical of DFT methods, mainly in charged systems with close molecular groups and opposite charges [56]. To simplify the calculations, they were carried out on the experimental geometry of $M_4$, but with the replacement of two paramagnetic $M^{II}$ ions with diamagnetic $Zn^{II}$ ions for each magnetic coupling. According to Ruiz et al. [57–60], magnetic exchange-coupling constants were determined using the *broken-symmetry* approach (BS-DFT). In all cases, a very tight energy convergence threshold was adopted ($\Delta E < 1.0 \times 10^{-9}$ au). Wavefunction stability was verified at each step through ORCA's stability test to ensure convergence to the most stable electronic configuration since the particular local electronic configuration of the $Co^{II}$ ion leads to energetically close excited states [61].

**Table 1.** Crystal data and structure refinement for compounds **1** and **2**.

| Compound | 1 | 2 |
|---|---|---|
| Empirical formula | $C_{72}H_{56}Co_4N_8O_{16}$ | $C_{72}H_{56}N_8Ni_4O_{16}$ |
| Formula weight | 1524.97 | 1524.01 |
| Temperature/K | 293(2) | 293(2) |
| Crystal system | Tetragonal | Tetragonal |
| Space group | I-42d | I-42d |
| $a$/Å | 14.322(1) | 14.172(3) |
| $b$/Å | 14.322(1) | 14.172(3) |
| $c$/Å | 38.300(1) | 37.900(8) |
| Volume/Å$^3$ | 7856.1(11) | 7612.1(3) |
| Z | 4 | 4 |
| $\rho_{calc}$/g cm$^{-3}$ | 1.289 | 1.330 |
| $\mu$/mm$^{-1}$ | 0.895 | 1.041 |
| $F$(000) | 3120.0 | 3136.0 |
| Radiation | Mo Kα (λ = 0.71073) | Mo Kα (λ = 0.71073) |
| 2$\theta$ range for data collection/° | 6.45 to 59.15 | 10.064 to 50.054 |
| Index ranges | $-18 \leq h \leq 18$<br>$-19 \leq k \leq 15$<br>$-51 \leq l \leq 51$ | $-16 \leq h \leq 16$<br>$-16 \leq k \leq 16$<br>$-42 \leq l \leq 45$ |
| Reflections collected | 38328 | 39546 |
| Independent reflections | 5178 | 3340 |
| $R_{int}$ | 0.058 | 0.081 |
| Data/restraints/parameters | 5178/0/215 | 3340/0/227 |
| Goodness-of-fit on $F^2$ | 0.998 | 1.095 |
| Final $R$ indexes [$I \geq 2\sigma(I)$] | $R_1$ = 0.0401, $wR_2$ = 0.1048 | $R_1$ = 0.0437 $wR_2$ = 0.1215 |
| Final $R$ indexes [all data] | $R_1$ = 0.0557, $wR_2$ = 0.1085 | $R_1$ = 0.0530, $wR_2$ = 0.1290 |
| Largest diff. peak/hole / e Å$^{-3}$ | 0.72/−0.32 | 0.81/−0.34 |
| Flack parameter | −0.011(8) | 0.06(4) |
| CCDC identification | 2417038 | 2417037 |

The parameters that determine the axial ($D$) and rhombic ($E$) components of the local zero-field splitting (zfs) were estimated from theoretical calculations based on the complete active space self-consistent field method (CASSCF) or through a subsequent second-order N-Electron Valence-State Perturbation Theory (CASSCF/NEVPT2) applied to CASSCF wavefunctions [62–64], which often provides accurate values of the energies of nearby excited states and for the zfs tensor of mononuclear first-row transition metal complexes. For this purpose, MZn3 models were built by replacing three $M^{II}$ ions with diamagnetic $Zn^{II}$ ones. Simple mononuclear models built from the experimental geometries were also studied (see Discussion). For these calculations, J, C and JK auxiliary basis sets were generated automatically [65]. The contributions to zfs of five triplet and ten doublet and ten quartet and twenty doublet excited states for nickel(II) and cobalt(II) models, respectively, generated from an active space with eight ($Ni^{II}$) or seven ($Co^{II}$) electrons in five d orbitals, were included through an effective Hamiltonian. The RIJCOSX method was applied by combining the resolution of the identity (RI) and "chain-of-spheres" COSX approximations for the Coulomb and exchange terms [54,66,67].

The magnetic behavior of 1 was analyzed using the XVPMAG package [68], a suite of computational tools designed to efficiently study large and complex systems with varying degrees of precision and methodological approaches. In this study, magnetic susceptibility per tetranuclear $M_4$ unit ($\chi_M$)-versus-temperature ($T$) plots were fitted using the Levenberg–Marquardt algorithm [69,70].

For the model incorporating exchange interactions, zero-field splitting (zfs) and the Zeeman effect (Equations (3), (6) and (7), see Section 3.2), the $\chi_{MT}$ product was simulated from the energy matrix. The energy matrix built in the basis of $Ms$ functions and of dimensions 256 × 256 was diagonalized, and the logarithm of the partition function ($Z$) was computed. Small positive and negative perturbations ($\Delta H$) were applied to derive the second derivative of log($Z$) with respect to $H$ (magnetic field), thus obtaining the magnetic susceptibility. Since measurements were conducted on a solid, unoriented sample, the results were averaged over 30 different orientations. Although fewer than 10 orientations can compromise accuracy, some tests have shown that 30 orientations provide sufficient precision compared to simulations performed using 5 × 105 randomly generated orientations. Considering the Thomson problem [71,72], care was taken to ensure that the chosen orientations eliminated any preferential bias by minimizing the electrostatic energy of 30 positive charges positioned on the surface of a sphere.

Additionally, a model incorporating spin-orbit coupling (SOC) for the $Co^{II}$ ion was applied, as described in Equations (2) and (3) (see Section 3.2). This more precise approach involved the diagonalization of a 20,736 × 20,736 energy matrix. While XVPMAG can handle such a large matrix, a simplified theory approach was employed to optimize computational efficiency, which is particularly suitable when the magnetic coupling is not excessively strong. In this method, the SOC effect was resolved using a reduced 12 × 12 energy matrix associating results with local effective spins ($S_{eff}$ = 1/2) and then incorporated into the Hamiltonian governing the magnetic interactions (Equation (3)) for these coupled local effective spins. The most efficient implementation of this approach involved constructing and diagonalizing an energy matrix for each total spin ($S_T$) using $S$ local basis functions and the irreducible tensor operator (ITO) method [73]. The largest matrix in this process had a dimension of only 3 × 3. Nevertheless, this final diagonalization step could be omitted, as the energy levels of the resulting states for this model are already well established.

The experimental magnetic data for 2 were analyzed using the software package PHI v3.1.6 [74].

## 3. Results

Compounds **1** and **2** were obtained by reaction of $M(NO_3)_2 \cdot 6H_2O$, $(py)_2CO$ and $NaO_2CPh$ in methanol. This process can be summarized by Equation (1):

$$4M(NO_3)_2 \cdot 6H_2O + 4(py)_2CO + 8NaO_2CPh \rightarrow [M_4\{(py)_2C(OH)O\}_4(O_2CPh)_4] + 8NaNO_3 + 4HO_2CPh + 20H_2O \quad (1)$$

Both complexes contain the monoanionic *gem*-diol form derived from di-2-pyridyl ketone. The formation of this species involves a nucleophilic attack by a water molecule on the carbonyl group of $py_2CO$ (see Scheme 1). Hydration water in metal nitrates and even the moisture in the solvent supply the required amount of water in the reaction mixture for the formation of $(py)_2C(OH)_2$, while the excess of $NaO_2Ph$ facilitates its deprotonation.

The IR spectra of **1** and **2** show medium intense bands in the region of 3500–3400 $cm^{-1}$ that are assigned to the $\nu$(OH) vibration of coordinated $[(py)2C(OH)O]^-$ groups. Other intense bands appear in the region 1600–1380 $cm^{-1}$, these may be due to contributions from the $\nu_{as}$(C=O) and $\nu_s$(C=O) of $PhCO_2^-$ (1601 $cm^{-1}$ and 1384/1387 $cm^{-1}$, respectively) and from the stretching vibrations of pyridyl rings (1560 $cm^{-1}$).

### 3.1. Structure Description

Compounds **1** and **2** crystallize in the tetragonal I-42d space group and consist of isomorphous systems. Selected bond lengths and angles are listed in Table 2. The asymmetric units are shown in Figure S1 and contain one $\{(py)_2C(OH)O\}^-$ ligand, one benzoate and one $M^{2+}$ cation (M = $Co^{2+}$ (**1**) and $Ni^{2+}$ (**2**)).

The structures consist of neutral tetranuclear $[M_4\{(py)_2C(O)(OH)\}_4(O_2CPh)_4]$ molecules, as can be seen in Figure 1. In both compounds, the metal ions and four oxygen atoms occupy alternated vertices of an $S_4$ local symmetry $[M_4(\mu_3\text{-}O)_4]$ cubane-like core. Each $\{(py)_2C(OH)O\}^-$ ligand is coordinated to three different metal ions by one oxygen (O1) and two pyridyl nitrogen atoms (N1 and N2). Furthermore, the coordination sphere is filled with one $\eta^1$-monodentate benzoate ligand, leading to distorted octahedral geometry around the metal ions. The M–O1 bond lengths are in the range 2.032(2)–2.233(3) Å and 2.005–2.166(4) Å, whereas the M–$O3_{benzoate}$ bond distances are 2.028(3) Å and 2.003(5) Å for **1** and **2**, respectively. As can be seen, the M–O1 lengths are shorter in **2** than in **1**, as expected when comparing nickel(II) and cobalt(II) complexes [75]. These bond lengths fall within the typical values found for similar compounds featuring the cubane-like motif [21–27]. Moreover, the average M···M distance in the $M_4O_4$ moiety is shorter for the nickel derivative [3.123(1) Å for **2** vs. 3.2283(8) Å for **1**]. In accordance with the point symmetry exhibited by the $M_4O_4$ units, the two opposite faces of the cubane that are perpendicular to the $S_4$ axis are equivalent to each other and distinct from the remaining four side faces. Both sets of faces are characterized by different M···M distances and M−O1−M bond angles. The four side faces show M−O1−M bond angles of 96.17(9)° and 102.61(10)° for **1** and 97.48(16)° and 102.78(17)° for **2**, while bond angles of 97.83(9)° and 95.47(16)° are found for the faces perpendicular to S4 in **1** and **2**, respectively.

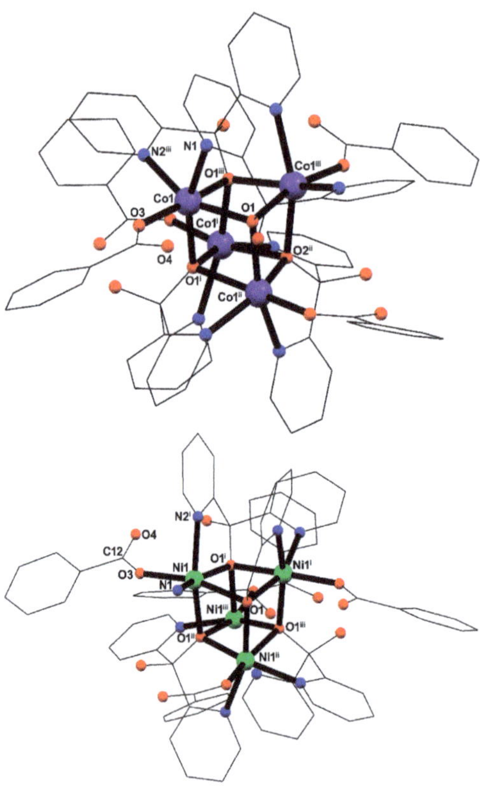


**Figure 1.** Molecular unit of compounds **1** (**top**) and **2** (**bottom**) with atom numbering. Symmetry operations—for **1**: (i) $3/2 - y, -1/2 + x, 3/2 - z$; (ii) $1/2 + y, 3/2 - x, 3/2 - z$; (iii) $2 - x, 1 - y, +z$; for **2**: (i) $-y, x, 2 - z$; (ii) $y, -x, 2 - z$; (iii) $-x, -y, z$.

**Table 2.** Selected bond lengths (Å), distances (Å) and bond angles (°) for **1** and **2**.

| (1) | | (2) | |
|---|---|---|---|
| Co1–O1$^{i}$ | 2.111(2) | Ni1–O1$^{i}$ | 2.005(4) |
| Co1–O1 | 2.032(2) | Ni1–O1 | 2.079(3) |
| Co1–O1$^{iii}$ | 2.233(3) | Ni1–O1$^{ii}$ | 2.166(4) |
| Co1–O3 | 2.028(3) | Ni1–O3 | 2.003(5) |
| Co1–N1 | 2.166(3) | Ni1–N1 | 2.032(6) |
| Co1–N2$^{iii}$ | 2.101(4) | Ni1–N2$^{i}$ | 2.088(5) |
| O1–Co1$^{ii}$ | 2.111(2) | O1–Ni1$^{i}$ | 2.166(4) |
| O1–Co1$^{iii}$ | 2.233(2) | O1–Ni1$^{ii}$ | 2.005(4) |
| O1–C1 | 1.395(4) | O1–C1 | 1.380(7) |
| C1–O2 | 1.390(5) | O2–C1 | 1.389(8) |
| O4–C12 | 1.233(6) | O4–C12 | 1.225(9) |
| C12–O3 | 1.258(5) | O3–C12 | 1.258(9) |
| Co1···Co1$^{i}$ | 3.2175(6) | Ni1···Ni1$^{i}$ | 3.089(1) |
| Co1···Co1$^{ii}$ | 3.2337(9) | Ni1···Ni1$^{ii}$ | 3.191(1) |
| Co1···Co1$^{iii}$ | 3.2337(9) | Ni1···Ni1$^{iii}$ | 3.089(1) |
| Co1–O1–Co1$^{i}$ | 102.61(10) | Ni1–O1–Ni1$^{ii}$ | 97.48(16) |
| Co1$^{i}$–O1–Co1$^{iii}$ | 96.17(9) | Ni1$^{i}$–O1–Ni1$^{ii}$ | 95.47(16) |
| Co1–O1–Co1$^{iii}$ | 97.83(9) | Ni1$^{i}$–O1–Ni1 | 102.78(17) |

Symmetry operations—for **1**: (i) $3/2 - y, -1/2 + x, 3/2 - z$; (ii) $1/2 + y, 3/2 - x, 3/2 - z$; (iii) $2 - x, 1 - y, +z$; for **2**: (i) $-y, x, 2 - z$; (ii) $y, -x, 2 - z$; (iii) $-x, -y, z$.

Intramolecular hydrogen bonds involving the $\{(py)_2C(OH)O\}^-$ alcohol group (O2) and the O4 atom from the benzoate moiety are present in the cubane-like core of both compounds, with the following geometric parameters for **1**: O2···O4 = 2.636(4) Å and ∠O2–H2···O4 = 167° and for **2**: O2···O4$^{ii}$ = 2.579(7) Å and ∠O2–H2···O4$^{ii}$ = 167° (ii = $y, -x, -z + 2$). Finally, the crystal packing is further stabilized by a series of weak intermolecular interactions of type Csp2–H···Csp2 and Csp2–H···O in both complexes.

### 3.2. Magnetic Properties

The static magnetic properties of compounds **1** and **2** under the form $\chi_M T$ versus $T$ plots are shown in Figures 2 and 3. Reduced magnetization plots at 2 and 4 K for both compounds were also obtained, with an additional measurement at 8 K for **1**. At room temperature, the $\chi_M T$ values were approximately 11.7 and 4.6 cm$^3$ K mol$^{-1}$ for **1** and **2**, respectively. These values are slightly higher than the theoretical estimates of 7.5 or 4.4 cm$^3$ K mol$^{-1}$ expected for four non-interacting Co$^{II}$ ($S = 3/2$, $g = 2.0$) or Ni$^{II}$ ($S = 1$, $g \approx 2.1$) ions. For the Co$^{II}$ case, the deviation and the first observed decrease upon cooling are attributed to first-order spin-orbit coupling (SOC), leading to a $g$-factor exceeding the free-electron value and the depopulation of higher excited $J$ states, respectively [76]. Later, both compounds exhibited a progressive increase in $\chi_M T$ upon cooling, reaching maxima of 12.4 and 8.7 cm$^3$ K mol$^{-1}$ at 5.3 K and 7 K for **1** and **2**, respectively. Subsequently, $\chi_M T$ decreased to 11.5 (**1**) and 7.6 (**2**) cm$^3$ K mol$^{-1}$ at 2.4 and 2.75 K, respectively. No maximum in magnetic susceptibility was observed within the measured temperature range (Figure S2). This behavior suggests the coexistence of ferromagnetic interactions and zfs, typical for Co$^{II}$ and Ni$^{II}$ ions in octahedral ligand fields, or dipolar magnetic couplings between the $S = 6$ (**1**) or 4 (**2**) ground states of neighboring tetranuclear units. The latter explanation is the most plausible in the case of the $Co_4$ complex with half-integer local spin momenta. Specifically, a positive axial zfs component ($D$) for the Ni$^{II}$ ion in **2** could partially explain the low-temperature $\chi_M T$ decay, as it would result in a non-magnetic $Ms$ ground state. For **2**, the magnetization value at 2 K and 60 kOe (7.4 μB) is lower than the expected saturation value of $\approx 8.4\ \mu_B$ for an $S = 4$ ground state with $g = 2.1$. Furthermore, the non-superposition of the $M$-versus-$H/T$ plots (insets in Figures 2 and 3) indicates strong magnetic anisotropy and magnetic couplings. While the reduced magnetization curves of **1** are not expected to superimpose due to the known zfs in octahedral cobalt(II) complexes, the expected effect should be less pronounced due to the large $D$ value typical of such systems. Therefore, the observed magnetic susceptibility behavior of **1** suggests the presence of magnetic interactions on the order of several $cm^{-1}$ between the $Co^{II}$ ions. The magnetic behavior of compounds **1** and **2** is intricate, primarily due to the interplay of first-order SOC or zfs with both intra- ferro- and antiferromagnetic interactions and possible additional intermolecular antiferromagnetic interactions. Although the four $M^{II}$ ions in these compounds are crystallographically equivalent, the molecular symmetry leads to a spin topology that includes multiple magnetic coupling pathways, complicating the analysis.

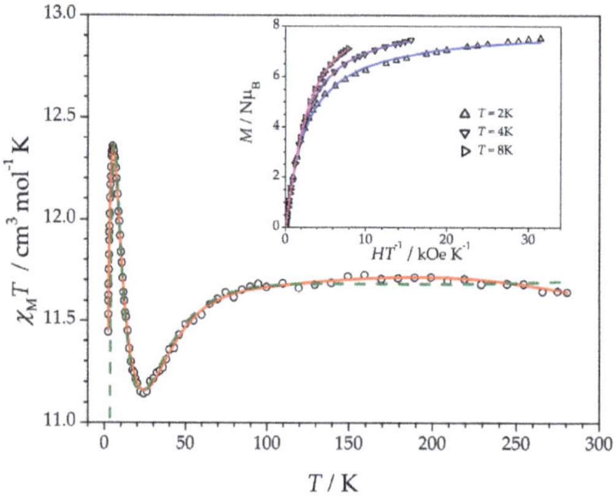


**Figure 2.** Temperature dependence of $\chi_M T$ (empty circles) for compound **1**. Solid and dashed lines represent the best-fit curves (see text) obtained using a model incorporating SOC (red line, Equation (2)) or zfs (green line, Equation (6)) applied to $Co^{II}$ ions, combined with an exchange-coupling Hamiltonian (Equation (3)). Inset: isothermal reduced magnetization curves (empty triangles) for **1** (solid lines are visual guides).

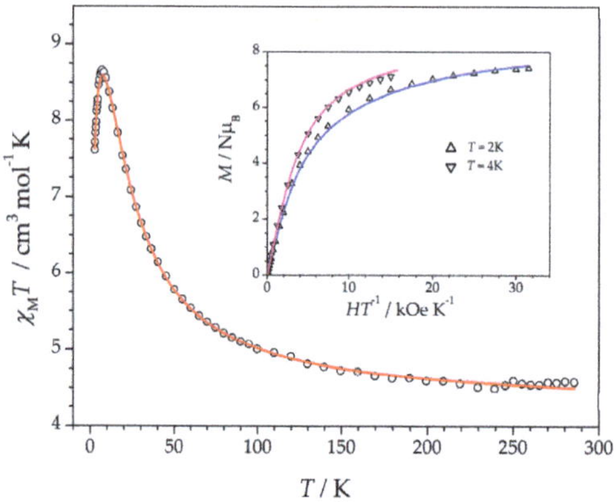


**Figure 3.** Temperature dependence of $\chi_M T$ (empty circles) for compound **2**. Inset: isothermal reduced magnetization curves (empty triangles). Solid lines represent the best-fit curves (see text) obtained using a model incorporating zfs (Equation (6)) applied to $Ni^{II}$ ions, combined with an exchange-coupling Hamiltonian (Equation (3)).

While simplified models offer some preliminary insights, a comprehensive theoretical study is necessary to quantify these magnetic parameters accurately. The presence of two distinct intramolecular M···M distances in the crystal structure implies the existence of at least two different magnetic coupling constants $J_i$ (Figure 4), which are mediated by two oxygen atoms from the alkoxo groups within the M–O–M exchange pathway, forming a $M_2O_2$ magnetic unit. As mentioned earlier, previous studies have highlighted the strong dependence of the magnitude and nature of these interactions on the M–O–M angle ($\alpha$, Figure 5) [19,31,32,59,77,78]. Specifically, $\alpha$ values below a critical "magic angle" favor ferromagnetic coupling, whereas larger angles promote antiferromagnetic interactions. This critical angle is also influenced by the specific metal ion involved, as observed in bis-μ-azido or bis-μ-thiocyanate dinuclear complexes [58]. While well established in copper(II) or nickel(II) dinuclear compounds, the correlation is more nuanced in cobalt(II) systems due to the additional magnetic orbitals involved, including one of $t_{2g}$ symmetry. Nevertheless, recent studies have indicated a tendency toward stronger ferromagnetic interactions in cobalt(II) complexes for smaller $\alpha$ values, consistent with observations in other transition metal complexes [78,79]. Furthermore, the $M_2O_2$ coupling units in these systems can be symmetrical or asymmetrical with one or two different $\alpha$ angles, respectively, adding another layer of complexity to the magnetic analysis. Other geometric distortions, such as the butterfly distortion of the $M_2O_2$ unit described by the $\delta$ angle (Figure 5), can affect the magnetic coupling, though in compounds **1** and **2**, this distortion is only moderate (Table 3). In dinuclear complexes, $\alpha$ and $\gamma$ (the angle describing the deviation of the O–R bond from the $M_2O$ plane) are often correlated due to the inherent geometric flexibility of the system [76,77]. However, in $M_4O_4$ cubane structures, some of the OR bridging groups connect more than two metal ions ($\mu_3$-OR), imposing a specific $\gamma$ angle and preventing a direct correlation between $\delta$ or $\gamma$ and $\alpha$ [79].

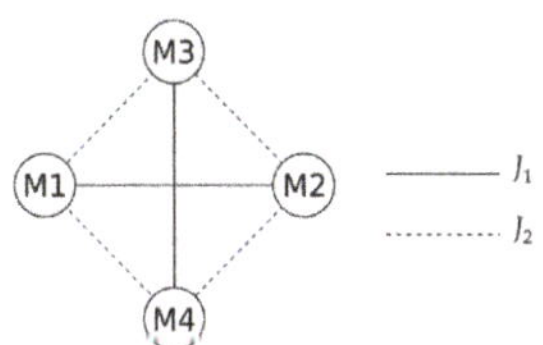


**Figure 4.** Schematic view of the topology of the spin coupling in **1** and **2**. The numbering system for the metal centers in this scheme is the one used in the definition of the Hamiltonians described in the text, and it relates to the structural labels as M1 = Co1; M2 = Co1$^{iii}$; M3 = Co1$^{i}$, M4 = Co1$^{ii}$ for **1** and M1 = Ni1, M2 = Ni1$^{i}$, M3 = Ni1$^{iii}$, M4 = Ni1$^{ii}$ for **2**.

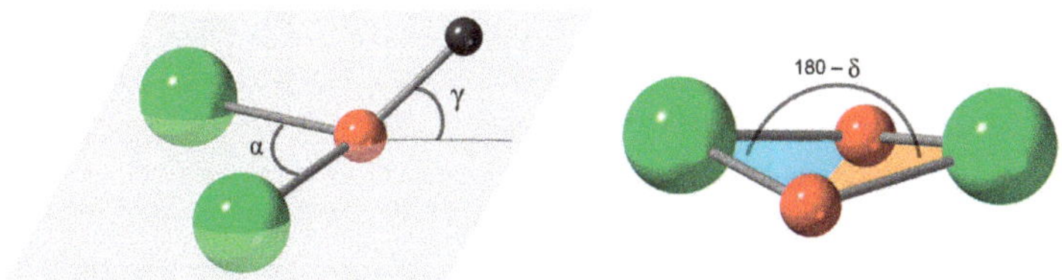


**Figure 5.** Depiction of structural parameters of the $M_2(OR)_2$ core tuning the magnetic couplings in **1** and **2**. Color codes: green, metal; red, oxygen; and gray, carbon.

**Table 3.** Most relevant structural parameters and DFT-calculated exchange-coupling constants ($J$) for the magnetic exchange pathways in **1** and **2**.

| | **1** | | **2** | |
|---|---|---|---|---|
| | $J_1$ | $J_2$ | $J_1$ | $J_2$ |
| $d_{M...M}$ [a] | 3.217 | 3.234 | 3.089 | 3.191 |
| $d_{M...O}$ [a] | 2.032; 2.233 | 2.032; 2.111 | 2.006; 2.166 | 2.166; 2.079 |
| $\alpha$ [b] | 97.8; 97.8 | 102.6; 96.2 | 95.5; 95.5 | 102.8; 97.5 |
| $\delta$ [b] | 14.5 | 11.0 | 19.0 | 8.3 |
| $\gamma$ [b] | 39.1; 39.1 | 57.7; 46.8 | 35.3; 35.3 | 48.6; 60.8 |
| $J_{DFT}$ [c] | +0.3 | +2.6 | +14.6 | +3.2 |
| $J_{Fit}$ [c] | +0.12/−0.6 | +3.4/+4.8 | +14.5/+15.1 | +2.3/+2.4 |

[a] In Å. [b] In degrees. [c] In $cm^{-1}$.

To streamline the computational estimation of magnetic coupling constants in **1** and **2**, a strategic simplification was employed: two $Ni^{II}$ or $Co^{II}$ ions in the tetranuclear complex were replaced with diamagnetic $Zn^{II}$ ions, maintaining the experimental geometry of the system. This substitution ensures that only two paramagnetic ions are involved in each magnetic coupling interaction at a time, significantly reducing the complexity of the calculations. Importantly, this approach preserves the electron configuration of the system and avoids electronic artifacts that could arise from geometric truncation or the construction of oversimplified models. The nature and magnitude of $J$ for the two systems were estimated using the broken-symmetry approximation, a well-established method described in prior studies. The results, summarized in Table 3, reveal that the magnetic interactions in both systems are predominantly ferromagnetic. Furthermore, the strength of these interactions is inversely correlated with the $\alpha$ angle for **2**, with lower $\alpha$ values corresponding to stronger ferromagnetic coupling, aligning with previously established correlations in the literature, mainly with those results found for similar $M_4$ cubanes [19,31,32]. It is worth noting that while the magnetic coupling constants ($J_1$ and $J_2$) in compound **1** generally agree with the predicted values from a previous correlation in $Co_4$ truncated dicubanes [79], some discrepancies exist. The observed $J_1$ value is lower than expected, likely due to one of the M–O bond lengths being significantly longer than usual, which can weaken the magnetic interaction (Table 3). However, it is important to remember that other geometric parameters can also influence the strength and nature of magnetic couplings. The interplay of these factors should be considered for a more in-depth analysis of the magnetic behavior of this compound.

A further contributing factor to the magnetic behavior of compounds **1** and **2** is the zfs tensor. This tensor is characterized by two components: an axial ($D$) and an equatorial ($E$) component, with their ratio ($E/D$) representing the rhombicity. These parameters originate from first-order SOC in $Co^{II}$ and second-order SOC in $Ni^{II}$, which also determine the $g$-factor components of these ions. To evaluate these parameters for $M^{II}$ ions, CASSCF calculations were performed on $M_4$ experimental geometries, wherein diamagnetic $Zn^{II}$ ions replaced three paramagnetic $M^{II}$ ions. The results in Table 4 reveal typical $D$ and $g$ values for cobalt(II) and nickel(II) complexes. The high rhombicity observed is consistent with the significant distortion of the coordination sphere around both metal ions. While CASSCF/NEVPT2 calculations offer a more accurate approach for determining zfs parameters, their computational cost becomes prohibitive when applied to the **MZn$_3$** models. Therefore, a reasonable approximation involves reducing the system to mononuclear $M_1$ models. The results obtained from these models must be compared with those from the more extensive **MZn$_3$** systems at the same calculation level to evaluate the effect of molecular modeling on the electronic structure and its zfs parameters. Two mononuclear models were employed: one in which the complete coordination sphere ligands were retained (**$M_1$** model), and another in which the peripheral pyridine or phenyl groups were replaced with methyl groups (**trimmed $M_1$** model). Given the consistency of results between the simplified mononuclear models and the more complex **$MZn_3$** system (Table 3), the simplified model is suitable for evaluating zfs parameters, enabling higher-level computational studies. Consequently, CASSCF/NEVPT2 calculations performed on the trimmed $M_1$ model yielded results comparable to those obtained with simpler, less accurate methodologies, but showing a weaker zfs. This agreement reinforces the reliability of the calculated order of magnitude and sign of $D$, the magnetic rhombicity ($E/D$) and the parallel ($g_{||}$) and perpendicular ($g_{\perp}$) components of the $g$-factor.

**Table 4.** Calculated zfs parameters, $g$-factor average ($g_{avg}$) and perpendicular ($g_{\perp}$) and parallel ($g_{||}$) components for **$MZn_3$**, **$M_1$** and **trimmed $M_1$** models of **1** and **2** derived from CASSCF and CASSCF/NEVPT2 (in parentheses) calculations.

| | $CoZn_3$ | $Co_1$ | Trimmed $Co_1$ | $NiZn_3$ | $Ni_1$ | Trimmed $Ni_1$ |
|---|---|---|---|---|---|---|
| $D$ [a] | +44.3 | +46.5 | +44.8 (+34.1) | +7.6 | +9.3 | +9.3 (+6.8) |
| $E/D$ | 0.216 | 0.250 | 0.255 (0.253) | 0.080 | 0.098 | 0.089 (0.080) |
| $g_{||}$ | 2.170 | 2.175 | 2.178 (2.089) | 2.300 | 2.291 | 2.290 (2.206) |
| $g_{\perp}$ | 2.666 | 2.706 | 2.690 (2.449) | 2.358 | 2.358 | 2.357 (2.253) |
| $g_{avg}$ | 2.512 | 2.541 | 2.531 (2.336) | 2.339 | 2.336 | 2.335 (2.238) |

[a] In $cm^{-1}$.

Following the conclusions derived from the theoretical study, a rational analysis of the experimental thermal dependence of $\chi_M T$ of **1** was conducted. This analysis initially involved examining the entire temperature range by applying a first-order SOC Hamiltonian ($\hat{H}_{SOC}$) through the $T$-$P$ isomorphism formalism, as described by Equation (2) [76]. This equation characterizes the coupling between orbital ($L$) and spin ($S$) angular momenta. The parameters $\lambda$, $g_e$ and $\alpha$ represent the SOC parameter, the electron $g$-factor and the orbital reduction factor ($\alpha = A\kappa$), respectively. Here, the $\kappa$ parameter accounts for the reduction in orbital momentum due to electron delocalization, while the $A$ parameter reflects the contribution of the upper $^4T_{1g}(^4P)$ state to the $^4T_{1g}(^4F)$ ground state, with values ranging from 1.5 to 1.0 under weak and strong crystal-field limits. Under axial distortion of the ideal $O_h$ symmetry in the cobalt coordination sphere, the ground triplet level $^4T_{1g}$ splits into the singlet $^4A_2$ and doublet $^4E$ levels. This splitting, driven by second-order SOC, further generates two and four Kramers doublets, separated by an energy gap defined by the $\Delta$ parameter. This model effectively reproduces the magnetic behavior of each $Co^{II}$ ion when isolated. However, to incorporate magnetic couplings between paramagnetic centers via a spin Hamiltonian, it was assumed that these magnetic interactions were sufficiently weak to act primarily when the ground Kramers doublet of each $Co^{II}$ ion was predominantly populated. Although this assumption is not strictly true, it is not far from reality, and therefore, it is unnecessary to use other, more sophisticated methods as we have done in the past. Consequently, each paramagnetic center was treated as an effective spin moment ($S_{eff} = 1/2$) with a temperature-dependent $g_{eff}$, obtained through the application of the $\hat{H}_{SOC}$ Hamiltonian. Under these assumptions, the magnetic interactions were also considered effective ($J_{eff}$), related to real interactions by a factor of 25/9 [76]. The exchange-coupling Hamiltonian in Equation (2) is defined based on the spin-interaction topology illustrated in Figure 4. Finally, the dipolar interactions between the ferromagnetic ground states of neighboring $Co_4$ units were included through $\theta$ under a mean-field model via Equation (5).

By combining the $\hat{H}_{Exchange}$ and $\hat{H}_{zeem-eff}$ Hamiltonians [Equations (3) and (4)], the best-fit parameters determined using the XVPMag software [68] were $A$ = 1.29, $\lambda = -103.3$ cm$^{-1}$, $\Delta$ = 454 cm$^{-1}$, $J_1 = -0.6$ cm$^{-1}$, $J_2$ = +4.8 cm$^{-1}$ and $\theta = -0.42$ K. The agreement between experimental and simulated data was remarkable, as indicated by the agreement factor $F = 7.8 \times 10^{-6}$, defined as $\sum(\chi_M T_{\text{exp}} - \chi_M T_{\text{calcd}}]^2/\sum(\chi_M T_{\text{exp}})^2$.

$$\hat{H}_{SOC} = -\alpha\lambda\hat{L}\cdot\hat{S} + \Delta\left[\hat{L}_z^2 - \frac{1}{3}L(L+1)\right] + \beta H\left[g_e\hat{S} - \alpha\hat{L}\right] \quad (2)$$

$$\hat{H}_{Exchange} = -J_1\left(\hat{S}_{M1}\cdot\hat{S}_{M2} + \hat{S}_{M3}\cdot\hat{S}_{M4}\right) - J_2\left(\hat{S}_{M1}\cdot\hat{S}_{M3} + \hat{S}_{M1}\cdot\hat{S}_{M4} + \hat{S}_{M2}\cdot\hat{S}_{M3} + \hat{S}_{M2}\cdot\hat{S}_{M4}\right) \quad (3)$$

$$\hat{H}_{zeem-eff} = g_{eff}\beta H\sum_{i=1}^{4}\hat{S}_{Co_i} \quad (4)$$

$$\chi_M T_{exp} = \chi_M T_{M_4} \times T/(T-\theta) \quad (5)$$

Further refinement was undertaken by incorporating second-order SOC, which explicitly accounts for the zfs experienced by the local $S = 3/2$ ground state of each $Co^{II}$ ion, as described by Equation (6), where $D$ represents the axial magnetic component of the zfs. A temperature-independent paramagnetism (TIP) term was introduced in the model to address the depopulation of the higher-energy Kramers doublet occurring at elevated temperatures. In order to minimize overparameterization, the zfs rhombic magnetic component ($E$) was considered null. As before, the dipolar magnetic interaction along the crystal lattice between neighboring $Co_4$ units was also considered. The combined model, incorporating zfs (Equation 6), exchange (Equation (2)) and Zeeman Hamiltonians (Equation (7)), yielded an excellent fit to the experimental data, as evidenced by the agreement factor $F = 6.6 \times 10^{-5}$. The best-fit parameters obtained were $g$ = 2.450, $D$ = +58.1 cm$^{-1}$, $J_1$ = +0.12 cm$^{-1}$, $J_2$ = +3.40 cm$^{-1}$, TIP = $797.1 \times 10^{-6}$ cm$^3$ mol$^{-1}$ and $\theta = -1.8$ K. The large positive $D$ value obtained is consistent with those reported in the literature for other octahedral high-spin cobalt(II) complexes. However, it is slightly higher than the local $D$ values derived from theoretical calculations (Table 4), even if the local zfs tensors are not aligned (Figure S3). Despite some minor discrepancies, this correspondence further validates the employed model and highlights the accuracy of both the experimental and theoretical approaches in describing the magnetic behavior of the cobalt(II) complex.

$$\hat{H}_{zfs} = D\sum_{i=1}^{4}\left[\hat{S}_{z,i}^2 - \frac{1}{3}S_i(S_i+1)\right] \quad (6)$$

$$\hat{H}_{Zeem} = g\beta H\sum_{i=1}^{4}\hat{S}_{M_i} \quad (7)$$

The situation is more straightforward for compound **2**. In this case, the corresponding Hamiltonian is given by the sum of terms $\hat{H}_{Exchange} + \hat{H}_{zfs} + \hat{H}_{Zeem}$ (Equations (3), (5) and (6)). Typically, the most frequently adopted approach involves assuming that the local $z$ axes are collinear and, consequently, parallel to the $S_4$ axis in this context [32,33]. In some reported cases, taking into account the distortions in the coordination environment of the nickel atoms, it has been deemed more appropriate to assume that the local anisotropy axes are orthogonal to the $S_4$ axis [21,25,80,81]. However, in our case, the distortion of the octahedral geometry is significant enough that it precludes a priori confidence in selecting the most suitable canting angle between $z$ and $S_4$. In fact, CASSCF calculations have revealed a canting angle of 67.7° in this compound. Therefore, we decided to investigate the effect on the quality of the simultaneous magnetization and susceptibility fit when the canting angle ($\beta$) is varied between 0 and 90 degrees for positive starting $D$ values. The results, obtained using the PHI package [74], are presented in Figure 6, which shows the optimal outcomes in terms of lower error residuals $R$ ($R = [\Sigma(\chi_{Mexp} - \chi_{Mcalcd})^2][\Sigma(M_{exp} - M_{calcd})^2]$. The best fits to the experimental data are attained with two distinct sets of values, as detailed in Table 5, corresponding to $\beta$ angles of 43 and 69°. An excellent fit was achieved without accounting for possible intermolecular interactions. No improvement was observed when an $E$ parameter was included to account for potential rhombic distortion of the $Ni^{II}$ ions. Remarkably, the values of $J_1$, $J_2$, $D$ and $\beta$ determined through the fitting procedure are in close agreement with those derived from computational calculations.

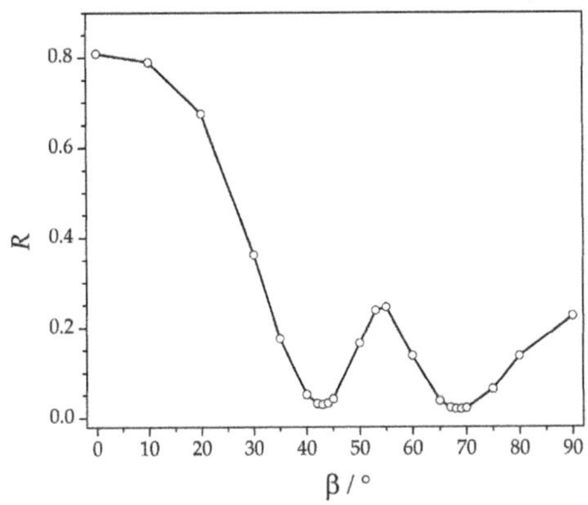


**Figure 6.** Residual error in the simultaneous fitting of $\chi_M T$ and M for different values of the $\beta$ angle and $D > 0$ for compound **2**.

**Table 5.** Best-fit parameters obtained for complex **2** (see text).

| Parameters | Fit | |
|---|---|---|
| $\beta$ [a] | 43 | 69 |
| $J_1$ [b] | 15.1(2) | 14.5(2) |
| $J_2$ [b] | 2.42(2) | 2.34(2) |
| $D$ [b] | 11.76(3) | 10.44(3) |
| $g$ | 2.050(1) | 2.056(1) |
| $R$ | 0.02966 | 0.01952 |

[a] In degrees. [b] In $cm^{-1}$.

## 4. Conclusions

New tetranuclear complexes $[M_4\{(py)_2C(OH)O\}_4(O_2CPh)_4]$ (M = Co, Ni) were synthesized by reacting $M(NO_3)_2{\cdot}6H_2O$, di-2-pyridyl ketone and sodium benzoate in methanol. X-ray crystallography revealed that compounds **1** and **2** feature a cubane-like $[M_4O_4]$ core with $S_4$ symmetry. Experimental magnetic data analyses for the nickel compound, conducted using simultaneous $\chi_M(T)$ and $M(H,T)$ fitting, confirmed ferromagnetic interactions among the four Ni ions in the cubane cluster, resulting in an $S = 4$ ground state. The obtained exchange-coupling $J$ values correlate well with previously established magnetostructural relationships, where stronger ferromagnetic coupling is associated with a smaller Ni–O–Ni angle. The analysis also indicated significant zfs ($D = + 10 - 12$ $cm^{-1}$) and enabled the assessment of local anisotropy axes orientation.

For the cobalt complex, the high magnetic anisotropy was analyzed using two different models for the fitting procedures, considering first-order SOC through *T-P* isomorphism formalism or local zfs effects. In both approaches, predominant ferromagnetic intracluster interactions were observed, yielding an $S = 6$ ground state. In contrast to the Ni compound, a low negative $J$ value was observed for the smallest Co–O–Co angle, suggesting that additional structural factors influence Co···Co interactions.

Theoretical calculations, including DFT and CASSCF/NEVPT2 methods, provided valuable insights into the magnetic coupling constants and zfs parameters, aiding in the fitting processes. The remarkable agreement between experimental data and theoretical models highlights the reliability of the methodologies employed. In highly complex systems such as these, computational approaches have once again proven to be essential tools for conducting comprehensive analyses of experimental results.

**Supplementary Materials:** The following supporting information can be downloaded at: https://www.mdpi.com/article/10.3390/magnetochemistry11040034/s1, Figure S1: Asymmetric units; Figure S2: Magnetic susceptibility versus $T$ curves; Figure S3: Orientation of the calculated local $D$ tensors.

**Author Contributions:** Conceptualization, M.G.F.V. and R.G.; investigation, C.P. and C.R.-D.; crystallography, G.P.G. and M.G.F.V.; magnetic study and analysis, C.P., J.C., M.A.N., R.C. and R.G.; theoretical calculations, S.V. and J.C.; formal analysis, J.C., R.C. and R.G.; resources, M.A.N., J.C., M.G.F.V. and R.G.; writing—original draft preparation, G.P.G., J.C., R.C. and R.G.; writing—review and editing, R.C. and R.G.; visualization, G.P.G. and R.C.; supervision, M.G.F.V. and R.G.; funding acquisition, J.C., G.P.G., M.G.F.V. and R.G. All authors have read and agreed to the published version of the manuscript.

**Funding:** Financial support from the Comisión Sectorial de Investigación Científica (Uruguay, Project 453), Agencia Nacional de Investigación e Innovación (Uruguay, FCE_1_2021_1_167507) and PEDECIBA is gratefully acknowledged. G.P.G., M.G.F.V. and M.A.N thank the Brazilian agencies FAPERJ and CNPq for financial support. C.P. thanks Agencia Nacional de Investigación e Innovación and S.V thanks Comisión Académica de Posgrado (Universidad de la República) for graduate scholarships.

**Institutional Review Board Statement:** Not applicable.

**Informed Consent Statement:** Not applicable.

**Data Availability Statement:** The original contributions presented in this study are included in the article and Supplementary Materials. Further inquiries can be directed to the corresponding authors.

**Acknowledgments:** We thank LABCri-UFMG and LDRX-UFF for crystallographic facility access.

**Conflicts of Interest:** The authors declare no conflicts of interest.

## References

1. Sessoli, R.; Gatteschi, D.; Caneschi, A.; Novak, M. Magnetic bistability in a metal-ion cluster. *Nature* **1993**, *365*, 141–143. [CrossRef]
2. Aromí, G.; Brechin, E.K. Synthesis of 3*d* Metallic Single-Molecule Magnets. In *Single-Molecule Magnets and Related Phenomena*; Winpenny, R., Ed.; Springer: Berlin/Heidelberg, Germany, 2006; Volume 122, pp. 1–67.
3. Sessoli, R.; Powell, A.K. Strategies towards single molecule magnets based on lanthanide ions. *Coord. Chem. Rev.* **2009**, *253*, 2328–2341. [CrossRef]
4. Christou, G. Single-molecule magnets: A molecular approach to nanoscale magnetic materials. *Polyhedron* **2005**, *24*, 2065–2075. [CrossRef]
5. Tasiopoulos, A.J.; Vinslava, A.; Wernsdorfer, W.; Abboud, K.A.; Christou, G. Giant Single-Molecule Magnets: A $\{Mn_{84}\}$ Torus and Its Supramolecular Nanotubes. *Angew. Chem. Int. Ed.* **2004**, *43*, 2117–2121. [CrossRef]
6. Manoli, M.; Inglis, R.; Manos, M.J.; Nastopoulos, V.; Wernsdorfer, W.; Brechin, E.K.; Tasiopoulos, A.J. A $[Mn_{32}]$ Double-Decker Wheel. *Angew. Chem. Int. Ed.* **2011**, *50*, 4441–4444. [CrossRef] [PubMed]

7. Powell, G.W.; Lancashire, H.N.; Brechin, E.K.; Collison, D.; Heath, S.L.; Mallah, T.; Wernsdorfer, W. Building Molecular Minerals: All Ferric Pieces of Molecular Magnetite. *Angew. Chem. Int. Ed.* **2004**, *43*, 5772–5775. [CrossRef] [PubMed]
8. Kong, X.J.; Wu, Y.; Long, L.S.; Zheng, L.S.; Zheng, Z. A Chiral 60-Metal Sodalite Cage Featuring 24 Vertex-Sharing $[Er_4(\mu_3\text{-}OH)_4]$ Cubanes. *J. Am. Chem. Soc.* **2009**, *131*, 6918–6919. [CrossRef]
9. Wang, W.G.; Zhou, A.J.; Zhang, W.X.; Tong, M.L.; Chen, X.M.; Nakano, M.; Beedle, C.C.; Hendrickson, D.N. Giant Heterometallic $Cu_{17}Mn_{28}$ Cluster with $T_d$ Symmetry and High-Spin Ground State. *J. Am. Chem. Soc.* **2007**, *129*, 1014–1015. [CrossRef]
10. Liu, T.; Zhang, Y.J.; Wang, Z.M.; Gao, S. A 64-Nuclear Cubic Cage Incorporating Propeller-like $Fe^{III}{}_8$ Apices and $HCOO^-$ Edges. *J. Am. Chem. Soc.* **2008**, *130*, 10500–10501. [CrossRef]
11. Zhang, Z.M.; Yao, S.; Li, Y.G.; Clerac, R.; Lu, Y.; Su, Z.M.; Wang, E.B. Protein-Sized Chiral $Fe_{168}$ Cages with $NbO^-$ Type Topology. *J. Am. Chem. Soc.* **2009**, *131*, 14600–14601. [CrossRef]
12. Müller, A.; Krickemeyer, E.; Das, S.K.; Kögerler, P.; Sarkar, S.; Bögge, H.; Schmidtmann, M.; Sarkar, S. Linking Icosahedral, Strong Molecular Magnets {Mo} to Layers—A Solid-State Reaction at Room Temperature. *Angew. Chem. Int. Ed.* **2000**, *39*, 1612–1614. [CrossRef]
13. Moushi, E.E.; Lampropoulos, C.; Wernsdorfer, W.; Nastopoulos, V.; Christou, G.; Tasiopoulos, A.J. Inducing Single-Molecule Magnetism in a Family of Loop-of-Loops Aggregates: Heterometallic $Mn_{40}Na_4$ Clusters and the Homometallic $Mn_{44}$ Analogue. *J. Am. Chem. Soc.* **2010**, *132*, 16146–16155. [CrossRef] [PubMed]
14. Kong, J.; Ren, Y.P.; Chen, W.X.; Long, L.S.; Zheng, Z.; Huang, R.B.; Zheng, L.S. A Four-Shell, Nesting Doll-like *3d–4f* Cluster Containing 108 Metal Ions. *Angew. Chem. Int. Ed.* **2008**, *47*, 2398–2401. [CrossRef] [PubMed]
15. Moushi, E.E.; Lampropoulos, C.; Wernsdorfer, W.; Nastopoulos, V.; Christou, G.; Tasiopoulos, A. A Large $[Mn_{10}Na]_4$ Loop of Four Linked $Mn_{10}$ Loops. *J. Inorg. Chem.* **2007**, *46*, 3795–3797. [CrossRef] [PubMed]
16. Stamatatos, T.C.; Abboud, K.A.; Wernsdorfer, W.; Christou, G. "Spin Tweaking" of a High-Spin Molecule: An $Mn_{25}$ Single-Molecule Magnet with an $S=61/2$ Ground State. *Angew. Chem. Int. Ed.* **2007**, *46*, 884–888. [CrossRef]
17. Ponomaryov, A.N.; Kim, N.; Hwang, J.; Nojiri, H.; Van Tol, J.; Ozarowski, A.; Park, J.; Jang, Z.; Suh, B.; Yoon, S.; et al. Structural Tailoring Effects on the Magnetic Behavior of Symmetric and Asymmetric Cubane-Type Ni Complexes. *Chem. Asian J.* **2013**, *8*, 1152–1159. [CrossRef]
18. Aromí, G.; Bouwman, E.; Burzurí, E.; Carbonera, C.; Krzystek, J.; Luis, F.; Schlegel, C.; Van Slageren, J.; Tanase, S.; Teat, S.J. A Novel $Ni_4$ Complex Exhibiting Microsecond Quantum Tunneling of the Magnetization. *Chem. Eur. J.* **2008**, *14*, 11158–11166. [CrossRef]
19. Herchel, R.; Nemec, I.; Machata, M.; Trávníček, Z. Solvent- Induced Structural Diversity in Tetranuclear Ni (II) Schiff-Base Complexes: The First $Ni_4$ Single-Molecule Magnet with a Defective Dicubane-like Topology. *Dalton Trans.* **2016**, *45*, 18622–18634. [CrossRef]
20. Ferguson, A.; Lawrence, J.; Parkin, A.; Sanchez-Benitez, J.; Kamenev, K.V.; Brechin, E.K.; Wernsdorfer, W.; Hill, S.; Murrie, M. Synthesis and Characterisation of a $Ni_4$ Single-Molecule Magnet with $S_4$ Symmetry. *Dalton Trans.* **2008**, *45*, 6409–6414. [CrossRef]
21. Escobar, L.B.L.; Guedes, G.P.; Soriano, S.; Marbey, J.; Hill, S.; Novak, M.A.; Vaz, M.G.F. Synthesis, Magnetic and High-Field EPR Investigation of Two Tetranuclear $Ni^{II}$-Based Complexes. *Inorg. Chem.* **2019**, *58*, 11120–11128. [CrossRef]
22. Yang, E.-C.; Wernsdorfer, W.; Hill, S.; Edwards, R.S.; Nakano, M.; Maccagnano, S.; Zakharov, L.N.; Rheingold, A.L.; Christou, G.; Hendrickson, D.N. Exchange bias in $Ni_4$ single-molecule magnets. *Polyhedron* **2003**, *22*, 1727–1733. [CrossRef]
23. Yang, E.-C.; Wernsdorfer, W.; Zakharov, L.N.; Karaki, Y.; Yamaguchi, A.; Isidro, R.M.; Lu, G.D.; Wilson, S.A.; Rheingold, A.L.; Ishimoto, H.; et al. Fast Magnetization Tunneling in Tetranickel (II) Single-Molecule Magnets. *Inorg. Chem.* **2006**, *45*, 529–546. [CrossRef]
24. Moragues-Canovas, M.; Helliwell, M.; Ricard, L.; Riviere, E.; Wernsdorfer, W.; Brechin, E.; Mallah, T. An $Ni_4$ Single-Molecule Magnet: Synthesis, Structure and Low-Temperature Magnetic Behavior. *Eur. J. Inorg. Chem.* **2004**, *2004*, 2219–2222. [CrossRef]
25. Guedes, G.P.; Soriano, S.; Comerlato, N.M.; Speziali, N.L.; Lahti, P.M.; Novak, M.A.; Vaz, M.G.F. Two cobalt (II) cubane compounds: The key role of small ligand changes on the crystal packing and magnetic properties. *Eur. J. Inorg.* **2012**, *2012*, 5642–5648. [CrossRef]
26. Yang, E.-C.; Hendrickson, D.N.; Wernsdorfer, W.; Nakano, M.; Zakharov, L.N.; Sommer, R.D.; Rheingold, A.L.; Ledezma-Gairaud, M.; Christou, G. Cobalt single-molecule magnet. *J. Appl. Phys.* **2002**, *91*, 7382–7384. [CrossRef]
27. Zhang, S.-H.; Zhang, Y.D.; Zou, H.H.; Guo, J.J.; Li, H.P.; Song, Y.; Liang, H. A family of cubane cobalt and nickel clusters: Syntheses, structures and magnetic properties. *Inorg. Chim. Acta* **2013**, *396*, 119–125. [CrossRef]
28. Ma, X.-F.; Wang, Z.; Chen, X.-L.; Kurmoo, M.; Zeng, M.-H. Ligand effect on the single-molecule magnetism of tetranuclear Co (II) cubane. *Inorg. Chem.* **2017**, *56*, 15178–15186. [CrossRef]
29. Wu, Y.; Xi, J.; Xiao, T.; Ferrando-Soria, J.; Ouyang, Z.; Wang, Z.; Luo, S.; Liu, X.; Pardo, E. Switching of easy-axis to easy-plane anisotropy in cobalt (II) complexes. *Inorg. Chem. Front.* **2021**, *8*, 5158–5168. [CrossRef]

30. Galloway, K.W.; Whyte, A.M.; Wernsdorfer, W.; Sanchez-Benitez, J.; Kamenev, K.V.; Parkin, A.; Peacock, R.D.; Murrie, M. Cobalt (II) Citrate Cubane Single-Molecule Magnet. *Inorg. Chem.* **2008**, *47*, 7438–7442. [CrossRef]
31. Palacios, M.A.; Mota, A.J.; Perea-Buceta, J.E.; White, F.J.; Brechin, E.K.; Colacio, E. Antiferromagnetic versus Ferromagnetic Exchange Interactions in Bis($\mu$-$O_{oximate}$) dinickel (II) Units for a Series of Closely Related Cube Shaped Carboxamideoximate-Bridged $Ni_4$ Complexes. A Combined Experimental and Theoretical Magneto-Structural Study. *Inorg. Chem.* **2010**, *49*, 10156–10165. [CrossRef]
32. Das, A.; Klinke, F.J.; Demeshko, S.; Meyer, S.; Dechert, S.; Meyer, F. Reversible Solvatomagnetic Effect in Novel Tetranuclear Cubane-Type $Ni_4$ Complexes and Magnetostructural Correlations for the $[Ni_4(\mu_3\text{-}O)_4]$ Core. *Inorg. Chem.* **2012**, *51*, 8141–8149. [CrossRef] [PubMed]
33. Aryaeifar, M.; Rudbari, H.A.; Moreno-Pineda, E.; Cuevas-Vicario, J.V.; Paul, S.; Schulze, M.; Wernsdorfer, W.; Lloret, F.; Moinii, N.; Blacque, O. Synthesis, characterization and magnetic properties of halogenated tetranuclear cubane-like nickel(II) complexes. *New J. Chem.* **2024**, *48*, 3603–3613. [CrossRef]
34. Zaleski, C.M.; Depperman, E.C.; Dendrinou-Samara, C.; Alexiou, M.; Kampf, J.W.; Kessissoglou, D.; Kirk, M.L.; Pecoraro, V.L. Metallacryptate Single-Molecule Magnets: Effect of Lower Molecular Symmetry on Blocking Temperature. *J. Am. Chem. Soc.* **2005**, *127*, 12862–12872. [CrossRef] [PubMed]
35. Papaefstathiou, G.S.; Perlepes, S.P. Families of Polynuclear Manganese, Cobalt, Nickel and Copper Complexes Stabilized by Various Forms of Di-2-pyridyl Ketone. *Comments Inorg. Chem.* **2002**, *23*, 249–274. [CrossRef]
36. Efthymiou, C.G.; Mylonas-Margaritis, I.; Raptopoulou, C.P.; Psycharis, V.; Escuer, A.; Papatriantafyllopoulou, C.; Perlepes, S.P. A Ni11 Coordination Cluster from the Use of the Di-2-Pyridyl Ketone/Acetate Ligand Combination: Synthetic, Structural and Magnetic Studies. *Magnetochemistry* **2016**, *2*, 30. [CrossRef]
37. Tsohos, A.; Dionyssopoulou, S.; Raptopoulou, C.P.; Terzis, A.; Bakalbassis, E.G.; Perlepes, S.P. The gem-Diol Form of $(py)_2$ CO as a Ligand in Cobalt(II) Carboxylate Clusters: A Cubane Complex and a Novel Nonanuclear Species with a Vertex-Sharing Double Square Pyramidal Structure. *Angew. Chem. Int. Ed.* **1999**, *38*, 983–985. [CrossRef]
38. Efthymiou, C.G.; Raptopoulou, C.P.; Terzis, A.; Boča, R.; Korabic, M.; Mrozinski, J.; Perlepes, S.P.; Bakalbassis, E.G. A Systematic Exploration of Nickel(II)/Acetate/Di-2-pyridyl Ketone Chemistry: Neutral and Cationic Clusters, and a Novel Mononuclear Complex. *Eur. J. Inorg. Chem.* **2006**, *2016*, 2236–2252. [CrossRef]
39. Tong, M.-L.; Zheng, S.-L.; Shi, J.-X.; Tong, Y.-X.; Lee, H.K.; Chen, X.-M. Synthesis, crystal structures and properties of six cubane-like transition metal complexes of di-2-pyridyl ketone in gem-diol form. *J. Chem. Soc. Dalton Trans.* **2002**, *2002*, 1727–1734. [CrossRef]
40. Papaefstathiou, G.S.; Escuer, A.; Mautner, F.A.; Raptopoulou, C.; Terzis, A.; Perlepes, S.P.; Vicente, R. Use of the Di-2-pyridyl Ketone/Acetate/Dicyanamide "Blend" in Manganese (II), Cobalt (II) and Nickel (II) Chemistry: Neutral Cubane Complexes. *Eur. J. Inorg. Chem.* **2005**, *2005*, 879–893. [CrossRef]
41. Papatriantafyllopoulou, C.; Efthymiou, C.G.; Raptopoulou, C.P.; Vicente, R.; Manessi-Zoupa, E.; Psycharis, V.; Escuer, A.; Perlepes, S.P. Initial use of the di-2-pyridyl ketone/sulfate "blend" in 3*d*-metal cluster chemistry: Preparation, X-ray structures and physical studies of zinc (II) and nickel (II) cubanes. *J. Mol. Struct.* **2007**, *829*, 176–188. [CrossRef]
42. Wang, H.-S.; Song, Y. Conversion of tetranuclear Ni complexes from a defect dicubane core to a $[Ni_4O_4]$ cubane-like core via addition of 2-hydroxymethylpyridine: Synthesis, crystal structures, and magnetic properties. *Inorg. Chem. Commun.* **2013**, *35*, 86–88. [CrossRef]
43. Katsenis, A.D.; Inglis, R.; Slawin, A.M.Z.; Kessler, V.G.; Brechin, E.K.; Papaefstathiou, G.S. Transforming the cube: A tetranuclear cobalt(II) cubane cluster and its transformation to a dimer of dimers. *Cryst. Eng. Comm.* **2009**, *11*, 2117–2120. [CrossRef]
44. Padhi, S.K.; Sahu, R. Co (II/III) coordinated pyridine alcoholate ligand generated through metal assisted nucleophilic addition to a C=O function: Temperature dependent synthesis of a mononuclear complex and a neutral cubane cluster. *Polyhedron* **2008**, *27*, 2662–2666. [CrossRef]
45. *CRYSALISPRO*; Version 1.171.33.55, release 5 January 2010; CrysAlis171.NET; Oxford Diffraction Ltd.: Abingdon, UK, 2010.
46. Sheldrick, G.M. A short history of SHELX. *Acta Cryst. A* **2008**, *64*, 112–122. [CrossRef]
47. Sheldrick, G.M. *SADABS, Program for Empirical Absorption Correction of Area Detector Data*; University of Göttingen: Göttingen, Germany, 1996.
48. Duisenberg, A.J.M. Indexing in single-crystal diffractometry with an obstinate list of reflections. *J. Appl. Crystallogr.* **1992**, *25*, 92–96. [CrossRef]
49. Duisenberg, A.J.M.; Kroon-Batenburg, L.M.J.; Schreurs, A.M.M. An intensity evaluation method: EVAL-14. *J. Appl. Crystallogr.* **2003**, *36*, 220–229. [CrossRef]
50. Neese, F. Software update: The ORCA program system—Version 5.0. *Wiley Interdiscip. Rev. Comput. Mol. Sci.* **2022**, *12*, e1606. [CrossRef]

51. Weigend, F.; Ahlrichs, R. Balanced basis sets of split valence, triple zeta valence and quadruple zeta valence quality for H to Rn: Design and assessment of accuracy. *Phys. Chem. Chem. Phys.* **2005**, *7*, 3297–3305. [CrossRef]
52. Becke, A.D. Density-functional exchange-energy approximation with correct asymptotic behavior. *Phys. Rev. A* **1988**, *38*, 3098–3100. [CrossRef]
53. Lee, C.; Yang, W.; Parr, R.G. Development of the Colle-Salvetti correlation-energy formula into a functional of the electron density. *Phys. Rev. B* **1988**, *37*, 785–789. [CrossRef]
54. Izsák, R.; Hansen, A.; Neese, F. The resolution of identity and chain of spheres approximations for the LPNO-CCSD singles Fock term. *J. Chem. Phys.* **2012**, *110*, 2413–2417. [CrossRef]
55. Weigend, F. Accurate Coulomb-fitting basis sets for H to Rn. *Phys. Chem. Chem. Phys.* **2006**, *8*, 1057–1065. [CrossRef]
56. Tomasi, J.; Mennucci, B.; Cammi, R. Quantum Mechanical Continuum Solvation Models. *Chem. Rev.* **2005**, *105*, 2999–3094. [CrossRef]
57. Ruiz, E.; Rodríguez-Fortea, A.; Cano, J.; Alvarez, S.; Alemany, P. About the calculation of exchange coupling constants in polynuclear transition metal complexes. *J. Comput. Chem.* **2003**, *24*, 982–989. [CrossRef]
58. Ruiz, E.; Cano, J.; Alvarez, S.; Alemany, P. Magnetic Coupling in End-On Azido-Bridged Transition Metal Complexes: A Density Functional Study. *J. Am. Chem. Soc.* **1998**, *120*, 11122–11129. [CrossRef]
59. Ruiz, E.; Alemany, P.; Alvarez, S.; Cano, J. Toward the Prediction of Magnetic Coupling in Molecular Systems: Hydroxo- and Alkoxo-Bridged Cu(II) Binuclear Complexes. *J. Am. Chem. Soc.* **1997**, *119*, 1297–1303. [CrossRef]
60. Ruiz, E.; Alvarez, S.; Cano, J.; Polo, V. About the calculation of exchange coupling constants using density-functional theory: The role of the self-interaction error. *J. Chem. Phys.* **2005**, *123*, 164110. [CrossRef]
61. Segger, R.; People, J.A. Self-consistent molecular orbital methods. XVIII. Constraints and stability in Hartree–Fock theory. *J. Chem. Phys.* **1977**, *66*, 3045–3050. [CrossRef]
62. Angeli, C.; Cimiraglia, R.; Malrieu, J.P. N-electron valence state perturbation theory: A fast implementation of the strongly contracted variant. *Chem. Phys. Lett.* **2001**, *350*, 297–305. [CrossRef]
63. Angeli, C.; Cimiraglia, R.; Malrieu, J.P. n-electron valence state perturbation theory: A spinless formulation and an efficient implementation of the strongly contracted and of the partially contracted variants. *J. Chem. Phys.* **2002**, *117*, 9138–9153. [CrossRef]
64. Angeli, C.; Cimiraglia, R.; Evangelisti, S.; Leininger, T.; Malrieu, J.-P. Introduction of n-electron valence states for multireference perturbation theory. *J. Chem. Phys.* **2001**, *114*, 10252–10264. [CrossRef]
65. Stoychev, G.L.; Auer, A.A.; Neese, F. Automatic Generation of Auxiliary Basis Sets. *J. Chem. Theory Comput.* **2017**, *13*, 554–562. [CrossRef]
66. Neese, F.; Wenmohs, A.; Hansen, A.; Becker, U. Efficient, approximate and parallel Hartree–Fock and hybrid DFT calculations. A 'chain-of-spheres' algorithm for the Hartree–Fock exchange. *Chem. Phys.* **2009**, *356*, 98–109. [CrossRef]
67. Izsák, R.; Neese, F. An overlap fitted chain of spheres exchange method. *J. Chem. Phys.* **2011**, *135*, 144105. [CrossRef]
68. Cano, J. *XVPMag: X-Virtual Package for Magnetism*; Universidad de Valencia: Valencia, Spain, 2010.
69. Levenberg, K. A method for the solution of certain non-linear problems in least squares. *Q. Appl. Math.* **1944**, *2*, 164–168. [CrossRef]
70. Marquardt, W.D. An Algorithm for Least-Squares Estimation of Nonlinear Parameters. *SIAM J. Appl. Math.* **1963**, *11*, 431–441. [CrossRef]
71. Thomson, J.J. On the structure of the atom: An investigation of the stability and periods of oscillation of a number of corpuscles arranged at equal intervals around the circumference of a circle; with application of the results to the theory of atomic structure. *Lond. Edinb. Dubl. Phil. Mag.* **1904**, *39*, 237–265. [CrossRef]
72. Moscato, P.; Haque, N.M.; Moscato, A. Continued fractions and the Thomson problem. *Sci. Rep.* **2023**, *13*, 7272. [CrossRef]
73. König, E.; Kremer, S. Irreducible tensor operator methods in intermediate-field coupling. *Int. J. Quantum Chem.* **1974**, *8*, 347–362. [CrossRef]
74. Chilton, N.F.; Anderson, R.P.; Turner, L.D.; Soncini, A.; Murray, K.S. PHI: A powerful new program for the analysis of anisotropic monomeric and exchange-coupled polynuclear *d*-and *f*-block complexes. *J. Comput. Chem.* **2013**, *34*, 1164–1175. [CrossRef]
75. Isele, K.; Gigon, F.; Williams, A.F.; Bernardinelli, G.; Franz, P.; Decurtins, S. Synthesis, structure and properties of $\{M_4O_4\}$ cubanes containing nickel (ii) and cobalt (ii). *Dalton Trans.* **2007**, *3*, 332–341. [CrossRef] [PubMed]
76. Lloret, F.; Julve, M.; Cano, J.; Ruiz-García, R.; Pardo, R. Magnetic properties of six-coordinated high-spin cobalt (II) complexes: Theoretical background and its application. *Inorg. Chim. Acta* **2008**, *361*, 3432–3445. [CrossRef]
77. Ruiz, E.; Alemany, P.; Alvarez, S.; Cano, J. Structural Modeling and Magneto-Structural Correlations for Hydroxo-Bridged Copper (II) Binuclear Complexes. *Inorg. Chem.* **1997**, *36*, 3683–3688. [CrossRef]
78. Song, X.; Xue, X. Study on the Magneto-Structural Correlation of a New Dinuclear Cobalt (II) Complex with Double μ-Phenoxo Bridges. *ACS Omega* **2020**, *5*, 8347–8354. [CrossRef]

79. Kant, S.; Saha, S.; Lloret, F.; Cano, J.; Mukherjee, R. A tetracobalt (II) cluster with a two vertex truncated dicubane topology endogenously supported by carboxylate-based (2-pyridyl) methylamine ligands: Magneto-structural and DFT studies. *Dalton Trans.* **2023**, *52*, 11922–11933. [CrossRef] [PubMed]
80. Sen, R.; Mondal, K.; dos Santos, M.; Escobar, L.B.L.; Brandão, P.; Reis, M.S.; Lin, Z. A chiral alkali metal capped Ni 4 cubane complex: Synthesis, structure, magnetic and catalytic bromination studies. *J. Mol. Struct.* **2023**, *1274*, 13441. [CrossRef]
81. Scheurer, A.; Ako, A.M.; Saalfrank, R.W.; Heinemann, F.; Hampel, F.; Petukhov, K.; Gieb, K.; Stocker, M.; Müller, P. Synthesis, Magnetic Properties, and STM Spectroscopy of Cobalt (II) Cubanes $[Co^{II}_4(Cl)_4(HL)_4]$. *Chem. Eur. J.* **2010**, *16*, 4784–4792. [CrossRef]

*magnetochemistry*

*Article*

# A Heptacobalt(II/III) Dicubane Cluster with Polyoxometalate and Acetato Ligands: Synthesis, Crystal Structure, and Magnetic Properties †

**Gonzalo Abellán-Dumont [1], Juan Modesto Clemente-Juan [2,*] and Carlos Giménez-Saiz [1,*]**

1 Departament de Química Inorgánica, Universitat de València, Dr Moliner 50, 46100 Burjassot, Spain; gonzalo.abellan-dumont@uv.es
2 Instituto de Ciencia Molecular (ICMol), Universitat de València, 46980 València, Spain
* Correspondence: juan.m.clemente@uv.es (J.M.C.-J.); carlos.gimenez@uv.es (C.G.-S.)
† This contribution is dedicated to the outstanding scientific careers of Miguel Julve and Paco Lloret, a great source of inspiration, both scientifically and personally, for their colleagues at the Inorganic Chemistry Department of the University of Valencia. Sadly, Miguel recently passed away. We will miss his friendship and scientific collaboration.

**Abstract:** The new polyoxometalate $[Co_7(OH)_6(H_2O)_2(CH_3COO)_4(PW_9O_{34})_2]^{13-}$ (**1**) has been synthesized and characterized by IR, UV-Vis-NIR, TGA-TDA, X-ray single crystal analysis, and magnetic studies; **1** consists of two trilacunary heptadentate B-α-$[PW_9O_{34}]^{9-}$ ligands encapsulating a heptacobalt dicubane-like $\{Co^{II}_6Co^{III}O_8\}$ core, in which the $Co^{2+}$ ions are further coordinated by two water molecules and four acetate anions acting as monodentate ligands. The magnetic properties of **1** have been fitted according to an anisotropic exchange model in the low-temperature regime and discussed on the basis of ferromagnetic interactions between $Co^{2+}$ ions with angles Co–L–Co (L = O, OH) close to orthogonality and weakly antiferromagnetic interactions between $Co^{2+}$ ions connected through a central diamagnetic $Co^{3+}$ ion.

**Keywords:** polyoxometalate; cobalt cluster; magnetic properties; acetate ligands; ligand substitution; crystal structure; anisotropic exchange; spin–orbit coupling

## 1. Introduction

Polyoxometalates (POMs) are discrete, anionic metal–oxo clusters containing early *d*-block metals such as W, Mo, or V in their highest oxidation states [1,2]. These metals form $MO_6$ octahedra, which share corners and/or edges while keeping one or more terminal M=O bonds, avoiding the formation of infinite solid-state compounds. Other metal or non-metal containing polyhedra (e.g., $PO_4$, $SiO_4$) can also be combined with the constituent $MO_6$ building blocks, giving rise to an enormous structural variety of POMs. The most prototypical POM structures are the Keggin (e.g., $[PW_{12}O_{40}]^{3-}$) and the Dawson (e.g., $[P_2W_{18}O_{62}]^{6-}$) anions, in which one or more $WO_6$ octahedra can be removed by controlled hydrolysis producing lacunary derivatives that are capable of acting as polydentate ligands of almost any kind of 3*d* or 4*f* metal. This fact, together with their ability to be covalently functionalized with organic components [3], combined with metal–organic frameworks [4], or accept electrons [5], allows POMs to have applications in many fields, such as catalysis [3,4,6,7], medicine [8,9], material science [10,11], sensor science [12,13], or magnetism.

In molecular magnetism, POMs play an important role because they allow the assembly of large magnetic clusters encapsulated by bulky diamagnetic POM ligands that assure the effective insulation of the magnetic cores and, therefore, avoid intermolecular magnetic interactions. Given this key feature, the usefulness of the effective Hamiltonians in magnetic clusters of high nuclearities and high symmetry has been examined [14], and POMs with single-molecule magnet behavior based on 3*d* or 4*f* metal ions have been obtained [15,16]. In addition, the ability of POMs to be reduced by a variable number of electrons has been used to study the exchange and transfer interaction in mixed-valence POM systems (heteropolyblues) [17].

Cobalt-containing POMs represent an excellent opportunity to conduct detailed studies on the magnetic exchange interactions in multicobalt clusters. The $Co^{2+}$ ion ($d^7$) exhibits a high spin S = 3/2, when octahedrally coordinated by POM ligands, and an unquenched orbital momentum. However, the extensive splitting of the $^4T_{2g}$ ground term, due to the first-order spin–orbit coupling and the octahedral distortion, leads to an effective ground spin doublet (S = $\frac{1}{2}$), which has a large anisotropy [18]. Insights in the anisotropic nature of the exchange interactions in $Co^{2+}$ clusters have been provided by inelastic neutron scattering (INS) studies on several POMs containing ferromagnetically coupled $Co^{2+}$ clusters, i.e., $[Co_4(H_2O)_2(PW_9O_{34})_2]^{10-}$, $[Co_3W(H_2O)_2(ZnW_9O_{34})_2]^{12-}$, and $[Co_3Na(H_2O)_2(P_2W_{15}O_{62})_2]^{17-}$ [19]. This knowledge has been applied to characterize the magnetic interactions in other POMs containing $Co^{2+}$ clusters with different topologies and coexisting ferromagnetic and antiferromagnetic interactions [20]. In particular, we have reported a series of POM compounds containing isolated dicubane $\{Co_7O_8\}$ cobalt clusters [21] and Keggin (B-α-$[PW_9O_{34}]^{9-}$) or Dawson (α-$[P_2W_{15}O_{56}]^{9-}$) trilacunary ligands, namely, $[Co_7(OH)_6(H_2O)_6(PW_9O_{34})_2]^{9-}$ (**2**), $[Co_7(OH)_6(H_2O)_4(PW_9O_{34})_2]_n{}^{9n-}$ (**3**), and $[Co_7(OH)_6(H_2O)_6(P_2W_{15}O_{62})_2]^{15-}$ (**4**) (see Figure 1). All these dicubane clusters contain six $Co^{2+}$ ions arranged in two triads separated by one central $Co^{3+}$ ion and represent rare examples of cobalt clusters encapsulated by POMs having cobalt in both oxidation states (see Figure 1, left). These compounds still remain the only examples of POMs containing heptanuclear cobalt clusters with a dicubane topology, although examples of this topology with nickel and manganese do exist in $[Ni_7(OH)_6(H_2O)_6(P_2W_{15}O_{56})_2]^{16-}$ [22], $[Ni_7(H_2O)_4(OH)_6(SiW_8O_{31})_2]^{12-}$ [23], and $[Mn_7O_6(H_2O)_6(P_2W_{15}O_{56})_2]^{14-}$ [24].

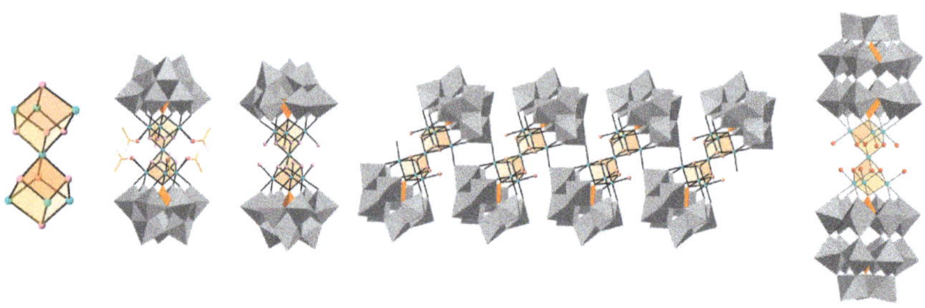


**Figure 1.** Heptacobalt dicubane $\{Co^{II}{}_6Co^{III}O_8\}$ cluster (left) encapsulated in POMs **1–4**. POM **1**, described in this work for the first time, has two water molecules and four acetate anions coordinating the six $Co^{2+}$ ions of the cluster. POMs **2–4** are previously reported members of this series. The formulas of **1–4** are mentioned in the main text.

Here we report on the synthesis, IR and UV-Vis-NIR spectra, thermogravimetric and differential thermal analysis, crystal structure, and magnetic properties of a new POM containing a heptacobalt dicubane cluster, in which four of the six water molecules that coordinate the $Co^{2+}$ ions in the previously reported POM **2** have been substituted by acetate anions that act as monovalent ligands: $[Co_7(OH)_6(H_2O)_2(CH_3COO)_4(PW_9O_{34})_2]^{13-}$ (**1**).

## 2. Materials and Methods

### 2.1. Materials

All reagents were of high-purity grade quality, obtained from commercial sources, and used without further purification. Pure water ($\rho > 18$ MΩ·cm) was used throughout. Reactions were performed in conventional round-bottomed flasks heated using a Heidolph magnetic hotplate stirrer fitted with aluminum alloy heating blocks suitable for round-bottom flasks. The temperature was controlled using a Pt-1000 sensor (Heidolph, Schwabach, Germany) connected to the hotplate.

### 2.2. Physical Techniques

IR spectra were recorded on an Agilent Cary 630 FTIR spectrometer (Agilent, Santa Clara, CA, USA) with diamond ATR sampling. Microanalysis of W, Co, P, and Na was carried out by energy-dispersive X-ray spectroscopy (EDX) using a SCIOS 2 field emission scanning electron microscope coupled with a microanalysis detector, Oxford Ultim Max 170. Carbon and hydrogen contents were determined using a FlashSmart elemental analyzer from Thermo Fisher Scientific (Waltham, MA, USA), both from the Central Support Service for Experimental Research of the University of Valencia (SCSIE). TGA/TDA measurements were carried out with a TA Instruments TGA550 (New Castle, DE, USA), heating from room temperature to 680 °C in a dry air atmosphere, using the Hi-Res mode, in which the heating rate varies depending on the mass loss of the sample. Diffuse reflectance spectra were registered on a Cary 4000 Spectrophotometer (Agilent), using $BaSO_4$ as reference.

### 2.3. Single-Crystal X-Ray Diffraction

A suitable crystal of $Na_{13}$-**1**·$37H_2O$, obtained as described in the synthesis section, was coated with Paratone-N oil, suspended on a small fiber loop, and placed in a stream of cooled nitrogen (150 K) on a Bruker APEX-II CCD diffractometer equipped with a graphite-monochromated (Mo) X-ray Source ($\lambda$ = 0.71073 Å). The data collection routines, unit cell refinements, and data processing were carried out using the APEX3 v2017.3-0 software package [25], and structure solution and refinement were carried out using SHELXT 2018/2 and SHELXL-2019/2 [26]. A multi-scan absorption correction was applied. The solution and refinement of the structure reveal that the terminal ligands of the six $Co^{2+}$ ions (Co2 and Co3) of the heptanuclear $\{Co_7O_8\}$ cluster correspond to four acetate anions and two water molecules, which are statistically disordered over the six positions occupied by the terminal ligands. Each of the two crystallographically independent acetate ligands is also disordered in several components that share the coordinating oxygen atom. EXYZ and EADP constraints commands were used to equate the xyz coordinates and Uij thermal parameters of O23a and O23b (the coordinating oxygen atoms of the disordered components of the first acetate ligand) and of O25a and O25b (belonging to the second acetate). Therefore, the refinement of acetate and water ligands involved partial occupancies, and appropriate similarity and local geometry restraints (SAME and FLAT commands). The components of their anisotropic displacement parameters were also restrained using a combination of hard and soft restraints (DELU and SIMU). The solvent/cation region of the structure also presents a high degree of disorder, as is commonly found in most crystal structures of large polyoxometalates. Therefore, some of the cations (Na4) and solvent water molecules were refined using partial occupancies, and their anisotropic components were restrained to have approximate isotropic behavior using the ISOR command. Even so, the disorder in this region could not be treated completely, and some large residual densities remain near the sodium cations and water molecules. H atoms on carbon atoms were included at calculated positions and refined with a riding model with relative isotropic displacement parameters. H atoms of water molecules or hydroxyl anions were not located. All non-H atoms were refined anisotropically. Crystal data and details of the data collection and refinement for $Na_{13}$-**1**$\cdot 37H_2O$ are listed in Table 1. CCDC 2419765 contains the supplementary crystallographic data for $Na_{13}$-**1**$\cdot 37H_2O$. These data are provided free of charge by the Cambridge Crystallographic Data Centre.

**Table 1.** Summary of the crystal data and refinement details for $Na_{13}[Co_7(OH)_6(H_2O)_2(CH_3COO)_4(PW_9O_{34})_2]\cdot 37H_2O$ ($Na_{13}$-**1**$\cdot 37H_2O$).

| **Formula** | $\mathbf{C_8H_{96}Co_7Na_{13}O_{121}P_2W_{18}}$ |
|---|---|
| $M_r$ (g/mol) | 6211.46 |
| Crystal system | Orthorhombic |
| Space group | *Cmca* (No. 64) |
| $T$ (K) | 150 (2) |
| $\lambda$ (Å) | 0.71073 |
| $a$ (Å) | 22.2275 (8) |
| $b$ (Å) | 20.0193 (8) |
| $c$ (Å) | 29.0406 (11) |
| $\alpha$ (deg) | 90 |
| $\beta$ (deg) | 90 |
| $\gamma$ (deg) | 90 |
| $V$ (Å$^3$) | 12,922.5 (9) |
| $Z$ | 4 |
| $\rho$ (mg/m$^3$) | 3.193 |
| $\mu$ (Mo K$\alpha$) (mm$^{-1}$) | 17.002 |
| $F$(000) | 11,224 |
| Crystal size (mm) | 0.063 × 0.179 × 0.199 |
| $\theta$ range (deg) | 2.308–39.681 |
| Index ranges for $h, k, l$ | −38/31, −35/34, −52/51 |
| Reflns. with $I > 2\sigma(I)$ | 16,291 |
| Independent reflns. | 18,871 |
| $R_{int}$ | 0.0527 |
| GOF on $F^2$ | 1.074 |
| Reflns. collected | 244,378 |
| Data/restraints/parameters | 18,871/328/551 |
| $R_1$, $wR_2$ [$I > 2\sigma(I)$] [a] | 0.0384, 0.0481 |
| $R_1$, $wR_2$ (all data) [a] | 0.0977, 0.1040 |
| Largest diff. peak/hole (e Å$^{-3}$) | 5.216/−6.043 |

[a] $R_1 = \sum(|F_0| - |F_c|)/\sum|F_0|$; $wR_2 = \sum[w(F_0^2 - F_c^2)^2]/\sum[w(F_0^2)^2]^{1/2}$; $w = 1/[\sigma^2(F_0^2) + (0.0431P)^2 + 308.57P]$, where $P = (F_0^2 + 2F_c^2)/3$.

### 2.4. *Powder X-Ray Diffraction (PXRD)*

PXRD measurements of compound **1** were collected on a D8 ADVANCE A25 CONTROLLER powder diffractometer from Bruker AXS using Cu K$\alpha$ radiation ($\lambda$ = 1.54056 Å) at room temperature and in a $2\theta$ range from 2 to 40°. Simulated diffractograms were obtained from single-crystal X-ray data using the Mercury 2024.1.0 software from the CCDC.

### 2.5. *Synthesis of $Na_{13}[Co_7(OH)_6(H_2O)_2(CH_3COO)_4(PW_9O_{34})_2]\cdot 37H_2O$ ($Na_{13}$-**1**$\cdot 37H_2O$)*

2.050 g (25 mmol) of $CH_3COONa$ and 1.012 g (3.6 mmol) of $CoSO_4\cdot 7H_2O$ were dissolved in 50 mL of water using a 100 mL round-bottomed flask. The resulting aqueous solution had a pH of ~7.9. Then, 2.42 g (1 mmol) of solid $Na_8H[PW_9O_{34}]$ (synthesized according to a literature method [27]) were added in small portions to the previous solution over a period of 15 min, with vigorous stirring. When the addition was completed, the stirring was continued for 10 min, and then, 0.072 g (0.3 mmol) of solid $Na_2S_2O_8$ was also added. After this, the round-bottomed flask was covered with a watch glass, the suspension heated to 80 °C for 2 h, and then allowed to cool back to room temperature. The resulting red-violet solution was transferred to a flat-bottomed crystallizing dish, and the solvent was allowed to evaporate at room temperature. After 8 days of standing in this open container, orange-pink crystals of **1** were obtained, collected by filtration, washed with a small amount of cold water, and air-dried (yield: 0.521 g; 17% based on B-α-$[PW_9O_{34}]^{9-}$). Elemental ratio Na–P–W–Co calculated for $Na_{13}[Co_7(OH)_6(H_2O)_2(CH_3COO)_4(PW_9O_{34})_2]\cdot 37H_2O$ (**1**): 6.5:1:9:3.5; found by EDX: 6.5:1:8.5:3.5. Anal. calculated for carbon and hydrogen: C, 1.55%; H, 1.56%. Found: C, 1.56%; H, 1.55%.

*2.6. Magnetic Measurements*

A crystalline sample of **1** was ground, and the powder was compacted to be used in the magnetic measurements. Variable-temperature susceptibility measurements were performed in the temperature range 2−300 K on a magnetometer equipped with a SQUID sensor (Quantum Design MPMS-XL-5, San Diego, CA, USA). The data were corrected for the diamagnetic contribution of the polyoxoanion using Pascal's constant tables. Isothermal magnetization measurements at 2 and 5 K were carried out up to a field of 7 T using the same equipment.

## 3. Results

*3.1. Synthesis*

The synthetic procedure leading to POMs encapsulating dicubane $\{Co_7O_8\}$ clusters has been previously reported by us [21]. Two approaches can be followed: (i) the direct reaction of $Co^{2+}$ ions with a previously prepared trilacunary ligand (B-α-$[PW_9O_{34}]^{9-}$ or α-$[P_2W_{15}O_{56}]^{12-}$), or (ii) the one-pot condensation of $WO_4^{2-}$, $PO_4^{3-}$, and $Co^{2+}$ ions. Both approaches require the appropriate conditions of temperature and pH, and also the use of $S_2O_8^{2-}$ as oxidant, and have led to three different POM compounds containing the heptacobalt cluster: $[Co_7(OH)_6(H_2O)_6(PW_9O_{34})_2]^{9-}$ (**2**), $[Co_7(OH)_6(H_2O)_4(PW_9O_{34})_2]_n^{9n-}$ (**3**), and $[Co_7(OH)_6(H_2O)_6(P_2W_{15}O_{62})_2]^{15-}$ (**4**).

POM $[Co_7(OH)_6(H_2O)_2(CH_3COO)_4(PW_9O_{34})_2]^{13-}$ (**1**), reported for the first time in this work, has been obtained following the first type of approach (see the experimental section for the details). Interestingly, we have found that POM **1** can also be obtained (although with a much lower yield) following the synthetic procedure to obtain the related POM **2** [21] (in which no acetate ligands are present). In that alternative procedure, POM **1** crystallizes from the mother liquor that results when POM **2** is filtered out from the solution. The polyoxoanions of **2** that have not been crystallized remain in solution as the solvent keeps evaporating and the acetate concentration increases, promoting the substitution of the terminal water molecules in **2** by acetate ligands. As the acetate ligands are incorporated into the POM, its anionic charge increases, presumably reducing its solubility. Therefore, the most likely reason for not obtaining a fully substituted compound (with six acetate ligands) is that the partially substituted one reaches its solubility limit before the full substitution takes place in significant amounts.

### 3.2. *Crystal Structure*

The polyoxoanion $[Co_7(OH)_6(H_2O)_2(CH_3COO)_4(PW_9O_{34})_2]^{13-}$ (**1**) consists of a heptanuclear cobalt cluster in which the cobalt atoms are coordinated by two heptadentate B-α-$[PW_9O_{34}]^{9-}$ ligands, six $\mu_3$ hydroxyls, and six terminal ligands (four acetate anions and two water molecules), see Figure 2a. The heptanuclear cluster can be topologically described as a $\{Co_7O_8\}$ di-cubane having two chemically different cobalt atoms (see Figure 2b): one central cobalt atom (Co1) at an inversion center (not directly coordinated by the trivacant B-α-$[PW_9O_{34}]^{9-}$ unit) and six cobalt atoms (Co2 and Co3 in the asymmetric unit) which are arranged in two separated triads, each of them sharing a common oxygen (O3 in Figure 2b) with the phosphorus atom of a trivacant POM unit.

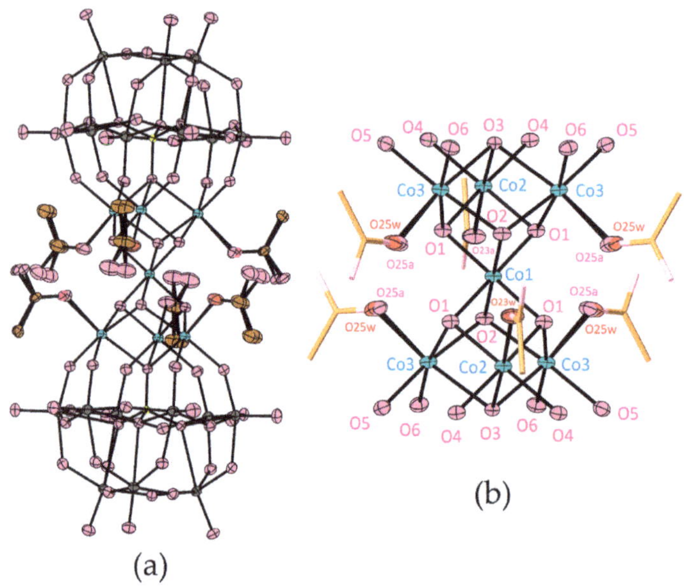


**Figure 2.** (**a**) Ellipsoid representation of POM **1** showing the disorder in the acetate groups and water molecules; (**b**) ellipsoid representation and atom labeling of the heptacobalt dicubane cluster contained in **1**, showing the disorder between the water molecules and acetate groups in the coordinating positions of the six $Co^{2+}$ ions (only one component of the disordered acetate groups is shown). Grey ellipsoids: W; yellow ellipsoids: P; red ellipsoids: O from coordinating water molecules; pink ellipsoids: other O; blue ellipsoids: Co; brown ellipsoids: C; acetate groups in (**a**) are shown as stick representations in (**a**). Hydrogen atoms from methyl groups have been omitted.

BVS calculations [28–30] clearly indicate that Co1 exhibits an oxidation state of +3, while all other cobalt atoms are divalent. In addition, BVS calculations confirm that the six $\mu_3$-O ligands (O1 and O2 in the asymmetric unit) that coordinate the $Co^{3+}$ ion and two $Co^{2+}$ ions correspond to hydroxyl groups. Therefore, each of the six $Co^{2+}$ ions is coordinated by three oxygen atoms belonging to a B-α-$[PW_9O_{34}]^{9-}$ trivacant unit, two hydroxyl groups, and one terminal ligand (water or acetate) that completes the octahedral coordination. The four acetate anions (acting as unidentate ligands) and the two water molecules are statistically disordered on the six positions that coordinate the six $Co^{2+}$ ions. We previously reported the synthesis and characterization of $[Co_7(OH)_6(H_2O)_6(PW_9O_{34})_2]^{9-}$ (**2**) and $[Co_7(OH)_6(H_2O)_4(PW_9O_{34})_2]_n{}^{9n-}$ (**3**), both containing the same $\{Co_7O_8\}$ dicubane cluster as **1**, but **2** has six water molecules (and no acetate ligands) coordinating the six $Co^{2+}$ ions [21], while in **3** two aquo ligands have been substituted by oxygen atoms of neighboring POMs, giving rise to a polymeric chain of polyoxoanions [21]. All Co–O distances in **1**, **2**, and **3** are very similar (see Table 2), suggesting that the substitution of four of the six water molecules in **2** to give rise to **1** has a negligible effect on the geometry of the heptanuclear cobalt cluster. Only a slight difference can be appreciated in the Co–O (terminal) bond distances when water molecules are substituted by acetate ligands: As expected, the charged acetate ligands are slightly closer to the $Co^{2+}$ ions (2.043(18) and 2.07(3) in (**1**)) than the water molecules (2.17(4) and 2.16(5) in **1**, and 2.13(3)–2.21(3) in **2**).

**Table 2.** Co–O bond distances (Å) involved in the heptanuclear dicubane cluster contained in POM **1** (atom labels are shown in Figure 2b), and ranges of the same distances in POMs **2** and **3**. The Co–O distances are classified according to the different types of oxygen atoms: $O^H$ (hydroxyl oxygen), $O^P$ (oxygen bonded to phosphorus), $O^W$ (oxygen bonded to tungsten), $O^{2H}$ (water oxygen), and $O^{Ac}$ (acetate oxygen).

| Type of Distance | 1 [1] | 2 | 3 |
|---|---|---|---|
| Co(III)–$O^H$ | Co1–O1 1.927(3) × 4<br>Co1–O2 1.919(5) × 2 | 1.91(3)–1.93(3) | 1.912(11)–1.917(12) |
| Co(II)–$O^H$ | Co2–O1 2.091(3) × 4<br>Co3–O1 2.092(3) × 4<br>Co3–O2 2.090(3) × 4 | 2.06(3)–2.11(3) | 2.069(10)–2.096(11) |
| Co(II)–$O^P$ | Co2–O3 2.282(4) × 2<br>Co3–O3 2.273(3) × 4 | 2.24(2)–2.29(2) | 2.232(10)–2.263(10) |
| Co(II)–$O^W$ | Co2–O4 2.038(3) × 4<br>Co3–O5 2.026(4) × 4<br>Co3–O6 2.034(3) × 4 | 1.97(3)–2.09(3) | 1.984(11)–2.030(11) |
| Co(II)–$O^{2H}$ | Co2–$O23_{water}$ 2.17(4) × 2<br>Co3–$O25_{water}$ 2.16(5) × 4 | 2.13(3)–2.21(3) | 2.100(10)–2.163(11) |
| Co(II)–$O^{Ac}$ | Co2–O23a 2.043(18) × 2<br>Co3–O25a 2.07(3) × 4 | — | — |

[1] The multiplier indicates the number of times that a particular distance appears in the dicubane cluster.

### 3.3. *IR Spectroscopy, TGA-DTA, UV-Vis-NIR Diffuse Reflectance Spectroscopy, and PXRD*

The IR spectra of POMs **1** and **2** in the fingerprint region are displayed together in Figure 3 for comparison. The IR bands attributed to the polyoxometalate skeleton [31] are very similar in both compounds (in $cm^{-1}$): P-O bands are at 1025s for **1** and **2**; W-O (terminal) at 950sh and 927s for **1**, and at 949sh and 928s for **2**; W-O-W (corner-sharing $WO_6$ octahedra) at 868s for **1**, and at 867s for **2**; W-O-W (edge-sharing octahedra) at 791s and 705s for **1**, and at 792 and 707 for **2**. The IR spectrum of **1** exhibits additional bands that can be attributed to the acetate ligands [32]: 1540m ($\nu_8$, $CO_2$ asym. stretching), 1410m ($\nu_2$, $CO_2$ sym. stretching), 1350sh ($\nu_3$, $CH_3$ sym. bending), 1054m ($\nu_{14}$, $CH_3$ rocking), and 662w ($\nu_5$, $CO_2$ sym. bending). The IR bands can be used to distinguish between different coordination modes of the acetate ligand in metal complexes [32], according to the following ranges of the difference $\Delta = \nu_8 - \nu_2$: 105–140 $cm^{-1}$ for monodentate bonding, 145–185 $cm^{-1}$ for bidentate chelate bonding, and 180–190 $cm^{-1}$ for bidentate bridging bonding. Applying these to the IR spectrum of **1**, we obtain $1540 - 1410 = 130\ cm^{-1}$, which clearly associates the acetate ligands in **1** with monodentate bonding, in agreement with the crystal structure of **1**.

The TGA–DTA curves of $Na_{13}$-**1**·37$H_2O$ are shown in Figure 4. The TGA curve suggests that a partial dehydration process takes place between 30 and 100 °C, with a mass loss of 9.15%, corresponding to the evolution of 32 water molecules (calcd. 9.28%). The slight slope change observed at 40 °C could indicate a different rate of release of water molecules. Two exothermic peaks in the DTA curve, at 37 and 73 °C, presumably correspond to different solvation water molecules. A gradual mass loss takes place from 100 to 353 °C, where an abrupt mass loss starts, finishing at 382 °C. The observed total mass loss between 100 and 382 °C is 6.21%, which can be related to the loss of the remaining seven (coordinated and uncoordinated) water molecules and four acetate ligands (calcd. 5.83%). The exothermic peak in the DTA curve at 381 °C can be attributed to the oxidative degradation of the acetate ligands. At higher temperatures, the collapse of the POM framework must take place. Cobalt may end as a mixture of different cobalt oxides (e.g., CoO, $Co_3O_4$, or $Co_2O_3$) or, when tungsten is present, even $CoWO_4$ [33]. If, for simplicity, we assume that the final decomposition product contains only $CoWO_4$, $WO_3$, $Na_2O$, and $P_2O_5$, the total calculated mass loss is 15.60%, while the experimental loss at 678 °C is 16.19%.

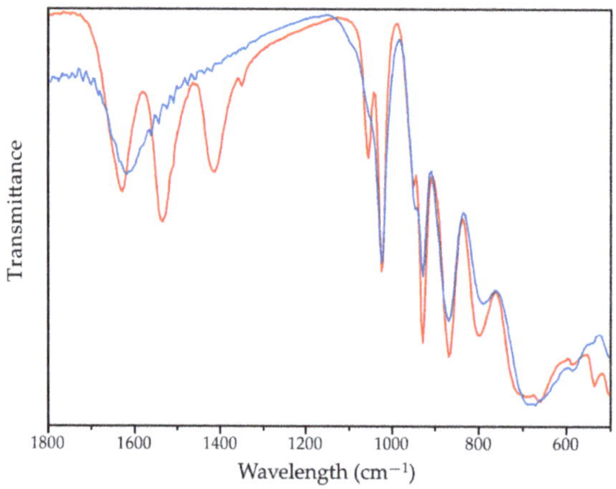


**Figure 3.** FTIR spectra of the salts of POMs **1** (red line) and **2** (blue line).

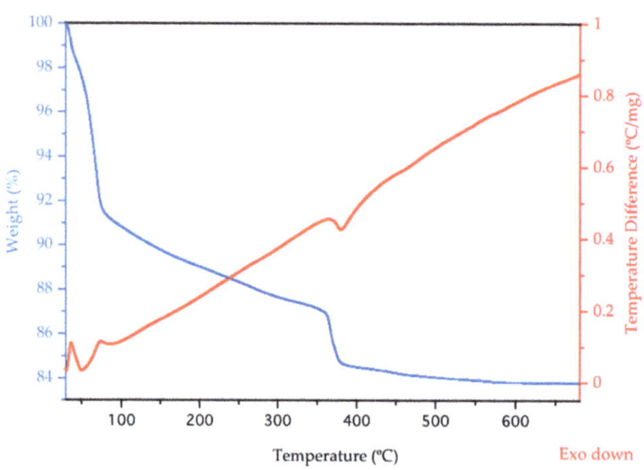


**Figure 4.** TGA–TDA curves of $Na_{13}$-**1**·37$H_2O$ (blue and red, respectively).

The solid UV-Vis-NIR spectrum of $Na_{13}$-**1**·$37H_2O$ is displayed in Figure S3. The strong absorption in the higher energy region (λ < 250 nm), tailing well into the visible, can be attributed to ligand-to-metal (LMCT) $O_{2p} \rightarrow W_{5p}$ charge transfer electronic transitions, common to all polyoxotungstates [34]. The band in the visible region, having a maximum at 543 nm, is very similar to the bands of monosubstituted Co(II) Keggin polyoxoanions and, therefore, can be ascribed to the octahedral Co(II) d–d electronic transition $^4T_{1g}$ (P) ← $^4T_{1g}$ (F) [35]. The splitting of this transition to $^4T_{1g}$ (P) has been attributed to spin–orbit coupling.

The experimental PXRD pattern of a bulk sample of $Na_{13}$-**1**·$37H_2O$ (see Figure S2) shows a very good agreement with the simulated pattern generated from the atomic coordinates of the single-crystal structure solution, although with some preferred orientation effects resulting from the high crystallinity of the material.

### 3.4. Magnetic Properties

The magnetic susceptibility on a ground polycrystalline sample of $Na_{13}$-**1**·$37H_2O$ is shown in Figure 5 as a plot of $\chi_m T$ vs. $T$, and Figure 6 shows a plot of the isothermal magnetization vs. H measured from 0 to 5 T at 2 and 5 K. The $\chi_m T$ vs. $T$ curves exhibit a smooth and continuous decrease from room temperature ($\chi_m T$ = 25.8 emuKmol$^{-1}$), and a more abrupt decrease below 20 K, indicating strong antiferromagnetic interactions within the heptacobalt units. This continuous decrease in the $\chi_m T$ value is due to the spin–orbit coupling of the $Co^{2+}$ ions. High-spin octahedral $Co^{2+}$ is an orbitally degenerate ion with a $^4T_1$ ground electronic term. Due to spin–orbit coupling and the low-symmetry crystal field, this ground term splits into six Kramers doublets. The smooth decrease from room temperature in all measurements can be attributed to the depopulation of the higher Kramers doublets. At lower temperatures (below 30 K), only the lowest Kramers doublet is significantly populated so that the exchange interaction between two octahedral $Co^{2+}$ ions can be conveniently described by assuming a coupling between these fully anisotropic Kramers doublets with effective spins $\frac{1}{2}$.

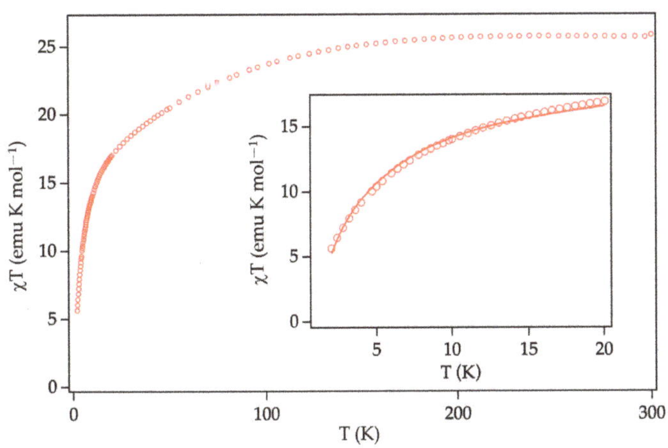


**Figure 5.** Experimental $\chi_M$T of the heptanuclear cluster in POM **1** at 0.1 T in the range 2–300 K. Inset: Thermal behavior of $\chi_M$T at 0.1 T in the range 2–20 K, and the solid line represents the theoretical simulation from the best fit.

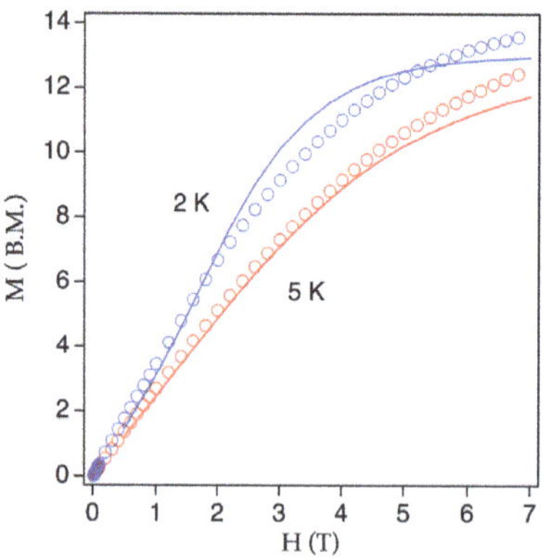


**Figure 6.** Experimental magnetization of the heptanuclear cluster in POM **1** at 2 and 5 K in the range 0–7 T. Solid lines represent the theoretical simulation from the best fit.

The Co-O distances for the central cobalt atom of the heptacobalt cluster are typical of the oxidation state +3. As already described for the other POMs containing these dicubane units [21], this central cobalt should be diamagnetic due to the strong field associated with the hydroxyl groups and only plays the role of an exchange pathway between the other spins. Typically, magnetic interactions between paramagnetic sites through diamagnetic metal ions are negligible or weakly antiferromagnetic.

Taking into account all aspects mentioned above, the effective exchange Hamiltonian for compound $Na_{13}$-**1**·$37H_2O$ can be written as Equation (1):

$$\begin{aligned}\hat{H} = -2\sum_{i=x,y,z} \{ & J_i(\hat{S}_i^1\hat{S}_i^2 + \hat{S}_i^1\hat{S}_i^3 + \hat{S}_i^2\hat{S}_i^3 + \hat{S}_i^4\hat{S}_i^5 + \hat{S}_i^4\hat{S}_i^6 \\ & + \hat{S}_i^5\hat{S}_i^6) + 2J'(\hat{S}^1 + \hat{S}^2 + \hat{S}^3)(\hat{S}^4 + \hat{S}^5 + \hat{S}^6)\}\end{aligned} \quad (1)$$

where $J_i$ represents the exchange components associated with the pathway between the three magnetically equivalent cobalt ions on each trinuclear unit. On the other hand, $J'$ is associated with the exchange interaction between cobalt ions from different trinuclear units through the diamagnetic central $Co^{3+}$. In order to simplify the number of parameters, an axial coupling has been considered between the cobalt ions inside each trinuclear unit ($J_x = J_y$), and the inter-trinuclear interaction has been considered isotropic and equal between all of them. Due to the disorder of the acetate groups and water molecules on the terminal coordinating positions, these parameters will correspond to an average value of interactions. A fit of the susceptibility times temperature product at 0.1 T was performed by numerical diagonalization of the full eigenmatrix, giving the following sets of parameters: $J_x = J_y = 6.2$ cm$^{-1}$, $J_z = 14.8$ cm$^{-1}$, $J' = -0.226$ cm$^{-1}$, $g_{av} = 4.40$ ($R = 3.6 \times 10^{-4}$. Other fits of similar quality can be obtained with $J_z/J_{xy}$ ratios closer to 1. This indicates that the fit is not very sensitive to the anisotropy of this interaction. In fact, when the magnetization adjustment is introduced, the best solution is obtained with an almost isotropic $J$ with a value of 11.6 cm$^{-1}$. At low temperature, where the antiferromagnetic interaction between the two trimeric units is dominant and the effective spin 1/2 model is valid, much of the sensitivity to the anisotropy is lost. For this reason, different $J_z/J_{xy}$ ratios can be obtained with adjustments of similar quality. It should be noted that the average value of this interaction (9.05 cm$^{-1}$) is very similar to the rest of the heptameric systems [21] and other previously studied cobalt systems [19]. This is also so because the presence of the acetate groups does not considerably alter the Co–O–Co angles.

## 4. Conclusions

In this work, we have described the synthesis of a new member of a series of POM compounds ($[Co_7(OH)_6(H_2O)_2(CH_3COO)_4(PW_9O_{34})_2]^{13-}$ (**1**)), in which two heptadentate lacunary POM ligands encapsulate a heptanuclear cobalt cluster having a dicubane topology. This dicubane cluster contains six $Co^{2+}$ ions and one central $Co^{3+}$ ion coordinated by six $\mu_3$-hydroxyl ligands, which justify the occurrence of this oxidation state in the cluster. Every $Co^{2+}$ ion is coordinated by two of these $\mu_3$-OH groups, three oxo ligands from the lacunary POM units, and one terminal ligand, which can be a water molecule or an acetate anion acting as a monodentate ligand. In the previously reported POM $[Co_7(OH)_6(H_2O)_6(PW_9O_{34})_2]^{9-}$ (**2**), all six terminal ligands were water molecules, while in **1**, four of them have been replaced by acetate groups. The magnetic properties of **1** are interpreted based on an antiferromagnetic coupling between two identical ferromagnetically coupled tricobalt units. A complex level scheme with a diamagnetic ground state (S = 0), quasidegenerate with an antisymmetric combination of ± M functions, is obtained. The substitution of some of the water molecules by acetate ligands in **1** does not significantly affect the geometry of the cluster, and then, the exchange parameters are similar to those obtained for the previously reported heptameric cobalt clusters having six water molecules as terminal ligands of the $Co^{2+}$ ions [21]. In any case, spectroscopic studies, for example, with inelastic neutron scattering, which determine energy levels directly, can help determine anisotropy. Furthermore, they will allow the use of models that explicitly introduce the orbital contribution and not just consider the fundamental Kramer doublet.

The preparation of **1** suggests that the substitution of the coordinating water molecules by other ligands is relatively easy in this dicubane cluster, allowing the functionalization of this magnetic POM, even with different degrees of ligand substitution. In this sense, an interesting research direction would be the use of dicarboxylate ligands that could allow the connection of the heptanuclear clusters along different directions of space and, then, the formation of MOFs based on POMs (POMOFs).

**Supplementary Materials:** The following supporting information can be downloaded at https://www.mdpi.com/article/10.3390/magnetochemistry11060048/s1, Table S1: O–Co–O angles in the heptanuclear cluster of $Na_{13}[Co_7(OH)_6(H_2O)_2(CH_3COO)_4(PW_9O_{34})_2]\cdot 37H_2O$ (**1**); Figure S1: Representation and atom labeling of the heptanuclear cluster in **1**; Table S2: Valence bond sum calculations for selected atoms in **1**; Figure S2: Powder X-ray diffraction pattern of **1**; Figure S3: UV-vis-NIR diffuse reflectance spectrum of $Na_{13}$-**1**$\cdot 37H_2O$; Figure S4: Additional crystal structure images of **1**. See Refs. [28–30].

**Author Contributions:** Conceptualization, C.G.-S.; methodology, C.G.-S. and J.M.C.-J.; software, J.M.C.-J.; validation, G.A.-D.; formal analysis, C.G.-S., J.M.C.-J., and G.A.-D.; investigation, C.G.-S., J.M.C.-J., and G.A.-D.; resources, C.G.-S.; data curation, C.G.-S., J.M.C.-J., and G.A.-D.; writing—original draft preparation, C.G.-S. and J.M.C.-J.; writing—review and editing, C.G.-S. and J.M.C.-J.; visualization, C.G.-S., J.M.C.-J., and G.A.-D.; supervision, C.G.-S.; project administration, J.M.C.-J. and C.G.-S.; funding acquisition, J.M.C.-J. and C.G.-S. All authors have read and agreed to the published version of the manuscript.

**Funding:** This study forms part of the Advanced Materials program and was supported by the Spanish MCIN with funding from European Union NextGeneration EU (PRTR-C17.I1), the Spanish MICIN grant (No. PID2020-117177GB-I00), and the Generalitat Valenciana (project MFA-2022-057). We also thank the project CIPROM-2022-060 from the Generalitat Valenciana for financial support. This project was also developed within the framework of the University of Valencia Vice-Rector's Office for Research program, call for special actions, reference UV-INV_AE-2667242.

**Institutional Review Board Statement:** Not applicable.

**Informed Consent Statement:** Not applicable.

**Data Availability Statement:** The original contributions presented in this study are included in the article/Supplementary Materials. Further inquiries can be directed to the corresponding authors.

**Acknowledgments:** We acknowledge J. M. Martínez-Agudo for technical support.

**Conflicts of Interest:** The authors declare no conflicts of interest.

## Abbreviations

The following abbreviations are used in this manuscript:

| | |
|---|---|
| ATR | Attenuated total reflectance |
| BVS | Bond valence sum |
| CCD | Charge-coupled device |
| DTA | Differential thermal analysis |
| EDX | Energy-dispersive X-ray spectroscopy |
| FTIR | Fourier transform infrared spectroscopy |
| IR | Infrared |
| POM | Polyoxometalate |
| PXRD | Powder X-ray diffraction |
| TGA | Thermogravimetric analysis |
| UV-Vis-NIR | Ultraviolet–visible–near infrared |

## References


1. Pope, M.T. *Heteropoly and Isopoly Oxometalates*; Springer: New York, NY, USA; Berlin/Heidelberg, Germany; Tokyo, Japan, 1983; ISBN 0387118896.
2. Pope, M.T.; Müller, A. Polyoxometalate Chemistry: An Old Field with New Dimensions in Several Disciplines. *Angew. Chem. Int. Ed. Engl.* **1991**, *30*, 34–48. [CrossRef]
3. Iftikhar, T.; Rosnes, M.H. Covalent Organic-Inorganic Polyoxometalate Hybrids in Catalysis. *Front. Chem.* **2024**, *12*, 1447623. [CrossRef]
4. Sun, Z.; Wang, R.; Kozhevnikov, I.V. Versatile POMOF-Based Materials: Synthesis, Mechanism, Topology and Catalytic Applications. *Coord. Chem. Rev.* **2025**, *524*, 216304. [CrossRef]
5. Gumerova, N.I.; Rompel, A. Synthesis, Structures and Applications of Electron-Rich Polyoxometalates. *Nat. Rev. Chem.* **2018**, *2*, 0112. [CrossRef]
6. Absalan, Y.; Gholizadeh, M.; Kim, E.-B.; Ameen, S.; Wang, Y.; Wang, Y.; He, H. Recent Progress on Organic Metal Compound/MOF Hybrids: From Controllable Synthesis to Potential Catalytic Applications. *Coord. Chem. Rev.* **2024**, *515*, 215972. [CrossRef]
7. Ren, Y.; Wang, M.; Chen, X.; Yue, B.; He, H. Heterogeneous Catalysis of Polyoxometalate Based Organic-Inorganic Hybrids. *Materials* **2015**, *8*, 1545–1567. [CrossRef] [PubMed]
8. Ma, M.; Chen, J.; Dong, L.; Su, Y.; Tian, S.; Zhou, Y.; Li, M. Polyoxometalates and Their Composites for Antimicrobial Applications: Advances, Mechanisms and Future Prospects. *J. Inorg. Biochem.* **2025**, *262*, 112739. [CrossRef]
9. Liu, J.; Huang, M.; Zhang, X.; Hua, Z.; Feng, Z.; Dong, Y.; Sun, T.; Sun, X.; Chen, C. Polyoxometalate Nanomaterials for Enhanced Reactive Oxygen Species Theranostics. *Coord. Chem. Rev.* **2022**, *472*, 214785. [CrossRef]
10. Hu, Y.; Wang, Y.; Zhao, J.; Chen, L. Recent Advances of Polyoxometalate-Based Materials Applied for Electron-Related Devices. *Coord. Chem. Rev.* **2024**, *506*, 215724. [CrossRef]
11. Kruse, J.-H.; Langer, M.; Romanenko, I.; Trentin, I.; Hernandez-Castillo, D.; Gonzalez, L.; Schacher, F.H.; Streb, C. Polyoxometalate-Soft Matter Composite Materials: Design Strategies, Applications, and Future Directions. *Adv. Funct. Mater.* **2022**, *32*, 2208428. [CrossRef]
12. Verissimo, M.I.S.; Evtuguin, D.V.; Gomes, M.T.S.R. Polyoxometalate Functionalized Sensors: A Review. *Front. Chem.* **2022**, *10*, 840657. [CrossRef] [PubMed]
13. Housaindokht, M.R.; Jamshidi, A.; Janati-Fard, F. Recent Advances in Polyoxometalates for Spectroscopic Sensors: A Review. *J. Mater. Sci.* **2022**, *57*, 13871–13902. [CrossRef]

14. Clemente-Juan, J.M.; Coronado, E.; Gaita-Arino, A. Magnetic Polyoxometalates: From Molecular Magnetism to Molecular Spintronics and Quantum Computing. *Chem. Soc. Rev.* **2012**, *41*, 7464–7478. [CrossRef]
15. Yang, Z.-X.; Gong, F.; Lin, D.; Huo, Y. Recent Advances in Polyoxometalate-Based Single-Molecule Magnets. *Coord. Chem. Rev.* **2023**, *492*, 215205. [CrossRef]
16. Baldoví, J.J.; Cardona-Serra, S.; Gaita-Ariño, A.; Coronado, E. Design of Magnetic Polyoxometalates for Molecular Spintronics and as Spin Qubits. In *Advances in Inorganic Chemistry, Vol 69: Polyoxometalate Chemistry*; VanEldik, R., Cronin, L., Eds.; Univ Valencia, Inst Ciencia Mol ICMol: Paterna, Spain, 2017; Volume 69, pp. 213–249. ISBN 0898-8838.
17. Tsukerblat, B.; Palii, A.; Clemente-Juan, J.M.; Gaita-Arino, A.; Coronado, E. A Symmetry Adapted Approach to the Dynamic Jahn-Teller Problem: Application to Mixed-Valence Polyoxometalate Clusters with Keggin Structure. *Int. J. Quantum Chem.* **2012**, *112*, 2957–2964. [CrossRef]
18. Palii, A.; Tsukerblat, B.; Klokishner, S.; Dunbar, K.R.; Clemente-Juan, J.M.; Coronado, E. Beyond the Spin Model: Exchange Coupling in Molecular Magnets with Unquenched Orbital Angular Momenta. *Chem. Soc. Rev.* **2011**, *40*, 3130–3156. [CrossRef]
19. Clemente-Juan, J.M.; Coronado, E.; Gaita-Ariño, A.; Giménez-Saiz, C.; Güdel, H.-U.; Sieber, A.; Bircher, R.; Mutka, H. Magnetic Polyoxometalates: Anisotropic Exchange Interactions in the $Co_3{}^{II}$ Moiety of $[(NaOH_2)Co_3(H_2O)(P_2W_{15}O_{56})_2]^{17-}$. *Inorg. Chem.* **2005**, *44*, 3389–3395. [CrossRef]
20. Galán-Mascarós, J.R.; Gómez-Garcia, C.J.; Borrás-Almenar, J.J.; Coronado, E. High Nuclearity Magnetic Clusters: Magnetic Properties of a Nine Cobalt Cluster Encapsulated in a Polyoxometalate, $[Co_9(OH)_3(H_2O)_6(HPO_4)_2(PW_9O_{34})_3]^{16-}$. *Adv. Mater.* **1994**, *6*, 221–223. [CrossRef]
21. Duan, Y.; Clemente-Juan, J.M.; Giménez-Saiz, C.; Coronado, E. Cobalt Clusters with Cubane-Type Topologies Based on Trivacant Polyoxometalate Ligands. *Inorg. Chem.* **2016**, *55*, 925–938. [CrossRef]
22. Bassil, B.S.; Xiang, Y.; Haider, A.; Hurtado, J.; Novitchi, G.; Powell, A.K.; Bossoh, A.M.; Mbomekallé, I.M.; de Oliveira, P.; Kortz, U. Heptanickel(II) Double-Cubane Core in Wells-Dawson Heteropolytungstate, $[Ni_7(OH)_6(H_2O)_6(P_2W_{15}O_{56})_2]^{16-}$. *Chem. Commun.* **2016**, *52*, 2601–2604. [CrossRef]
23. Zhang, Z.; Wang, E.; Qi, Y.; Li, Y.; Mao, B.; Su, Z. Synthesis, Characterization, and Crystal Structures of Double-Cubane-Substituted and Asymmetric Penta-Ni-Substituted Dimeric Polyoxometalates. *Cryst. Growth Des.* **2007**, *7*, 1305–1311. [CrossRef]
24. Fang, X.; Kögerler, P.; Speldrich, M.; Schilder, H.; Luban, M. A Polyoxometalate-Based Single-Molecule Magnet with an S = 21/2 Ground State. *Chem. Commun.* **2012**, *48*, 1218–1220. [CrossRef] [PubMed]
25. *APEX3*, version v2017.3-0; Bruker AXS Apex3 Inc.: Madison, WI, USA, 2017.
26. Sheldrick, G.M. Crystal Structure Refinement with SHELXL. *Acta Crystallogr. Sect. C Struct. Chem.* **2015**, *71*, 3–8. [CrossRef] [PubMed]
27. Domaille, P.J.; Hervé, G.; Tézé, A. Vanadium(V) Substituted Dodecatungstophosphates. In *Inorganic Syntheses, Volume 27*; Ginsberg, A.P., Ed.; Wiley: Hoboken, NJ, USA, 1990; pp. 96–104.
28. Brown, I.D.; Altermatt, D. Bond-Valence Parameters Obtained from a Systematic Analysis of the Inorganic Crystal Structure Database. *Acta Crystallogr. Sect. B Struct. Sci.* **1985**, *41*, 244–247. [CrossRef]
29. Wood, R.M.; Palenik, G.J. Bond Valence Sums in Coordination Chemistry. A Simple Method for Calculating the Oxidation State of Cobalt in Complexes Containing Only Co—O Bonds. *Inorg. Chem.* **1998**, *37*, 4149–4151. [CrossRef]
30. Brese, N.E.; O'Keeffe, M. Bond-Valence Parameters for Solids. *Acta Crystallogr. Sect. B Struct. Sci.* **1991**, *47*, 192–197. [CrossRef]
31. Finke, R.G.; Droege, M.W.; Domaille, P.J. Trivacant Heteropolytungstate Derivatives. 3. Rational Syntheses, Characterization, Two-Dimensional Tungsten-183 NMR, and Properties of Tungstometallophosphates $P_2W_{18}M_4(H_2O)_2O_{68}{}^{10-}$ and $P_4W_{30}M_4(H_2O)_2O_{112}{}^{16-}$ (M = Cobalt, Copper, Zinc). *Inorg. Chem.* **1987**, *26*, 3886–3896. [CrossRef]
32. Nickolov, Z.; Georgiev, G.; Stoilova, D.; Ivanov, I. Raman and IR Study of Cobalt Acetate Dihydrate. *J. Mol. Struct.* **1995**, *354*, 119–125. [CrossRef]
33. Zhang, Z.; Muhammed, M. Thermochemical Decomposition of Cobalt Doped Ammonium Paratungstate Precursor. *Thermochim. Acta* **2003**, *400*, 235–245. [CrossRef]
34. Yamase, T. Photo- and Electrochromism of Polyoxometalates and Related Materials. *Chem. Rev.* **1998**, *98*, 307–326. [CrossRef]
35. Weakley, T.J.R.; Malik, S.A. Heteropolyanions Containing Two Different Heteroatoms—I. *J. Inorg. Nucl. Chem.* **1967**, *29*, 2935–2944. [CrossRef]

MDPI AG
Grosspeteranlage 5
4052 Basel
Switzerland
Tel.: +41 61 683 77 34

*Magnetochemistry* Editorial Office
E-mail: magnetochemistry@mdpi.com
www.mdpi.com/journal/magnetochemistry

www.ingramcontent.com/pod-product-compliance
Lightning Source LLC
Chambersburg PA
CBHW061547220926
51850CB00005B/272
* 9 7 8 3 7 2 5 8 4 6 4 1 2 *